HONEY BEE PESTS, PREDATORS, AND DISEASES
THIRD EDITION

HONEY BEE PESTS, PREDATORS, AND DISEASES, THIRD EDITION

Edited by ROGER A. MORSE
and KIM FLOTTUM

PUBLISHED BY THE A. I. ROOT COMPANY
Medina, Ohio, U.S.A.

First Edition published in 1978 by Cornell University Press.
Second Edition published in 1990 by Cornell University Press.
Third Edition published in 1997 by The A.I. Root Company.

Printed in The United States of America.

Honey Bee Pests, Predators, and Diseases
Includes Bibliographical References
ISBN 0-936028-10-6
1997

This edition produced in 2013 under licence from the A.I.Root Company
by Northern Bee Books, Scout Bottom Farm,
Mytholmroyd, Hebden Bridge HX7 5JS (UK)

Available from A.I. Root Co. Medina, Ohio
Northern Bee Books, UK
And all book stores

With Special Acknowledgements to:

The Authors

Kathy Summers, Design

Susan Pohl, Copy Editing

Mary Gannon, Copy Editing

From The Publisher

Thirty two authors have come together to produce this book. These people are specialists in their respective field(s) and their contributions are to be considered the latest information available in honey bee science at publication.

The Third Edition has added significantly to the body of knowledge offered in the Second Edition, published only seven years previous. Especially important are the chapters on Varroa mites, tracheal mites, disease resistance, subspecies of *Apis mellifera*, and the International Trade Agreements and Country Record For Honey Bee Diseases.

Differing from the First and Second Editions format, the scientific names index has been incorporated directly into the General Index.

The Editors and the Publishers express their appreciation to all those who have contributed to this work.

CONTENTS

Authors

John T. Ambrose, B.A., M.S., Ph.D.

Professor of Entomology and Extension Apiculturist, Department of Entomology, North Carolina State University, Raleigh, NC 27695

Leslie Bailey, M.A., Ph.D., D.Sc.

Retired. Insect Pathologist, Entomology and Nematology Department, Rothamsted Experiment Station, Harpenden, Hertfordshire AL5 2JQ, UK

Brenda V. Ball, B.Sc., M L Biol, C.Biol.

Insect Pathologist, Entomology and Nematology Department, Crop and Environment Protection Division, Rothamsted Experiment Station, Harpenden, Hertfordshire AL5 2JQ, UK

D. Michael Burgett, B.S., M.S., Ph.D.

Professor of Entomology, Department of Entomology, Oregon State University, Corvallis, OR 97331

Nicholas W. Calderone, B.S., Ph.D.

Assistant Professor of Apiculture, Department of Entomology, Comstock Hall, Cornell University, Ithaca, NY 14853

Dewey M. Caron, B.S., M.S., Ph.D.

Professor of Entomology, Department of Entomology and Applied Ecology, University of Delaware, Newark, DE 19717

David De Jong, B.S., M.S., Ph.D.

Professor of Ecology, Department of Genetics, School of Medicine, University of Sao Paulo, 14.049 Ribeirao Preto, SP, Brazil

Keith S. Delaplane, B.S., M.S., Ph.D.

Associate Professor of Entomology, Department of Entomology, University of Georgia, Athens, GA 30602

George C. Eickwort, B.S., M.S., Ph.D.

Deceased, former Professor of Entomology and Chairman, Department of Entomology, Cornell University, Ithaca, NY 14853

Richard D. Fell, B.S., M.S., Ph.D.

Professor of Entomology, Department of Entomology, Virginia Polytechnic Institute and State University, Blacksburg, VA 24061

Ingemar Fries, B.S., Ph.D.

Research Leader, Department of Entomology, Swedish University of Agricultural Sciences, S-750 07 Uppsala, Sweden

Martha Gilliam, B.A., M.S., Ph.D.

Research Microbiologist, Carl Hayden Bee Research Center, Agricultural Research Service, United States Department of Agriculture, 2000 East Allen Road, Tucson, AZ 85719

Ernesto Guzman-Novoa, M.S., D.V.M., Ph.D.

Research Entomologist, Secretariat of Agriculture and Hydrologic Resources, Santa Cruz 29-B, Fracc. Las Haciendas, 52140 Metepec, Edo. de Mex., Mexico

Carol H. Hoffman, M.S., D.V.M., Ph.D.

Instructor, San Jose City College, San Jose, CA

Wm. Michael Hood, B.S., M.S., Ph.D.

Associate Professor, Department of Entomology, Clemson University, Clemson, SC 29634-0365

David A. Knox, B.S., M.S.

Entomologist, Bee Research Laboratory, Agricultural Research Service, United States Department of Agriculture, BARC-East, Beltsville, MD 20705

Phillip J. Lima, B.S.

Entomologist, Retired, Animal and Plant Health Inspection Service. Plant Protection and Quarantine. Hyattsville, MD 20787

Andrew G. Matheson, B.S., M.S.

National Advisor SPS, Ministry of Agriculture, P.O. Box 2526 Wellington, New Zealand

Roger A. Morse, B.S., M.S., Ph.D.

Professor of Apiculture, Department of Entomology, Cornell University, Ithaca, NY 14850

Don Nelson, B.S.A., M.S., Ph.D.

Research Scientist, Northern Agricultural Research Center, Agriculture and Agri-food Canada, Box 29, Beaverlodge, AB T0H 0C0, Canada

Richard Nowogrodzki, B.A., M.S., Ph.D.

Former Research Specialist in Apiculture, Trumansburg, NY 14847

Robert E. Page, Jr., B.S., Ph.D.

Professor of Entomology, Department of Entomology, University of California, Davis, CA 95616

Jeffrey S. Pettis, B.S., M.S., Ph.D.

Bee Research Laboratory, Agricultural Research Service, United States Department of Agriculture, BARC-East, Beltsville, MD 20705

Diana Sammataro, B.S., M.S., Ph.D.

Research Associate, Department of Entomology, The Ohio State University, OARDC, Wooster, OH 44691

Walter S. Sheppard, B.S., M.S., Ph.D.

Associate Professor of Entomology, Department of Entomology, Washington State University, Pullman, WA 99164

Hachiro Shimanuki, B.A., Ph.D.

Microbiologist and Research Leader, Bee Research Laboratory, Agricultural Research Service, United States Department of Agriculture, BARC-East, Beltsville, MD 20705

John A. Skinner, B.S., M.S., Ph.D.

Assistant Professor of Entomology, Department of Entomology and Plant Pathology, University of Tennessee, Knoxville, TN 37901

Kenneth W. Tucker, B.S., Ph.D.

Research Entomologist (retired), Honey Bee Breeding, Genetics and Physiology Research Laboratory, United States Department of Agriculture, Baton Rouge, LA 70820

John D. Vandenberg, B.S., M.S., Ph.D.

Research Entomologist, Agricultural Research Service, United States Department of Agriculture, Plant, Soil and Nutrition Laboratory, Cornell University, Ithaca, NY 14853

Allen W. Vaughan, B.S., M.S.

Ecological Effects Branch, Environmental Fate and Effects Division, Office of Pesticide Programs, Environmental Protection Agency, Washington, DC 20460

Jon L. Williams, B.S., M.S.

Research Entomologist, Honey Bee Breeding, Genetics and Physiology Laboratory, Agricultural Research Service, United States Department of Agriculture, 1157 Ben Hur Rod, Baton Rouge, LA 70820

William T. Wilson, B.S., Ph.D.

Research Entomologist, Honey Bee Research Laboratory, Agricultural Research Service, United States Department of Agriculture, 2413 East Highway 83, Weslaco, Texas 78596

CHAPTER ONE

Introduction

Roger A. Morse

Introduction
Chalkbrood
Tracheal mites
Varroa mites
The sources of the new diseases
American foulbrood
Other diseases
Pests and predators
Pesticides
Other species and races of honey bees as a threat
Honey bee biology
Legislation pertaining to the beekeeping industry
Diagnosis
The future

CHAPTER ONE
Introduction

Two striking changes have occurred in the beekeeping industry in recent years. One is the increasing movement of colonies around the country for honey production and especially pollination. Over one million colonies are moved for the pollination of agricultural crops alone; most of these bees are rented for use on two crops, occasionally three (Robinson et al. 1989). Because most pollination is done in the spring, the same colonies may be used for honey production in the summer and fall. Nearly one million colonies of bees are transported north in the spring for honey production alone. As an example of the changes that are taking place, one beekeeper trucked bees from Florida to California for the pollination of almonds, to Maine for blueberry pollination, to Massachusetts for cranberry pollination, and to New York for honey production, after which time the bees were returned to Florida. This mass cross-country movement of colonies of bees has also resulted in the movement of several of their pests, predators and diseases.

The second big change is the discovery since 1972 of three diseases of honey bees new to North America: chalkbrood, tracheal mites and varroa mites. At the same time, Africanized honey bees, which introduce new genetic material into the country, have been found in several states since 1990. In some areas, and in extreme cases, commercial beekeepers have lost up to 80 percent of their colonies in a single year because of one or both mites. In many states, especially in the North, there are half as many hobby beekeepers as there were 1984 to 1987. In many areas, feral honey bee populations, those that live in hollow trees and buildings, have almost disappeared.

One salvation for the industry has been the ease with which one may grow new queens and colonies, especially in the Southern states in the spring. Another is the fact that stock resistant to tracheal mites in Europe was available and has been imported. While not perfect, this stock has been of great value in restoring the industry in some areas. In the case of varroa mites, we had the benefit of over 10 years of European research, since they had accidentally carried varroa mites into that part of the world in the late 1960s.

It is the spread and additional information about these three new diseases that prompts the 1997 revision of this textbook, which was last updated in 1990.

Chalkbrood

Chalkbrood is a fungal disease of honey bee larvae that was found on leafcutter bees by Baker and Torchio in Utah in 1968. It was reported infecting honey bees in California in 1972 (Thomas and Luce 1972). By 1974, the disease had been found in honey bee colonies in at least 35 states (De Jong and Morse 1976); very soon thereafter it was everywhere in North America. Large numbers of colonies died of chalkbrood, and no chemical has been developed to control it. The impact of chalkbrood has been overshadowed by losses to mites, but it is not uncommon to find white, gray or black larval mummies on the bottomboards that project in front of colonies as a result of chalkbrood.

Tracheal mites

On July 3, 1984, tracheal mites (*Acarapis woodi*) were found in Southern Texas. They had been found four years earlier in Mexico (Wilson and Nunamaker 1982), and their movement north should not have been a surprise. Detecting tracheal mites, especially a low-level infestation, is difficult. Nevertheless, within a few months of the Texas discovery, the mites were found as far away as New York and South Dakota. The history of the spread of the mites is recorded by Henderson and Morse (1990).

Varroa mites

On September 25,1987, a migratory beekeeper and an apiary inspector in Wisconsin found varroa mites (*Varroa jacobsoni*) while they were in the process of checking colonies for American foulbrood preparatory to moving the bees back to Florida. Within a few weeks, varroa mites had been found in Florida, Illinois, Ohio and Pennsylvania followed by finding infestations in several other states (Graham 1987a,b). The widespread movement of colonies for pollination and honey production was obviously responsible for the rapid spread of the varroa mites.

The sources of the new diseases

One might logically ask when and how these pests entered the country and how we might prevent the spread of further problems. People in the beekeeping industry generally believe that chalkbrood was imported in pollen brought into the United States from Europe for bee food. In retrospect, it appears that the bees in North America had not been exposed to the disease in decades and had lost any resistance they might have had to it.

Tracheal mites probably came into the United States by way of Mexico or perhaps via an illegal importation into Florida. For many years they were thought to be a danger to North American beekeeping. The United States and Canada had, and still have, strong legislation prohibiting the importation of honey bees from abroad except under special circumstances, but Mexico did not, and this was the source of the tracheal mites. There is no proof, but it is suspected that varroa mites entered the country, probably in Florida, in an illegal shipment of bees (queens) from Brazil. It is clear that the source of the varroa mites in the United States is

South America (Delfinado-Baker and Houck 1989).

American foulbrood
The chief concern of state apiary inspection departments, which are usually found in state departments of plant industry, has been American foulbrood, a bacterial disease of larval honey bees that is almost always fatal. The chief problem in coping with this disease is that the bacteria may form a spore that remains viable for at least 60 years. Spores may lie dormant under cocoons or in scales in comb in active hives and be exposed and ready to infect new larvae if the comb is broken and the opportunity presents itself. A great problem in American foulbrood control is reinfection from old equipment that has been in storage for many years. It was found that antibiotics would give good control of American foulbrood in 1944 (Haseman and Childers, 1944). However, the industry was reluctant to endorse this recommendation since the material proposed had a long life, care was needed to make effective treatments, and the drug did nothing to negate the long-term danger from the spores. The discovery that Terramycin, a short-lived drug, and more specifically the use of antibiotic extender patties (grease patties with Terramycin), gave a more uniform dose and has gained widespread acceptance. State regulatory agencies, through thorough inspection programs, and chemotherapy, used by beekeepers, have contributed to keep levels of this disease low in many areas.

Other diseases
European foulbrood (a bacterium affecting larval honey bees), nosema (a microspordian affecting adult bees), sacbrood (a virus affecting larvae), and a host of lesser diseases that have plagued the North American beekeeping industry for years have been reasonably well-studied. The key to reducing the impact of most of these diseases appears to be the maintenance of strong (populous) colonies of bees. A general recommendation for the control of these diseases is to requeen the affected colonies, combine weak colonies, or both.

Pests and predators
The seriousness of pests and predators varies from place to place and year to year. An easy way to provoke an argument among beekeepers is to open the subject of bears; they are a serious pest in some parts of North America and cherished wildlife elsewhere. The electrical fences designed in recent years deliver a strong shock to an animal that touches an electrified wire and will usually drive away naive bears.

Skunks are an irritation for bees and beekeepers. They move up and down rows of hives attacking the weaker colonies that are less well-defended. A skunk working in an apiary will keep colonies in a state of alert for long periods of time.

One result of their activity is that beekeepers are stung more frequently in such apiaries because of the agitated state of the guard bees in skunk-molested colonies.

Wax moths destroy great quantities of honeycomb every year but are also of benefit in destroying comb that may contain American foulbrood spores in feral or let-alone colonies that died from the disease. Mice probably cause more physical damage to beekeeping equipment than any other pest. Mice will enter a colony in the late fall and chew out an area six to eight inches in diameter where they build a nest and remain for the winter. Birds, squirrels and a small number of other pests are also of concern.

Pesticides

Pesticides are not within the scope of this book, but because pesticide losses are sometimes mistaken for disease problems, a new appendix reviewing the use and problems for honey bees created by pesticides has been written. The first recorded loss of honey bees in the United States because of insecticide use took place in 1881 when a farmer noted that his bees were being killed as a result of visiting a flowering pear tree that he had sprayed with Paris green (an arsenical). However, it was during the present century that serious problems occurred. The creation of the Environmental Protection Agency, in 1970, has done much to alleviate the problem.

Other species and races of honey bees as a threat

From time to time there have been suggestions that *Apis cerana*, the small Indian honey bee, which is less than half the size of the European honey bee, should be imported into the United States for experimental purposes, or perhaps as a bee for hobbyists. Like our own bees, there are several races of the Indian bee, most of which have been little studied. We are aware that one viral disease (Thai sacbrood virus) is a severe threat to these bees, and we have little information concerning how this virus might affect European or African honey bees.

During the past several years we have watched as the dwarf honey bee, *Apis florea*, has moved west from Iran, once the westernmost limit of its natural habitat, into Oman and most recently, Sudan. It is now in a position to invade the whole of Africa. There is no way to determine what its effect might be.

The cape honey bee, the bee that until recently has been restricted to living in the southernmost areas of South Africa, has found its way north into the Transvaal, Natal and the Orange Free State in that country. It was unwittingly transported there by commercial migratory beekeepers. The cape bees are different in that their workers may produce diploid eggs through parthenogenesis that develop into workers. These cape workers produce chemicals (pheromones) that cause them to be recognized as queens. These false or pseudo-queens dominate, and the populations dwindle. The result has been the devastation of several commercial beekeeping operations in the invaded areas in South Africa. This cape bee activity outside of its native range has been seen only since 1992, and it is still too early to know the long-range effects. However, our present knowledge is such that we can state that cape bees are definitely pests insofar as at least one other African race is concerned (Allsopp and Crewe 1993, Cooke 1993).

Honey bee biology

The biology of the European honey bee is well-known. More has been written about honey bees than about most animals except for dogs, horses, cows, poultry, pigs, sheep and man. There are thousands of bee books and hundreds of bee journals. However, a few salient points about honey bee biology are worth mentioning as they bear heavily on the subject of honey bee pests, predators and diseases.

Honey bees are social insects; no honey bee can live alone. There is no normal social order until a colony has a queen and several hundred workers, and even then, such a small population must be given special protection. A minimum number for a colony to survive, depending on the climate, is about 5,000 to 10,000 bees. Colonies in Canada and the northern tier of states need more bees. Food exchange, presumably accompanied by the exchange of pheromones, alerts the colony to changes, for example, the loss of a queen. An alarm odor produced only by workers is used to alert others to danger. Honey bees control their brood nest temperature and are the only insects to do so. The closeness in which honey bees live may also be a detriment and can cause contagious disease organisms to be exchanged quickly between individuals. A high brood nest temperature may also give some protection against disease, but the role of elevated brood nest temperatures, analogous to fevers in mammals, has been little studied.

A honey bee colony may have a reservoir of as much as 80 pounds of honey, 12 pounds of pollen and thousands of bees, which represents a great hoard of food for an animal such as a bear. A beehive is usually a warm, dry cavity that may be an attractive home for a mouse, snake or other small animal. Knowledge of these and related facts about honey bee biology aid the beekeeper in controlling many of the unwanted problems that may plague a colony.

Legislation pertaining to the beekeeping industry

At a time when the honey bee disease situation was probably never worse, we are seeing a number of states reducing their support for bee disease inspection. California, for example, has given up bee disease inspection at the state level. New York state is spending less than half the amount of money on apiary inspection that it was devoting to the problem five years earlier. The reasons are several: A need for tax money for other purposes; the complexity and indecision on problems relating to tracheal and varroa mites; the difficulty and cost of inspections for tracheal and varroa mites; the fact that in an extreme example, a colony of bees may be transported and used in as many as five states (three is not unusual) for pollination and honey production in a single year; the fact that some states have no apiary inspectors; that good over-the-counter chemicals are available for some diseases; and last, that the chief problem is one of education, not destruction of offending colonies.

Despite these developments, there has been little change in legislation except for Canada, where a blanket prohibition against the importation of United States bees has been enacted. Bees, queens and beekeeping equipment can still flow freely into the United States from Canada. Since there are no federal laws pertaining to apiary inspection, only a federal prohibition of bees from abroad (other than Canada), we have tabulated the existing state legislation.

Diagnosis

The U. S. Department of Agriculture Bee Research Laboratory in Beltsville, Maryland, will examine honey bee disease samples from anywhere at no charge. The purpose of this service is to provide a continuing surveillance of the worldwide picture. As a result, hopefully, the diseases that pose a threat to North American beekeeping will be known before they pose a threat here. For information on how to prepare samples for shipment, see the final pages of "Summary of Control Methods," Chapter 22.

The future

Of immediate interest in bee disease control is the selection of resistant strains of honey bees. Bringing potentially resistant honey bees into North America appears to be a reasonable course of action. At the same time, researchers are probing the genetics of many plants and animals, including honey bees, hoping to isolate and eliminate or enhance genetic material that aids in their growth and development.

CHAPTER TWO

Viruses

Brenda V. Ball and Leslie Bailey

CHAPTER TWO
Introduction

All forms of life are attacked by viruses, and there is a wide variety of virus types. Each type is either host-specific or has a narrow host range. Virus particles are little more than genetic material (DNA or RNA) enclosed in a protective coat of protein, and they multiply only within the living cells of their host. Most are so small that they can be seen only by special techniques, notably electron micros-copy, which only became feasible about the middle of this century. Even under an electron microscope, the particles of many unrelated viruses that cause different diseases look alike (Figure 2.1) and can be distinguished only by special methods, such as serology, or occasionally by the diseases they cause.

Many viruses that cause severe and often fatal diseases multiply within and spread between individual hosts for prolonged periods, even indefinitely, with-out causing obvious sickness. This feature is characteristic of many if not all known bee viruses. Some bee viruses depend on but do not invariably accompany the common microsporidian midgut parasite *Nosema apis*, and some are aggravated and transmitted by the mite *Varroa jacobsoni*. Most multiply and spread independent-dently, sometimes causing severe diseases, but more often they occur in bee colo-nies that continue to appear healthy even when infected with several different viruses. Many viruses of bees are extremely common, and there are more different types than of the other known bee parasites. Accordingly, viruses have probably always been prime sources of confusion and error in the diagnosis and manage-ment of bee diseases. This is well-exemplified by the history of paralysis.

Paralysis
History

Although paralysis was first described more than a century ago, the cause was not identified until 1963 (Bailey et al. 1963). During the first 20 years of this century, several common parasites, visible by ordinary microscopy, were discov-ered in adult bees. The last to be identified, the tracheal mite *Acarapis woodi*, was quickly accepted as the cause of paralysis, or "Isle of Wight disease," as it was then known in Britain (Bailey and Ball 1991). However, the evidence for this was largely anecdotal, entirely circumstantial, and frequently conflicting. Viruses were rarely considered until Burnside (1933) showed that bacteria-free filtrates of bees with paralysis in the United States were infective. The infectivity was destroyed by heating, and it seems likely that he was working with the same virus identified as the cause of paralysis in Britain (Bailey et al. 1963; Figure 2.2b), in the United States (Rinderer and Green 1976), and then throughout the rest of the world (Bailey 1976).

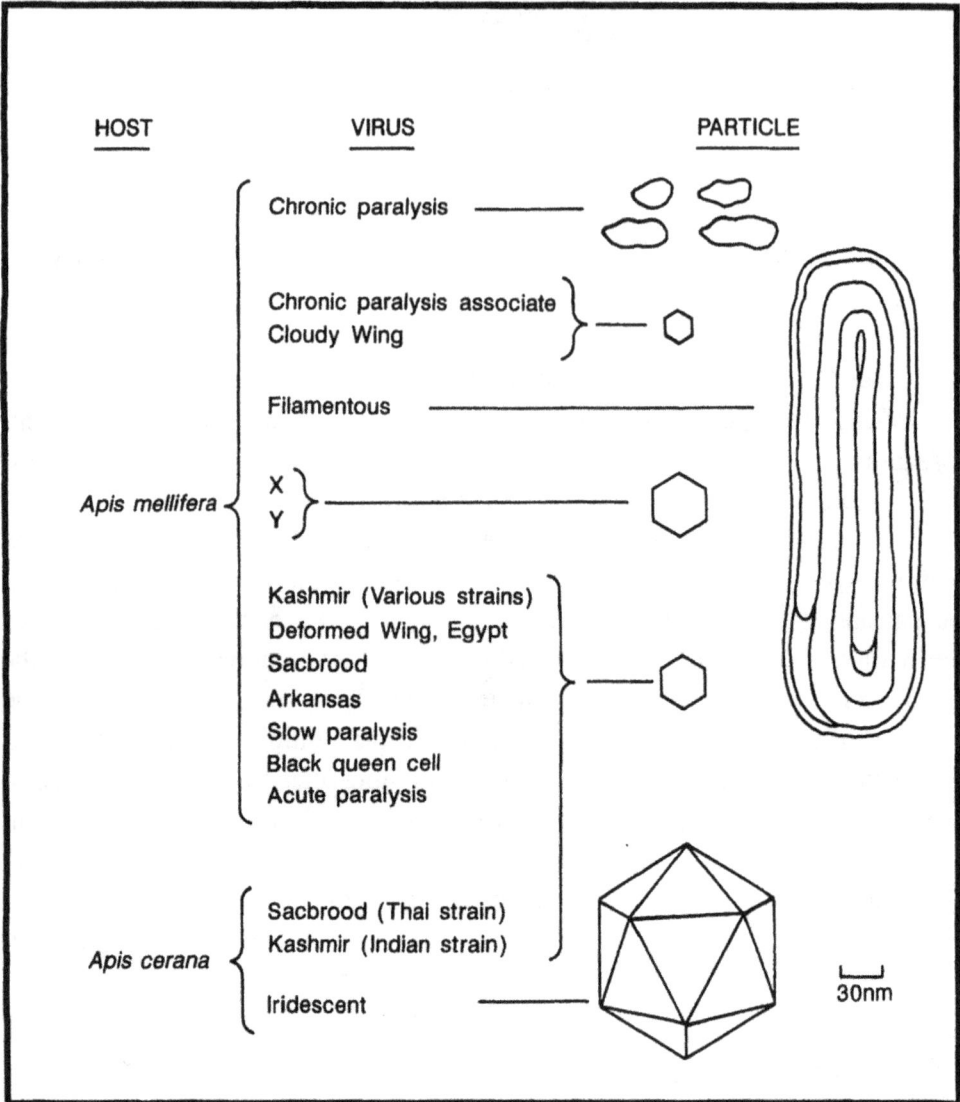

Figure 2.1. Viruses of honey bees

The virus was named chronic paralysis virus to distinguish it from acute paralysis virus, an unrelated virus, also found in some of the bees during the investigations. Chronic paralysis virus is frequently, perhaps always, accompanied by a "satellite" virus—chronic paralysis virus associate (Figure 2.2a)—that is of unknown significance, variable in amount, and especially numerous in the relatively few queens that develop paralysis. It may have a role in defense mechanisms because the associate particle interferes with the multiplication of chronic paralysis virus in individual bees, inhibiting particularly the replication of the longest, most infective particles (Ball et al. 1985).

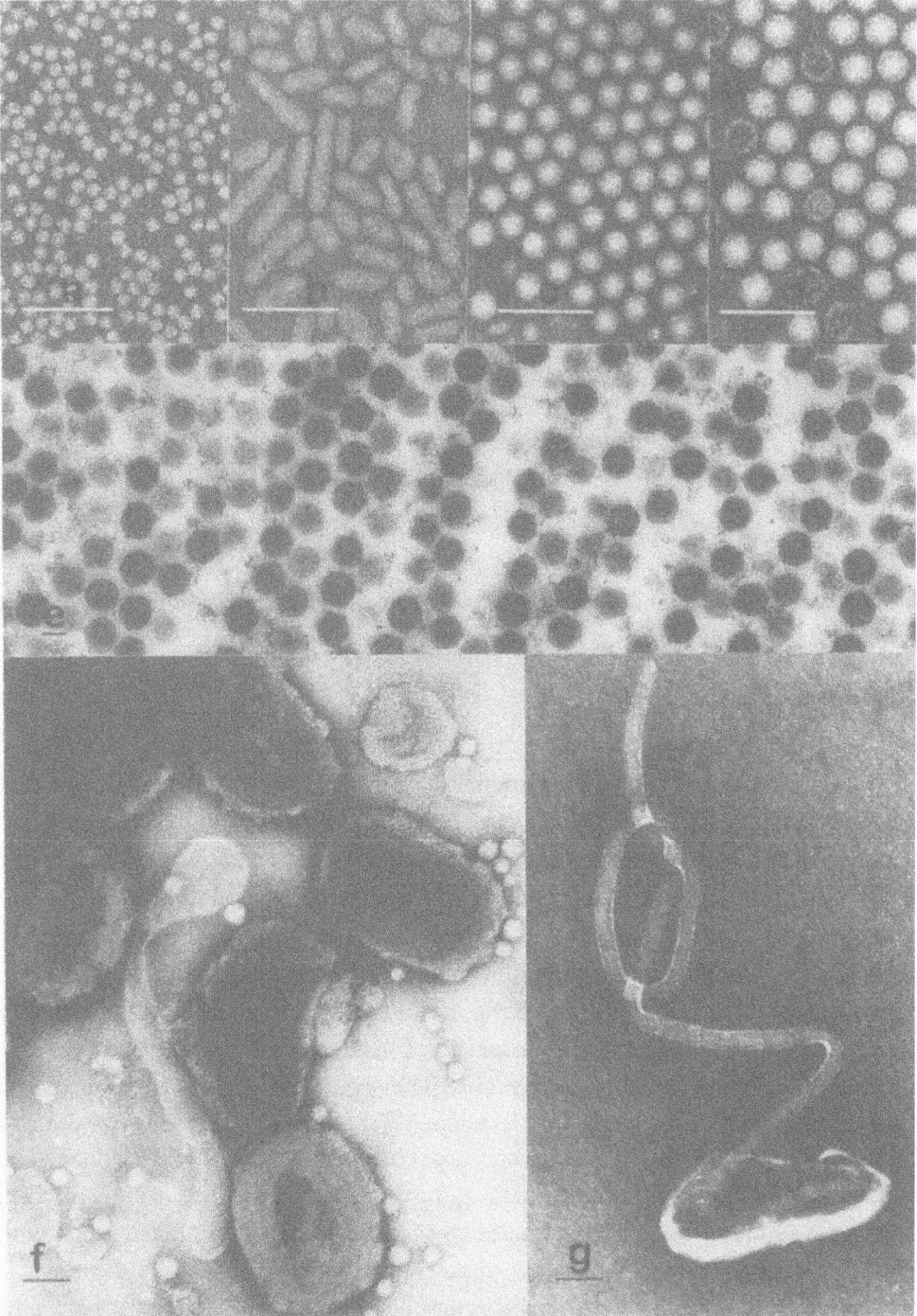

Figure 2.2. Electron micrographs of representative types of virus particles from bees: (a) particles 17 nm in diameter (cloudy wing virus, chronic paralysis virus associate); (b) chronic paralysis virus particles 20 x 30 to 60 nm; (c) particles 30 nm in diameter (sacbrood virus, acute paralysis virus, Kashmir bee virus, deformed wing virus, Egypt bee virus, slow paralysis virus, black queen cell virus, and Arkansas bee virus); (d) particles 35nm in diameter (bee viruses X and Y); (e) *Apis* iridescent virus in ultra thin section of the cytoplasm of adult bee fat body cell; (f) filamentous virus particle; (g) filamentous virus particle with ruptured envelope releasing the single flexuous rod, or nucleocapsid, which contains DNA. Scale bars = 100 nm.

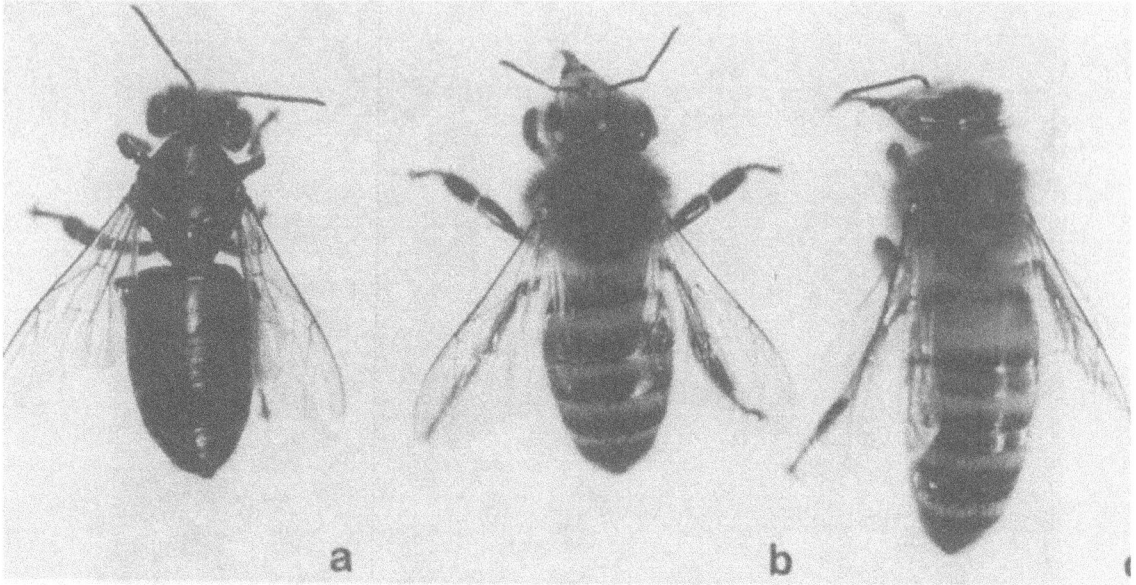

Figure 2.3. Bees infected with chronic paralysis virus: (a) Type 2 syndrome; (b) healthy individual; (c) Type 1 syndrome.

Symptoms

Paralysis is the only common disease of adult bees that has striking, well-defined symptoms (Figure 2.3). It is serially transmissible experimentally and indefinitely to adult bees by inoculating them with purified preparations of the virus by injecting it, rubbing it on their bodies, or adding it to their food. It has two sets of symptoms (syndromes). One (Type 1, Figure 2.3c), seemingly the most common in Britain, includes an abnormal trembling motion of the wings and bodies of affected bees. These bees fail to fly and often crawl on the ground and up the stems of grass, sometimes in masses of thousands of individuals. Frequently they huddle together on top of the cluster in the hive. They often have bloated abdomens and partially spread, dislocated wings. The bloated abdomen is caused by distention of the honey sac with fluid: This accelerates the onset of so-called "dysentery," and sick individuals die within a few days of the onset of symptoms. Severely affected colonies suddenly collapse, particularly at the height of the summer season, typically leaving the queen with a few workers on neglected combs. All these symptoms are identical to those attributed to the "Isle of Wight disease" in Britain at the beginning of the century.

The other syndrome (Type 2, Figure 2.3a) has been given a variety of names: "black robbers" and "little blacks" in Britain; *Schwarzsucht, mal noir* or *mal nero* in continental Europe and "hairless black syndrome" in the United States. It could well have been the condition described by Aristotle of black bees that he called "thieves." At first the affected bees can fly, but they become almost hairless, appearing dark or almost black, which makes them seem smaller than healthy bees, with a relatively broader abdomen. They are shiny, appearing greasy in bright light, and they suffer nibbling attacks by healthy bees of their colony, which makes them seem like robber bees (Drum and Rothenbuhler 1983). In a few days they become flightless, tremble, and soon die. Both syndromes can occur in one colony, but one or the other soon predominates.

Susceptibility of bees

The difference between the two syndromes is probably caused by genetic differences between individual bees. There is considerable evidence that susceptibility of bees to paralysis is influenced by several inherited qualities, so some variation in these qualities may well lead to variations in the symptoms. Rinderer et al. (1975) and Kulincevic and Rothenbuhler (1975) selected strains of bees more susceptible than usual to the Type 2 syndrome; and circumstantial evidence (Bailey 1965, 1967) indicated that susceptibility to paralysis in Britain was much influenced by hereditary factors. Inbreeding with colonies showing much paralysis, or allowing them to rear their own queens that are likely to mate with drones from local colonies with paralysis, maintains a higher incidence of paralysis than when colonies are supplied with queens from elsewhere.

In sensitive infectivity tests with extracts of live bees collected from normal colonies in Britain, paralysis virus was detected among many colonies throughout the year, with no obvious seasonal variation. Nevertheless, less than two percent of colonies in Britain are noticeably affected by paralysis today, so bees have a considerable degree of resistance to the spread and multiplication of the virus under ordinary circumstances.

Natural spread of paralysis

When adult bees from seemingly healthy colonies in Britain are kept crowded in cages at 35°C (normal brood nest temperature), paralysis often spreads quickly among them (Bailey et al. 1983a). The virus may spread contagiously and most readily when individuals are active but in unusually prolonged bodily contact. They do not transmit sufficient virus to cause paralysis by food exchange because many millions of virus particles are required to cause paralysis when given to a bee in food. Very few particles, however, will do so when they enter the body via a wound, probably because they can easily infect vital tissues. In nature, this is most likely to occur when adult bees are brushing close against each other, which breaks many hairs or bristles from the cuticle to expose live tissue. This can be simulated in the laboratory by rubbing virus preparations on the freshly "shaved" body surfaces of bees. Furthermore, the sensitivity of bees to ingested virus is

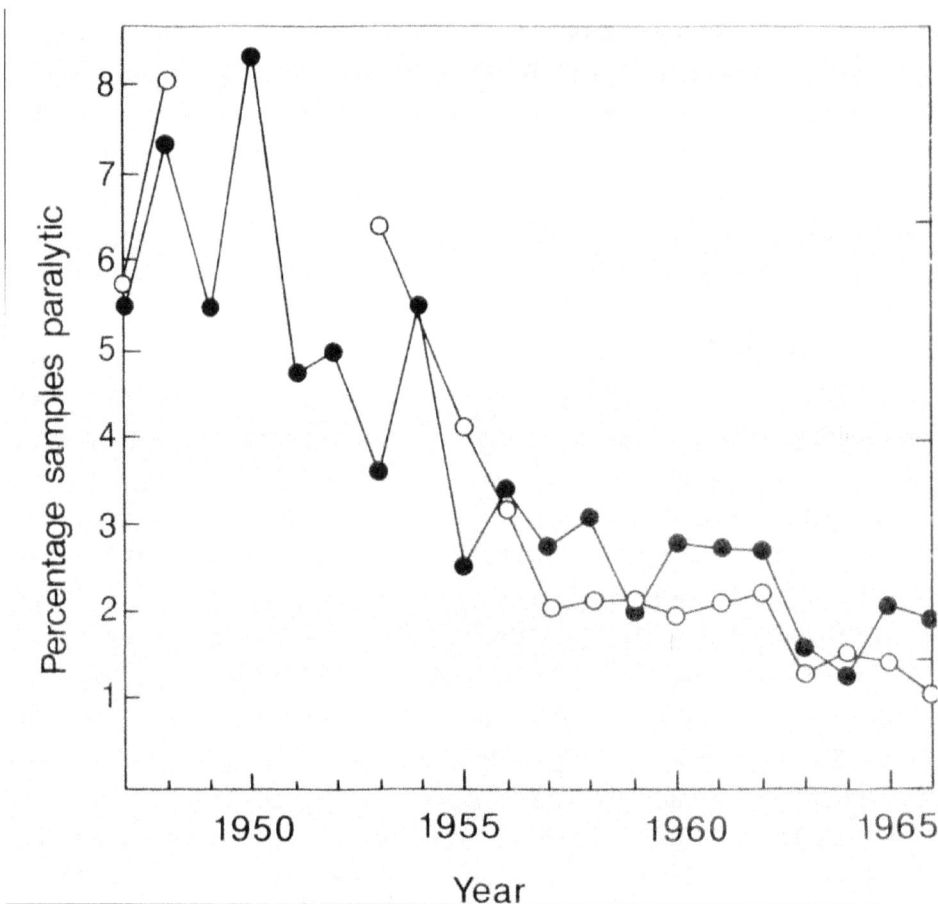

Figure 2.4. The percentage of samples of adult bees diagnosed as suffering from paralysis (●) in England and Wales and the estimated total number of bee colonies (○).

increased by the admixture of broken hairs with the food (Rinderer and Rothenbuhler 1975), perhaps because the hairs damage the intestinal tract, again enabling the virus to invade exposed tissue. Accordingly, paralyzed bees may be expected in the field mostly when local events confine adult bees to their colonies at times of the year when they would normally be foraging. These events may occur at irregular intervals during the usually active season for bees for a variety of reasons, both natural (e.g., sudden failures of nectar flows) and artificial (e.g., when too many colonies are kept for the available nectar). These circumstances are the same as those that aggravate infestation by *Acarapis woodi.*

Changing incidence in Britain

The incidence of colonies showing signs of paralysis in Britain has declined significantly from about seven percent to less than two percent since official records began in 1947 (Bailey et al. 1983a; Figure 2.4). This decline was significantly correlated with the number of colonies in the country and corresponds exactly with the decline of *Acarapis woodi* (see Chapter 13), which is transmitted by close bodily contact between live bees. It contrasts with the somewhat increasing incidences, during the same period, of common pathogens that do not require such

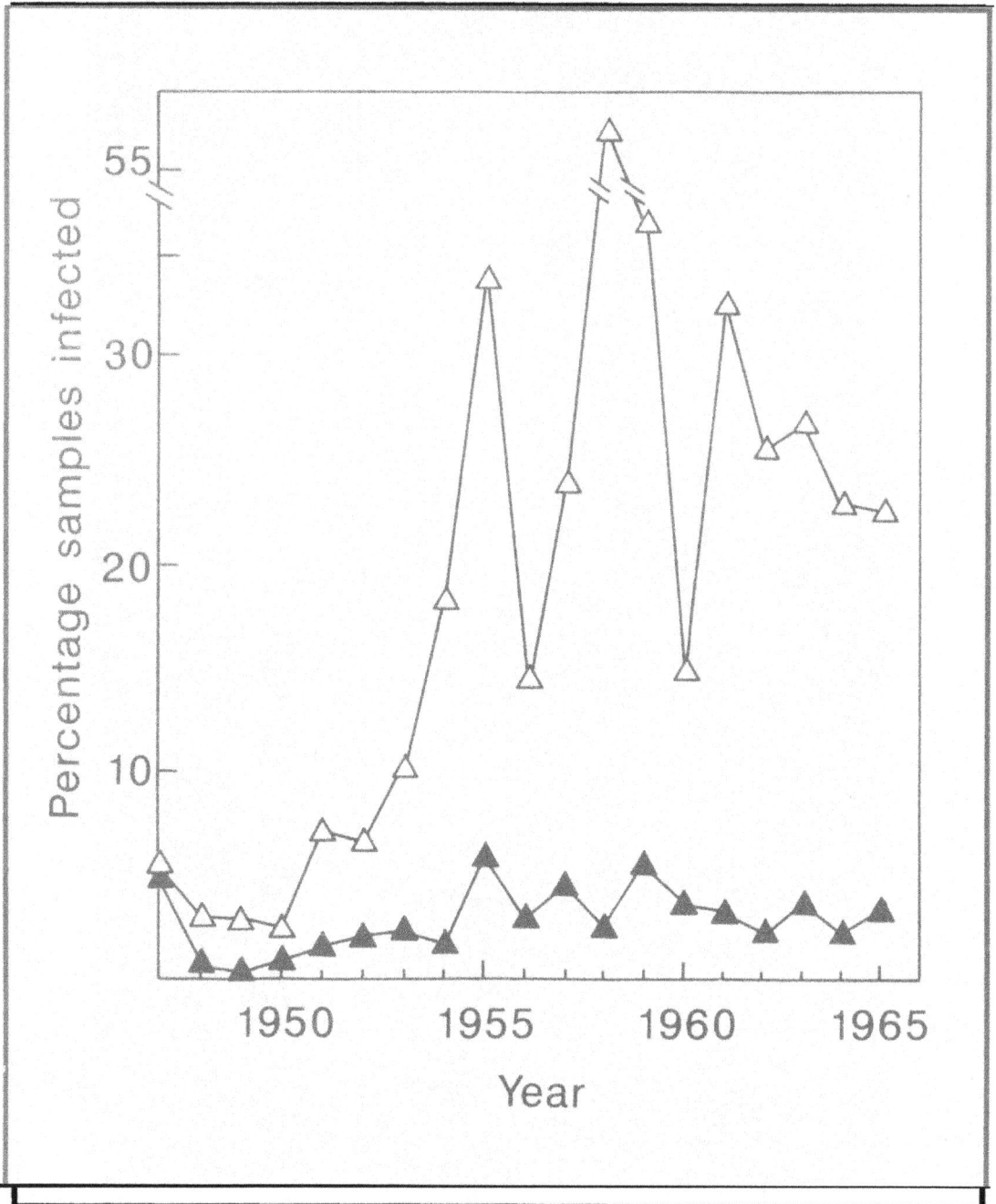

Figure 2.5. The percentage of samples of adult bees infected with *Nosema apis* (△) or with *Malpighamoeba mellificae* (▲) in England and Wales.

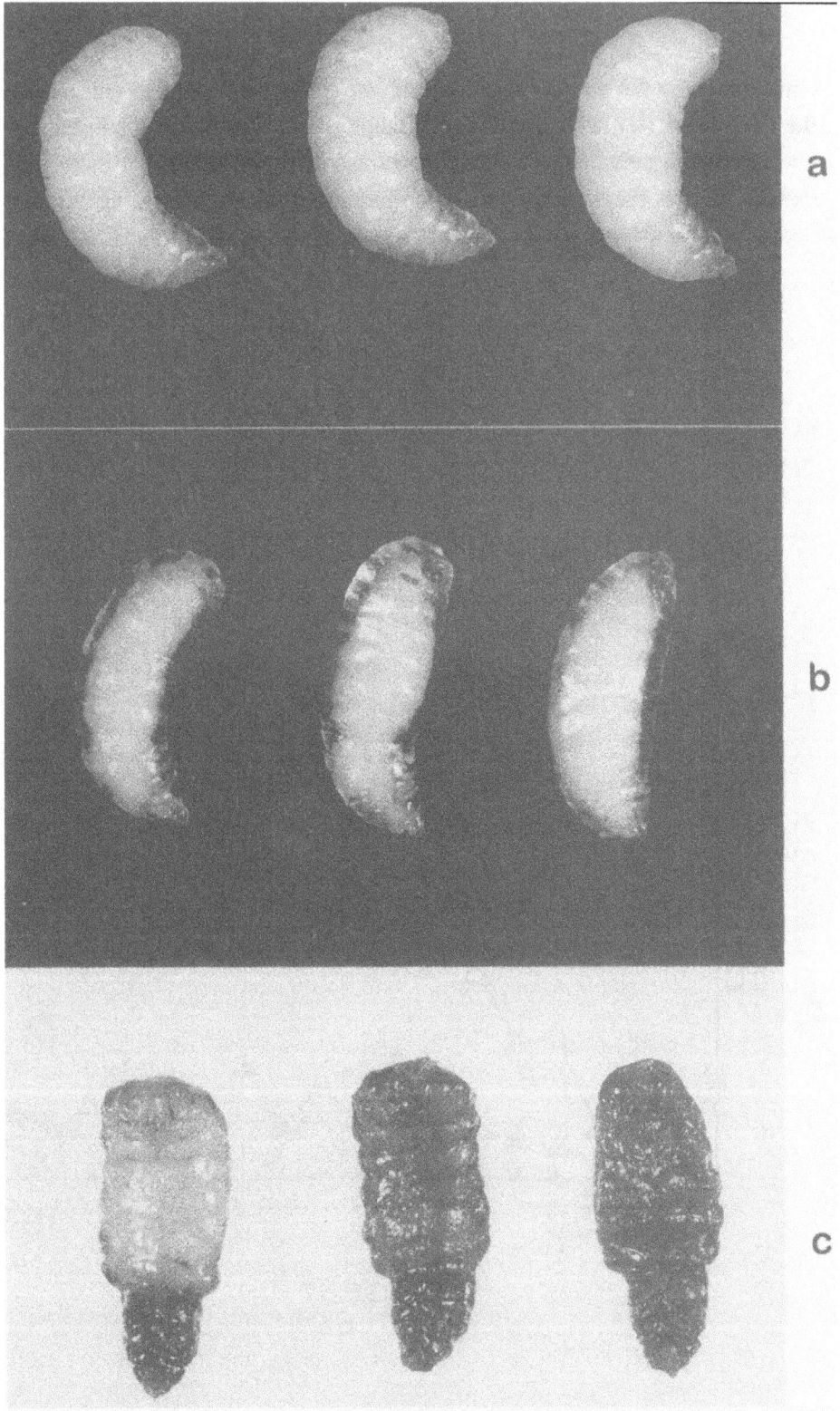

Figure 2.6. Sacbrood: (a) healthy individuals; (b) saclike appearance of diseased larvae showing the typical accumulation of fluid between the body of the larva and its unshed skin; (c) dead larvae start to darken from the head region and eventually dry down to a flattened gondola-shaped scale. (Continued next page Fig. 2.6d)

contact (Figure 2.5). Paralysis virus and *A. woodi* are independent, but their incidences in Britain, along with the number of colonies, were probably even greater early in the century than later (Bailey and Perry 1982). One-quarter—more in poor seasons—of colonies in those early days could well have shown signs of paralysis. This would have been impressive and is in accord with the accounts of bee diseases at the time. The inevitable coincidence of newly discovered but commonly occurring parasites, especially *A. woodi,* with paralysis in many of the diseased colonies almost certainly led to popular misapprehensions about their effects.

Sacbrood

History

G. F. White (1917) in the United States caused sacbrood by inoculating larvae with heat-sensitive filtrates of watery extracts of diseased larvae, thus showing that the agent was a virus. This was long before virus particles could be seen by electron microscopy, and sacbrood virus was one of the earliest insect viruses to be detected. Sacbrood, although suspected to be present elsewhere in the world, was not regarded as a common disease and was long considered to be absent from Britain. There, it used to be regarded as a noninfectious, hereditary fault called "addled brood" because experimenters had failed to spread the disease by placing combs of affected larvae into healthy colonies. Sacbrood virus was not identified until almost half a century after White's experiments (Figure 2.2c) (Bailey et al. 1964).

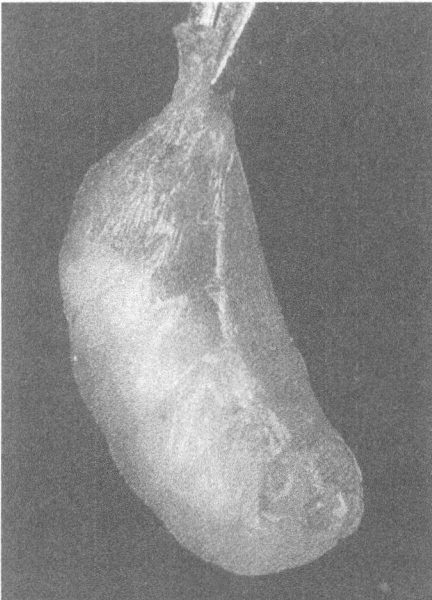

Fig. 2.6d A sacbrood-infected larva lifted out of a cell. (Hansen photo)

Symptoms

Sacbrood virus multiplies extensively in several larval tissues, but larvae continue to appear normal until after they are sealed in their cells. They then turn pale yellow and fail to pupate because they cannot dissolve the thick, tough endocuticle of the final larval skin; they remain stretched on their backs with their heads toward the cell cappings. Ecdysial fluid containing millions of virus particles collects between the body and the unshed, saclike skin of the larvae (Figure 2.6b). The larvae then soon die, desiccate and turn dark brown (Figure 2.6c).

Natural history

Larvae are most susceptible to sacbrood at about two days old, and they become infected by virus in their food. The virus in dead larvae quickly loses infectivity, and continuity of infection, especially over winter, is provided by adult bees in which sacbrood virus multiplies without causing obvious sickness. The youngest worker bees are the most susceptible and almost certainly become infected when they are engaged in their earliest duties of cell cleaning. When young bees remove larvae killed by sacbrood, they ingest a little of the virus-rich ecdysial fluid and readily become infected, which is easily demonstrated experimentally. Within a day, much sacbrood virus begins to accumulate in the hypopharyngeal glands of these susceptible adult bees (Bailey 1969a), and they probably transmit the virus when they become nurse bees and feed larvae with their glandular secretions. The virus may also infect larvae that ingest pollen freshly collected by infected foragers, which add much virus from glandular secretions to their pollen loads. However, infection with the virus accelerates the normal sequential duties of adult bees and substantially shortens their lives; infected bees stop eating pollen, start foraging much earlier than usual, and collect much less pollen than healthy individuals (Bailey and Fernando 1972). Thus several behavioral changes are invoked by infection which, together with the lability of the virus, limit its spread within the colony.

Interestingly, the effects of sacbrood virus on worker bees are exactly the same as those caused by brief anesthesia with CO_2; the lives and behavior of drones, which never eat pollen, are seemingly unaffected by the virus, even though it multiplies in remarkable quantities in the brains of infected individuals.

Sacbrood is most evident in spring and early summer, when the division of labor of bees is least well-developed. Transmission of sacbrood virus from infected adults to larvae decreases when foraging increases later during the major nectar flows, at which time most colonies recover spontaneously. The virus then has only infected adult bees in which to overwinter. Many of these bees die prematurely, further decreasing the amount of infection for the next season.

Incidence

Surveys in England and Wales showed that most colonies carry some infection, and up to 30 percent usually contain a few larvae that are visibly affected (Bailey and Ball 1991). The virus has also been detected in seemingly healthy pupae in South Australia and New South Wales (Dall 1985), so it is probably much more common than has previously been supposed. It almost certainly causes more damage by its inapparent but harmful effects on adult bees than would appear from the incidence of diseased larvae. A strain of sacbrood virus has been isolated from larvae of *Apis cerana* in Thailand (Bailey et al. 1982), and it seems widely distributed in this bee species in Southeast Asia. It is closely related to sacbrood virus from *A. mellifera* but has distinctive properties.

Acute paralysis and Kashmir bee viruses

These two viruses are considered together because they are serologically related and affect adult bees similarly.

Acute paralysis virus commonly occurs in seemingly healthy adult bees, especially during the summer, but until relatively recently, it has never been associated with disease or mortality in nature. In continental Europe and North America, strains of acute paralysis virus have now been shown to kill both adult bees and brood in colonies infested with the parasitic mite *Varroa jacobsoni* (Ball and Allen 1988, Hung et al. 1996). The mite induces the virus to multiply when it feeds on apparently healthy but virus-infected bees, and the virus rapidly becomes systemic and lethal; the mechanism of virus induction is unknown. Adult female mites acquire virus by feeding on an overtly infected host and can then act as virus vectors, transmitting acute paralysis virus to other adult bees or developing pupae within the brood cell. Adult bees in which the virus is actively multiplying can also infect young larvae by secreting large amounts of virus into their food. In severely affected colonies, the adult bee population rapidly diminishes, and symptoms in diseased brood may resemble those of American or European foulbrood. Acute paralysis virus isolated from sick bees from Central America in sufficient quantity to have caused their sickness and death (Bailey et al. 1979) suggests that *V. jacobsoni* was also in the colonies.

The above seems analogous to recent reports in the United States of unusual adult bee and brood mortality in colonies infested with *V. jacobsoni* (Shimanuki et al. 1994) and of acute paralysis virus and strains of Kashmir bee virus in individual dead adult bees from at least some of the affected colonies (Hung et al. 1996).

Kashmir bee virus was first detected in *Apis cerana* from Kashmir (Bailey and Woods 1977). Several closely related but serologically distinguishable strains of the virus were later identified in *A. mellifera* from different regions of Australia (Bailey et al. 1979), where the virus appeared to be responsible for several local disease outbreaks, causing the mortality of both adult bees and brood. At first the virus seemed likely to be confined to Southeast Asia and Australia and to be unrelated to other bee viruses. However, strains of Kashmir bee virus recently characterized from Canada and Spain more closely resemble acute paralysis virus with respect to their coat protein profiles and their serological reactions, than those strains of Kashmir bee virus already known (Allen and Ball 1995).

Like acute paralysis virus, strains of Kashmir bee virus persist as inapparent infections of adult bees and pupae (Dall 1985, Anderson and Gibbs 1988); thus the virus may also be activated to multiply to lethal levels by *V. jacobsoni*. All strains of Kashmir bee virus are highly virulent; few particles are required for infection, and the virus multiplies rapidly when introduced into the hemolymph of adult bees or pupae, causing death within three days. Strains of Kashmir bee virus are widely distributed in the United States (Bruce et al. 1995), and there is some evidence of mortality caused by this virus in colonies infested with *V. jacobsoni* (Hung et al. 1996). This mite acts as a vector of several unrelated honey bee viruses (Ball 1989, Ball and Allen 1988), and in areas where strains of Kashmir bee virus are endemic, the combination of the mite and the virus could well have severe effects on beekeeping.

The occurrence of strains of Kashmir bee virus in *A. mellifera* on the continents of North America, Europe and Australasia area now renders obscure its origins in this bee species. The virus may be more widely distributed than previously thought but, similar to acute paralysis virus, it may have remained undetected in some areas of the world until the advent of *V. jacobsoni* and the consequent increased interest in honey bee viruses.

Deformed wing and Egypt bee viruses

Deformed wing virus was first isolated from diseased adult *A. mellifera* from colonies infested with the parasitic mite *V. jacobsoni* in Japan (Figure 2.2c). Newly emerged bees from affected colonies had deformed or poorly developed wings (Figure 2.7), and extracts of these individuals all contained particles of a similar, unstable virus that had a distant serological relationship to Egypt bee virus. The occurrence of wing malformation in newly emerged bees has been reported

Figure 2.7. Newly emerged bees killed by deformed wing virus.

from several other countries and in each instance, the damage has been attributed to the feeding activities of mites.

The appearance of symptoms of wing deformity probably depends upon the stage at which individual bees become infected. The virus multiplies slowly, and pupae infected at the white-eyed stage of development survive to emergence but are malformed and soon die. Brood may die earlier in development, and bees infected as adults appear normal until death. The virus can cause mortality in colonies not infested with *V. jacobsoni*, but in laboratory and field studies the mite transmits the virus in the same way as acute paralysis virus (Ball 1989).

Deformed wing virus has been detected in *A. mellifera* from most European countries, Saudi Arabia, Iran, Vietnam and Argentina and in *A. mellifera* and *A. cerana* from China. Isolates of this virus differ in their coat proteins, but all are serologically closely related to each other and distantly related to Egypt bee virus. Egypt bee virus has been detected only in adult bees from Egypt, and nothing is known of its natural history.

Slow paralysis virus

This virus (Figure 2.2c) was first isolated during laboratory experiments on bee virus X. It causes mortality of bees after about 12 days when injected into the hemolyph, and infected individuals typically suffer a paralysis of the anterior two pairs of legs for a day or two before death.

Slow paralysis virus had not been found as a cause of mortality in nature until recently, when large amounts of virus were detected in both diseased brood and dead adult bees from colonies infested with *V. jacobsoni* in Britain (Ball, unpublished observations). The virus presumably persists as an inapparent infection in the same way as acute paralysis virus and may be activated to multiply by the feeding activities of mites. Nothing more is known of its natural history or distribution.

Black queen cell virus, Filamentous virus and Y viruses

These three viruses (Figure 2.2c, d, f and g, see Figure, pg. 15) are considered together because they are intimately associated with *N. apis*; otherwise they are entirely disparate (Bailey et al. 1981, 1983b).

Black queen cell virus was first identified as the cause of death of queen larvae or prepupae after they had been sealed in their cells, which developed dark brown to black cell walls. Diseased larvae have a pale yellow appearance and tough saclike skin, resembling those killed by sacbrood virus, but unlike this virus, black queen cell virus does not multiply when ingested by young worker larvae or by young adult bees. Like sacbrood virus, however, it multiplies abundantly and fatally when injected into pupae.

Filamentous virus, first identified in the United States (Clark 1978), multiplies in the fat body and ovarian tissues of adult bees, causing the hemolymph of severely infected individuals to become milky-white with its large particles. There are no other known symptoms.

Bee virus Y multiplies when ingested by adult bees but not when injected into them, so it may be restricted to the gut. There are no known symptoms of infection.

All three of these viruses multiply greatly in adult bees only when the bees are also infected with *N. apis*. Laboratory tests and field observations on some 200 normal, undisturbed colonies for four years in Britain showed that black queen cell or Y virus shortened the lives of individual bees and that the mortalities of colonies in winter were significantly greater when they were infected with either virus and *N. apis* than with *N. apis* alone. The same trend occurred with filamentous virus but to a less significant degree (Bailey et al. 1983b).

These associations of the viruses with *N. apis* are reflected in the coincidences of their annual cycles of incidence in undisturbed, naturally infected colonies (Figure 2.8 a and b).

All three viruses have been found in North America, Australia and Europe; filamentous virus has also been found in Russia and Japan.

Bee virus X

Bee virus X (Figure 2.2d) is distantly related serologically to bee virus Y (Bailey et al. 1980b) and is almost indistinguishable from it both physically and chemically. It also multiplies only in adult bees when ingested by them but not when injected into their hemolymph. It may, therefore, be restricted to the gut, like the Y virus. Yet in contrast to this virus, it has no relationship with *N. apis*, is most prevalent in bees in midwinter, and is more virulent.

In Britain, bee virus X has been found associated with the protozoan *Malpighamoeba mellificae* in dead bees in late winter (Bailey et al. 1983b), but it is not dependent on this parasite in the way that other viruses depend on *N. apis* because it multiplies equally well in bees in the presence or absence of *M. mellificae*. The virus may spread especially when fecal contamination ("dysentery") is unusually severe, the same as for *M. mellificae*. Bee virus X shortens the lives of bees at least as much as does *M. mellificae* and, during its winter peak, accelerates the death of bees already infected with the protozoan. This is particularly harmful because young bees are not produced in winter to replace those that die. *Malpighamoeba mellificae* is invariably blamed when found in colonies that die in late winter, but the prime cause is usually the virus.

Bee virus X has been found in bees from the continents of Europe and

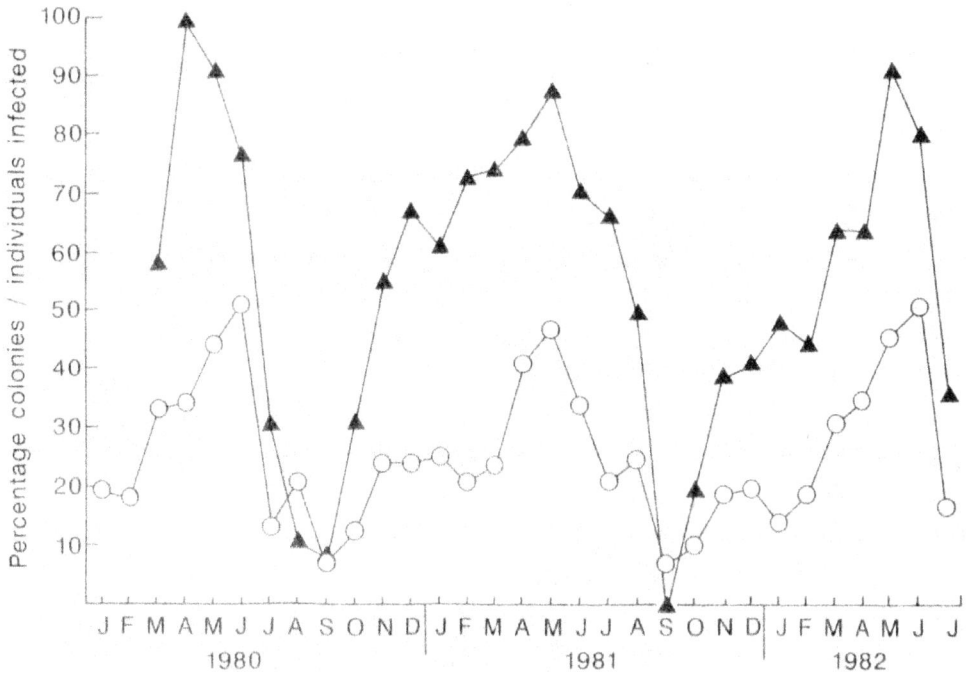

Figure 2.8. Mean percentage of 25 undisturbed bee colonies infected with (a) filamentous virus (▲) and of individuals infected with *Nosema apis* (◯) (b) black queen cell virus (△) and bee virus Y (●).

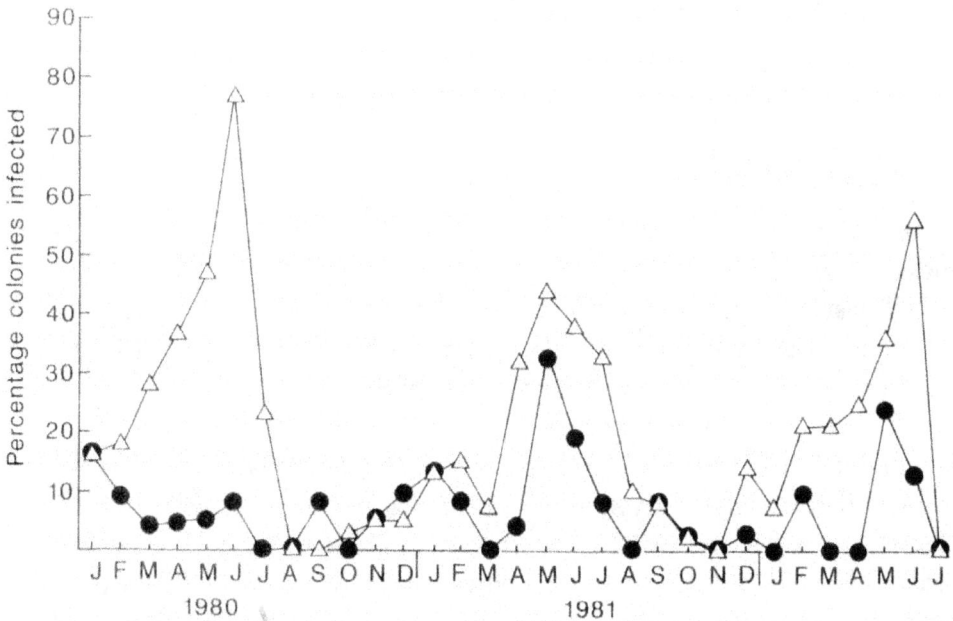

Australasia area and from Argentina, Canada and Iran.

Cloudy wing virus

This is a common virus of bees that sometimes causes their wings to become opaque, although this is not a reliable symptom of infection. It can be diagnosed with certainty only by serology; sufficient virus multiplies in the head and thorax of infected individuals to be detected readily by immunodiffusion. The lives of infected individuals are shortened, and colonies in which most individuals are infected soon become inactive and die.

The virus particles are small (Figure 2.2a), and infection is probably air-borne over short distances because it spreads between cages of bees that are incubated together in the laboratory. Crystalline arrays of particles occur in the region of the tracheoles between the muscle fibrils of the thorax where the main inhalatory spiracles are located (Bailey et al. 1980a). Laboratory infection experiments gave equivocal results, but cloudy wing virus appears not to multiply when fed to adult bees or injected into their hemolymph. However, the virus has recently been detected in dead, newly emerged bees from colonies infested with V. jacobsoni, which indicates that it may be infective by injection into pupae.

There is no seasonal cycle of incidence, which suggests that irregular events determine the multiplication and spread of the virus within bee colonies, similar to those controlling the spread of chronic paralysis virus and other contagiously transmitted pathogens. About 15 percent of colonies in Britain have been found to be infected by cloudy wing virus.

The virus has been detected in honey bees from the continents of Europe, North America and Australasia area and in bees from Egypt.

Apis iridescent virus

This virus (Figure 2.2e) is associated with "clustering disease" of *Apis cerana* in India. The signs alleged to be the most striking and consistent are the unusual activity, especially in summer, of colonies that frequently form small detached clusters of flightless individuals from which many bees are lost, crawling on the ground. Large colonies are said to perish within two months of becoming visibly affected, although the symptoms may abate when bee foraging increases. These signs are reminiscent of those that used to be known as the "Isle of Wight disease" in Britain (see section on paralysis above), and they were first attributed to tracheal mites, apparently *Acarapis woodi*, that were found in the affected colonies. However, the mites, which had already been known in *A. cerana* for some 20 years (Milne 1957), did not occur or were rare in many samples of sick bees, every individual of which was infected with *Apis* iridescent virus (Bailey and Ball 1978). This was exactly analogous to the finding of chronic paralysis virus in sick individuals only of colonies severely affected by paralysis in Britain, whereas about the same percentage of both sick and apparently healthy bees from the same colonies were infested with *A. woodi* (Bailey 1969b).

Apis iridescent virus is the only example of an iridovirus from the Hymenoptera. These viruses are so called because the crystalline masses they form when purified, and the tissues where they multiply, appear blue-violet or green when illuminated with bright white light. These infected cells can easily be seen by microscopic examination, even from bees preserved in formalin or alcohol. Infected cells appear bright blue in contrast to the surrounding creamy-white tissues.

The virus multiplies in many different tissues: fat-body, alimentary tract, hypopharyngeal glands and ovaries. It has been found only in Kashmir and Northern India and may be limited in nature to *A. cerana*, although it can multiply in *A. mellifera*. Curiously, it does not multiply when injected into larvae of the greater wax moth, *Galleria mellonella*, whereas many of the iridoviruses from other insects multiply readily in this host.

Arkansas bee virus

This virus was originally detected in Arkansas by injecting apparently healthy local adult bees with extracts of pollen loads trapped from foragers of the same colonies. Individuals injected with the virus die after about 14 days but otherwise show no sign of disease (Bailey and Woods 1974). The virus has been detected, together with chronic paralysis virus, in dead bees from declining colonies in California, but it has not been found outside the United States.

Lommel et al. (1985) found a viruslike particle of the same size and shape as that of Arkansas bee virus. It was shown to be in the original isolate of Arkansas bee virus and in some Californian bees but was unrelated to it or any other known bee virus. They named it "Berkeley bee virus," but whether it multiplies independently of Arkansas bee virus and what its affects are on bees, remains unknown.

Outlook

Most known bee viruses shorten the lives of individual bees. All of them are harmful, and most are very common. Severe outbreaks of overt disease are relatively infrequent, but the constant attrition by such a variety of background viral infections is probably great. The presence or absence of viruses linked with well-known parasites (*Nosema apis, Varroa jacobsoni*) could well account for the considerable variations reported of the apparent effects of these parasites. Some honey bee viruses may still be localized, but their ability to persist as inapparent infections, together with the continued and increasingly rapid international trade in bees, seems to make their ultimate worldwide distribution inevitable.

As yet there are no known direct methods of controlling bee viruses, and few viruses are easily diagnosed, which may persuade many beekeepers and their advisers to continue to disregard them. However, past ignorance of these viruses has led to mistaken or seriously inadequate diagnoses of diseases, which in turn have caused confusion and much wasted effort. Moreover, knowledge of the natural history of a virus, as of any pathogen, helps to identify the circumstances that lead to its multiplication and spread within the colony and hence enables countermeasures to be devised. The traditional priority for controlling any bee disease has always been to find some direct treatment, preferably chemotherapy. Even if such a treatment becomes available, knowledge of the natural history of the viruses would be central to devising the most efficient methods of application.

Chemotherapeutic treatments now available against *N. apis* and *V. jacobsoni* should, if used correctly, indirectly suppress the viruses associated with these parasites. Because selecting virus-free or resistant bees is unlikely, the best hope for controlling the independent viruses is appropriate management of bees. This requires knowledge of the natural history of the viruses. Viruses transmitted by unusually close contact between individual bees—notably chronic paralysis virus, but others may be similar—are likely to spread more than usual in circumstances that suppress normal foraging activity. In nature, these would be the failure of adequate nectar flows, or beekeepers overpopulating an area with bees, although other beekeeping practices such as transportation of colonies, queen rearing, or periods of feeding much sugar may also assist the spread of contagiously transmitted viruses. Sacbrood may be largely controlled in nature by division of labor among adult bees, which is best established during good nectar flows. Interestingly, the virus appears largely self-limiting in the way its effects accentuate this division of labor; young infected bees cease to feed and tend larvae and start foraging much earlier in life than usual, which reduces the risk of transmission of virus to other larvae during feeding. Other viruses, such as bee virus X, probably spread by fecal contamination of combs, which is aggravated by prolonged confinement of bees in their hives.

Though little is known on the subject, factors that resist the multiplication and spread of bee viruses seem to depend greatly on the normal behavior of bees in favorable circumstances. Therefore, the best general management will minimize any beekeeping practice that hinders or diverts the natural development and field behavior of free-living colonies, a principle that also favors the natural resistance of bees to their well-known pathogens (Bailey and Ball 1991). This management is entirely compatible with the traditional needs of beekeepers for maximum yields of honey, but it is an ideal with which some aspects of modern beekeeping, particularly migratory and pollination work and international trade in live bees, are increasingly at odds.

CHAPTER THREE

———⊶⊷⊷⊷⊷⊷—⊷———

Bacteria
Hachiro Shimanuki

CHAPTER THREE
Bacteria

Before the discovery of *Varroa jacobsoni* and *Acarapis woodi*, the two most important diseases of honey bees were European foulbrood and American foulbrood, both caused by bacteria; they have been studied more than any other bee disease. Septicemia and powdery scale, two other bacterial diseases of bees, have been given less consideration because they cause little if any economic loss to beekeepers. Until 1906, the two foulbrood diseases were not differentiated and were merely called "foulbrood." Then Phillips, in the introduction to White's publication (1906) on bacteria in the apiary, used the terms "European foulbrood" and "American foulbrood." He made it clear that the designations did not refer to the geographical distribution of the diseases but to the areas where they were first investigated scientifically.

American foulbrood disease

American foulbrood disease is one of the most dreaded bee diseases in the world. It is highly contagious, and if unchecked, can kill a colony or spread to other colonies within an apiary or to another apiary nearby. Although the spores of *Bacillus larvae*, the causative agent, remain viable indefinitely on beekeeping equipment, the disease is usually diagnosed by examining for diseased brood. Because the spores can remain dormant for years and still cause an outbreak, the advent of movable-frame hives probably introduced a major method of transmission. American foulbrood disease is one of the most economically significant diseases of honey bees, causing annual losses in the United States of $5 million or more. Concern about American foulbrood disease in the United States resulted in the establishment of today's apiary inspection programs.

History

Despite the early interest in foulbrood disease, White (1907) was the first to demonstrate conclusively that *B. larvae* was the cause of American foulbrood disease. White fulfilled Koch's postulates by rearing *B. larvae* in pure culture, reinfecting the larvae, reproducing the symptoms, and reisolating the causative organism.

Distribution

American foulbrood disease has been reported by beekeepers on every continent. But it does not appear to be present in every country, although statistics on its distribution are sometimes difficult to obtain. In the United States, from

38 reporting states in 1992, less than one percent of all colonies inspected had American foulbrood disease. This is a reduction of the 1.8 percent reported in 1984 and approximates the magnitude of American foulbrood incidence in the United States over the last 25 years.

Etiology

The disease is spread from colony to colony by robbing and drifting bees. In addition, beekeepers can spread American foulbrood by inadvertently feeding honey or pollen from diseased colonies or interchanging brood combs between diseased and healthy colonies. Pankiw and Corner (1966) demonstrated that American foulbrood disease can also be transmitted by package bees. Subsequently, Bitner et al. (1972) demonstrated that viable spores of *B. larvae* fed to queens could ultimately be isolated from workers. This finding implies that both queens and workers are potential carriers of the causal agent. In a later study, however, Wilson and Alzubaidy (1975) reported that queens from American foulbrood – diseased colonies contained only occasional *B. larvae* spores, whereas workers from the

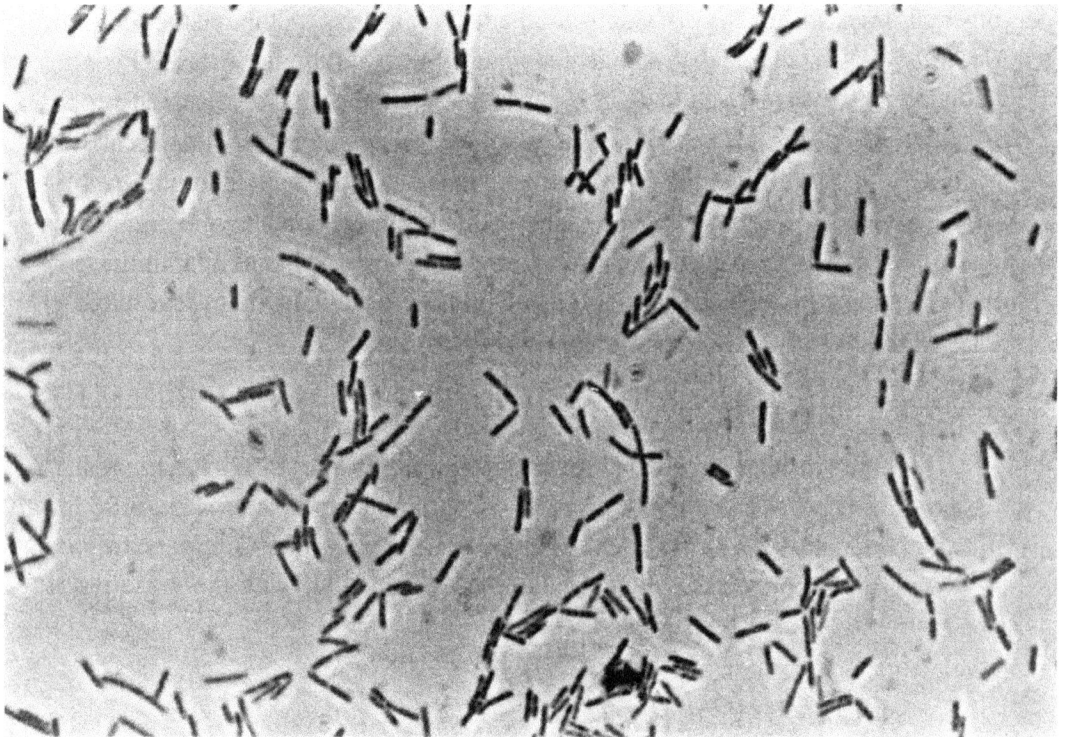

Figure 3.1. Rods, the vegetative stage of *Bacillus larvae*, from a honey bee larva that died of American foulbrood disease. Magnification about 440. (Shimanuki photo).

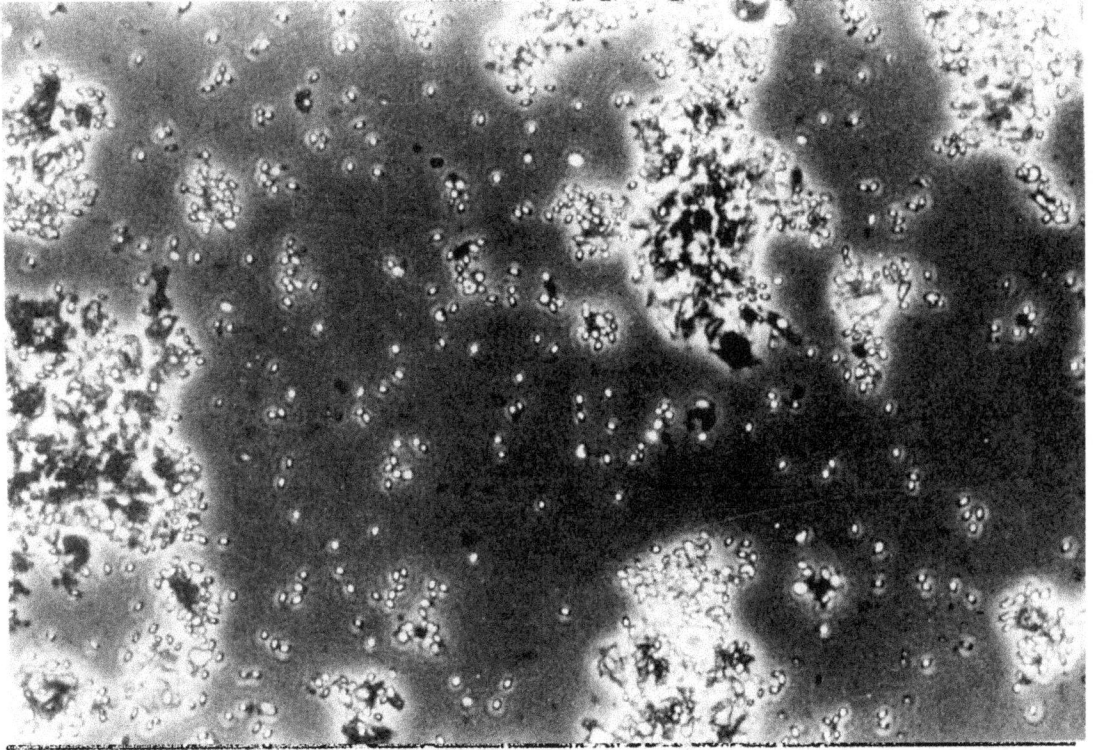

Figure 3.2. Spores of *Bacillus larvae,* from a honey bee larva that died of American foul-brood disease. Magnification about 500. (Wilson photo).

same colonies had larger numbers of spores in their guts. The authors reported that even though the queens from diseased colonies were spore carriers, the queens were unable to incite the disease in healthy colonies.

White (1920a) demonstrated that both *Bacillus brandenburgiensis* and *B. burrii* were synonyms of *B. larvae*. Tarr (1937) found that normally, only the spores of *B. larvae* are capable of producing the disease. Subsequently, *B. larvae* was found to be a microaerophilic, gram-positive, spore-forming rod that does not grow on routine laboratory media. Several culture media, however, can be employed for its cultivation (Foster et al. 1950, Bailey and Lee 1962, Shimanuki et al. 1965). Sporulation of *B. larvae* in vitro has always been considered a problem, but Dingman and Stahly (1983) obtained good sporulation (5×10^8 spores/ml) using their new broth medium. Spores of *B. larvae* survive for years in the dried remains (scales) of larvae that have died from American foulbrood disease. Haseman (1961) showed that the spores were viable after 35 years in the scale stage. Using scales from a Haseman sample 69 years after they were collected, Shimanuki and Knox (1994) were still able to recover a few surviving *B. larvae* spores.

Woodrow (1941a) found that larvae became less susceptible to American foulbrood disease with age; thus he was unable to infect larvae later than 53 hours

Figure 3.3. Widely scattered and partially open cells are the easily recognized symptoms of American foulbrood, caused by *Bacillus larvae*, found on honey bee brood combs.

after egg hatch. Later, Woodrow (1942) concluded that one spore is sufficient to infect a larva one day after egg hatch. Using Woodrow's data, Bucher (1958) determined the LD_{50}[1] of *B. larvae* to be 35 spores in one-day-old honey bee larvae.

Spores of *B. larvae* germinate approximately one day after ingestion by the larva; after germination, the bacteria multiply in the midgut and penetrate to the body cavity through the gut wall (Bamrick 1964). It is not known what mechanism(s) enable *B. larvae* to penetrate the gut wall. Bamrick (1964) believed that the larva ultimately dies from a septicemic condition.

Symptoms and identification

Larvae affected with American foulbrood, unlike those with European foulbrood disease, usually die in an upright position. Consequently, the diseased larvae may be capped and not visible. In heavily infected colonies, a scattered brood pattern and punctured and sunken cappings on brood cells are typical.

---

[1]LD_{50} is an abbreviation for "lethal dose, 50 percent," the level at which 50 percent of the organisms are killed by a toxic material.

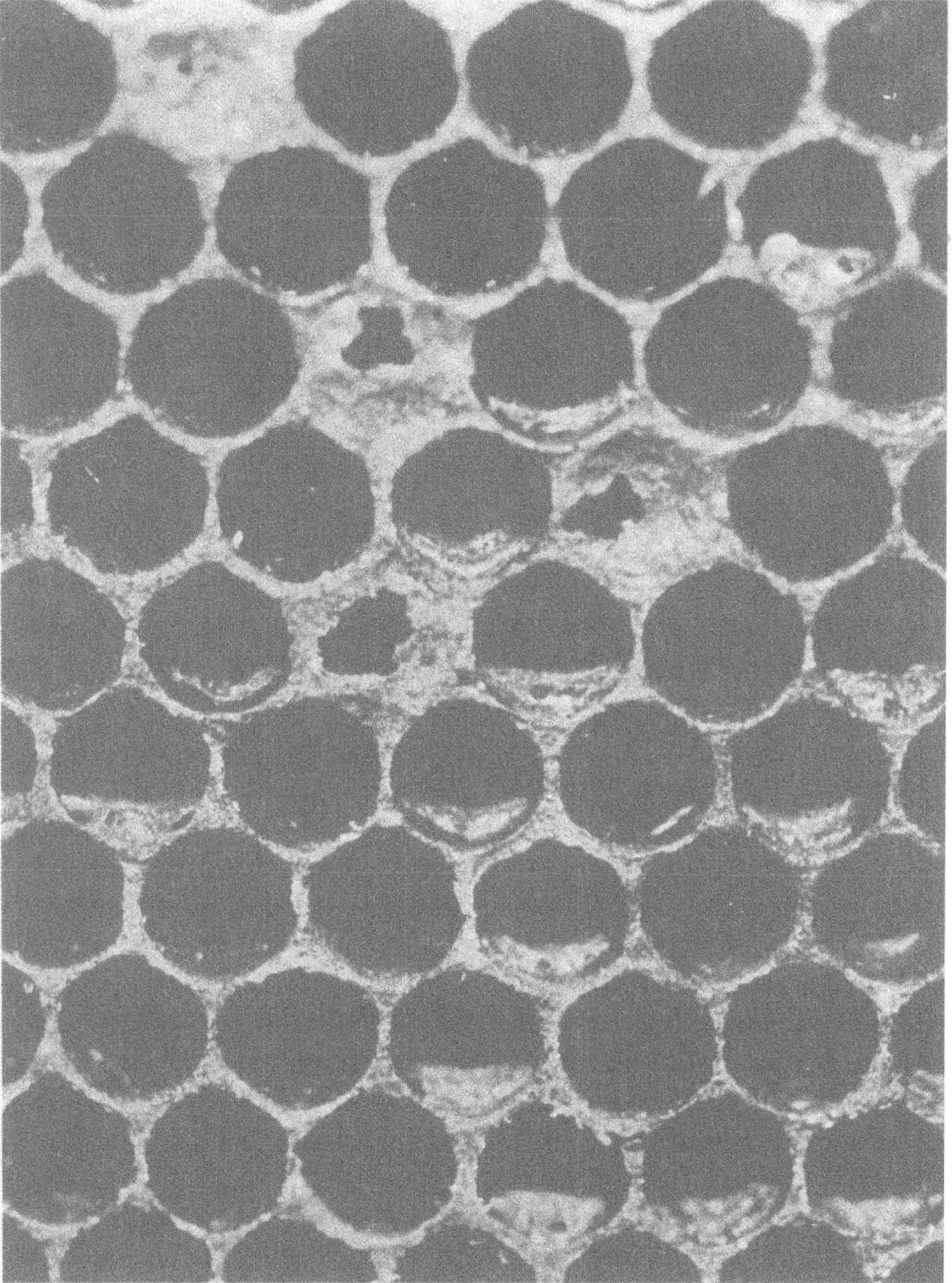

Figure 3.3a. Close up of the diseased brood cells, showing sunken and punctured cappings. Also visible are the American foulbrood scales, formed from diseased larva that died and dried out. Scales are flat, and extend the entire length of the cell. They adhere tightly to the cell wall and are difficult to remove, a characteristic of the disease. Scales are visible when the comb is held such that sunlight shines to the bottom of the cell. Scales are dark black. (Shimanuki photo).

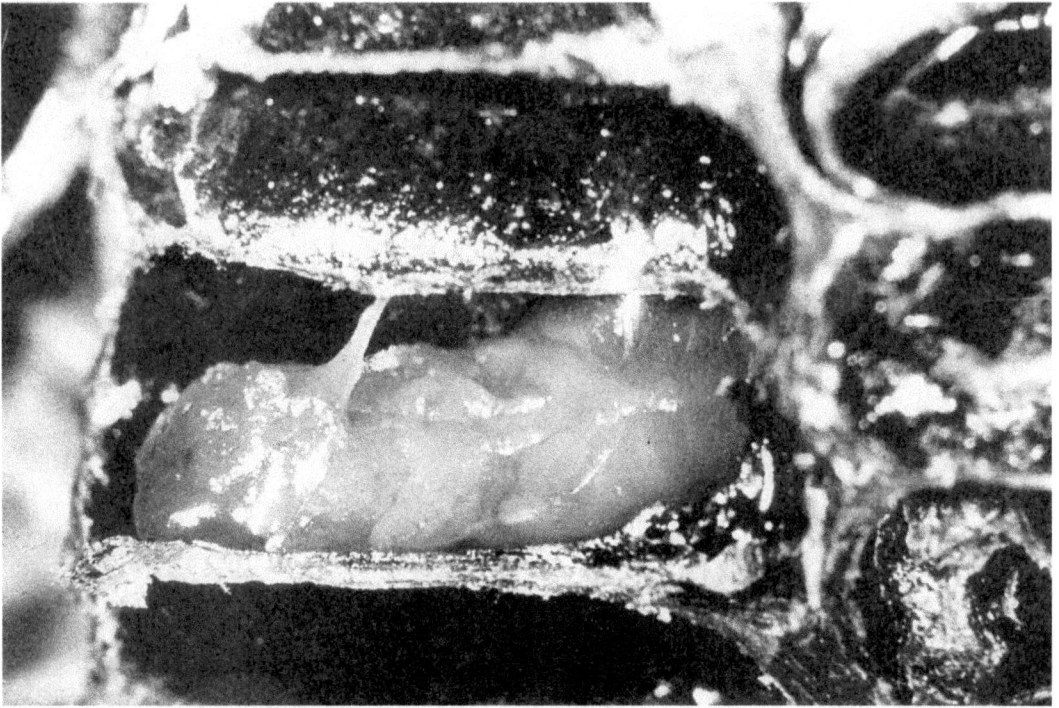

Figure 3.4. Cross section of a cell with a pupa within that has died of American foulbrood disease. Note pupa is full length in cell, and died with its tongue extended, indicating approximate age at death. (Smith photo).

Larvae that have died of American foulbrood disease exhibit a "ropy" condition that can be demonstrated by inserting a matchstick or similar implement into the dead mass and drawing out the material into a threadlike projection longer than 2.5 centimeters. The color of the larva also changes from its healthy pearly-white to dark brown. Finally, after a month or more, the dead larva dries out and typically becomes a brittle scale that adheres tightly to the cell wall. The most characteristic symptom of the disease is the pupal tongue protruding upward from the scale to the center of the cell.

In most cases, American foulbrood disease can be diagnosed in the field by experienced beekeepers or by personnel specially trained for disease inspection. When the field diagnosis is questionable, several tests are available to make a more accurate diagnosis.

The Holst milk test (Holst 1946) is a simple test based on the proteolytic enzymes produced by *B. larvae*. The test is conducted by suspending a suspect scale, or a smear of a diseased larva, in a tube with 3 to 4 ml of one percent powdered skim milk in water. The tube is then incubated at 37°C (99°F). If *B. larvae* spores are present, the suspensions clear in 10 to 20 minutes. No clearing is observed with diseased material from either European foulbrood or sacbrood disease.

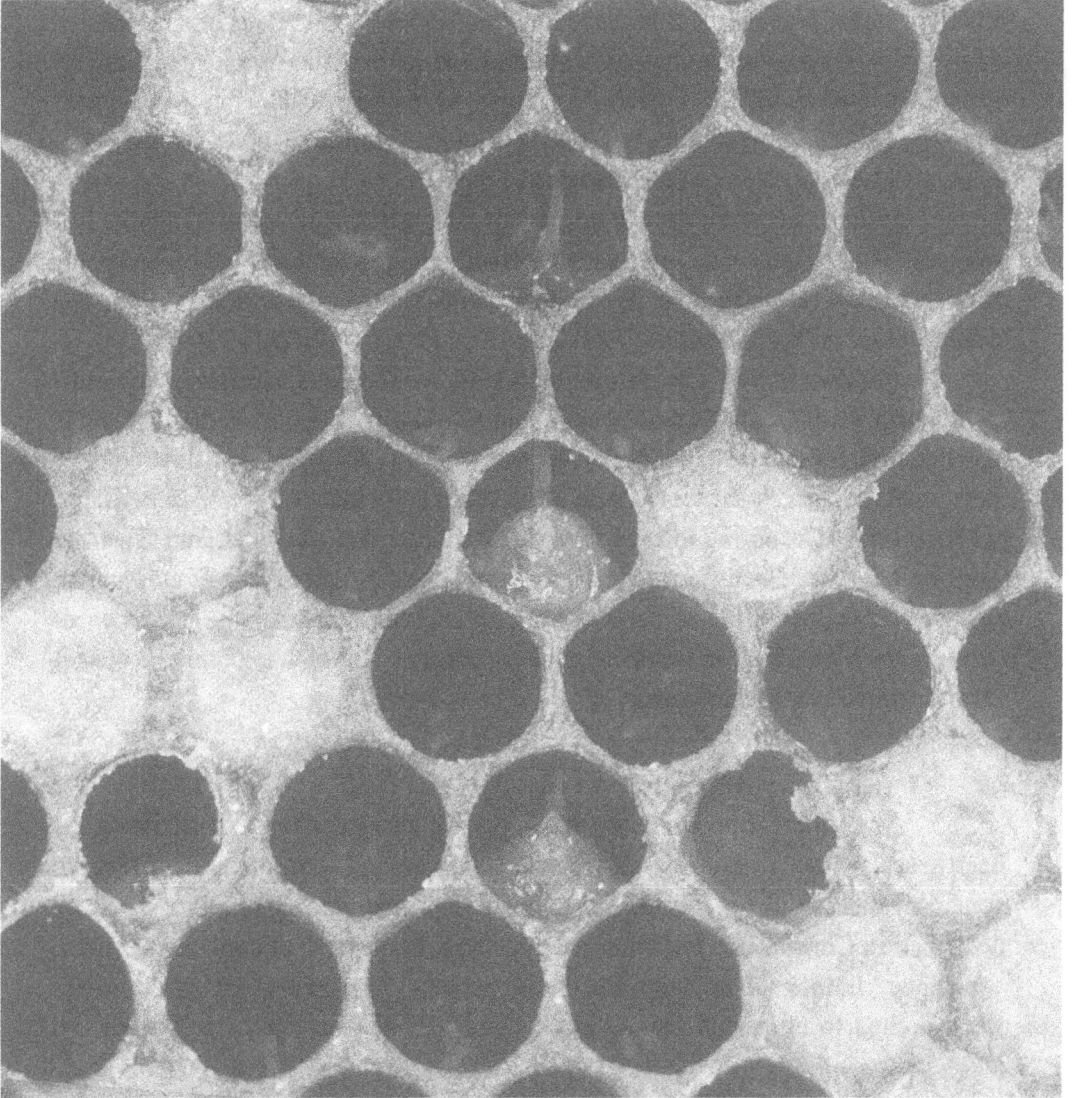

Figure 3.4a. Looking into comb cells, showing three dead larvae (center) with tongues extended. Typical characteristic for identification. When a bee pupa dies of this disease, the proboscis often remains extended, while the rest of the pupa melts (top cell), then dries to form a scale. In addition, one or more legs may protrude and point upward. Larvae that have died of European foulbrood do not exhibit extended tongues or legs. (Hansen photo).

      The modified hanging drop method (Michael 1957) is another test that can be useful for differentiating American foulbrood from other brood diseases. A bacterial smear is made on a cover glass from either a scale or a diseased larva. After the smear is dried and fixed with a heat lamp or open flame, the smear is stained with carbol fuchsin or a suitable spore stain for 30 seconds; the excess stain is then washed off with water. While the smear is still wet, the cover glass is

inverted, smear side down, and placed onto a droplet of oil on a standard microscope slide. The slide is blotted dry and examined with a microscope using the oil
immersion lens. In the areas where oil droplets form pockets of water, the spores of
B. larvae exhibit Brownian movement. The spores formed by other *Bacillus* spp.
associated with known bee diseases remain fixed.

In some laboratories, a gram stain or a simple stain is used to diagnose bee
diseases. This method relies solely on differentiating the bacteria by their morphology. A further refinement of this method is the fluorescent antibody technique
(Toskov et al. 1970, Zhavnenko 1971, Otte 1973, Peng and Peng 1979).

Bacillus larvae can also be identified biochemically from other spore-
forming bacilli. *B. larvae* usually reduces nitrate to nitrite (Lochhead 1937) and is
also negative for catalase production (Haynes 1972). Positive identification of *B.
larvae* can be made by the biochemical tests and growth media requirements. Lloyd
(1986) proposed a combination of cultural, morphological and biochemical tests to
identify *B. larvae*. His procedure calls for growth on semi-solid blood agar and two
smears, one stained with 10 percent nigrosin and another using the gram procedure. If the identification is still in doubt, a catalase test can be conducted using
growth from a sub-culture on agar.

Control

The general approach to controlling this disease in the United States has
been to seal off and then destroy all infected colonies and equipment to prevent the
spread of disease to neighboring colonies. The hives are generally burned, as
described in Chapter 22, and care must be taken to destroy all adult bees, brood,
wax, honey, frames and inner covers; the insides of all hive bodies and bottomboards
are scorched to eliminate any spores lodged on their surface or in cracks and
crevices.

Selective breeding for disease resistance has long been a goal of researchers. Park (1936) first definitively demonstrated differences in colony resistance to
American foulbrood disease. Various mechanisms of resistance were then studied
by other investigators: The role of the proventricular valve (Sturtevant and Revell
1953); removal of diseased larvae (Woodrow 1941b); protection of larvae by adults
(Thompson and Rothenbuhler 1957); and larval resistance (Rothenbuhler and Thompson 1956). The role of the peritrophic membrane was dismissed as a factor in
the resistance or susceptibility to American foulbrood disease (Davidson 1970).
Yet another means of resistance was proposed by Rose and Briggs (1969): Different
levels of bacterial inhibitors were found in the brood food provided by different
nurse bees. Similarly, Rinderer and Rothenbuhler (1969) reported that drone larvae
were more resistant than queen or worker larvae and suggested that this was caused
by a difference in diet of the different castes.

One of the earliest methods for treating foulbrood disease was the "shaking" method. The possibility of salvaging adult bees from combs containing diseased brood was first proposed by Schirach in 1769. Later McEvoy rediscovered this method, which is sometimes referred to as the McEvoy method (Howard 1907). Simply, the method involves transferring the adult bees to a disease-free hive with-

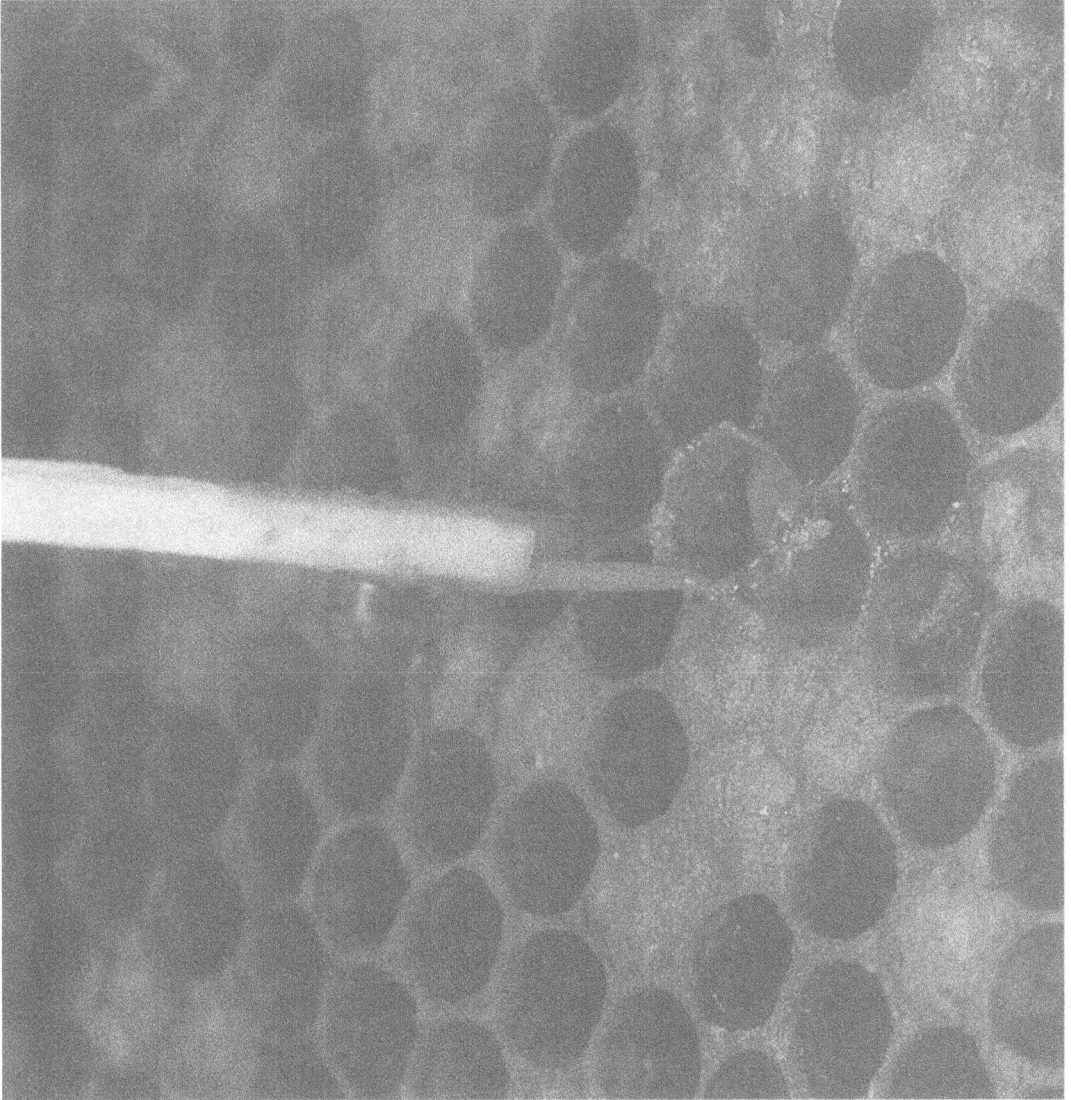

Figure 3.5. Foulbrood 'ropy,' or 'stretch' test. A small, dry stick (wooden match size) is inserted into a diseased larva's cell (at least half-way down), twisted one rotation and removed slowly, straight up. The remains of a larva that has died of American foulbrood disease will be light coffee-colored and will adhere to the match as it is withdrawn. It will be the consistency of glue and will stretch into a fine thread about 2.5 cm (1 inch) long. Remains that have dried into a scale will not stretch. When inserting the stick into the cell, a noticeable 'foul' odor should be detectable. (Hansen photo).

out drawn comb, well away from the infection source; the potentially contaminated honey carried by the bees is consumed in comb building.

Haseman and Childers (1944) began a new era when they successfully cured colonies of American foulbrood disease and protected them from the disease by feeding them sodium sulfathiazole. Later, Gochnauer (1951) reported that oxytetracycline (Terramycin®) was also effective in the treatment of the disease. In the United States, only oxytetracycline can be used for the prevention and control of American foulbrood disease because sodium sulfathiazole is no longer approved for use on honey bees.

Oxytetracycline can be fed in a number of formulations. The most commonly used carrier is powdered sugar (see Chapter 22). Recently, the use of the antibiotic extender patty was approved by the U.S. Food and Drug Administration. (It is currently available as a premix from a bee supply company.) However, no chemical should be fed to any colony without first consulting the regional apiary inspector or some other local official. There are specific state and federal regulations governing the use of chemotherapeutic agents in the United States and most other countries.

Wilson et al. (1970) proposed antibiotic "extender patties" as a new method of introducing a chemotherapeutic agent to a colony. This method incorporates oxytetracycline in a soft patty made of hydrogenated cooking oil and sugar. Because no water is used, the drug is stable over a long period (Gilliam and Argauer 1975a,b). Drug residues in colonies fed antibiotic extender patties, however, have been found in honey stored in the brood area, although not in the surplus honey (Wilson 1974). The greatest risk of contaminating the honey crop occurs when feeding oxytetracycline in sugar syrup, even though the oxytetracycline is reportedly unstable in aqueous solutions. Argauer and Herbert (1992) found detectable levels of oxytetracycline in the pollen cells, even when oxytetracycline was fed in sugar syrup.

Michael (1964) established the efficacy of ethylene oxide against *B. larvae* and certain other infective organisms and pests of the honey bee. Shimanuki (1967) made the first field trials of bee equipment fumigated with ethylene oxide. Knox et al. (1976) combined ethylene oxide fumigation with the feeding of oxytetracycline. In their studies, they showed that it was feasible to combine the shaking method, ethylene oxide fumigation and drug feeding to save bees and equipment from American foulbrood-diseased hives.

Studier (1958) investigated sterilization techniques for the control of American foulbrood disease, using cobalt-60 to irradiate combs with diseased larvae. Shimanuki et al. (1984) demonstrated that high-velocity electron beams can also be used not only to decontaminate bee equipment from American foulbrood -diseased colonies but also to kill *Nosema apis* spores and destroy all stages of the greater wax moth, *Galleria mellonella*. The authors reported that this procedure reduced neither the biological efficacy nor the palatability of pollen for adult honey bees.

Another experimental approach to the treatment of American foulbrood disease is the use of bacteriophages specific for *B. larvae*, a technique proposed and used by Smirnova (1954). Dingman et al. (1984) described the DNA of one such bacteriophage. These procedures are promising alternatives to burning bees and bee equipment.

Shimanuki et al. (1992) found that antibacterial material from chalkbrood mummies inhibited the growth of *Melissococcus pluton* and *B. larvae*. This material was subsequently identified as the fatty acid, linoleic acid (Feldlaufer et al. 1993a). In their next paper, Feldlaufer (1993b) showed that related fatty acids also had antibacterial activity.

Comments

The appearance of large numbers of hobby beekeepers in recent years has made it imperative that recognition of bee diseases be taught as a first lesson. Beekeepers must learn to distinguish American foulbrood disease from other, less important diseases.

American foulbrood disease is a real threat, and beekeepers may be tempted to feed oxytetracycline indiscriminately. But if oxytetracycline is used incorrectly, honey can become contaminated. Furthermore, the development of resistance of *B. larvae* to oxytetracycline is always a possibility. For these reasons, other new controls, including new chemicals, must be investigated.

European foulbrood disease

European foulbrood disease is not considered a serious disease by most beekeepers. The disease is still poorly understood, although it was described before American foulbrood disease. It is seasonal in nature and is more serious in some instances than in others; for example, colonies used for pollination are known to be more susceptible to the disease.

Schirach, as early as 1771, described a foulbrood disease that he believed was caused by "improper food which was consumed by the larvae" (Phillips and White 1912). The causative agent is now known to be *Melissococcus pluton*, a bacterium that has had the names *Bacillus pluton* and *Streptococcus pluton* in the past.

History

Cheshire and Cheyne (1885) published the first comprehensive work on European foulbrood disease. They isolated and described *Bacillus alvei*, which they believed to be the etiological agent of the disease. White (1912, 1920b)

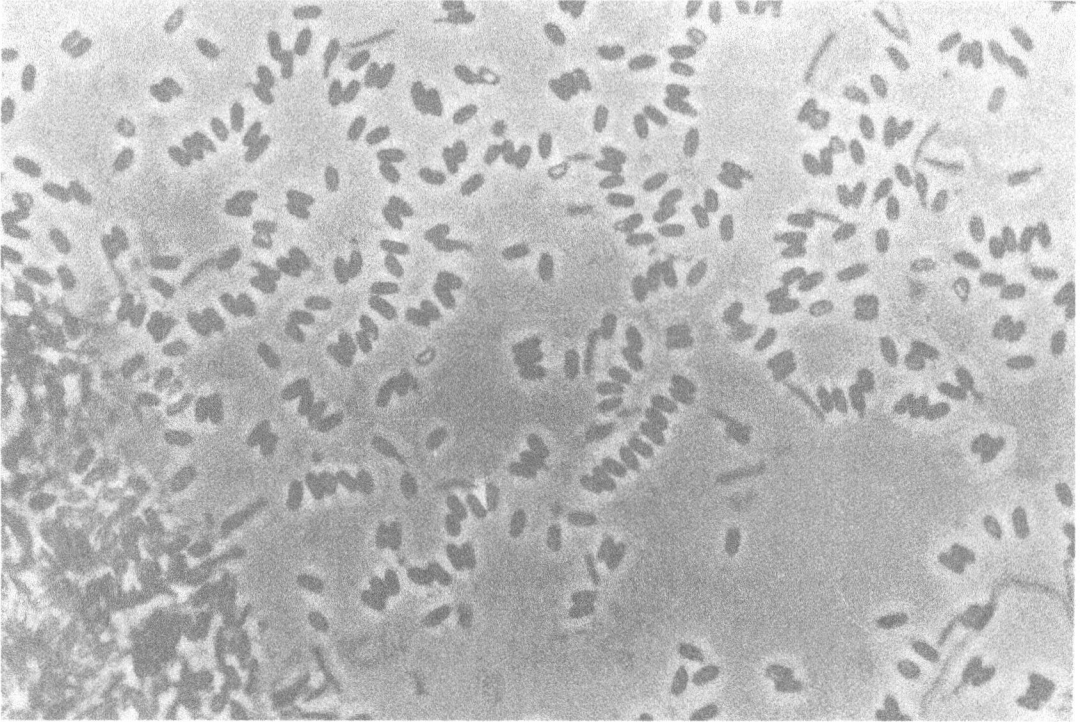

Figure 3.6. *Bacillus alvei* spores. Magnification 1,000. (Shimanuki photo).

refuted their claim and named a new bacterium, *Bacillus pluton*, as the causative agent. However, the distinctions between *B. pluton* and *B. alvei* have been questioned over the years. Lochhead (1928) believed that the two names were synonyms and represented different stages in the life cycle. Burnside (1934b) suggested that *Streptococcus apis* and *B. pluton* were identical, whereas *B. alvei* and *Achromobacter (Bacterium) eurydice* were synonyms. Burri (1941) concluded that *A. eurydice* dissociated from the rod-shaped forms to the streptococcuslike cells of *B. pluton*.

The question was resolved by Aleksandrova (1949), who proved that *B. pluton* was the causal agent of European foulbrood. She successfully grew *B. pluton* and reinfected larvae with the organism. Bailey (1956) suggested that *Bacillus pluton* be renamed *Streptococcus pluton* on the basis of the gram reaction and morphology. Subsequently, Bailey and Collins (1982a) observed that *S. pluton* differs sufficiently from the type species, *S. pyogenes*, on the basis of cultural and biochemical tests, to merit the creation of a new genus, *Melissococcus* (Bailey and Collins 1982b). Thus the new proposed scientific name for the causative agent of European foulbrood disease is *Melissococcus (=Streptococcus) pluton*.

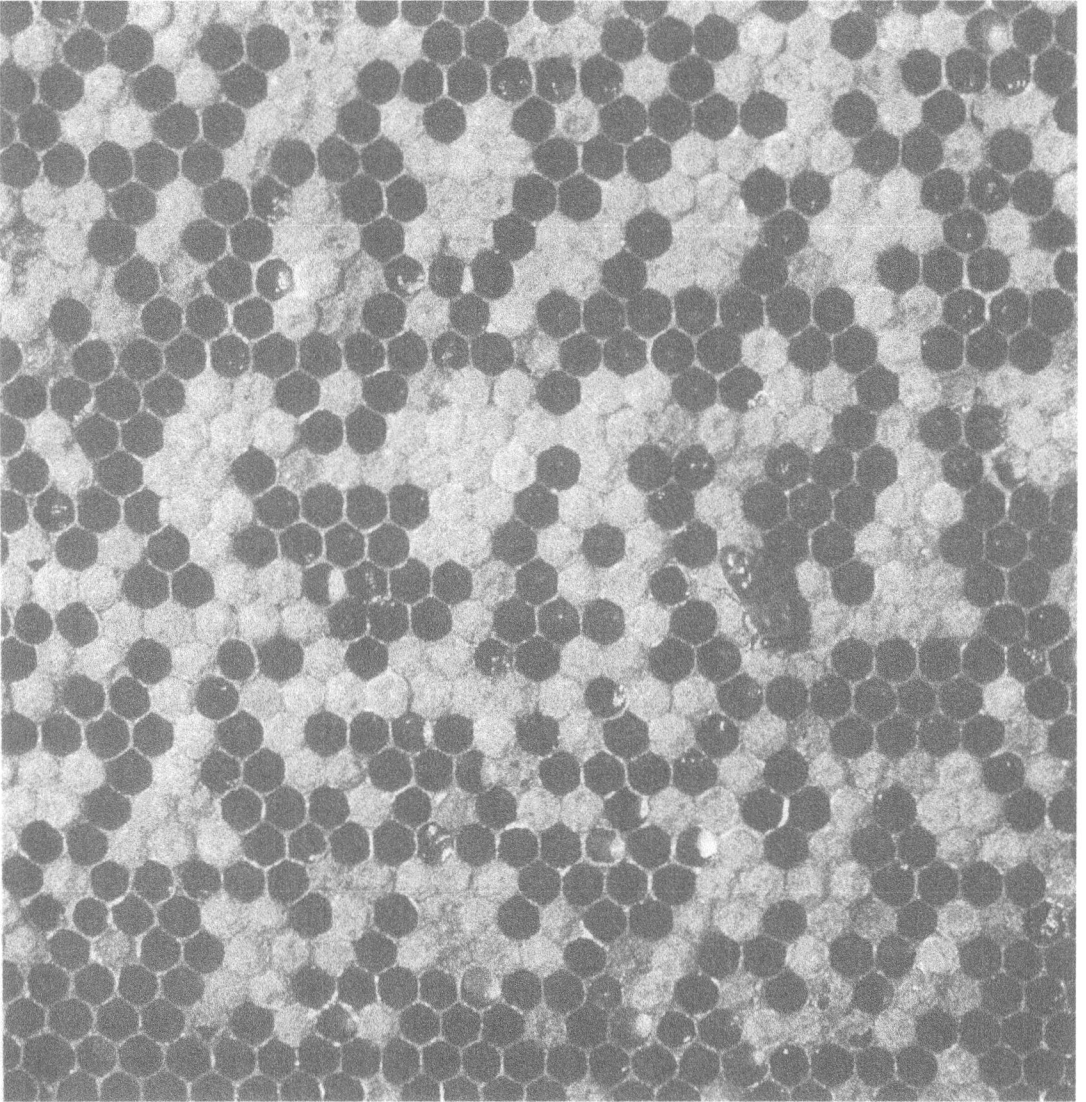

Figure 3.7. The pepperbox appearance of a comb heavily infested with European foulbrood. Often called shotgun appearing. There are many uncapped cells, and many normal, healthy capped cells in no apparent order. (Hansen photo).

Distribution

European foulbrood disease is found on every continent where honey bees are kept. Because European foulbrood is not considered a serious disease, however, no precise records are available concerning its occurrence. In the United States, the disease is enzootic in some states.

Etiology

European foulbrood disease affects only young larvae (those less than 48 hours old) and strikes primarily in mid- to late spring (May and June, in Northern temperate areas), the time when colonies are building up to maximum population. Because diseased colonies fail to increase normally, they may not provide the beekeeper with surplus honey. European foulbrood is also found in some colonies in the fall, but it is less widespread than in the spring.

Larvae infected with European foulbrood disease have a varied microflora including *Melissococcus pluton, Bacillus alvei, Achromobacter (Bacterium) eurydice* and *Bacillus laterosporus*. However, *Achromobacter (Bacterium) eurydice*, one of the organisms associated with European foulbrood disease, is sometimes found in healthy-appearing larvae (Bailey 1960), and organisms such as *B. alvei* and *B. laterosporus* are most likely saprophytes and play no part in the infectious cycle. Bailey (1963b) tested the pathogenicity of these organisms: *B. alvei* spores showed no pathogenicity; *A. eurydice* showed pathogenicity only when mixed with *M. pluton*; and although one isolate of *Enterococcus faecalis* did show some pathogenicity, in general, the number of larvae ejected after the bees were fed *M. pluton* was always higher than the number ejected after the bees were fed *E. faecalis*. He concluded that *M. pluton* was the cause of European foulbrood disease. Serological tests by Bailey and Gibbs (1962) indicated that *E. faecalis* and *M. pluton* were related but not identical and that the pleomorphic nature of *M. pluton* was especially evident when culture media were stored for a few weeks. Under such conditions, Bailey and Gibbs observed almost rodlike forms instead of the typical lanceolate cocci shapes associated with *M. pluton*.

The infectious cycle of European foulbrood disease begins when a larva consumes brood food contaminated with *M. pluton*. Once the larva is infected, *M. pluton* establishes and multiplies in the midgut. According to Tarr (1938), the pathogen is localized in two to three-day-old larvae between the peritrophic membrane and the food in the midgut. By the time the larva is five days old, the area in the midgut that should be occupied by the food mass is occupied by the bacteria. *Melissococcus pluton* first destroys the peritrophic membrane and then, as the disease progresses, invades the intestinal epithelium (Tarr 1938). The larva therefore competes for food with the rapidly multiplying bacteria, creating an abnormal demand for larval food. The nurse bees eject larvae that require more than the usual amount of food (Bailey 1960). If populations of nurse bees are sufficient, or if there is time to eliminate infected larvae, most colonies can keep the infection contained. The onset of the disease usually coincides with the first nectar flow.

European foulbrood disease appears to be readily transmitted within a colony by nurse bees that inadvertently infect the larvae while feeding them. The bacteria may overwinter on the sides of the cell walls or in feces and wax debris on the bottom of the hive.

Symptoms and identification

Larvae infected with European foulbrood usually die while still in the coiled stage. The larvae turn yellow and then brown, at which time the tracheal system becomes visible. Sometimes, diseased larvae apparently die in the upright stage, but at other times they appear to collapse and seem to be twisted or melted

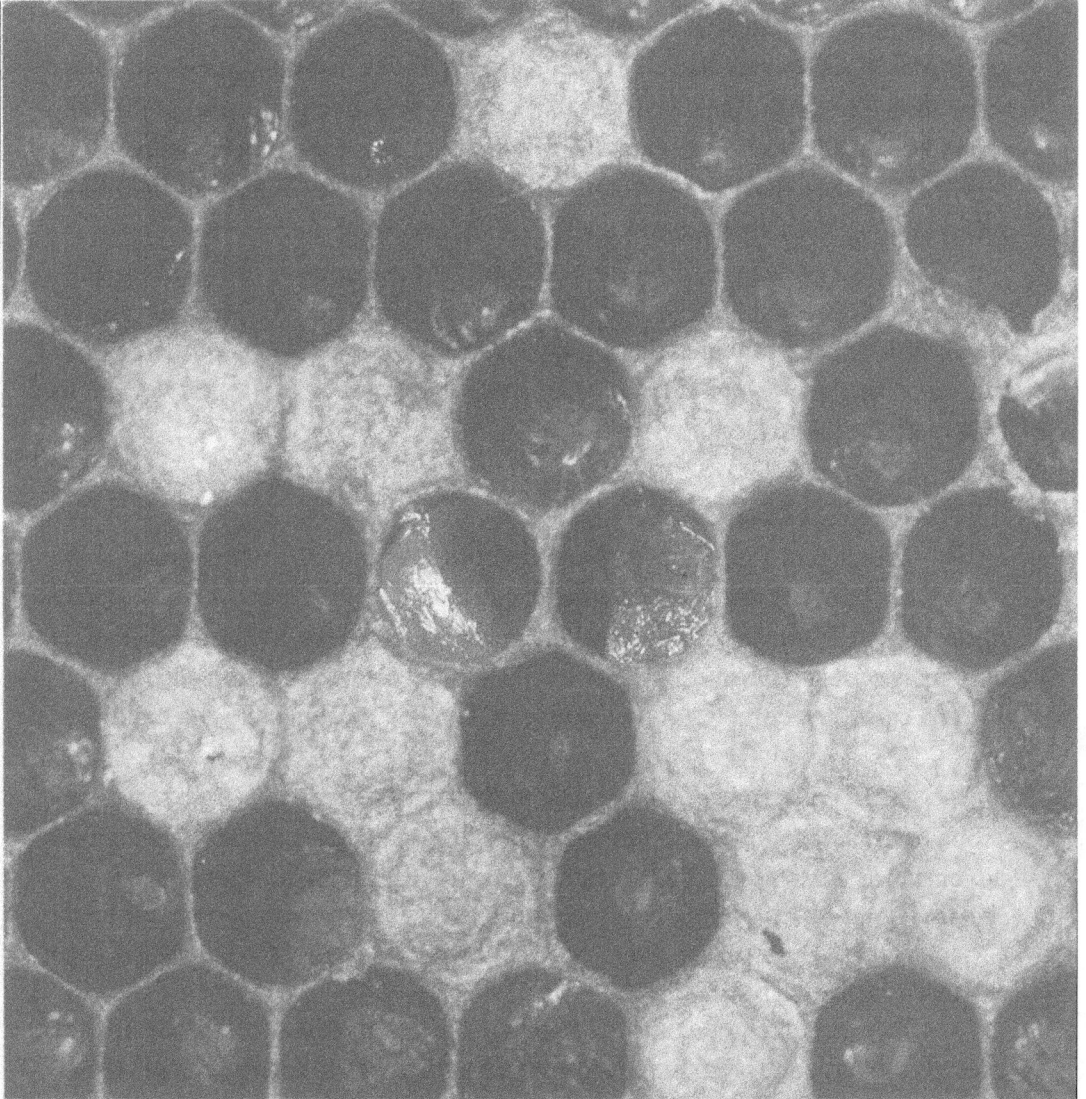

Figure 3.8. Cappings over diseased brood are concave and sometimes punctured. Healthy brood have convex, unpunctured cappings. Larvae usually die in the coiled stage, turn yellow, then brown. A ropiness test can be performed when the remains appear melted in the bottom of the cell. The rope will be granular and difficult to remove. Scales of dead larvae, unlike those of AFB, are rubbery, and easy to remove. (Hansen photo).

in the bottom of the cell. At this stage, the larval remains can be tested for ropiness, which should be slight, with threads less than 1.5 cm long when the material is drawn out with a matchstick; the remains appear granular when drawn out.

If the disease is widespread in a colony, the combs take on a so-called pepperbox appearance, the term used to describe combs with many uncapped cells mixed with normal, capped cells. Another useful symptom is the appearance of the cappings: Cell cappings of healthy brood are convex, whereas those over diseased brood are concave and sometimes punctured. The odor of larvae infected with European foulbrood disease varies with the presence of saprophytes. Typically it can be described as sour, hence the German term for the disease, *Sauerbrut*. After some time, the larval remains dry out and form a scalelike protrusion in the cell. Characteristically, scales of European foulbrood disease are rubbery rather than brittle as are those caused by American foulbrood disease. European foulbrood scales are much easier to remove than are American foulbrood scales.

A more precise and sensitive method for the diagnosis of European foul-brood disease was reported by Pinnock and Featherstone (1984). They developed an enzyme-linked immunosorbent assay (ELISA) for detecting *M. pluton*. Using this technique, they were able to demonstrate the presence of *M. pluton* even in apparently healthy colonies.

Control

Treatment for European foulbrood disease is usually not required if the infection is light because most colonies can overcome a light case. A good, steady nectar flow will sometimes eliminate the disease. European foulbrood disease is extremely serious, however, in colonies whose populations are drastically reduced. In such cases, the colonies do not gather sufficient stores to overwinter and may die.

Different strains of bees vary in their susceptibility to European foul-brood of bees, but no strain is immune to the disease. Originally, requeening was advocated as a treatment. The success of this procedure can be attributed partly to the introduction of a potentially more prolific queen, but, more important, the break in the brood cycle provides nurse bees with the opportunity to remove affected larvae and polish the cells.

Several chemotherapeutic agents have been used successfully for the prevention and control of European foulbrood disease (Katznelson et al. 1952). The most widely used antibiotic today is oxytetracycline (Terramycin®), discussed above in connection with American foulbrood disease.

More recently, ethylene oxide fumigation has been shown to be of value for the prevention and control of European foulbrood disease. Shimanuki et al. (1969) found that the level of the disease was reduced and the populations were much higher in colonies that were established on equipment fumigated with ethylene oxide than in colonies on unfumigated equipment or colonies fed oxytetracycline. Nevertheless, they found some European foulbrood disease in colonies where the equipment had been fumigated.

The ultimate responsibility for the prevention and control of European foulbrood disease rests with the beekeeper, who must exercise judgment when feeding pollen from unknown sources, restrict drifting between colonies, and inspect colonies closely before exchanging brood combs.

Comments

European foulbrood disease occurs quite frequently, and more often than not is unobserved by the beekeeper. However, in most areas, the level of infection is usually quite low, and the colonies can cope with the disease without assistance.

European foulbrood may have more than one causal agent, and several bacteria could produce symptoms that mimic the disease. Organisms such as *Enterococcus faecalis* and *Achromobacter eurydice* could be opportunistic pathogens. Further research is necessary to establish the pathology of European foulbrood. Specifically, histological sections should be prepared to study the normal and diseased larvae to determine the tissues that are affected and also the role played by *Bacillus alvei* and *A. eurydice* in the series of events that lead to European foulbrood.

In certain sections of the United States, European foulbrood is a major problem each year. That it appears to be aggravated when colonies are used for pollination suggests that nutrition may play a major predisposing role in the disease cycle.

Septicemia

Septicemia refers to any of a wide variety of diseases caused by the presence in the blood (or, in insects, the hemolymph) of pathogenic microorganisms or their toxic products. In honey bees, the term is generally used to designate a particular bacterial disease of adult bees that was first described by Burnside (1928). He identified the pathogen as *Bacillus apisepticus* and fulfilled Koch's postulates to prove that it was the cause of the disease. Septicemia is not considered serious, and data on its distribution are lacking, although the disease is probably distributed worldwide.

Etiology

Landerkin and Katznelson (1959) reclassified *B. apisepticus* as *Pseudomo-*

nas apiseptica. They concluded that *P. apiseptica* was similar to *Pseudonomas aueruginosa*; the major difference was that *P. aueruginosa* did not produce septicemia in bees. *Pseudomonas apiseptica* is a gram-negative, non-spore-forming bacterium that Colwell (1970) attempted to fit into the genus *Vibrio.* She concluded that the classification of *P. apiseptica* as a separate entity is questionable and in need of further study.

Burnside (1928) believed that moisture aided in the transmission of septicemia and found in cage tests that the disease was not transmitted from diseased to healthy bees. Wille and Pinter (1961) reported that the disease occurs when the stress on a colony increases; stresses include massive artificial feeding, comb building in a swarm, or the mobilization of the colony population to counteract adverse weather.

Infections of septicemia, once established, kill the bees rapidly; the highest rate of kill occurs 20 to 36 hours after inoculation. Although the hemolymph of infected bees changes in color from colorless to chalky white (Burnside 1928), probably the most striking symptom is the rapid degeneration of muscles. For example, it is impossible to pick up dead affected bees by the appendages because the wings, legs, head, thorax and abdomen fall off when they are handled. Affected bees in colonies appear restless, are not observed to feed, and appear unable to fly.

It is believed that the bacteria invade via the spiracles. How the bacteria winter in the colonies is not clear, but *P. apiseptica* probably survive in adult bees. The disease erupts when the predisposing factors are present. Wille (1962) reported that septicemia disease can be found mixed with nosema disease, mite infestation or other unknown conditions and that not all bees with the gross symptoms of septicemia yield *P. apiseptica.*

Control
No strain or caste of honey bees is known to be resistant to septicemia. Wille and Pinter (1961), however, observed differences in the blood cells and suggested that not all bees are equally susceptible.

Streptomycin has been used successfully to control septicemia in Switzerland, but the development of strains of *P. apiseptica* that are resistant to streptomycin limits the usefulness of this antibiotic (Wille 1962). Feeding bees citric acid has also been advocated for septicemia, but Heinz-Gerhard (1968) states that it merely acidifies the sugar syrup to bring the pH closer to that of nectar and does not affect *P. apiseptica.*

Comments
The name septicemia should be changed because it describes a symptom that could be caused by several organisms. In addition, further research on the

taxonomy of *P. apiseptica* is indicated. An investigation of the presence or absence of toxins could prove useful in establishing the cause of mortality and also aid in placing *P. apiseptica* in the proper taxon.

Powdery scale disease

Powdery scale disease is rarely reported, perhaps because the average beekeeper is unable to identify it. The disease was first described by Katznelson (1950), who identified the bacterium associated with the disease as *Bacillus pulvifaciens*. The organism is gram-positive and produces spores; when it is first isolated it produces a reddish-brown pigment that can be lost by subculturing. Characteristically, maximum growth occurs at 45°C (113°F), which caused Katznelson (1950) to conclude that *B. pulvifaciens* was a new species and not *B. laterosporus* or *B. larvae*. Powdery scale disease affects only the larvae of honey bees. The scale that results from the dead larvae is powdery rather than brittle, as in American foulbrood disease, or rubbery, as in European foulbrood disease. It is also light brown to yellow and extends from the base to the top of the cell. When handled, the scales crumble into a dust, hence the name.

Control

Kirkor and Czyz (1958) and Katznelson (1950) conducted tests of the infectivity of powdery scale disease. Katznelson had success with pure cultures of *B. pulvifaciens*. Kirkor and Czyz sprinkled the powder on open and sealed brood and found that within 48 hours, the nurse bees were removing larvae from the cells and that other cells contained yellowing larvae. Later, punctured cappings were seen over affected larvae. Cappings of diseased larvae were punctured as in American and European foulbrood disease. The symptoms of the dead larvae appear somewhat similar to those of larvae that have died from stonebrood (see Chapter 5) (Kirkor and Czyz 1958).

Comments

The etiology of powdery scale disease is still in question. Katznelson (1950) was able to infect larvae only when he used an unheated suspension of spores from powdery scale disease. No disease was produced when he fed larvae either a heat-shocked suspension of material from larvae that had died of powdery scale disease or an unheated suspension from pure cultures of *B. pulvifaciens*. It is likely that *B. pulvifaciens* is a bacterium that may be commonly associated with honey bees and that under certain stress conditions, this organism could become pathogenic.

Spiroplasmas

Spiroplasmas are prokaryotes in the class Mollicutes. A spiroplasma species lethal to honey bees was discovered by Clark (1977b, 1978b) in Maryland, U.S.A.; this pathogen was found in 36 percent of the foraging bees in one of the apiaries studied and was also detected in bee samples sent from New York and Hawaii. It has since been found in honey bees from California, Florida, Louisiana, Texas and Wisconsin in the United States, as well as in France and Peru (Clark and Whitcomb 1984). Clark (1978b) isolated the spiroplasma from nectar collected from the blossoms of tulip poplar (*Liriodendron tulipifera*) and Southern magnolia (*Magnolia grandiflora*) trees.

Rickettsia disease (filamentous virus)

Rickettsiae are small, gram-negative bacteria that are intracellular parasites. Krieg (1963), in a general discussion of rickettsiae of insects, described several means of identifying rickettsial forms, including microscopic morphology, the use of immunofluorescent stains and infectivity studies; Machiavello's stain makes rickettsial bodies look red, whereas other bacteria stain blue.

There have been a few reports of rickettsiae in bees: From the former Soviet Union (Poltev and Grobov 1967), Switzerland (Wille 1964b, 1966, 1967) and Italy and Germany (Wille 1964b). Poutiers et al. (1969) were able to grow a culture of *Rickettsia grylli*, isolated from crickets, in cells from honey bee embryonic tissue in hanging drop slides.

Wille (1964b, 1966, 1967) observed rickettsial infections in up to 25 percent of the diseased adult bees examined in Switzerland; the typical site of infection was the fat body tissue. In acute infections, the hemolymph became milky, and high numbers of rickettsial cells measured 0.1×0.3 µm, as determined by electron micrographs. Wille believed that larvae and adult bees could be infected.

Wille's identification of these organisms as rickettsiae, however, has been challenged by Clark (1977a, 1978a), who determined that the rickettsiae were actually a virus that he named F-virus for filamentous virus. Bailey (1981) suggested that they be described as "resembling rickettsiae." Thus, the existence of rickettsial diseases of honey bees remains in question.

CHAPTER FOUR

---⬤⬤⬤---

Protozoa
Ingemar Fries

CHAPTER FOUR

Protozoa

Protozoans are small, unicellular organisms. Each cell possesses organelles whose functions are similar to organs in more complex organisms. The common view of protozoans as one-celled animals may be erroneous because animals, as well as plants, always develop from embryos. In this sense, one-celled animals do not exist (Margulis et al. 1990).

Most protozoans that are pathogenic to insects cause chronic infections that may or may not kill the infected host. Clear clinical symptoms of infection are absent in protozoan infections in honey bees, and thus diagnosis is dependent on microscopic examinations. Infection may shorten the life span, alter behavior, and modify specific body functions. No protozoans infecting insects are known to produce toxins, although Tanada and Kaya (1993) report tumorlike growth and inflammatory response in infected insect tissue. Simple as well as complex life cycles are represented, and both sexual and asexual reproduction occur. Although the protozoans infecting honey bees have been studied for many years, large gaps remain in the knowledge of their genetics, phylogenetics and general life cycles.

Nosema disease

In 1909, the German scientist Enoch Zander classified a parasite infecting the midgut epithelium of honey bees as a microsporidium and named it *Nosema apis*. Zander published his findings simultaneously in several bee journals, using the term "Nosema-seuche" for the conditions caused by the parasite (Zander 1909a, b, c). The first recorded observation, however, of *N. apis* causing a severe disease in adult honey bees was made by Dönhoff and Leuckart (1857), before the parasite had been named and classified. White (1919) studied the effects of *N. apis* on the

---

The classification of the large and heterogeneous group of microorganisms called protozoans ("first animals") is changing as knowledge of their biology and phylogenetics increases. More than a decade ago (Levin et al. 1980), the Committee on Systematics and Evolution of the Society of Protozoologists established seven phyla in the elevated subkingdom Protozoa under the kingdom Protista. Recently, it has been suggested that the protists be reorganized into the broader category protoctists. The new category would include fungi and algae in addition to the small eukaryotes consisting of one or a few cells (Margulis et al. 1990). In the new taxonomy, used in this chapter, the former subkingdom Protozoa is distributed among 36 phyla. The new taxonomy divides protozoa known to be associated with honey bees into four different phyla: (1) Zoomastigina, which move by using flagella (e.g., flagellates); (2) Rhizopoda, motile by means of pseudopods (e.g., amoebas); (3) Apicomplexa, which usually move by body flexion (e.g., gregarines); and (4) Microspora, obligatory intracellular parasites without means of locomotion.

honey bee, calling it "nosema disease," a term that has persisted.

Nosema disease is caused by the microsporidian parasite *N. apis* that infects the epithelial cells of the honey bee ventriculus. Like all members of the phylum Microspora, *N. apis* is an intracellular parasite that multiplies only within

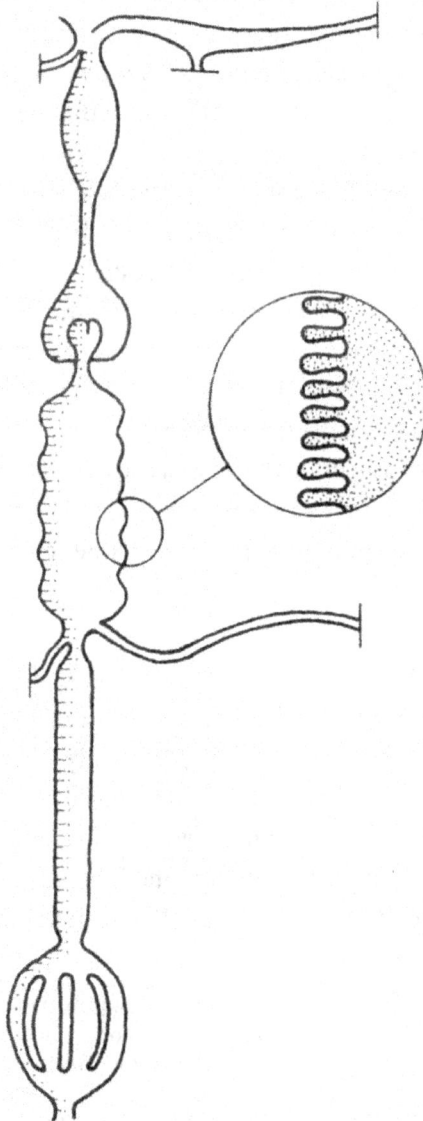

Figure 4.1. A simplified diagram of several internal organs of the honey bee showing the location of the ventriculus and Malpighian tubules. From Dade. (Yatcko drawing).

living cells. Members of the Microsporidia are the most important protozoan pathogen of insects and have a large negative economic impact on the breeding of beneficial insects. The most striking characteristic of Microsporidia is its infection mechanism. After germination of the spores, the parasites enter their host cells through a hollow polar filament coiled in the spores (Figure 4.2). The polar filament penetrates the host cell membrane upon germination of the spore.

Although Zander (1909a) was convinced that *N. apis* caused the most important disorder of honey bees, other influential scientists have claimed that *N. apis* infection is not serious (White 1919) and that natural infections rarely cause major damage to infected colonies (Bailey and Ball 1991). It is clear, however, that infected colonies are less productive (Farrar 1942, Hammer and Karmo 1947, Moeller 1962, L'Arrivée 1965b, Cantwell and Shimanuki 1969, Fries et al. 1984). Infected colonies also show an increased rate of winter loss from the bee cluster (Jeffree and Allen 1956), and winter loss of honey bee colonies and nosema disease are related (Farrar 1942, Fries 1988a), although the direct cause of colony death during winter is often unknown. Supersedure of infected queens increases the economic damage

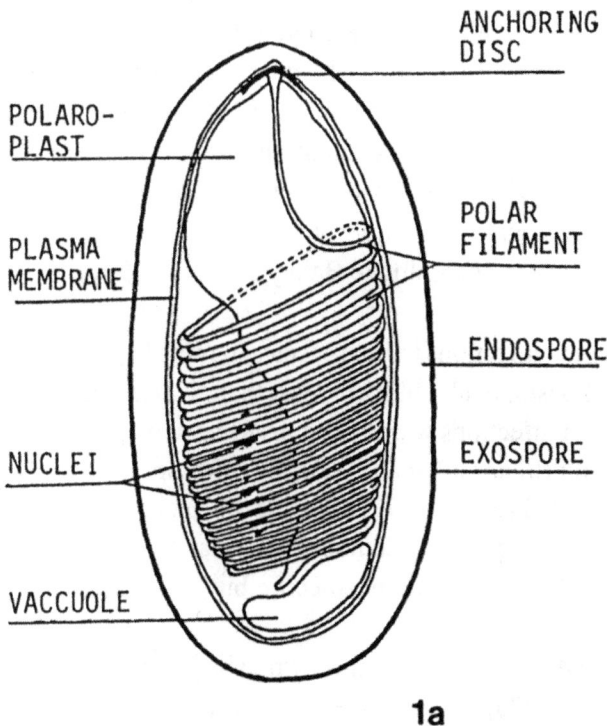

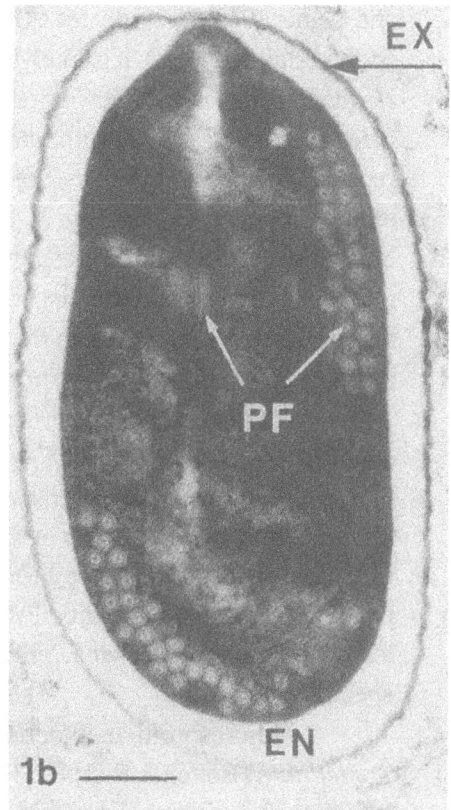

Figure 4.2. General Microsporidia morphology, showing polar filament used to penetrate host membrane. (Fries drawing).

caused by the parasite (Farrar 1947). Thus, nosema disease must be regarded as a threat to profitable beekeeping. Diagnosis of the disease, as well as control measures, should be an integrated practice in beekeeping operations.

Distribution

Nosema disease occurs worldwide wherever bees are kept (Matheson 1993). Although present in a variety of climates, the parasite is believed to have the largest negative impact on beekeeping in temperate climates where the bees are confined inside the hive for longer periods. A survey for nosema disease in Mexico indicates that the disease probably is not a serious problem for beekeepers under Mexican (warm) conditions (Wilson and Nunamaker 1983). It should be pointed out, however, that information on the distribution and impact of *N. apis* in tropical and subtropical climates is scarce.

In temperate climates, the incidence of the disease varies greatly between years as well as between colonies and apiaries (Michailoff 1928, Doull and Eckert 1962, Furgala and Hyser 1969). The seasonal variation in intensity of infection is similar in both the Northern (Michailoff 1928, Bailey 1955b) and Southern (Doull and Cellier 1961) hemispheres. Although emerging bees are always free from nosema spores (Bailey 1955c), it is likely that the parasite is present wherever bees are kept. *N. apis* is believed to infect only the epithelial cells of the ventriculus (Bailey 1972a, Fries 1988b, Graaf 1991), although other tissues as well as hemolymph have been reported to become infected (Steche 1965, Gilliam and Shimanuki 1967). Several authors report infections by *N. apis* in other hosts (Fantham and Porter 1912a, Showers et al. 1967, Dörntlein and Reng 1970), but only Showers and co-workers (1967) have infected other insects with spores of *N. apis* and then isolated new spores that were infectious to the honey bee. According to Bailey and Ball (1991), these results require confirmation.

Observations of microsporidian infections in both species in mixed apiaries of *A. cerana* and *A. mellifera* (Yakobson et al. 1992) suggest that *N. apis* may have multiple hosts. No microsporidian infections were found in surveys of apiaries containing only *A. cerana* in Kashmir (Shah and Shah 1986), but microsporidian infections in *A. cerana* have been reported (Singh 1975, Lian 1980). Recently, in cross infections experiments, Fries and Feng (1995) demonstrated that *N. apis* is also infective for *A. cerana*. Thus, *N. apis* may be tissue-specific but is not host-specific.

Microsporidian infections in *A. florea* have been shown to be infective to *A. mellifera* (Böttcher et al. 1973, 1974, 1975). A newly described microsporidian infection in *A. cerana* (Fries et al. 1996) is also infective for *A. mellifera* and multiplies more readily in *A. mellifera* than *N. apis* does in *A. cerana* (Fries and Feng

1995). Whether or not this new species of microsporidia, infective for *A. mellifera*, is capable of becoming a problem for beekeeping with the European honey bee, if transferred to populations of these bees, needs to be investigated. Reports in the past of damage in *A. cerana* colonies attributed to *N. apis* infections (Lian 1980) may actually be reports on this new species of microsporidia, since differences to *N. apis* are small when investigated under the light microscope (Fries et al. 1996). An unidentified microsporidian infection of honey bee brood has been reported in South Africa (Buys 1972, 1977), along with a report of ventricular cells of honey bees infected with a microsporidian other than *N. apis* (Clark 1980).

Etiology

The honey bee becomes infected by *N. apis* upon ingesting spores from the parasite. After the spores pass into the midgut, they soon germinate under the influence of the gut juices. Several chemical stimuli cause germination of nosema spores in vitro (Laere 1977), but what triggers germination in vivo is not known.

When the spores germinate, the coiled polar filament is discharged, everts like the finger of a glove (Lom and Vávra 1963), and may penetrate an epithelial cell wall. The infective cell, the sporoplasm, enters the host cell through the hollow polar filament and receives a plasma membrane during this process (Weidner et al. 1984). Sporoplasms that do not become injected into host cells are probably soon destroyed by the gut juices (Avery and Anthony 1983). The initial infection is usually found in the posterior part of the ventriculus from which it spreads to the anterior part (Kellner 1980, Fries 1988b), even though most spores germinate shortly after entering the ventriculus (Bailey 1955c). The reason for this location of the initial infection is probably the shorter distance of spores in the gut lumen to the ventricular epithelium in the posterior part of the gut (Fries 1988b). A negative correlation between the distance from spores in lumen to the epithelium and the site of initial infection has been demonstrated (Graaf et al. 1994b).

The sporoplasms that enter a host cell soon mature to mother cells, so-called meronts. Within 24 hours after infection, the parasite is already dividing in the host cytoplasm (Fries et al. 1992), producing merozoite stages (Figure 4.3). After a series of divisions, the merozoites mature to sporonts, distinguished under TEM (transmission electron microscopy) from earlier stages by electron-dense depositions externally to the plasma membrane (Youssef and Hammond 1971, Fries 1989). The sporonts divide once and produce two sporoblasts, which is typical for the genus *Nosema* (Larsson 1986). The sporoblasts then mature into the final stage of development, the spore. Under optimal temperature conditions, around 30°C (Lotmar 1943), the parasite may complete the development within 48 to 60 hours (Kellner 1980). Throughout development, each stage has two nuclei in close arrangement, so-called diplokaryon (Fries 1989). There is no evidence of sexual activ-

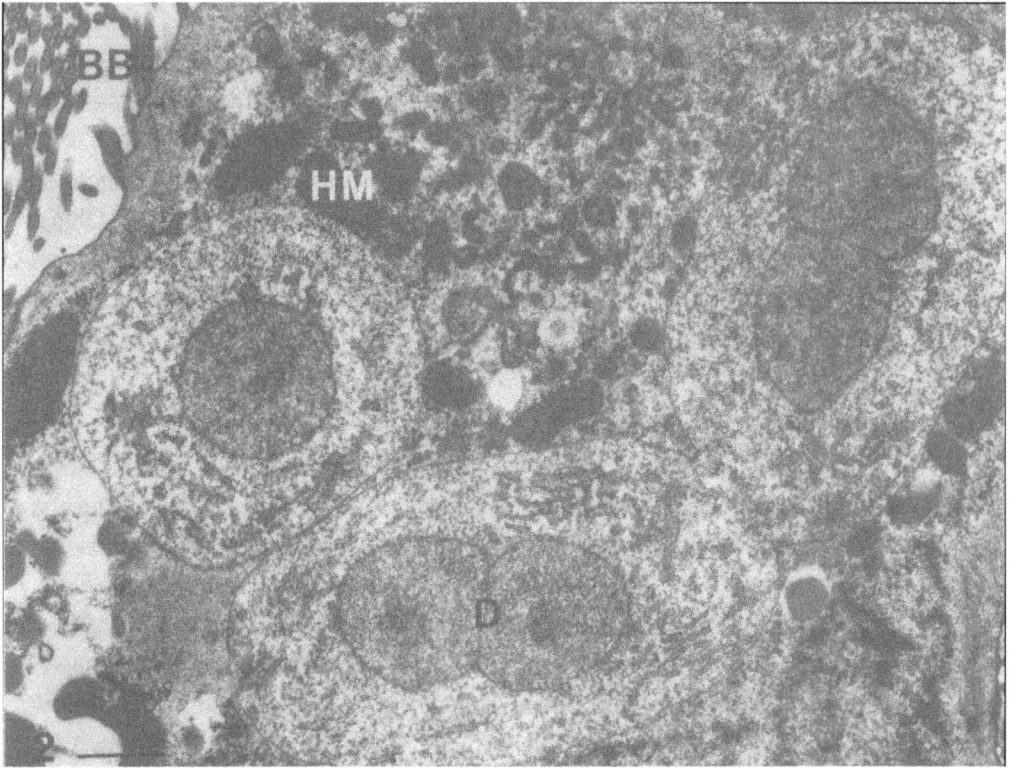

Figure 4.3. Transmission electon microscopy (TEM) of host cytoplasm showing merozoites (p). (Fries photo).

ity such as meiosis or nuclear fusion in *N. apis*, and Fries (1989) suggested that the
life cycle is asexual, which may be true for the entire genus *Nosema* (Loubès 1979).

      Once the parasite becomes established in the epithelium, the entire ven-
triculus becomes infected within two weeks under favorable temperature conditions (Fries 1988c), at which time 30 to 50 million spores can be found in the midgut
(Bailey and Ball 1991); more than 200 million spores form in the intestinal system if
the bees do not deposit their feces (Lotmar 1943). Clearly, new cells may be
reinfected from newly formed spores released in the gut lumen when old infected
epithelial cells are shed (Liu 1990b). The rapid spread of the parasite in the epithelium, however, makes an intercellular route of transmission inevitable, and spores
that germinate within the gut epithelium are probably infective agents (Fries 1989).
Two types of spores appear to be produced, one durable with a thick endospore
primarily responsible for transmitting the disease between hosts, and one with a
thin endospore that germinates upon formation inside the host cytoplasm (Fries et
al. 1992, Graaf et al. 1994a). The latter type, that may be young spores of the durable
type that germinate inside the host cell, probably is important for the horizontal

spread of the parasite in the epithelium. Although in vitro studies on other *Nosema* species have indicated that vegetative stages may invade adjacent cells (Ishihara 1969, Kawarabata and Ishihara 1984), no *in vivo* observations support this hypothesis in *N. apis* or any other member of Nosematidae. A suggested life cycle for *N. apis* is shown in Figure 4.4.

The durable spore type voided with the feces from infected bees transmits the parasite between hosts and may remain viable for longer periods. They resist refrigeration (Revell 1960), freezing, exposure to microwaves (Cantwell et al. 1971) and lyophilization (Bailey 1972b). The spores may remain viable in fecal deposits inside the beehive for more than a year (Bailey and Ball, 1991).

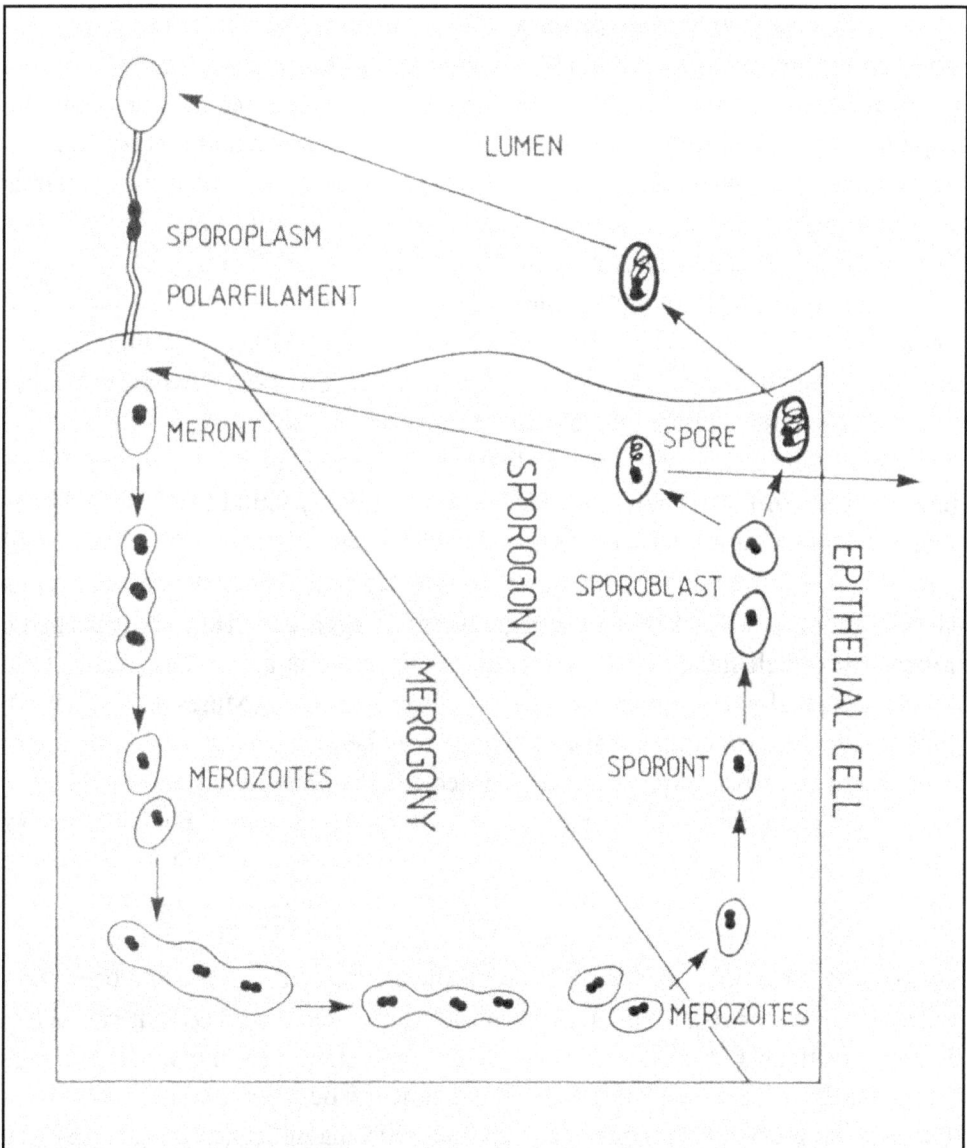

Figure 4.4. Suggested life cycle for *N. apis*. (from Fries 1993).

Symptoms and identification

Adults of all honey bee castes are susceptible to *N. apis*, whereas larvae do not become infected (Zander 1911, White 1919). A greater proportion of worker bees becomes infected than drones or queens, probably because of the cleaning activities of young house bees in which drones and queens do not participate (Fyg 1945, Bailey 1972a). Nosema-infected bees have a reduced life span compared to noninfected bees (Hassanein 1953, Wang and Moeller 1969, 1970). When the queen becomes infected, her egg-laying capacity is reduced as her ovary degenerates (Fyg 1945, Hassanein 1951, Liu 1992), and she is soon superseded by the bees (Farrar 1947).

Lotmar (1936, 1939) demonstrated the detrimental effect of *N. apis* infections on the hypopharyngeal glands of infected bees and made the assumption (Lotmar 1936) that poor brood development in heavily infected colonies could be explained by the inability of the bees to produce a sufficient amount of brood food. The effect of nosema infections on the development of the hypopharyngeal glands is now well-documented (Hassanein 1951, Wang and Moeller 1969, Wang 1971). Physiological effects of nosema infections include reduced ability of the bee to store protein in the fat body (Lotmar 1939). The fatty acid composition of the hemolymph (Tomaszewska 1979), as well as that of the whole bee, is altered (Roberts 1968), and the *corpora allata* cells are affected (Liu 1986) by nosema infections. Hormones regulating development as well as behavior are produced in the *corpus allatum*, and thus behavioral differences can be expected between infected and noninfected individuals. Indeed, Hassanein (1953) found that infected bees began to forage at an earlier age than healthy bees; Wang and Moeller (1970) demonstrated faster physiological aging in infected bees. Infected ventricular cells show signs of lysis. Lack of zymogen granules and aggregation of ribosomes in the invaded host cell indicate that secretion of digestive enzymes has ceased (Liu 1984). Eventually the infected bee may die from starvation (Muresan et al. 1975). Pathological effects connected with *N. apis* may be partly caused by virus infections that usually occur only in nosema-infected bees (Bailey and Ball 1991).

No specific clinical symptoms are connected with *N. apis* infections. Heavy infections are often correlated with dysentery, not because nosema infections cause dysentery, but rather that when dysentery occurs, the disease is aggravated and effectively spread in the honey bee colony (Bailey 1967d). The simplest way to diagnose *N. apis* infections is by microscopic examination of crushed bees, ventricular content or fecal matter. A visual test has been described that can detect heavy infections, but it requires special test tubes, staining and overnight storage of the material (Malone et al. 1993). The characteristic *N. apis* spore is easily detected in fluids under a light microscope at 250-500X magnification (Figure 4.5). The detection level is simply a statistical consideration. Provided each infected bee is

NOSEMA FIELD TEST

Visual symptoms are not specific but may include bees unable to fly; quivering bees; spotting of top bars, bottomboard, or outside of the hive; swollen abdomens; misangled wings; crawling bees. But the most distinctive *visual* symptom is internal – you need to take a close look at the midgut. You will need a fine-pointed forceps, a sheet of white paper, at least a couple dozen live bees, and maybe a hand lens.

Select an adult worker from inner cover or entrance. Avoid young house bees. Grasp the live bee by the thorax between thumb and forefinger. With a fine-pointed forceps, remove the head at the base. This action severs the midgut from the head, making removal possible.

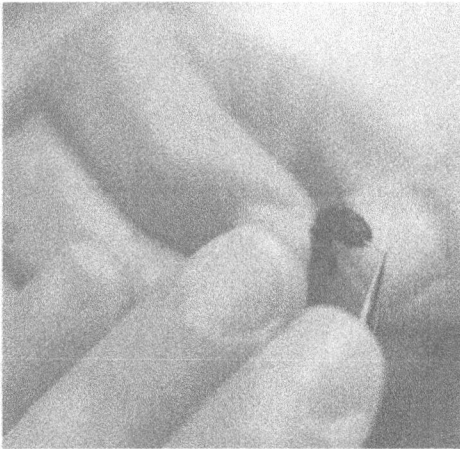

Top: Turn the bee around and, holding the forceps at right angles to the bee's body, carefully and firmly grasp the very tip of the abdomen – *only* the last segment and the sting. Holding firmly, gently pull straight out. When done correctly the sting mechanism, rectum, and hindgut should follow. Don't squeeze the thorax so tightly when pulling that you crush the insides. **Middle**: Continue to pull as straight as pos-

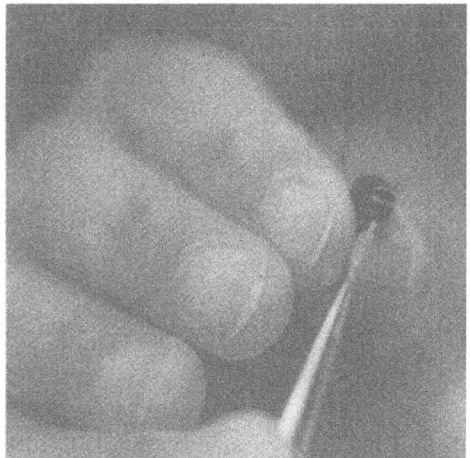

sible, with steady but gentle pressure. First the hindgut will be removed, followed by the midgut, and finally the honey stomach. If you fail the first time, try again. Diagnosing nosema is very important.

Bottom: When the midgut emerges, hold the bee over the sheet of white paper so the entire system is laid out when completely removed.

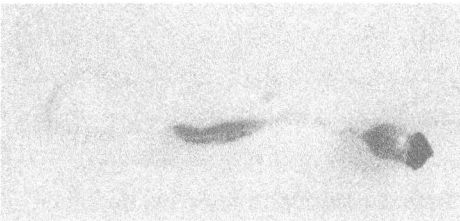

Observe the midgut. If healthy, it should be tan, with the constrictions (wrinkles) obvious. If a heavy nosema infection is present, it may be white, bloated, and without apparent constrictions. The color is the best *visual* indicator. Confirm with a lab test.

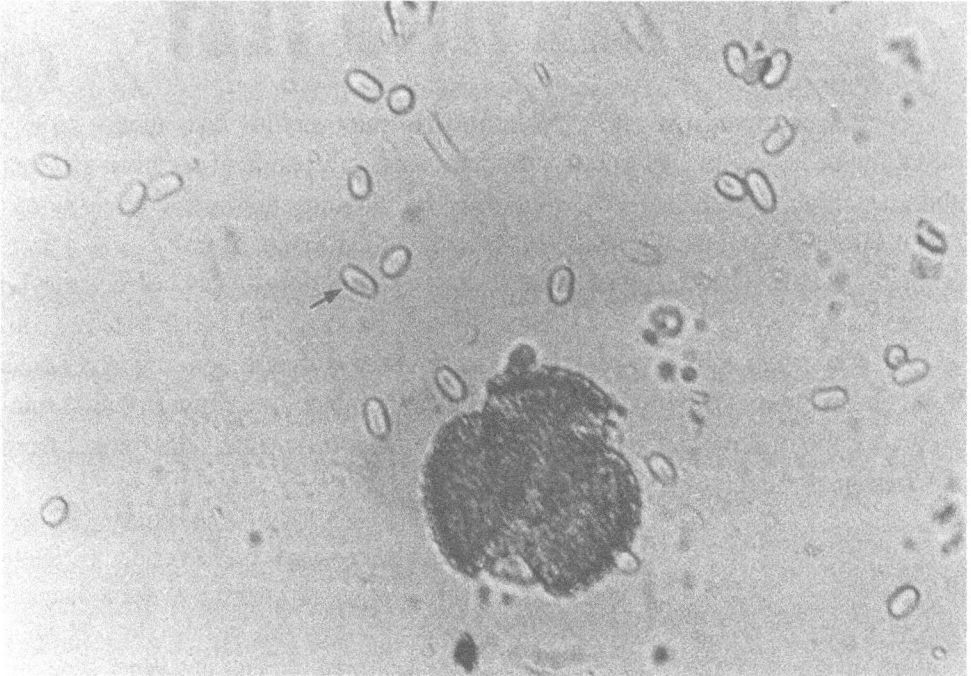

Figure 4.5. *N.apis* spores (arrow) are easily detected in fluids under a light microscope at 250-500X. (Press photo).

detected, the number of bees necessary in a composite sample to detect a specified proportion of infected bees with a certain probability can be inferred from the following formula:

[1-(Proportion of diseased bees)] x [Number of bees sampled] =
[1-(Probability of detecting the disease)]

Thus if a five percent infection level is to be detected with a probability of 95 percent, we must examine 59 bees. Similarly, if a one percent infection level is to be detected with a probability of 99 percent, we need to examine 459 bees.

The infection level found is heavily dependent on season. A typical infection has low or undetectable levels in the summer, a small peak in the fall and a slow rise in infection level during the winter. As brood rearing starts in the spring, although flight possibilities are still limited, the level of infection increases rapidly (Borchert 1928, Michalloff 1928, Bailey 1955b). The sampling site also affects the level of infection found. Infected bees may congregate in the warmer parts of the hive (Moeller 1956). Flight bees are always more infected than house bees (L'Arrivée 1963). Investigating individual bees to determine the proportion of infected bees is laborious. The average number of spores per bee in composite samples, however, is closely correlated to the percentage infected of bees (Fingler et al. 1982) and is preferable. Lotmar (1940) used haemacytometer counts to describe the infection level using the method described in detail by Cantwell (1970). For practical purposes, more subjective estimates of infection level where the number of spores is put roughly into a few categories are quite sufficient (Gross and Ruttner 1970). Although there is a significant negative correlation between the nosema disease level in composite samples of dead bees from the bottomboard in the spring and the honey yield (Pechhacker 1980), live bees sampled at the same time and investigated for nosema will explain much more of the variation in honey yield (Fries et al. 1984).

Figure 4.6. Colonies appearing heavily stained, especially in the spring, may be suffering from dysentery rather than nosema. However, the two often go hand-in-hand. (Fries photo).

Figure 4.7. The most important source of infection is spore-contaminated comb. Defecation inside the hive, such as on these top bars, is a prime source of spores.

Control

Nosema disease reduces the honey yield and contributes to winter losses and losses of queens despite the lack of clear clinical symptoms. The infective agent, the spore, can be transmitted naturally between colonies through drifting and robbing but also through water collection sites where bees may deposit their feces (Borchert 1928, L'Arrivée 1965a). The most important source of infection within the colony is comb contaminated with feces (Zander 1911, Bailey 1953). Zander (1911) believed that nosema infection caused bees to defecate inside the colonies, but Bailey (1967d) demonstrated that nosema disease is not likely to cause dysentery in infected colonies. Dysentery, however, may aid the spread of the disease within colonies where the bees defecate inside the hive.

To cure infected colonies, Zander (1911) suggested they be transferred to clean comb and the hive be cleaned out. Infected colonies should also be stimulated to produce young bees because hatching bees are free from infection; reinfection through contaminated water-collecting sites should be avoided if possible (Zander 1911). These recommendations are still valid.

The most important source of infection is spore-contaminated comb. Colonies transferred to contaminated combs in the spring are likely to develop the disease at detectable levels the following year (Bailey 1955d). Renewing the old comb in the beehive lowers the probability of nosema disease reaching detrimental levels. Putting bees on comb foundation before winter feeding reduces the incidence of nosema disease (Villumstad 1970, 1980, Fries 1988a). Even in colonies wintered on spore-contaminated combs, however, disease levels are not always detectable the following spring. Availability of infectious material, thus, is only one factor, albeit an essential one, that creates epizootic disease development in infected colonies (Fries 1988a).

All conditions that may cause bees to defecate inside the hive, such as unsuitable winter feed or a long winter without flight possibilities, increase the risk of detectable levels of infection (Moeller 1972). The disease development is aggravated by stress conditions (Doull 1961), and colonies seem to be sensitive to disturbance during both the summer and the winter (Oertel 1967, Bretschko 1979). Brood rearing in the early season increases the risk of nosema disease in temperate climates if flight possibilities are limited (Fries 1988a). This may be because the bees become infected as they clean the combs for the expanding brood nest. The increased temperature in the hive as brood rearing increases also favors the development of the parasite in the individual bee (Lotmar 1943).

The nutritional status of honey bees has perhaps not received enough attention when nosema disease is considered. Crowding of colonies where the individual hive has difficulty in supporting its need for pollen may increase the risk of nosema disease (Bailey and Ball 1991). Supplying honey bee colonies with protein, either by supplementary feeding or by using apiary sites with rich pollen supplies, reduces the level of infection (Kleinschmidt and Furguson 1989).

In addition to recommendations for disease control or prevention, based on the biology of the parasite and the parasite-host relationship, it is also possible to use medication in the bee colonies and to decontaminate combs and hive parts. Although several hundred compounds have been tested for control of nosema disease, only the antibiotic fumagillin combines excellent control with low bee toxicity (Moffet et al. 1969). Fumagillin was isolated from the fungus *Aspergillus fumigatus* (Hanson and Eble 1949) and shown to be effective against *N. apis* in cage experiments in the early 1950s (Katznelson and Jamieson 1952). Since that time, treatment with fumagillin has become part of many beekeeping operations. Fumagillin does not affect the spores or the germination process, as formerly believed by Katznelson and Jamieson (1952). The drug affects the intracellular development of the parasite (Liu 1973), possibly by inhibiting DNA replication in the parasite, without a similar effect on the host cell (Hartwig and Przelecka 1971). Despite the extensive use of this drug, the parasite has shown no signs of resistance (Gross and Ruttner 1970).

Fumagillin fed to package bees or to wintering colonies represses disease development (Kulikov 1961, Furgala and Boch 1970). If the drug is present in the food, it effectively prevents transmission of the disease to queens in mating nuclei (Girardeau 1972) or in queen cages used for mailing (Lehnert et al. 1973). Feeding fumagillin in sugar syrup is more reliable than using fumagillin dust or extender patties containing the drug (Furgala and Gochnauer 1969, Williams 1973). Fumagillin retains some amoebicidal activity after storage in sugar syrup in the beehive throughout the winter (Furgala 1962b). The spore is not affected, however, and fecal deposits still provide an inoculum for housecleaning bees in the spring (McDonald 1978). A detailed description of fumagillin treatment is found in Chapter 22.

To reduce the need of changing combs for control of nosema disease transmission, acetic acid fumigation has been used to kill spores of *N. apis* on wax combs (Bailey 1957a, Jordan 1957). The acid may be diluted to 60 percent, and approximately 2 ml solution/liter volume to be treated is used. For treatment, solution is placed on a plate in an empty super on the top bars above a stack of supers. To increase the evaporation area, a crumbled paper or ball of cotton is placed in the acid. The stack of supers is sealed and left undisturbed for at least a week and may be reused after being sufficiently aired.

Fumigation with ethylene oxide effectively kills honey bee pathogens, including *N. apis* (Michael 1964). Several studies have demonstrated that fumigation with ethylene oxide kills nosema disease (Grobov et al. 1970, Shimanuki et al. 1970, Cantwell et al. 1975). Because of the dangers of handling the gas, fumigation must be done only in central plants under strict control.

Cantwell and Shimanuki (1969) demonstrated that heat treatment, 49°C for 24 hours, could be used to kill nosema spores. Because colonies with treated combs produced more honey than colonies on untreated equipment, their results imply that spore transmission on comb is also important for reducing the honey yield in colonies where no signs of disease appear. Besides fumigation techniques and heat treatment, gamma radiation may be used to kill spores of *N. apis* (Hornitzky 1986).

It seems clear, at least in temperate climates, that profitable beekeeping requires control of nosema disease. Management methods, including wax renewal, comb fumigation, heat treatment, radiation and chemotherapeutic treatment with fumagillin, may protect colonies from the ravages of nosema disease. None of these remedies, however, can ensure complete control of the disease under all circumstances.

Tolerance
Several observations indicate variations in susceptibility to nosema infections in individual honey bees (Furgala 1962a, L'Arrivée 1965b, Bornus et al.

1977). These variations have been related to differences in the activity of the enzyme chimozine in the ventriculus (Zherebkin 1976) and to the nutritional status of the bee (Bornus et al. 1977, Rinderer and Elliott 1977). In selection experiments, Rinderer and Sylvester (1978) and Rinderer et al. (1983) found a positive correlation between response to *N. apis* infections and the longevity of honey bees, but the direct effect of the parasite in the ventriculus was not studied. In a study with free-flying colonies in New Zealand, little variation in response to *N. apis* infections could be shown, indicating a lack of variability in susceptibility to the infections in these bees (Malone et al. 1992). Further work is needed to confirm genetic variation in resistance to *N. apis* infections in both individual bees and at the colony level.

Amoeba disease

Maassen (1916) observed what he considered to be amoebic cysts in the Malpighian tubules of honey bees infected with *N. apis* and described the transmission of the parasite from combs stained with feces from infected bees. The amoebas infecting the Malpighian tubules of honey bees were named *Vahlkampfia (Malpighamoeba) mellificae* by Prell (1926b). According to the taxonomy presented by Margulis et al. (1990), the amoebas belong to the phylum Rhizopoda. The systematic position on family level may very well change in the future as more details about the biology and morphology of this parasite become known.

M. mellificae has a worldwide distribution (Matheson 1993). It occurs on all continents where *A. mellifera* is kept, but it is less prevalent than *N. apis*. Only recently was the parasite detected in Sweden (Fries 1994), and it has never been found in Norway, Spain (Bailey and Ball 1991) or Finland (Varis et al. 1992).

Etiology

Adult bees of all castes may become infected with *M. mellificae*. In nature, however, drones and queens rarely, if ever, become infected (Örösi-Pál 1963). Cysts from *M. mellificae* in fecal deposits will transmit the parasite to new hosts as the cysts are ingested by other bees. As for *N. apis*, this may occur as the bees clean the combs for the expanding brood nest in the early season (Maassen 1916, Bailey 1955a). Although ingested cysts may transmit the parasite, vegetative stages, so-called trophycytes, are also infective (Schulz-Langer 1964). Whether vegetative stages of the parasite contribute to the spread between colonies or between bees within colonies is not known.

Schulz-Langer (1964), who cultivated the amoebas in vitro (Schulz-Langer 1961) suggested that the cysts germinate in the rectum and that the amoebas move back through the alimentary canal to infect the Malpighian tubules. She proposed a life cycle of the parasite based on in vivo as well as in vitro observations. In this

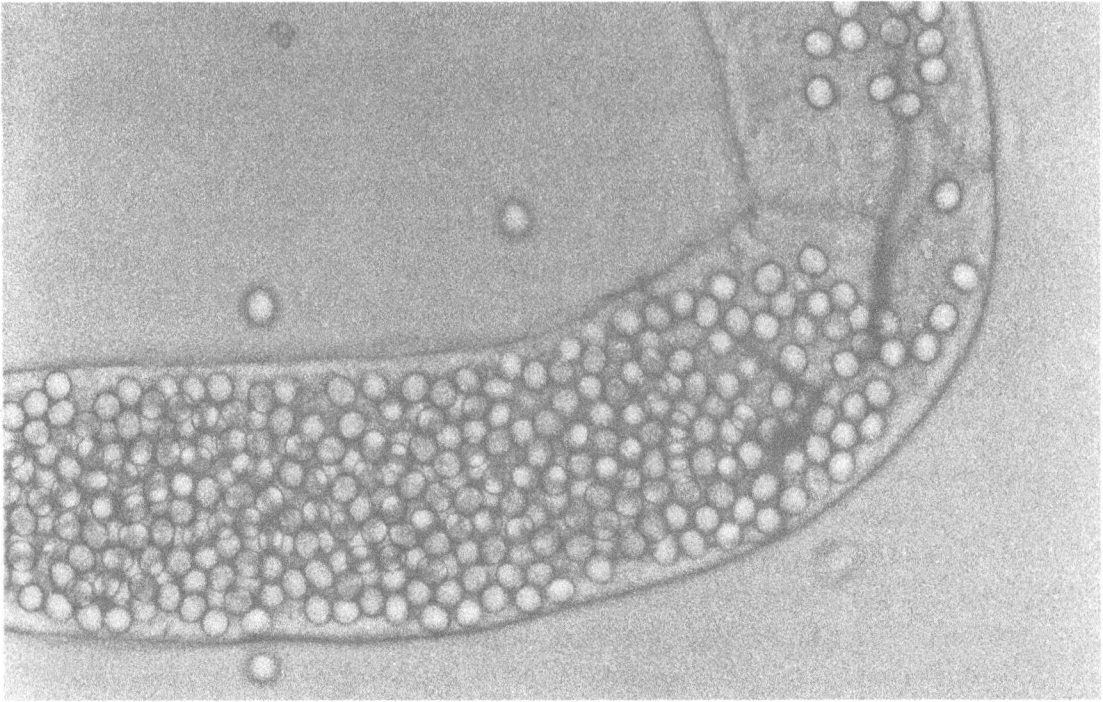

Figure 4.8. Cysts in an infected Malpighian tubule. (Fries photo).

description, *M. mellificae* is an extracellular parasite. Other authors claim, however, that the primary trophocytes later infect the Malpighian tubules (Liu 1985, Tanada 1993). Although such development would correspond to the development of *Malpighamoeba locustae* infecting locusts (Harry and Finlayson 1976), intracellular stages of *M. mellificae* have yet to be described. Giordani (1959) found many protozoa in the rectums of honey bees where the Malpighian tubules contained cysts of *M. mellificae* and suggested that they may be vegetative stages of the parasite. According to Schulz-Langer (1964), a flagellated form makes its way into the Malpighian tubules after the amoeba has existed in the rectal part of the intestine. In the Malpighian tubules, the amoeba changes into the trophic amoeba and later encysts. New cysts are formed 22 to 24 days after infection in bees kept at 30°C (Fyg, 1932) and as many as 500,000 cysts may be found in one infected bee (Bailey and Ball 1991). In contrast to infections with *N. apis*, few attempts to infect honey bees in the laboratory with *M. mellificae* using cysts from Malpighian tubules have been successful (Bailey 1968b).

Symptoms and identification
No clear clinical symptoms are associated with infections by *M. mellificae*. The epithelial cells of infected tubules appear flattened (Fyg 1932, Schwantes and

Eichelberg 1984) and are consumed by the invading parasite (Liu 1985). Secretory transport vesicles, present in healthy organs, are absent in the cells of infected tubules, indicating that the parasite causes an osmoregulatory malfunction in the infected insect (Schwantes and Eichelberg 1984).

Several authors have connected infections by *M. mellificae* with spring dwindling (Prell 1926a, Fyg 1932) and even colony deaths (Schiller 1937). But the overall effect of the amoebic infection is uncertain. There is no experimental evidence that the infections shorten the lives of infected individuals, nor does the parasite cause dysentery in infected colonies, although the incidence of infection is correlated with dysentery (Bailey 1968b). A mixed infection of *N. apis* and *M. mellificae* may be more detrimental to the individual bee than separate infections (Fyg 1932).

A certain diagnosis of *M. mellificae* can be made only by microscopic examination of the Malpighian tubules for the presence of amoebic cysts. The Malpighian tubules originate in the junction of the midgut and hindgut. The long, threadlike tubes project into the hemocoel and may be removed for examination with a pair of fine tweezers. The cysts measure approximately 5 to 8 mm in diameter and can be seen in an infected tubule in Figure 4.8.

Control

Control of amoebic infections is based on hygiene and decontamination of equipment. The antibiotic fumagillin, commonly used to control nosema disease, has no effect on *M. mellificae* (Bailey 1955a), and at present there is no chemotherapeutic treatment available that will control the infections. Combs stained with feces that contain cysts from the amoebas can be sterilized by fumigation with acetic acid, as described for nosema (Bailey 1955a). If infected bees are transferred to clean equipment in the early summer, the procedure is more likely to remove detectable levels of *M. mellificae* infection the following year than of *N. apis*.

Gregarines

The class Gregarinia, commonly referred to as gregarines, forms a large group of nearly 1,500 named species divided into three orders (Levine 1982). Gregarines (at least some *Leidyana* species) are found in the intestinal tract of the honey bee where they attach to the gut epithelium through organelles. They may encyst in the honey bee, but it is not known if they can complete their life cycle there to produce spores infective to other individuals. The species found in honey bees may not be specific to this host (Wallace 1966) but could originate from nectar gathering, water collection or even other insects that may enter the beehive.

Gregarines were first found in honey bees in Europe (Morgenthaler 1926), but they also occur in North (Fantham et al. 1941) and South America (Stejskal 1955). Warm climates are probably most favorable because spores are killed by freezing (Stejskal 1955). Although gregarines may cause pathological changes in the cells where they attach (Stejskal 1965), there is little evidence that they cause any measurable damage to infected bee colonies (Oertel 1965). Stejskal (1965), however, states that these parasites cause serious losses in apiculture. The number of parasites in an individual bee is generally low. Stejskal (1955) reported up to 3,000 gregarines in single bees in Venezuela, whereas Hitchcock (1948) found only 12 to 300 gregarines per bee in the United States. There seems to be little reason to control gregarine infections, at least in temperate climates; Stejskal (1965) reports that fumagillin treatments for control of N. apis would also prevent gregarine infections.

Flagellates

Flagellates are non-spore-forming protozoa in the phylum Zoomastigina (Margulis et al. 1990) that generally possess flagella, pseudopodia or both for means of locomotion. The flagellates in the class Kinetoplastida (trypanosomatids) have a long, slender shape and are equipped with a single flagellum (Tanada and Kaya 1993); flagellates found in honey bees belong to this group.

Fantham and Porter (1912b) described two trypanosomatids from the bee intestine, but Lotmar (1946) believed that they described a single species for which she gave the name *Leptomonas apis*. The parasites produce visible symptoms of their presence as they attach to the intestinal wall in the pyloric region. A dark spot develops at the point of attachment on the dorsal side of the pylorus, but there is no evidence that the presence of flagellates in the bee's intestine causes a pathological condition (Lotmar 1946). The number of flagellates that can be found in the bee is dependent on the temperature and nutritional status of the bee. In the winter, flagellates are often scarce or completely absent (Lotmar 1946). It is not known if flagellates found in honey bees are specific to this host (Wallace 1966).

Later, Lom (1964) reidentified *L. apis* as a *Crithidia* spp., and in Australia, *Crithidia mellificae* has been cultivated in vitro and described in some detail (Langridge 1966, Langridge and McGhee 1967). Experiments have indicated that this flagellate is not a pathogen (Langridge and McGhee 1967), so there seems to be no need to control flagellates in honey bees.

CHAPTER FIVE

Fungi

Martha Gilliam and John D. Vandenberg

CHAPTER FIVE
Fungi

Fungi are common saprophytes of bees and combs. Chalkbrood, the important fungal disease of bees, is caused by species of Ascosphaerales, the spore cyst fungi. Some 20 species of these Ascomycetes associated with 50 species of bees have been described. Some are pathogenic and cause chalkbrood; others are saprophytes thriving on nectar and pollen collected by bees. When various species of bees are managed by humans, chalkbrood often becomes a problem. Stonebrood and other fungal diseases of bees are considered to be of minor importance.

Chalkbrood

Chalkbrood is a fungal disease of honey bee larvae. It is caused by *Ascosphaera apis*, a heterothallic organism that sporulates only when mycelia of opposite sex (designated + and -) come together. After interaction of the two mating types, spores form within dark, brownish-green fruiting bodies known as spore cysts or ascocarps.

Diseased larvae become mummified. The mummies are white, owing to the mycelium of the fungus. If large numbers of fruiting bodies are formed, however, the mummies become dark gray or black.

The disease is sometimes transient and not considered serious, although endemic infection can be persistent and damaging (Bailey 1981). Estimates of losses from chalkbrood vary and obviously depend on the extent of infection in colonies and apiaries. Some that have been reported are a one to five percent reduction in the honey crop and a 49 percent reduction in foraging capacity (Heath 1982a). Yacobson et al. (1991) reported a 37 percent reduction in honey production.

Chalkbrood was not found in the United States until 1968 (Hitchcock and Christensen 1972, Thomas and Luce 1972). Reviews include those of Heath (1982a, b), Gilliam (1990), and Bailey and Ball (1991).

History

Maassen (1913) in Germany published the first observations on chalkbrood. He later described the associated fungus and named it *Pericystis apis* (Maassen 1916). In 1912, Priess gave Claussen a moldy honey comb and a slant culture of a fungus. He wanted to know whether the fungus was a pathogen. Claussen (1921) published a detailed paper on the morphology of the fungus and retained the name *P. apis*.

In England, Betts (1912a) described a hive fungus and named it *Pericystis alvei.* Later, she examined the chalkbrood fungus and noted that it was not identical to *P. alvei* (Betts 1919). Subsequently, detailed mycological observations on *P. apis* were reported by Varitchak (1933) in France, Maurizio (1934, 1935) in Switzerland and Prökschl (1953) in Austria. Moreover, Maurizio (1934, 1935) demonstrated the existence of two morphological types of *P. apis*. Each was heterothallic and capable of causing chalkbrood. The two varieties could not be crossed (see also Christensen and Gilliam 1983). One variety, the usual one in primary cases of chalkbrood (cases of actual outbreaks of the disease in the bee colony), has small cysts. The other, more commonly found in secondary cases (where the fungus develops on combs that have been kept outside the hive), has much larger cysts and is stalked. This large-fruited variety prefers low temperatures, especially during cyst formation. The optimal temperature for cyst formation in the large-fruited variety is 20°C (68°F), and in the small-fruited variety, 30°C (86°F).

Prökschl (1953) designated the small-fruited form, originally named *P. apis*, as *Pericystis apis* variety *minor*, and the large-fruited form described by Maurizio (1934, 1935) as *Pericystis apis* variety *major*. In the United States, Spiltoir (1955) studied the life cycle, and Spiltoir and Olive (1955), noting that the name *Pericystis* had previously been used as a generic name for red algae, reclassified the fungus and established a new genus and family, *Ascosphaera* and Ascosphaeraceae. Spiltoir and Olive (1955) validated the variety under *Ascosphaera apis* variety *major* without seeing any material. They also established the type variety as *Ascosphaera apis* variety *apis*.

Skou (1972) compared cultures of members of the Ascosphaeraceae and designated Ascosphaeraceae as the only family in Ascosphaerales under the series Plectomycetes in the class Ascomycetes. For the first time, all members of the family Ascosphaeraceae were grown side by side. He raised *A. apis* variety *major* to a separate species as *A. major; A. apis* was retained for the small-fruited form; and *Ascosphaera proliperda* was erected for a new *Ascosphaera* species found associated with a solitary bee, *Megachile centuncularis. Bettsia alvei* (= *Pericystis alvei*), a saprophytic pollen mold, was the sole member of the only other genus in the family Ascosphaeraceae. Skou (1985, 1988a) further demonstrated the unique nature of the Ascosphaerales by reviewing their habitats, hyphal formation, spore cyst structure, and the membranous nature of spore ball coverings.

In 1975, Skou described other new species: *Ascosphaera aggregata* from the alfalfa leafcutting bee, *Megachile rotundata*, and from *Megachile centuncularis* and *Osmia rufa*; and *Ascosphaera fimicola* from feces of *O. rufa*. He named the conidial state of *B. alvei* as *Chrysosporium farinaecola*. Skou and Hackett (1979) described the homothallic *Ascosphaera atra*, and Kish et al. (1988) detailed its development in cells of *M. rotundata*. Skou (1982) described the heterothallic

Ascosphaera asterophora, also associated with *M. rotundata* larvae. In 1984, Skou and King described *Ascosphaera osmophila* associated with *Megachile* species and with a mason bee, *Chalicodoma mystaceana*, in Australia. In 1988, Bissett presented a key to the 11 *Ascosphaera* spp. described to that time and included detailed examinations of each. He described three new species, each associated with larvae or nests of *M. rotundata*. Pinnock et al. (1988), Skou (1988b), and Skou and Holm (1989) described new *Ascosphaera* species from various species of solitary bees. Youssef and McManus (1991) developed a culture method that allowed limited sporulation of *A. aggregata* in vitro for the first time.

In 1974, Stejskal described *Arrhenosphaera cranei* as a new member of the Ascosphaeraceae. This organism was said to cause chalkbrood in honey bees in Venezuela. Skou (1985) tentatively named a new species of Ascosphaerales, *Chrysosporium* "hispanicum," from pollen in nests of *Osmia cornuta*. Later, he described six additional *Chrysosporium* species, all closely related to *Chrysosporium farinaecola*, from the nests of species of *Osmia* (Skou 1992).

Rose et al. (1984) developed a key to several *Ascosphaera* species that cause chalkbrood. They found no pair of species to be interfertile.

Stephen et al. (1981) made measurements and observations of spores, spore balls, and cysts of several *Ascosphaera* species. They compared these with observations of similar characters made by other authors and found some differences among strains of *Ascosphaera aggregata* and other species. Gilliam and Lorenz (1994, 1995) pointed out the continuing problems of misidentification and use of invalid taxonomic names of Ascosphaerales and of incorrect assignment of mating types. Based on measurements of 2,800 spore cysts from 23 strains of *A. apis* from feral and managed colonies of honey bees and from a carpenter bee, *Xylocopa californica arizonensis*, Gilliam and Lorenz (1995) concluded that more variability exists in diameters of spore cysts than had been reported previously. Maghrabi and Kish (1985a, b, 1986, 1987a, b) conducted electrophoretic studies of six *Ascosphaera* species. Their findings were consistent with previous descriptions that were based on morphology. Results of assays for 19 enzymes produced by *A. apis* strains isolated from honey bees from the United States (Gilliam and Lorenz 1993) and from Spain (Alonso et al. 1993) were remarkably similar with the major exception that none of the strains from the United States produced μ-glucosidase, whereas most of the Spanish strains did.

Three species of spore cyst fungi are of concern to researchers who work with honey bees. *Bettsia alvei* is the pollen mold, *A. apis* is the chalkbrood pathogen, and *Ascosphaera major* has been reported to cause chalkbrood in Europe (Maurizio 1934, 1935), but reports of its isolation from mummies are rare. Under normal conditions, spores of *A. major* are usually evacuated by larvae without germinating (Puerta et al. 1989).

Distribution

Chalkbrood in honey bees has been reported from Europe, including the British Isles, for many years (Maassen 1913, Betts 1919, 1932, 1951, Zander 1919, Anderson 1934, 1938, Maurizio 1934, Dreher 1938, Deans 1940, Morgenthaler 1944, Tabarly and Monteira 1961, Roussy 1962, Bailey 1963a, 1968a, McLellan 1964, Giauffret and Taliercio 1967, Barthel 1971, Lunder 1972, Matus and Sarbak 1974). Seal (1957) reported chalkbrood in New Zealand, and Reid (1984) surveyed its distribution. Furuya et al. (1981) and Takatori and Tanaka (1982) reported on the occurrence and identification of chalkbrood in Japan. Rossi and Carranza (1980) first reported chalkbrood in Argentina. Chalkbrood has also been reported from Asia, including countries of the Middle East and countries east of the Indian sub-continent (see Matheson 1993). White (1993) reported chalkbrood in Australia, which until that time had remained the only major beekeeping country without the disease. Nixon (1983), Heath (1985), Bradbear (1988), and Matheson (1993) have reviewed other reports of chalkbrood from throughout the world.

Baker and Torchio (1968) reported the first record of *A. apis* from the United States. Later, Thomas and Luce (1972) reported that chalkbrood had been found in honey bees in California in 1968, and Hitchcock and Christensen (1972) found the disease in honey bee larvae in Nebraska and Wyoming and noted that other occurrences of the disease in California, Minnesota, North Dakota, and Montana had recently come to their attention. Gilliam and Taber (1973) reported chalkbrood in Arizona, and Connor (1974a) found chalkbrood in Ohio. By 1974, the disease had been found in at least 35 states (De Jong 1976).

Gochnauer et al. (1972) reported the first record of chalkbrood in Canada in honey bees from British Columbia and Saskatchewan. Nelson et al. (1976) surveyed 5,374 colonies in five Canadian provinces in 1975 and found that 32 percent of these colonies had mummies in the frames. However, 75 percent of the infected colonies each had fewer than 10 cells with chalkbrood. In a similar survey conducted in Western Canada in 1976, 33 percent of the package bee colonies and 20 percent of the wintered colonies had chalkbrood (Nelson et al. 1977). It is possible that the disease was unreported for some years in Canada and the United States (Gochnauer et al. 1975), since commercial beekeepers indicated that the disease had been present for decades but was considered insignificant in comparison with American foul-brood (Connor 1974b).

In only 12 years after the initial discovery of chalkbrood in honey bees in California, the disease had spread throughout the North American continent. It continues to cause problems, although losses are often sporadic because of the cyclic nature of the disease. Thus, at various times, individual beekeepers may have a range of opinions on chalkbrood from regarding it as a curiosity when only a few mummies are occasionally found to considering it a serious economic problem when colonies die from the disease. A survey in the United States revealed that more beekeepers have had colonies with chalkbrood than with nosema disease, American foulbrood, European foulbrood, tracheal mites, or varroa (Flottum 1991b).

In countries such as New Zealand (Heath 1985) and Japan (Takeuchi et al. 1989) that had reported chalkbrood earlier, increases occurred in the incidence of the disease in honey bee colonies in the early 1980s. At the same time, chalkbrood problems and research efforts increased in Europe. By 1983, the increased incidence and severity of chalkbrood in honey bees throughout the world merited the organization of a special session on the disease to be held at the XXXth International Apicultural Congress of Apimondia in 1985 in Nagoya, Japan (Gilliam 1986a).

When and how *A. apis* was introduced into the United States and the methods by which it spread have been discussed by Herbert et al. (1977), Gilliam (1978a, b, 1989, 1990), and Heath (1985). In summary, chalkbrood was either present before 1968 and was considered unimportant, or it was introduced around that time. The first view seems unlikely because numerous competent microbiologists and apiary inspectors have examined diseased bees in the United States for many years and would have observed and reported chalkbrood if it were present before 1968. For example, Burnside (1927, 1930) of the United States Department of Agriculture conducted classic studies on saprophytic and pathogenic fungi of honey bees and was familiar with the European work on chalkbrood. Beekeepers were probably seeing another disease, stonebrood, which also produces mummies.

Another theory is that *A. apis* may have been present in solitary or wild bees that served as a reservoir of inoculum for honey bees. However, many early reports of *A. apis* from solitary bees may have been other species of *Ascosphaera* that had not been described at that time.

It seems probable that *A. apis* was introduced into honey bees in the United States in the 1960s on imported pollen, bees, or queens. The high incidence of infection and the rapid spread of chalkbrood throughout North America are indicative of a pathogen new to an area where the host has little natural resistance. The spread would have been accelerated by susceptible bee stock, migratory beekeeping, queens, and package bees. Pollen is a likely source of the introduction since *A. apis*, pulverized chalkbrood mummies and even large numbers of whole mummies were found in several shipments of pollen from foreign sources where were received shortly after chalkbrood had been found in the United States and were the only shipments that had been examined (Gilliam 1989). Numerous prior shipments to various locations were never checked. In fact, the best way to collect chalkbrood mummies is from pollen traps or dead bee traps placed on infected hives.

Chalkbrood caused by *A. aggregata* in solitary bees, including the alfalfa leafcutting bee, has been reported from Europe (Skou 1972, 1975), North America (Thomas and Poinar 1973, Richards 1985), and in South America where *M. rotundata* is cultivated (W. P. Stephen, personal communication). It has not been reported from New Zealand (Read and Donovan 1984) or Australia, although *M. rotundata* is managed for alfalfa pollination in those countries.

Etiology

According to Bailey (1967a), honey bee larvae are most susceptible to chalkbrood if they ingest spores of *A. apis* when they are three to four days old and then are chilled briefly two days later, immediately after they are sealed in their cells to pupate. Because of the chilling factor, the affected larvae are usually found on the periphery of the brood area. For this reason, it was once believed that only

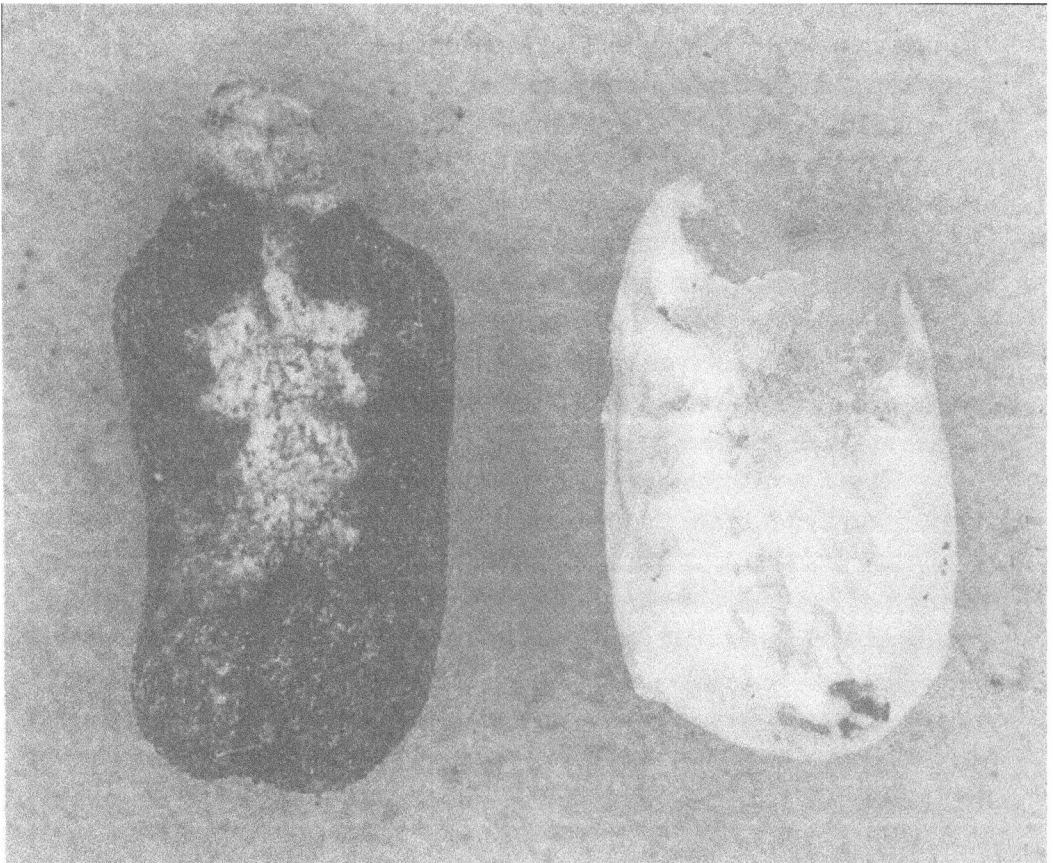

Figure 5.1. Larva killed by chalkbrood. The white mycelia grow through the surface of a dead larva (right); later the larva dries into a chalklike lump, sometimes called a mummy. Mummies may remain white, or, if large numbers of fruiting bodies are formed, become gray to black (left). (Hansen photo).

drone larvae are affected, since they are frequently on the periphery of the brood nest.

The spores then germinate in vivo in the gut of the larva, which is almost anaerobic. The mycelium, however, is aerobic and therefore survives in a three to four-day-old larva longer than in a younger one because it is deprived of oxygen for a shorter time (Bailey 1967a). The spores germinate particularly in the hind end of the gut. When the larva becomes sealed in its cell to pupate, the mycelium develops rapidly in the abdomen (Bailey 1976). Bignell and Heath (1985), however, found that the gut of a fifth-instar honey bee larva is not anaerobic and suggested that other conditions must stimulate spore germination. They suggested a possible role of carbon dioxide in the infection process. Heath and Gaze (1987) showed the importance of carbon dioxide for spore activation (the first stage of germination) of *A. apis*. Kish (1980) demonstrated the stimulative effect of carbon dioxide on spore germination of *A. aggregata*.

Gochnauer and Margetts (1979) showed the production of phosphatase by cultures of *A. apis* but did not observe the production of several other enzymes for which they tested. Because of the absence of enzymes that are found in many other pathogens, they concluded that *A. apis* is a noninvasive pathogen that kills its host by direct competition for primary nutrients. Kowalska (1984a) reported a similar lack of various enzymes.

At first, the dead larvae are covered with a fluffy white growth of mycelia and are swollen to the size of the cell. Later, the dead larvae dry into hard, shrunken, chalklike lumps, which may become gray to black if sufficient fruiting bodies are formed. Many of the cells in heavily infected colonies may remain sealed, and the mummies will rattle if the comb is shaken. But larval remains can be found in sealed or unsealed cells. Most larvae die in the upright stage. Adult bees remove many of the mummies, which can be found at the hive entrance and on the bottomboard.

The majority of insect entomopathogenic fungi infect their hosts through the external cuticle; infection from ingested spores occurs rarely (Roberts and Humber 1984). In this regard, *A. apis* and other *Ascosphaera* species, including *A. aggregata* (Vandenberg and Stephen 1982), may differ. Barthel (1971) and Matus and Sarbak (1974) stated that natural infection by *A. apis* can occur in two ways, either by ingestion of spores with food or via the body surface from spores adhering to combs and cell walls. Maurizio (1934), however, thought that infection takes place through the mouth and not through the cuticle or spiracles, whereas Roussy (1962) found that the spores germinate on the surface of larvae. Gilliam et al. (1978) and Glinski (1981) obtained infected larvae following inoculations of the cuticular surface as well as by feeding. Heath and Gaze (1987) conceded that cuticular infections could occur because of the carbon dioxide levels in colonies and at the larval surface because spore activation is rapid when enhanced carbon dioxide is in the

Figure 5.2. Chalkbrood mummies removed from cells and discarded. This sight is common on bottomboards and at hive entrances of infected colonies. The black mummies are covered with fruiting bodies. (Killion photo).

vicinity. Based on tests of enzyme production by mycelia and on scanning electron microscopy of larvae fed spores of *A. apis*, Alonso et al. (1993) postulated that N-acetyl-ß-glucosaminidase may break down chitin in the cuticle whereas hyphae inside the larval body exert pressure on the cuticle. These actions may explain how the fungus pierces the cuticle. Alonso et al. contended that it is more likely that the conditions conducive to penetration of the cuticle exist inside the larvae rather than on the surface. In addition, these authors found production by mycelia of enzyme activity previously unreported or undetected including DNAase, RNAase, amylase, and cellulase. Bamford and Heath (1989) conducted histological studies of larvae fed spores of *A. apis* and observed that the cuticular elements of the larval integument were not a major barrier to hyphal development of *A. apis*. Penetration of the integument by hyphae appeared random, without preferential use of intersegmental regions or spiracles. Bamford and Heath (1989) postulated that bound enzymes might be involved.

Entomopathogenic fungal mutants lacking enzymes responsible for degradation of cuticular layers and penetration through the insect cuticle have been shown to have a longer LT_{50} value than the wild type or to have a variable effect as far as pathogenicity is concerned (see Khachatourians 1991). Such studies on *Ascosphaera* species have not been conducted but may provide clues for two possible routes of infection and for the increased incidence and severity of chalkbrood. If *A. apis* initially infected larvae via the cuticle but mutated and lost this ability, infection through ingestion could be more severe and death of the host more rapid. Some strains with the delayed ability to penetrate the cuticle may survive. Thus, a mixture of original wild-type strains and strains with defects in the production of one or more enzymes could result and account for different routes of infection. In addition, some strains of *A. apis* have been reported to vary in virulence for honey bee larvae (Glinski and Chmielewski 1989, Sawathum and Ritter 1995).

Gochnauer et al. (1975) postulated that once a colony is infected, the spores can remain viable on the combs and eventually germinate when conditions become favorable, and the disease can then reappear. They also suggested that *A. apis* might survive in soil, find its way into the food chain of honey bees, and be transmitted to larvae via contaminated brood food.

Although the ability of infection to spread is probably low (Bailey 1963a), the spores are resistant and remain infective for at least 15 years (Toumanoff 1951). Hale and Menapace (1980) demonstrated that *A. apis* remains viable for at least one year at temperatures below 27°C (81°F). Maurizio (1934) and Betts (1951) thought that the disease was probably carried by honey and that, therefore, one should avoid feeding honey from infected colonies. Maurizio (1934) found the fungus in the intestines of bees from an infected colony and stated that the organism wintered in bees and in honey. De Jong and Morse (1976) found *A. apis* in the honey sac contents of adult worker bees from infected colonies and showed that spores are passed from bee to bee in food exchange. They also reported that queen bees can transmit the disease. Nelson and Gochnauer (1982) found viable spores in adults but not in honey. Herbert et al. (1977) demonstrated transmission of chalkbrood by contaminated bees and queens. Stejskal (1974) reported that honey bee larvae could be infected with *Arrhenosphaera cranei* when fed nectar that contains spores or when fed by worker bees that have cleaned infected combs.

Pollen has been shown to be a carrier for spores and to lead to disease transmission (Mehr et al. 1976, 1978, Menapace 1978, Moffett et al. 1978), although De Jong (1977) was able to obtain infection by feeding contaminated pollen only during a pollen dearth. Hale and Menapace (1980) found that spores remain viable in pollen for at least 12 months.

Further spread of the disease occurs mostly through drifting of bees from infected colonies. Transfer of spores by the beekeeper also occurs, however, when contaminated tools and combs are interchanged in infected and healthy colonies

(Barthel 1971, Herbert et al. 1977). Nelson and Gochnauer (1982) found viable spores in adult bees from a diseased colony, and Gilliam (1986b) isolated the pathogen from worker bees from colonies that had never had chalkbrood.

Several factors seem to contribute to the development of the disease in bee colonies. Bailey (1967a) noted that chilling of larvae immediately after they are sealed in their cells, which increases susceptibility to the disease, is most likely in spring (when chalkbrood is most evident) because colonies are then growing rapidly, and the ratio of brood to adult bees is the largest. He postulated that chilling enables oxygen to diffuse into the gut and reactivate the mycelium, which then continues to grow when the brood temperature is restored to 35°C (95°F). Puerta et al. (1994) suggested that chilling is an important predisposing factor for chalkbrood if and when it takes place 24 hours before or after the sealing of cells.

Seal (1957) found that a hive that is poorly ventilated and only partly occupied during winter provides an excellent resting place for spores of *A. apis*. Then, following expansion of the brood nest in spring or early summer until it includes the combs harboring the spores, the larvae can become infected. He suggested that races of bees with an excessive tendency to swarm are the most susceptible to chalkbrood, because the departure of swarming bees leaves too large a brood chamber for the remaining bees to care for properly. De Jong (1977) was able to cause an increase in infection rates by increasing the amount of brood or by decreasing the number of foraging workers. Befus-Nogel et al. (1992) increased the incidence of chalkbrood by inserting combs of honey between combs of brood.

Dallmann (1966) found that chalkbrood occurs particularly during rainy summers in apiaries that are located in moist, cool places. Gochnauer (1963) stated that fungal infections of bees appear in colonies that have excessive hive moisture. Also, Lunder (1972) attributed the occurrence of chalkbrood to wet weather, poor foraging conditions for long periods, weak colonies, open hives, and genetic factors. Inbred lines of honey bees seem to be particularly susceptible to infection (Moeller and Williams 1976). Lunder (1972) wondered whether an acid-ambient environment resulting from increased air pollution might promote chalkbrood development. Dreher (1938) thought the frequent and persistent occurrence of chalkbrood in heather regions in Germany was primarily caused by the predominant practice of skep beekeeping and by the damp oceanic climate. However, Deans (1940) found no connection between chalkbrood and heather locations in Scotland.

Barthel (1971) thought that chalkbrood had become partially epidemic in East Germany because of the high relative humidity and high temperatures. Cury (1951) believed that chalkbrood occurred mainly in hot months and only in poorly ventilated hives. Highly hydrated honey was found to create dampness and permit development of chalkbrood; the administration of antibiotics in syrup also seems to favor the growth of fungi (Tabarly 1962). However, Menapace and Wilson (1979) reported that feeding Terramycin in extender patties did not enhance chalkbrood infections.

Chalkbrood also seems to occur in colonies that are first weakened by another disease. Deans (1940), in Scotland, found that in almost every serious case of chalkbrood, very few foraging bees were present because of high levels of infestation with tracheal mites. Once a colony was weakened by other causes, the bees did not seem to clean out the cells to keep chalkbrood in check or to eradicate it. He suggested that in severe cases of chalkbrood, some other circumstance has first weakened the colony.

Matus and Sarbak (1974) found that chalkbrood occurred in apiaries where secondary factors, such as unsatisfactory beekeeping or another disease, had occurred. Maurizio (1934) found secondary cases of chalkbrood on combs with European foulbrood cadavers. Wille (1964a) noted that chalkbrood seemed to occur jointly with nosema disease, bacterial septicemias, and rickettsioses rather than alone; and a possible relationship between sacbrood and chalkbrood has been suggested (Mehr et al. 1976, Moeller and Williams 1976). In addition, Dreher (1938) observed that the fungus seemed to grow first in injured brood.

Giauffret and Taliercio (1967) attributed the spread of fungal disease of bees to the use of antibiotics that upset the equilibrium of the intestinal flora of bees and therefore allowed the fungi to flourish. Humidity, lack of ventilation, watery food, and genetic factors were suggested as other aids in the spread of infection.

Gilliam and Taber (1973) reported chalkbrood in colonies that had been fed excessive amounts of pollen and extensively manipulated. Some of these colonies also had other diseases, such as American foulbrood. In addition, Hitchcock and Christensen (1972) found chalkbrood in a colony severely weakened by American foulbrood. The old queen had become a drone layer, and the adult bee population was very small.

Bailey (1963a) suggested that other insects, primarily solitary or wild bees, play a part in spreading chalkbrood. There have been several reports of *A. apis* associated with various solitary bees (Melville and Dade 1944, Clout 1956, Baker and Torchio 1968, Torchio 1971, Batra et al. 1973, Thomas and Poinar 1973). However, most of these are unconfirmed and suspect or were later shown to be other *Ascosphaera* species now known to occur in association with non-*apis* bees (see Stephen et al. 1981, Bissett 1988). Recent reliable data are those of Rose et al. (1984), who isolated *A. apis* from the alkali bee, *Nomia melanderi*, and both *A. apis* and *A. aggregata* from the alfalfa leafcutting bee, *M. rotundata*, and of Gilliam et al. (1994), who reported *A. apis* from a giant carpenter bee, *X.c. arizonensis*.

In Denmark, Skou (1972) reported *Ascosphaera proliperda* causing chalkbrood in *M. centuncularis*. *Ascosphaera major* was also found to cause chalkbrood in these bees, although it seemed to be only a facultative parasite (Holm and Skou 1972). They suggested that the ascospores of the fungus stick to the bodies of the bees as they pass through one cell after another when emerging. Therefore, this is one method of transfer of the organism from one generation to the next. Vandenberg et al. (1980) demonstrated this phenomenon for *A. aggregata* and the alfalfa leafcutting bee.

In recent years, many studies have been conducted on chalkbrood in solitary bees. They have focused on the domesticated alfalfa leafcutting bee, a commercial pollinator of alfalfa in Western North America, and on other bees with potential as useful pollinators. Studies of etiology and pathogenesis of *A. aggregata* for larvae of *M. rotundata* (Hackett 1980, Vandenberg and Stephen 1982, 1983a, b, McManus and Youssef 1984) were followed by quantitative assays revealing various biotic and abiotic influences on larval susceptibility to fungal infection (Vandenberg 1992a, b, 1994, Goettel et al. 1993b, Inglis et al. 1993, Vandenberg and Goettel 1995). Related studies of other *Ascosphaera* spp. infections in *M. rotundata* and other solitary bees have been done by Youssef et al. (1984, 1985), Rust and Torchio (1991, 1992), and Torchio (1992).

Control

In the United States, no chemotherapeutic agent is registered for use against chalkbrood, although some are available in other countries. In the past, the losses caused by the disease were not considered serious enough to justify research on treatment (Bailey 1963a). Because the adult bees generally carry the dead brood out of the hive, the disease often disappears without intervention by the beekeeper. Dreher (1938) emphasized the importance of burning mummies and not allowing them to accumulate; Deans (1940) stated that the disease is not a problem in strong colonies.

For severe cases, destruction of affected combs or sections of combs has been recommended (Kenward 1932, Betts 1951, Gochnauer et al. 1975). Nelson and Gochnauer (1982) and Koenig et al. (1986) found that providing new comb reduced the incidence of chalkbrood. Kenward (1932) suggested rendering good combs safe, after removal of shrunken larvae, by exposing them to sun rays or by fumigating with sulfur fumes or 40 percent formalin. Giauffret et al. (1969), Thomas and Luce (1972), Cantwell et al. (1975), Gochnauer and Margetts (1980), and Mabuchi (1982) found that ethylene oxide fumigation killed *A. apis* in infected combs. Faucon et al. (1982) used methyl bromide to fumigate hive equipment. They found that spores of *A. apis* were killed, but chemical residue was detected in wood and wax. Shimanuki et al. (1984) demonstrated the efficacy of high-velocity electron beams for killing spores associated with chalkbrood mummies.

Moisture accumulation and poor ventilation in hives should be prevented because dampness encourages the development of chalkbrood. Zander (1919) emphasized the importance of wrapping colonies during winter because cold temperatures increase the likelihood of chalkbrood, as well as keeping them dry. Seal (1957) stated that serious spread of the disease could be prevented by closing the hives for winter and keeping them clear of long grass to prevent dampness and to allow adequate ventilation. One can also enlarge the entrance to a colony to aid ventilation (Gochnauer et al. 1975).

Seal (1957) recommended strengthening badly diseased colonies by adding young adult bees and hatching brood and by feeding sugar syrup. He also stated that chalkbrood can be prevented by not allowing the bees to winter in too large a brood chamber. Tabarly (1962) reported that chalkbrood decreases in affected colonies when honey with a water content above 19 percent is replaced with a more concentrated honey containing less than 17 percent water.

Many experiments have concerned the use of chemicals to control chalkbrood. Cardinale et al. (1994) list many of the papers published from 1982 to 1990 that concern various chemicals tested in vitro or in vivo for control of chalkbrood in honey bees. Elbe and Weide (1961) found that 0.7 percent thymol prevented the growth of *A. apis* in culture. When the thymol solution was sprayed on infected combs, the disease disappeared. However, it was necessary to spray every comb and the inner walls of the brood chamber. The bees would not accept 0.7 percent thymol in syrup. These authors also reported that chlorides of potassium, calcium and sodium, as well as fumes of various chemical substances, did not inhibit the growth of the fungus. Tanaka et al. (1984) found that trichloroisocyanuric acid in water placed in the hive controlled chalkbrood. Presumably, the vapors of this solution acted on the fungus. Vapors of propionic acid proved efficacious when tested by Kajikawa and Nakane (1986).

Kowalska (1984b) reported that thymol inhibited *A. apis* growth in vitro. Barthel (1971) found that thymol in a 2 percent solution had a fungistatic effect in 20 minutes. He stated that stimulation of the cleaning instinct of bees is the principal control measure. Herbert et al. (1985, 1986) have shown that certain alkyl amines not only stimulate the removal of chalkbrood cadavers but also inhibit the growth of the fungus in vitro.

Dallmann (1966) tested the disinfectant Fesia-Form (formaldehyde base) on bee colonies infected with chalkbrood. He sprayed the solution on brood combs, the inside parts of the hive, and the flight board. Within a week after spraying, the adult bees from severely infected colonies had removed the mummies, and no reinfection was observed throughout the year. He found that a 4 percent solution of Fesia-Form killed the fungus. Barthel (1971) performed in vitro tests on the activity of fungicides against *A. apis* and reported that 4 percent Fesia-Form killed the spores after 30 minutes. Samsináková et al. (1977) also showed that formalin vapors inhibit the growth of the fungus.

Many antimycotics have been tested for possible use in the treatment of chalkbrood. Giauffret and Taliercio (1967) reported that amphotericin B was the most effective agent of those they tested against *A. apis*. However, it was not stable. Actidione exhibited a high toxicity for bees, and griseofulvin did not inhibit the growth of *A. apis*. Glinski and Chmielewski (1979, 1981) and Glinski and Rzedzicki (1980a, b, c) conducted a series of studies on some polyene antibiotics, among them amphotericin B. They found these compounds had a low minimum inhibitory concentration toward *A. apis* in vitro, low toxicity to adult bees and no detectable residues in honey following hive treatments. The effectiveness in vitro of amphotericin B, actidione, and nystatin was shown by Kowalska (1984b). Samsináková et al. (1977) also found nystatin effective at low concentrations; other antimycotics were also efficacious. Moeller and Williams (1976) found that feeding infected colonies 250 parts per million benomyl in sugar syrup reduced infection. Hartwig (1983) observed the recovery of 80 to 87 percent of chalkbrood-infected colonies following spray application of ascocidin in sugar syrup. Glinski (1986) reported that ascocidin was more effective than either nystatin or amphotericin B. It had to be sprayed in sugar syrup on brood and on bees on combs four times at five-day intervals after removal of combs with infected brood or after shaking bees from diseased colonies onto new combs.

The antiseptics tested by Giauffret and Taliercio (1967) were quite stable but were more toxic for bees than the antimycotics. For example, cetyl-trimethyl ammonium was toxic at levels of 0.5 g per hive. Glinski et al. (1981) tested various antiseptics and disinfectants in vitro and found chloramine T and abacin most effective. Kowalska (1984b) determined the minimum inhibitory concentrations of merthiolate and brilliant green to be less than 5 ug/ml. Samsináková et al. (1977) observed limited growth of *A. apis* in vitro after treatment with salicylic acid.

Various preservatives and other chemicals have been tested against *A. apis* in vitro and against chalkbrood in honey bee colonies. Thomas and Luce (1972) reported that sorbic acid and methyl parahydroxybenzoate inhibited *A. apis* in culture. Kowalska (1984b) also reported the efficacy in vitro of sorbic acid and of thiabendazole. Samsináková et al. (1977) observed the inhibition of fungal growth following treatment with methyl parahydroxybenzoate, allyl isothiocyanate, or pro-polis. Taber et al. (1975) fed sorbic acid and sodium propionate in pollen-sugar patties to badly infected bee colonies; chalkbrood disappeared seven days after the treatment started, there were no symptoms of toxicity in the bees, and no reduction in sealed brood occurred. Nelson and Gochnauer (1982) observed a 50 percent reduction in chalkbrood following treatment with sorbic acid and sodium propionate. However, Gochnauer et al. (1979) found that neither potassium sorbate nor sodium propionate prevented fungal growth. Furthermore, Menapace and Hale (1981) reported that treatment with a combination of potassium sorbate and sodium propionate (up to 0.1 percent concentration in pollen cakes) did not prevent chalkbrood. De Jong (1977) observed a decrease but not elimination of chalkbrood following treatment with 1 percent thiabendazole.

Gochnauer et al. (1979) used vapors of citral and geraniol to inhibit the growth of cultures of *A. apis*. Later, however, Menapace and Hale (1981) were not able to prevent chalkbrood in honey bee colonies treated with pollen cakes containing citral at 10 or 1,000 parts per million. There are a few reports of encouraging results obtained with plant extracts for chalkbrood control in bee colonies (Colin et al. 1989, Liu 1995a).

A recent interesting discovery about the interaction of bee diseases concerns the effect of chalkbrood on European and American foulbrood. Shimanuki et al. (1992) noted that a decline in the incidence of European foulbrood in the United States was accompanied by an increase in chalkbrood. Ethanol extracts of mixtures of mycelia and spores of *A. apis* were shown to contain an antimicrobial substance active against the bacterial pathogens of both European foulbrood and American foulbrood diseases. This substance was identified as linoleic acid (Feldlaufer et al. 1993b).

Although a variety of studies, both in the laboratory and in the field, have demonstrated the potential of fungicidal compounds and other treatments for controlling chalkbrood in the alfalfa leafcutting bee (Hackett 1980, Stephen et al. 1982, Kish 1983, Parker 1984, 1985, 1987, 1988, Kish and Panlasigui 1985, Youssef and McManus 1985, Fichter and Stephen 1987, Youssef and Brindley 1989, Mayer et al. 1990, Goettel et al. 1991, 1992, 1993a), only a few have led to registration or widespread adoption in controlling chalkbrood. Stephen et al. (1982) showed that halogenated compounds (including hypochlorites) were effective in killing the fungal spores, and Mayer et al. (1988) demonstrated the efficacy of dipping nest materials in calcium hypochlorite for controlling chalkbrood. Goerzen and Watts (1991) used paraformaldehyde fumigation for mold control in leafcutting bee nests and later registered this treatment for use in Canada. Goettel et al. (1993a) showed that this was effective in controlling chalkbrood. Because paraformaldehyde is not registered for use on leafcutting bee nest materials in the United States, only certain prophylactic management techniques combined with hypochlorite dipping of nesting materials and overwinter cocoons have been widely adopted.

Variation in susceptibility of bee colonies to chalkbrood is an important factor in expression of the disease, and the pathogen is often present in colonies that never show symptoms of chalkbrood (Gilliam 1986b). Also, some colonies are more adept than others at removing chalkbrood mummies (Gilliam et al. 1983). Thus, resistance or tolerance to *A. apis* exists in some bee colonies. Lunder (1972) suggested the use of resistant queens, and Mraz (1973) recommended requeening with nonsusceptible strains. Nelson (1975) made crosses between New Zealand and California stocks of bees and found that the resulting stock was less affected by chalkbrood than the California stock. Gilliam et al. (1983) showed that it is possible to select and breed honey bees for resistance or tolerance to the disease. Increased resistance is evidenced by elevated hygienic behavior (uncapping and removal of dead and dying brood) on the part of the nurse bees; by decreased longevity of the fungal spores in resistant colonies; and by reduced pathogen contamination of bees, brood, and stored food in resistant colonies.

Hygienic behavior is considered the primary mechanism of resistance to American foulbrood (Rothenbuhler 1964a) and to chalkbrood (Gilliam et al. 1988). The behavior was explained as being controlled by two recessive genes, one for uncapping and one for removal (Rothenbuhler 1964a). However, the underlying genetic mechanism may be more complex (Moritz 1988), and resistance to chalkbrood involves other factors in addition to hygienic behavior (Milne 1983, Spivak and Gilliam 1993). Hygienic bees also uncap and remove pupae infested with the mite, *Varroa jacobsoni* (Boecking and Drescher 1992).

Simple tests using comb inserts of frozen brood have been devised for use in testing hygienic behavior of bee colonies (Taber 1982, Gilliam et al. 1983). Bees from a naturally mated queen that remove freeze-killed brood within 48 hours are considered hygienic; bees that take more than a week to remove dead brood are considered nonhygienic (Taber and Gilliam 1987). In studies of the relation between hygienic behavior and chalkbrood, most but not all, colonies showed a good correlation between uncapping and removal of freeze-killed brood and resistance to the fungus (Gilliam et al. 1983, 1988, Milne 1983). Taber (1986) also noted similar exceptions and stated that research is required to determine different mechanisms of resistance to chalkbrood.

Research has defined some secondary mechanisms of resistance to chalkbrood, and no doubt others exist. These include faster filtering of *A. apis* spores by the proventriculus of bees from hygienic colonies than from nonhygienic colonies and increased resistance to *A. apis* spores of larvae from hygienic colonies (Rath 1985). Also, molds present in bee bread (stored pollen) inhibit the growth of *A. apis* (Gilliam et al. 1988). Colonies that were resistant to chalkbrood had more of these molds, which were apparently introduced by the bees, than colonies that were susceptible.

In Denmark, lines of honey bees were tested for hygienic behavior, and after four generations of selecting and breeding, queens were offered to Danish beekeepers (Holm 1986). Colonies headed by these queens had 9.1 percent chalkbrood compared to 71.4 percent in the original stock of the beekeepers.

Queens heading colonies that are highly susceptible to chalkbrood should be replaced. Elimination of susceptible strains can often be accomplished with routine requeening, which has the added benefit of producing a break in the brood cycle to allow the bees to remove mummies and clean the cells. Queens heading colonies in an apiary that are always free of chalkbrood when adjacent colonies have the disease are good candidates to serve as breeders.

Stephen and Fichter (1990a, b) successfully selected lines of the alfalfa leafcutting bees that had resistance to chalkbrood and concluded, after appropriate backcrosses, that resistance was polygenic. Rank et al. (1990) documented the greater productivity of a univoltine strain of leafcutting bees and concluded that restricting bees to univoltinism would minimize chalkbrood and facilitate its control.

Differences have been reported in the relative hygienic behavior between subspecies of *Apis mellifera*, *Apis cerana*, and Africanized honey bees. For example, Kefuss (1995) found that *A. mellifera intermissa* colonies from Tunisia had the highest hygienic behavior of several subspecies of *A. mellifera* that he tested from France, Tunisia, and Chile. Boecking et al. (1993) reported that *A. cerana* is more effective than *A. mellifera* in removing mite-infested sealed brood. In some studies, Africanized honey bees had better hygienic behavior than European bees (Cosenza and Silva 1972, Loper 1995); in another, there were no differences (Lengler 1977); and in a recent study, European colonies were found to be superior to Africanized colonies in this regard (Danka and Villa 1994). Thus, the issue of hygienic behavior in Africanized bees is confusing and requires more research.

Authors' Opinions

Chalkbrood is a stress-related disease of honey bees. Most of the numerous management procedures that have been recommended to control the disease are aimed at reducing stress and pathogen load and at maintaining strong colonies. Strong, healthy colonies receiving optimal nutrition are less likely to have chalkbrood than weaker ones.

It is unlikely that an antifungal chemical to control chalkbrood in honey bees will be approved in the near future in the United States. Registration is a time-consuming process, even after completion of the required expensive tests to show efficacy without residues. The market for a product must justify this expense, and the beekeeping industry is too small for most companies to invest the large amounts of money required to steer a product from development to final approval.

Even though most of the research on chalkbrood control has concerned chemicals, this research has been plagued by problems and difficulties. Many chemicals used by beekeepers may not be beneficial in any way other than stimulating the cleaning instinct of the bees. Moreover, many antifungal compounds are toxic to bees and humans. Often the compounds are not accepted by the bees, and many are unstable in the carriers used. Publications on chemical control include field results that were not repeatable at other locations, observational rather than solid experimental evidence, and laboratory results that were not verified using bee colonies in properly designed experiments. Chemicals that inhibit the growth of pathogens in culture often do not control diseases in bee colonies and vice versa.

Researchers must recognize that genetically determined hygienic behavior of honey bees affects test results of chemical trials. For example, use of bee colonies exhibiting good hygienic behavior can give positive test results that are due to the genetics of bees rather than to the chemical. Thus, test results and recommendations for treatment are frequently questionable because researchers generally use randomly selected colonies about which nothing is known of the hygienic behavior. Proper control colonies are often not included, and often trials report chalkbrood in controls that were supposed to be free of the disease. Therefore, controlled field experiments using colonies whose hygienic behavior characteristics have been determined are required to assess whether the test chemical reduces or eliminates the disease. Sister queens should be used to head all of the colonies to reduce variability. The test colonies should be inoculated with a standard dose of *A. apis*. Natural infection levels are too variable to be used. And finally, residues in bee products must be determined.

Many chemical control methods require laborious procedures, such as frequent spraying of the chemical on all of the inner hive components, which cannot be used by beekeepers with large numbers of colonies. Many antifungal compounds are too expensive for beekeeper use. Thus, effective chemicals that can be easily applied by methods such as dusting or feeding and are cost-effective are necessary if this route of control is to be taken. However, chemical control may be difficult to achieve because *A. apis* is long-lived and has highly resistant spores that are well-protected by their thick walls and by their location within spore balls within spore cysts.

The selection and breeding of hygienic and resistant or tolerant bees seems the most promising method for control of chalkbrood in all affected bee species. We need to identify and understand other natural mechanisms that bees use to cope with chalkbrood. These can then be employed in control programs. For example, the results of Spivak and Gilliam (1991, 1993) indicate that physiological resistance of honey bees to chalkbrood may exist and that nonhygienic colonies, at least those with no physiological resistance, may cope with chalkbrood by sealing larvae in their cells to avoid further spread of the pathogen by nurse bees to larvae during feeding. What are the mechanisms of physiological resistance? One may be the antifungal products of molds associated with stored pollen (Gilliam et al. 1988). These molds are now being tested for control in field tests with honey bee colonies (Gilliam and Taber unpublished). No doubt other mechanisms exist. In the field of plant pathology, research on disease control with antagonistic microorganisms has been conducted for some years (Becker and Schwinn 1993), and honey bees have been used to transfer these antagonistic biocontrol agents to flowers of some crop plants (Peng et al. 1992, Thomson et al. 1992). Becker and Schwinn (1993) pointed out that development costs for an indigenous nonmanipulated microorganism in the United States are much less than those for a pesticide. Also, the reduced requirement for toxicological studies shortens the time required for development.

Bettsia alvei (Pollen Mold)

Bettsia alvei, formerly called *Ascosphaera alvei* and *Pericystis alvei*, is a fungus that is saprophytic on pollen stored by honey bees in cells of the comb. Therefore, it is commonly referred to as pollen mold. It does not attack brood (Betts 1932).

Pollen that is moldy with *B. alvei* may be mistaken for chalkbrood. However, it breaks up easily into fragments representing the original pollen loads (Betts 1951). *A. apis* and *B. alvei* are similar, but the latter has a low optimum temperature for spore germination and growth and therefore usually becomes evident in early spring after a wet winter (Bailey 1963a), whereas chalkbrood generally occurs later in the season. Moreover the fruiting bodies of *B. alvei* are small compared with those of *A. apis*, and the spores within the former are single.

History

Betts (1912a) first described the pollen mold and named it *Pericystis alvei*. She observed that some combs on which the fungus was present contained cells that were no longer white but had become gray, owing to the formation of cysts. She also gave a detailed morphological description of the organism.

In 1919, Betts compared *B. alvei* with the chalkbrood fungus and noted that the two organisms, although similar, were distinct. *B. alvei* did not grow at brood nest temperature, and the cysts contained simple, spherical, transparent spores. In contrast, the chalkbrood fungus grew well at brood nest temperature, and the cysts contained spherical bodies, each of which contained several small spores. Zander (1919) and Claussen (1921) confirmed Betts's observations.

Burnside (1927) described a pollen fungus and named it *Ovularia farinaecola*. Betts called his attention to the similarity between this fungus and *B*.

alvei. Subsequently, Burnside (1934a) compared cultures received from Maurizio in Switzerland and Betts in England with isolates from the United States and agreed that *O. farinaecola* was indeed *B. alvei*. Therefore, he withdrew the name he had given to the fungus in 1927.

Maurizio (1929) pointed out that as early as 1919, Morgenthaler in Switzerland had suspected that *B. alvei* was heterothallic, and she confirmed that the cysts arise only when the two opposite mating types come together. In 1934, she reported that *A. apis* and *B. alvei* each exist in two forms based on the difference in size of the cysts.

Spiltoir (1955) established that *A. apis* and *A. alvei* were ascomycetes. Spiltoir and Olive (1955) reclassified the genus *Pericystis* and gave *Pericystis alvei* the name *A. alvei* without examining fertile cultures of the organism. Then Skou (1972) separated *A. alvei* from the genus *Ascosphaera* and founded a new genus *Bettsia* with the genus type *B. alvei*. He also reported that *B. alvei* grew poorly, if at all, on ordinary media but grew well and produced spore cysts on a medium containing honey, yeast extract, and pollen. The organism did not start growth at room temperature but grew fairly well at 18°C (64°F).

Distribution

B. alvei has been found on pollen in hives of honey bees in Great Britain (Betts 1912a, b, 1919, 1932, 1951), Switzerland (Maurizio 1929), the United States (Burnside 1927, 1934a, Batra et al. 1973), France (Giauffret and Taliercio 1967), and Denmark (Skou 1972). *B. alvei* was also identified as the cause of a mycosis in a pupa of an unidentified wild bee from New Zealand (Thomas and Poinar 1973).

Etiology

B. alvei is common in hives during winter and early spring and grows on pollen stored in combs (Betts 1912b), mainly those removed from the hive (Skou 1972). The pollen supplies on which it develops are rendered useless for bees (Giauffret and Taliercio 1967).

The organism does not grow in cells that are filled with pollen and finished with a layer of honey on top before they are sealed with wax. However, it grows well in cells that are not filled with honey or where the honey on top has been removed (Skou 1972).

Because the spores do not die during the heat of summer, they are transported to new nests by swarming bees and germinate within one to five days at 15° to 18°C (59° to 64°F) (Zander 1919). According to Giauffret and Taliercio (1967), *B. alvei* develops on pollen more readily at a temperature below 30°C (86°F) and is found particularly in the cold season in weakened hives.

The contents of the cells attacked by the fungus dry into hard plugs that often split into layers. These plugs are permeated by mycelia. Combs on which the fungus is present may contain cells in which the fungus is gray, owing to the formation of cysts.

Control

B. alvei is undoubtedly a normal inhabitant of a healthy bee colony (Betts 1912b). Evidently, pollen mold has not been considered a serious problem, and therefore control measures are deemed unnecessary. Betts (1951), however, recommended soaking moldy combs in water for 24 hours and then propping them up and syringing them to wash out the moldy pollen. The combs should then be allowed to dry and, if desired, syringed again. Afterwards the cells containing very hard lumps of pollen must be cut away, leaving the midrib on which the bees can build more cells. Glinski and Rzedzicki (1980c) demonstrated the efficacy of polyene antibiotics to control growth of *B. alvei*. In addition to their potential use for chalkbrood control, these compounds may also control pollen mold.

Zander (1919) reported that with the arrival of warmer weather, the spoiled masses of pollen are always removed from the hive. This cleaning is not easily accomplished, however, because of the hard pluglike masses caused by the growth of the fungus through the pollen. Therefore, the bees gnaw the cells to remove the pollen masses from the combs. Zander recommended keeping hives dry and wrapping them during winter to avoid the growth of *B. alvei* on pollen.

Stonebrood

Stonebrood in honey bees is rare and is considered to be of minor importance to beekeepers. Both brood and adult bees are susceptible to attack by several species of fungi belonging to the genus *Aspergillus*. *Aspergillus flavus* causes stonebrood most frequently, although *A. fumigatus* is encountered occasionally.

Infection of larvae is more often recognized than infection in adult bees, and infected larvae and pupae are transformed into hard, stonelike mummies after death. The abdomens of infected adult bees also become mummified.

History

Stonebrood was first described by Maassen in 1906 in Germany. He isolated and identified *A. flavus* from dead bees and noted that affected larvae and pupae had been transformed into hard, leathery mummies.

Bahr (1916) in Denmark reported finding stonebrood on rare occasions. Betts (1919) speculated that the disease might have been imported to Great Britain from continental Europe. Zander (1919) described stonebrood as appearing suddenly and disappearing again. In 1923, Vincens reported stonebrood in France. Morgenthaler (1927) described stonebrood caused by *Aspergillus niger.*.

Toumanoff (1928) concluded that in some cases, mortality resulted from toxic products elaborated by the fungus in the bee's intestine and liberated by the digestive juices of the gut. Burnside (1930) found that the action of the pathogenic fungi on bees is both physical and chemical in nature. Tissues are penetrated by the mycelia and are digested by enzymes produced by the fungus. Burnside demonstrated that a transient toxic substance in a strain of *A. flavus* was the cause of fatal poisoning in bees. In addition, he found that *A. flavus, A. oryzae, A. effusus* (= *A. oryzae* variety *effusus*), *A. parasiticus, A. flavus-oryzae* (= *A. oryzae?*), *A. fumigatus, A. nidulans, A. ochraceus,* and members of the group *A. glaucus* killed bees when inoculated experimentally. *A. flavus* attacks bees more frequently than other Aspergilli, but *A. fumigatus* is highly pathogenic. Also, *A. nidulans, A. niger,* members of the *A. glaucus* group, and *A. ochraceus* attack bees in nature.

Dreher (1953) observed several spontaneous outbreaks of stonebrood in Germany. He stated that a spontaneous recovery was impossible; the bees could not remove the mummies from the cells because the mummies were connected to the cell wall by mycelia. He observed no toxic effect of *A. flavus* and concluded that an epidemic outbreak of stonebrood probably occurs when a highly virulent strain of the fungus attacks colonies that have a low resistance.

Bailey (1968a) stated there is no experimental evidence to show that *A. flavus* and *A. fumigatus* are primary pathogens even though these organisms occasionally multiply in bees. McLellan (1964), however, found an outbreak of stonebrood (*A. flavus*) in Scotland that was the first reported occurrence of the disease in the United Kingdom in 13 years.

Mitroiu et al. (1966) and Giauffret and Taliercio (1967) tested antimycotic chemicals against the causative agents of stonebrood. However, these chemicals were not tested against the fungal diseases in vivo.

Distribution

Stonebrood has been reported from Europe (Maassen 1906, Bahr 1916, Zander 1919, Vincens 1923, Toumanoff 1928, Dade 1949, Dreher 1953, Bailey 1963a, McLellan 1964, Giauffret and Taliercio 1967, Tomac et al. 1983), North America (Burnside 1930, Steinhaus 1949a, Bailey 1963a), Venezuela (Stejskal 1958), Australia (Hornitzky et al. 1989), Egypt (Shoreit and Bagy 1995), and in *Apis florea* in Iran (Alizadeh and Mossadegh 1994).

Etiology

Aspergillus flavus and occasionally *A. fumigatus* and other species of

Aspergillus infect and kill both larval and adult honey bees. These fungi are ubiquitous and are commonly found in soil. They are pathogenic for other insects and can cause respiratory diseases in humans and other animals.

Larvae in sealed and unsealed cells may be affected as well as pupae. Capped pupae are less susceptible (Cury 1951). Most infected larvae die in the sealed stage before pupation, and adult summer bees seem particularly susceptible, although they may die at any age (Dreher 1953).

Bees are attacked when the fungal spores are ingested (Burnside 1930). After the spores germinate within the alimentary canal, the resulting mycelia may attack all the softer tissues. Spores can, however, germinate on the cuticle, and the mycelia will then penetrate the tissues.

As the fungus invades the tissues, the bodies of the larvae and the abdomens of the adult bees become hardened. In infected larvae, the fungus develops rapidly, passing through the cuticle and forming a characteristic whitish-yellow ring or collar behind the head. Within one to three days, it envelops the whole larva as a false skin. The fungus then produces spores on the external part of the dead larva, and the color changes to green. This growth is powdery. *A. flavus* is yellow-green, and *A. fumigatus* is gray-green. The disease causes a mummification of brood, and the mummies are solid and hard. At times, the fungal spores are formed in such large numbers that they completely fill the comb cells that contain mummified larvae.

Adult bees usually allow any brood killed by stonebrood to remain in combs for some time, or they only partially remove it because destruction of the cell walls is often necessary for complete removal (Burnside 1930). Dreher (1953) found that the mummies were not loose in the cells but were attached to the cell walls by aerial mycelia.

The first noticeable symptoms of stonebrood in adult bees are abnormal restlessness, feebleness, and paralysis. The abdomen is generally dilated. Spores form earliest and most abundantly near the head. The abdomen of a dead adult bee frequently shows mummification similar to that of the entire body of the larva. It does not decay, and the interior often becomes hard as a result of the fungal growth. The fungus forms spores on dead adult bees, especially on the transition parts of the thorax and abdomen.

Usually the colony is not seriously affected because only a small percentage of adult bees or larvae is infected. Death of naturally infected colonies has been observed, however (Dreher 1953). Cases are known where scarcely any brood disease occurred, yet adult bees died in large numbers (Betts 1919).

Bailey (1963a) suggested that because of the rarity of the disease and the fact that the causative fungi are common, the spores may cause disease only when they are present in subnormal (stressed) larvae or prepupae. If much of the brood dies, then the colony may die because it will be weakened, and the remaining brood and aging bees may be susceptible to attack by the fungus.

The method of natural spread of the disease is not known. Betts (1919) stated that stonebrood was undoubtedly disseminated by beekeepers by the interchange of combs from diseased to healthy colonies. She also believed that honey from infected combs fed to bees would cause the disease. Giauffret and Taliercio (1967) stated that the spread of fungal diseases was probably related to the use of antibiotics, which upset the equilibrium of the normal intestinal flora of bees. They also listed humidity, lack of ventilation, food or supplies containing too much water, and genetic factors as predisposing factors in the development of fungal diseases. Maassen (1906) suggested that the fungus probably first develops in comb cells filled with pollen, but he noted that this had not been proved. Cury (1951) believed that larvae and adult bees are attacked more frequently when the humidity is high and that stonebrood is transmitted in ingested pollen or honey containing fungal spores. Grigortsovskaya and Borodai (1972) fed spores of *A. fumigatus* and *A. niger* in sugar syrup to bees of various ages. After three to four days, the bees became hairless and were unable to fly. The younger bees died first.

Attempts to infect colonies artificially have been unsuccessful (Bailey 1963a). Therefore, stressed or abnormal bees may be more susceptible, and toxins (Toumanoff 1928, Burnside 1930), perhaps aflatoxins, produced by the fungi might be responsible for the disease. Indeed, Hilldrup and Llewellyn (1979) observed increased mortality of caged adult bees fed aflatoxins in concentrations as low as 5 parts per million. Hilldrup et al. (1977) and Gunst et al. (1978) showed that aflatoxins are produced by *A. flavus* when it is grown on a variety of apiarian substrates, including dead bees. Burnside (1930) concluded that the pathogenicity of the fungi that attack bees through the gut wall appears to be determined by the ability of the spores and mycelia to resist the action of the intestinal fluids.

Control

There is no known treatment for stonebrood (Bailey 1963a, Giauffret and Taliercio 1967). In severe cases, however, Betts (1919, 1951) recommended burning the bees, combs, and all contents of the hive and then disinfecting the hive. To save a mildly diseased colony, she suggested shaking the bees onto new equipment, disinfecting the old hive, and burning all combs. While doing this work, one should protect the eyes, mouth, and nose to reduce the risk of human infection. Betts emphasized that the honey from colonies infected with stonebrood is not safe for human consumption because *A. flavus* has been known to grow in human nasal passages.

Dreher (1953) fumigated badly infected colonies with sulfur, sterilized the hives and rendered the combs. Colonies in which only the brood was affected were treated by brushing the bees into artificial swarm boxes and feeding the bees in these boxes for two days in a cool, darkened room. Hives and accessories were sterilized and the frames provided with foundation. The bees were then returned to the old hives and regularly fed until comb was built. He noted no relapse.

Giauffret et al. (1969) found that fumigating combs with ethylene oxide for 15 hours at 22°C (72°F) killed *A. flavus*. Cantwell et al. (1975) found that the fungus could be killed within one-half hour with ethylene oxide at a concentration of 100 mg/liter of fumigant. Giauffret and Taliercio (1967) studied the possibility of treating fungal diseases with antimycotics and antiseptics. They determined the in vitro sensitivity of two strains of *A. flavus* to antimycotics and compared the toxicity of these products for adult bees and brood. They found that nystatin and thiabendazole were most effective against *A. flavus*. Actidione was the most toxic test substance to adult bees and larvae. They noted, however, that studies on the activity of these antimycotics against fungal diseases should be done in vivo. Mitroiu et al. (1966) found that stamycin, a mycostatic agent, inhibited development of cultures of *A. niger* and *A. fumigatus*. Glinski and Rzedzicki (1980c) and Glinski et al. (1981) also tested antimycotics and disinfectants against *A. flavus* in vitro and found that polyene antimycotics, including amphotericin B and nystatin, have very low minimum inhibitory concentrations.

Gochnauer et al. (1975) stated that no treatment for stonebrood is required because bees remove diseased brood, and the colony appears to recover spontaneously. Zander (1919), however, found that when bees attempted to remove larvae affected by stonebrood, they gnawed away the cell walls and reached the mummies only with difficulty. Therefore, they were only partly successful in their attempts. Larvae that could not be removed were covered with propolis so that only the heads remained visible.

Maassen (1906) observed that bees were not very susceptible to the disease. The susceptibility increased if the hive temperature was high, the hive was poorly ventilated, or the bees were confined for a long time.

Other Fungal Diseases

Melanosis

Fyg (1934) discovered a disease that affected the reproductive systems of queen bees, rendering them sterile. The disease was designated H-melanosis (from the German *Hefe*, meaning yeast) to distinguish it from another melanosis (B-melanosis) caused by a bacterium. The systematic position of the yeastlike microorganism presumed to cause H-melanosis is still undetermined.

According to Fyg (1964), the organism probably enters the reproductive organs through the sting chamber and vaginal orifice. It produces a melanosis, or black coloration, in the oviducts and ovaries. The poison sac and poison gland of the queen are also affected and contain large black swellings. These swellings exert pressure on the oviduct, and egg laying ceases. Because infected queens become sterile, Cury (1951) recommended that they be replaced, although such a queen will soon be superseded by a new queen in any case (Bailey 1963a).

Healthy queens can be infected by vaginal inoculations in the laboratory (Fyg 1964), and adult workers and drones have been infected with cultures injected into the thorax (Fyg 1936). Direct feeding of the pathogen to bees has not produced infection, however. Unfortunately, nothing is known about the natural incidence of the disease (Bailey 1968a).

Skou and Holm (1980) observed symptoms similar to H-melanosis in both queens and workers. They were not able to correlate the occurrence of such symptoms with infection with *Nosema apis*, insemination methods, or other factors. Furthermore, they were not able to induce the symptoms by inoculating bees with a variety of yeast species. They concluded that the real cause of melanosis remains undetermined.

The pathogen described by Fyg (1964) produces white, smooth, shiny colonies on bee-wort agar at 30°C (86°F). These colonies never darken, even in old cultures. Fyg (1964) contends that this organism is not identical with the organism *Melanosella mors apis* described by Örösi-Pál (1936, 1939b). Although Skou and Holm (1980) isolated many yeasts from queen bees with melanosis, they were unable to isolate any fungi (or other microorganisms) from the black knots or other affected tissues.

Örösi-Pál (1936) found a living mated queen that had ceased to lay eggs. The genital orifice was plugged with a brown substance in which the sting was embedded. This "anal wad" was lodged solidly between the last two abdominal segments and the sting. He cultivated spores from the wad and ovaries of this queen and also found the organism in hindguts of infected worker bees. The most striking sign of the disease in worker bees was an evagination of the hindgut. He gave the name of *Melanosella mors apis* to the pathogenic agent and was able to cause mortality in worker bees within 24 hours after feeding them pure cultures of the organism.

Gontarski (1937/38) found a fungiform organism between the ovarioles of a queen suffering from ovarian atrophy. Fyg (1964), however, examined 224 cases of ovarian atrophy of queens and did not encounter this condition.

Poltev and Neshatayeva (1969) stated that the agent *M. mors apis* corresponds morphologically and culturally to *Aureobasidium pullulans* (=*Pullularia pullulans*). Skou and Holm (1980) concur that these two fungi are synonymous.

Poltev and Neshatayeva (1969) injected worker bees in the thorax and abdomen with *A. pullulans* obtained from culture collections and from queens with melanosis. Infection began in the regions injected, and no pathological differences were observed between strains of the fungus. They postulated that the melanosis organism is introduced into the hive with honeydew honey.

Skou and Holm (1980) isolated *Paecilomyces lilacinus* and induced a lethal disease upon vaginal inoculation of queens. They did not name the disease but concluded it is not a problem among honey bee colonies.

Gontarski (1950) found melanized hypopharyngeal glands in about 15 percent of the pollen-collecting bees from a colony. He thought that queens became infected when they were fed the hypopharyngeal gland secretion by diseased workers. Nectar-collecting bees from the same colony were not infected. Therefore, pollen was suggested as the source of infection.

Liu (1991a) reported black spots on the surface of the ovaries of a few queens and noted that melanization is a defensive mechanism of insects in which melanin deposits are formed around invading microorganisms or wounds inflicted by microorganisms or other pests.

Other Molds

Fungi are commonly found as saprophytes on and inside bees and on brood combs. Most of the fungi collected by the widespread foraging of bees are probably unable to become established within the bee or the beehive (Burnside 1927). However, after the death of the bee, some of the fungi germinate and mummify the softer tissues inside the exoskeleton. Others do not germinate within the bee either before or after death.

Betts (1912b), Burnside (1927), and Maurizio (1931) described fungi associated with hives. Combs that have been used for brood rearing and for the storage of honey and pollen are susceptible to overgrowth by fungi if sufficient moisture and proper temperatures for the growth of the fungi are present (Burnside 1927). Damage is generally limited to poorly ventilated hives, weak colonies, and combs in unoccupied hives. Through the proper regulation of humidity within the hive, the bees are able to protect themselves and their stores from attack by fungi.

Burnside (1927) noted that the Penicillia are the most common group within the beehive. He frequently found entire combs and, at times after the death of the bees, all the combs of a hive overgrown with these organisms. Badly infected combs are not readily accepted by bees. He added that actual parasitism of bees has not been established for any species of *Penicillium*. The Aspergilli occur far less abundantly in the hive, and species of *Mucor* do not grow well on brood combs.

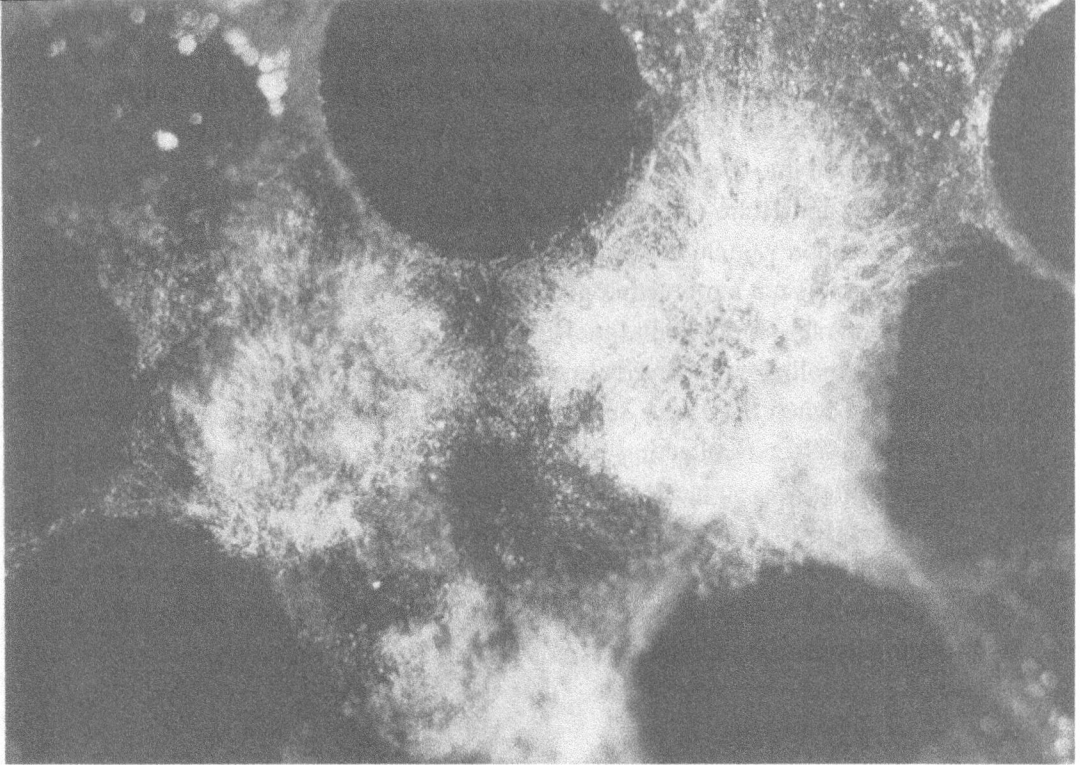

Figure 5.3. Fungi are commonly found as saprophytes on and inside bees and on brood combs. Some fungi will germinate in larval remains and dead bees if temperature and humidity conditions are right. (Killion photo).

Fielitz (1925) reported that *Trichoderma lignorum* and *Mucor mucedo* found on mummified bees were pathogenic to brood and adult bees when introduced into healthy colonies on brood combs. Nicholls (1934) and Chowdhury (1953) also found that *T. lignorum* caused disease in bees. In Tasmania, the fungus appeared to be associated with *Nosema apis* (Nicholls 1934). Bees fed *T. lignorum* spores in sugar syrup died (Grigortsovskaya and Borodai 1972). Mycelium was found in the epithelium of the midgut and in muscles, and the gut contained many fungal spores.

Mucor hiemalis was reported to be virulent for young adult bees when they were exposed to a temperature of 20°C (68°F) (Burnside 1935a). However, the normal temperature of the brood nest is above the tolerance of *M. hiemalis*. Moreover, when bees are old enough to leave the hive, they are no longer susceptible. Therefore, this disease is of little importance.

Queen larvae in sealed cells have been reported to be attacked by *Aspergillus niger* (Burnside 1939). Prest et al. (1974) isolated the fungus from dead queen larvae and found other species of fungi in discolored worker and drone brood. Gilliam and Prest (1977) isolated *A. niger* and various other fungi from healthy adult queens.

In Denmark, a case of a disease similar to stonebrood was described and attributed to a fungus of the genus *Claviceps* (Cowan 1881). Drone brood was attacked first, then worker brood, and finally adult bees.

Cury (1951) described a disease of larvae and adult bees caused by *Rhizopus equinus* that was obtained via the pollen of flowers contaminated with horse manure. In addition, he stated that in cold countries, *Scopulariopsis brevicaulis* caused a disease called yellow brood or black brood. The affected larvae died and turned yellow or black according to the age of the fungus and strongly adhered to the walls of the cell. The disease does not affect protected brood and generally disappears when the temperature rises.

Shabanov and Georgiev (1979) reported that moldy honey and nectar were responsible for extensive colony mortality in Bulgaria. Kunchev et al. (1983) found *Geotrichum candidum utilis* (=*G. candidum*) and *Aspergillus niger* in dead bees and concluded that fungal contamination of pollen was responsible for the mortality.

Stejskal (1976a, b) described *Labyrinthula apis* and *Endomycopsis apis* from the hemolymph of adult honey bees. He suggested that both may produce toxins and cause premature mortality.

Shaw and Robertson (1980) observed honey bees collecting spores of *Neurospora intermedia* on their corbiculae. The collecting occurred during a pollen dearth, and the authors suggested that the bees may employ the spores as a pollen substitute. Shaw (1990) reviewed reports on the incidental collection of fungal spores by bees and on the collection of spores in lieu of pollen. Spores of rust fungi, powdery mildews, and *Neurospora* have been collected by bees in lieu of pollen.

Yeasts

Yeasts have been found in nectar (Betts 1920, Lochhead and Heron 1929, Batra et al. 1973), honey (Fabian and Quinet 1928, Lochhead and Farrell 1931, Lochhead and McMaster 1931, Tysset and Rautlin De La Roy 1974), bee bread or stored pollen (Wilson and Marvin 1929, Pain and Maugenet 1966, Egorova and Bab'eva 1967, Egorova 1971, Gilliam 1979), hives (Betts 1928a, b, c, 1929a), apiary soils (Lochhead and Farrell 1930), and bees (see Batra et al. 1973, Gilliam et al. 1974).

Yeasts are not considered to be pathogens of honey bees, although some osmophilic species cause fermentation of honey. Giordani (1952), however, isolated a yeast belonging to the genus *Torulopsis* from the digestive tracts of bees suffering from a disease of unknown nature. When this yeast was fed to healthy bee colonies, it developed in the digestive tracts of the bees and caused death. Injection of some species of yeasts into bees causes death (Burnside 1930, Batra et al. 1973). However, some yeasts may ferment the provisions of honey bees (Betts 1920, Moreaux 1949, Vecchi 1959) and cause sickness or death of the larvae or adult bees that ingest these provisions.

In fact, yeasts may be beneficial to honey bees by playing a role in the production and preservation of bee bread (Pain and Maugenet 1966, Egorova 1971) and by supplying vitamins and other growth factors (Egorova 1971, Batra et al. 1973). Giordani (1957) fed caged bees *Saccharomyces*, Torula yeast (*Candida utilis*), or bee-collected pollen in sugar syrup. From these experiments, she concluded that the yeasts had approximately the same nutritional value for young bees as pollen. Moreover, the Torula yeast gave better growth results than pollen. Hajsig (1959) cites the results of other researchers to point out that yeasts have a favorable effect on the larval development of bees. Peng et al. (1984) followed the course of the Torula yeast consumed by bees and used histological techniques to demonstrate digestion of the yeast. Much more research is needed to assess the effect of yeasts (and other microorganisms) on the nutrition and physiology of honey bees.

Yeasts are more readily isolated from adult honey bees from colonies that have been (1) treated with antibiotics such as Terramycin (Tomasec 1957, Hajsig 1959, Mitroiu et al. 1966, Gilliam 1973, Gilliam et al. 1974), (2) treated with herbicides (Gilliam et al. 1974, 1977), (3) caged (Gilliam 1973), (4) suffering from diseases not caused by yeasts (Betts 1928c, Lavie 1951, Kamburov and Hajsig 1963, Gilliam 1973), or (5) receiving deficient diets (Gilliam 1973). Thus, their presence in bees may be indicative of stress conditions (Gilliam 1973).

CHAPTER SIX

Nematodes
Roger A. Morse

Parasitic and nonparasitic associations
Entomopathogenic nematodes

CHAPTER SIX

Nematodes

Nematodes have never been reported to be of economic importance as far as honey bees are concerned. But when you make an anatomical examination of a worker bee, it is a shock to find a three-inch-long nematode neatly curled up in her abdomen. This has occurred several times in the Ithaca, New York, area. I once found two such nematodes when examining the honey stomachs of 2,111 bees, which attests to their rarity. Chitwood assigned these nematodes to the genus *Agamomermis*, a group of agricultural importance, which were found in the larval stage and that he could not identify further (Morse 1955). Larvaform nematodes are difficult to identify.

Parasitic and nonparasitic associations

Vandenberg (1990) divided nematodes into three groups: parasitic, nonparasitic and artificial. Among the parasitic associations, he cited reports and reviews by Kramer (1902), Milum (1938), Fyg (1939), Paillot et al. (1949), Toumanoff (1951), Morse (1955), Welch (1963), Borchert (1966), Poinar (1975) and Bailey (1981). Since the publication of Vandenberg's paper, I find no papers on nematodes reviewed in *Apicultural Abstracts*, a journal that abstracts nearly 1,500 papers on bees and beekeeping annually; this, too, testifies to their rarity.

The identification of many parasitic nematodes is in question according to Vandenberg, and their biologies are even more vague. It is generally thought that nematodes that infest honey bees must be general feeders that are capable of surviving in a variety of insects. It is suggested that the bees become infested when they ingest an egg or early stage larvaform while feeding on nectar or pollen on a flower. It appears unlikely that even under the most opportunistic situation, an adult nematode would be able to contact a honey bee to deposit an egg or larva directly on its body. There are no records of nematodes that live free in bee hives and attack the honey bees therein.

Regarding the nonparasitic nematodes, Vandenberg (1990) cites papers by Batra (1980) and Giblin and Kaya (1984). During the dispersal stage, these nematodes inhabit the reproductive tracts or Dufour's glands of honey bees. The nematodes are left in the cells of developing bees where they feed on the cell linings and the microorganisms they find there. They attach to the bees as they leave their cells and are thereby transported to a more favorable site. Vandenberg states it is not clear if the bees are harmed by the nematodes.

Entomopathogenic nematodes

The use of nematodes as biological control agents has aroused considerable interest in recent years. Vandenberg (1990) cites several researchers (Cantwell et al. 1972b, Hackett and Poinar 1973 and Kaya et al. 1982) that have tested one nematode, *Neoaplectana carpocapsae*, for its effect on honey bees. The effect was minimal, and there is no indication that this nematode species poses any direct threat to honey bees or their young. It is postulated that the high temperature in the honey bee's brood nest may protect the brood and bees in it against this species of nematode.

The subject of entomopathogenic nematodes was reviewed in detail by Kaya and Gaugler (1993) as follows. Several species of nematodes in the families Steinernematidae and Heterorhabditidae are of the greatest interest and of value in biological control programs. They have a broad host range and kill almost every insect against which they have been tested. Nematodes in these two families are always mutualistically associated with bacteria in the genus *Xenorhabdus*, of which five species have been described, though probably more exist. The third-stage juvenile nematodes carry the bacteria in their intestinal tracts. The nematodes enter the host insect through a natural opening such as the mouth, anus, spiracles or any wounds that may exist. It is possible that some species enter through the thin, intersegmental membranes, but this is considered rare. The bacterial cells are released into the insect's hemolymph along with the nematode feces. The bacteria multiply rapidly and kill the insect, usually within two days, through septicemia. The nematodes then pass through two or three generations, feeding on both the bacteria and tissue of their hosts before producing infective stages that seek a new host. It is a curiosity that each species of nematode is associated with only one species of bacteria but that one bacterial species may be associated with more than one nematode species. Whereas it is unlikely that these pathogenic nematodes pose any serious threat to honey bees, interest in them is growing exponentially according to Kaya and Gaugler (1993); the use of nematodes in biological control programs should be watched closely.

CHAPTER SEVEN

Insects: Lepidoptera (Moths)
Jon L. Williams

Greater wax moth
    Distribution
    Economic importance
    Morphology
    Biology
        Eggs
        Larvae
    Galleria-caused absconding
    Galleriasis
    Control
        Natural control
        Applied control
        Chemical control
        Microbial control
        Physical control
        Control in comb honey
        Control in queen rearing materials
    Future control techniques
Driedfruit moth
    Biology
    Control
Lesser wax moth
    Biology
    Bald brood
Bumble bee wax moth
Indianmeal moth
Mediterranean flour moth
Death's-head moth
Moths in bee-collected pollen

CHAPTER SEVEN
Insects: Lepidoptera

Several moth species attack the products of honey bees — honey, pollen, and wax. Included are the greater and lesser wax moths, *Galleria mellonella* and *Achroia grisella*, and several species that are primarily pests of stored products: Indianmeal moth *(Plodia interpunctella)*, the Mediterranean flour moth *(Anagasta kuehniella)*, and the driedfruit moth *(Vitula edmandsae serratilineella)* (Milum 1940a, Smith 1960, Singh 1962). Death's-head moths *(Acherontia* spp.) occasionally feed on honey and nectar (Brügger 1946, Raw 1954, Smith 1960). The bumble bee wax moth, *Aphomia sociella*, is a rare pest of honey bee colonies (Toumanoff 1939).

Greater wax moth

Many people outside the beekeeping arena consider the greater wax moth a useful insect. It is raised commercially in the United States and Europe for larvae, which are used as fish bait and live food for insectivorous zoo animals. The larvae are also extensively used for studies of the physiology, pathology and biological control of insects (Dutky et al. 1962, Bedding 1984, Schaefer et al. 1989, Schaupp and Kulman 1992). The greater wax moth, however, is a perennial pest of honey combs, causing major losses to commercial beekeepers (Oertel 1969, Williams 1976). The greater wax moth infests colonies of *Apis cerana, A. dorsata,* and *A. mellifera* (Smith 1953, Singh 1962, Morse and Laigo 1968, Shimanuki 1981, Sihag 1982).

The destructive activities of the greater wax moth are most severe in the tropics and subtropics, presumably because, having evolved in Southern Asia with honey bees (Morse 1975b), the species is best adapted to warm climates. Unprotected combs in honey bee colonies that become weakened or die during hot weather may be quickly reduced to a mass of webbing and debris by *Galleria* larvae. Colony conditions that enable wax moths to flourish include a lack of food, failing queens, queenlessness, and disease, especially pesticide- and *Varroa*-induced collapse of the worker bee population. Beekeepers of the subtropical areas of the United States reported that large numbers of combs were destroyed in 1993 by the greater wax moth as a result of severe colony losses in July and August from the mite *Varroa jacobsoni*. The capability of *V. jacobsoni* in warm climates to rebound to lethal population levels within eight to 10 months following the late-season application of fluvalinate was unknown until that time.

Large numbers of combs are fumigated at the end of each beekeeping season in warm climates so as to protect them against wax moth damage. A typical

Figure 7.1. Wax moth adults. (Flottum photo).

1,000-colony beekeeping operation in the United States averages about 57,000 drawn combs (Anderson 1969), half or more of which are in storage for about six months out of each year. In the warmer sections of the country, protection of combs against moth pests and other elements is an essential part of the beekeeper's management program. In addition to *Apis* colonies, the greater wax moth infests the nests of meliponids (so-called stingless bees) and bumble bees (Noguiero-Neto 1953, Oertel 1963).

Distribution

The greater wax moth occurs throughout the world, almost everywhere honey bees are kept. Its distribution is limited mainly by an inability of the species to tolerate prolonged subfreezing temperatures. *Galleria mellonella* causes little or no damage at high elevations (Paddock 1926). For example, natural infestations of beekeeping equipment have not occurred for over 40 years in Laramie, Wyoming, U.S.A. (elevation, 2,184 m).

Figure 7.2. Greater wax moth larva in a typical hiding place beneath the top bar of a frame. (Flottum photo).

Figure 7.3. Combs heavily damaged by the greater wax moth. Honey bees do not rebuild comb over areas in which damage has exposed plastic foundation (lower comb), unless some beeswax covering remains (Perrone, USDA).

Economic importance

Several assessments of the economic significance of the greater wax moth to the United States beekeeping industry have been made. Paddock (1918) assumed that *Galleria* destroyed approximately five percent of hives of comb in Texas annually (14,000 units), and Oertel (1969) estimated losses in Louisiana to be approximately $31,000 a year at that time. The greater wax moth causes extensive losses to the major U.S. producers of queens and package bees in the Gulf Coast states, Georgia, and California. A survey of over 100 Southern commercial and semicommercial beekeepers concluded that *Galleria*-caused losses in the United States were estimated to be about $3 million or more during 1973 and about $4 million in 1976 (Williams 1976). Annual losses from the greater wax moth were roughly comparable to those from American foulbrood disease in the United States, which, including operational costs for state apiary inspection programs, probably reach $5 million or more.

Morphology

A newly hatched *Galleria* larva has a creamy white body that becomes gray to dark gray on the dorsal and lateral surfaces as the larva ages; strains of larvae that are white at maturity have been bred commercially in the United States by fish-bait producers. *G. mellonella* larvae, the largest of the comb-infesting Lepidoptera, can reach as much as 28 mm (1.25 in.) in length and weigh up to 240 mg (Hase 1926).

Greater wax moth cocoons are usually bare and white, but some are almost completely covered with dark fecal pellets and frass; they are usually 12 to 20 mm long and about 5 to 7 mm in diameter. Often the last-instar larvae migrate from the feeding site and spin cocoons on hive bodies or on the inner cover. Sheets of as many as 10,000 cocoons may be found covering the inner surface of a two-story, 10-frame Langstroth hive, but normally only about 250 normal-sized larvae are able to develop on one dark brood comb (frame size: 447.7 x 231.8 mm).

Adults are heavy-bodied, fairly small moths. The females range in length up to about 20 mm and can weigh as much as 169 mg (Marston and Campbell 1973); the males are considerably smaller. Male moths may also be distinguished from females by their lighter color and the margins of the forewings. In addition, the female's labial palps extend forward, giving the head a beaklike appearance. The anterior two-thirds of the forewing of normal adults of both sexes is rather dark, but the posterior one-third has irregular light and dark areas interspersed with darker streaks and uneven spots. Dorsally, the thorax and head are light-colored. The size and color of both sexes vary considerably, however, depending on the diet of the larvae. Silver-white adults have been reared from wax foundation, whereas those reared on a diet of brood comb are predominantly brown to dark gray to almost black (Milum 1940a). *G. mellonella* adults may be smaller than some adult lesser wax moths (see below) if the larvae have developed slowly because of poor diet and low temperatures.

Biology

The life cycle of the greater wax moth lasts from four weeks to six months; in the longer cycle, dormancy occurs in the prepupal stage (Marston et al. 1975). The adults require neither food nor water. The female begins laying eggs about four to 10 days after emergence; she produces about 300 to 600 eggs, though individual moths may lay up to 1,800 eggs (El-Sawaf 1950). Adults live from three to more than 30 days, but most mated females die within seven days when held at 30° to 32°C (86°-90°F) (El-Sawaf 1950, Nielsen 1971). Greater wax moths live longer at lower temperatures; females average about 3.8 days at 40°C (104°F) and 19.6 days at 20°C (68°F) (Marston et al. 1975).

The reproductive behavior of *Galleria* is not typical of lepidopterans in that the female searches for the male. The males release a two-component phero-mone (nonanal: undecanal, 7: 3) and short pulses of ultrasonic sound to attract the females (Leyrer and Monroe 1973, Spangler 1984, 1985).

Nielsen and Brister (1977, 1979) observed that greater wax moths usually emerged as adults in the evening during warm weather, left the bee colonies, and flew into trees to mate. One day later, usually soon after dark, mated female moths returned to bee colonies to lay eggs. The female moths approached the colony entrance, became motionless for a few minutes, then entered by running past the guard bees. After leaving bee colonies, male moths seldom appeared again, though a few were observed 13 meters above ground on trees, fanning their wings to attract females. Mated females did not remain in bee colonies, but flew to trees each morning shortly before daylight.

Although mated female greater wax moths were attracted to and laid eggs in only a few bee colonies, nearly all of the colonies in an apiary were infested with

Figure 7.4. A super with frames removed to expose hundreds of *Galleria mellonella* cocoons attached inside. With infestations this heavy, wooden frames and supers become weakened when the moth larvae chew away wood to create an area for cocoon attachment. (Killion photo).

Galleria larvae because newly hatched larvae dispersed to neighboring colonies after feeding on honey. Laboratory studies demonstrated that newly hatched *Galleria* larvae can travel 50 meters or more between feedings. The few moths that were observed during the winter remained and mated inside bee colonies that contained brood, and they flew from the colonies on warm days (Nielsen and Brister 1977, 1979).

Apparently, honey bees respond to a greater wax moth only when it is near or actually touched by the bees. Female wax moths are attacked preferentially over males by both European and Africanized honey bees (Dickens et al. 1986, Eischen et al. 1986).

Eggs

The eggs of the greater wax moth vary from pinkish to cream to whitish. They are difficult to detect with the naked eye, though most are glued together in sheets of 50 to 150 eggs. Undoubtedly, the female moth's habit of depositing eggs in small cracks and crevices that barely admit the ovipositor (Milum and Geuther 1935, Makings 1958) affords considerable protection against removal by worker bees.

Eggs develop rapidly at warm temperatures (29°-35°C)(84°-95°F) and begin to hatch three to five days after oviposition (Dutky et al. 1962, Nielsen 1971). Egg hatch is extended to about 30 days at 18°C (64°F). Short exposure to temperature extremes (at or above 46° [115°F] for 70 minutes, or at or below 0°C [32°F] for 270 minutes) will cause 100 percent mortality of eggs (Cantwell and Smith 1970).

Larvae

For their first meal, *Galleria* larvae eat honey, nectar, or pollen, if available, and beeswax otherwise (Nielsen and Brister 1979). Typically, the newly hatched larvae then burrow into the outer edges of cell walls or into pollen stored in cells. Developing larvae extend their tunnels to the midrib of the comb; there they continue to feed and grow, protected from worker bees.

Larvae of the greater wax moth grow rapidly. If diet and temperatures are favorable, larvae can double in body weight daily for the first 10 days after hatching, and they begin to spin cocoons as early as day 18 or 19 (Beck 1960). Thus, all of the combs in a honey bee colony are often destroyed within 10 to 15 days after the adult bee population has collapsed. Aggregations of *Galleria* larvae produce substantial quantities of metabolic heat; temperatures as high as 25°C (77°F) above the environmental temperature are produced in the center of aggregations of larvae (Smith 1941, Roubaud 1954). Such heat production probably helps to protect the massed larvae against fungal, microsporidial and viral diseases (Allen and Brunson

1957, Stairs 1978). Exposure to cool temperatures, on the other hand, adversely affects the physiology of wax moth larvae. For example, many of the mature larvae that are subjected to cooling are unable to spin cocoons (Cymborowski and Bogus 1976).

Rapid growth occurs under optimal conditions, but *Galleria* larvae can survive even if food is available intermittently or if they feed continuously on a marginal diet. Under such conditions, the total developmental period (egg to adult) is greatly extended — up to six months or so. The size of adults decreases with increasing developmental period (Milum and Geuther 1935, Marston et al. 1975). Developing larvae may feed on practically all of the honey bee products in the colony. Dark combs are especially preferred whether they contain honey, pollen or both. Bee brood (larvae and pupae) will also be attacked if *Galleria* larvae are short of food (Paddock 1918, Milum 1935). In warm climates, larvae are often abundant in the wax, pollen, and debris accumulated on the bottomboard of the hive.

In warm weather, combs of harvested honey or combs freshly extracted and in storage may suffer rapid and heavy damage and even complete destruction unless properly handled. In subtropical and tropical climates, *Galleria* larvae often infest impure cakes of beeswax, slumgum, and wax cappings. In addition, beekeepers rearing queens in warm climates must protect their wax cell cups, waxed wooden cell cups, and waxed cell bars from infestation; otherwise, small *Galleria* larvae may consume pupal and adult queens in finished cells. Although their natural diet usually includes beeswax, *Galleria* larvae rarely complete their development on refined wax, such as foundation, or on the new wax of comb honey (Milum 1940a). In the laboratory, the greater wax moth readily adapts to a variety of beeswax-free diets that contain cereal products (Dutky et al. 1962, Marston and Campbell 1973, Marston et al. 1975). Lack of food causes *Galleria* larvae to become highly cannibalistic, and the larger larvae will devour smaller larvae, prepupae, and pupae.

Galleria-caused absconding

Wax moth depredation, along with overheating, lack of water, and attack by other pests, is one of the common causes of absconding by *Apis mellifera* colonies in South Africa. When colony populations decrease sufficiently to expose combs, wax moth larvae develop unhindered and gradually attack combs occupied by bees, causing the host colony to abscond (Fletcher 1976).

According to Singh (1962), *Galleria* is a pest of nearly all colonies of *A. cerana*, *A. dorsata*, and *A. florea* in India. In that country, larval *Galleria* populations often become so damaging during dearth periods or monsoon seasons that infested honey bee colonies abscond. Wild colonies and deserted combs of *A. dorsata* and *A. florea*, in addition to managed colonies of *A. cerana*, stored combs, and improperly cleaned wax, are constant sources of wax moths there. Greater wax moths are also common residents of active *A. dorsata* colonies in the Philippines; populous colonies are lightly infested, whereas weaker colonies and abandoned nests contain many larvae (Morse and Laigo 1968, Morse 1975).

Galleriasis

During the warmest months of the year, when *Galleria* populations are largest, honey bee queen breeders in the Southern United States commonly observe patches of apparently normal emerging workers and drones that cannot leave the cells of combs in queen-mating nuclei. After they have chewed away the cell cappings, the bees remain trapped in their cells by silken threads that *Galleria* larvae have spun at the bases of the cells (Nielsen and Brister 1977). Entire combs of worker bees that have developed from brood of nearly the same age may be observed trapped in this way. Dissection of combs that contain affected bees reveals one to three *Galleria* larvae near the bottom of most cells. This condition, called galleriasis, can also occur in frames of brood in Langstroth colonies, especially if the first generation is reared in newly drawn combs.

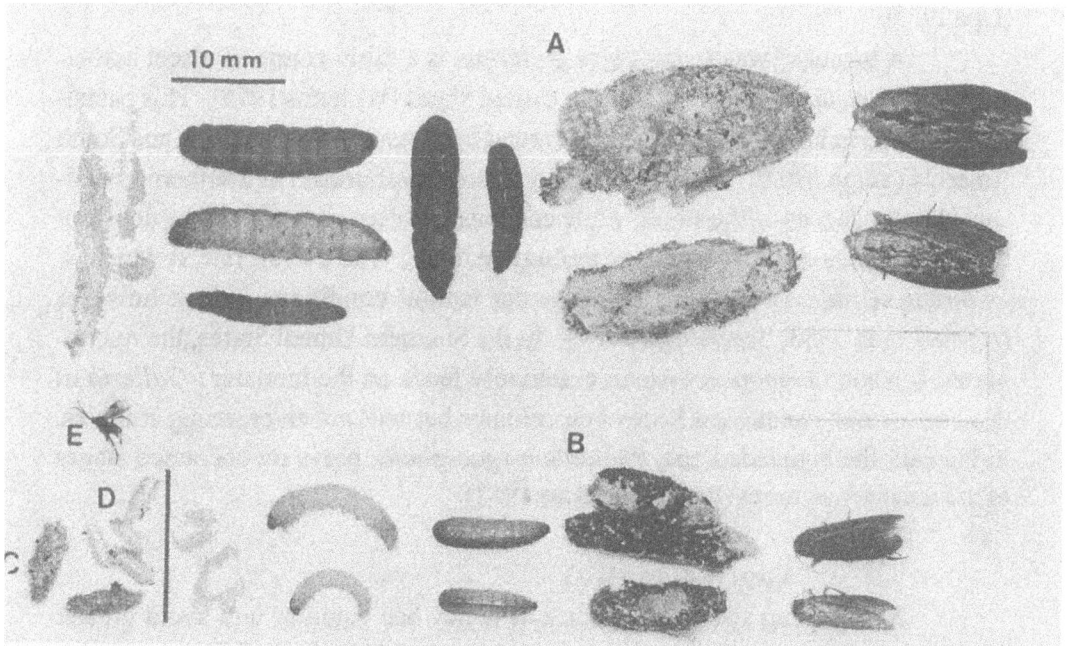

Figure 7.5. (A) Eggs, larvae, cocoons and adults of the greater wax moth, *Galleria mellonella*. (B) Same, for the lesser wax moth *Achroia grisella*. (C) Cocoons spun prematurely and parasitized by the wasp *Apanteles galleriae*. (D) White cocoons and (E) an adult of *Apanteles galleriae*. (Burks photo).

Control

Since wax moths are found everywhere honey bees are kept, a variety of control methods have been devised. Their cost and effectiveness vary greatly.

Natural control

Galleria are the natural hosts for certain viruses, bacteria, protozoa, and parasitoid insects. In Europe, a nonoccluded virus occasionally destroys fish-bait cultures of *Galleria* (Vago 1968). A nuclear polyhedrosis virus and various kinds of bacteria cause problems in commercial fish-bait cultures of *Galleria* in the United States. The larvae also are the natural host for a virus known as the densovirus of *Galleria mellonella* (*Gm*DNV)(Tal and Attathom 1993). The bacterium *Bacillus thuringiensis* (commonly referred to as "Bt") apparently has limited natural effect on local populations of the greater wax moth, though many beekeepers use commercial Bt preparations for wax moth control, as described below. The protozoan *Coelogregarina* is said to cause fatal infections of *Galleria* in Europe (Raw 1954). Two microsporidians are known from laboratory cultures of *G. mellonella* larvae; spores of both, *Nosema galleriae* and *Nosema* sp., are distinguished from those of *Nosema apis* by their smaller size and infectivity to larvae of the greater wax moth (Lipa 1977).

A braconid wasp, *Apanteles galleriae*, is a fairly common insect associated with *G. mellonella* in the Southern United States (Williams 1976). This parasitoid of *Galleria* larvae has also been recorded from Asia, Africa, Europe and South America (Singh 1962). It is common and seasonally abundant in Southern Louisiana, U.S.A.; dozens of the small, white cocoons are often seen among the debris of damaged combs from midsummer through autumn. The overall rate at which *A. galleriae* parasitizes *Galleria* larvae under natural conditions is low, however (Ahmad et al. 1983, Shimamori 1987). In the Southern United States, the red imported fire ant, *Solenopsis invicta*, commonly feeds on the immature *Galleria* of dead or severely weakened honey bee colonies but will not enter strong colonies. In Hawaii, the bigheaded ant, *Pheidole megacephala*, preys on cocooned stages of the greater wax moth (Eckert and Bess 1952).

Applied control

Maintaining strong and populous honey bee colonies will avoid greater wax moth losses to equipment in use. Unfortunately, the logistics of many beekeeping enterprises prevent the frequent inspections of outapiaries (Anderson 1969) that enable the operator to combine weak colonies or to collect and treat dead units before *Galleria* larvae have destroyed the combs. Fumigant chemicals and temperature control are used to prevent the greater wax moth and other moth pests

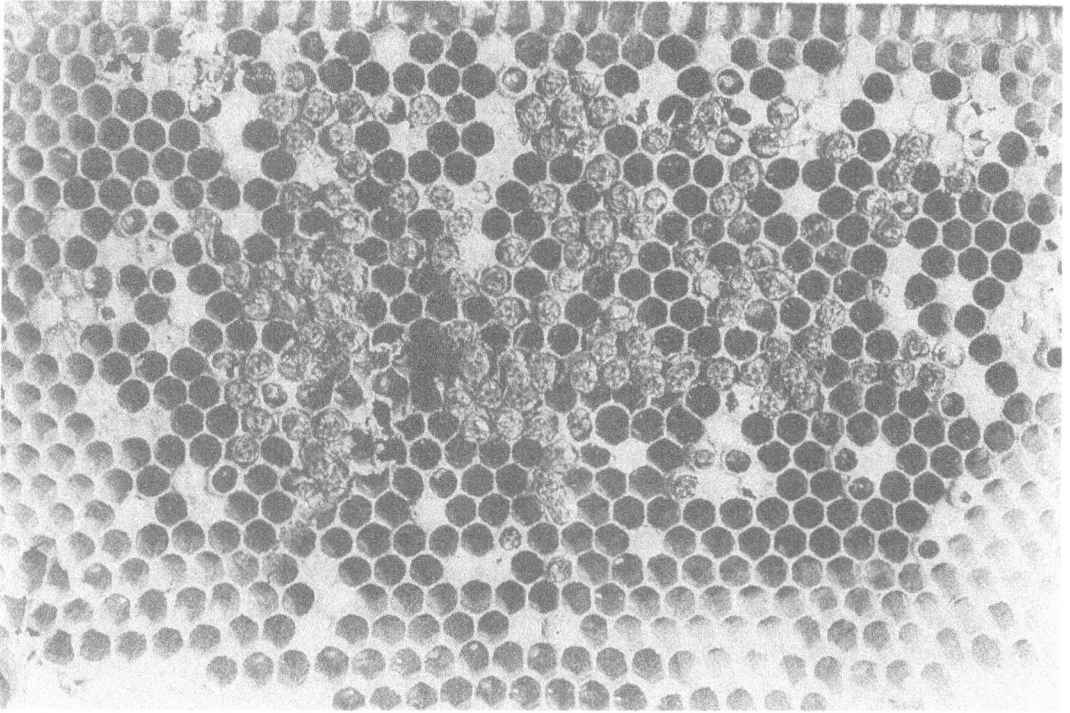

Figure 7.6. Honey bees that died of galleriasis during emergence from brood cells. They were unable to leave the cells after completing development because they were tethered to the comb by their abdomens with silken threads spun by small *Galleria mellonella* larvae. (Burks photo).

from damaging honey bee products, including combs, comb honey, wax, and bee-collected pollen, after their removal from the colony. Presently, only heating, freezing, carbon dioxide fumigation, and the insecticidal bacterium *Bacillus thuringiensis* are approved by the U.S. Environmental Protection Agency for control of the greater wax moth in bee products to be used for human consumption.

Chemical control

Fumigation has been used to prevent destruction of combs by wax moths for over 200 years in the United States (Paddock 1930). Fumigants such as acetic acid, calcium cyanide, ethylene dibromide, methyl bromide, paradichlorobenzene, and phosphine have been used in different countries to protect honey bee products, especially combs, from moths and other insects during storage (Burges 1978, Cantwell 1980, Bailey 1981, Shimanuki 1981, Goodman et al. 1990). As of 1994, Phostoxin™ was approved in the United States for fumigation of empty combs and combs containing honey to be used as bee food. Phostoxin (aluminum phosphide) pellets liberate gaseous phosphine when exposed to air. Supers of combs that are treated with phosphine gas inside gas-tight plastic bags or under plastic tarps are effectively protected against stages of the greater wax moth (Goodman et al. 1990). The U.S. Environmental Protection Agency, however, restricts the use of Phostoxin to state-certified individuals because phosphine is highly dangerous to humans and may be explosive under certain conditions. Paradichlorobenzene (PDB), another fumigant that is used in some countries to protect stored combs, does not kill wax moth eggs, and treated combs must be aired before use. More importantly, the safety of PDB for honey is questionable (Wallner 1992b).

An alternative is fumigation with carbon dioxide. A modified atmosphere of 73.4 percent carbon dioxide and 20.9 percent nitrogen will kill all greater wax moth larvae, the most resistant stage, at 37.8°C (100°F) in 28 hours (Cantwell et al. 1972). The use of carbon dioxide for wax moth control is discussed further under Control in comb honey, below.

Various enclosures, such as second-hand house trailers, are used by beekeepers in the United States to fumigate and store supers of combs during the off-season. Some operators fumigate combs in small rooms or chambers before storing them in larger mothproof rooms.

Microbial control

A water-dispersible concentrate of the *B. thuringiensis* (Serotype 7) (Cantwell and Shieh 1981) is no longer available commercially in the United States for use as a stored-comb protectant. This formulation provided excellent protection of stored combs against *Galleria* larvae without affecting the taste of honey later produced in the treated combs. Development of spray methods for treatment of an entire stack of supers of combs had largely overcome a primary impediment to widespread use of *B. thuringiensis* (Vandenberg and Shimanuki 1990b). Combs treated with *B. thuringiensis* and stored at 30°C for two to four months had slight to moderate damage from *Galleria* larvae (Vandenberg and Shimanuki 1990a). Unfortunately, the rapid loss of viability of Bt spores observed on treated combs at elevated temperatures may be more problematic in the Southern United States. Daytime air temperatures of comb storage facilities there may be considerably above 30°C during five or more months of the year. At the time of this writing, an effort is being made to formulate a Bt unit that may be marketed in the future.

Physical control

Exposure of beekeeping equipment to temperatures above or below the range tolerated by *Galleria* is a safe, relatively rapid method of eliminating or preventing infestations. Temperature manipulation, like carbon dioxide fumigation, eliminates the hazard of contaminating bee products with chemical residues. If artificial cold is used, the following minimum treatments are required to kill all stages of the greater wax moth: minus 7°C (20°F) for 4.5 hours; minus 12°C (10°F) for three hours; or minus 15°C (5°F) for two hours (Cantwell and Smith 1970). (Bulky materials — containers of bee-collected pollen, combs containing honey, and so on may require considerably longer exposure times to reach killing temperatures.) Good sanitation practices alleviate wax moth damage. In South America, concerns about potential chemical hazards and the need for inexpensive controls have fostered the use of light and open-air storage to protect combs from *Galleria* damage (Popolizio et al. 1973a, b).

Control in comb honey

Producers and packers of comb honey risk losing a high-value commodity if it is not protected constantly from the greater wax moth and certain other moth pests (Cantwell and Smith 1970, Morse 1978). Large quantities of comb honey may be rendered unsalable by warm storage temperatures, which accelerate moth development (Roberts and Smaelli 1958).

Infestations of *Galleria* and other moths must be prevented from the time comb honey is removed from colonies until it is packaged because a single larva can destroy its market value. LeMaistre's report (1934) that almost all beekeepers in Southwestern Ontario, Canada, had *Galleria* larvae in their stores of comb honey indicates that suitable precautions should be taken to protect comb honey from wax moth damage even in northern climates. A large-scale carbon dioxide fumigation procedure (Cantwell et al. 1972, Jay et al. 1972) was used for three years at the Waycross, Georgia, plant of the Sioux Bee Honey Association. Included were a gasification system consisting of a 3.628-metric-ton liquid carbon dioxide holding vessel, vaporizers, automatic application controls, and three large treatment rooms. Each room held 7,000 supers of comb honey arranged in stacks 3.66 meters (12 feet) high. All of the supers were palletized (40 supers per pallet) to reduce handling and to help prevent damage to the fragile combs. The original atmosphere was purged through vent tubes near the ceiling while the carbon dioxide gradually filled the treatment room. A five-day treatment period followed, during which an atmosphere of 80 percent carbon dioxide was maintained automatically. Supers of honey thus treated and stored remained free of wax worms for two months or longer and were removed from the treatment room as honey was required for packing.

During 1974-'75, the cost of controlling wax moths in one room for 10 months (three treatments of 11,818.2 kg of carbon dioxide each) ranged from 91 cents to $1.14 for each kilogram of honey for each treatment. Over 680,000 kg of gallberry comb honey were fumigated with carbon dioxide at the Sioux Bee packing plant during the period in which this method of control was used.

Section comb honey or comb honey destined for cut comb packs may be frozen to destroy any eggs or young larvae present (Morse 1975, 1978). Care should be taken to allow extra time for the honey to reach the freezing point before timing the exposure. Both empty and honey-filled brood combs are routinely held in a freezer for 24 hours before storage at the USDA-ARS, Honey-Bee Breeding, Genetics, and Physiology Research Laboratory at Baton Rouge, Louisiana, U.S.A.

Control in queen rearing materials

Honey bee breeders occasionally contaminate stocks of rendered beeswax or queen rearing supplies made of beeswax, such as queen cell cups, with insecticides or other chemicals and subsequently experience great difficulties in queen production. Storage or periodical treatment in a home freezer is a safe, nonchemical method of protecting wax cell cups and other queen rearing materials from wax moth damage.

Future control techniques

Whether available wax moth control measures are acceptable will depend on the size, economics, and geographic location of a given beekeeping operation. Some approaches to the development of control methods based on toxic chemicals, biological agents, genetic manipulation (sterile insect release method), insect growth regulators, and traps containing attractants have been discussed previously (Morse and Nowogrodzki 1990). An ideal method of protecting stored combs should be inexpensive; easy and safe to apply; safe for bees that receive treated combs; and have no effect on honey. Potential advances in wax moth control technology are impeded by the limitations imposed by these objectives. For example, spraying entire stacks of stored combs with the *Galleria*-specific nuclear polyhedrosis virus should be feasible and should provide long-term protection of stored combs (Doughtery et al 1982, Vandenberg and Shimanuki 1990a), but commercialization of this virus is currently not possible in the United States. Mass production of the virus in wax moth larvae and the extensive registration testing required are prohibitively expensive.

The lack of safe, low-cost wax moth management strategies for beekeeping operations of the subtropics and tropics indicates that different research and development approaches may be beneficial. Additional research might include methods of decontaminating combs (e.g., improved comb handling techniques for freezing), exploitation of endemic predatory ants, such as the red imported fire ant (*Solenopsis invicta*) of the Southeastern United States, low-hazard fumigants (acetic acid, mixtures of gases and propylene oxide); barriers to prevent reinfestation of decontaminated combs (vegetable oil, repellents, and light and moving air), and materials applied directly to combs (nonhazardous chemicals and biological agents). Techniques to suppress wax moth populations near apiaries might include periodic release of biological control agents. Organisms that might be exploited include *Bacillus thuringiensis*, the *Galleria*-specific nuclear polyhedrosis virus, *Nosema galleriae* that develop in *Galleria* larvae, and wasp parasitoids that develop in *Galleria* eggs, larvae, or pupae.

Driedfruit Moth

The moth species *Vitula edmandsae* has a Western U.S. form (designated *V. edmandsae serratilineella*) that has acquired the common name driedfruit moth. The driedfruit moth has been reported from nearly all of the Rocky Mountain states of the United States and from Western Canada (Cockle 1920, Okumura 1966). It was also inadvertently introduced into Europe (Tiedemann 1958). In the Eastern United States, a separate race is recognized (designated *Vitula edmandsae edmandsae*) and has been referred to as the bumble bee wax moth (Weatherston and Percy 1968) because it has been reported as a common resident of *Bombus* nests (Milum 1940a, 1953). This name, however, can cause confusion with the European bumble bee wax moth, *Aphomia sociella* (described below).

Biology

Driedfruit moths are mottled gray and are about 20 mm long. They are attracted to black lights (Okumura 1966, Simmons and Nelson 1975). Development from egg to adult requires about 88 days. The species overwinters in the larval stage. Larvae of the Indianmeal moth, the Mediterranean flour moth, and mature larvae of the driedfruit moth are similar in size (about 15 mm long) and color (either white or light pink). A taxonomic key is available for identification of moth larvae that infest honey bee combs (Okumura 1966).

In addition to a wide variety of dried fruits, the larvae feed rather commonly on pollen and honey in unprotected stored combs, tunneling through the cell walls in the process. Occasionally, they also feed on pollen and honey in the combs of active colonies (Okumura 1966). Grant (1976) reared this species on one of the beeswax-free diets used for the greater wax moth.

Large numbers of driedfruit moth larvae can complete their development on combs without destroying the midrib or the entire comb, unlike *G. mellonella*. Larvae of the driedfruit moth form a dense mass of silk webbing over the comb face (Okumura 1966, Wilson and Brewer 1974).

The driedfruit moth has also been recorded from nests of bumble bees, carpenter bees and alfalfa leafcutting bees (Linsley 1943, Free and Butler 1959, Bohart 1972). From 1977 to 1983, the driedfruit moth increased from a rare to an important pest of alfalfa leafcutting bees, *Megachile rotundata*, managed for alfalfa seed production in Western Canada (Richards 1983, 1984). Larvae of the driedfruit moth enter uncapped tunnels in artificial nests of the alfalfa leafcutting bee and consume pollen and nectar provisions of cells; they also are capable of feeding on live, immature stages of the alfalfa leafcutting bee. After artificial nests of *M. rotundata* are stored for winter, most driedfruit moth larvae leave the bee tunnels at maturity and aggregate in spaces between nests (Richards 1984). Increased driedfruit moth populations associated with managed alfalfa leafcutting bees can be expected to foster additional infestations in nearby honey bee operations. In British Columbia, Canada, larval and adult driedfruit moths have been found in virtually all types of honey bee beekeeping equipment and facilities, heated and unheated, and in overwintered colonies (Scott et al. 1984).

Control

Extensive studies indicate that a sex pheromone-based management program for the driedfruit moth is feasible in both field and storage areas of beekeeping operations that use honey bees or alfalfa leafcutting bees (Struble and Richards 1983, Scott et al. 1984). Increased catches of moths in pheromone traps near honey bee colonies may result from colony odors chemically guiding female driedfruit moths to the source of larval food (Scott et al. 1984). Gravid females also are attracted to pheromone-baited traps that capture numerous male driedfruit moths, possibly in response to male secretions (Struble and Richards 1983). In addition, fumigation with sulfur dioxide appears to be a promising control measure for driedfruit moth infestations of stored honey bee combs (Szabo and Heikel 1987).

Lesser wax moth

The lesser wax moth, *Achroia grisella*, has a scattered distribution throughout the temperate and tropical climates of the world (Smith 1953, Singh 1962, Hassanein et al. 1969); it occurs in most, if not all, of the United States (Vansell 1936, Eckert and Bess 1952). Although generally of minor importance to beekeepers, this species can destroy neglected or inadequately protected combs (Milum 1940a).

Biology

Lesser wax moths are easily distinguished from greater wax moths by their appearance and behavior, even when the latter are much reduced in size as a result of inferior diet. Adult *A. grisella* are small, slender-bodied, silver-gray to buff-colored, and have a conspicuously yellow head. Adult males reach a maximum length of about 10 mm, whereas the female moths are usually slightly larger, about 13 mm maximum length (Kunike 1930). Lesser wax moths weigh only about one-

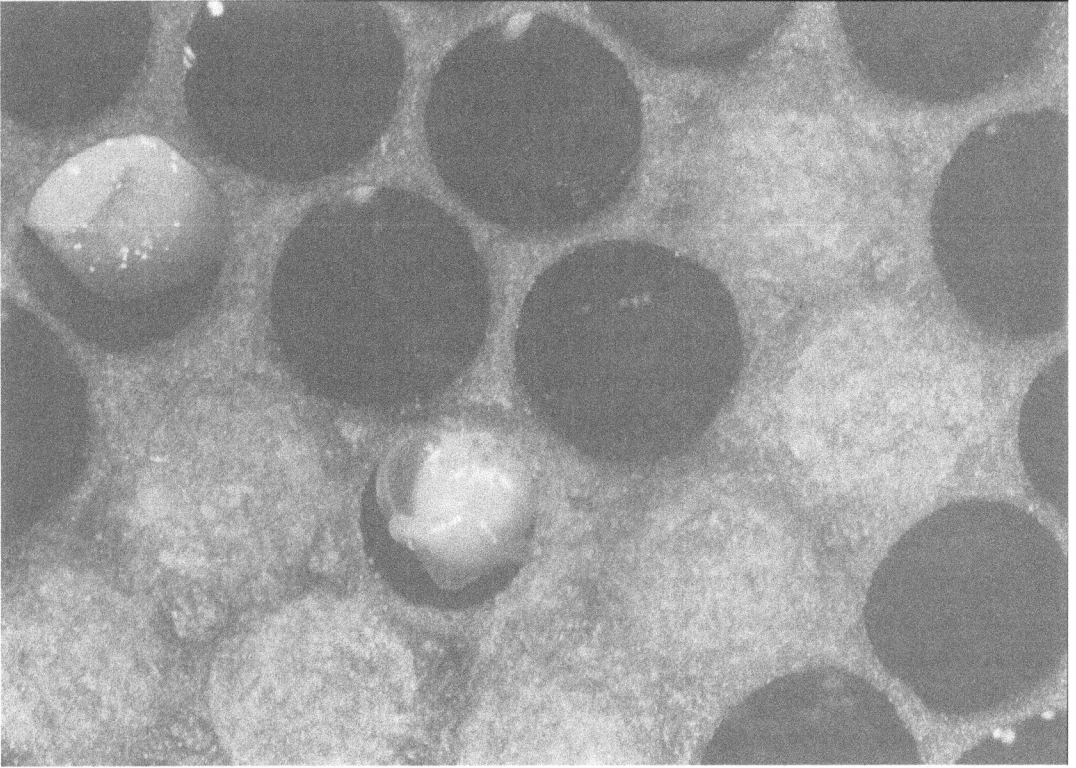

Figure 7.7. A light case of bald brood. The cell is uncapped with a normal, healthy pupa inside. This is usually attributed to the lesser wax moth, and is uncommon in the United States (Killion photo).

sixth to one-tenth as much as greater wax moths. The average weights of 100 adult *Achroia* of each sex reared on a diet of dark combs containing fresh pollen, within 24 hours of occlusion, were as follows: males, 11.3 mg (range 3.0-17.5 mg); females, 20.3 mg (range: 9.9-33.8 mg) (personal observation 1973).

When disturbed, the moths run rapidly across combs and throughout the hive for 30 seconds or more. If lesser wax moths are on a comb when it is removed from the hive, they often run to one edge of the comb and quickly fly down to the ground among vegetation. The mate-locating phase of the reproductive behavior of the lesser wax moth, as with the greater wax moth, is the reverse of the female-signaling-male searching system common to most moths. Male *A. grisella* also release a sex pheromone (undecanal and cis-11-octadecanal) from wing glands and use ultrasonic sound to signal or call female moths (Greenfield and Coffelt 1983, Spangler and Takessian 1983, Spangler 1984). The possible adaptive value of such a sex-role reversal is unknown but may relate to the ectosymbiotic relationship of these moths to honey bees.

The 250 to 300 eggs produced during the seven-day lifespan of female moths (Kunike 1930) are usually laid in crevices (Makings 1958). The larvae have narrow white bodies and brown heads and pronotal shields; they reach a maximum length of about 20 mm (Milum 1940a). Each larva tends to live segregated in a silken tunnel covered with frass while growing, whereas maturing *Galleria* larvae tend to congregate (Singh 1962).

Studies by Hassanein et al. (1969) indicate that the optimal natural diet of *A. grisella* larvae is dark comb that contains either pollen or brood. Larvae are most often found, however, on the bottomboard among debris, in partially rendered cakes of beeswax, and in debris remaining from combs destroyed by *G. mellonella* larvae; they also develop in slumgum, foundation, and bee-collected pollen (Paddock 1930, Vansell 1936, Milum 1940a). Thriving populations of *Achroia* are found only in very weak honey bee colonies.

Larvae of the lesser and greater wax moths frequently infest beekeeping materials simultaneously in warm climates, but the *Galleria* larvae will consume the smaller, less active *Achroia* larvae under crowded conditions. Larvae of *A. grisella* may be easily reared on the beeswax-free diet developed for the greater wax moth by Dutky et al. (1962). Certain natural enemies of the greater wax moth, including the nonoccluded and polyhedrosis viruses and the wasp *Apanteles galleriae,* also attack the lesser wax moth (Raw 1954, Bailey 1966, Ahmad et al. 1983).

Bald brood

Single cells or patches of honey bee brood with uncapped cells that expose the heads of live, apparently normal pupae are called bald brood. This condition usually occurs when lesser wax moth larvae perforate the outer cell walls and the cappings; worker bees then chew away the remainder of the cell caps, thereby exposing the developing pupae. A few malformed adult bees result from the honey bee pupae that have fecal pellets deposited on them by lesser wax moth larvae in uncapped cells. In Europe, bald brood is usually caused by the lesser wax moth, infrequently by the greater wax moth (Milne 1942, Bailey 1963a). Bald brood is probably uncommon in the United States, though it has been observed in widely separated areas.

A condition resembling bald brood, but not caused by moths, has been reported. Occasionally, colonies in Europe do not cap their brood normally, but either turn the cell edges slightly inward or leave a small hole in the center of the capping (Anonymous 1959). Requeening the affected colonies has eliminated the condition in the latter case.

Bumble bee wax moth

The bumble bee wax moth, *Aphomia sociella,* occurs in Europe and Asia. In some European countries, it is common in the nests of various species of *Bombus* (Free and Butler 1959, Pouvreau 1967). It is rarely a pest of honey bee colonies (Toumanoff 1939). Adult *A. sociella* are heavy-bodied and are similar to, but slightly smaller than, greater wax moths. The body and forewing are reddish brown; females have a prominent dark spot on each forewing (Milum 1940a). Mature larvae are pale yellow and reach a length of about 22 to 30 mm. As in the greater and lesser wax moths, larvae construct dense tunnels of silk, in which they feed. According to Pouvreau (1967, 1973), the larvae readily consume brood (eggs, larvae, and pupae), as well as pollen and honey stored in the wax cells within bumble bee nests, and they frequently cause the adult bees to desert the colony. These moths are attracted to the odors of active bumble bee colonies.

Indianmeal moth

The Indianmeal moth, *Plodia interpunctella,* is one of the worst moth pests of whole grains and cereal products. The larvae feed on a wide variety of other foods, including dried fruits, dried milk, candy, dried roots, herbs, nuts, seeds, dead insects, and unprotected bee-collected pollen (Metcalf et al. 1962). The species is a minor pest of nests of alfalfa leafcutting bees, *Megachile rotundata* (Bohart 1972). It also occurs in the nests of bumble bees *(Bombus),* mining bees *(Anthophora* and *Osmia*), and paper wasps *(Polistes)* (Linsley 1943). This moth has spread throughout the world from Europe (Metcalf et al. 1962).

The moths are up to 9 mm long; the forewing is grayish and the remaining one-half to two-thirds, as well as the head and thorax, are reddish brown with darker markings. This coloration gives the moth a banded appearance (Silacek and Miller 1972, Simmons and Nelson 1975). At rest, the moth holds its wings together along the body line (Metcalf et al. 1962). At maturity, the brown-headed larvae average about 13 mm in length and are usually dirty white, but their body color varies to pinkish or greenish (Hamlin et al. 1931). In warm climates, Indianmeal moth larvae develop on pollen and cocoons or dead brood in stored combs (Eckert and Shaw 1960), combs containing clusters of dead bees, accumulations of dead bees (Eckert and Bess 1952, Wilson and Brewer 1974), or debris from these materials. An advanced infestation can be identified by the loose, flimsy webbing that is constructed by larvae across the face of combs. Fortunately, this pest does not feed on the beeswax foundation or processed wax. The life cycle may be completed in four to six weeks during summer or in heated buildings.

Cold storage at subfreezing temperatures in airtight containers is probably the best method of protecting large quantities of pollen against deterioration and insect attack. (Freezer storage requires nearly 1,440 cm³ of space for each kilogram of pollen.) Pollen intended for beekeeping use may also be dried and stored in airtight containers, or it may be mixed with expeller-type soybean flour (1 : 1) and then stored at room temperature (Whitefoot and Detroy 1968).

Mediterranean flour moth

Milled cereal products are most often damaged by the Mediterranean flour moth, *Anagasta kuehniella*, although whole grain is also attacked. Before the development of modern fumigation techniques, it was considered the most serious insect pest of flour mills in the United States. The Mediterranean flour moth occurs throughout most of the world (Metcalf et al. 1962).

Stored honey bee combs that contain pollen are sometimes infested, but this moth cannot develop on empty brood combs or dead insects (Essig 1940, Milum 1940a, Eckert and Bess 1952). *A. kuehniella* is an occasional resident of bumble bee nests (Essig 1940, Milum 1940b). The pale gray moths are about 6 to 13 mm long and have two dark zigzag markings on the forewings. As soon as they have hatched, larvae begin to spin small silken tubes in which they live and feed. Flimsy cocoons are constructed by mature larvae within the food material, their webbing or nearby cracks and crevices.

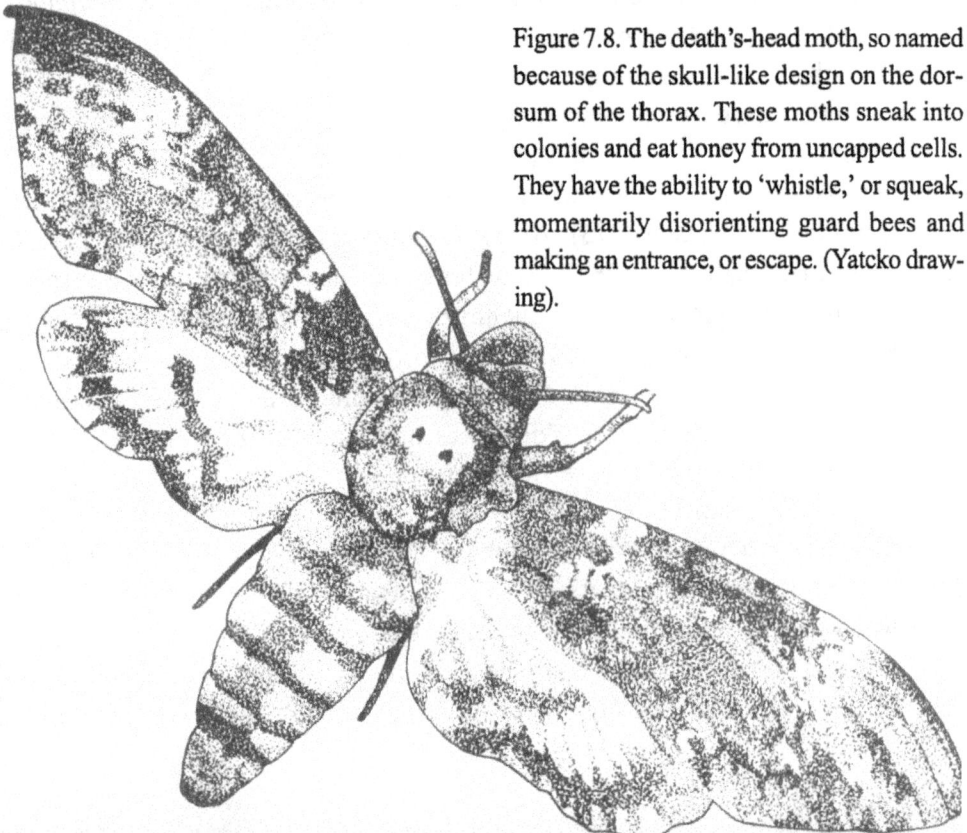

Figure 7.8. The death's-head moth, so named because of the skull-like design on the dorsum of the thorax. These moths sneak into colonies and eat honey from uncapped cells. They have the ability to 'whistle,' or squeak, momentarily disorienting guard bees and making an entrance, or escape. (Yatcko drawing).

Death's-head moth

Honey bee colonies in some regions of the world are subject to occasional predation by unique sap-, nectar- and honey-feeding hawk moths (Sphingidae) of the genus *Acherontia*. Their common name is derived from the pattern on the dorsum of the thorax, which resembles a death's-head design. *Acherontia atropos* is indigenous to parts of Europe and Africa (Brügger 1946, Lundie 1952, Budel and Herold 1960, Smith 1960). Other species of *Acherontia* frequent honey bee colonies in Asia (Smith 1953, Singh 1962).

A. atropos is a medium-sized, heavy-bodied moth with dark gray forewings and yellow hindwings that have two heavy dark bands toward the apex. Moths of this species can produce a whistling sound by expelling air from the pharynx. Although the adults feed mainly on sap from tree wounds, they may resort to robbing nectar and honey from the hives of honey bees, taking a teaspoonful or so from open cells with each visit. Nightly attacks may cause the bees to become agitated and remain so during the daytime. The *Acherontia* cadavers sometimes found in hives indicate that these moths are not always successful thieves (Brügger 1946).

Moths in bee-collected pollen

Commercial-scale harvesting of pollen collected by honey bees has received considerable attention in many countries, principally to supply the pharmacy, health food, and beekeeping markets (Anonymous 1975, 1976b). Pollen trapped from honey bee colonies must be handled properly to prevent its destruction by fermentation, molds, and heat, as well as attack by a variety of pollen-feeding insects, including moths. The Indianmeal moth and the driedfruit moth may infest spilled bee-collected pollen in handling and storage facilities.

Acknowledgments

Details of the biology of the greater wax moth and the fumigation of comb honey were provided by R. A. Nielsen, former USDA-ARS entomologist, and the Sioux Bee Honey Association, respectively, at the writing of the first edition. This chapter was prepared in cooperation with the Louisiana Agricultural Experiment Station, Baton Rouge.

CHAPTER EIGHT

---∞∞∞---

Insects: Dipteran Pests and Predators of Honey Bees
Diana Sammataro

CHAPTER EIGHT

Insects: Diptera

Diptera (*di*, two; *ptera*, wings), or true flies, make up the fourth largest order of insects. Although they are generally small in size, many flies are of great economic importance either as pests on humans and animals or as beneficial predators and parasites. True flies (not to be confused with dragonflies or butterflies) have two wings rather than four; their hind wings are reduced to small, knobby structures called *halteres*, which help keep them balanced in flight. Flies undergo complete metamorphosis — egg, larva (maggot), pupa, and adult. Those discussed here are coarctate, meaning that the pupa is encased inside the last larval skin (*puparium*) in which the larva forms a pupa. The puparium is often dark brown, oval in shape, and hard. Many fly larvae are aquatic and feed on diverse matter such as the tissue of living and dead plants, animals, or insects.

Although many flies are not major enemies of bees, they can become important pests in some areas or at certain times of the year, especially if their normal prey or host is missing or other natural population controls are not working. Many species are mimics of bees, including the bee flies (Bombyliidae), flower flies (Syrphidae), and soldier flies (Stratiomyiidae), but only a few trouble honey bees. Several species of the first two families are predaceous on or live in the nests of social insects (Borror et al. 1989).

Ectoparasites

The most common dipterous parasite of honey bees in the world is *Braula coeca*, Nitzsh 1818, sometimes written as *B. coeca* also known as bee louse or blind louse. Its common name is braula.

Braula coeca

Description

Braula belongs to the infraorder Muscomorpha and the family Braulidae and is known only from honey bees. It is not a true louse but a highly modified, reddish-brown, wingless fly about 1.2 to 1.5 mm long and 0.75 mm wide (males are smaller). The body is covered with stiff, spinelike hairs that are long on the terminal abdominal segments (Grimaldi and Underwood 1986). The abdomen is composed

---

Records early in this century may have confused flies with other pests or diseases not known then. Such inconsistencies have been noted here and, it is hoped, clarified.

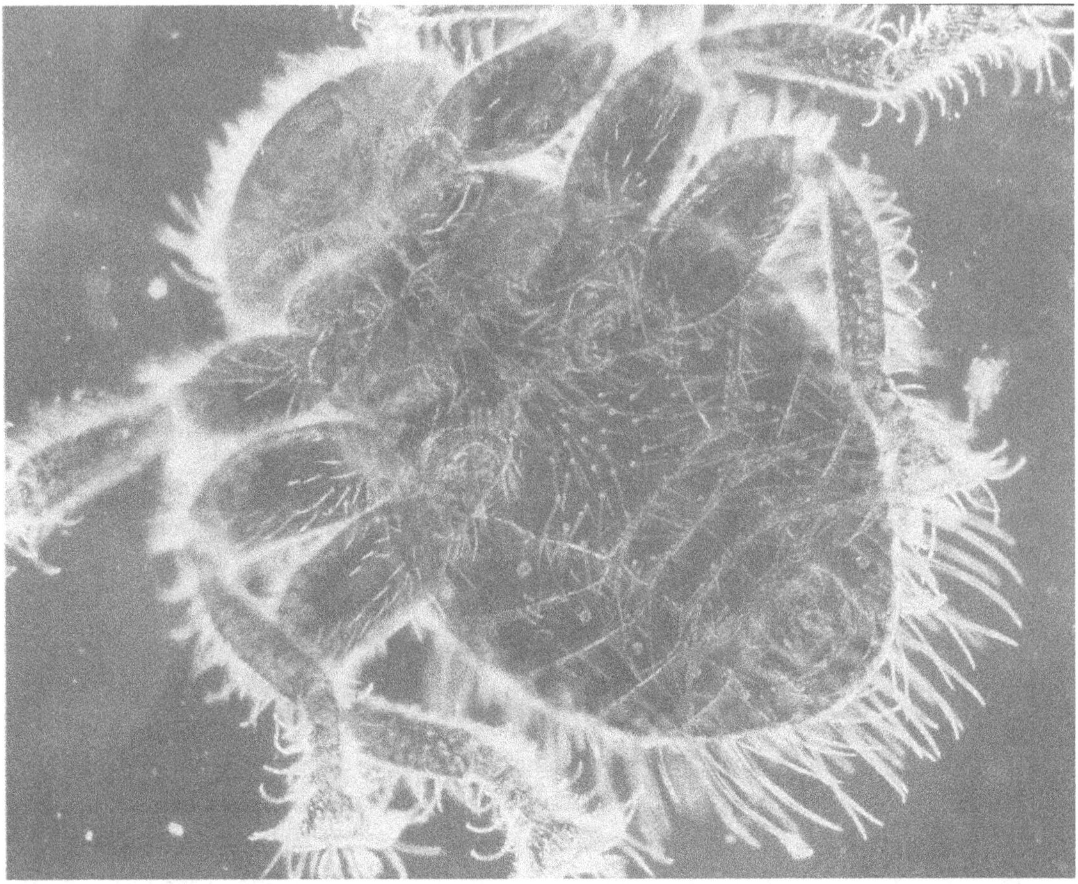

Figure 8.1. *Braula coeca* adult. It has a compressed head, held upright, rudimentary compound eyes and no ocelli. The legs are equal in length and long in proportion to the rest of the body. (Sammataro photo).

of five segments and occupies 60 percent of the total body length. The head is compressed and held upright, with rudimentary compound eyes but no ocelli; the legs, equal in length, are long in proportion to the rest of the body.

The pretarsal claw of each leg is modified into a comb or *ungues*, with 12 to 16 small, spinelike teeth on each side of the middle line of the foot. The fly attaches to the host bee by drawing the ungues through the bee's hairs. Each claw has an apical, brushlike *pulvillus* (lobe associated with the claw) that may have sensory functions (Ramirez and Malavasi 1992). Superficially similar in appearance, braula and varroa are often confused because they may be found in the same hive (Perret-Maisonneuve 1925, Accorti 1986, Calatayud and Feuerriegel 1987, Morse 1987). Braula is longer than varroa, however, and has six legs (Shimanuki and Knox 1991).

Braula may be found on mated queens, drones, and workers (Markosyan et al. 1973, Weems 1983). One or two braula are often found per worker, but the louse never leaves the hive with foraging bees (Grimaldi and Underwood 1986). They are rarely found on drones. The number of lice reported on queens ranges from 75 (Benton 1896) to 187 (Grimaldi and Underwood 1986); a daily collection from one queen totaled 371 (Bailey and Ball 1991). Flies are found mostly on the queen because she is fed more often and lives longer (Manley 1958, Crane 1990). Excessive numbers of flies may interfere with the queen's work and can cause her death (Eckert and Shaw 1960).

Range

Braula is distributed worldwide. It is abundant in regions with a moderate or Mediterranean-type climate (Eckert and Shaw 1960) and may die in cold climates (Wedmore 1932) or overwinter as eggs or pupae (Manley 1958). It was first reported

Figure 8.2. The pretarsal claw of *Braula coeca*. It is modified into a comb, or ungues with 12 to 16 small, spurlike teeth used to attach itself to the host.

Figure 8.3. A queen honey bee with a moderately large population of *Braula coeca* attached. (Morse photo).

in the United States on queens imported from Italy (Surface 1913, Root and Root 1919, Örösi-Pál 1980) and was well-established by the 1920s (Aldrich 1924, Phillips 1925, Argo 1926a, b, Paddock 1926). It is now recorded in 14 states east of the Mississippi, ranging from Maine to Florida (Morse and Flottum 1990).

Braula was first found in France from imported Italian queens (Perret-Maisonneuve 1925, Robineau-Desvoidy 1936), and also in Scotland (MacDougall 1914), Spain (Gil Collado 1931), Bulgaria (Dryenski 1933), Russia (Beliavsky 1929, Zivihinovic 1936), South Africa (Skaife 1921b, Taylor 1934, Anderson et al. 1983, Clauss 1983), Mauritius (Crane 1982), Tasmania (Nicholls 1932), and Sicily (Morse 1987).

Braula has been in Brazil since 1902 (Van Emden 1921, De Azevedo 1924, Hasselman 1926, Ronna 1936b), in Central America (Belize 1987), and in England by the 1930s (Herrod-Hempsall 1931, Wedmore 1932, Manley 1958). Testimony to its wide distribution comes from a great number of reports (Grassi and Parona 1881, Felizardo 1925, Porter 1936, Hassanein and El-Salaam 1962, Papadopoulo 1964, Stejskal 1967, Atakishiev 1971, Buys 1975, Laurence and Mohammed 1975, Smith 1978, Martins and Rosa 1982, Mwale 1982, Nixon 1982, Ahmadi 1984, Faveaux 1984, Patetta 1984, Smith and Caron 1984a,b, Bradbear 1988, Hussein 1989, Shimanuki et al. 1992, Matheson 1993). Usually considered a minor pest, this fly has been re-ported in some countries to cause losses in honey production (Oytun 1963), and it attacks weakened hives (Markosyan et al. 1973). Braula is not found in Australia, New Zealand, or Papua New Guinea (Crane 1990), and although recorded on *Apis cerana*, Morse (1987) proposes that this report confused the flies with varroa mites.

Biology

The female braula lays her eggs beneath the wax cappings on the walls of honey cells (Wedmore 1932, Morse and Hooper 1985, Bailey and Ball 1991, Jean-Prost and Médori 1994) or at random on brood (Weems 1983). If eggs are laid on brood cells, Suire (1931) relates that the larvae live on food brought by nurse bees, but this is disputed by Beliavsky (1929). Weems (1983) reports that eggs on brood cells die. The eggs are firmly stuck to the top of the wax cappings of the honey cells (Beliavsky 1929). Eggs are elliptical with wing-shaped, netlike flanges on each side but pinched in at the top and bottom. They are white, 0.76 mm long and 0.43 mm wide (Beliavsky 1929).

Emerging larvae have no anterior spiracles (Cantwell 1975, Örösi-Pál 1980, Grimaldi and Underwood 1986) and are about 2.0 mm long, white, and transparent (Beliavsky 1929, Wedmore 1932). While eating, they make tunnels 0.75 mm in diameter (Eckert and Shaw 1960), which appear as raised lines of wax debris and can be up to 10 cm (four inches) long. Larvae eat honey, pollen, or beeswax (Örösi-Pál

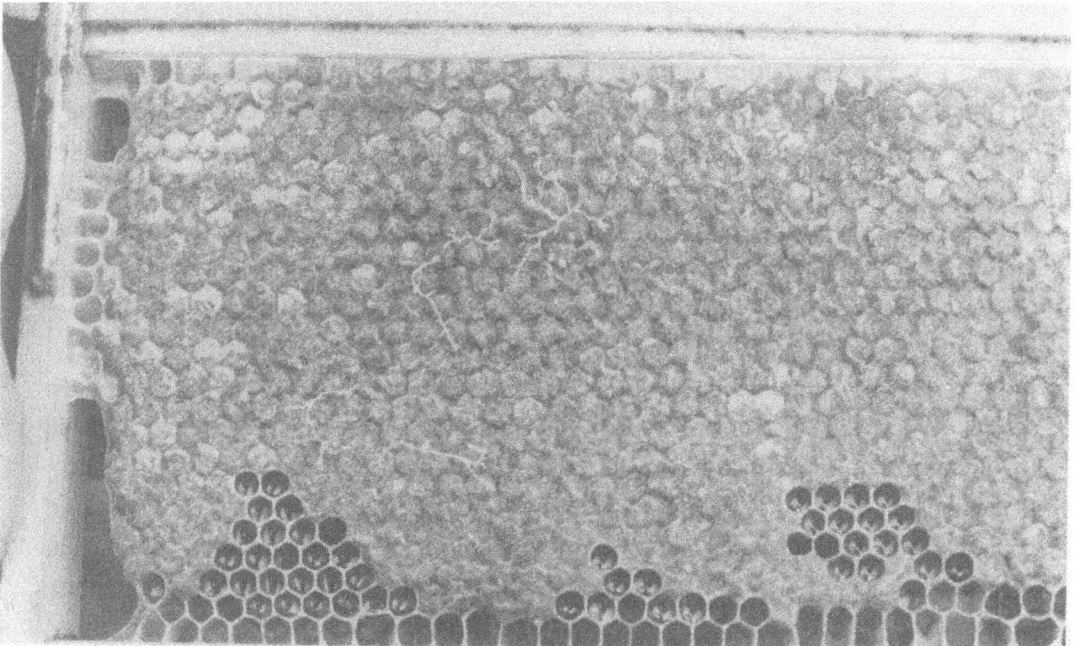

Figure 8.4. Tunnels made by *Braula coeca* larvae in cappings of honeycomb. Tunnels are 0.75 mm wide and can be up to 10 cm long. These tunnels can make comb honey unsaleable. (Smith photo).

1980, Meyer 1983). Bailey and Ball (1991) suggest that microorganisms in the larva's intestine may help digest wax.

Once they mature, the larvae pupariate in the tunnel areas, which are approximately 1.5 mm wide. An adult fly emerges in three weeks, starting out white, transparent, and soft and then hardening and turning darker. Adults probably mate on the comb, but this is not recorded; little information is available about the males. It is not clear when the louse adjoins the bees, but this may occur when new bees eclose or emerge (Wedmore 1932). The time from egg to adult is 21 days.

Adults feed on honey and pollen fed to bees (Borror et al. 1989) and are especially attracted to the queen, stealing food from her outstretched mouth parts. Adult flies scratch with their claws near the edge of the labrum (upper lip) until the bee extends its tongue (Phillips 1925). They move quickly to the base of the bee's mouth parts near the opening of the salivary gland where they feed on food passed to the bee or possibly on salivary secretions (Imms 1942). After feeding, flies move to the thorax to rest. Adults die within six hours of hatching if they do not attach themselves to a bee (Herrod-Hempsall 1931). Brown (1988) reports braula can hop like fleas. Populations of braula are high in the fall and low in spring, but Argo (1926b) reports large numbers in spring and again in the fall. Although braula are more irritating than harmful to bees, their tunnels can disfigure comb honey (Morse and Flottum 1990), making it unsalable.

Control

The mechanical removal of honey cappings during regular honey extraction controls braula under normal circumstances (Morse 1981). Comb honey may be protected by freezing after being harvested. In areas where braula seems to be a pest, the smoke of phenothiazine from smoldering strips is recommended (Tsivilev 1968, Grobov et al. 1983, Sulimanovic et al. 1983). Markosyan et al. (1973) report that using 1.5 g phenothiazine per hive, treating in the morning or evening, kills adult flies only. Basan (1977) suggests burning phenothiazine tablets, and Danielyan et al. (1975) maintain that this material is effective against both varroa and braula.

The smoke from rolled-up corrugated cardboard containing cigarette ends (not filters) and burned in a smoker will cause the flies to drop off the bees. A sheet of sticky paper on the bottomboard is used to catch flies (Brown 1988). Tobacco smoke kills overwintering flies in one to two minutes (Herrod-Hempsall 1931, Eckert and Shaw 1960) or makes them drop off onto the paper on the hive floor, which later can be burned. Fumes of naphthalene (Ronna 1936b, Volcinschi 1969) and camphor placed on the hive floor kill braula (Manley 1958). Nicholls (1932) suggests sprinkling bees with dilute honey in the evening to dislodge the flies, which can be collected on paper and burned. Beliavsky (1929) reports good control with fumes of carbon bisulphide (sulfated carbon) and formalin which also kills the braula larvae.

To incite bees to pick the lice off a queen, Eckert and Shaw (1960) suggest dabbing her with a matchstick dipped in honey or putting the queen in a matchbox and blowing in tobacco smoke until all the braula drop off. This is a temporary measure, though, because if the queen is returned to the hive she can be reinfested quickly. Perret-Maisonneuve (1925) achieved complete control by covering the queen's thorax with paint. The lice can be detected using a flotation method in edible oil (Vyslouzil 1984) which is probably similar to the ether roll used to detect varroa mites.

Perizin (coumaphos), a systemic insecticide fed in syrup, is registered for controlling varroa in some parts of Europe (Accorti 1986, Marchetti and D'Agaro 1986, Lahitte 1986, 1987, Abeille de France 1987, Kessler 1987). Perizin also kills braula, but this chemical is highly lipophilic and can contaminate beeswax for several months. Kulincevic et al. (1991) report that braula was killed in reasonably large numbers when exposed to fluvalinate aerosol or fumigants but not amitraz.

Recently, Apistan strips, normally used to control varroa mites, have been successful in killing braula as well (I. B. Smith, personal communication), especially when used in conjunction with a sticky board, which will capture the adults as they fall off the bees.

Other species of Braula

Range

Örösi-Pál (1939, 1963a, b, 1966, 1976, 1980) and Crane (1990) have made the following observations about braula species. *B. angulata*, originally from Asia and Southern Europe, is now in South America. *B. kohli*, perhaps from South America, is found also in Belgium (Coorman 1966) and Zaire and on *A.m. carnica* and Middle Eastern races of bees. *B. orientalis* is from the Pacific coast of Russia, Turkey, Israel, and Egypt. *B. pretoriensis* is from Africa south of the equator (Mwale 1992) and is now in South America and on *A.m. carnica* and Middle Eastern races of bees. *B. schmitzi* Örösi is from Russia, Asia, Turkey (Oguz 1976), Israel, Southern Europe, Brazil, and Argentina and is found on *A.m. ligustica* bees (see Örösi-Pál 1980 for differences between *B. schmitzi* and *B. coeca*). It is probably also in Central and South America. Crane (1990) reports on two other unidentified species from Africa.

Several species may cohabit a single colony. The appearance of *Braula* species in different parts of the New World was correlated directly with the Old World countries from which queens were originally imported.

New species of Braulidae
Two new species of Braulidae were discovered on *Apis laboriosa* from Nepal that were 3.5 to 11 times, respectively, the size of *B. coeca*. (Grimaldi and Underwood 1986). They are *Megabraula antecessor* and *Megabraula onerosa*.

Predators

Asilidae (robber flies)
Description
Robber flies include over 5,000 species worldwide and about 1,000 in North America. Found in a variety of habitats, the adults have a bearded face with the top of the head hollowed out between the eyes. They also have long legs, elongated abdomens, and are relatively large (0.6 to 4.0 cm long). All robber flies are predaceous on insects; some even resemble bees. The asilid larvae live in soil, decaying wood, or other places where insects would be found (Borror et al. 1989).

Range
The following species are listed as eating honey bees:

Old World: *Molobratia teutonus, Dasypogon diadema, D. teutonus, Stenopogon sabaudus, Echthistus cognatus, Machimus annulipes, M. rusticus,* and *M. fimbriatus* all eat honey bees (Adamovíc 1949, 1950, 1963a,b, 1971). Adamovíc (1972) reports that *Dasypogon diadema* and *Molobratia teutonus* ate honey bees about 66 to 78.5 percent of the time, respectively. Unknown species are reported feeding on bees in Africa (Clauss 1983), and Londt (1993) reports four subfamilies as being apivorous: Apocleinae, Dasypogoninae, Laphriinae, and Stenoponinae. Some other references include Poulton (1906), Séguy (1927), Brügger (1935), Bromley (1945), and Delgado and Afonso (1986).

New World: *Diogmites missouriensis* caught bees frequently on flowers in Arkansas (Scarbrough 1979), and Bromley (1936) reports that *D. misellus* occasionally takes honey bees. Bullington (1978) records *Dasylechia atrox* in Michigan, and *Mallophora ruficauda* Wiedemann is recorded in Argentina (Ciencia y Abehas 1976). Copello (1922, 1925, 1927) reports that the eggs of *M. ruficauda* are laid in posts, and the larvae burrow in the ground feeding on dynastid beetles; the adults prey on honey bees. Rinaldi et al. (1970, 1971) record *Eicheraz ricnotes* from Argentina as well.

O'Neill (1992) lists *Megaphorus willistoni* and the following genera as apivorous: *Stenopogon, Deromyia, Promachus, Mallophora,* and *Proctacanthus. Mallophora orcina* (Wiedemann) and *M. bomboides* (Wiedemann) are known as the southern bee killers, and *M. nigra* Williston is called the black bee killer

(Knutson and Murphy 1990). *Promachus vertebratus* was a honey bee killer introduced into Puerto Rico (Wolcott 1922) to control other insects. *P. fitchi* is also
called the Nebraska bee killer and is found in Argentina (Bromley 1930). *Stenopogon
inquinatus* takes honey bees on flowers (O'Neill and Kemp 1991) in Montana, and
S. engelhardti hunted bees opportunistically (Powell and Stage 1962), whereas
other asilids had distinct bee appetites (Linsley 1960). Additional species include
Proctacanthus milbertii (or the Missouri bee killer) and *Saropogon dispar*,
Diogmites angustipennis, and *D. symmachus*, all of which are called the Texas bee
killers. *Bombomima* and *Erax* occasionally prey on bees (Bromley 1930). Fattig
(1945) recorded 30 asilid species caught with honey bee prey in Georgia. See Shelly
(1979) and Bromley (1946 and 1948) and references therein.

Australia: *Asilus rufirentris*, *Blepharotes flavus* (Tryon 1919), and
Dakinomyia spp. (Daniels 1978) are recorded as eating honey bees.

Control
While not a serious problem overall, Londt (1993) and Bromley (1930)
have suggested that robber flies could be a threat to bees if they are numerous or
their natural prey is scarce. Bromley suggests plowing fields known to contain the
fly larvae.

Phoridae (humpbacked flies)
Description
The Phoridae are small flies (2.0 to 4.0 mm long) that are easily recognized
by their humpbacked appearance and flattened hind femur. The males and females
have a similar appearance with a shiny head, brown thorax, yellow-brown legs, and
black abdomen. About 3,000 species occur worldwide, but only one
(*Pseudohypocera kerteszi* (= *P. nigrofascipes*) is listed as a pest of honey bees
(Reyes 1983). They are found in many habitats, especially decaying material, and
may be internal parasites (see below) of insects, whereas others are commensals
(cohabiting nests of other insects) and are thus discussed as both predators and
endoparasites.

Biology
The larvae feed on pollen first, then on prepupae and pupae of bees as the
fly population increases. Chaud-Netto (1980), who successfully reared these phorids
in Brazil, recorded that each female can produce up to 57 viable offspring. The fly
has been reported infesting 8 to 10 percent of bees in colonies in Mexico and
Colombia (Robinson 1981, Pérez Gómez 1975). When populations were high enough
the flies caused considerable colony losses in Colombia (Robinson 1982). Flies are

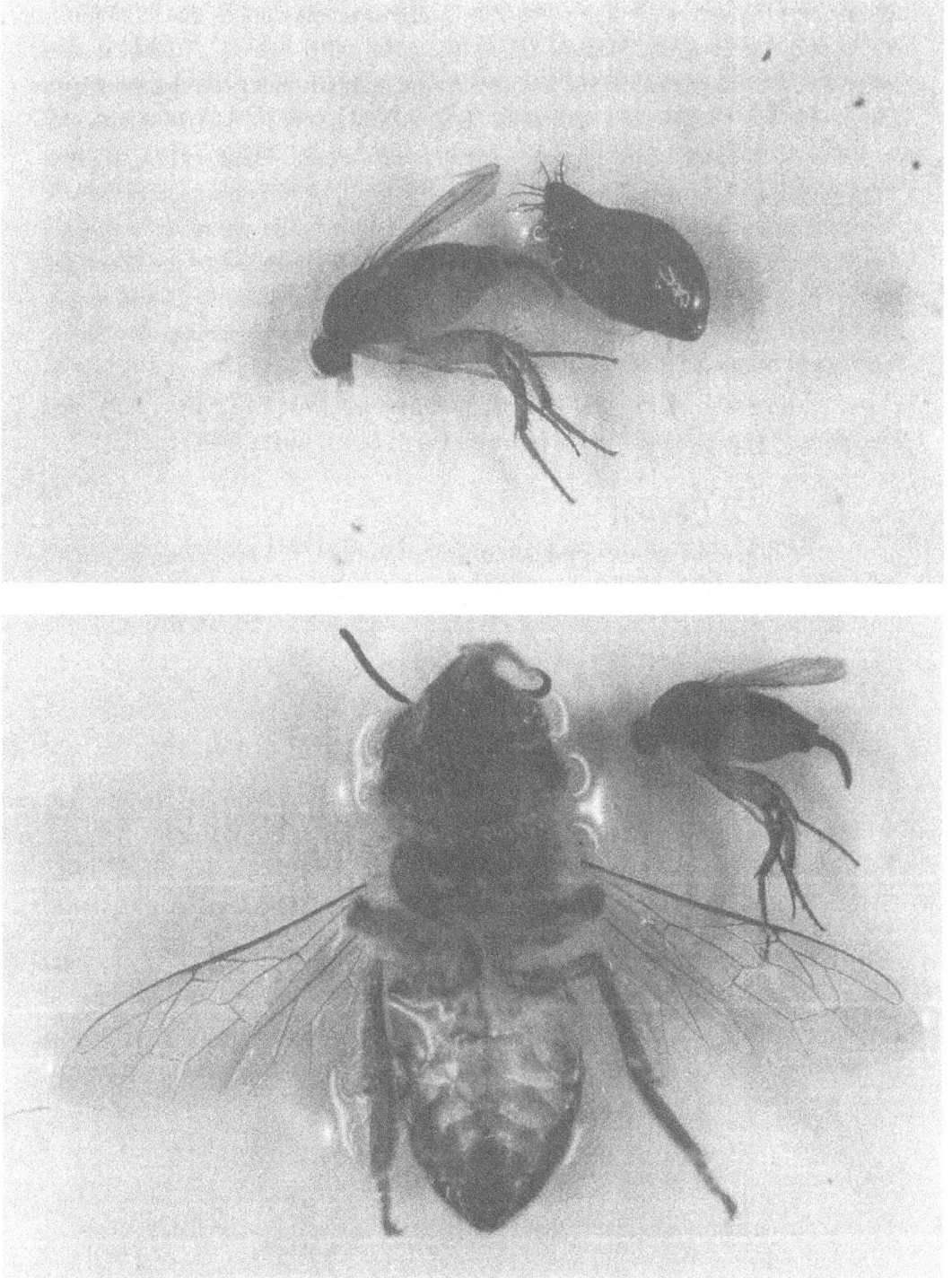

Figure 8.5. Top: Adult and pupa of the phorid fly *Melaloncha rommai*, which parasitizes honey bees in Brazil. Bottom: An adult fly next to a worker honey bee for size comparison. (Van de Sande photo).

found in weak or abandoned hives, attracted to the odor of fermenting pollen, and can survive several months in stored combs, presumably living off the pollen that is there. The flies are also associated with the stingless bees in the genera *Melipona* and *Trigona* (Robinson 1981). Örösi-Pál (1938) also mentions *P. clypeata*, which is reported to feed on honey, wax, and pollen. Other species of these flies attack the nests of social insects (Borgmeier 1924).

Control
Reyes (1983) recommends reducing hive entrances, relocating apiaries into half shade or full sun to reduce humidity, and keeping moth crystals in stored equipment.

Endoparasites
An endoparasite is an insect that lives within the host, most often killing it. When fly larvae become parasites in animals or people, the condition is referred to *myiasis*; when they infest bees, the term used is *apimyiasis*. The following species are endoparasites of honey bees.

Calliphoridae (blow flies)
These flies are found worldwide, and many species, such as the screw worm fly, are economically important. The size of house flies, many are metallic green or blue. Most are scavengers, and the larvae feed on feces or carrion. Only one species in this family is reported to infest bees.

Pollenia
Pollenia sp. is recorded in Egypt (Ibrahim 1984). Its larvae feed first on soft tissues of the thorax for about nine days, after which the bee dies. The maggot then enters the abdomen and feeds until it is mature. These flies pupate outside the host, and the adults emerge in about 12 days. Adult flies live only three days.

Control
No control measures for blow flies or *Pollenia* are recorded.

Conopidae (thick-headed flies)
Conopidae are medium-sized, brownish flies that resemble small thread-waisted wasps or syrphid flies with long antennae (Borror et al. 1989). Adults are usually found on flowers, but the larvae are endoparasites on honey bees, bumble bees (Schmid-Hempell 1991), and wasps. The following species are reported on honey bees.

Range

In the New World, Van Duzee (1934) records *Physocephala marginata* and *P. sagittaria* in bee hives, as well as *Zodion fulvifrons* (Severin 1937) and *P. texana* (Riedel and Shimanuki 1965, and see Hüttinger 1974).

Old World species include *Thecophora apivora* and *T. longirostris* reported by Zimina (1968). *T. occidensis* (Smith 1966), *Zodion notatum* (U.S.S.R. 1988, Zimina 1968), *Physocephala truncata* (Zimina 1973), *P. vittata,* and *Sicus sp.* (Sychevskaya 1956) are also noted. Unidentified conopids from Africa are recorded by Lundie (1965) and Hargreaves (1934). Zimina (1968) also documents that *Physocephala rufipes* parasitized *Bombus* species.

Biology

The parasitizing behavior is rarely seen. The female fly lays a newly hatched (first instar) larva on the host's abdomen, which then eats into the soft intersegmental membranes to enter the insect, feeding on the soft tissues therein (Hüttinger 1974). The larvae of *Zodion fulvifrons* are 0.3 to 1.5 cm long. When the maggot grows large enough to fill the abdominal cavity, it starts to feed on the thoracic muscles, paralyzing the bee, which drags itself around by its legs. The adults emerge two to three weeks after the larvae pupariate. The incidence of parasitism ranges from 50 percent in South Dakota (Severin 1937) to 7.8 percent in Wyoming (Riedel and Shimanuki 1965).

Control

No control measures are recorded.

Phoridae

Eight species in the genus *Melaloncha* are recorded as pests on neotropical honey bees: *Melaloncha ronnai*, in Brazil and Costa Rica, causes considerable mortality at certain seasons as recorded by Medina Solis (1980), Ronna (1936a, 1937a, b, c), Ramirez (1984), and Van de Sande et al. (1989).

Biology

Females oviposit on the host's abdomen where young larvae develop and enter through the intersegmental membranes. Örösi-Pál (1938), however, states that the female has a hard ovipositor and pierces the bee's body to lay eggs. In one week, when the egg hatches, the larva moves to the thorax to feed on muscle tissue, eventually killing the host. The dying bee lies on its side, trembling and shaking until death. The flies pupariate in three days within the host or outside in the soil. The intrapuparial stage lasts 21 to 31 days, depending on weather, heat, or humidity. Colonies have been reported with up to 59 percent infestation (Örösi-Pál 1938). The flies are rarely seen in hot, dry months such as April in Costa Rica. Eighty percent of bees examined were parasitized in December and 49 percent in September and October when it is humid (Solis 1980). Before the arrival of the honey bee, most species of *Melaloncha* parasitized meliponid bees (Borgmeier 1935) and *Bombus mexicanus* (Ramirez 1984). The infestation may be most severe in colonies placed in shady, wet areas (Ronna 1936a).

Control

Ronna (1936a) suggests burying dead and dying bees and spraying the apiary grounds with insecticides.

Other species of phorids

Additional phorid species are minor pests, but there is considerable confusion over the names of these humpbacked flies. Listed here are the names corrected by Örösi-Pál (1938 and references therein): *Borophaga* (= *Peromitra*) *incrassata* (Bailey and Ball 1991) or *Phora incrassata* (= *Hypocera incrassata*) was reported incorrectly by Packard (1869) to cause foulbrood disease in Europe and was reported feeding on dead bees in Algeria by Faveaux (1965). A megaselid was reported but not identified in Russia (Vyslouzil 1984). *Megaselis rufipes* (= *Aphiochaeta rufipes*) was reported widely in Europe as feeding on dead bees and brood (Clout 1956). Maciera et al. (1983) report that *M. scalaris* can parasitize stingless bees and honey bees in South America, Africa and India (Örösi-Pál 1938). *M. preacuta* (= *Aphiochaeta preacuta*) was found in brood from a colony attacked by a fungus and was reported with *M. rata* (= *Aphiochaeta rata*) in one colony. These scavengers are often found on dead bees or brood, especially those killed by American foulbrood disease, but do not affect healthy hives.

Sarcophagidae (flesh flies)

Flesh flies are similar to blow flies but are gray with black thoracic stripes, never metallic, and are commonly found around sweet material. Many are saprophytes (fungus feeders), and nearly all larvae feed on animal material or insects.

Range

There are over 2,600 species worldwide, and the subfamily Miltogramminae invades bee nests (Borror et al. 1989). The most important sarcophagid pest is *Senotainia tricuspis* meigen (= *Myiapis angellozi* or *Apimiase angellozi*), found in the Old World and Australia. Rocha and Delgado (1986) record it for the first time in Portugal, and it is widely spread across Russia (Smirnov and Luganskii 1987).

Biology

The flies parasitize bees on flowers or may swoop down on bees at the entrance of the hive, depositing newly hatched larvae on a bee's back every six to 10 seconds (Séguy 1930, Bailey and Ball 1991). One adult female is capable of producing up to 700 larvae (Boyko 1958). Wedmore (1932) notes that the fly lays an egg on the bee's body and the larva enters the spiracles. The larva lives in the thorax until it is 2.0 mm in length, after which the bee dies and the maggot moves to the abdomen. It grows to 8.0 mm long, leaves to pupariate in the ground, and emerges as an adult in six to eight weeks. The larvae may also eat their way into the bee host via the membranes and develop in eight to 11 days (Boyko 1939, 1971, Simintzis 1949, Giordani 1956) into the second instar. The maggots feed first on soft tissues and hemolymph in the abdomen. When the bee dies, the fly larva eats the muscles of the thorax and molts into the third instar. Some larvae will find another dead bee to consume, growing to 8 to 9 mm before pupariating. This is done outside the host in the soil in front of the hives; the fly emerges in seven to 18 days, or after overwintering there (Bailey and Ball 1991). In Southern France, August is the peak month for *Senotainia*, which attack only hives that are in full sun (Simintzis and Fiasson 1951).

Damage

Reports of damage are conflicting. Symptoms of sick bees resemble descriptions of bees infested with the tracheal mite, *Acarapis woodi* (Guilhon 1945). These flies may be the cause of "disappearing disease" (called "May disease" in France), which is caused by a spiroplasma (Mathis 1975, Jean-Prost and Médori 1994). In the early 1930s, devastating losses occurred, including paralyzed bees unable to fly and crawling out of colonies (Bailey and Ball 1991). Others report that no bees were lost, and it now appears that these symptoms may have been confused with other diseases or parasites.

Control

Various control methods have been proposed (Boyko 1949, 1959, Meged 1960, Smirnov and Luganskii 1987). These include drenching the soil in front of the hives with insecticides. Hives with artificially infested bees sustained no apparent damage (Simintzis 1958).

Another species, *Sarcaphaga surrubea*, was reported to damage honey bees in the New World (Braun 1957), and *S. nigriventris* is recorded by Ronna (1936a) and Séguy (1965) but Knutson and Murphy (1990) relate they are detritus and fungus feeders and do not damage bees.

Tachinidae

The second largest family in the order Diptera has about 1,300 species in North America and 9,000 worldwide. Members of this valuable group are endoparasitic on economically important insect pests such as butterflies and moths, sawflies, and beetles. Resembling muscids and flesh flies, the tachinid's abdomen is covered with large and small bristles; many are beelike mimics. *Rondaniooestrus apivorus* is the only tachinid known to be associated with honey bees and was the first fly known to cause apimyiasis (Villeneuve 1917). It is recorded only on honey bees in Africa (Skaife 1921a, Hargreaves 1934, Van Emden 1944, Buys 1975).

Biology

The female fly rests on the hive for 10 minutes, then hovers in front of its entrance and deposits newly hatched larvae on about a dozen or more bees as they enter the hive. She then rests for another 10 minutes and starts again. The larvae are glued onto a bee with a drop of gummy fluid but soon penetrate the bee via the intersegmental membranes. Once inside, they feed on body fluids and tissues until they are large enough to fill the abdomen, usually within four weeks. Finally the bee dies outside the hive where the maggot exits and pupariates in the soil (Skaife 1930).

Control

The fly does no economic damage to hives according to Hepburn (1991). The slow-moving insects are easily killed with a fly swatter (Taylor 1934, and Skaife 1930).

Minor fly pests

Drosophilidae (pomace flies) are common, small (3.0 to 4.0 mm) flies that are normally found around decaying fruit or vegetables. *Drosophila busckii* was found in the thoraces of bees in France (Mages 1956), and *D. ampelopilos* (Wedmore 1932) was reported to lay its eggs on honey cappings. The larvae feed on the honey and yeasts and regurgitate liquids on top of the wax cells. This action wets the comb cappings, making the appearance of the honey unpleasant. In addition, this causes the honey to ferment and may give rise to dysentery in bees. *D. flavohirta* is reported to compete with honey bees for eucalyptus nectar (Tribe et al. 1989) in South Africa. *Cacoxenus indagator* Loew is a "social parasite" or scavenger in other bees' nests (Knutson and Murphy 1990).

A soldier fly (Stratiomyidae), *Hermetia illucens,* is reported by Copello

(1926) in Argentina. It has a wasplike shape, and the female fly lays her eggs on decomposing organic matter or in cracks in a beehive. The larvae, which hatch in four to six days, feed on honey and pollen and in two to 20 days the adults emerge. They may destroy weak hives.

An Old World syrphid (flower fly or bee mimic) is recorded in bee nests by Kharizanov and Radeva (1979); from Australia, Goebel (1981) reports that *Tenthredomyia australis* maggots are found in honey and can be persistent pests even in strong hives, where they enter to lay eggs in the honeycombs. He recommended covering stored honey and burying or burning infested frames.

Zaitsev (1976a, b) reports a parasitic bombyliid fly in Iraq and another, in the genus *Anthrax*, is reported to parasitize bee larvae especially in arid regions in Algeria (Faveaux 1965). *Anthrax* is in turn parastized by a Hymenopteran parasitoid, *Bembex handlirschella* Ferton.

Summary

Aside from *Braula* and certain endoparasites, most flies do not pose a real problem to honey bees. In most cases, flies bother bees when the colony is dead or dying and abundant fungi, yeasts, tissues, or pollen are exposed to scavengers and saprophytes.

CHAPTER NINE

———⊶⊷⊷———

Insects: Hymenoptera (Ants, Wasps, and Bees)
Richard D. Fell

Formicidae (Ants)
> *Army ants*
> *Argentine ants*
> *Carpenter ants*
> *Formica ants*
> *Other ant species*
> *Honey bee defenses against ants*
> *Control*
> > Chemical control
> > Hive stands
> > Repellents

Wasps
> *Vespa (Hornets)*
> > Behavior
> > Bee defenses against hornets
> > Control procedures
> *Vespula and Dolichovespula (Yellowjackets)*
> > Control procedures
> *Polistes (Subfamily Polistinae)*
> *Philanthus and Palarus (Beewolves and bee pirates)*
> > Control
> *Mutillidae*
> *Other parasitic wasps*

Bees
> *Bombus (Bumble Bees – Subfamily Bombinae)*
> *Melipona, Trigona, and Lestrimelitta (Stingless bees – tribe Meliponini)*
> > Control
> *Apis (Honey Bees)*
> > Robbing
> > Reproductive Parasitism

CHAPTER NINE
Insects: Hymenoptera (Ants, Wasps, and Bees)

The order Hymenoptera[1] is one of the largest insect orders, containing over 100,000 described species worldwide. The members of this order are a diverse group and display a wide variety of lifestyles, habits, and behaviors. Many are important parasites or predators of insects, some of which cause problems to honey bees or honey bee colonies.

Several of the Hymenoptera discussed in this chapter have acquired common names because of their frequent interactions with humans. Names such as wasp, hornet, or yellowjacket may have different meanings to people in different areas of the world. In order to help avoid confusion and better define the terminology used in the chapter, the following definitions apply. The term *wasp* is used in a general sense for any hymenopteran, other than a bee or ant (or sawfly). Use of the term *hornet* is restricted to wasps in the genus *Vespa*. The name *yellowjacket* is used for members of the genera *Vespula and Dolichovespula*, all of which are social and are typically yellow and black in coloration.

Formicidae (Ants)

Ants are among the most successful and common insects and can be found throughout the world from the tropics to the Arctic. Despite their small size, the impact of ants upon the terrestrial environment is considerable; they are leading predators of insects and other small invertebrates in most terrestrial environments and among the primary movers of soil in both temperate and tropical regions. The diversity of ants exceeds that of all other social insects, and they are most abundant and ecologically dominant in the tropics. Over 8,800 species of ants have

---

[1]The order Hymenoptera is divided into two main suborders, the Symphyta and the Apocrita. The Symphyta is composed primarily of the sawflies and wood wasps, most of which are phytophagous or xylophagous. The members of this suborder do not interact with honey bees and are not considered further. The other suborder, the Apocrita, includes the ants, bees, and wasps. A number of the species within the suborder possess an ovipositor that has been modified into a sting, forming a group that is sometimes referred to as the aculeate Hymenoptera. The primary function of the sting is to inject venom, either for the paralysis of prey or for use in defense. The suborder Apocrita contains the largest group of insects exhibiting high levels of social behavior. All of the ants (family Formicidae) are eusocial, as are the wasps in the subfamilies Polistinae and Vespinae (family Vespidae). Among the bees, several families contain species that exhibit social behavior. The most commonly observed social bees are those in the family Apidae, including the bumble bees (Bombinae), the stingless bees (Meliponinae), and the honey bees (Apinae).

been described (Holldobler and Wilson 1990), although numerous undescribed species undoubtedly exist. Given their diversity, abundance, and frequent pest status, it is not surprising that ants can cause severe problems with honey bee colonies.

Ants are frequently mentioned as a serious pest of honey bees and interact with them in many ways. Ants, especially the well-known carpenter ants, may attack the wood in beehives. Other ant species may invade beehives and consume brood, honey, pollen, and even adult bees. Many early texts cite a need to control ants in apiaries (Benton 1896, Scholl 1912, Root and Root 1908). Buys (1975) reported results from a survey sent to South African beekeepers in which 62 percent of those responding felt that ants were a rather serious to serious problem; 33 percent reported badly weakened or destroyed colonies. Hussein (1989) reported that ants can be a serious problem in Oman, although infestation levels were low (1.4 to 1.7 percent). Shahid (1992) noted that ants were a problem in parts of Pakistan, and Veeresh (1990) reported that ants frequently damage hives in India. Ibrahim and Yusoh (1989) stated that weaver ants were a common pest of *Apis cerana* apiaries in Western Malaysia; Seeley et al. (1982) found that weaver ants were a major pest of both *A. cerana* and *A. florea* in Thailand. In South America, several species of ants have been reported as pests of honey bees (Winston et al. 1979), and Nunamaker et al. (1986) stated that ants were one of the most serious pests to beehives in Belize, Central America.

Army ants
Legionary or army ants in the subfamilies Dorylinae and Ecitoninae have been reported as serious pests of honey bee colonies in parts of Africa and the Neotropics. Army ants forage in groups of tens or hundreds of thousands and can take over and destroy an entire apiary within a few hours. Dubois and Collart (1950) described the destruction of brood (both larvae and pupae) by army ants in Zaire (formerly the Belgian Congo), and Mwale (1992) noted that 7 percent of the colonies examined in Malawi had army ant damage (*Dorylus* sp.). These ants may also be a problem in the Amazon forest and other parts of South America. Hellmich and Rinderer (1991) stated that doryline ants destroy more honey bee colonies in Venezuela than all other pests and diseases combined. Winston et al. (1979) also noted that army ants (*Eciton* sp.) were a frequent cause of absconding by Africanized bees in French Guiana. Honey bees are unable to defend themselves against such attacks and may abandon their nest when overrun by raiding ants.

Argentine ants
The Argentine ant, *Iridomyrmex humilis*, is native to Brazil and Argentina

but has spread from South America throughout much of the world. It is frequently reported as an important agricultural and urban pest in the United States, South Africa, and Australia, but is a common pest of honey bee colonies wherever it is found. Workers of this ant species are small, 2.2 to 2.8 mm long, and light to medium brown in color with a darker-colored gaster. Argentine ants nest in exposed soil and soil under cover but may also be found in rotten wood, refuse piles, and in or under beehives (Smith 1965). Colonies may be large and have hundreds of queens (Holldobler and Wilson 1990, Thompson 1990).

Argentine ants are capable of destroying strong hives, often attacking colonies over a period of several days. May (1961), for example, described the loss of 160 colonies by a beekeeper in South Africa over six years. These ants, however, more frequently attack weak colonies. Newly established hives and colonies that have been manipulated or moved are more susceptible (Buys 1990). Ants often initiate colony attack by entering a hive and stealing honey from combs. As more workers are recruited, the ants attack and eat brood. They eventually attack any adult bees that have not abandoned the nest. The ants (and robbing bees from other hives) then remove any remaining honey from the comb. Recently manipulated hives are often susceptible because propolis seals between hive bodies break during inspection, leaving cracks for the ants to enter. The presence of crushed bees between hive bodies may also serve as an attractive food source for the ants (Buys 1990). Colony movement can make bees more susceptible to attack because of the bees' disorientation and a disruption of normal guarding activities. In addition, Spangler and Taber (1970) found that *I. humilis* workers have low odor levels and no alarm pheromones. This lack of detectable odors reduces the responses of the bees to the ants, allowing the latter easier entry to hives.

In the United States, Argentine ants are found throughout the South, along the West Coast, and in Arizona, Illinois, Maryland, and Missouri (Thompson 1990). Problems with *I. humilis* attacking honey bee colonies have been reported from California (Newell and Barber 1913), Hawaii (Messing 1991), Louisiana (Cook 1908, Cockerham and Oertel 1954), and Florida (Robinson and Oertel 1950) but undoubtedly occur in other areas across the Southern United States. Similar problems have been reported in Colombia (Robinson 1982), Bermuda (De Jong 1990a), and Zimbabwe (Papadopoulo 1965). Buys (1987) has also reported that Argentine ants can compete with honey bees for nectar. His studies in South Africa indicate that the ants can collect as much as 66 percent of the nectar produced by flowers of the black ironbark tree, *Eucalyptus sideroxylon*.

Table 9.1. Additional records of ant species entering or nesting in honey bee colonies.

Species	Common Name	Geographic Area	Reference
Camponotus noveboracensis	carpenter ant	New York, U.S.A	Morse and Gary 1961
C. pennsylvanicus	black carpenter ant	New York, U.S.A	Morse and Gary 1961
Diacamma rugosum		Thailand	Seeley et al. 1982
Dolichoderus bituberculatus		Thailand	Seeley et al. 1982
Dorylus labiatus		India	Singh 1962
Formica fusca	silky ant	New York, U.S.A.	Morse and Gary 1961
Iridomyrmex pruinosus		Arizona, U.S.A.	Spangler and Taber 1970
Lasius niger		England	Mace 1927
	corn louse ant	Missouri, U.S.A.	Burrill 1926
Monomorium destructor		India	Singh 1962
M. indicum		India	Singh 1962
M. pharaonis	Pharaoh ant	South Dakota, U.S.A.	Worden 1917
Odontoponera transversa		Thailand	Seeley et al. 1982
Oecophylla smaragdina	weaver ant	Thailand	Wongsiri et al. 1986
Pheidole megacephala		Hawaii, U.S.A.	Messing 1991
Pheidologeton diversus	marauder ant	Thailand	Wongsiri et al. 1986 Seeley et al. 1982
Solenopsis (probably *geminata*)	fire ant	Louisana, U.S.A.	Cockerham and Oertel 1954
S. invicta	imported fire ant	Louisana, U.S.A.	Cockerham and Oertel 1954

Carpenter ants

Several species of ants in the genus *Camponotus* will attack honey bee colonies on occasion. Most of these species do not cause serious problems, but some can destroy colonies or damage hive materials. Smith (1965) reported that *C. abdominalis floridanus*, the Florida carpenter ant, can seriously disrupt honey bee colonies, plundering hives for both food and living quarters. Walshaw (1967) observed that this ant could destroy small or weak colonies overnight and that in some areas of Florida special precautions are needed to protect small colonies such as mating nucs. Burrill (1926) reported damage to hives and honey stores by *C. herculeanus pennsylvanicus* in Missouri. The carpenter ant *C. herculeanus* typically nests in the trunks of living or dead trees (Holldobler and Wilson 1990) and may raid nearby bee colonies. In one case reported by Burrill, the ants raided two hives, destroying much of the comb and woodenware. In the other case, the ants raided honey stores in supers above a bee escape, damaging comb honey sections. Carpenter ants occasionally cause minor structural damage to hives but rarely cause serious problems, especially to sound, dry wood. Carpenter ants potentially involved in structural damage to hives in the United States include *C. pennsylvanicus* in the East and *C. modoc* and *C. vicinus* in the West. In Europe, *C. herculeanus* is a common pest that could damage beehives (Fowler 1990).

Other species of *Camponotus* may cause problems. In India, *Camponotus compressus* has been reported as a serious pest of beehives (Singh 1962). Winston et al. (1979) reported that *Camponotus* ants were a major cause of absconding by Africanized honey bees in French Guiana. De Jong (1990a) observed at least two species of *Camponotus* ants killing and removing *Apis cerana* from the exterior of a recently transferred hive in South Korea.

In some areas, pollen traps must be protected from ants that would otherwise carry away much of the harvest. *Camponotus punctualis* and *Crematogaster jheringi* have been reported as common invaders of pollen traps in Argentina (Santis and de Regalia 1978).

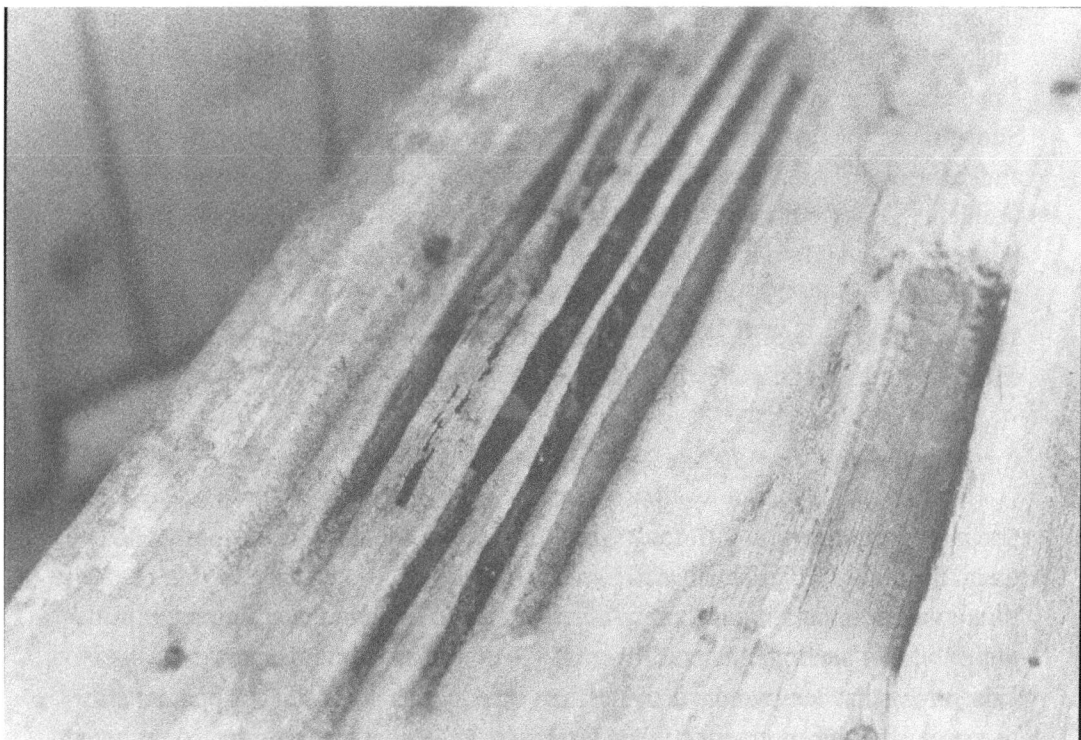

Figure 9.1. This super was structurally damaged by carpenter ants tunneling through the wood. (De Jong photo).

Formica ants

Formica is a common genus of ants found throughout much of the Northern regions of both the Old World and the New World. Many are soil nesters or mound builders, and some species have been reported as aggressive predators that may invade honey bee colonies. Corner (1960) recorded the destruction of several colonies in an apiary in British Columbia, Canada, by ants that were tentatively identified as *Formica integra*. Burrill (1926) described the invasion and honey collection by foragers of *Formica fusca subsericea* of a hive in Missouri, noting that the bees paid relatively little attention to the ants. In Europe, the red wood ant, *Formica rufa*, a mound builder, attacked and destroyed colonies placed near large nests. These ants also destroyed an apiary of 20 hives in Romania (Prosie 1959) and one of 26 hives in West Germany (Muthel 1955).

Other ant species

Several ant species have been described as occasional invaders of honey bee colonies, but most do not cause serious problems, entering hives in search of food or potential nest sites. The beehive not only provides a dry, warm, and protected environment, but it also contains spaces suitable for use as nest sites by several ant species. The space between the inner and outer covers may be used by ants, especially if the bees do not have access to the area. *Crematogaster lineolata* has been commonly observed nesting on the inner cover of hives (Burrill 1926). Such infestations are more a nuisance to the beekeeper than a problem to the bees, and control efforts are rarely warranted. Some species can cause more serious problems. Drees and Vinson (1986) stated that fire ants (*Solenopsis* sp.) will invade beehives to feed on developing brood and can destroy weak colonies. Fell (personal observation 1990) observed mating nucs destroyed by fire ants (*Solenopsis* sp.) in Alabama, U.S.A. Additional records of ant species that have been found entering or nesting in honey bee colonies are listed in Table 9.1.

Honey bee defenses against ants

Honey bees are normally capable of defending their nests against ants and may exhibit several different types of defensive behavior when approached by them. Sistrank (1956) described a characteristic behavior in which a worker bee will turn away from an ant and fan with her wings to dislodge the attempted intruder with a blast of air. Spangler and Taber (1970) described similar responses by bees to ants, noting that bees would also kick their hind legs backward in an apparent effort to remove the ants more effectively. Defensive behaviors by bees are also triggered by odors given off by some ants. The ant *Iridomyrmex pruinosus analis* gives off 2-heptanone as an alarm pheromone. This chemical, which is used by honey bees as an alarm odor (Boch et al. 1970), stimulated ant defensive behaviors (Spangler and Taber 1970). These authors found that other ant compounds would release defensive behaviors, including benzaldehyde, citral, and formic acid (secreted by both *Formica* and *Camponotus* sp.).

Apis cerana workers exhibit a pivoting and fanning behavior as an ant defense, similar to that observed with *A. mellifera* workers. The major ant pests of *A. cerana* in Thailand are weaver ants (*Oecophylla smaragdina*), and the bees use a blast of air in an attempt to dislodge ants trying to enter their nest. Entrance smoothing may increase the effectiveness of this fanning behavior (Seeley et al. 1982). Ants not repelled by fanning are grabbed by guard bees and pulled apart.

Because direct assaults on the hive entrance are rarely successful, weaver ants may prey on *A. cerana* by perching on vegetation near the nest and pulling down flying bees. This behavior is only partly successful and requires the cooperative effort of several ants. As a counterdefense, guard bees may hover a few millimeters above the ants to distract them, enabling foragers or drones to fly past safely. This distraction behavior is most commonly observed during daily periods of drone flight (Seeley et al. 1982). In Korea, *A. cerana* workers also have been observed to defend their nest by stinging (occasionally) and flying off with large *Camponotus* workers. This behavior reduces the likelihood that the ants will successfully invade a hive by removing them from the immediate area (De Jong 1990a).

The use of propolis forms an important part of the honey bee nest defense system. Cracks and other small openings can be filled or the entrance reduced to prevent the entry of ants. Some races, such as the Africanized honey bee, do this more than others. In Korea, beekeepers with *A. cerana* are careful to tape or put mud on the outside of all cracks and joints of traditional log or box hives. The bees also tend to guard any openings in the hive, even those too small for them to pass through. These bees use little, if any, propolis on the hive interior, although they sometimes seal cracks with beeswax (De Jong 1990a). The invasion of ants may cause a colony to abscond. Seeley et al. (1982) observed an *A. cerana* colony, which was kept in a crudely built hive with many openings, abscond when invaded by *Monomorium destructor*.

The dwarf honey bee, *A. florea*, also uses propolis to protect its nests by placing bands of sticky propolis on the branches or twigs that support the single comb nest (Akratanakul 1977, Seeley et al. 1982). Ants are a constant threat to *A. florea* colonies, especially the weaver ant, *Oecophylla smaragdina*. Seeley et al. (1982) noted that 37 percent of the colonies were under attack by weaver ants when first discovered. The use of a sticky resin band provides an effective defense, but the ants can invade a nest if the propolis dries out slightly (Akratanakul 1977). Ants unable to gain access to a nest may stand on surrounding vegetation and catch bees that fly too closely to their position. *A. florea* workers are no match for an ant and are quickly subdued if captured. If a weaver ant falls onto an *A. florea* nest, the bees at first withdraw, but then they mob the ant, kill it, cover it with gooey resins, and drop it from the nest (Seeley et al. 1982).

Colonies of *A. dorsata* are rarely bothered by ants and are successful at repelling them. Nests are covered by a protective curtain of bees that quickly responds to potential threats. Seeley et al. (1982) observed several nests that were approached by the arboreal ants, *Dolichoderus tuberculata* and *Polyrachis armata*. Guarding bees lunged at the ants that approached too closely, grabbing them or causing them to scurry away.

Control

Beekeepers in many areas of the world face the need to control ants in apiaries. The approach to such problems and the recommendations for control should be considered with respect to the specific country and its beekeeping industry. Ibrahim and Yusoh (1989) have suggested several factors to consider before making control recommendations. The chosen method should provide quick, effective control, should use locally available materials, and should suit the beekeeper's finances.

Chemical control

The use of chemical control procedures is often recommended in areas where ants are a persistent problem. Considerable care must be exercised, however, if insecticides are to be used in apiaries or near beehives because most of these materials are also highly toxic to bees. The beekeeper should carefully follow label directions and application instructions, being sure that both the insecticide and the formulation are legal and appropriate for the pest and the intended application. The effectiveness of control measures can be improved if apiary sites are kept free of debris and heavy vegetation (Cockerham and Oertel 1954). Rotten wood, boards, trash, and similar materials should be removed from the areas around hives. Underbrush, weeds, and grass should be cut close to the ground. Cleanup efforts will reduce the number of potential ant nesting sites and better allow insecticide applications to reach the soil.

Specific chemical recommendations will depend on the ant species and the area where the application is to be made. Beekeepers should check with extension agents or apiary inspection personnel in local or regional agriculture offices for current treatment recommendations. In the past, insecticides such as DDT (dichlorodiphenyltrichloroethane) or chlordane were used to control ants in apiaries (Robinson and Oertel 1950, Wright 1952, Dyce 1953). Spray applications using wettable powder formulations of DDT or chlordane were recommended for treating the ground area around hives (Cockerham and Oertel 1954, Eckert and Shaw 1960). These materials have been banned in many countries because of environmental persistence and effects on non-target organisms. The use of a diazinon granular formulation has been recommended for ground treatment around (sides and rear) and under hives in some areas (Williams 1985). Drees and Vinson (1986) have recommended the use of baits, insecticide barriers, or both around hives to control fire ants. The careful application of diazinon or chlorpyrifos to hive stands or the ground around hives can help reduce fire ant problems. Individual ant mounds near hives can be treated with insecticide drenches.

Hive stands

Many attempts have been made to construct or alter hive stands to prevent ant access to hives. Some beekeepers have even constructed moats around apiaries (Hefinger 1922). Selders (1933) claimed that hive stands made with the wood from sassafras (*Sassafras albidum*) repelled ants, but most efforts have been directed toward limiting ant access to hives by using barriers. Dalton (1931) and Koover (1949) built antproof hive stands by using containers of axle grease over which the ants had to pass. Koover stated that his barriers had been effective for five years and prevented Argentine ants from entering his hives. Eckert and Shaw (1960) suggested placing hive stand supports in containers of oil, and Clauss (1983) recommended tying rags soaked in oil around hive stand supports or on top of vertical stakes supporting a hive. Oil deters ants only temporarily, and reapplications may be required more than once a week (Ibrahim and Yusoh 1989). Western (1989) suggested that an effective "ant isolator" could be made with two tin cans, one placed inside of the other. The inner can is taller and narrower and is slightly overfilled with wet cement. A wider, shorter can is placed over the inner can and the two cans are pressed together until the cement has hardened. The outer surface of the inner can and the inner surface of the outer can are then smeared with grease or thick oil as an ant barrier. Four such isolators are placed under a hive or hive stand so that the wider can is in an inverted position.

Cockerham and Oertel (1954) and Eckert and Shaw (1960) suggested the use of tanglefoot bands as a barrier but noted that the bands must be renewed frequently. Ibrahim and Yusoh (1989) tested several sticky glues for effectiveness in weaver ant control in *A. cerana* apiaries in Malaysia. A glue mixture made from rubber latex and used engine oil in a 1:4 ratio and applied to hive stands had the potential to prevent ants from reaching hives for two to three weeks. As an alternative, insecticides can be used with hive stands. Dyce (1953) reported that this approach was effective in Costa Rica. Almost all beekeepers in Brazil use hive stands to reduce ant damage to colonies. They sometimes place inverted tin cans

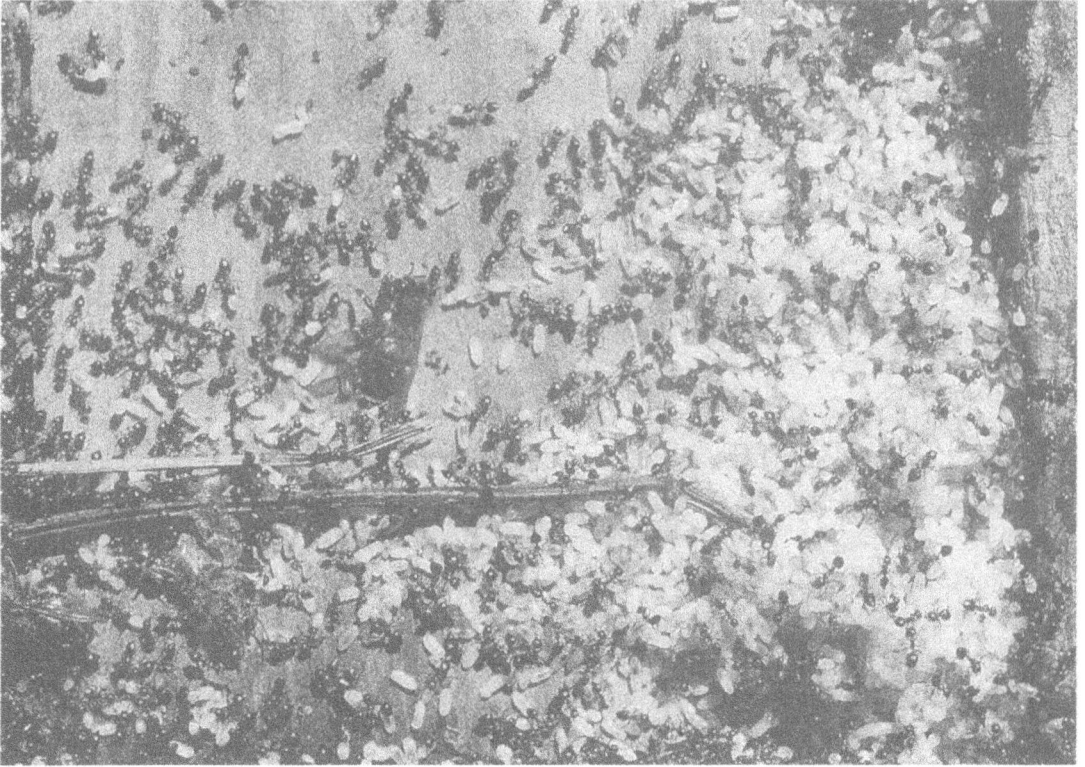

Figure 9.2. An ant colony living on an inner cover of a weak colony in Ohio. Eggs and pupae are stacked in one corner, but the ants covered the whole inner cover. Ants can bite, sting, pull arm hair and be a general nuisance when a beekeeper is working a colony. (Flottum photo).

filled with automotive grease over the vertical posts of the stand (De Jong 1990a). Other beekeepers suspend hives from wire supports. Walshaw (1967) described a queen breeder in Florida who eliminated ant problems by stapling his mating nucs to wires suspended from posts.

Repellents
Many beekeepers have noticed ant colonies between the inner and outer covers of hives or in insulating materials placed between the hive cover and the top bars of frames. Such colonies are often a nuisance for the beekeeper when handling a hive but usually do not present a serious problem to the bees. Nonetheless, beekeepers are often concerned about the presence of ants and have used various natural and synthetic materials as repellents in an effort to deter ants from nesting in hives. Natural repellents used for this purpose include catnip (*Neptea cataria*) (Starnes 1946, Bond 1947), tansy (*Chrysanthemum vulgare*) (Stepp 1935, Dunn 1938, Starnes 1946), and green leaves of black walnut (*Juglans nigra*) (Bond 1947). Common chemicals used for the same purpose include alcohol (Root 1919), sodium

fluoride (Ramsey 1946), borax powder (a household cleanser) (Nikiel 1972), and salt or powdered sulfur (Dadant 1957). Most of these materials have not been carefully tested and their effectiveness as ant repellents is questionable. A simpler solution is to avoid the use of insulation materials under the cover or to leave a hole in the inner cover so that the bees can keep the ants away.

Wasps

Vespa (Hornets)

The hornets in the genus *Vespa* are the most important wasp predators of honey bees because of their large size and behavior, and they can cause serious economic problems to beekeepers in some areas. These wasps have been recognized as an important threat to honey bee colonies since Roman times (Fraser 1951). Serious predation problems by hornets are most common in the tropical or subtropical areas of the Middle East, Southeast Asia, and Japan (Table 9.2), but not the United States.

Attacks by hornets on colonies of *A. mellifera*, *A. cerana*, and *A. florea* can lead to serious damage, complete colony destruction, or nest abandonment (Matsuura and Sakagami 1973, Seeley et al. 1982, Shah and Shah 1991). In some areas of Japan, hornets have led to the total elimination of wild *A. mellifera* colonies. They also cause serious damage to large numbers of managed colonies each year unless the hives are protected (Matsuura 1988). Similar reports come from India (Shah and Shah 1991) and Israel. The Israel Beekeeper's Association (1949) reported a loss of 3,000 of the association's 28,000 hives because of the depredations of hornets.

Behavior

The predatory behavior of many *Vespa* species on honey bees has been well-documented (Matsuura and Sakagami 1973, Seeley et al. 1982, Matsuura 1988, Shah and Shah 1991). Most hornets attack singly, preying on bees at or near a hive entrance. Some attacks on colonies, however, may involve several individuals or a mass attack with as many as 16 to 18 hornets (Matsuura 1988). Small numbers of *Vespa tropica* workers (two to three), for example, can cause an *A. florea* colony to abscond (Seeley et al. 1982). Mass attacks on *A. mellifera* colonies by *V. mandarinia* involve an average of 7.9 ± 3 hornets and lead to the complete destruction of an unprotected colony.

Hornets often initiate an attack by landing near a hive or hovering near the entrance. *V. velutina* workers typically hover in front of a hive and catch returning foragers. The hornets fly to a support and consume the bees' internal organs, discarding the exoskeleton (Shah and Shah 1991). In late October and November

Table 9.2. Additional records of *Vespa* species preying on honey bees or attacking honey bee colonies.

Hornet Species	Honey Bees Species Attached[1]	Geographic Area	Reference
V. affinis	A.m.	Taiwan	Sonan 1927
	A.m.	Thailand	Wongsiri et al. 1986
	A.c.	Malaysia	Ibrahim & Yusoh 1989
V. analis	A.m., A.c.	Indonesia	van der Vecht 1957
	A.m., A.c.	Japan	Matsuura &
			Yamane 1984,
			Matsuura 1988
V. basalis	A.m., A.c	India	Singh 1962,
			Kshirsagar & Mahindre 1975,
			Sharma et al. 1985
		Pakistan	Muzaffar & Ahmad 1986
V. crabro	A.m.	Azerbaijan	Atakishiev 1971a,
			Kadymov 1981
	A.m.	Germany	Olberg 1959
		Great Britain	Andrews 1882,
			Herrod-Hempsall 1937,
			Buzzard 1951, Bunn 1988
	A.m.	Italy	Alber 1953, Longo 1980
	A.m.	Japan	Okada 1960,
			Matsuura 1988
V. mandarinia[2]	A.c.	China	Kellogg 1941
	A.m., A.c., A.d.	India	Kshirsagar & Mahindre 1975,
			Sharma et al. 1979, 1985,
			Hussain & Ali 1980
	A.m., A.c.	Japan	Matsuura & Skagami 1973,
			Okada 1956, 1961, 1984,
			Ono et al. 1987
	A.m.	South Korea	Choi 1974
		U.S.S.R.	van der Vecht 1957
V. orientalis	A.c.	China	Kellogg 1941
	A.m.	Egypt	Wafa 1956, Wafa &
			Sharkawi 1972
	A.m.	India	Kshirsagar & Mahindre 1975,
			Sharma & Raj 1988
	A.m.	Israel	Rivnay & Bitinsky-Salz
			1949, Ishay et al. 1967
	A.m.	Jordan	W. Robinson 1981
	A.m., A.c.	Pakistan	Muzaffar & Ahmad 1986

V. simillima[3]	A.m., A.c.	Japan	Matsuura & Skagami 1973,
			Higo 1983, Okada 1984,
			Sakai 1989, Sakai &
			Ono 1990
V. tropica[4]	A.c.	China	Kellogg 1941
	A.m., A.c.	India	Kshiragar & Mahindre 1975,
			Sharma et al. 1985,
			Sharma & Raj 1988
	A.c., A.d.	Malaysia	Schmidt et al. 1985,
			Ibrahim & Yusoh 1989
	A.m., A.c.	Pakistan	Muzaffar & Ahmad 1986
	A.c.	Sri Lanka	Robinson 1982
	A.m., A.c., A.f.	Thailand	Burgett & Akratanakul 1982,
			Seeley et al. 1982,
			Wongsiri et al. 1986
V. velutina[5]	A.m., A.c.	India	Singh 1962, Sharma et al. 1985,
			Sharma & Raj 1988, Shah
			& Shah 1991
	A.m., A.c.	Indonesia	van der Vecht 1957
	A.m., A.c.	Pakistan	Muzaffar & Ahmad 1986

1. A.m. = *Apis mellifera;* A.c. = *A. cerana;* A.d. = *A. dorsata;* A.f. = *A. florea*
2. *V. mandarinia magnifica* is a subspecies, but has been cited as *V. magnifica* (see Edwards 1980).
3. Synonym *V. mongolica*
4. Synonyms *V. cincta, V. ducalis* (a subspecies of *V. tropica*) (Edwards 1980).
5. Synonym *V. auraria* (a subspecies of *V. velutina*) (Edwards 1980).

---

when temperatures are low and the bees are inactive, the hornets may enter bee colonies and scavenge dead bees, attack isolated workers, or steal honey from open cells. Several other *Vespa* species exhibit similar predatory behavior. In Japan, *V. simillima*, *V. analis*, *V. tropica*, and *V. crabro* will hover near the entrance of hives and catch foragers. *V. simillima* and *V. analis* will also attempt to enter hives; *V. tropica* does not (Matsuura 1988). Wasps that enter a hive may carry off larvae or pupae, preferring the latter. Colony attacks generally start in midsummer and continue until late fall, depending on the hornet species. The peak period for attack in Japan is August and September for *V. simillima* and *V. analis* but extends through October for *V. mandarinia*.

   Vespa crabro, one of the more widely distributed hornet species, is found throughout much of Europe, North America, and Japan, as well as other parts of Asia. The natural distribution of this wasp did not include North America, but it was introduced accidentally into the United States from Europe in about 1840 (Shaw and Weidhaas 1956). *V. crabro* generally does not cause problems for beekeepers in the United States and is rarely seen around hive entrances, although Caron and Schaefer (1986) reported European hornets capturing bees in an apiary in Maryland. In Germany, the wasp is also of minor importance as a bee predator (De Jong 1990a). This species often constructs its nests in protected locations such as tree cavities, the attics of houses, or even underground (Matsuura and Sakagami 1973). These hornets will attack yellowjacket nests (De Jong 1979) and occasionally prey on foraging honey bees in the field. They have a rather curious behavior of pouncing on any dark object on a flower, which may include the capture of honey bees (De Jong 1979). As do other hornet species, *V. crabro* prey upon bees as a protein source for the hornet larvae. After catching a bee, the wasp typically bites off the head and abdomen and flies back to the nest with the thorax. Other feeding behaviors may also be observed. Olberg (1959), for example, described attacks in which *V. crabro* bit open a bee's abdomen to feed on the contents of the honey stomach.

The oriental hornet, *V. orientalis*, is an important predator of honey bees in the Mediterranean area and has been known to destroy entire apiaries (Rivnay and Bitinsky-Salz 1949). These wasps forage regularly at the entrances of hives and can seriously deplete a colony's forager force. *V. orientalis* workers may also enter a hive to feed on brood and young bees after killing the guard bees. Brar et al. (1985) reported that this species visited *A. mellifera* colonies in the Punjab region of India from July to December with peak populations from August to October. The hornets visited colonies in the middle of the day and made few attacks in the morning and evening.

The most damaging attacks to honey bee colonies are made by *Vespa mandarinia*. They are the most common hornets attacking honey bee colonies in Japan and account for almost 74 percent of the *Vespa* species captured in hornet traps attached to hives (Matsuura 1988). (The second most common species is *V. simillima*, which accounts for approximately 18 percent of the trapped hornets.) This large wasp (*V. mandarinia* is 10 to 15 times the size of a honey bee) can inflict catastrophic damage on an *A. mellifera* colony by group predation. The mass attacks of *V. mandarinia* involve a unique strategy consisting of hunting, slaughter, and nest occupation phases (Matsuura and Sakagami 1973, Matsuura and Yamane 1984, Matsuura 1985). In the hunting phase, the hornets kill one bee at a time, make a ball of meat from the thorax, and carry it back to the nest. This phase may continue for an indefinite period of time or may shift to the next phase. The shift to the slaughter phase depends to a large extent on the distance between the hornet's nest and the apiary; the closer the nest, the more likely the shift (Matsuura and Sakagami 1973). The attacking hornets may mark the hive site with a pheromone from the van der Vecht gland (basal tuft of the terminal gastral sternite) before the mass attack starts (Ono et al. 1991). This pheromone helps to coordinate group predation such that several hornets may concentrate on a single hive, biting and killing counterattacking bees. All bees in a hive will be killed unless the hive is abandoned (this occurs less than 6 percent of the time). The mean time from the initiation of an attack to hive occupation in 36 mass attacks observed by Matsuura (1988) was 105.4 minutes. Individual hornets participating in one attack were estimated to kill more than 2,000 honey bees each, with total hive losses ranging from 12,550 bees to more than 23,000 bees in 22 hives monitored during attacks. During the occupation phase, the hornets carry stored honey, larvae, and pupae back to their nest. The hornets also post guards at the hive entrance and may occupy a hive for up to three weeks (mean 10.4 days) (Matsuura and Sakagami 1973, Matsuura 1988).

Mass attacks generally are not observed with other hornet species, although Burgett and Akratanakul (1982) reported a similar hunting, slaughter, and occupation sequence by *V. tropica* on an *A. mellifera* colony in Thailand. Hornet attacks on *A. mellifera* colonies apparently succeed because of the lack of an effective defense, especially against mass attacks (Matsuura 1988).

Hornets also attack other honey bee species. Seeley et al. (1982) reported on the attack of an *A. florea* colony by *V. tropica* in Thailand. The initial attack involved a single hornet that landed on vegetation near the nest. From this point, the attacking hornet caught and killed more than 80 bees. The following day, two wasps attacked the nest, and after resisting an attempt by the bees to mob one of the wasps, moved onto the comb, forcing the bees to retreat. Shortly after the arrival of a third wasp, the bees absconded. After occupying the bees' nest, the wasps removed brood and carried it back to their nest.

Hornets in the genus *Vespa* are a threat to *A. cerana* colonies but generally are not as successful in their attacks on this species. *V. simillima* and *V. mandarinia* are major natural enemies of *A. cerana* in Japan (Matsuura and Sakagami 1973, Ono et al. 1987), as is *V. tropica* (Seeley et al. 1982). Other reports of hornet predatory behavior on this honey bee species include *V. velutina*, *V. mandarinia magnifica*, and *V. basalis* in India (Singh 1962, Shah and Shah 1991, Kumar et al. 1993), *V. cincta* (=*tropica*) in Sri Lanka (Robinson 1988), and *V. orientalis*, *V. cincta*, and *V. mandarinia* in China (Kellogg 1941). Colony destruction by hornets is not common with *A. cerana* because of its defensive behaviors, but continued predation can seriously decimate a colony population. Ibrahim and Yusoh (1989) have reported that both *V. affinis indonesiensis* and *V. tropica tropica* will attack *A. cerana* colonies in Malaysia in large numbers and can weaken them and cause absconding. De Jong (1990a) reported that *A. cerana* beekeepers in Korea frequently suffered colony losses from attack by *V. mandarinia*.

If *A. cerana* colonies are kept in apiaries with *A. mellifera*, they are attacked less frequently by hornets. Adlakha (1975) and Kumar et al. (1993) have

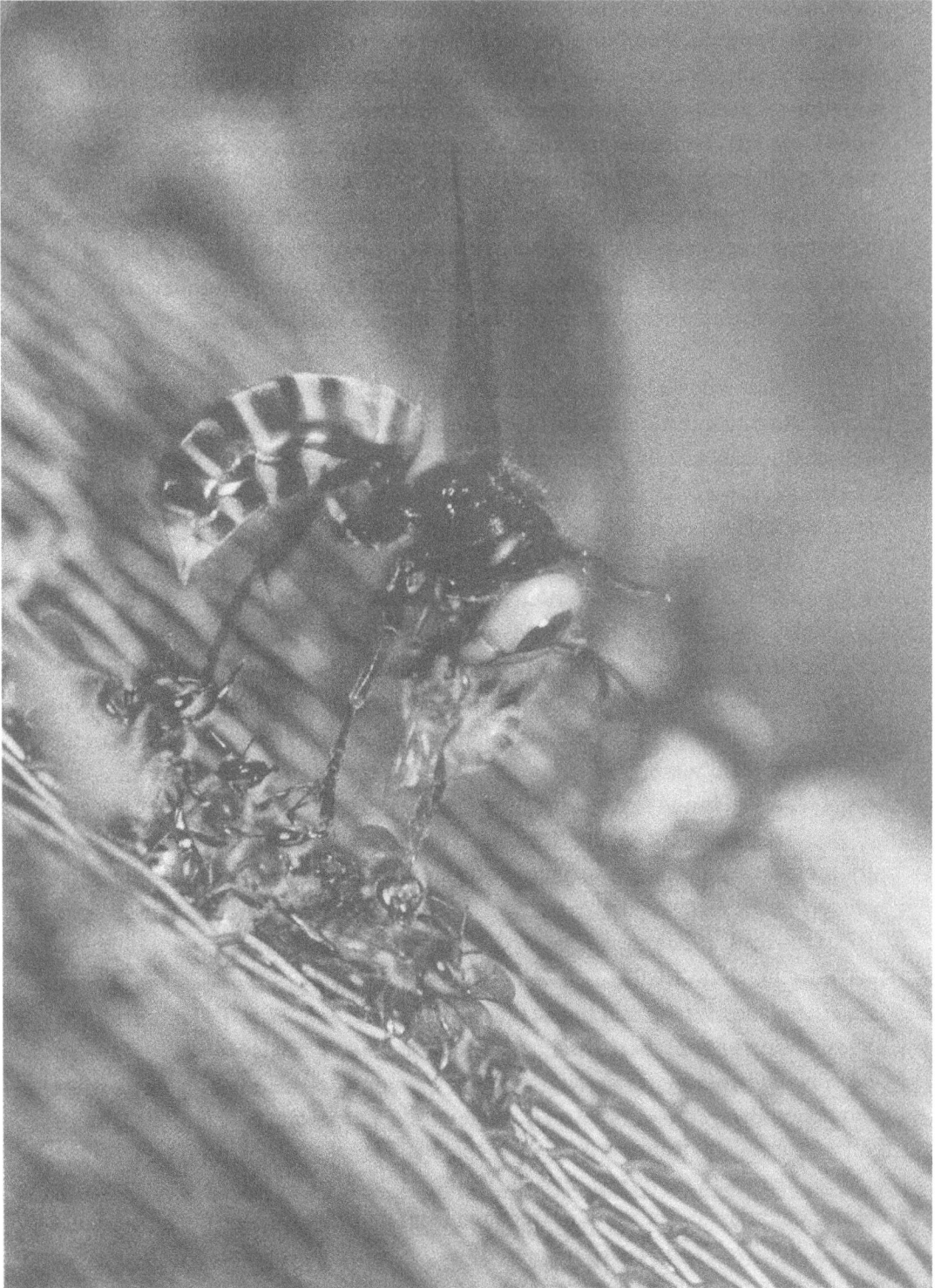

Figure 9.3. *Vespa mandarinia* is 10 to 15 times the size of a honey bee, and can inflict catastrophic damage on a honey bee colony by group predation. (Ono photo).

Figure 9.4. "Slaughter phase," or group attack by *Vespa mandarnia* on a honey bee colony in Japan. (Ono photo).

reported that *Vespa* species preferentially attack *A. mellifera* colonies in mixed apiaries in India, and Sakai (1989) noted that *V. simillima xanthoptera* attacked *A. mellifera* colonies more easily than *A. cerana* in Japan. If two species of hornets attack bees in a mixed apiary, the larger species generally will prey on the *A. mellifera* colonies and the smaller species on *A. cerana* (Adlakha 1975).

Apis dorsata colonies are not attacked as frequently as other honey bee species, and there are few reports of *Vespa* predation on this species. Hussain and Ali (1980) reported that these bees are attacked by *V. mandarinia* in Bhutan, and Schmidt et al. (1985) noted that *V. tropica* attacked *A. dorsata* in Sabah (Malaysia). Seeley et al. (1982) indicated that *A. dorsata* exhibit strong defensive behaviors to *Vespa* wasps, suggesting that such wasps could be a dangerous enemy.

Bee defenses against hornets

Several honey bee species exhibit behavioral defenses against *Vespa* wasps, but the responses by *A. mellifera* workers are rarely effective. Guards respond to an attacking hornet one or two at a time, allowing the hornet to seize individual bees (Shah and Shah 1991). Guards defending against attacks by the giant hornet *V. mandarinia* are easily overcome by the wasps; as few as five or six hornets have been reported to destroy colonies in Korea (De Jong 1990a). *A. mellifera* in most areas possess little counterattacking ability against hornet mass attacks and appear to have no effective means of repelling most hornets (Matsuura 1988). Ishay et al. (1967), however, have described a mass defense to the attacks of *V. orientalis*.

A. cerana, on the other hand, has evolved highly effective defenses against hornets. *A. cerana* workers may counter an attacking hornet by retreating inside the nest and then surrounding and stinging the hornet if she pursues her attack (Sakagami 1960, Singh 1972, Matsuura and Sakagami 1973). Workers may also gather at the colony entrance, facing toward any attacking hornets and shaking their bodies from side to side in a shimmering behavior. This behavior may involve 10 to 200 workers depending on the size of the colony and will keep hornets away from the entrance. If attempts are made to approach more closely, the guards will increase the intensity of shimmering (Shah and Shah 1991). Shimmering guards may hum loudly, raise their abdomens, and bend their heads downward as they face the attacking hornets (Kloft and Schneider 1969). If a hornet approaches closely, several guards may dart toward the wasp and pull it down, after which 30 to 60 bees immediately join the attack. They seize the hornet and form a ball of bees around it (Shah and Shah 1991). Ono et al. (1987) reported that balling of *Vespa simillima* by *A. cerana* in Japan involves 180 to 300 workers. The counterattacking bees do not attempt to sting the hornet but rather kill it by excessive heat production. The bees raise the temperature inside the ball to 46°C within minutes. The lethal temperature for the hornets is 45° to 47°C, whereas the upper lethal temperature for the Japanese honey bee is 48° to 50°C (Ono et al. 1987). These authors also reported that in Japan, *A. mellifera* workers will occasionally ball attacking hornets but do not raise the temperature inside of the cluster sufficiently to kill the hornet (~43°C). Shah and Shah (1991) did not observe similar behaviors with *A. mellifera* workers in Kashmir and noted that they did not appear capable of seizing or holding attacking hornets. Butler (1974), however, described a shimmering behavior by *A. mellifera* workers in Cyprus.

A. cerana workers may also adjust their flight behavior in response to predation pressures from hornets. Outgoing foragers fly off quickly after leaving the hive and do not make the circling movements around the hive common to *A. mellifera* foragers. On returning, the foragers use dodging tactics and land directly at the entrance (Shah and Shah 1991).

Dwarf honey bees, *A. florea*, also exhibit shimmering behavior in response to the approach of *Vespa* wasps (Lindauer 1956, Butler 1974). Bees in the outer curtain covering the nest flip their abdomens upward and outward, causing a synchronous ripple to spread over the surface, preventing an attacker from landing. Persistent approaches by hornets may be met with a more aggressive response. Bees nearest the hornet may turn, face the attacker, and attempt to pull it into the curtain to maul it (Seeley et al. 1982).

A. dorsata exhibit similar shimmering responses against hornets, except the responses are more forceful and frequent and of longer duration (Seeley et al. 1982). The larger size of the *A. dorsata* workers also makes them formidable defenders.

Control procedures

Matsuura and Sakagami (1973) summarized procedures used by Japanese beekeepers against hornets. These methods include killing individual hornets, destroying nests, trapping or mass poisoning with baits, the use of entrance traps, and the use of protective entrance screens. Shah and Shah (1991) described similar approaches used in India: killing queens in the spring, destroying nests, and swatting hornets at hive entrances. Mishra et al. (1989) mentioned additional recommended approaches, many of which relied on insecticide baits. Both groups of authors noted that the commonly recommended techniques were not particularly effective.

Colony destruction by burning, fumigation, or insecticide treatment depends on locating the hornets' nest, a relatively difficult task given their foraging range. In Japan, the giant hornet *V. mandarinia* often forages in apiaries 1 to 2 km from the nest and may fly as far as 8 km (Matsuura and Sakagami 1973). Sharma et al. (1979) were able to follow *V. auraria* (= *velutina*) foragers back to their nest after tying a bit of tissue paper onto their hind legs but could not follow *V. magnifica* (= *mandarinia*). Poison baits can be effective if they are sufficiently attractive to the *Vespa* species, but not to the honey bees. Relatively slow-acting, non-repellent insecticides that can be carried back to the nest and distributed to other colony members offer the greatest chances of success. Appropriate toxicants currently available include avermectin, sulfluramide, and hydromethylnon (Akre and Mayer 1994). Several researchers have proposed this type of control. Matsuura and Sakagami (1973) described an approach to mass poisoning that used prey items treated with an insecticide. Rivnay and Bitinsky-Salz (1949) used a bait against *V. orientalis* consisting of a 1:1:2 mixture of water, talcum powder, and chopped meat. The bait was poisoned with 1 percent thallium sulfate, lead arsenate, or benzene hexachloride (materials no longer in use). Longo (1980) recommended the use of a bait composed of a 10 percent solution of hydrolyzed protein mixed with pyrethrin to control *V. orientalis* and *V. crabro* in Sicily, and Lord (1994) indicated that beekeepers in Jericho place vicom (Ficam) poison bait in bee yards to control *V. orientalis*.

The most effective bait used in Jordan involves a mixture of 5 gm of Ficam 80W (active ingredient bendiocarb) with 250 gm of soft meat (Barry Girling, personal communication). Mishra et al. (1989) tested a series of baits consisting of overripe fruits and the feasibility of treating baits with fenitrothion. They also tested a new approach to baiting that involved attaching small gelatin capsule cups with the bait to the thoraxes of live hornets and allowing them to return to their nests where the food was consumed. Sixteen to 25 capsule loads were sufficient to kill colonies of *V. cincta* (=*tropica*) in the field.

Ibrahim and Yusoh (1989) tested several fermented sugar baits in *A. cerana* apiaries in Western Malaysia against *V. tropica*. The baits were used in wasp traps but were not particularly effective. Lim et al. (1989) recommended a similar system using fermented grape juice to control hornets in Korea. Shah and Shah (1991) developed a simple and effective baiting system for use against *V. velutina*. Jars half-filled with a mixture of fermented honey and water were placed on top of several hives in an apiary. Hornets entering the jars to feed drowned. In a 10-day period, more than 11,480 hornets were captured, significantly reducing the population of hornets observed at hive entrances. Higo (1983) developed a mass trapping technique for *V. xanthoptera* (= *simillima*) in Japan using a trap baited with fermented honey, and Okada (1980) trapped *V. mandarinia* on sticky paper by placing a hornet of that species on the paper as an attractant.

Hornet traps attached to hive entrances also reduce the success of hornet attacks. Matsuura and Sakagami (1973) developed traps that protect *A. mellifera* colonies against mass attacks by *V. mandarinia*. Matsuura (1988) observed that no colonies were exterminated by hornets in an apiary over two years, if they were protected by entrance traps. Hives still lost bees to hornet predation, but the mean numbers of honey bees killed were 283 and 325 per hive over the two years. This compared with entire colony losses, averaging approximately 15,000 bees per hive without traps. Caron and Schaefer (1986) presented plans for a modified hornet trap for use on hive entrances. The basic design of most traps is similar, allowing bees to enter and exit the hive but trapping hornets in a screen chamber. The screen mesh used in trap construction (ca. 5/16″) allows honey bees, but not the larger hornets, to pass through.

Other traps and screening systems have been used to reduce hornet populations or the effects of attacks. Kshirsagar (1971) described a trap for use in apiaries in India for the control of predatory wasps. Dave (1943) recommended the use of queen guards, and Subbiah and Mahadevan (1957b) suggested the elimination of alighting boards to reduce wasp attacks. The destruction of individual hornets with a stick or racket is practiced in several areas, including Egypt (Blake 1966) and South Korea (De Jong 1990a). An alternate technique used in South Korea to reduce the effects of *V. mandarinia* attacks involves the use of a 1-cm mesh net placed in front of the hive. The netting is attached to the top of the hive, stretched to a point about a half meter in front of the entrance, and anchored.

Attempts at using biological control to reduce hornet populations have been minimal. Muzaffar and Ahmad (1986) found that the parasitic mite *Pyemotes ventricosus* attacked the larvae and pupae of *V. velutina*. Infestation levels were low, but Ahmad et al. (1986) later reported the control of four species of *Vespa* in Pakistan by releasing adults artificially infested with this parasitic mite species.

Vespula and *Dolichovespula* (Yellowjackets)

Wasps in the subfamily Vespinae and belonging to the genera *Vespula* and *Dolichovespula* generally are not considered serious pests or predators of honey bees in North America, Europe, or Asia. A recent survey in the United States and Canada on the impact of wasps on beekeeping indicated that they caused minor problems in 28 states and four provinces; 17 states and five provinces reported that social wasps were not a problem (Akre and Mayer 1994). Several species may rob hives of honey, eat brood, or prey on individual bees, but serious predation and colony losses are rare. Two yellowjacket species introduced into New Zealand, *Vespula vulgaris* and *V. germanica*, have caused serious problems for beekeepers, however (Clapperton et al. 1989, Stringer 1989, Matheson 1991).

The species of yellowjackets most commonly associated with honey bees in the United States are *V. pensylvanica*, *V. vulgaris*, and *V. germanica* (Mayer et al. 1987). Other species listed as occasional predators of honey bees include *V. maculifrons* (Bromley 1948), *V. squamosa* (De Jong 1979), *D. arenaria* (Ma 1949), and *D. maculata* (Akre and Mayer 1994). Morse and Gary (1961) collected insects from dead bee traps and found several dead yellowjacket species, including *V. consobrina*, *V. vidua*, *D. arenaria*, and *V. maculifrons*. These latter yellowjacket species were probably incidental invaders that were attracted to honey stored in the hive.

Yellowjackets may forage on dead or dying bees at or near the entrances of hives. Free (1970) described the behavior of *V. germanica* and *V. vulgaris* foraging on dead bees, and Cotton (1842) observed that some yellowjacket species would prey on dying drones near hive entrances in the autumn. Attacks on live bees by yellowjackets are less common but have been observed. Rawson (1963) reported an attack on a foraging bee by *V. vulgaris* in England, and Amster (1940) described attacks on bees in his Maryland apiary by *V. germanica*. Other reports of predatory attacks by *V. vulgaris* include Mace (1927) and Herrod-Hempsall (1937) in England, Bromley (1948) in the United States, Atakishiev (1971a) in the U.S.S.R., and Stringer (1989) and Matheson (1991) in New Zealand. Similar reports can be found for *V. germanica* (Andrews 1882 and Mace 1927 in England; Stringer 1989 in New Zealand), *D. sylvestris* Scopoli (Watson 1922 in England, Atakishiev 1971a in the U.S.S.R.), and *D. arenaria* in Utah (Ma 1949).

When yellowjackets raid a beehive, they are usually repelled by the bees (Edwards 1980), although weak colonies such as mating nucs are susceptible to attack (Mayer et al. 1987). In general, wasps are most likely to cause problems and rob hives in late summer and fall. Yellowjacket colony populations usually peak in August or September, but some wasps remain active and forage well into autumn. *Vespula* colonies are typically small and have a fairly short colony duration, dying with the onset of cold weather, but colonies of some species (*V. vulgaris* species group) can attain a large size (5,000+ workers) and persist late into the season (Akre and Mayer 1994). Some species may also have perennial colonies, which pose an even greater threat to honey bees. During cold weather, yellowjackets may enter beehives to rob honey when the bees are inactive or clustered. *V. germanica* and *V. vulgaris* have robbed nucleus colonies in England in the autumn (Rawson 1963), and *V. vulgaris* robs colonies in Korea (Choi 1974). In the United States, several species, including *V. germanica*, *V. maculifrons*, *V. vulgaris*, and *V. squamosa* steal honey from hives during cold weather (De Jong 1979, Akre and Mayer 1994). *V. squamosa* also has been observed to form large, overwintering, multiqueen nests in Florida (Tissot and Robinson 1954). Heavy robbing of colonies by this species in the Lake Placid area of Florida occurs during the winter (January to April). Large numbers of bees can be killed defending their colonies, resulting in significant reductions in colony strength (De Jong 1979). Reducing colony entrances, however, can help the bees defend their hive and lessen the severity of such attacks.

The most serious damage to honey bee colonies from yellowjackets occurs in New Zealand where there were no vespid wasps until their first accidental introduction in the 1940s. The German yellowjacket (*V. germanica*) was introduced into New Zealand during the Second World War and the common yellowjacket (*V. vulgaris*) in the late 1970s (Matheson 1991, Donovan 1992). Both species have spread throughout the country with serious effects on the beekeeping industry. Changes in the life cycle have contributed to the problems. Many areas of the country have relatively mild winters, and colonies in some areas were able to survive the winter and renew worker production in the spring. Colonies contained multiple queens and grew to huge sizes (Thomas 1960); reproductives were produced in large numbers and at an earlier date. These factors led to extremely high wasp populations in many areas of New Zealand; mean nest numbers of 8.4 per hectare were reported in some study areas (Donovan 1992) with one report of 45 nests per hectare (Stringer 1989). Honey bee colony losses occurred as wasp populations increased; a survey for the 1974-1975 season showed that 3,900 colonies

(1.9 percent of hives) were totally destroyed by yellowjackets and another 10,100 colonies damaged (Walton and Reid 1976). By 1987 the figures for losses and serious damage had reached 2.6 percent and 6.6 percent, respectively (Matheson et al. 1989, Donovan 1992).

The colony attacks by yellowjackets were often initiated by foragers removing dead and dying bees in front of hives. The wasps then progressed to entering hives to steal honey and finally to preying on weakened bees and brood. Continued heavy predation led to colony destruction, after which other hives were attacked (Donovan 1992). Additional problems with yellowjackets occurred in many areas because of competition for honeydew. The wasps compete with the bees for honeydew in beech forests, reducing honey production significantly (Moller and Tilley 1989, Stringer 1989).

Control procedures

Several approaches to the control of yellowjackets have been advocated. The most obvious is to locate and destroy nests. Yellowjackets do not forage very far, and nests typically can be located within a few hundred meters of an apiary where they are causing problems. Studies by Akre et al. (1975) showed that 90 percent of *V. pensylvanica* workers foraged at distances of less than 400 meters. Flying wasps can be tracked back to their nest, especially if they have been dusted with flour or powdered sugar. Early morning or evening are often good times to attempt nest location because the number of other flying insects is often lower (Mayer et al. 1987). Matsuura and Sakagami (1973) described a simple technique used in Japan to help locate nests; bits of meat or fish are tied to a cotton ball so that wasps picking up the bait can be followed back to the nest more easily.

Located nests can be destroyed with an insecticide. Materials such as propoxur 1.5 EC can be mixed in water (following label directions) and poured or sprayed into an entrance hole. Carbaryl dust formulations (5 percent) can also be used. Nest entrances should not be plugged after treatment. Commercially available wasp and hornet sprays (in aerosol cans) can be used to treat aerial nests (Mayer et al. 1987).

The process of nest location and destruction can be difficult and only partly effective, especially when colony densities are high. Alternate methods involve the use of traps or poison baits. Yellowjackets can be baited to meat or fish suspended over water to which a small amount of detergent or wetting agent has been added. Wasps attempting to collect pieces of the meat fall into the water and drown (Mayer et al. 1987, Akre and Mayer 1994). Heptyl butyrate traps, available commercially, can also be used in bee yards to catch the western yellowjacket, *V. pensylvanica*. The traps are not attractive to other species such as *V. vulgaris* or *V.*

germanica (Edwards 1980). Several other traps have been suggested for reducing yellowjacket populations, including cone traps baited with fresh meat and bottle or jar traps with dilute sugar solutions in the bottom (Mayer et al. 1987).

Toxic baits offer the potential to control yellowjackets over a wide area. Wagner and Rierson (1969) developed a 1 percent mirex bait with heptyl crotonate as an attractant, and Davis et al. (1969) used heptyl butyrate as an attractant in a bait formulation. Beekeepers in New Zealand have used several mirex-based baits to control yellow jackets. Perott (1975) described a bait composed of 0.5 to 1.0 percent mirex in canned fish, and Caron and Schaefer (1986) mentioned a bait mixture of mirex and icing sugar. Akre and Mayer (1994) discussed some of the characteristics of a good bait and suggested several toxicants (see section on control of *Vespa*). They also suggested the potential use of chlordane in a wasp bait on a limited basis. The wettable powder formulation is non-repellent if mixed in a bait at levels of less than 0.5 percent but effective in killing both larval and adult wasps.

The potential use of biological control agents to control yellowjacket populations has been examined in New Zealand. The ichneumonid wasp *Sphecophaga vesparum* is a parasitoid of vespine wasps (MacDonald et al. 1975) that has been introduced into the country. Female parasitoids invade nests and lay their eggs on wasp prepupae or early pupae. The parasitoid larvae move into the wasp's abdomen, feeding on and killing the host before spinning a cocoon. Adult parasitoids emerge within the nest and lay more eggs on other wasp pupae (Stringer 1989, Donovan 1992).

The initial studies of the biological control potential of this parasitoid were made on *Sphecophaga vesparum burra*, a subspecies brought to New Zealand from the United States. Tests with this parasitoid showed that it was not well-adapted to the German yellowjackets in New Zealand, which were introduced from Europe. Subsequent tests were made with the European subspecies *S. v. vesparum* from Switzerland (Donovan 1992). This parasitoid showed a greater potential to attack the nests of both the German and common yellowjackets. Efforts were made to mass rear and release *S. v. vesparum* in several areas of New Zealand, starting in 1985 (Donovan et al. 1989). By 1992, as many as 16 percent of the nests in some areas had parasitized pupae (Donovan 1992). The long-term success of *S. vesparum* in reducing yellowjacket populations is questionable. Akre and Mayer (1994) stated that such parasitoid pressures are not sufficient to reduce yellowjacket populations significantly. Donovan (1992) suggested that the potential release of other biological control agents be examined. Beetles in the genus *Metoecus* parasitize wasp larvae and potentially could be released to increase the biological control pressure on yellowjackets. Hackett and Poinar (1973) examined the potential use of the nematode *Neoaplectana carpocapsae* for the control of yellowjackets but found that it would infect honey bees. Akre and Mayer (1994) also suggested that efforts be made to find, collect, and identify pathogens that could be used against vespine wasps.

Polistes (Subfamily Polistinae)

Polistes are medium-sized wasps in the vespid subfamily Polistinae. They are found throughout most of the world and are most diverse in the Old and New World tropics. *Polistes* colonies are generally small and they build fiber comb nests without an envelope. They commonly prey on caterpillars but will capture other prey. *Polistes* have been reported as minor pests of honey bees in some areas. Matheson (1984) reported that *Polistes humilis* and *Polistes chinensis* were occasional pests of honey bees in New Zealand. Al-Ghamdi (1990) listed *Polistes watti* as a pest in Saudi Arabia, and Singh (1962) noted that *Polistes hebraeus* will visit hives in India. Morse and Gary (1961) collected *Polistes fuscatus* from dead bee traps attached to the front of hives in New York. Other records of *Polistes* as an occasional honey bee predator or minor hive pest include *P. major castaneicolor* in Arizona (Schmidt and Schmidt 1986) and *P. gallicus* in Sicily (Longo 1980). Significant damage to honey bee colonies by *Polistes* wasps has not been reported.

Philanthus and Palarus (Beewolves and bee pirates)

Beewolves are ground-nesting, solitary wasps belonging to the genus *Philanthus* (family Sphecidae). Members of this genus prey on bees and other Hymenoptera, using the captured prey as a food source for their larvae. The genus *Philanthus* contains about 136 species and is found throughout much of the world (excluding Australia, South America, and Antarctica) (Evans and O'Neill 1988). Many species are unselective with regard to prey, whereas others confine themselves to bees of one or several families (such as the Halictidae). A few species occasionally prey on honey bees, and one Old World species specializes on *Apis mellifera* (Table 9.3). Buys (1975) reported that 59 percent of the beekeepers responding to a survey in South Africa considered these *Philanthus* wasps to be a moderately serious to serious problem.

The most important beewolf is the Old World species *Philanthus triangulum*, which is a specialized predator of honey bees. It is found throughout most of Europe, Africa, and areas of the Middle East (Table 9.3). *P. triangulum* nests in sandy soils and is often found in sandy areas bordering cultivated fields (El-Borollosy et al. 1972, Ozbek 1982, Evans and O'Neill 1988). They may construct nests in areas covered with coal cinders or ash (Thiem 1932) or even in the cracks of paved streets or along the foundation walls of buildings (Thiem 1935). Nests are often aggregated, but nest spacing within aggregations is usually maintained at distances of 10 cm (Simonthomas and Simonthomas 1972, Lomholdt 1975). The nesting behavior of *P. triangulum* has been described by Simonthomas and Simonthomas (1972, 1980) and reviewed by Evans and O'Neill (1988).

Hunting by *P. triangulum* occurs primarily on flowers (El-Borollosy et al. 1972, Simonthomas and Simonthomas 1972) but also occurs near hive entrances in Egypt when few flowers are present in the field (Simonthomas and Simonthomas 1980). Beewolves hunting near hives catch returning foragers. Female wasps foraging in the field typically seize bees on flowers and sting them in the thorax between the front legs (Rathmayer 1962). The venom injected by the beewolf causes a

Table 9.3. Records of *Philanthus* species preying on honey bees.

Species	Geographic Area	Reference
P. basalis basalis[1]	Sri Lanka	Krombein 1981
P. bicinctus	Western United States	Evans & O'Neill 1988
P. crabroniformis[2]	Western United States	Alcock 1975, Evans & O'Neill 1988
P. crotoniphilus	Southwestern United States	Evans & O'Neill 1988
P. fuscipennis	Zaire	Dubois & Collart 1950
P. pulcherrimus	India	Dutt 1914, Evans & O'Neill 1988
P. remakrishnae	India	Singh 1962
P. sanbornii[3]	United States, Southern Canada	Evans 1955, Kurczewski & Miller 1983, Evans & O'Neill 1988
P. sicarius	Zaire	Dubois & Collart 1950
P. triangulum[4]	Africa (records - Botswana, Egypt, South Africa & Zaire)	Callan 1954, Clauss 1983, Simonthomas & Simonthomas 1980, May 1961, Dubois & Collart 1950
	Europe (records - England, France, Germany, Netherlands, Scandinavia)	Betts 1936c, Nixon 1954, Thiem 1932, 1935; Olberg 1952, 1959; Haeseler 1977; Erlandson 1961; Simonthomas & Simonthomas 1972
	Saudi Arabia	Al-Ghamdi 1990
	Turkey	Ozbek 1982
	U.S.S.R.	Sedova 1959
Philanthus species - unidentified	Jordan	Robinson 1981
	Oman	Hussein 1988

1. Preys on *Apis cerana;* other records in the table involve *A. mellifera* as prey.
2. Synonym: *P. flavifrons*
3. Synonym: *P. eurynome*
4. May be referred to as *P. diadema*, or as an African subspecies *P. triangulum diadema*.

general paralysis, but time to overall paralysis is dependent on location of the sting (Rathmayer 1966). Paralyzed bees are carried back to the nest and used to provision brood cells. Piek and Spanjer (1986) reviewed the chemistry and pharmacology of *Philanthus* venom.

The female *P. triangulum* may provision a cell with as many as six bees. Simonthomas (1966) reported that 19 percent of the *P. triangulum* cells he examined contained one bee, 41 percent two bees, 22 percent three bees, 10 percent four bees, and 8 percent from five to six bees. Unfertilized eggs are often laid in cells provided with one or two bees for the production of males (Simonthomas and Veenendaal 1978). Individual nests average three to six cells (El-Borollosy et al. 1972) but may contain up to 14 cells (Evans and O'Neill 1988). Female wasps catch up to 10 honey bees per day in Egypt (Simonthomas and Simonthomas 1980), using them both as a food source and for provisioning the nest.

Predation by beewolves can weaken or seriously damage colonies, especially when populations are high. Simonthomas and Simonthomas (1980) noted that populations of 3,000 *P. triangulum* are not uncommon in some areas of Europe and can result in the capture of 30,000 honey bees daily. Dead bees are often found around nest entrances and can provide estimates of the predation potential of these wasps. Thiem (1932) recorded 150 to 250 dead or paralyzed bees per square meter in a *P. triangulum* nesting area in Germany; Betts (1936b) gave figures as high as 700 bees. Thiem (1935) estimated that a nesting area of 1.5 ha in an alkali dump in Germany contained 1.9 million captured honey bees. In the Dakhla oasis in Egypt, heavy predation of honey bees by *P. triangulum* made beekeeping almost impossible (Simonthomas and Simonthomas 1980).

In North America, at least four species of *Philanthus* prey on honey bees occasionally. Higher levels of predation may occur, however. Kurczewski and Miller (1983) found that honey bees could constitute 40 percent of the prey of *P. sanbornii* in Florida. The wasps foraged on bees from hives located close to a nesting area. The proximity of easy prey such as honey bees can influence the foraging behavior of the female wasps.

The other group of predatory, solitary wasps that attack honey bees are the bee pirates. These ground-nesting sphecids belong to the genus *Palarus*. Several species have been reported as serious pests of honey bees. *Palarus latifrons* can be a serious pest of *A. mellifera* in Southern Africa (Clauss 1983, 1985, May 1961), and *Palarus orientalis* is considered a problem in parts of India where it preys on *A. cerana* (Singh 1962). The general biology of the bee pirates is similar to that of other ground-nesting sphecids; the bees caught as prey are stung and paralyzed and then used to provision the nest.

Clauss (1985) described the activity and behavior of the banded bee pirate, *P. latifrons*, in Botswana. Male wasps first appear in late September and females one to two weeks later. High populations can be found from mid-November to mid-March, after which numbers decrease. Female banded bee pirates may prey on honey bees foraging on flowers or at water sources, but they also are strongly attracted to honey bee colony odors and forage at or near hive entrances. When colonies are active, *Palarus* typically captures foragers returning to the hive. During periods of low flight activity, the wasps may land at the entrance or even enter a hive to capture a bee. Predator activity by *Palarus* occurs only at higher temperatures (21°-40°C). Damage to strong colonies by this wasp species is minimal, and most well-managed colonies should be able to compensate for forager losses from *Palarus* and still produce good honey crops. Banded bee pirates may cause significant losses of young queens, however, capturing them on orientation or mating flights (Clauss 1985).

The behavior of *P. orientalis* is similar to that of *P. latifrons* with respect to foraging on honey bees. These wasps nest gregariously in sandy banks or at the edges of fields and forage at beehives during the middle of the day in hot weather. Flying bees are caught near hives, stung and paralyzed, and carried back to the nest. Four to five bees are used to provision a cell before an egg is laid on one bee (Singh 1962).

Control

In areas where beewolves constitute a serious problem to beekeepers, control measures have been proposed. El-Borollosy et al. (1972) suggested destroying nests and nest sites by plowing, but such practices may be impractical if nests are widely scattered (Simonthomas and Simonthomas 1980). The destruction of nesting areas by irrigation has also been suggested. In the Werra River region of Germany, nesting areas were eliminated by covering the ground with clay and sod. Adult wasps were able to emerge through the sod but did not construct new nests in the same area (Thiem 1935). Efforts to destroy *Philanthus* larvae with soil insecticide treatments have not been effective, but nests and adult females can be destroyed by pouring insecticides into individual burrows during the late afternoon (Thiem 1932). The treatment of individual nests is labor-intensive and not practical on a large scale.

Alternate approaches have involved efforts to capture bee wolves with nets or destroy them with paddles or flyswatters, but none of these methods has been effective. Bounties for dead beewolves were offered in Egypt, but although 24,000 wasps were captured in one season, they were almost as abundant the following year (Simonthomas and Simonthomas 1980). Biological control has also been proposed. Simonthomas and Simonthomas (1980) suggested that the cuckoo wasp, *Hedychrum intermedium* Dahlbohm, could be used for control of this beewolf species. This chrysidid wasp is the main parasite of *P. triangulum* and can eliminate or significantly reduce its populations within a year.

One of the most effective methods for reducing problems with bee wolves is to remove hives from the area when the wasps are active (Simonthomas and Simonthomas 1980). Because honey bees are the primary prey, removing hives limits reproductive success and beewolf populations decline significantly. It is also suggested that beekeepers maintain large colonies because they are better able to survive predation. Small colonies can be combined to reduce predation effects.

The use of strong, well-managed colonies is recommended for reducing the effects of predation by the banded bee pirate, *Palarus latifrons*. Clauss (1985) indicated that strong colonies can compensate for losses by *Palarus* and still produce good honey crops, especially if optimal conditions such as shade are provided for the colonies. Clauss also suggested that natural control agents may exist and possibly could be used to help reduce wasp populations. He described two parasites found around nests of *Palarus*, one a small wasp, the other a fly, but has not presented further information on their biology of potential use as biological control agents.

Mutillidae (Velvet ants)

The Mutillidae are a family of solitary, parasitic wasps, consisting of several thousand species worldwide. The females are wingless and antlike in appearance and are often covered with a bright pubescence that gives them their common name, the velvet ants. The males are winged and frequently larger than the females, an adaptation in many species for phoretic copulation (Deyrup 1988). Both sexes also have heavy exoskeletons. The mutillids are primarily parasitoids of solitary, ground-nesting wasps and bees, although several species will attack social bees or wasps (Mickel 1928). The females have a long sting that can be used for defense.

Only one species of mutillid wasp, *Mutilla europaea*, has been reported commonly as a pest of honey bee colonies. This antlike wasp is about 14 mm in length, black in color, and has an orange-red thorax. The abdomen is covered with long hairs and has three silver-white bands (two in the male). *M. europaea* is a parasite of bumble bee brood and lays its eggs inside bumble bee cocoons containing pupae or prepupae (Alford 1975). This velvet ant has been described as a minor pest of honey bees in England and is a more serious pest in parts of Southern Austria and Germany (Mickel 1928, Alfonsus 1930, Alford 1975). The first records of colony attack by this *Mutilla* species were described in a letter written by an Austrian beekeeper who indicated that the wasps were a frequent enemy of bees in his region of the country (Schoenfield 1878). The females entered colonies and killed as many as 200 bees in a 24-hour period (Schoenfield 1878). Others have made similar observations (Beljavsky 1935). Schoenfield suggested that velvet ants probably use their sting to kill the bees. A somewhat different account of the behavior of *Mutilla europaea* was presented by Scholz (1879 in Mickel 1928). His description indicates that the wasps can enter beehives without hindrance and will bite and feed on individual bees. Subsequent reports have indicated that *M. europaea* will kill both brood and adult bees in hives (Betts 1936a, Meyerhoff 1954) and that mutillids can be reared on the brood in honey bee colonies (Meyerhoff 1954). Mutillid adults will also feed on the honey stores of their host (Alford 1975), although such feeding would have little impact on a colony. Honey bees are relatively ineffective at repelling an invading mutillid and are rarely successful in stinging velvet ants because of their heavily sclerotized cuticle.

Alfonsus (1930) presented reports of a second mutillid species, *Mutilla differens* (*M. europaea?*) that occasionally caused problems in South Austria. Attacks on colonies by this species were reported to cause losses of several hundred bees and keep a colony disturbed for several days. Bees killed by this mutillid were described as having a contracted abdomen; the proboscis, legs, and wings were stretched forward.

Few beekeepers in the United States have reported problems with mutillids. C. V. Riley published a letter from a Texas beekeeper in 1870 describing an attack on a colony by large velvet ants (known by the local name 'Cowkillers') (Mickel 1928). These wasps were probably the large mutillid *Dasymutilla occidentalis*. A similar report was made by Cook (1902), who noted that *Mutilla coccinea* (= *D. occidentalis*) would enter hives and feed on the larvae. *D. occidentalis* has been reported as a parasite of bumble bees (*Bombus fraternus*) (Fattig 1943), thus the occasional entry of this species into honey bee hives would not be surprising.

Damage by mutillids to honey bee colonies is usually not significant and rarely requires control efforts. Borchert (1966) suggested eliminating bumble bee colonies near apiaries and keeping hives on stands at least 30 cm off the ground. Barriers and stand elevation are used to protect hives against ants and should also work against the wingless mutillids.

Other parasitic wasps

There are few reports of parasitic wasps attacking honey bees. Erickson and Medenwald (1979) reported on the parasitism of queen pupae by *Melittobia acasta* (Hymenoptera: Eulophidae) in Wisconsin. The wasp was discovered ovipositing on a queen pupa held in an incubator. Fell (unpublished data) made similar observations in New York, finding several parasitized queen pupae in an incubator used for holding queen cells until emergence. Female wasps were collected and tentatively identified as a *Melittobia* sp. Jelinsski and Wojtowski (1984) reported that *Melittobia acasta* was a parasite of honey bee brood in Poland. *M. acasta* and *M. hawaiiensis* are parasitoids of solitary bees and wasps and will attack several species of bumble bees (Alford 1975, MacFarlane and Donovan 1989). Parasitism of honey bees by *Melittobia* is probably opportunistic and relatively rare. *M. hawaiiensis* has not been reported as a parasitoid of honey bees but can be reared on honey bee larvae (Varanda et al. 1984).

The only other hymenopteran parasitoid reported from honey bees is the braconid *Syntretomorpha szaboi* (Verma and Joshi 1988, Walker et al. 1990). This parasite was described from India and oviposits on adult *Apis cerana*, presumably while they are foraging on flowers. The larvae develop internally, emerging from the anal aperture of the bees before spinning a cocoon and pupating. Infestations may occur between late April and November with overall parasitism rates between 1.7 percent and 5.2 percent. Parasitized bees were described as sluggish and darker in color than normal bees.

Bees

Bombus (bumble bees - subfamily Bombinae)

Bumble bees are commonly found dead at the entrance of honey bee colonies. They are of little importance as pests but may occasionally attempt to enter hives. Sakagami (1958) reported that *Bombus* species commonly enter hives during the autumn in Japan. He identified *Bombus speciosus* robbing honey from an exposed comb. Morse and Gary (1961) collected a total of 27 bumble bees of seven different species from dead bee traps attached to honey bee colonies in New York; the species included *Bombus affinis*, *B. auricomis*, *B. impatiens*, *B. perplexus*, *B. ternarius*, *B. terricola*, and *B. vagans*. Thoenes (1993) collected 147 dead workers of *B. pennsylvanicus sonorus* (and five dead female carpenter bees, *Xylocopa californica*) from dead bee traps over a one-month period in Arizona (U.S.A.). Thoenes suggested that the foraging bumble bees and carpenter bees were attracted to the odor of mesquite floral volatiles emitted from the honey bee colonies while the bees were concentrating nectar to honey. Similar responses to floral volatiles by foraging bees could explain observations of other native bee species in dead bee traps (Morse and Gary 1961).

Trigona, and Lestrimellita (Stingless bees - tribe Meliponini)

The stingless bees are a diverse group of tropical eusocial bees. Colonies live in permanent nests of often complex and elaborate structure and may contain hundreds to thousands of individuals. Stingless bees produce small amounts of honey and have been kept for centuries by humans in various areas of the New World tropics. Some species of meliponids may store up to two kilograms of honey in the nest. They may rob nests of other species for wax or edible brood (Michener 1974).

The name stingless bee refers to the non-functional sting possessed by females within this group. Many species, however, have alternate defense systems and react aggressively to colony disturbance. Defensive behaviors may include swarming over intruders and biting. Several species also have a chemical defense system that not only improves colony defense but allows for robbing the food reserves of other social bee species, including *A. mellifera*. This defense system is based on secretions from enlarged mandibular glands, some of which may cause painful lesions on human skin.

The earliest reports of stingless bees robbing honey bee nests were those of Moure (1946) and Kempff Mercado (1952). Moure (1946) observed that workers of *Trigona (Oxytrigona) tataira* successfully invaded honey bee colonies, with little if any resistance, to remove food. Kempff Mercado (1952) noted that colonies located near the edge of virgin forests were particularly susceptible to attacks by stingless bees. Entire colonies could be lost to *Melipona flavipennis*, one of the largest and most aggressive meliponids. Several species of *Trigona* also attacked and robbed honey bee colonies.

The robbing behavior of stingless bees in the genus *Trigona* has been described by Moure (1946), Kempff Mercado (1952), and Rinderer et al. (1988). During nest plundering, stingless bee workers produce cephalic secretions that disrupt normal honey bee colony defenses. Honey bee workers may become motionless, wander over the comb in a disoriented fashion, or cluster outside the entrance (Rinderer et al. 1988). The air outside attacked colonies has often been described as having a strong floral odor, a factor that has often been associated with the exocrine secretions of the mandibular glands. Bian et al. (1984) identified the major components of the mandibular gland secretion of *Trigona tataira*, and Rinderer et al. (1988) showed that the major responses of honey bees were caused by two diketones present in this secretion, (E)-3-nonene-2,5-dione and (E)-3-heptene-2,5-dione. The primary response to these chemicals was a negative chemotaxis with the two compounds probably functioning as irritants.

Roubik et al. (1987) examined the cephalic secretions of two other species of *Trigona (Oxytrigona daemoniaca* and *O. mellicolor)* and found that formic acid was present in the mandibular secretions, as well as the diketones. Formic acid is a well-known repellent (Blum 1981) and undoubtedly increases the effectiveness of the repellent secretions of these bees when robbing honey bee colonies.

Bees in the genus *Lestrimelitta* are highly social meliponids but obtain pollen and honey by robbing the nests of other social bee species. *Lestrimelitta limao* is a common Neotropical species and obligate robber of other social bee species such as *Trigona*. Weak colonies of *Apis mellifera* are occasionally attacked (Michener 1974); Stejskal (1962), for example, reported attacks on mating nuclei by *L. limao* in Venezuela. The mandibular gland secretion of *Lestrimelitta* contains citral, which is thought to act by disrupting the social organization of the colony being robbed. *Lestrimelitta* workers that successfully invade colonies carry off honey and pollen in their enlarged crops.

Control

Stingless bees should not be killed, whether they are from wild colonies or from managed hives. Most reports of stingless bee problems in apiaries of *A. mellifera* colonies refer only to occasional food robbing (Roubik et al. 1987). Many species of stingless bees are already endangered because of habitat destruction. In the past, beekeepers in some areas of the American tropics attempted to reduce stingless bee populations by exposing workers to insecticides that they could carry back to their nest (Kempff Mercado 1952, 1957). Such measures are no longer recommended.

Apis (Honey bees)

Intraspecific and interspecific parasitism may occur in honey bee populations. These interactions may include various types of robbing behavior between colonies of either the same or different species, and social reproductive parasitism, a phenomenon most frequently observed between Africanized and European honey bees (*A. mellifera*). Both types of behavior can cause serious problems for beekeepers (Danka et al. 1992).

Robbing

Robbing behavior is familiar to most beekeepers and involves the act or series of acts by which honey bees from one colony steal food from one or several other colonies (Atwal and Dhaliwal 1969). Nogueira-Neto (1949) divided robbing behavior in the social bees into two categories: "mild pillage" and "exterminating pillage." The mild form of robbing typically involves small numbers of bees and causes only minor damage to the colony being attacked. The more severe form of robbing can lead to complete destruction of the attacked colony. Honey bees involved in robbing often exhibit characteristic behaviors when approaching a colony, flying in a distinctive swarming pattern in front of the hive entrance and appearing hesitant to enter (Gary 1966, 1992; Sakagami 1959). Robbing is most likely to occur under poor nectar flow conditions and may continue for variable periods of time.

Interspecific robbing behavior is commonly observed between *A. cerana* and *A. mellifera* and has been reported from India (Atwal and Dhaliwal 1969), China (Kellogg 1959), Korea (De Jong 1990a), and Japan (Sakagami 1959). Both mild and exterminating behaviors have been observed between the two species, but the mild pillage of *A. mellifera* colonies on *A. cerana* frequently develops into the more severe kind. *A. cerana* guards pay less attention to robber bees than *A. mellifera* guards, frequently allowing their entry with little or no resistance (Sakagami 1959, Atwal and Dhaliwal 1969). Robbing can then intensify and lead to complete domination of the *A. cerana* colony by the *A. mellifera* workers. The robbers may occupy the victimized colony and completely remove all honey stores. Members of the *A. cerana* colony often abscond, if possible, in response to such an attack (Kellogg 1941, Sakagami 1959). Robbing by *A. mellifera* has led to the extermination of many *A. cerana* colonies in Japan and makes it very difficult to keep colonies of both species in the same apiary or even in the same area (Sakagami 1959).

A. cerana workers may rob *A. mellifera* colonies, especially if the colonies are weak. The effects of such robbing are generally mild, but the *A. cerana* robbers are often persistent. They return to the same colony and are not easily chased away. *A. cerana* workers are also more active in cooler seasons and will frequently rob *A. mellifera* colonies when they are less able to defend themselves (Kellogg 1941). Exterminating robbery of *A. mellifera* colonies by *A. cerana* is rare and has only been reported when the *A. mellifera* colonies are very weak (Sakagami 1959).

A. florea and *A. dorsata* have been reported to rob weak *A. mellifera* colonies in India (Chahal et al. 1991, Sihag 1991). Other reports of interspecific robbing include attacks on *A. florea* colonies by *A. cerana* and stingless bees (Seeley et al. 1982), robbing attacks on *A. dorsata* by *A. cerana* (Seeley et al. 1982), and robbing of *A. cerana* colonies by *A. dorsata* workers (Singh 1959). The intermingling of different species during robbing has offered the potential for the interspecific spread of diseases and parasites and has been suggested as a possible mechanism for the spread of tracheal mites from *A. mellifera* to *A. cerana* (Atwal and Dhaliwal 1969, De Jong 1990b).

Intraspecific robbing is a behavior familiar to most beekeepers and is commonly observed with both *A. mellifera* and *A. cerana*. It probably also occurs in other *Apis* species. Robbing activity is favored by management practices, especially when numbers of colonies of varying strengths are kept in the same apiary. Robbing occurs most frequently under dearth conditions and is often triggered by hive manipulations or practices that expose honey or other sugar sources to potential robbers. Weaker colonies are most susceptible to robbing and are typically less able to defend themselves. The tendency toward robbing may vary with different races of *A. mellifera*: The Italian bee (*A. mellifera ligustica*) and the Caucasian bee

(*A. mellifera caucasica*) are more likely to initiate robbing than are bees of other common races (Ruttner 1975). Africanized bees are more difficult and dangerous to handle under robbing conditions but do not appear significantly more inclined to initiate robbing than most of the European races (De Jong 1990a).

Once robbing has started in an apiary, it is difficult to control. Thus, preventive efforts are advisable if conditions suggest that robbing could occur. Hives should be manipulated as little as possible, and care should be taken not to expose or spill honey. Supers or other hive parts should be covered during colony inspections. If robbing starts, the best response is to improve the defensive position of the colonies being robbed. Entrances should be reduced and other openings in the hives sealed. Under severe conditions, weak colonies can be moved to a different location.

Reproductive parasitism

Social parasitism refers to a phenomenon of queen usurpation in which an invading queen takes over a nest from the resident reproductive female. The behavior is common to many wasps and bumble bee species and can be intra- or interspecific. The intraspecific usurpation found in vespine wasps and bumble bees often results from nest site competition and involves takeover efforts by individual queens. In honey bees, social parasitism is a phenomenon primarily associated with Africanized bees and involves takeovers by small swarms containing workers and a queen (Michener 1975, Rinderer and Hellmich 1991). Workers from the swarm enter the hive, kill the resident queen, and allow the parasitizing queen to assume the reproductive role in the nest (Rinderer and Hellmich 1991). Reports by South American beekeepers of the occurrence of this type of parasitism have ranged from rare (Camazine 1986) to frequent (Vergara et al. 1993). The takeover of established European honey bee colonies by Africanized bees has been observed in Argentina, Brazil, Venezuela, and in parts of Central America, including Mexico (Camazine 1986, Danka et al. 1992, Hellmich and Rinderer 1991, Vergara et al. 1993).

Female reproductive parasitism in honey bees is most common with colonies that are queenless or have failing queens. Small colonies or stressed colonies may also be more susceptible to usurpation (Rinderer and Hellmich 1991). Studies in Mexico have shown that queenless colonies are invaded much more often than are queenright colonies (Vergara et al. 1993). These authors have suggested that queenless colonies may play a role in attracting invading swarms to an area. The strength of queenright colonies did not appear to have a significant effect on colony invasion, however, because medium and strong colonies were invaded as often or more often than were weaker colonies. The sources of parasitizing queens are unknown and may result from absconding swarms or small afterswarms (Rinderer and Hellmich 1991).

Social parasitism in honey bees has had an impact on beekeeping, since it contributes to the Africanization process. Usurpation of a European colony by an Africanized queen leads to complete Africanization of the colony and ultimately advances the process when the invaded colony produces swarms and drones. However, the frequency of nest usurpation is relatively low. Danka et al. (1992) found a usurpation rate of 4.7 percent per year for commercially managed European honey bee colonies in Venezuela. Queen supersedure (and the associated mating events) and colony failures played a more significant role in the Africanization process.

CHAPTER TEN

Other Insects
Dewey M. Caron

Thysanura (bristletails)
Odonata (dragonflies, damselflies)
Orthoptera (grasshoppers, crickets, mantids, roaches)
     Mantids
     Cockroaches
Dermaptera (earwigs)
Isoptera (termites)
Psocoptera (barklice)
Hemiptera (true bugs)
Neuroptera (lacewings, fishflies, antlions)
Coleoptera (beetles)
     Carabidae (ground beetles)
     Cleridae (checkered beetles)
     Cryptophagidae (cryptophagid beetles)
     Dermestidae (dermestid beetles)
     Meloidae (blister beetles)
     Nitidulidae (sap beetles)
     Ptinidae (spider beetles)
     Scarabaeidae (scarab beetles)
     Silphidae (carrion beetles)
     Tenebrionidae (darkling beetles)
     Additional coleoptera
Strepsiptera (twisted-winged parasites)

CHAPTER TEN
Other Insects

Thirteen of the 27 insect orders include species that have been recorded as pests of honey bees or their comb, hives, or stored food. Insects in the orders Lepidoptera (moths and butterflies), Diptera (flies), and Hymenoptera (ants, wasps, and bees) are covered elsewhere in this book. This chapter treats the remaining 10 insect orders, which include pests, predators, or parasites of bees; the orders are arranged phylogenetically in accordance with Borror et al. (1989). Only a few of these insects are of more than minor importance to beekeepers, and most of the problems they cause are relatively limited seasonally or geographically.

A few simple measures will control the insect pests that enter beehives. Primarily, the bee colonies should be populous, and the bees should have access to all parts of the hive, especially the area above the inner cover. To control scavengers, the bottom of the hive should be kept relatively free of debris. To keep out larger insects, such as pollen-feeding beetles, the entrance opening can be reduced to a narrow slit just tall enough to accommodate a honey bee.

Thysanura (bristletails)

Members of the order Thysanura, or bristletails, are small, elongate insects, each with three taillike appendages at the posterior end. In most species, the body is covered with scales. The silverfish, *Lepisma saccharina*, is a species commonly found in human dwellings because it can feed on the starch in books, clothing, curtains, and stored food. Silverfish (Beljavsky 1927, Borchert 1974) and another bristletail species, *Ctenolepisma lineata*, cited in Borchert (1974), have been found in honey bee hives. They are probably attracted to the stored honey (Toumanoff 1939, Borchert 1974). Subhapradha (1961) considered an unidentified *Lepisma* species to be a pest of bees in India and believed that their droppings can contaminate honey. If they are numerous, he suggested transferring the infested combs into a populous colony, in which the bees will control the pests.

Odonata (dragonflies, damselflies)

The order Odonata includes the dragonflies and the damselflies. All species are predaceous, but only some species of dragonflies prey on honey bees. The adults capture flying insects on the wing. Because the immatures (called nymphs or naiads) must live in fresh water, dragonflies are rarely found far from lakes, rivers, or streams. Some of the larger dragonfly species feed on honey bees, and there have been occasional reports of dramatic depletion of honey bee populations. Dragon-fly predation on queens on their mating flights may be especially costly to bee colonies and beekeepers (Needham and Heywood 1929). The problems seem to occur only seasonally and under exceptional conditions, however, when dragonfly populations are unusually high (Örösi-Pál 1939a).

Wright (1944) described the damage done by the dragonflies *Anax junius* and *Coryphaeschna ingens* to an apiary in Louisiana, U.S.A., along the Missis-sippi River. Normally in this apiary 75 to 85 percent of queens returned to their hives from mating flights, but during a period of high dragonfly populations only 5 per-cent of the queens returned. On one day, 79 queens were recorded taking mating flights, and only three returned to their hives. Dragonflies of two species were observed feeding on drones and workers as well as on the queens. Dragonflies of two other species were present but did not feed on the bees. The destruction of worker bees was so extensive that the colonies could not be used to make up packages. Root (1966) stated that dragonflies caused $1,000 in damage in four or five days to a group of 300 to 400 colonies in Florida, U.S.A. He observed that in parts of Florida dragonflies were so numerous they could darken the sky.

Figure 10.1. Adult dragonflies are all predaceous, but only a few species cause problems, and then only on rare occasions. (Flottum photo).

Frey (1914) observed dragonflies, *Aeschna grandis*, eating bees in Germany. So many were resting in the trees around his apiary that their wings gave the trees a glassy appearance. By May, the colonies had very low worker bee populations, and they did not swarm or yield a honey crop. Beljavsky (1927) stated that one beekeeper in the Tomsk area of Russia had heavy predation from *A. grandis* in 1887: only 32 hives survived from a total of 92. In contrast, *Aeschna* predation on honey bees apparently does not occur in England (Betts 1939). In India, the unnamed species of dragonflies listed by Singh (1962) and Gandheker (1959) are a problem for a limited time during the rainy season and mainly for queens on their mating flights. Goodacre (1923) indicated that a red dragonfly species, *Hemianax papuensis*, created a problem for several Australian apiaries in 1921 and 1922, and McKay (1967) reported colonies that lost more than one-quarter of their bee population to an unidentified species of dragonfly in the Darwin area.

In addition to the dragonfly species already mentioned, *Epiaeschna heros*, *Cordulegaster diastatops*, and *Aeschna cyanea* have been observed feeding on honey bees (Byers 1930, Toumanoff 1939, Bromley 1948, Ma 1949).

No effective control measures are available to protect bees from dragonfly predation, although bird predation on dragonflies is occasionally helpful in India (Gandheker 1959) and Australia (Goodacre 1923). In areas where problems occur, they are localized and seasonal, so the best solution is to move the colonies, at least temporarily (such as during mating times). Populous colonies are able to tolerate the loss of some workers, and attention to colony management by the beekeeper can overcome the occasional loss of a queen.

Orthoptera[1] (grasshoppers, crickets, mantids, roaches)

The order Orthoptera includes grasshoppers, crickets, mantids, and roaches. Most species are plant feeders, but some may be omnivorous, and a few are predaceous. Orthopterans typically have an elongated body with long, prominent antennae and cerci (tails). Those that have wings have two pairs, with the first pair somewhat thickened. The wings are long and narrow and have many veins. Members of this order hatch from the egg as small versions of the adult but lack wings. They grow larger in a series of stages (molts), adding wings and reproductive ability with the last molt.

---

[1]This order is undergoing further division at press, and may be split into other orders.

Mantids

Several species of mantids have been observed to feed on honey bees on flowers, but none is an important bee predator. Bromley (1948) listed 28 records of Chinese mantids, *Tenodera aridifolia sinensis*, feeding on honey bees in Connecticut, U.S.A., in flower gardens or on goldenrod flowers. In the former U.S.S.R., *Mantis religiosa* eats honey bees (Beljavsky 1927). In India, the smallish (2.5 cm long) green mantis, *Odontomantis micans,* is a predator of *Apis cerana* (Subba Rao et al. 1972), and *Creobrator gemmatus* also preys on honey bees (Singh 1962). Okada (1959) listed *Tenodera aridifolia sinensis* (= *Paratenodera sinensis*) as a bee pest in Japan. Bregante (1972) stated that *Brauneria subaptera* and species of *Coptopteryx* eat honey bees in Argentina.

Cockroaches

Cockroaches may enter the hives of weak honey bee colonies and feed on honey, pollen, dead bees, and wax. They have been reported as minor pests in India (Gandheker 1959, Subhapradha 1961, Singh 1962), South Africa (Lundie 1951, 1952), South America (Stejskal 1955, Ordetx and Espina Perez 1966), and North America

Figure 10.2. Mantids will feed on honey bees, usually capturing foragers on flowers, but occasionally moving right to a colony. (Caron photo).

(Haydak 1963, Scott et al. 1970, Amos 1972). In contrast, Toumanoff (1939) and Johnsen (1953) labeled cockroaches, especially *Blatta orientalis* and *Periplaneta americana*, as among the worst insect enemies of beekeepers in Indochina.

Cockroaches may destroy comb and foundation (Toumanoff 1939, Haydak 1963), their excreta may contaminate honey (Subhapradha 1961), and they may impart an unpleasant odor to a hive (Subhapradha 1961). Stejskal (1955) found gregarines (see Chapter 4) in diseased bees and implicated cockroach feces as their source. Cockroaches are uncommon in healthy, populous colonies however, and in general the damage they do is negligible.

Dermaptera (earwigs)

The Dermaptera, or earwigs, are long, slender insects with large forceps-like tails. The adults have four wings but rarely fly. Earwigs are usually nocturnal and hide in cracks, crevices, under bark, and in similar places during the day. They are mainly scavengers. Some earwigs have special glands that secrete a foul-smelling, yellowish-brown fluid that probably serves as protection. The name earwig refers to an ancient superstition that these insects enter the ears of humans; they do not.

Du Buysson (1900) was apparently the first person to report that earwigs may be found in bee hives. Earwigs may frequent the area just beneath the hive cover (Toumanoff 1939), but they do little or no harm (Doupe 1921, Webster 1947). It is possible that earwigs help spread honey bee diseases: Danielyan (1963) demonstrated that washings from the earwig *Forficula fomis* contained European foulbrood bacteria.

The primary earwig species identified from beehives in Europe and North America has been *Forficula auricularia* (Beljavsky 1927, Borchert 1974). Kaczmarek (1984) reported *F. auricularia* as the fourth most common insect of over 6,000 insects in 11 orders associated with beehives in Poland. Morison (1941) stated that *F. auricularia* enter hives for shelter and food and that they may pierce honey cappings and spoil comb sections with excreta, food fragments, and the cast skins of the nymphs. Crumb et al. (1941) found plant materials and beeswax in the digestive tracts of earwigs taken from a hive, and they concluded that the earwigs had fed outside the hive as well as inside; when a poisoned bait was scattered on the ground around the hive, the earwigs disappeared. It has also been suggested that earwigs can eat honey, the soft parts of injured and dying bees (Haydak 1963, Borchert 1974), and perhaps larvae (Toumanoff 1939).

Control of earwigs is rarely, if ever, needed. Any damage they do is slight. A tachinid fly, *Digonochaeta spinipennis*, parasitizes earwigs, at least in Scotland (Morison 1941), and this may keep their populations in check. Nevertheless, Morison

(1941) recommended control efforts against earwigs, including mowing the grass in apiaries and eliminating areas that attract earwigs, such as piles of rubbish, wood, or stone. He also suggested trapping earwigs by stuffing flowerpots with hay or straw and inverting them on a stake or by putting bits of fruit between two boards left on the ground. Morison suggested periodically collecting the earwigs from the traps and feeding them to chickens.

Isoptera (termites)

The Isoptera, or termites, are soft-bodied, small to medium-sized insects that, like honey bees, are highly social. In some termite species, a colony can include over a million individuals. The principal component of the termite diet is cellulose, obtained from wood or other plant material.

Termites can be classified into two major groups according to their habitat. The subterranean termites nest in wood that is buried in, or in close contact with, moist soil. They always remain in the nest, or in galleries or exterior tunnels built from the nest, and avoid direct exposure to the air, which would dry them out. The second group, called the dry-wood termites, live in wood above ground, some-

Figure 10.3. Termites can cause extensive damage to wooden bee equipment, the stand colonies sit upon, and the buildings where beekeepers store equipment. (Caron photo).

times in human dwellings or other wooden structures. Virtually anywhere in the world, and especially in the tropics and subtropics, termites can damage wooden beekeeping equipment. Both subterranean and dry-wood termites infest hives in the United States, mostly in the southern states (Eckert and Shaw 1960). Singh (1962) listed the species *Termes obesus* as a pest of wooden wall hives and mud hives in India. Adamson (1943) listed three termite species as pests for beekeepers on the Caribbean island of Trinidad: *Heterotermes tenuis*, *Nasutitermes costalis*, and *Microcerotermes arboeus*.

Hive parts that come in contact with the ground need protection from termites as well as from ants (see Chapter 9) and fungi. In some areas, hives suspended from trees or human-made structures may also require protection. A barrier between hive and ground can protect hives from the soil-infesting termites, but it is not effective against dry-wood termites. Eckert and Shaw (1960) suggested hive stands of concrete, tile, or stone. Adamson (1943) recommended metal stands. Other suitable materials may be available locally.

Termite-resistant lumber may be available in some areas for constructing hive bodies, bottomboards, or hive stands. Eckert and Bess (1952) found that redwood is less susceptible than other woods to termite damage, but redwood is

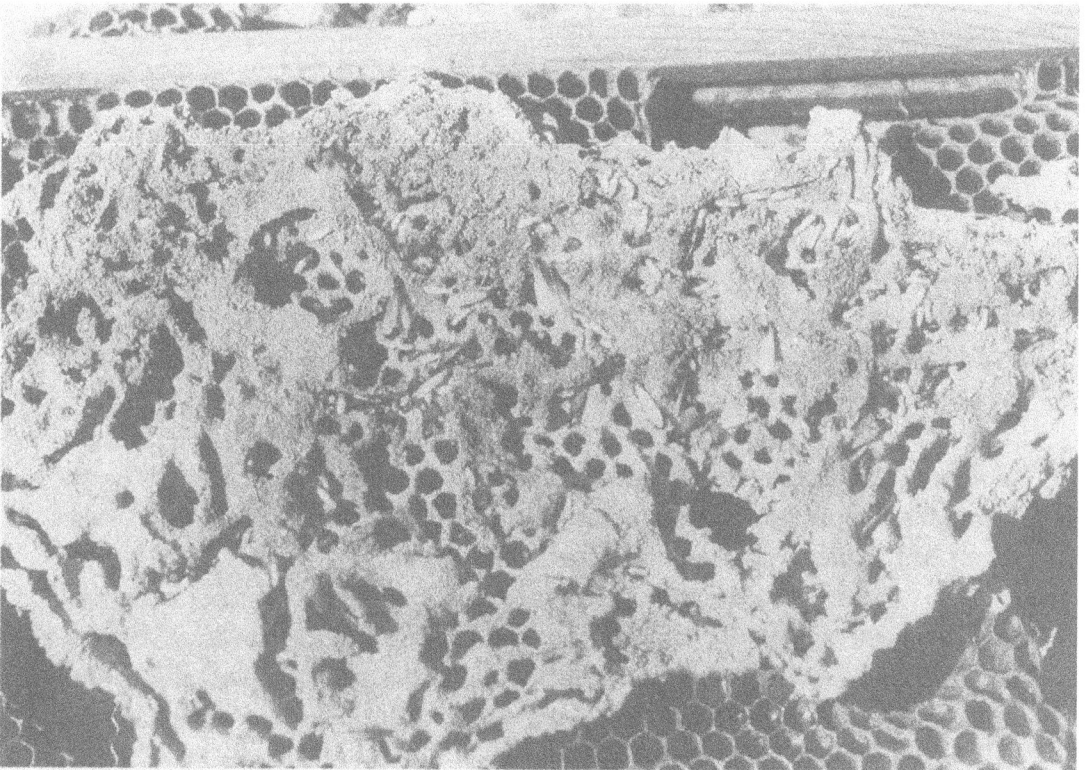

Figure 10.4. A termite (genus *Reticulotermes*) nest inside a beehive. (Caron photo).

prone to injury from carpenter bees. Rayment (1917) suggested placing beehives on wooden blocks made of "red gum" or "Jarrah" in the South Pacific because these woods survive 40 to 50 years of soil contact. Kent (1979) stated that in Costa Rica, wood from the tree pochote (*Bombacopsis quinatum*), native to the province of Guanacaste, is preferred for hive bodies because it is resistant to termites and rot and need not be painted; the wood is expensive, however, and highly acidic, which makes it less useful for frames. Plastic hive bodies and bottomboards might be a solution where termites are a special problem.

Alternatively, wooden hive stands and hive parts that come in contact with the ground can be treated with wood preservatives. Singh (1962) suggested solignum or pentachlorophenol in India. He also proposed coating the termite galleries with coal tar or spraying an insecticide on their mounds. Great care must be taken, and all appropriate local restrictions must be followed if an insecticide (such as chlordane [G. Morse 1979], the use of which is no longer permitted in the United States) is used in an apiary to control termites or ants.

Psocoptera (barklice)

The psocids, known commonly as booklice or barklice, are small, soft-bodied insects with chewing mouthparts. Most species live on or under bark, stones, or dead leaves, feeding on fungi, bits of pollen, dead insects, or other debris. A few species cause minor damage to books and papers in homes and libraries.

Toumanoff (1939) and Beljavsky (1927) stated that *Liposcelis divinatorius* (previously called *Troctes divinatorius*) can be numerous in hive-insulating materials and in hive inner covers made of cloth. Beljavsky (1927) collected 446 specimens of *L. divinatorius* from a single hive and believed they were attracted to the warmth in the hive. Knowlton (1951) found wingless forms of *Liposcelis* in crevices between hive parts in living honey bee colonies and huge numbers among the dead bees and debris at the bottom of a dead colony.

Haydak (1963) listed *L. divinatorius* and *Trogium pulsatorium* (= *Atropos pulsatoria*) as occasional invaders of bee colonies that feed mostly on pollen remnants and perhaps on wax moth eggs. Borchert (1974) and Most (1989) also described *T. pulsatorium* as a harmless hive occupant.

Although Toumanoff (1939) believed that psocids, like most insect species that come in contact with bee colonies, contribute to the spread of bee diseases, psocids are not very common in standard Langstroth hives, and control measures are not necessary. The packing material in insulated and double-walled hives may provide a hiding and nesting area for psocids; the use of such hives is not recommended.

Hemiptera (true bugs)

The insects for which entomologists reserve the name "bug" are generally classified in the order Hemiptera (e.g., Borror et al. 1989), although some authors prefer an alternative arrangement in which this group of insects is referred to as the Heteroptera. The name Hemiptera (meaning "half-winged") refers to the fact that only the outer half of each front wing looks like a typical membranous insect wing; the half closer to the body is leathery and thickened. The wings fold flat over the insect when not in use and are the major structure one sees when looking at these insects.

The mouthparts of a bug are formed into a beak that is used for piercing and sucking. In some species, the beak is large and can inflict a painful bite on humans. Many bugs eat plant matter, but some are predaceous. The few species recorded as preying on honey bees usually capture foraging bees on or around flowers, although the author found them in and on bee colonies in Panama (Caron 1993). None are of major importance for beekeepers.

Toumanoff (1939) described the family Reduviidae (called assassin bugs) and their close relatives (family Pyrrhocoridae) as honey bee predators. He stated that adults, eggs, and larvae of one unidentified species in the genus *Pyrrhocoris* have been found in beehives in the former U.S.S.R., especially in colonies weakened by mites. Toumanoff wondered whether the bugs aided in the spread of mites, but it seems more likely that colonies weakened by mites were unable to keep the bugs out. Toumanoff (1939) identified *Pyrrhocoris apterus* as a frequenter of hives in France. He also found that members of the genera *Triatoma* and *Conorhinus* attack honey bees in Brazil.

Borchert (1974) identified the beautiful red adults of *Rhinocornis iracundus* and the soil-encrusted larvae of *Reduvius personatus* as ambushers of honey bees and other bees on flowers in Germany. Lundie (1952) stated that assassin bugs are pests of honey bees in Africa. In Australia, Roff and Brimblecombe (1963) found the assassin bug *Pristhesancus papuensis* seizing bees visiting flowers. Toumanoff (1939) also listed this species. Mahadevan (1951) and Subbiah and Mahadevan (1957a) listed *Acanthaspis siva* as a predator of the Indian honey bee, *Apis cerana*, as do Ambrose and Livingstone (1987, 1990).

Swadener and Yonke (1973a,b) reported the honey bee as a new prey record for the assassin bug *Apiomerus crassipes*. Both Stejskal (1969) and Ma (1949), however, had earlier identified the same species as a minor pest of honey bees in Venezuela. Ma (1949) also listed *Apiomerus spissipes* as a bee pest; he gave its common name as the bee-catching bug. Riley and Howard (1892) reported that *Apiomerus flaviventris* was seen feeding on honey bees.

Blake (1946) reported that *Phymata wolfi* (family Phymatidae, ambush bugs) were killing hundreds of honey bees in West Virginia, U.S.A. Individual bees on sweet clover and other flowers were attacked by several bugs each. *Phymata erosa* and *Phymata pennsylvanica* have also been recorded as honey bee predators (Bromley 1948, Ma 1949).

In addition, a few other bugs have been observed to feed on honey bees. Bromley (1948) cited four records of the pale soldier bug, *Podisus placidus*, capturing honey bees foraging on sumac blossoms, and two records of immature spined soldier bugs, *Podisus maculiventris*, feeding on honey bees. Knowlton (1947) observed adult boxelder bugs, *Leptocoris trivittatus*, feeding on recently dead or nearly dead honey bees in Utah, U.S.A. He did not believe that this species would attack active honey bees. Morse and Gary (1961) recovered two single specimens of Hemiptera from dead bee traps on honey bee colonies: *Euschistus tristigmus* and *Sehirus cinctus*. These bugs were probably incidental visitors, however, and unlikely to have been feeding on honey bees.

In general, no control measures are needed for Hemiptera.

Neuroptera (lacewings, fishflies, antlions)

Neuropterans have four wings, each with a network of many veins, that fold rooflike over the body when at rest. The order includes the snakeflies, the lacewings, the dobsonflies, and the fishflies (the immatures of which are called hellgrammites).

Most neuropterans are predaceous, but records of feeding on honey bees are scarce. Banaszak (1980) reported the lacewing *Chrysopa vulgaris* as a common inhabitant of beehives in Poland. An antlion, *Myrmeleon januaris*, was observed eating bees in Brazil (Nascimento and de Souza 1970). Morse and Gary (1961) found a single fishfly, *Neohermes* sp., in one of their dead bee traps on honey bee colonies; its presence was probably the result of an incidental nocturnal visit.

Coleoptera (beetles)

The order Coleoptera is the largest of all the insect orders and includes at least 40 percent of all the described species of insects. Beetles vary in their habits and can be found nearly everywhere on earth. They are distinct because the elytra, as the front wings are called, are thickened, hard, and brittle and meet in a straight line down the middle of the body. The elytra do not aid in flight but serve as a very effective protective shield. Most beetles have mouthparts adapted for chewing. They have four developmental stages, and the larvae (called grubs or worms) frequently have a life history and food habits different from those of the adult.

Several beetles have been described as pests of honey bees, but most are only occasional hive visitors that feed on pollen and debris (see, e.g., Gentry 1982).

Many beetle species are pollen feeders; Leonard (1983) found one or more representatives of 22 beetle families in pollen traps on honey bee colonies or in pollen collected from the traps. In general, beekeepers find beetles only in weak hives, among the debris on bottomboards, or on combs in storage.

The following description of beetle pests is organized by families within the order Coleoptera; the families are arranged alphabetically.

Carabidae (ground beetles)

Borchert (1974) listed the carabids *Carabus auratus* and *Calosoma sycophanta* as occasional bee pests. These are active, predaceous species that eat a wide variety of insects. Haydak (1963) also noted that *Calosoma sycophanta* can capture and devour adult bees at the entrance of a hive. This species has been introduced into the Northeastern United States from Europe to help control the gypsy moth. Occasional predation by this and other ground beetle species should not be considered a serious problem. Elevating a hive on a hive stand might afford protection if ground beetles are very numerous in an area.

Cleridae (checkered beetles)

Toumanoff (1939), Paillot et al. (1949), and Borchert (1974) described *Trichodes apiarius* as a pest of brood in weak colonies. These small, black, red-banded beetles usually feed on dead or dying bees but may attack brood and cause comb damage with their galleries. They have one generation each year. The adults attack insects at flowers and may eat an occasional honey bee forager (Beljavsky 1927, Ma 1949, Borchert 1974). They may also attack solitary bees and their nests. Other checkered beetles listed as preying on honey bees are *T. ornatus* (Ma 1949) and *T. alvearium* (Beljavsky 1927), although Haydak (1963) said that the latter species does not attack honey bees. The best control against checkered beetles is to keep colonies strong and healthy and to keep bottomboards as clean and free of debris as possible.

Cryptophagidae (cryptophagid beetles)

Cryptophagus scanicus was listed by Banaszak (1980) as the most common coleopteran found in a survey of mite and insect fauna associated with honey bee hives. Hejtmanek (1951) found them in debris from hives that had wintered in cellars. These beetles live on hive refuse on the bottomboard and cause no damage to the bees. Related species occur in wasp and bumble bee nests.

Dermestidae (dermestid beetles)

Dermestids are small, oval-shaped beetles that are generally scavengers on plant and animal products. Some species, including the familiar carpet beetle (*Anthrenus scrophulariae*), are common household pests.

Adults and larvae of *Dermestes lardarius*, called the larder beetle or the bacon beetle, sometimes feed on pollen, brood remains, and other debris in honey bee colonies and may damage wooden hive parts and comb by tunneling (Beljavsky 1927, Haydak 1963, Borchert 1974). Toumanoff (1939) believed this species to be an important pest of stored comb. Borchert (1974) found that in frames made from spruce, the beetles attacked only the exterior, whereas they were more destructive in frames of basswood; however, in no instance was damage extensive. The beetles are found only in weak colonies or on stored comb (Haydak 1963). Paillot et al. (1949) found *Nosema apis* Zander spores in the intestines of *D. lardarius* and believed that larder beetles could aid in the spread of nosema disease.

Borchert (1974) found larvae of two other dermestids, *Dermestes vulpinus* and *D. frischii,* in beehives. *D. vulpinus* was also noted by Smith (1960) as a hive pest in Southeast Asia. Their tunnels can harbor other insect pests of the hive. Toumanoff (1939) listed yet another dermestid species from Southeast Asia, *D. cadaverinus.*

Shaw (1961) and Haydak (1963) reported that a cabinet beetle, *Trogoderma ornata*, bored through comb, causing the comb's disintegration. The combs had American foulbrood disease and were in storage. Iselin (1968) reported a heavy infestation of *Trogoderma glabrum* in comb honey from Kansas, U.S.A. The author stated that these beetles are commonly found in old honeycombs and in poorly managed hives, where they consume the remains of brood and pollen in addition to honey. The damage they do resembles that caused by wax moths, but the characteristic small, threadlike tunnels are distinct. Dermestids, unlike moths, do not make silken webbing, although they do leave behind considerable dustlike frass and the transparent skins from larval molts (Milum 1940b). Iselin (1968) noted that frass from beetle tunnels, along with the damage they do to the edges of comb honey sections, renders infested comb honey unmarketable.

Control measures for *Trogoderma* and *Dermestes* may be necessary for stored comb and weak hives. Fumigants used for wax moth control in stored equipment are effective in control of these beetles. Species of *Dermestes* can be a serious problem in museums, and insect collections must be continually fumigated to protect them against these pests. Modern methods of treating comb honey, such as fumigation with carbon dioxide or ethylene oxide or freezing, control these beetle pests in honey that will be consumed. For weak hives, the same precautions necessary for wax moth control work for these pest species. It is easier to prevent an infestation than to bring one under control.

Meloidae (blister beetles)

The beetle family of greatest consequence to beekeepers is undoubtedly the *Meloidae*, commonly called blister beetles; those in the genus *Meloe* are sometimes known as oil beetles. Blister beetles cause two types of damage to bees, corresponding to two different life histories of the beetles. The first type involves predation on bee eggs and larvae; the second involves parasitization of adult bees. In all cases, the damage is done by the first larval stage, called a triungulin, which differs greatly in appearance and behavior from the later larval stages, which are generally inactive (Borror et al. 1989).

An example of a species that preys on immature bees is *Meloe cicatricosus*. Triungulins of this species wait on flowers for foraging bees, on which they hitch rides. In the hive, the triungulin dismounts and enters a brood cell that contains an egg or a larva, which it eats. It remains in the cell and is fed by nurse bees, developing through the various larval stages and eventually emerging as an adult (Seltner 1950).

The second type of life history is that exemplified by *Meloe variegatus*, an inhabitant of North America, Europe, and Asia. The triungulin burrows into the abdominal segments (Bailey 1963a) or leg segments (Raw 1954) of an adult bee to feed on its hemolymph. The larval beetle eventually causes the death of the bee, then attacks another adult bee. The resulting losses in honey bee colonies can be extensive, and masses of dead bees can be found at hive entrances (Seltner 1950, Borchert 1974). In the former U.S.S.R., such damage occurs in late May and June (Minkov and Moiseev 1953).

Other meloid species have been described as pests on honey bees. Danielyan and Nalbandian (1971), in their study of blister beetles in Armenia, found that four of eight species were serious bee pests; these were *Meloe variegatus, M. proscarabaeus, M. cavensis,* and *M. hungarus*. Beljavsky (1927) listed *M. proscarabaeus* and *M. violaceus* as bee pests. Borchert (1974) listed *M. proscarabaeus, M. variegatus, M. faveolatus,* and *M. tuccius*. Quintero and Canalas (1987) listed *M. leavis* and another unidentified *Meloe* species parasitizing honey bees in Mexico.

Cros (1932), in a review of the life history of the genus *Meloe*, questioned the record of *M. cavensis* as a honey bee pest. He did not see this species on honey bees in Algeria, although it is common on other bee species there. Cros also stated that only meloid species in which the triungulin has a flat, triangular head attack adult bees; it uses its head to attach itself between abdominal segments of the bee and feed on its hemolymph.

Borchert (1974) reported one to three percent of adult honey bees in infested hives parasitized by meloids. An infested bee typically had one or two triungulins, but in one case 15 were found on one bee. The beetles usually attach to the thorax and adhere so tightly that it is difficult to remove them with forceps. Borchert observed grooming and shaking movements by bees that were apparently attempting to rid themselves of the triungulins.

According to Seltner (1950), triungulins also get on honey bee drones and queens. A *Meloe* larva, species not determined, was sent to the *American Bee Journal* by a beekeeper in Mexico. The specimen had been removed from a young queen, and the beekeeper stated that he saw them on young queens and workers "quite often" in Mexico (Park 1922). Whether they can kill queens is unknown.

Rayment (1926) gave an excellent account of the life history of *Meloe* species on wild bees, not using honey bees in his description. Composite flowers are preferred host-seeking locations. The beetles can and do make mistakes by attaching to *drone flies* (*Eristalis*) and to wasps such as *Ammophila*. Even pieces of velvet dangled near flowers will elicit attacking behavior from *Meloe* larvae. According to Paillot et al. (1949), an adult *Meloe* female can deposit as many as 4,000 eggs in the earth, and as many as 230 triungulins have been counted on a single foraging bee.

Beljavsky (1933) pointed out that *M. variegatus* has been misclassified as a parasitic fly in the family Braulidae. Early insect classifiers, including Linnaeus, thought blister beetle larvae were lice.

Toumanoff (1939) did not believe that *Meloe* damage justified special control measures. Paillot et al. (1949) believed that strong colonies can normally cope with the number of *Meloe* larvae usually found and that it would be difficult to implement any effective control measures. Minkov and Moiseev (1953), however, used naphthalene (5 g per hive) to control *M. variegatus* in the former U.S.S.R.; a beekeeper informed them that adult bees from colonies treated with naphthalene were not attacked in the field by *Meloe* larvae (apparently because of a residual odor). Naphthalene is an insecticide that will also kill bees, and its use in colonies is not legal in the United States.

Toumanoff (1939) also mentioned the meloid genus *Lytta* as infesting bee species, including perhaps the honey bee. Linsley and MacSwain (1952) described the biology and host relationships of the meloid genus *Nemognatha* in North America. Most species infest bees other than honey bees, mostly in the family Anthophoridae, but one species, *Nemognatha piezata*, was recorded from *Apis mellifera*. Individuals of *Nemognatha* and *Meloe* species have been found in pollen collected by honey bees (Leonard 1983).

Nitidulidae (sap beetles)

In Africa, a beetle in the family Nitidulidae, *Aethina tumida*, known as the small hive beetle, has been called as destructive as the wax moth *Galleria*. Lundie (1940, 1951, 1952) described the biology of the beetle, which May (1969) reviewed. The beetle is only about 0.5 cm long; both adult and larval stages cause damage in the hive.

According to Lundie (1940), *A. tumida* is active only during the summer, during which five generations are reared. The beetles spend egg, larval, and adult stages inside the beehive but pupate in the soil. Adults gather near the rear of the hive and feed on pollen that falls from the brood nest above them. The larvae rapidly multiply in small colonies, and they damage wax comb as they feed on honey and pollen. They also attack combs in storage. Larvae develop as rapidly on pollen alone as on a pollen and honey mixture, but they cannot survive on a diet of only honey.

The small hive beetle is widely distributed in tropical and subtropical Africa. Protective measures are apparently necessary in some areas. The best control is to maintain populous, active bee colonies, in which the bees keep the number of beetle larvae down. Smith (1960) recommended single-walled hives. It is also important to protect stored combs. The beetles and wax moths may be present at the same time. Lundie (1940) controlled the small hive beetle with carbon bisulphide fumigation of stored comb. Lundie (1951, 1952) and Smith (1960) also stated that the fumigants used for wax moth, such as paradichlorobenzene (PDB), provide sufficient control.

Adults of the dusky sap beetle, *Carpophilus lugubris*, were the most common invaders of pollen traps placed on honey bee colonies in Maryland, U.S.A. (Leonard 1983, Leonard et al. 1983). They were most prevalent in July and August. Pollen traps that protect pollen from moisture and thereby discourage invasion by fungus-feeding arthropods were recommended, as was keeping pollen well-ventilated and protected from insects during and after collection, as outlined by Chambers (1982).

Ptinidae (spider beetles)

The spider beetle *Ptinus fur* will attack and damage beeswax comb in weak colonies or combs in storage. Borchert (1974) described its life history. The beetle is an omnivore that attacks a large variety of food sources of both animal and vegetable origin, including other *Ptinus* individuals. It can attack stored comb and comb in the hive (Toumanoff 1939, Paillot et al. 1949). *Ptinus raptor* has also been found attacking stored pollen in bee hives (Brassler 1929, Haydak 1963). Rayment (1935) listed *P. exulans* as an occupant of a deserted hive in Australia. Ptinid specimens were the most common insect collected from bottomboard debris in Czechoslovakia (Vyslouzil 1984). Linsley and MacSwain (1941) listed another spider beetle,

Ptinus californicus, as a pest of wild bees in the United States and reviewed the related literature. No control measures are mentioned for *Ptinus*.

Scarabaeidae (scarab beetles)

The scarab beetles are a large family with diverse habits. Several members of this family have been recorded as bee pests. The adults of these species are fairly large and thick-bodied, so honey bees cannot sting them. Beekeepers can keep the beetles out of hives by reducing the entrances to narrow slits.

Rinaldi et al. (1967) detailed the damage caused by *Euphoria lurida* in Argentina. They reported that 52.6 percent of the hives had at least one *E. lurida* present, and 33 of the beetles were found in one hive. The beetles attack the wax combs, apparently seeking pollen and honey. They may feed on debris from brood rearing, and are frequently found in the brood chamber of the hive.

Adult *Euphoria sepulchralis* are common in Gainesville, Florida, U.S.A. (F. Robinson, personal communication 1981); they can be found in virtually every honey bee colony there in May and June, spending the nights in the grass but entering the hives each morning. The beetles can damage combs in their quest for stored pollen to eat. They are also common around Baton Rouge, Louisiana, U.S.A. (J. Harbo, personal communication 1987), and probably in much of the Southeastern United States.

Rangarajan (1964) found *Protaetia aurichalcea* in India in *Apis cerana* colonies during the winter of 1961-'62. The beetles were feeding on stored pollen; there were four to five beetles per hive. The author recommended removing the beetles by hand. Hameed and Adlakha (1973) stated that the scarabs *Protaetia impavida* and *Anomala dimidiata* infest combs of both *Apis cerana* and *A. mellifera* in the Kulu Valley, India. *P. impavida*, commonly called the cherry blossom beetle, feeds extensively on stored pollen. As many as 15 to 20 beetles were found on the frames in a super. The authors believed that the beetles might contribute to the spread of adult bee diseases. Atakishiev (1971b) stated that unidentified scarab beetles were carriers of the wingless fly *Braula coeca* (the bee louse; see Chapter 8) in the former U.S.S.R.; braulid-carrying beetles introduced into beehives emerged minus the fly.

Adlakha and Verma (1976) reported that in India *Torynorrhina opalina* were observed on the alighting boards of *Apis mellifera* hives but not on those of *A. cerana* hives, although the latter were much more numerous in the area. Narrow hive entrances kept the beetles out of the hives proper. Bees tried unsuccessfully to sting the beetles, which were apparently seeking pollen stores.

Vidano and Onore (1971) labeled *Potosia opaca*, a scarab beetle found in the Mediterranean region, the "black chafer of hives." The adult bees eat honey, but the larvae feed on decaying organic matter. Kashkovskii (1958) listed *Potosia hungarica* as a bee pest in the former U.S.S.R. Beekeepers find the beetle on combs of pollen and open brood. Lecq (1933) reported that a dung beetle, *Copris lunaris*, was observed consuming honey in a hive. In the Caribbean, adults of *Phileurus didymus*, a large scarab beetle, have occasionally been found chewing through comb in beehives (Laurence 1973).

Lundie (1951, 1952) and Smith (1960) identified two large scarab beetles, *Diplognatha gagates* and *Oplostoma fuligineus*, as invaders of beehives in South Africa. Captured individuals lived quite well in captivity for over a year on honey and pollen. Donaldson (1989) said *O. fulgineus* preferentially feeds on open honey bee brood as well as honey and pollen in beehives. Johannsmeier (1980) provides additional information about the life history of these two species and about their occurrence in beehives, and mentions another scarab, *Pachonoda rufa*, that enters beehives in South Africa. No estimates of the extent of the problems caused by these three species are given by any of these authors. The control methods mentioned are narrowing hive entrances to 8 mm or less or killing the beetles by hand (Phokedi 1985).

Lundie (1952) and Smith (1960) mention two additional scarabs as inhabitants of hives in Africa. These are *Coenochilus bicolor*, which has been reported from Zimbabwe and Zambia, and *Rhizoplatys trituberculatus*, a horned beetle. Evans and Nel (1989) reported *Macrocyphonistes kilbeanus* and *Rhizoplatys auriculatus* feeding on honey bee brood.

Johnsen (1953) found individuals of *Cetonia cupria* in mating nuclei but not in sufficient numbers to cause damage. Toumanoff (1939) listed *Cetonia cardui* (= *Cetonia apaca*) as a possible bee pest. Individuals of *Diplotaxis tristis*, *Gnorimella maculosa*, and *Phyllophaga fraterna* were collected from dead bee traps on honey bee colonies in New York, U.S.A. (Morse and Gary 1961).

Silphidae (carrion beetles)

Healthy colonies should not be attractive to carrion beetles. Johnsen (1952a) stated that a burying beetle, *Necrophorus humator*, may be attracted to hives that contain decomposing dead bees. Probably other species of carrion-eating beetles are attracted as well.

Tenebrionidae (darkling beetles)

Singh (1962) described two species in the family Tenebrionidae as occupants of the debris on the bottomboards of weak hives in India: *Platybolium alvearium* and a species of *Bradymerus*. Cleaning of the bottomboard and inspection of empty combs will keep beehives free of these beetles. Johnsen (1953) found both adults and larvae of the mealworm, *Tenebrio molitor*, in hives packed for winter. They were especially abundant in the straw packed around the colonies.

Haragsim (1965) identified *Tribolium madens* as a permanent beehive inhabitant that is capable of causing considerable damage to pollen stores, and *Tribolium confusum* as a species frequently spread inadvertently by beekeepers in pollen substitutes fed to colonies. *T. confusum* was commonly found in pollen samples by Leonard (1983) and, along with *T. castaneum*, was the most common insect pest of stored pollen pellets (Ibrahim 1972) *T. myrmecophilum* has been found associated with stingless bees (genus *Trigona*) in Australia (Rayment 1935); Neboiss (1962) described a new species, *Tribolium apiculum*, from a *Trigona* nest.

Additional Coleoptera

Larvae of the drugstore beetle, *Stegobium paniceum* (family Anobiidae), were reported infesting chalkbrood mummies in beehives in Great Britain (Heath and Smithers 1986). This species was not found on healthy bees, but it might infest foodstuffs in the hive, according to these authors.

Oryzaephilus surinamensis (family Cucujidae) was commonly found in pollen traps on bee colonies (Leonard 1983). Morse and Gary (1961) found nine species of beetles in dead bee traps on bee colonies in New York, U.S.A.: *Adalia bipunctata* (family Coccinellidae), *Agriotes pubescens* (family Elateridae), *Cicindela sexguttata* (family Cicindelidae), *Dicerca divaricata* (family Buprestidae), *Hister* spp. (family Histeridae), *Staphylinus maculosus* (family Staphylinidae), and the three scarab species mentioned above. It is probable that all these beetles became entrapped accidentally. One or two specimens of each were collected, except that four *Cicindela sexguttata* were taken.

Caron (1996) found the recently introduced multicolored Asian *Harmonia axyridis* overwintering in some beehives in apiaries of the Eastern United States Although not considered serious, the beetle adults may be a nuisance and perhaps result in less successful overwintering.

Rayment (1935) listed two beetles in the family Nitidulidae associated with stingless bees (*Trigona* spp.) in Australia: *Brachypeplus planus* and *B. meyricki*. When Rayment placed specimens of *B. planus* on beeswax combs of *A. mellifera*, they were carried off by the bees. Apparently the beetles are not serious pests.

Strepsiptera (twisted-winged parasites)

Members of the order Strepsiptera (twisted-winged parasites) are tiny, unusual insects. Most species are parasites of other insects. In these species, the female lacks wings, eyes, and legs, and her head is fused to the rest of the body, so that her overall shape is that of an oval or pellet. The female never leaves the host insect; she spends her entire adult life protruding from the abdominal segments of the host. Hosts of the parasitic Strepsiptera include insects in the orders Orthoptera, Hemiptera, Homoptera, Thysanura, and Hymenoptera.

In the Hymenoptera, bees of the families Andrenidae, Halictidae, and Colletidae and wasps in several families, particularly paper wasps of the subfamily Polistinae, are known to serve as hosts. All of the Strepsiptera that parasitize the bees and wasps are classified in the family Stylopidae and are commonly called stylops.

Stylops as parasites of honey bees are apparently extremely rare. Adlakha and Sharma (1976) found *Apis cerana* parasitized by stylops in the Kangra Valley and Jamalpur, India. They reported that infested colonies have a tendency to desert combs and form small clusters on the bottomboard or hive walls; infested individuals could not fly or sting. By destroying the parasitized bees, further spread of the stylops was checked.

Johnsen (1953) identified stylops on *A. mellifera* in Denmark. All other references in the literature can be traced to a single secondhand report by Beljavsky (1936) that *Stylops melittae*, a common parasite of Andrena bees, parasitized honey bees in one mountain apiary in the former U.S.S.R. This record was disputed by Betts (1937) and Borchert (1974), but it has been cited by Toumanoff (1939), Bailey (1963a), and Paillot et al. (1949); the latter authors provided an illustration of *S. melittae* and gave life history information.

Leonard (1983) found first-instar stylops larvae to be common in samples of honey bee-collected pollen from many locations around the world. Considering the number of times that honey bees have been examined, however, the scarcity of records of stylops parasitization indicates that these insects rarely survive in honey bee colonies.

CHAPTER ELEVEN

Spiders and Pseudoscorpions

Dewey M. Caron

CHAPTER ELEVEN
Spiders and Pseudoscorpions

Introduction

Spiders and pseudoscorpions, along with mites (covered in Chapters 12 through 14), are arachnids. Arachnids are related to insects, but they lack antennae, have four pairs of legs, a body of one or two segments, and special mouth appendages termed chelicerae (Savory 1977).

Numerous spiders and some pseudoscorpion species are honey bee predators. Several familiar web-building spiders, crab spiders, and others that hunt prey at flowers where foraging honey bees are common may capture and consume bees. Good ecological studies, however, demonstrate that spiders and pseudoscorpions have a very small effect on honey bees and have other balancing positive influences. Nyffeler and Breene (1991) believe that orb-weaving spiders "should not be addressed as honey bee pests." Studies by Morse (1984) show that crab spiders are also not significant as foraging bee predators. Pseudoscorpions (reviewed herein and Caron 1992) certainly are not of significance. They may hitch a ride on a bee but feed on other occupants of the hive.

Spiders

All spiders are predaceous and feed mainly on insects (Gertsch 1949, Foelix 1982). The prey is killed by poison injected through the fangs of the spider. Spiders capture their prey in a variety of ways. Some actively forage, whereas others such as crab spiders lie in camouflage on flowers and pounce on visiting insects such as foraging bees, flower-frequenting flies, or beetles. Most spiders build a web of silk to trap terrestrial or flying prey.

Combfooted spiders

Several of the larger species of spiders in the family Theridiidae (combfooted spiders) are potential predators of honey bees. These spiders build irregular webs in which they hang in an inverted position (Kaston 1978, Nyffeler and Benz 1978). Rayment (1917) listed an unidentified species of black widow spider (*Latrodectus*) as a pest of Australian honey bees. The spiders apparently live under the colony and around the hive cover.

Botha (1970) and Smith (1960) described the black widow *Latrodectus indistinctus* as a bee pest in Africa. All species of black widow spiders produce venom that is poisonous to insects and vertebrates. The two most common North

American species, *L. hesperus* (the western black widow) and *L. mactans* (the eastern black widow), may eat honey bees and represent some danger because they may accidentally bite a beekeeper as they nest under and around a bee hive. Both species can be locally abundant and often prey on large bees, such as bumble

Figure 11.1. A black widow spider, *Latrodectus hesperus*. They are common in warm areas of the United States and will feed on honey bees when they are available. They can be a nuisance to beekeepers when they live on or near hives. (Ross photo).

bees, carpenter bees, and honey bees.

Botha (1970) indicated that another black widow spider, known locally as the gutter spider, *L. geometricus*, can be numerous around hives near farm houses in South Africa. The spiders live beneath the hive or between the outer and inner covers of the hive. They catch insects, including honey bees, in webs constructed near barns, farm buildings, and other structures. This spider also occurs in the southernmost United States (Kaston 1978). Toumanoff (1939) listed not only the genus *Latrodectus* as predators of honey bees but also spiders in the closely related genera *Theridion* and *Achaearanea*. He stated, however, that none of these is of prime importance as a honey bee pest.

The domestic spider, *Achaearanea tepidariorum*, is familiar to nearly everyone. Its tangled webs can be found in virtually any neglected room or honey house around the world (Gertsch 1949, Kaston 1978). These spiders usually build their webs beneath some object; window corners, door frames, and projecting eaves of buildings will often harbor one or more webs. Many beekeepers have undoubtedly seen bees trapped in the web of this spider. But the webs are usually placed inside buildings or other structures and thus do not frequently trap honey bees.

Langstroth believed that spiders were not necessarily harmful to bees and that spiders should have free access to combs in storage where they would destroy wax moths (Root 1966). Langstroth was probably referring to the black widow or the domestic spider, but it seems unlikely that spiders could effectively protect stored equipment to the degree stated by Langstroth.

Hunting spiders

Toumanoff (1939) stated that hunting spiders probably capture honey bees because of their large size. This group of relatively primitive spiders, which do not use silk to capture prey, includes tarantulas (family Theraphosidae) and trapdoor spiders (family Ctenizidae). Although many species in this group are exceptionally large and fierce predators, it is unlikely that they use bees as prey, simply because their habits do not bring them into contact with bees. Many species spend their lives hidden under leaf litter or in the soil.

Lynx spiders

The green lynx spider, *Peucetia viridars* (family Oxyopidae), was observed by Quicke (1988) to prey on honey bees "on numerous occasions." The captured bees were held just behind the head by the chelicera of the spider, which is about the same body size as the bee's body. The spiders hunt in flowering ground vegetation. Whitcomb et al. (1963, 1966), Turner (1979), Randall (1982), and Nyffeler et al.

(1987), all identify honey bees as prey for the same lynx spider. All agree that this species is a food generalist. Nyffeler et al. (1987) estimated that the spiders consumed less than one prey individual per day.

Turner (1979) reported honey bees to be the single most important prey of the green lynx spider. Randall (1982) found a 44:12 ratio of beneficial to harmful prey. Nyffeler et al. (1987) found more than half of the prey to be honey bees and other beneficials. In their study, more beneficials (10 percent honey bees) were taken by spiders in overgrown fields adjacent to cotton fields than within the crop. A leafhopper pest of cotton was a 30 percent prey selection.

Jumping spiders

Subhapradha (1961) listed jumping spiders (*Salticus* spp., family Salticidae) as bee pests. These spiders do not build webs; rather, they use their keen vision to stalk their prey. They build silken cases as retreats, frequently using the insides of the covers of beehives or the ventilation holes of hives. Subhapradha watched spiders seize bees twice their own size, bite them, and hold on until the venom took

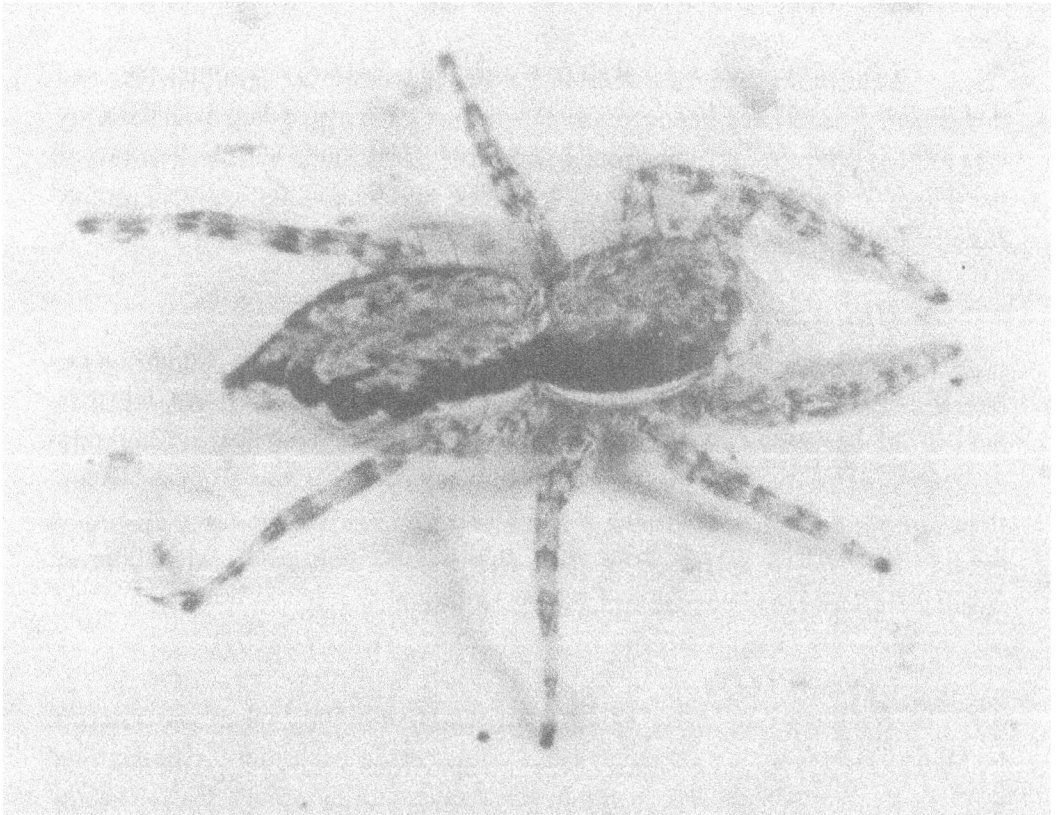

Figure 11.2. A jumping spider (family Salticidae). These spiders are aggressive and pursue their prey, characteristically making quick pouncy movements. (Carrel photo).

effect. Subhapradha suggested that examining beehive exteriors and scraping off the spider retreats when they are found would be an effective control measure.

Jackson (1978) and Quicke (1988) identify jumping spiders, *Phidippus audax* and *P. johnsoni* (family Salticidae) as honey bee predators. These species wait at flowers for prey. Quicke (1988) observed that honey bees constituted 90 percent of the prey of *P. aurdax* in a study site where honey bees were common as they foraged on flowering *Baccharis*. Bees were typically grabbed and held by the spiders' chelicerae.

Toumanoff (1939) also listed the Salticidae as bee pests. They are widely distributed, and some are fairly large. Larger members of the genus *Salticus* and other genera undoubtedly will capture honey bees if given the opportunity. These spiders "jump" onto the backs of prey such as honey bees and paralyze them rapidly with their venom. Honey bees may be frequent victims as seen in the observations of Quicke (1988), but their numbers and extent of distribution are limited.

Orb weavers
The orb weavers (families Tetragnathidae and Araneidae) are identified in

Figure 11.3. An orb weaver, *Argiope aurantia,* with a drone honey bee in its web. (De Jong photo).

the literature as the worst enemies of honey bees. The family Tetragnathidae, or four-jawed spiders, includes several species of large brown spiders that frequent meadows and marshy areas. Toumanoff (1939) identified *Tetragnatha extensa* as the most serious bee enemy. He also listed *Araneus diadematus, A. angulatus, Nuctenea cornuta, N. patagiata,* and *Metepeira salpetaria* as pests.

Orb weavers are among the largest and most familiar spiders. Their large and beautiful orb webs are often found around houses and gardens. Any of the larger-bodied species of orb weavers that build webs where honey bees are active might be expected to use bees as prey. Bromley (1948) provided the results of a survey of orb weavers that captured honey bees in Connecticut. He listed the black-and-yellow garden spider, *Argiope aurantia,* as a common predator of honey bees. The banded garden spider, *A. trifasciata,* the large white orb weaver, *Metepeira obesa,* the red-and-yellow orb weaver, *M. raji,* and the dusky orb weaver, *M. domiciliorum,* were also identified with honey bee captures.

Borchert (1974) also stated that orb weavers capture and consume honey bees. In apiaries where hive bodies are close, the spiders may build their webs between two hives. Borchert lists *Araneus quadratus, A. marmoreus, A. angulatus,* and *A. diadematus* as potential bee pests.

Latham (1939) believed that spiders played a major role in preventing him from producing a honey crop in Connecticut for 31 years! He did not identify the six types of spiders implicated, but his description of their bright body markings and round geometric webs makes it clear that he was speaking of orb weavers. In the one year of a good honey crop, spiders appeared to be scarce, and colonies built up in numbers and stayed strong. In other years, the field force would be gone after 10 days or so despite normal buildup. In some webs, bee fragments in a clump as big as a marble were described. Latham (1939) estimated that spiders took seven million field bees in 10 days, a phenomenal rate of predation.

One unusual case of spider predation was reported by Tribe (1989). He observed 22 drones, flying in a known drone congregation area, become entangled in a spider web (unidentified by name) 10 meters high between two trees. Comets of drones formed below the trapped drones until the web became completely destroyed and the suspended drones fell from their position, after which the drone comets also disappeared.

Large-bodied web-spinning spiders have the capacity to kill honey bees. They have been reported to constitute 23 percent of the prey of *Agelena labyrinthica,* 22 percent of *Aranus trifolium* plus 15 and 14 percent, respectively, of *Argiope bruennichi* and *A. aurantia* (Bilsing 1920, Nyffeler and Benz 1978). In good bee foraging areas, honey bees were the most frequent prey (greater than 60 percent), but in less favorable terrain (identified as "grasshopper-rich habitats") they accounted for less than 20 percent of the prey of orb-weaving spiders (Nyffeler 1982).

In an abandoned grassland meadow near Zurich, Nyffeler and Breene (1991) further quantified and analyzed the prey of orb weavers *Argiope bruennichi*, *Araneus quadratus*, and *Araneus diadematus*. Honey bees were a mere 17 percent of 7,480 prey items identified as food of these three species during the summer (May to September). Virtually all (99 percent) of the honey bees were captured in the last two months, which coincided with the spiders reaching full size and building strong, large webs.

The authors counted the web density and analyzed the pattern of bee deaths. They concluded that the web-weaver species killed a honey bee only every fourth day (August) or fifth day (September) per square meter. In another grassland with more thistle, one species, *A. bruennichi*, was estimated to kill a honey bee every fifth day in August. Both estimates discounted rainy days.

The two most efficient bee killers in the study of Nyffeler and Breene (1991), *A. bruennichi* and *A. quadratus*, both spin vertical orb webs fully capable of multiple bee captures per day but that are probably barely visible to the insect eye (Craig 1986). In the favorable bee-foraging habitat with flowering thistle, the estimated bee kill was higher (0.61 vs. 0.04-0.45 bees killed/spider/day). Nyffeler and Breene (1991) concluded that orb-weaving spiders could only negatively affect bee colonies in small, localized areas during short time periods.

Funnelweb weavers

Another group of web-building spiders, the funnelweb weavers (family Agelenidae), may also prey on honey bees. Toumanoff (1939) identified *Tegenaria domestica* as one species observed feeding on bees. Flies and mosquitoes are also included in its prey because it frequently builds its webs in cellars and neglected buildings. Bromley (1948) observed four captured honey bees in webs of a highly common grass funnelweb weaver, *Agelenopsis naevia*, in Connecticut, U.S.A. Many people are familiar with this genus and recognize its abundant webs among the grass or in shrubs on dewy fall mornings. Borchert (1974) identified another grass funnelweb spider, *Agelenopsis labyrinthica*, as a bee pest in France. Large numbers occur in areas of heath where they build their webs. Up to 30 dead bees were counted in a single web (Olberg 1960). Olberg recommended driving sheep through the area of the webs to protect bees because the webs are not immediately repaired.

Haplodrassus lapidosus, a hunting spider related to the funnelweb weavers, was listed as a bee pest by Toumanoff (1939). These spiders, however, generally make silk only under stones or in debris, and it is difficult to imagine honey bees coming into contact with such spiders.

Crab spiders

Another group of spiders that prey on honey bees are the crab spiders (family Thomisidae). These short, broad spiders somewhat resemble crabs, holding their legs in a crablike posture and walking sideways and backward most of the time (Gertsch 1949). They do not spin webs to capture prey. Some forage for prey, but most lie in ambush, especially on or near flowers. There they capture bees and flies, which often are much larger than they are, and quickly disable them with their venom. Toumanoff (1939) specified *Thomisius onustus* and *T. rotundatus* as bee predators. Both rapidly kill the honey bees they capture and return to capture additional bees as soon as they have finished feeding, though the feeding process sometimes takes hours. The spiders feed on the liquid contents of the bee.

Another common crab spider that captures and consumes honey bees is *Misumena vatia*, the goldenrod spider. This spider can change color over several

Figure 11.4. A crab spider female has constructed its egg case on the screened side of a package of honey bees. She is feeding on an escaped bee. (De Jong photo).

days to allow it to blend in with the flowers it frequents (Borror et al. 1976, Kaston 1978). The goldenrod spider's feeding habits are typical of the crab spiders. The spider positions itself on a flower, with its two pairs of long front legs grasping the flower for support. The spider closes its long front legs quickly when they are touched by a fly or a bee. A bee captured by the goldenrod spider will struggle for only a few seconds before death.

Makarov (1966) contended that crab spiders are the most dangerous spiders for bees in the former U.S.S.R. Of more than 100 species described there, he listed three unidentified species as especially troublesome. The spiders seized bees foraging on flowers and injected venom between the head and the thorax. It takes a spider about 24 hours to finish each bee meal. Grozdanic and Vasic (1966) listed honey bees as the prey of unidentified crab spiders in Eastern Europe. Bromley (1948) watched a yellow crab spider, *Misumena aleatoria*, capture 48 bees. The spider waited for and captured prey on goldenrod blossoms.

Morse (1979, 1984, Fritz and Morse 1985, Morse and Fritz 1989) has used crab spider foraging to study optimal foraging theory. The species *Misumena vatia* is a set-and-wait forager that feeds on bumble bees, honey bees, and flower-visiting syrphid flies. Although bumble bees represent the most favorable energy intake, the crab spider does not specialize on them.

Honey bees foraging on common milkweed had the probability of capture only once every 33 days (Morse 1985). This flower has a pollinia that entangles foragers. Although honey bees had a 4.7 percent entanglement rate, this did not increase their risk of being captured by crab spiders, which represent the only predation risk to foragers on this flower.

The numbers presented from studies over numerous seasons by Morse clearly demonstrate that crab spiders, although they capture and eat foraging honey bees, should not be considered a major bee pest. The probability of an encounter or capture is remote under all circumstances.

Spider control

Control measures against spiders are mostly unnecessary. Pellet (1916) and Root (1966) suggested keeping spider webs brushed away from active colonies and other areas around apiaries, especially where heavily laden forager bees fly. Because spiders are seldom present in large numbers and usually build webs among foliage or in buildings where bees do not forage, they probably take relatively few bees as prey. In any case, effective spider control would be difficult to achieve. Beekeepers should not consider spiders a major problem.

Pseudoscorpions

Pseudoscorpions (sometimes called book scorpions) resemble miniature scorpions; they have greatly elongated mouthparts (pedipalps) that end in strong pincers. They are not true scorpions (order Scorpiones). They lack a sting, and the abdomen is short and oval. They eat small insects and other arthropods and are harmless to humans.

Many species of pseudoscorpions use large insects for transportation, termed phoresy, holding on with one or both pedipalps. *Ellingsenius indicus* is such a species, perhaps the same one not identified to species by Singh and Venkataraman (1948). Murthy and Venkataraman (1986) even believe this pseudoscorpion can be used as a bioindicator as to whether swarms of the Indian honey bee *Apis cerana* have settled into a new homesite. This pseudoscorpion moves by attaching to the legs of bees. It does not feed on the bees themselves but eats other arthropods such as mites and moths (Murthy and Venkataraman 1985). Two swarms that adopted their new homesite had the pseudoscorpions move off bee legs and settle into crevices in the nest. In two other instances, the pseudoscorpions remained attached to the bees, and the swarms did not remain in the location they had first selected.

The pseudoscorpion *Paratemnus minor* is a common, though relatively unimportant, predator of honey bees, in the state of Sao Paulo, Brazil (Marques and

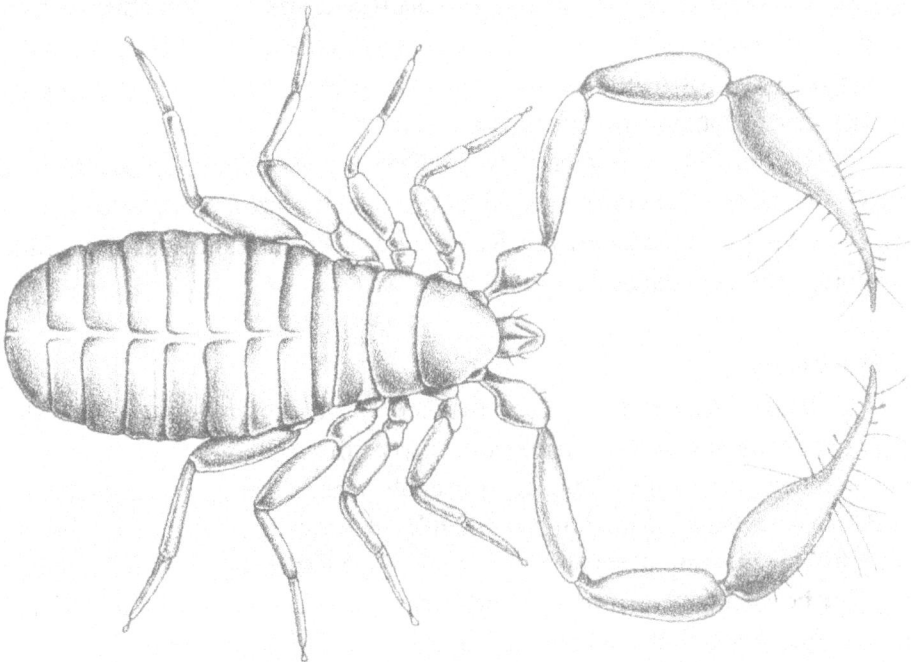

Figure 11.5. A pseudoscorpion (order *Pseudoscorpiones*). About 4 mm long, excluding its pedipalps. (Lay drawing).

Nascimento 1981). Typically, groups of 25 to 35 individuals were found living in cracks or joints in an outside wall or bottomboard of a wooden beehive. An individual pseudoscorpion grabs a passing bee with its pedipalps and holds on for 40 to 50 minutes until the bee is paralyzed, then feeds on the body fluids.

Borchert (1974) reported that the pseudoscorpion *Chelifer cancroides* has been found on bees and inside bee colonies in Europe. It is rarely found on combs but may be common on frames or on hive walls. It shuns light and flees to cracks and crevices when the hive is opened. According to Borchert, *C. cancroides* eats silverfish, mites, moth larvae, and bee lice but does not harm honey bees.

Borchert (1974) listed other European pseudoscorpions from bee hives as *Cheridium museorum* and *Chernes cimicoides*. He listed four species from Africa: *Ellingsenius sculpturatus, E. fulleri, E. ugandanus,* and *E. somalicus.* One species, *E. indicus.* was listed from India. Subbiah et al. (1957) also identified this last species as a pest in India. Ahmadi (1984) collected pseudoscorpions (no species identified) from two weak colonies in Southern Iran and characterized them as uncommon pests.

In a footnote to an article by Hirst (1921), Whyte noted that he had observed young pseudoscorpions sucking the body fluids of artificially injured honey bee larvae. Singh and Venkataraman (1948) fed captive pseudoscorpions wax moth larvae, and Örösi-Pál (1938) also observed pseudoscorpions eating wax moth larvae. These observations suggest that pseudoscorpions could feed on bee larvae if they gained entry into brood cells.

On the other hand, pseudoscorpions that kill wax moth larvae benefit honey bee colonies. Furthermore, their consumption of mites in beehives has been reported in Southern and Eastern Africa (Hirst 1921, May 1969) and in Germany (Alfonsus, cited in Örösi-Pál 1938). In India, colonies of bees that had pseudoscorpions were remarkably free of wax moths and mites (Singh and Venkataraman 1948). In general, the question of whether the benefits of pseudoscorpions to honey bees are outweighed by the harm they cause is unresolved.

I observed and reported the bee-eating behavior of pseudoscorpions in bee colonies in Belize (Caron 1992). The species have yet to be identified. Other unidentified species have been collected over several years in Southern Mexico by Francisco Reyes (Caron 1994).

Pseudoscorpion control

If pseudoscorpions cause problems for beekeepers, adequate control can probably be achieved by eliminating their hiding places within a hive. Cracks can be sealed, joints nailed more tightly, and hives painted on the outside.

CHAPTER TWELVE

Mites: An Overview

George C. Eickwort

CHAPTER TWELVE
Mites: An Overview

Mites, like the spiders and pseudoscorpions considered in Chapter 11, belong to the class of arthropods called the Arachnida. Although most mites are very small—barely visible to the naked eye at best—a few (including the ticks and the important honey bee parasite Varroa) are as large as some insects. Members of the subclass Acari (the technical name for mites) are characterized by the following anatomical features visible with a microscope: four pairs of segmented legs (three pairs in the larval stage that hatches from the egg); an unsegmented body, not divided into two parts as is that of a spider or a pseudoscorpion; and two pairs of mouthparts, the sensory palpi and the piercing or cutting chelicerae.

Unlike other arachnids, mites exhibit a wide diversity of feeding habits and habitats. Although many mites are predators in terrestrial habitats, others are saprophagous, fungus feeders, feeders on green plants, aquatic in freshwater or marine habitats, parasites of vertebrates, and parasites of other arthropods, especially insects. Indeed, among the arthropods, the mites are second only to the insects in terms of numbers of species.

A honey bee hive provides an acceptable habitat for a wide diversity of mites. More than 86 species of mites have been recorded in association with the different species of *Apis* and their nests (De Jong et al. 1982b, Eickwort 1988). These mites fall into the following ecological categories, modified from Eickwort (1988): (1) mites that are scavengers in the hive debris and may invade the stored provisions, (2) predatory mites that feed on the scavengers, (3) mites that "hitch-hike" (are phoretic) on foraging honey bees to reach the places where they feed (flowers, hives), and (4) mites that are parasitic on adult honey bees and their brood.

The mite species include incidental associates of honey bees, frequent but facultative associates, and obligate associates, which are known only from honey bees or their hives. Most of the incidental associates are of little importance to beekeepers and are mentioned briefly, if at all, in this chapter. The more frequent facultative acarine associates vary widely in their economic importance and are also covered in this chapter; some of the most conspicuous mites have little impor-

---

Professor Eickwort died in an accident on July 11, 1994. His chapter is reprinted here with little change from the second edition.

tance or are even beneficial because they prey on harmful mites. The most important mites are parasites of the bees themselves and are the subjects of Chapters 13 and 14. A list of most of the mites known from honey bees can be found in De Jong et al. 1982b, and a technical account of the evolutionary relationships of the more important acarine associates of honey bees is given in Eickwort 1988.

Scavenger mites

Honey bees efficiently care for their brood cells and food-storage combs, and most of the interior wood surfaces of their nests are coated with propolis. However, the hive bottom is not given the same care. Debris consisting of bits of old honeycomb and dead bees and fungi and other microbes that infest the detritus collect there and attract a wide variety of saprophagous mites, most belonging to the suborder Astigmata. These scavenger mites are typically the most abundant mites in a beehive and may reach very high densities in the hive bottom; 350,000 mites per kilogram of litter have been recorded in beehives in the U.S.S.R. (Grobov 1975).

These scavenger mites also occur in many other habitats, especially those that contain food resources concentrated by humans, so they are often called *stored-product mites*. Almost every stored-product mite species occasionally can be found in beehives; this chapter discusses only the more common species. Their evolution from more specialized insect or vertebrate associates is discussed in OConnor 1979, 1982; their taxonomy, distribution, and identification are *reviewed* in Hughes 1976.

The astigmatid scavenger mites are slow-moving globular mites, with clear bodies and (sometimes) darker-pigmented legs. Their bodies sometimes bear long hairs. The chelicerae form stout pincers, and, unlike the other mites that occur in beehives, the astigmatid mites can ingest solid food.

Glycyphagus species (family Glycyphagidae), especially *Glycyphagus domesticus*, are typically the most common scavenger mites in beehives. The body setae are branched and exceptionally long, and the legs terminate in a long, slender segment (tarsus) (Figure 12.1a). Strictly mycophagous, *Glycyphagus* feed on fungal contaminants in the hive.

Several genera of the family Acaridae have species that commonly invade beehives, rivaling in numbers the Glycyphagidae. These acarid mites are similar in morphology and biology, and identification of the genera is best left to a specialist. The acarid mites lack the elongated tarsi of the glycyphagids (Figure 12.1b). Acarids also feed on fungi, but, more catholic in their food choices, they consume other organic remains as well. *Acarus* is typically the most abundant acarid genus in beehives, especially *Acarus siro* and *A. immobilis*. Second in importance is the

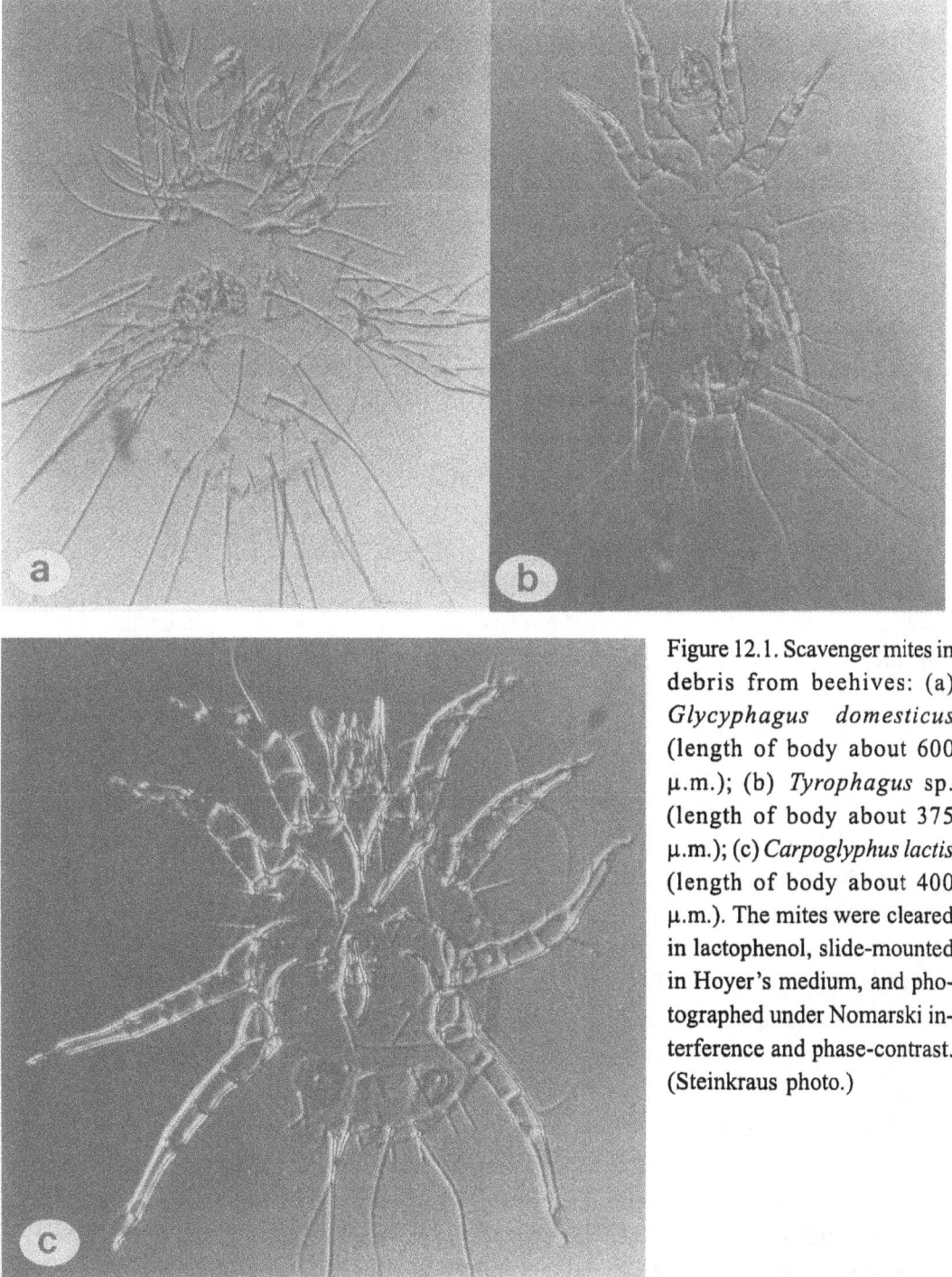

Figure 12.1. Scavenger mites in debris from beehives: (a) *Glycyphagus domesticus* (length of body about 600 μ.m.); (b) *Tyrophagus* sp. (length of body about 375 μ.m.); (c) *Carpoglyphus lactis* (length of body about 400 μ.m.). The mites were cleared in lactophenol, slide-mounted in Hoyer's medium, and photographed under Nomarski interference and phase-contrast. (Steinkraus photo.)

genus *Tyrophagus*, represented by *Tyrophagus putrescentiae*, *T. longior*, and *T. palmarum*. The closely related *Tyrolichus casei* Oudemans also commonly occurs in beehives.

The astigmatid mite *Carpoglyphus lactis* (family Carpoglyphidae) occurs in foods with a high sugar content that have begun to ferment. In beehives, *C. lactis* occurs principally in stored pollen and old honeycomb, and can be a pest in stored honey (Chmielewski 1971, Baker and Delfinado 1978). Morphologically, it resembles the acarids; it can be distinguished from them by the apodemes (thickened rods) on the venter. The apodemes from legs 1 and 2 both join the midventral apodeme; in the acarids the apodemes from legs 2 do not join the midventral apodeme (Figure 12.1c).

Suidasia pontifica (family Suidasiidae), a species of stored-product mite commonly found in the tropics, has been recorded in colonies of *Apis cerana* and *A. florea* (Koeniger et al. 1983) and can be expected in tropical *A. mellifera* hives. Also, the house-dust mites, *Dermatophagoides* spp. (family Pyroglyphidae), are recorded occasionally from beehives (e.g., Haragsim et al. 1978), as are *Aeroglyphus robustus* (Banks) (family Aeroglyphidae) (Delfinado-Baker and Baker 1982a) and *Forcellinia galleriella* Womersley (family Acaridae) (Bowman and Ferguson 1985, Delfinado-Baker and Baker 1987). The latter species, like other species of *Forcellinia*, occurs in ant nests, where it is saprophagous.

In healthy colonies, scavenger mites do little or no damage, despite their abundance. Although they are capable of invading stored honey and pollen, rendering the food less suitable to the bees because of metabolic wastes and the mechanical transmission of fungi and bacteria (Chmielewski 1975), their impact is slight. They are potential allergens for humans (*Dermatophagoides* is the allergen involved in allergies to house dust), and prolonged exposure may lead to allergic responses, including dermatitis, in beekeepers. No control measures are recommended except for cleaning out the debris that collects in old beehives, especially after the winter.

In addition to the Astigmata, fungus-feeding mites belonging to the suborder Prostigmata, families Scutacaridae (*Scutacarus, Imparipes*) and Tarsonemidae (*Tarsonemus*), occur in small numbers in the hive debris. These minute mites appear to be unimportant to the honey bees, even when they invade the provisions.

Predatory mites

Many species of mites that prey on the scavenger mites occur in the hive debris. With few exceptions, these mites are not specialized for living in beehives, but occur in the habitats where their prey, the stored-product mites, occur. Most predatory mites in beehives belong to the suborder Mesostigmata; they are fast-moving mites with dark plates on the dorsum and venter of their relatively flattened bodies. Many are large enough to be seen with the naked eye and may therefore be the most readily observed mites in a beehive. Although they are relatives of the important honey bee parasites *Tropilaelaps* and *Varroa*, the predatory mesostigmatid mites do not harm honey bees or their brood.

As with the scavenger mites, many species of predatory mites may incidentally occur in beehives. Among the more frequently encountered predatory mites are species in the families Ascidae (*Blattisocius, Proctolaelaps, Lasioseius,* and *Melichares*), Macrochelidae (*Macrocheles*), and Parasitidae (*Parasitus*) (Eickwort 1988). In contrast to these generalist predators, *Melittiphis alvearius* (Berlese) (family Laelapidae) (Figure 12.2) is known only from hives of *Apis mellifera*. It has been collected in small numbers in Europe and New Zealand and has recently been found in California, U.S.A. It is presumed, on the basis of its morphology, to be a predator on other arthropods in beehives, although its feeding behavior has never been observed directly (Samsinak et al. 1978, Cook and Bowman 1983).

The other predatory mites found in beehives belong to the suborder

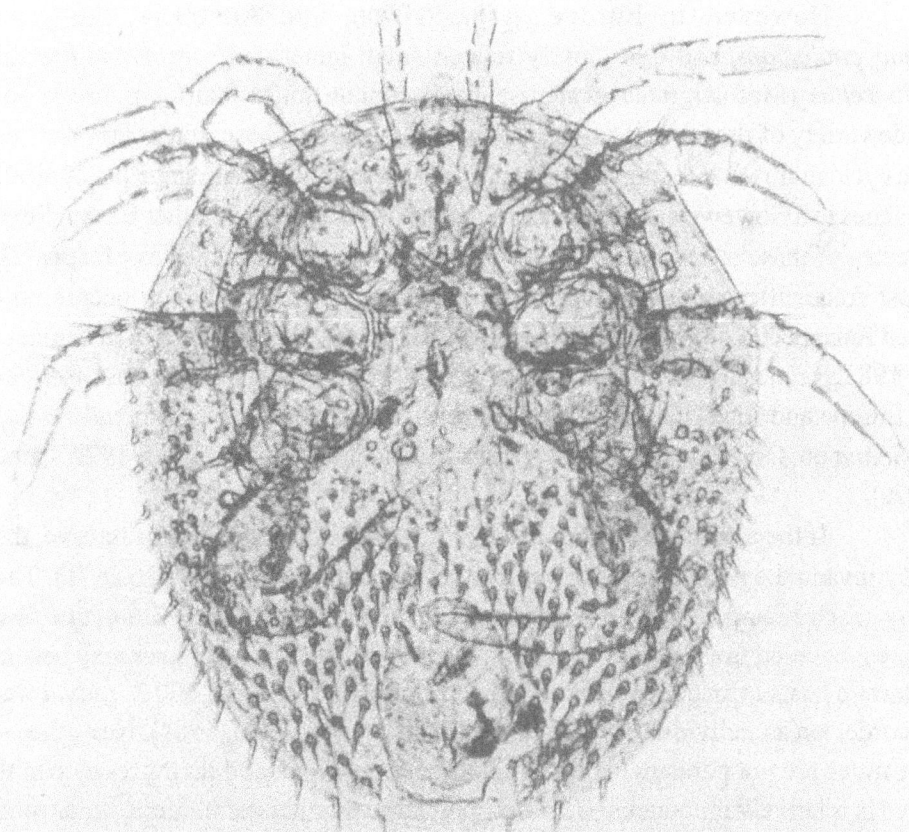

Figure 12.2. *Melittiphis alvearius*, a predatory mite, has a body about 800 μ.m. long. (Steinkraus photo).

Prostigmata. Most abundant are species of *Cheyletus* (family Cheyletidae), which are also important predators of Astigmata in stored grain.

Phoretic mites

Mites that inhabit the nests of insects frequently have one developmental stage specialized for phoresy. This phoretic stage attaches to a newly emerged adult insect and uses it as a winged vehicle to gain transport to a new nest. Such phoretic stages occur in the many specialized astigmatid mite species that inhabit the nests of most wild bees; many, if not most, individual bumble bees, megachilid bees, and other ground-nesting bees carry phoretic mites. However, the stored-product astigmatid mites that are scavengers in honey bee colonies rarely form such phoretic stages, and most of the other acarine hive inhabitants similarly do not use adult bees as phoretic hosts. Consequently, worker honey bees are remarkably free of phoretic mites.

However, in Europe, Asia, Africa, and Australia, the genus *Neocypholaelaps* and two closely related small genera, *Afrocypholaelaps* and *Edbarellus* (Mesostigmata: Ameroseiidae), contain mites that are phoretic on a wide variety of flower-visiting insects, including bees. These mites carry out their life cycle in flowers, feeding on pollen, and it is the adult female stage that typically attaches to a flower-visiting insect as a means of transport to another flower. Seven species of these mites are known to be phoretic on various species of *Apis*. The most frequently encountered are *Neocypholaelaps indica*, which occurs on at least four species of *Apis* in Asia (Mo 1971, Baker and Delfinado 1976, Koeniger et al. 1983, Ramanan and Ghai 1984); *Neocypholaelaps favus* Ishikawa on *A. mellifera* in Europe and Asia (Ishikawa 1968, Haragsim et al. 1978); and *Afrocypholaelaps africana* on *A. mellifera* in Africa, Australia, and New Guinea (Elsen 1972, Clinch 1979).

If the ameroseiid mites disembark from a honey bee within its hive, they may invade the pollen stores, in which they feed and reproduce (Mo 1971). They may reach remarkably high numbers there; up to 3,000 *Neocypholaelaps favus* were recovered from one hive by Ishikawa (1968). Bees in such hives may become phoretic hosts to large numbers of mites (Figure 12.3b): Up to 400 *N. indica* were recorded on an individual *Apis cerana* by Ramanan and Ghai (1984). Nevertheless, the mites are not parasitic on either adult bees or brood, and their presence in the hive is relatively innocuous. Bees carrying large numbers of phoretic mites might exhibit "discomfort" and attempt to remove the mites (Underwood, cited in Baker and Delfinado 1976), and presumably their ability to fly is impaired.

Neocypholaelaps and the related genera can be confused with *Tropilaelaps*, which occurs on adult bees and on combs in Asia and is a serious

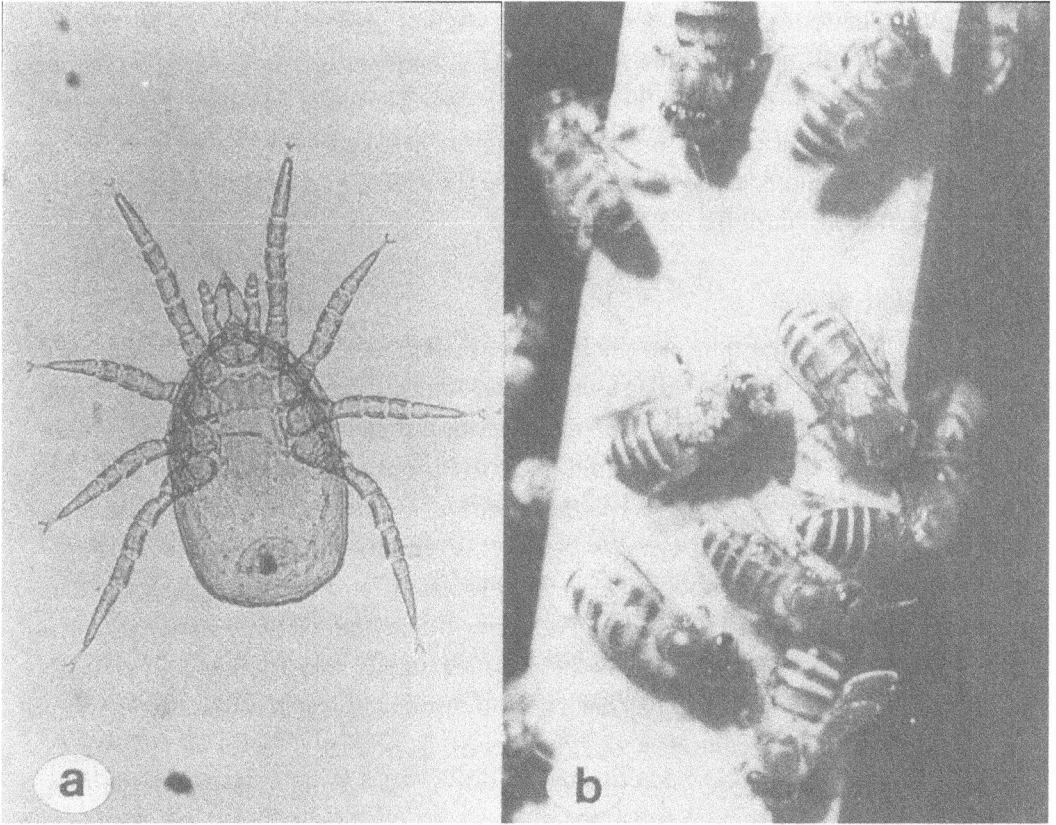

Figure 12.3(a) An undescribed species of *Neocyphalaelaps* about 500 μ.m. long (Steinkraus photo). (b) *Neocypholaelps indica* phoretic on *Apis cerana* (Underwood photo).

honey bee pest. *Neocypholaelaps* and relatives are about half the size of *Tropilaelaps* (500 vs. 1,000 μm long, respectively); they are more weakly sclerotized (thus the dorsal and ventral plates are less obvious) and have far fewer body setae and an oval plate between the hind legs in the female (Figure 12.3a).

Tarsonemus species (family Tarsonemidae) are occasionally encountered as phoretic females on honey bees. One species, *Tarsonemus apis*, has been described as a specific associate of honey bees and other bees (see Lindquist 1968). It has been recovered from pollen stores in hives, and Homann (1933) hypothesized that it is a pollen feeder. However, other species of *Tarsonemus* are fungivorous, and fungus feeding should be investigated for *T. apis*. Given the difficulty of identifying species of *Tarsonemus* accurately, it is likely that many records of *T. apis* actually refer to widespread, unspecialized species of *Tarsonemus* (such as *T. fusarii*) that incidentally invade beehives and feed on fungi. These mites are not known to cause problems in beehives.

A specialized tarsonemid, *Pseudacarapis indoapis*, is known only as phoretic females on worker *Apis cerana* in India (Lindquist 1968, 1986). While Lindquist (1968) suggested that this may be a commensal mite, feeding on pollen that collects on the adult bee, the absence of other life stages of the mite on the bee indicates that its life cycle is carried out elsewhere, probably in the hive or on flowers. The mite is evidently not parasitic, but there is no evidence to indicate whether it feeds on pollen or on fungus.

Parasitic mites

The most serious acarine pests of honey bees are those mites that feed directly on the bee brood (see Chapter 13) and adults (Chapter 14). *Varroa jacobsoni* Oudemans is a large mesostigmatid mite that was originally restricted to *Apis cerana*. In colonies of that species, the mite parasitizes only drone brood and does not significantly harm the colony's health. However, *V. jacobsoni* was able to successfully infest colonies of *Apis mellifera* when European honey bees were brought into contact with Asian honey bees. In colonies of *A. mellifera*, the mites also parasitize worker brood. The mites feed on the blood of the larval and pupal brood, and the adult females feed on the blood of the adult bees, which also serve as phoretic hosts. *V. jacobsoni* has been spread throughout most of the world and is the most serious acarine pest of honey bees. A second species of *Varroa*, *V. underwoodi* has recently been discovered in *A. cerana* colonies in Nepal and Korea. The closely related *Euvarroa sinhai* is so far restricted to *Apis florea* in Asia. It is also a parasite of drone brood and causes little harm to the colony.

Tropilaelaps is an Asian genus of mesostigmatid mites (family Laelapidae) restricted to honey bees. Both described species, *Tropilaelaps clareae* and *T. koenigerum*, were originally restricted to *Apis dorsata* F. When *A. mellifera* came into contact with *A. dorsata*, *T. clareae* switched to it and has since become a devastating pest of honey bees in Asia. The mites feed on the blood of bee larvae and pupae, and the adult females are phoretic on adult bees.

The other specialized parasites belong to the genus *Acarapis* in the family Tarsonemidae. The three *Acarapis* species (*A. externus*, *A. dorsalis*, and *A. woodi*) are restricted to honey bees as hosts (Eckert 1961a, Delfinado-Baker and Baker 1982b); all are tiny mites. Their chelicerae are modified to form sucking stylets that are used to pierce the cuticle of adult bees and suck blood. *A. externus* and *A. dorsalis* occur worldwide and feed externally on bees. *A. externus* occurs principally on the neck of the bee and in pits on the back of the head, while *A. dorsalis* occurs principally in a groove across the top of the bee's thorax. These two species do not appear to affect the bees adversely. The third species, *Acarapis woodi*, occurs within the first thoracic tracheae of honey bees. The mites pierce the lining of the tracheae and feed on blood. Only the adult females leave the tracheae, in order to transfer to other bees. The significance of these mites to the health of a colony has been a matter of considerable debate. *A. woodi* is the subject of Chapter 14.

Other parasites that attack honey bees are incidental associates that usually attack other hosts. Species of the *Pyemotes ventricosus* group (family Pyemotidae), polyphagous parasites of insect larvae, occasionally invade beehives (Homann 1933, Grobov 1975). While early records of *Pyemotes* from beehives refer to *P. ventricosus*, it is now apparent that several very similar species have been included under that name, and the actual species in beehives may be *Pyemotes anobii* (Cross and Moser 1975) or *P. tritici*.

An adult female *Pyemotes* inserts her chelicerae through the cuticle of an insect host and injects a potent salivary venom, which paralyzes the host within two to four hours. The mite then ingests the fluids of its host, and its abdomen swells greatly so that the mite resembles a 1-mm-long light bulb. Within the mother mite's abdomen, her young pass through all immature stages and are born as adults. The males (which make up only about three to five percent of the offspring) seldom leave the mother's abdomen; they mate with their sisters while within their mother. The first adult progeny may emerge within seven days after the mother initially attacks a host, and about 250 offspring are produced by each mother. The newly emerged and mated females then seek another host insect to parasitize. Attacks of *Pyemotes* are usually fatal to the host insect (Cross 1965, Bruce and LeCato 1980).

Pyemotes kill both bee larvae and pupae and can wipe out numerous hives within an apiary (Veitch 1936). However, because the mites are not phoretic on the adult bees, they are not readily spread to distant apiaries, and infestations are usually localized. Consequently, *Pyemotes* are rarely cited as pests in beehives. The usual hosts of these mites are not honey bees but grain-infesting moth larvae, as well as many other insects (Cross et al. 1975). *Pyemotes* are called straw itch mites because their bites cause a severe dermatitis in humans. An infested colony should be destroyed so that the mites do not spread to other hives in an apiary.

The larvae of red velvet mites (cohort Parasitengona of the suborder Prostigmata) are parasites, and some of the insect-infesting species occasionally attach to adult honey bees. They insert their chelicerae through the cuticle and suck blood, swelling to such an extent that they become visible to the naked eye. When they have completed feeding, they detach and molt, and then become free-living predators on arthropods. No species is known to be specific to honey bees, or to any other kind of bee. However, species of the polyphagous genus *Leptus* (family Erythraeidae) commonly do attach to honey bees as well as to many other insect hosts (Fletchmann 1980). These conspicuous parasites may well alarm a beekeeper, but there is no evidence that they do serious harm to their bee hosts.

The mites do not invade the hives. They apparently attach to foraging workers when the bees visit flowers or land on vegetation.

Conclusions

Only a very few of the many species of mites that invade beehives or occur on adult honey bees are important causes of mortality to the bees. A beekeeper should be prepared to recognize *Varroa jacobsoni*, the tracheal mite *Acarapis woodi*, and, in Asia, *Tropilaelaps clareae*. Other mites are unlikely to require control measures other than routine hive hygiene.

Acknowledgments

I thank Carol Henderson and Byron Alexander for helpful comments on this chapter, Benjamin Underwood for Figure 12.3b. Donald Steinkraus photographed the remainder of the mites, and F. Robert Wesley prepared the prints. Mercedes Delfinado-Baker of the Beneficial Insects Laboratory (retired), USDA-ARS, provided a specimen for Figure 12.2 and also reviewed this chapter.

CHAPTER THIRTEEN

—⊶⊷—

Tracheal Mites

William T. Wilson, Jeffrey S. Pettis,
Carol E. Henderson and Roger A. Morse

CHAPTER THIRTEEN

Tracheal Mites

Introduction

Beekeepers worldwide have suffered colony losses from honey bee diseases for as long as anyone can remember. But nothing prepared North American beekeepers for the devastation caused by the tracheal mite (*Acarapis woodi*) when it invaded colonies in Mexico in the early 1980s and was found in the United States in 1984. Since then, North American beekeepers have lost tens of thousands of colonies and millions of dollars as a result of this microscopic parasite. In 1987 when researchers and beekeepers thought that they were making reasonable progress in controlling this mite, the bee industry suffered another serious setback with the discovery of *Varroa jacobsoni* in the United States. The combination of these two parasitic mites has had catastrophic consequences that will undoubtedly continue for many years. With current technology, we are unable to exterminate either species of mite. Thus we must find ways to combat their destructive behavior for as long as we keep honey bees.

Acarapis woodi is referred to as the tracheal mite because it feeds and reproduces in the tracheae (respiratory passages) of adult honey bees (*Apis mellifera*). In the past, colonies infested with this parasite were said to have "acarine disease," but this term is too general to be useful because it can refer to problems caused by any mite species (Delfinado–Baker 1984a).

The tracheal mite was first described by Rennie (1921), who gave it the name *Tarsonemus woodi*; this name was later changed to *Acarapis woodi*. Rennie believed it to be the causative agent of "Isle of Wight disease." This disease, or complex of disease conditions, was first noted around 1905 on the Isle of Wight in the United Kingdom; in succeeding years it was reported throughout Great Britain and continental Europe, sometimes in epidemic proportions (Bailey 1963a, 1981). Bees with disjointed wings were seen crawling outside affected hives, partially paralyzed, and unable to fly. Many colonies were lost, and some reportedly died within a month after disease symptoms were detected. When Rennie found tracheal mites in bees from the affected colonies, he made the association between the mites, paralysis, and the increased colony mortality. However, the etiological agent or physiological condition that caused Isle of Wight disease still remains unknown. Some pathologists maintain that several diseases and diverse factors caused the observed symptoms (Bailey 1963a, 1981). The history of the infestation and the discovery of tracheal mites were reviewed earlier by Phillips (1922, 1923).

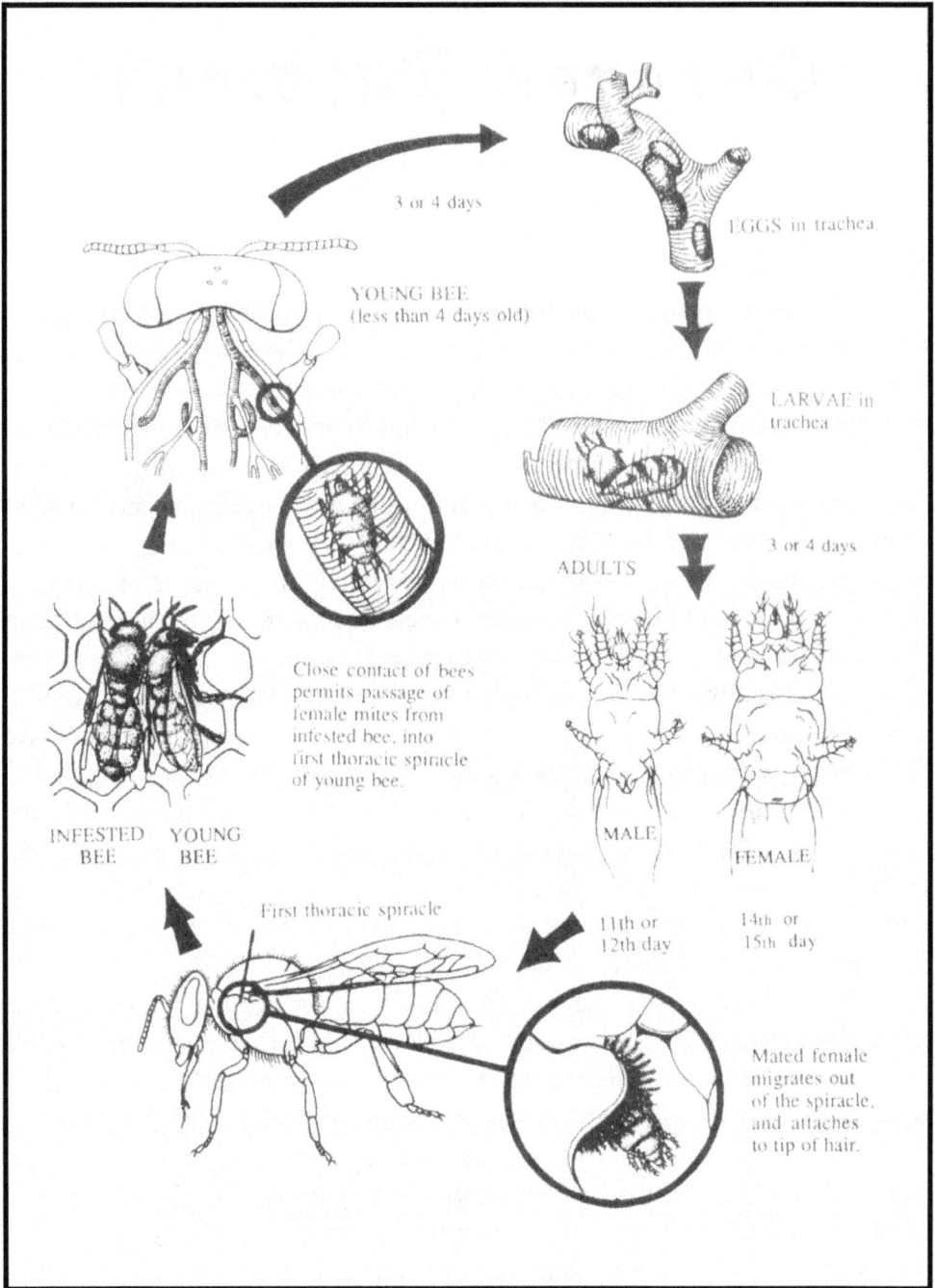

Figure 13.1. Life cycle of the honey bee tracheal mite, *Acarapis woodi*. (From B. Alexander; Sammataro drawing).

Beekeepers in many parts of the United States and Canada have reported unusually high rates of mortality in honey bee colonies during the winters between 1986 and 1994 and attributed them to tracheal mites. Studies by Furgala et al. (1989) and Otis and Scott-Dupree (1992) documented the situation in Minnesota and New York, respectively. These and other losses support the belief that *A. woodi* was indeed the cause of the serious problems experienced by British beekeepers in the years following 1905 (Adam 1987a, Gruszka 1987, Kjer et al. 1989). The effects of *A. woodi* on host honey bees continues to be the subject of research and debate (Bailey 1989). Much of the research, however, has shifted to the applied arena where chemical controls and breeding for resistance to the mite have taken center stage.

Biology

Delfinado-Baker and Baker (1982b) provided a detailed morphological description of the tracheal mite. Female mites measure 143 to 174 μm (length) by 77 to 81 μm (width), whereas males are somewhat smaller: 125 to 136 μm (length) by 60 to 77 μm (width). Males otherwise resemble females except for differences in the external sexual organs.

Adult male and female mites, along with larvae and eggs, are found in the honey bee respiratory system. This system is composed of ten spiracles (external openings), internal branched trunks, and the tubes called tracheae (Snodgrass 1956). All stages of *A. woodi* are found in the large tracheal trunks branching from the first pair of thoracic spiracles (see Figure 13.1). Mites are occasionally found in the air sacks of the head and thorax. Female mites enter the first thoracic spiracle of an adult bee that is generally less than three days of age (Lee 1963, Gary et al. 1989) but older bees may become infested (Smith et al. 1991a), especially in the winter cluster. Dispersing females appear to use cuticular hydrocarbons to discriminate between old and young bees (Phelan et al. 1991). Additionally, they are attracted to the current of expired air coming from the first thoracic spiracle and seem to ignore the second and third spiracles, which are used for inspiration (Hirschfelder and Sachs 1952). Pettis and Wilson (1996) wrote an extensive treatise on the biology of the tracheal mite.

Scientists have wondered why young bees are more frequently infested, while older ones appear somewhat resistant. It was initially believed that the stiffening of the thoracic hairs in aging bees prevented mites from entering the spiracles (Örösi–Pál 1935, Sachs 1952). Even old bees that had been experimentally shaven were not as readily infested as young bees (Lee 1963). A second hypothesis is that the spiracular closing mechanism and grooming behavior become more efficient in older bees, thus more effectively excluding the mites from the tracheae.

In self-grooming, bees comb their pleural hairs with their second pair of legs and may dislodge mites. Lee (1963) tested this hypothesis, however, by removing the second pair of legs from bees exposed to mites and found that young bees were still more readily infested than older ones. Additional grooming by nestmates may also take place. Interestingly, female mites have evolved an efficient means of locating young host bees. Mites that select an older host would produce fewer offspring, thus providing selection pressure for preferential infestation of younger bees.

Once inside the host, the female mite begins to lay eggs within one to two days. Morgenthaler (1931) reported that the females lay from five to seven eggs over a period of three to four days, and that the eggs hatch after three to four days. Pettis (1991) obtained similar results on oviposition: 0.85 eggs were laid per day over the initial 12 days of oviposition.

Recent studies show that females frequently produce eight to 10 offspring, but the idea of as many as 21 offspring per female has been proposed (Pettis 1991, Pettis and Wilson 1995). One generation per host is common, but a second generation from eggs laid by daughter mites is possible in longer-lived bees in the fall and winter (Pettis and Wilson 1989). The mite goes through a six-legged larval stage, followed by a pharate nymphal stage (Lindquist 1986), and finally develops into an adult male in 11 to 12 days or an adult female in 14 to 15 days (Delfinado-Baker and Baker 1982, Royce et al. 1988, Pettis 1991). Unfortunately, the destructive sampling techniques required to study *A. woodi* within an adult bee have limited our understanding of basic mite biology.

Mating occurs shortly after an adult female mite completes ecdysis. Mated females migrate to new host bees by exiting the trachea through the spiracular opening. Adult males apparently do not disperse (Lindquist 1986). There is debate about the sex ratio in *A. woodi*. Delfinado-Baker (1988) reports the female:male ratio of 3:1 or 4:1, yet males sometimes outnumber females. Lozano et al. (1989) reported the ratio at 2.4:1 after examining more than 40,000 adult bees. The observed ratio can vary widely and lead to reporting of a disproportionate number of males because newly mated females migrate out of a trachea whereas males do not (Lindquist 1986).

Honey bee tracheal mites will infest queens, drones, and workers (Morgenthaler 1933, Bailey 1981, Pettis et al. 1989). Drones have larger tracheal trunks, and there is evidence that they are preferentially infested over workers (Royce and Rossignol 1991, Dawicke 1991). Drifting drones may be important in dispersing mites between distant colonies. Workers, however, being much more numerous, are of prime importance when examining the effects of tracheal mites on bee colonies. Queens, because of their longevity, may serve as a reservoir for mites, but more than two mite generations within a queen seem unlikely because tracheae become increasingly damaged and brittle from mite feeding.

Attempts to rear *A. woodi* in vitro have thus far been only partially successful, with no complete generations reared (Giordani 1967, W. Bruce, personal communication 1995). Studies on the other two species of *Acarapis* (i.e. *A. dorsalis* and *A. externus*) have shed some light on reproductive patterns, but these two species live externally on bees and may differ from the internal *A. woodi* (Delfinado-Baker and Baker 1982, Lindquist 1986, Ibay and Burgett 1989).

Intracolony dispersal

Mated adult female tracheal mites disperse from older host bees to young bees (less than three days of age) (Lee 1963, Gary et al. 1989). This transfer, which is an essential component in the life cycle of *A. woodi*, is the only period when

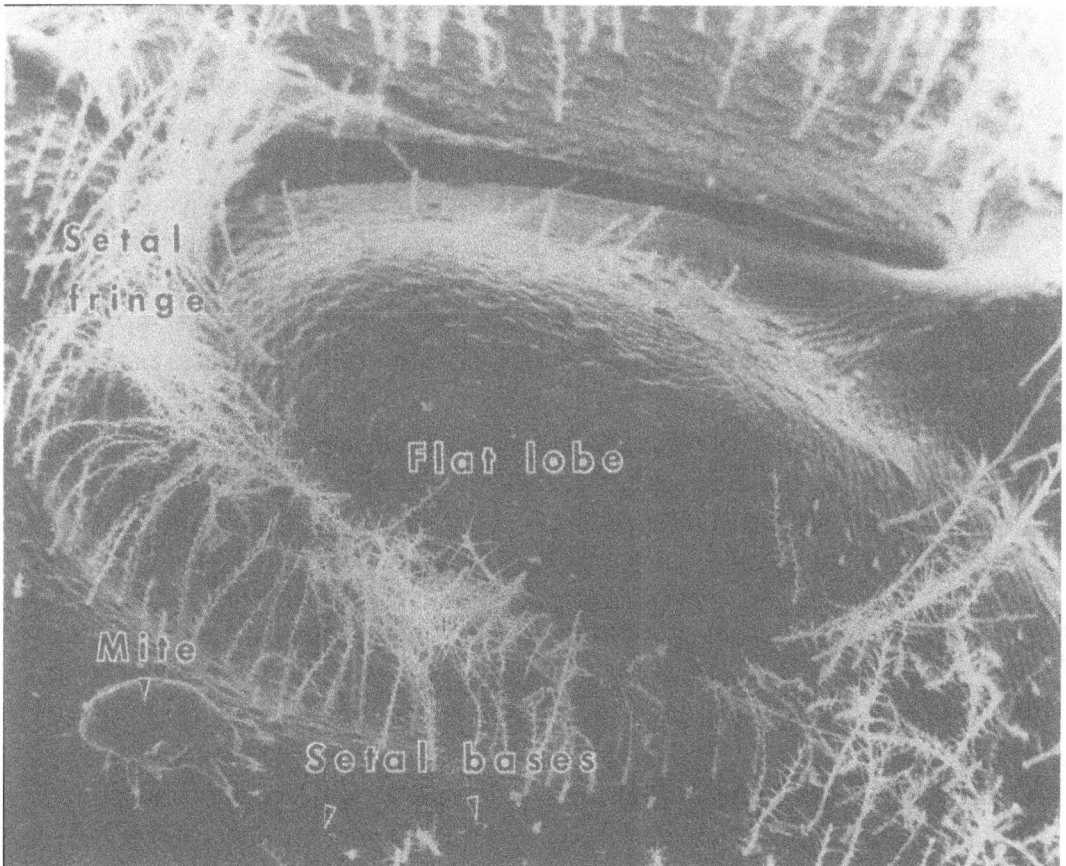

Figure 13.2. Scanning electron micrograph of the first spiracle of an adult honey bee, concealed under the flat lobe with its dense fringe of hairs. These hairs, and the lobe, present no obstacle to mites entering the spiracle. (Styer photo).

mites are known to be outside of the tracheal system. Female mites transfer to a new host by climbing to the tip of a seta on the host bee and then using the forelegs to grasp the seta of a passing bee (Hirschfelder and Sachs 1952). *A. woodi* females and eggs have been found frequently at the base of the wings of honey bees, especially during winter (Royce et al. 1988). The relatively short life span of the host (less than 30 days in spring) has probably been one selective pressure shaping the timing of parasite dispersal from the host.

By examining the rate of mite dispersal between host bees, researchers have determined some factors that influence the dispersal process. Bailey (1958) demonstrated that rising temperatures resulted in increased numbers of mite transfers among bees held in an incubator. Pettis et al. (1989) found that mites continued to disperse across a single screen barrier to young bees. Presumably, close contact occurs during trophallaxis between the older infested and younger confined bees as they feed venter to venter through the screen wire. Hirschfelder and Sachs (1952) observed that mites do not readily leave bees to go onto wax foundation but rather transfer directly to another bee. The same authors believe that although "under natural conditions inside the hive, any mite which happens to get onto the comb surface may reach a new host," this is not a common form of transmission. Hirschfelder and Sachs (1952) examined mite responses to puffs of air, vibrations, and other stimuli that might serve as cues to host location. Intermittent air puffs and vibration were both positive stimuli to the mites. The same authors believed that the vibration would attract the mites to the wing bases; this has been borne out by Royce et al. (1988), who showed that mites move first to the wing base of a new host and then into the tracheal system. Gary and Page (1987) and Page and Gary (1990) demonstrated phenotypic and genotypic variation in susceptibility of young bees to mite infestation. Recent work has shown that dispersing mites use cuticular hydrocarbons as important cues in host location (Phelan et al. 1991). Mites in a laboratory bioassay responded positively to extracts of young bee cuticle as opposed to controls and old bee extract (Smith 1990).

Only female tracheal mites move outside the host and begin dispersing from bees that are approximately 13 days old (Pettis 1991, Pettis and Wilson 1995). No cues have been identified that initiate dispersal behavior from a bee; perhaps mating may serve as a cue. On the colony level, a diurnal cycle in mite dispersal has been demonstrated. Approximately 85 percent of all mite transfers to young bees occur at night (Pettis et al. 1992). A reduction in overall host activity was associated with the increased nocturnal dispersal of mites (Pettis et al. 1992). From these findings, the authors proposed that mites may detect movement of the host and wait to initiate dispersal until the host is motionless.

Transfer is perhaps a vulnerable period in a mite's life cycle, as it is the

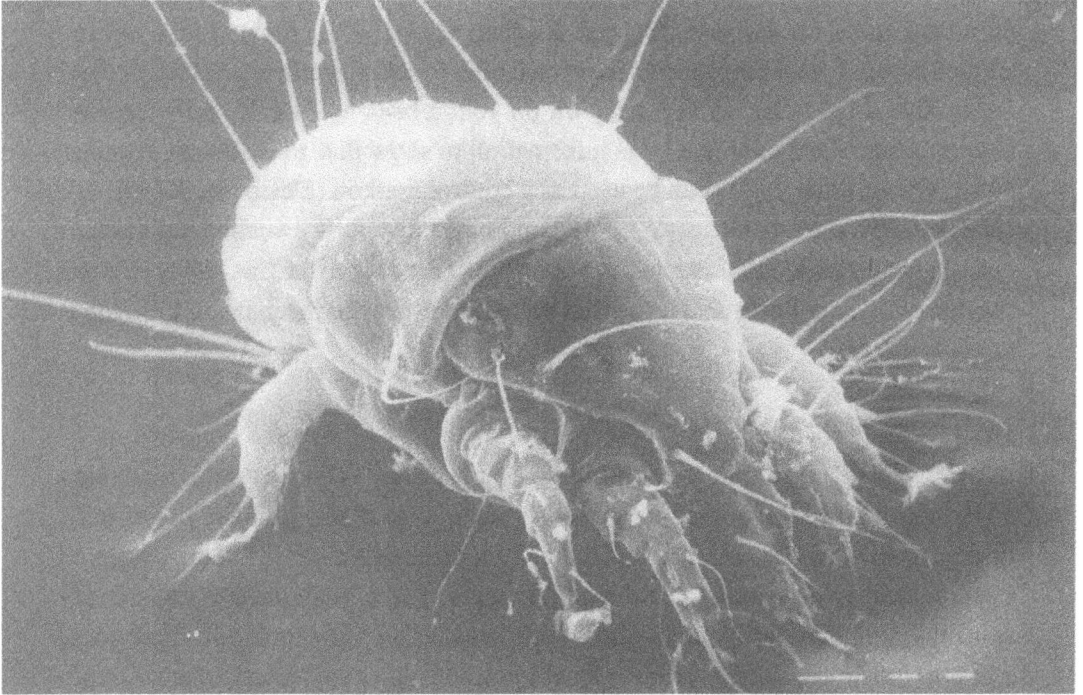

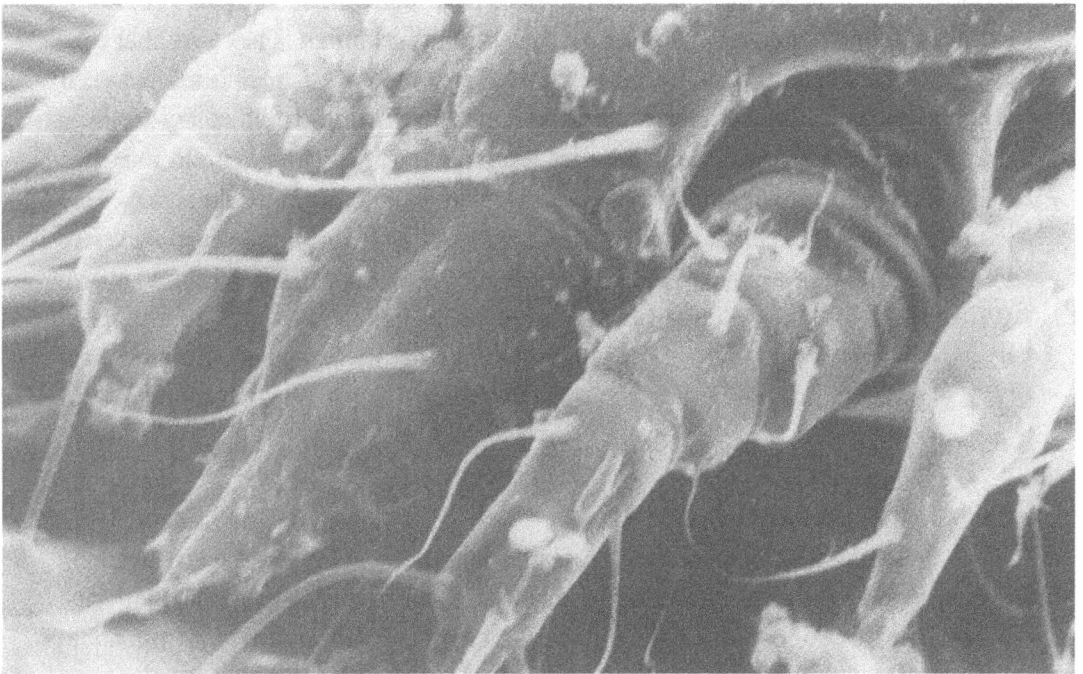

Figures 13.3, 13.4. Scanning electron micrographs of adult tracheal mites. They have no eyes, but many sensory hairs for determining position and location. Bottom: close-up of mouth parts. (Styer photos).

only time mites are outside the bees. Eischen et al. (1986) showed that amitraz applied topically to young queens prevented mite transfer. Smith et al. (1991b) have shown that a light film of vegetable oil on young bees will inhibit mite transfer. Several researchers have used this information to show that an oil-based extender patty (Wilson et al. 1971) can be used as a control method (Delaplane 1992b, Liu and Nasr 1993, Sammataro et al. 1994). Grooming behavior by worker bees has also been shown to limit mite dispersal to young bees (Pettis and Pankiw 1994). Grooming behavior is heritable and is the first potential resistance mechanism to be identified.

Intercolony dispersal

The tracheal mite is a true obligate parasite and cannot exist for more than a few hours off of the adult honey bee. As such, it must rely on host mobility for dispersal between bee colonies. In nature, feral colonies are more widely dispersed than those in apiaries maintained by beekeepers. Three methods are probably responsible for intercolony movement of the parasite: robbing, drifting of individual workers and drones, and swarming. Robbing could play a role in mite dispersal, but the biological and behavioral mechanisms are unknown. Atwal and Dhaliwal (1969) reported that *A. woodi* could be transmitted from parasitized *A. indica (cerana)* colonies to uninfested *A. mellifera* colonies through robbing. They said that robbing was the single most important factor in the spread of tracheal mites between colonies in India. Infested drones and workers, by drifting, could easily serve as vectors to move mites to new colonies. Perhaps mite parasitism could alter bee behavior such that bees drift more readily, thus enhancing mite dispersal. This hypothesis has not been demonstrated experimentally. Swarming bees move mites to new areas, as the parasite is harbored inside infested workers within a swarm (Wilson et al. 1988a). Although swarming certainly disperses tracheal mites to new areas, it has also been proposed as a natural regulatory mechanism to limit tracheal mite population growth within a colony (Royce et al. 1991, see natural resistance section).

Seasonality of infestations

Bailey (1981), reporting on research done in Great Britain, states that among overwintered bees, mite-infested individuals die considerably sooner than uninfested bees of the same age. "The difference becomes statistically significant from about March onward," but the difference in life span between infested and uninfested bees is only slight in summer. The prevalence of mites in individual colonies can vary widely from month to month but generally increases in the late summer and fall and is lowest in late spring and summer (Killion and Lindenfelser 1988, Otis et al.

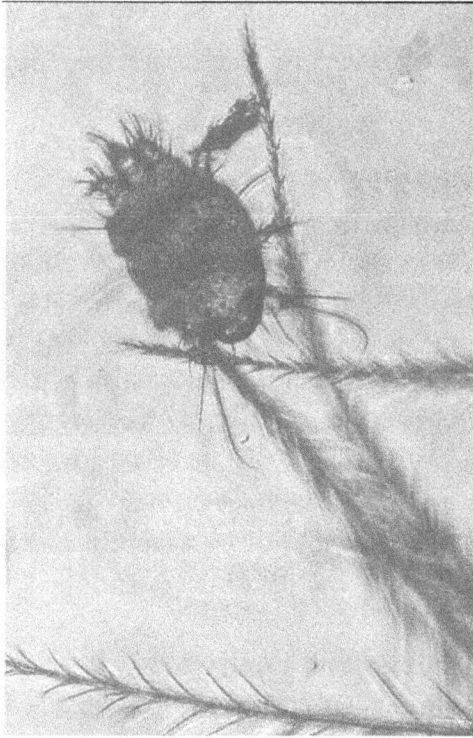

Figure 13.5. Mite bee-to-bee dispersal occurs when a mated female exits a tracheal tube and crawls to the tip of a seta. There she waits for an appropriately aged bee (one to four days old) to come in contact. She then transfers to her new host and seeks a spiracle opening to begin the process of reproduction. (Pettis photo).

1987). Often the spring decline in mite prevalence is associated with a nectar flow and the "dilution factor" caused by the rapidly expanding population of young bees, many of which are uninfested.

Distribution

Acarapis woodi is known to be present on every major continent except Australia, yet its distribution is apparently somewhat patchy (Bradbear 1988, Matheson 1993). Australia, New Zealand, the Hawaiian Islands, and Vancouver Island (Canada) are known sources of certified tracheal mite-free bee stocks. Strict laws governing the importation of live bees into Australia and New Zealand have kept these areas free of tracheal mites. Once mites become established within an area, detection of low mite levels is economically not feasible. Consequently, the difficulty of detecting low mite levels has allowed tracheal mites to move into many new areas despite regulatory efforts. Any transport of live bees, including queens, from an area known to have tracheal mites will carry with it some risk of spreading *A. woodi,* regardless of the level of sampling. Because of the mite's small size, most beekeepers find it very difficult to monitor the level of tracheal mites within their colonies. This can lead to unintentional spreading of the mite or chemical treatment at inappropriate times.

Distribution among *Apis* species

The most frequently recorded host for *A. woodi* is the European honey bee, *A. mellifera*, but two other host species are also reported, the Asian honey bee, *Apis cerana* (Milne 1957, Atwal 1971, Delfinado-Baker and Baker 1982), and the giant honey bee, *Apis dorsata* (Delfinado-Baker 1988a). Reports from India indicate that tracheal mites have caused serious losses in *A. cerana* colonies similar to the losses in European colonies in North America (Singh 1957, Atwal and Sharma 1970, Adlakha 1976, Shah 1987). A more complete search of potential hosts within the genus *Apis* could help determine the host specificity of this parasite and help resolve the issue of how old the host-parasite relationship is between tracheal mites and *A. mellifera* (DeJong et al. 1982b).

Distribution in Africanized honey bees

In 1971, Nascimento (1971) reported finding *A. woodi* in honey bees in Brazil. This was 15 years after the 1956 introduction of the African bee into that country (Kerr 1957), so we can be rather sure that the parasitized bees were Africanized. Lehnert et al. (1974) and Wiese (1977) firmly established that Africanized honey bees (AHB) in Brazil harbored the tracheal mite. Flechtmann (1977) believed that Africanized bees in feral colonies were responsible for the spread of the mite in Brazil. *Acarapis woodi* was discovered in Colombia in 1979 (Menapace and Wilson 1980) before the AHB arrived, but Roubik and Reyes (1984) believed that the mite was spread later within Colombia by feral AHB colonies. The transport of *A. woodi* from Colombia into Panama apparently did not occur, however, since Roubik and Reyes (1984) conducted an extensive survey of the AHB population and *A. woodi* was not found. They suggested that the life span of the AHB was too short during periods of intense foraging for the mite to complete its life cycle, and therefore mite populations would not build up. However, this concept of the AHB not being a suitable host for either *A. woodi* or *Varroa jacobsoni* is not supported by the current buildup of populations of both parasites in Mexico (Wilson et al. 1995). The feral Africanized bees in South Texas also harbor both species of mite (W. Rubink, personal communication 1995).

Diagnosis

There are at least three species of *Acarapis* mites (see Chapter 12). They are clearly distinct from other honey bee mites but closely resemble each other.

Delfinado-Baker and Baker (1982b) describe the *Acarapis* species as "extremely difficult to detect and identify owing to their small size and the similarity between species." In most cases identification is made by observing where the mites are found on the bee's body. *Acarapis woodi* lives only within the tracheae, whereas the others are found exclusively on the bee's external surface: *A. dorsalis* in the dorsal groove on the thorax, and *A. externus* on the underside of the back of a bee's head. Though the location of mites on a bee is usually a good indication of the species, it is not infallible (Eckert 1961a). A redescription with detailed illustrations of the mites is given by Delfinado-Baker and Baker (1982) and Delfinado-Baker (1984b).

Shimanuki and Knox (1991) described several methods of diagnosis for *A.woodi*. Most involve a physical search for and examination of the first pair of tracheae. Apiculturists generally recommend that bees be collected from the honey supers or entrance, but Calderone and Shimanuki (1992) caution that not all areas of the hive are representative of the actual mite prevalence. A common dissection method involves separating the thorax from the rest of the body and then removing the prothorax. The remaining part of the thorax is placed in 7.5 percent potassium hydroxide at 37°C for 24 hours to dissolve muscle, after which the exposed tracheae are examined under a microscope. Peng and Nasr (1985) stained the cleared tubes to enhance mite detection. Mites preferentially infest the area between the spiracular opening and the first bifurcation of the first pair of tracheal tubes, with less than 5 percent of mites moving into the air sacs and head (Robinson et al. 1986). Tamasko et al. (1993) describe a sequential sampling method that can be used by apiculturists to determine the amount of sampling needed to detect a particular mite infestation level.

A more detailed method of dissection can be employed to examine live mites, count life stages, or determine the extent of an infestation (mite load) within individual bees. The head of a live bee is removed, and the fresh tracheal tubes extracted with jeweler's forceps. Tracheae are dissected under magnification (ca. 40X) with fine needles, and the number of live or dead adult mites and immature stages can be determined (Eischen et al. 1987, Cox et al. 1989). An alternative method involves immobilizing a bee and, using fine forceps, pulling off the flat lobe covering the first thoracic spiracle. This extracts internal tissues, including the main tracheal trunk, which can then be examined for mites (Smith et al. 1987). The criterion for determining if a mite is alive is to observe movement of a body appendage (e.g., a leg). This may first require gentle prodding with a needle. Recently, Liu (1995b) described a staining technique to determine whether a mite is alive. During the late 1980s, bee regulatory agencies within many U.S. states provided free diagnostic service to beekeepers so they would know if their bees were infested. Unfortunately, most state and private bee diagnostic services are no longer available because of a lack of funding.

Alternative methods of diagnosis

Because the methods described above are slow and cumbersome, several people have sought alternatives. Camazine (1985) found he could break honey bee thoraxes apart in water with a household blender and that the air-filled tracheae would float to the surface. Three research teams (Fichter et al. 1986, Ragsdale and Furgala 1986, Grant et al. 1993) have used enzyme-linked immunosorbent assays (ELISA) to detect *A. woodi*. The first and third groups used mice, while the second used rabbits for antibody production. The second team reported that the assay detects *A. woodi* at any life stage and does not respond "to equal numbers of the other two *Acarapis* species." Thus, the test is species specific since no cross reaction took place. The third team reports that bulk bee samples can be analyzed with mite levels detectable down to 5 percent of infested bees (Grant et al. 1993).

Effects on individual bees

There has been much discussion as to what damage *A. woodi* causes in

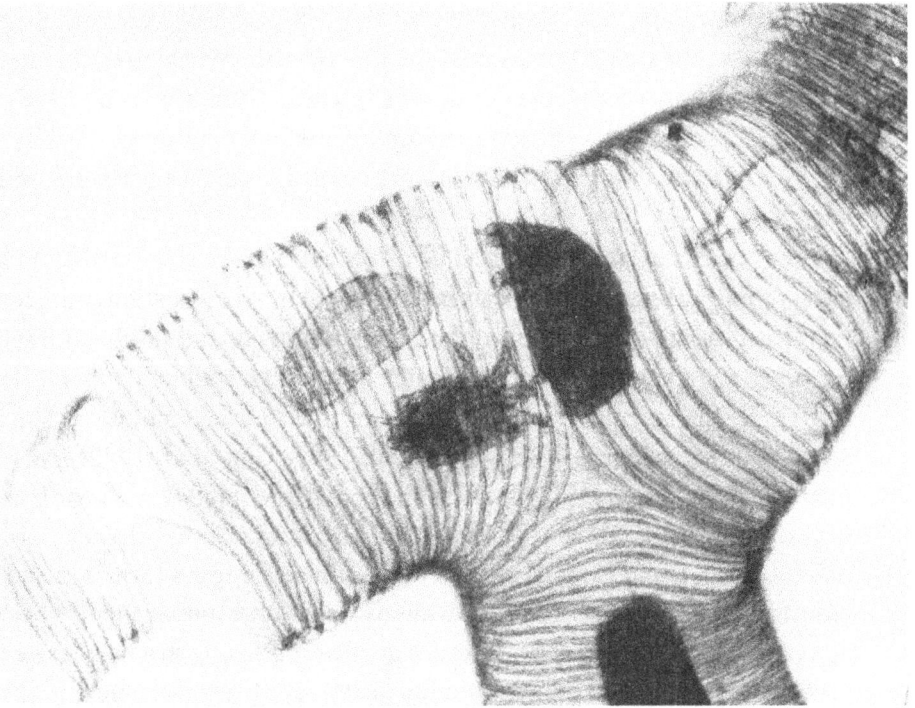

Figure 13.6. Tracheal tube removed showing egg, larva, and adult tracheal mite. (USDA, Wilson photo).

honey bees. When observing a tracheal trunk that is literally packed with various stages of mites, it is hard to believe that these organisms do not cause at least mechanical obstruction of the air passageway, but there are no data supporting this assumption. It has been assumed that the flight ability of bees would be impaired because of the reduced oxygen supply to flight muscles. White (1921) found that bees could not fly if the anterior spiracles were plugged with wax. Two studies have shown that tracheal mites reduce the longevity of adult bees (Bailey 1958, 1961, Maki et al. 1988). In addition, data from Royce and Rossignol (1990a) indicated that infested bees should die sooner than uninfested bees. Komeili and Ambrose (1990) showed that infested colonies had higher bee mortality and lower production. In contrast, studies on foraging behavior have shown no significant differences between lightly infested and non-infested bees in the number of foraging trips, the frequency of trips, round-trip times, the frequency of pollen collection, or the time between foraging trips (Gary and Page 1989). Likewise, the nectar loads of infested foragers and controls were not statistically significantly different in size (Gary and Page 1989). The foraging studies warrant repeating with more heavily infested worker bees.

Acarapis woodi feeds by piercing the bee's tracheal wall with its stylets, placing its mouth opening near the puncture, and sucking up hemolymph. Congo

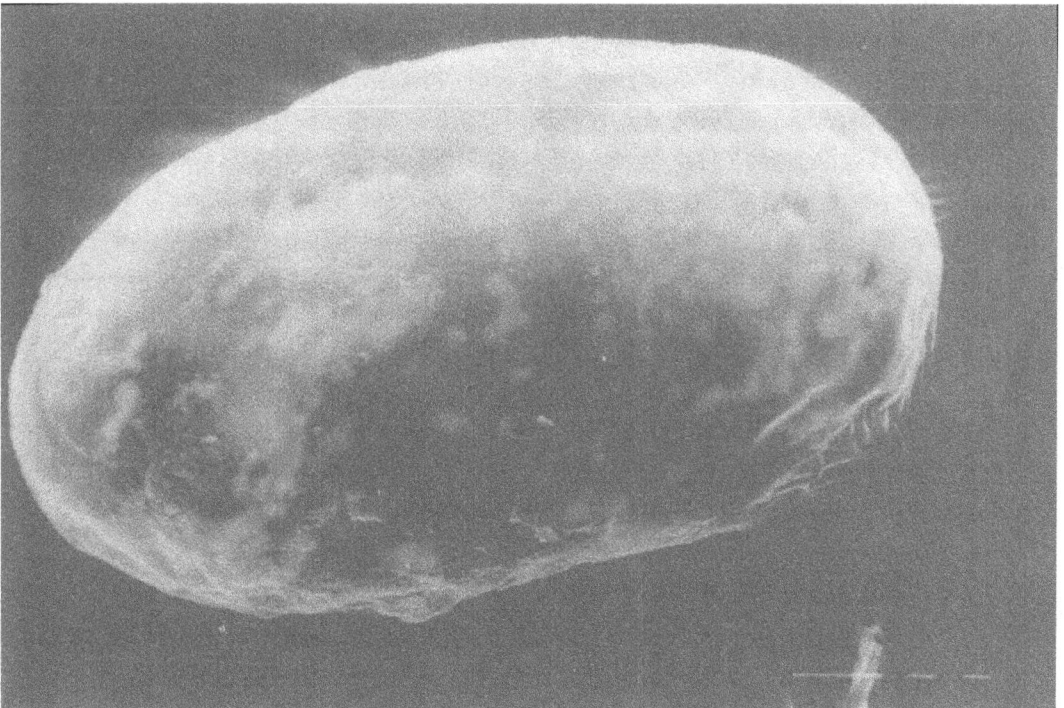

Figure 13.7. Close-up of tracheal mite egg. (Styer photo).

Red stain solution injected into host bees was subsequently found in mites in the tracheae of those bees (Örösi-Pál 1934). Larval mites also feed in the same manner as adults but are less mobile within the tracheae (Hirschfelder and Sachs 1952).

Infested honey bee tracheae will often show brown blotches with brown scabs or even appear entirely blackened with dark bands, depending on the number of mites infesting the host (Delfinado-Baker 1984b, Liu et al. 1989a). In comparison, the healthy tracheae are clear, translucent, showing no signs of stains. Liu et al. (1989a) showed that a film forms on the inside walls of tracheal tubes as mites feed, but the nature and composition of this film are not known. Infested bees seem to have higher bacterial counts than uninfested bees, but this does not necessarily mean that these bees are more susceptible to disease (Fekl 1956). Bailey (1965b) states that spraying infested bees with suspensions of pathogenic bacteria did not increase the susceptibility to disease, but more data would be necessary to affirm this conclusively. Mite feeding may also affect the surrounding flight muscles or hypopharyngeal glands (Liu et al. 1989b, Liu 1990b). Komeili and Ambrose (1991) used an electron microscope to study the tracheae and flight muscles of non-infested, infested, and crawling honey bees. They found that vertical flight muscles had degenerated in crawling bees.

Colony-level effects

Perhaps it is more realistic to discuss the pathogenicity to colonies as a whole rather than to individual bees. The economic impact of the mite is still not resolved (Morgenthaler 1929, Bailey 1964, 1985, Giordani 1967, Adam 1968, 1985). Bailey (1958) demonstrated decreased longevity for infested bees, and increased susceptibility to winter loss for colonies that showed infestation levels greater than 30 percent. A significant loss in bee populations has been correlated with increased mite populations (Eischen et al. 1989). Additionally, a reduction in honey production was recorded in colonies heavily infested with *A. woodi* (Eischen et al. 1989). Studies using mite-infested package bees (King 1987) and studies on winter survival of infested and uninfested colonies indicate that tracheal mites are a problem at the colony level (Eischen 1987, Furgala et al. 1989, Otis and Scott-Dupree 1992). Research in this area continues. Wiese (1977) said that *A. woodi* presented the greatest threat to bee colonies in some parts of Brazil.

Signs and symptoms

When the tracheal mite invaded beekeeping operations, many owners saw drastic increases in winter mortality and crawling worker bees in the spring. These symptoms of tracheal mite parasitism were in sharp contrast with anything they had previously experienced in bee management. Those who checked clusters in midwinter often found the bee population split into several smaller clusters with little chance for survival. Abnormal winter clusters had been reported earlier in infested colonies in places such as India (Shah 1987). U.S. beekeepers also reported that in early spring, colonies had large quantities of stored honey, but all of the bees were dead or the surviving population was reduced to a "handful" of workers and a queen. When bees were checked more closely, a large percentage often had the "K-wing" condition where the two wings on one side of the thorax had become unattached. Frequently in early spring, there were not only large numbers of crawling bees but also fecal spots at the colony entrance. Bees that were unable to fly often crawled up grass stems and formed small clusters. We believe that the K-wing and the inability to fly result from damage to major flight muscles, especially from the lack of oxygen. Crawling bees from heavily infested colonies in Mexico often had distended abdomens (Eischen, personal communication 1985). Liu et al. (1989a, b) showed that tracheal mite infestations caused degenerative damage to the hypopharyngeal glands and resulted in brittle and abnormal tracheal walls in adult bees. Komeili and Ambrose (1991) support the belief that tracheal mites contribute to the occurrence of crawling bees. Not all authors agree, however, that severe dwindling and crawling bees are caused by tracheal mite infestations (Bailey and Ball 1991).

Canadian scientists demonstrated that colonies in New York with a high proportion of mite-infested bees became weakened and reared less brood (Scott–Dupree and Otis 1988). This resulted in the slow buildup of colonies in spring and less honey stored during summer.

Thoenes and Buchmann (1992) reported an interesting phenomenon in which a mite-infested colony absconded from a hive over a five-hour period leaving honey, sealed brood, and dead workers behind. Royce et al. (1991) have proposed that swarming reduces mite density within a colony and that modern methods of curbing swarming in movable-frame hives may increase the spread and impact of some parasites (Royce and Rossignol 1990b).

It is true that no one symptom or sign characterizes this disease and that their absence does not always imply freedom from the mite (Shimanuki et al. 1992). However, beekeepers soon found that if all or part of these symptoms were seen in many colonies at the same time, the bees were invariably infested with *A. woodi*. Positive diagnosis can only be made by microscopic examination of the tracheae (Bailey and Ball 1991).

History of the North American infestation

The first confirmation of *A. woodi* in North America was from bees collected in an apiary about 20 miles southeast of Guadalajara, Mexico, in 1980 (Wilson and Nunamaker 1982). The mite spread rapidly across Mexico and within five years was found in nearly all states (Zozaya et al. 1982, Guzman-N. and Zozaya-R. 1984, Wilson and Nunamaker 1985). By 1988, all states including Baja, California, were infested (Clark et al. 1986, W. Wilson, unpublished data 1987, Eischen et al. 1990). Tracheal mites were first found in the United States in honey bees collected on July 3, 1984, near Weslaco, Texas (Delfinado-Baker 1984b). The chronology of what followed is reported in 21 APHIS reports (numbered 1 through 22, but report number 18 was not issued) dated July 13 to November 2, 1984 (Henderson and Morse 1990). Canada closed its eastern border to bee imports from the United States following the·1984 discovery of tracheal mites. In 1987, the entire Canadian border was closed because of the tracheal mite and the discovery of *Varroa jacobsoni* in the United States (Anonymous 1987). This action slowed the spread of tracheal mites into and across Canada but did not stop them, and tracheal mites have become a widespread problem there. For a more detailed description of the initial regulatory efforts in the United States and geographical movement of the tracheal mite, see the following: Anonymous (1984a,b,c,d,e,f, 1985) and the second edition of *Honey Bee Pests, Predators, and Diseases* by Morse and Nowogrodzki (1990).

In North America, a pattern of increasing colony loss appeared shortly after tracheal mites were discovered in a new area (Henderson and Morse 1990). When bee populations were first exposed to the mite, winter losses in some apiaries were as high as 75 to 100 percent of colonies (H. Shimanuki, personal communication 1987). During the second winter, colony loss was usually down to 50 percent, and by the third winter, the loss was often 25 percent or less (M. Ellis, personal communication 1989). Although Northern winter losses were most severe during the late 1980s, every year still brings reports of heavy loss (greater than 75 percent) of colonies in some Northern apiaries. Most commercial beekeepers in the United States and Canada could not afford to stay in business with such losses, and therefore they treat with miticides. A compounding problem is that many of the queens used in Northern colonies are raised in the subtropic regions of the Southern United States where selection pressures from the mite are not as severe.

Resistant (tolerant) stocks of bees

Great variability among U.S. colonies in their susceptibility to *A. woodi* has been demonstrated (Gary and Page 1987, Page and Gary 1990). Such variability suggests that this mite species has not previously been in contact with North American honey bee stocks. Indeed, the North American bee population appears to have been isolated from tracheal mites based on numerous surveys that were negative for tracheal mites. In the 1980s, approximately 200,000 bees from 4,000 apiaries in the United States and Canada were examined in a search conducted by agencies within the U.S. Department of Agriculture: APHIS (Animal and Plant Health Inspection Service) and the ARS Beneficial Insects Laboratory in Beltsville, Maryland. No *A. woodi* were found in these surveys (Shimanuki et al. 1983, Shimanuki and Knox 1989). Previous North American surveys also proved negative (Eckert 1961a).

Numerous bee breeding programs have been launched to develop resistant stock. In England, Brother Adam started selecting for bee stocks that had heritable resistance (Adam 1968, 1985, 1987b), and over a period of many years, he developed the Buckfast line, which is currently available in many countries. Buckfast queens were first introduced into the United States and Canada in 1961 (Smith 1964). One queen breeder has had exclusive rights to the propagation and distribution of Buckfast queens in North America since the late 1960s (Morse 1990b). Using only North American stock, Page and Gary (1990) demonstrated successful selection for tracheal mite resistance in two generations.

Dr. R. Morse of Cornell University began a selection program by importing 26 mother queens from six different beekeepers in England, Scotland, and Wales (Morse 1990a). However, a bioassay test that compared the relative susceptibility of the British stocks to commercial stocks available in the United States indicated the British bees were no more resistant than the U.S. bees (Gary et al. 1990). An Arizona beekeeper, L. Hines, suffered heavy colony losses from *A. woodi* during the winters of 1988-'90. Using the surviving colonies, the beekeeper and Dr. G. Waller developed and marketed a stock of bees that appears to have considerable resistance to mites (Waller and Hines 1990, Hines 1992, Loper et al. 1992). A former California queen breeder, Steve Taber (1990), made selections and developed a stock of bees with apparent resistance. Mite resistance in Buckfast bees was compared with a commercial stock from California in field tests in the state of Washington (Milne et al. 1991) and in Ontario with Canadian standard stock (Lin et al. 1992, Nasr et al. 1992). The Buckfast bees proved to be superior. The USDA ARS Bee Research laboratory in Baton Rouge, Louisiana, imported queens in 1989 from Yugoslavia onto an isolation island off the coast of Louisiana (Rinderer et al. 1993). After the bees were determined to be safe (pathologically and with no African genes) progeny queens were released to three U.S. queen breeders for propagation and sale of queens (Anonymous 1993c). Yugoslavian queens are reported to have some resistance to both the varroa and tracheal mites (Guzman et al. 1993, Rinderer et al. 1993). Clark et al. (1990) selected a large number of queens from beekeepers in Western Canada to breed for resistance to *A. woodi* and reported considerable success. Unfortunately, some of these programs lack funding for long-term selection of mite-resistant stocks.

Many of these stocks are exhibiting tolerance to tracheal mite infestations, but it remains to be demonstrated whether the stocks are commercially useful. However, colonies may still need periodic chemical treatments to maintain low mite populations. Unfortunately, these selection programs are proceeding without knowledge of the resistance mechanisms involved. Recently, grooming behavior by host bees has been proposed as a resistance mechanism within colonies (Pettis and Pankiw 1994). Earlier Peng (1988) reported a series of behavior responses including grooming as important resistance mechanisms used by *Apis cerana* to control *V. jacobsoni.*

Royce et al. (1991) have proposed swarming as a means of limiting mite population growth at the colony level. They hypothesize that the break in the brood cycle associated with swarming will act to reduce mite population growth. If this is true, then unmanaged rustic and feral colonies should have lower mite levels than managed colonies, which do not swarm as often. Two studies found that indeed rustic and feral colonies have lower mite prevalence than managed colonies in the same area (Pettis et al. 1987, Royce et al. 1993). Additional support for the swarm hypothesis is gained indirectly from studies that demonstrate that requeening or making colony divisions will result in lower mite levels (Eischen et al. 1988).

Bailey and Perry (1982) proposed that the reduced number of mite-infested colonies in England in recent years is the result of a decrease in the numbers of colonies kept in the British Isles. Fewer colonies reduces competition and thus improves forage for bee populations in the remaining colonies. Older, mite-infested bees would then forage in the field more often and consequently have less contact in the hive with young bees, leading to reduced mite transmission. Reduced bee contact time is the basis for Bailey and Perry's hypothesis. A test of this hypothesis found that approximately 85 percent of mite dispersal occurred at night when older foragers are in the colony (Pettis et al. 1992). Additionally, when foragers were held within the colony during the day, no increase in mite dispersal was observed. This evidence contradicts Bailey's reduced-contact-time hypothesis.

Magnitude of the problem and economic loss

Prior to the discovery of *A. woodi* in Mexico in 1980 (Wilson and Nunamaker 1982), there was little concern among beekeepers over the biological impact that this parasite might have on American beekeeping. The leading British expert said that the mite was not a serious problem in England (Bailey 1963a) and that it would not be a problem in the United States (Bailey 1986). In the early 1980s, however, some Mexican beekeepers (D. Cardoso, personal communication 1984) were expressing concern over an increase in colony losses and weak bee populations in tracheal mite-infested apiaries. There were nationwide surveys on mite distribution in Mexico (Zozaya-R. et al. 1982, Wilson and Nunamaker 1985) and a study to find suitable miticides (Guzman-N. and Zozaya–R. 1984). With the exception of a USDA ARS-sponsored study on miticides in Italy in the early 1960s, U.S.-led mite control studies did not start until late 1984 (Anonymous 1984e).

When tracheal mites were found in the United States in 1984 (Delfinado-Baker 1984b), state and federal agencies attempted to contain the spread of the mite through quarantines, but were ineffective. As the mite spread across the United States (Delfinado-Baker 1985), the customary winter loss of 5 to 10 percent of colonies per year in the Northern states quickly increased to 50 percent or greater colony loss (Furgala et al. 1989) and in some apiaries reached 100 percent (Wilson and Collins 1994). Within a year of mites being detected in a new area, there was always a drastic increase in bee mortality, and surviving colonies were often weak. In India, after 20 years of infestation, winter losses from acariosis still exceed 50 percent in untreated colonies (Shah 1987).

Before 1984, the queen and package bee industry was rather stable and profitable in the warmer, subtropical parts of the United States. Finding the tracheal mite in South Texas in July 1984 and three months later in Florida (Anonymous 1984e), however, caused a serious disruption in the placement of orders for both queens and packages for the spring of 1985. The highly respected Florida queen industry basically vanished in 1985 from the lack of customers. Beekeepers did not want to purchase queens from states where colonies were infested. The purchase

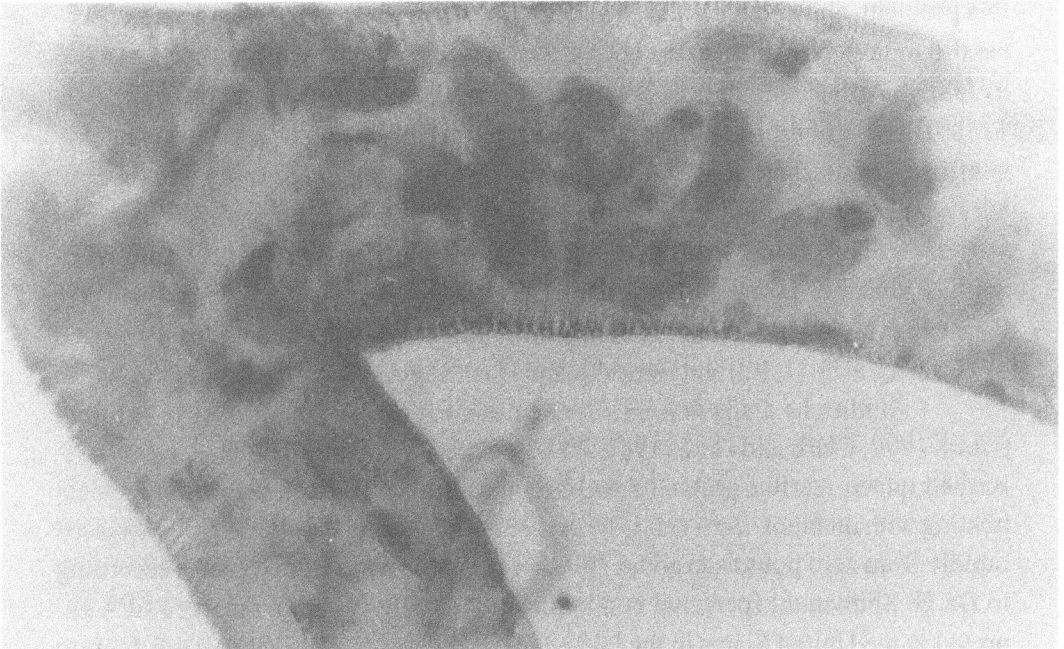

Figure 13.8. Tracheal tube removed from honey bee showing serious infestation of adults, larvae, and eggs, plus scarring of the tracheae from feeding. (USDA photo).

of uninfested California queens increased dramatically. Regrettably, the monetary loss to Florida queen breeders was in the millions of dollars, and many family businesses were forced to close (Wilson and Collins 1994).

In the late 1980s, tens of thousands of *A. woodi*-infested colonies died each year. When *V. jacobsoni* was discovered in the United States in 1987 (Anonymous 1987a), the losses increased because bees (and beekeepers) found it difficult to deal with the stress of two parasites. In the 1990s, thousands of mite-infested colonies are still dying each year, and this loss will continue until more effective chemical treatments are found or genetically resistant stocks of honey bees become readily available and widely used. Delaplane (1992a) estimates an annual loss of more than $5 million in bees and bee products for the state of Georgia alone when colonies are left untreated. Bee mortality across the United States seriously decreases honey production and results in fewer colonies that are suitable for crop pollination. Tracheal mite infestations in United States have cost the beekeeping industry millions of dollars in recent years. The addition of the varroa mite has greatly increased this cost.

Chemical control

Following heavy winter losses in tracheal mite-infested colonies during 1986-87, many U.S. beekeepers suddenly became interested in finding a solution to this problem, and chemical control seemed to be the easiest answer. The first successful mite-control tests in the United States were conducted in the spring of 1986 by USDA ARS scientists near Weslaco, Texas (Cox et al. 1986, Wilson et al. 1986, 1988b). These findings on the miticidal properties of menthol fumes were quickly confirmed by Herbert et al. (1988). During this same period of time, mite-control tests were being conducted in Northeastern Mexico by Dr. Frank Eischen and associates. They found that amitraz was effective against the tracheal mite when applied topically to queen bees, and Apitol was effective when fed to workers (Eischen et al. 1986, 1987). Additional miticide studies were conducted in Mexico by Garza-Q. et al. (1990) and Garza-Q. and Dustmann (1993).

Studies by Canadian scientists further supported the efficacy of menthol (Clark 1990, Clark and Gates 1991, Nelson et al. 1993), but Clark et al. (1989) described queen rearing problems and brood mortality (Nelson et al. 1993) when treating with menthol. Between 1987 and 1990, beekeeper use of menthol increased rapidly from zero pounds to about 100 tons per year in the United States according to Dr. H. Shimanuki (personal communication 1990). Menthol received EPA approval in the United States in the late 1980s. Beekeepers reported varying degrees of success when using menthol for tracheal mite control mainly caused by a lack of vaporization. Menthol fumes must penetrate a bee's tracheal tubes to be effective, and then only the adult mite is killed. Unfortunately, cool weather in spring and fall reduced the effectiveness of menthol and often resulted in poor control. Vaporization did improve with menthol-grease boards (Wilson et al. 1990). Poor control, combined with the high cost of menthol, however, soon encouraged beekeepers to search for additional miticides such as formic acid.

Fumes of formic acid kill adult tracheal mites, but repeated treatments on a four-day or weekly schedule over a two-week period are needed (Hoppe et al. 1989, Garza-Q. et al. 1990, Wilson and Collins 1993, Wilson et al. 1993) since the life cycle of the mite is about two weeks. Liu and Nasr (1992) reported that similar treatments in Canada controlled adult mites, resulting in fewer mite eggs and immatures. Even with such intensive treatment, some bees still harbor live mites that reinfest other bees within a hive. Some apiculturists believe that 95 percent or more of the mites must be killed during treatments or rapid reinfestation will occur. A level this high is difficult and expensive to achieve. When a colony becomes infested, it will always be infested to some degree because eradication of the mite is not possible with our current technology and available miticides.

Formic acid and other miticides can be toxic to workers and queens. Fumes from both formic acid and menthol cause worker bees to fan excessively and disrupt bee behavior through repellency. Though chemical treatments disrupt bees and cause them to produce less honey (Guzman-N. and Zozaya-R. 1984, Cox et al. 1989), the long-term benefits from treating outweigh the problems. Untreated colonies frequently die during cold winters. Care should be taken by the applicator to avoid skin contact or inhalation of fumes. Clark (1992) tested a gel strip containing formic acid that provided greater safety for the applicator.

The need for an effective mite treatment often exists long before research and product registration have taken place. This situation can lead to the illegal use of miticides, but often it encourages research scientists to establish practical projects. North American scientists have relied heavily on the screening of varroicides in Europe to provide a new list of potential control compounds for the tracheal mite. Amitraz is an example of one that works well for both species of mites provided that the formulation and method of application deliver the toxicant in a suitable manner (Henderson 1988, Moffett et al. 1988, Wilson and Collins 1993). However, some varroicides, e.g., fluvalinate and Perizin, are not effective against the tracheal mite, (Cox et al. 1986, Pettis et al. 1988).

Several new miticides have been tested in the United States for tracheal mite control but none have been highly effective except for amitraz smoke (J. Baxter, unpublished data 1994). Citral fumigation and feeding the systemic Apitol killed about 50 percent of the adult mites, but this does not meet the needs of commercial beekeepers (Wilson et al. 1994). Plastic strips that are impregnated with a miticide (viz, amitraz, flumethrin, or fluvalinate) do not control tracheal mites because the contact poison only kills mites that are on the exterior of the bee, such as *V. jacobsoni*. Apistan strips (fluvalinate) for varroa control did cause some adult bee mortality when the bees were left in the shipping cages for periods longer (four to five days) than recommended by the manufacturer (Pettis et al. 1991). Miticur® (amitraz in plastic) was advertised as a control for tracheal mites, but research proved it was ineffective (Wilson and Collins 1993).

With increased use of chemicals for mite control, the possibility of contamination of both honey and beeswax increases. Consequently, companies that bottle honey for the retail market have increased their testing of honey for possible miticide residues. U.S. beekeepers are careful when applying chemicals to avoid contaminating honey, but this level of care may not be exercised by some foreign beekeepers if residue testing is not enforced before the honey is exported to the United States (R. Adee, personal communication 1994).

Use of naturally occurring miticides, and especially botanicals, has a lot of appeal for beekeepers and honey consumers because there is less chance for product contamination with objectionable residues. The successful testing of menthol and other botanicals by Vecchi and Giordani (1968) set the stage for future miticide studies. Menthol provides good control of tracheal mites (Cox et al. 1986, Herbert et al. 1988). Mazeed (1987) reported success with five natural substances, and Calderone et al. (1991) studied botanicals such as clove oil for mite control. Ellis (1994) found certain essential oils, especially citral, to be effective in laboratory studies, but Wilson et al. (1994) demonstrated limited control of *A. woodi* with citral under field conditions. Calderone and Shimanuki (1995) evaluated four seed-derived oils for mite control and showed a reduction in mite prevalence.

Because tracheal mite control is essential for successful bee management (Wilson and Collins 1994), studies on chemotherapeutic treatments will continue for the next several years or longer depending on the progress in developing a tracheal-mite-resistant stock(s) that possesses the heritable characteristics essential for successful commercial beekeeping.

Non-chemical control

Researchers have dissected thousands (perhaps millions) of tracheal trunks to count live and dead adult mites, but they have not reported abnormalities in mites except for yellowing (Eischen 1987) and shrinking of the body, possibly because of water loss. Dr. F. Eischen (personal communication 1995) observed filamentous growth in tracheal tubes from bees in an infested colony in Mexico in the 1980s. The mite population diminished rapidly in this colony, but the possible pathogen was not identified. Several years ago, Lavie (1952) reported that a yeast

(*Acaromyces*) controlled *A. woodi*, but it was not a successful pathogen in field tests in Mexico several years later (Francisco Reyes, personal communication 1984). Liu (1991b) found virus-like particles in tracheal mites from Scotland and suggested that viral diseases could be a part of the resistance mechanism in British honey bees (Liu 1992c). Although pathogens have not yet provided successful mite control, hopefully new research will discover an effective biocontrol agent.

Heat can kill tracheal mites inside live adult honey bees. Adult bees exposed to 39°C for 48 hours had many dead adult and immature mites with no obvious harm to the bees. However, at 42°C for six hours, the mites die faster but so do the bees (Harbo 1993).

Antibiotic extender patties were originally developed by Wilson et al. (1971) for control of American foulbrood. However, cursory observations by beekeepers and scientists suggested that the patties might aid in control of tracheal mites. Delaplane (1992b), Liu and Nasr (1993), and Sammataro et al. (1994) demonstrated that extender patties containing vegetable oil prevented or reduced tracheal mite infestations to an economically acceptable level.

The presence of a young, highly productive queen is beneficial during spring to rebuild bee populations in mite-damaged colonies (Eischen et al. 1988). When large numbers of young worker bees are present, many of them escape infestation. Dustmann (1993) recommends starting new colonies from packages each year as a means of developing productive colonies with low mite counts, but this does not fit the management plan of most commercial beekeepers in North America.

Establishing mite-free colonies

Bailey (1981) notes that several researchers have found that sealed brood, which is not infested by tracheal mites, may be taken from affected colonies and used to start mite-free colonies or to strengthen others without mites. All adhering adult bees must be removed from sealed brood combs. The brood is then allowed to emerge in an incubator or in a warm climate above a double screen over a colony being used as an incubator. New colonies started in this manner should be given queens from an uninfested source. M. Mazeed (personal communication 1984) used this method in Egypt after *A. woodi* was introduced there in 1977. Because this method requires additional labor it should be used only under special circumstances.

CHAPTER FOURTEEN

———— ✸✸✸ ————

Mites: Varroa and Other Parasites of Brood
David De Jong

Introduction

CHAPTER FOURTEEN
Mites: Varroa and Other Parasites of Brood

Introduction

All of the known honey bee brood mites are large, obligate, external parasites of Asian honey bees that cannot survive more than a few days away from the host. They feed and reproduce on honey bee brood, principally during the bee's late larval and pupal stages, inside the sealed brood cell. Newly developed adults, along with the original mother mites(s), leave the cell with the emerging bee. The bee usually survives, though it is weakened and has a shortened life span. Adult female mites are typically phoretic on adult bees, that is, they use the bees for transportation. In some species the adult mites also feed on adult bee hemolymph.

The brood mites were little known until man brought *Apis mellifera*, the most important bee species for honey production, into areas of Asia that previously had been occupied only by other honey bee species. Unfortunately, *A. mellifera* has proved to be a highly susceptible host species for at least two of the Asian bee mites, *Varroa jacobsoni* and *Tropilaelaps clareae*.

The best known of the brood mites, *V. jacobsoni,* was originally restricted to its natural host, *Apis cerana*, in Asia but is now found throughout the world as a serious pest of *A. mellifera*. In temperate areas of Europe, Asia, the Middle East, Northern Africa, and North and South America, it has become impossible to keep bees without annual chemical treatments to control *V. jacobsoni* infestations.

Another Asian bee species, *Apis dorsata*, is the natural host of the mite *T. clareae*. This mite is probably the cause of the primary disease problem for *A. mellifera*-based beekeeping in tropical and subtropical regions of Asia. It has also been carried by beekeepers to new regions, having reached the African continent.

Table I. Natural distribution of brood mites on their honey bee hosts.

Mite species

Bee host	Varroa underwoodi	Varroa jacobsoni	Euvarroa sinhai	Euvarroa wongsiri	Tropilaelaps clareae	Tropilaelaps koenigerum
Apis andreniformis				+		
Apis florea			+			
Apis cerana	+	+				
Apis koschenikovi		+				
Apis dorsata					+	+
Apis laboriosa					+	+

Varroa jacobsoni

The importance of *V. jacobsoni* has provoked numerous publications and, unfortunately, a variety of common names for the mite and for the disease it provokes (Matheson 1994). Correct common names are designated according to international rules. The technical name for the parasitism caused by *V. jacobsoni* infestation of a colony is "varoosis," and the common name "varroa disease." "Varroasis" and "Varroatosis" are incorrect, though widely used. The mite can be called varroa, but not Varroa, because the latter is the genus name. Normally, we designate the presence of varroa in the colony as an "infestation," but it is also correct to call this an "infection" with parasites.

V. jacobsoni apparently began to infest its new host, *A. mellifera*, during this century, as a result of the introduction of this bee species into regions of Asia where *A. cerana* was native. These two bee species were in close proximity for many years before it became evident that *V. jacobsoni* was a problem for *A. mellifera* (Crane 1978b). This implies that there was an adaptation phase by this mite to the new alternative host. Recently *V. jacobsoni* has been found in colonies of *A. mellifera* in Papua New Guinea and Indonesia (Anderson 1994). Unlike *V. jacobsoni* in other regions of the world, the mites were found not reproducing in these colonies, though they reproduced normally in *A. cerana* drone brood. Despite the lack of reproduction, small populations of *V. jacobsoni* persist in the *A. mellifera* colonies, apparently from mites spreading from *A. cerana* colonies in the vicinity. This peculiar situation may represent the initial adaptation phase found earlier in parts of Asia.

Biology

Adult female *Varroa jacobsoni* are large (1.1 mm long x 1.6 mm wide), hard, reddish-brown, flattened, phoretic external parasites of worker and drone bees. On adult bees they are often found on the abdomen, especially between the overlapping abdominal sternites; there they are physically protected and can reach the intersegmental membranes, which they pierce with their chelicerae to feed on hemolymph.

Typically, the mites only leave the adult bees to enter the brood cells of last-stage bee larvae, where they begin their reproductive phase. They strongly prefer drone larvae over workers. Generally, they do not invade cells with queen larvae, although varroa has been observed in queen cells in heavily infested colonies (Harizanis 1991).

V. jacobsoni appears to be highly specialized to survive and reproduce on its honey bee host. Physiological adaptations include low proteolytic enzyme activity to allow nearly direct uptake and utilization of bee hemolymph proteins (Tewarson and Engels 1982) and the production of few, large, rapidly developing offspring during the short time that the bee cell remains sealed.

Reproduction

Varroa jacobsoni has a haplo-diploid or arrhenotokous sex-determination system. Males are produced from unfertilized eggs (as in bees). They are haploid, with three submetacentric and four acrocentric chromosomes. Females develop from fertilized eggs, and are diploid, with 14 chromosomes (Steiner et al. 1982). Females that produce only males in one reproduction, continue to do so in subsequent cycles, demonstrating that only young females can mate, and that without sperm, only male eggs are produced (Ruijter 1987).

Varroa females initiate reproduction by entering the brood cells of last-stage worker or drone larvae, normally within about 20 or 40 hours, respectively, before the cell is sealed (Boot et al. 1992). Immediately upon entering the cell, the mites go underneath the bee larva to the cell bottom and enter the larval food. They

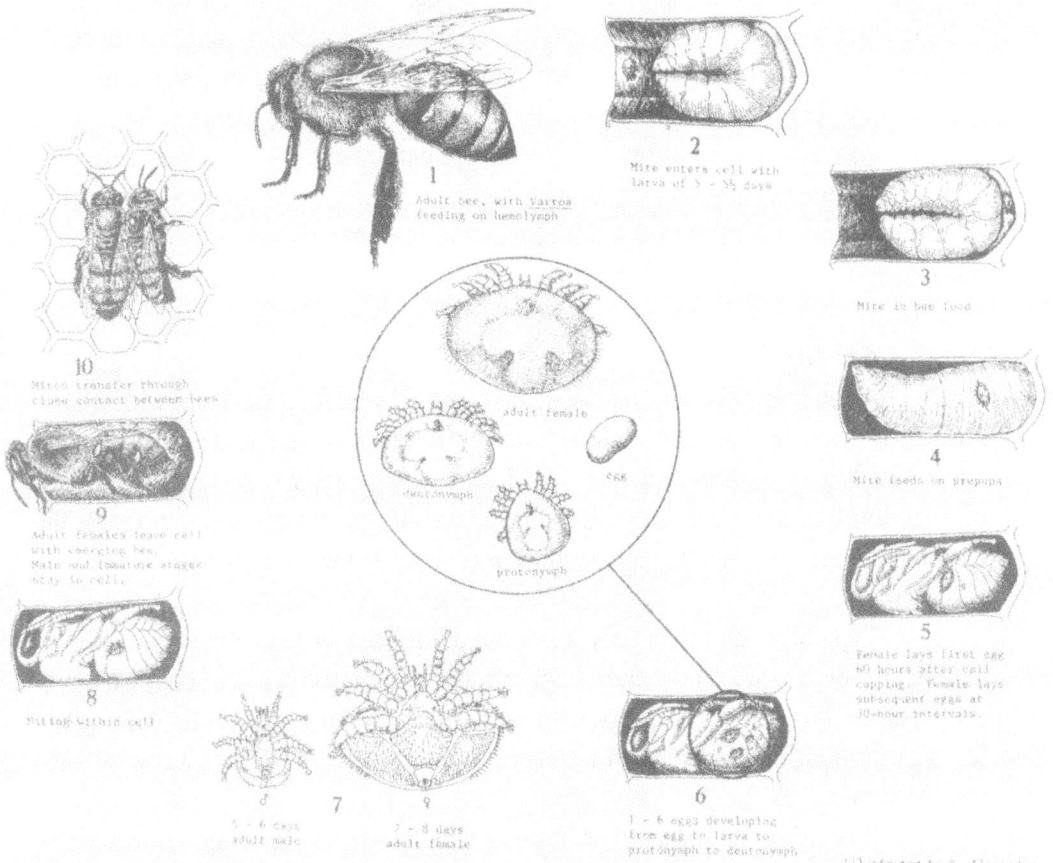

Figure 14.1. Life cycle of *Varroa jocobsoni*. (Alexander drawing).

take a characteristic position, upside down, with the venter facing the opening of the cell, and the entire body submerged, except for the peritremes (two respiration tubes), which are rotated away from the body to protrude out of the liquid food, perpendicular to the body of the mite. These tubes act as "periscopes" to allow the mites to breath while submersed in the "worker jelly" or "drone jelly," which has a consistency similar to royal jelly. The mites enter the food by themselves, and do so in the laboratory, even when no larva is present. Without the feeding activity of the larva, however, they cannot leave this position, otherwise they would remain imprisoned and die. Some of this larval food is consumed, but its value to the mite is unknown. Mites remain in this position for a short time, up to about four hours after the cell is sealed in worker brood. As many as 21 have been found "stuck" in a single drone cell (De Jong et al. 1982b, De Jong 1984, Ifantidis 1988).

After leaving the food, the mites begin feeding on the hemolymph of the prepupa. Originally the mite weighs about 0.3 mg, increasing to nearly 0.5 mg, engorged by the hemolymph taken from the developing bee (Steiner et al. 1994). A feeding female mite becomes substantially thicker. About 60 to 70 hours after the brood cell is sealed, the first egg is laid (Ifantidis 1983, Steiner et al. 1994). They are laid singly, usually on the cell wall, at intervals of about 30 hours (26-32). Eggs are oval and white, about 0.30 mm long by 0.23 mm wide. The first egg is generally male, and the rest female. Males develop from egg to adult in about 5.5 to 6.2 days and females in 6.5 to 6.9 days. Females deposit five or six eggs, of which four (one male and three females) would have time to reach maturity. Mortality of the nymphs, however, reduces the effective reproduction to about 1.5 new females per original female mite (Rehm and Ritter 1989, Martin 1994). When numerous original females invade the same worker brood cell, reproduction is reduced and resorption of developing eggs occurs (Steiner et al. 1995).

When the egg is laid, the eight-legged protonymph is visible within. It hatches within half a day of being laid. Male and female protonymphs have pointed chelicerae, with one or two "teeth." Sex at this stage can only be distinguished by dissection or by examining the chromosomes (Steiner et al. 1982). The protonymph feeds, grows, and then molts to a deutonymph stage. The final molt is directly to the adult stage (Steiner 1988, 1995). Normally, mites have a six-legged larval phase before the protonymph and a tritonymph stage after the deutonymph. The abbreviation of the reproductive cycle in varroa demonstrates a high degree of specialization to adapt to the short reproductive cycle imposed by the necessity to complete reproduction during the 10 to 15 days that the brood cells are sealed.

Each of the two nymphal stages (protonymph and deutonymph) is divided into an earlier ambulatory/feeding phase, and a later immobile/non-feeding phase. This second phase has been called a "chrysalis." In this immobile phase, the legs are held out straight-ahead. The egg and the nymphal stages have an opaque, creamy-white color. It is easy to distinguish the female deutonymph because of its size and shape, which are similar to that of the adult female; the female protonymph and all the stages of the male are all about the same size (ovoid and about 500 µm wide), however, and are difficult to distinguish without magnification (Steiner 1988).

Adult males have a more or less spherical body and are very lightly tanned, with slightly darker legs. The male chelicerae, instead of being pointed as in the female, are highly modified and have an open hollow tip that is used to transfer sperm packets from the genital opening in the ventrally located sternogenital shield of the male to the genital opening of the female (Akrantanakul 1976). This same tubelike structure is found in the male deutonymph (Delfinado-Baker 1984b). Nevertheless, males still are able to feed on bee pupae at the feeding site prepared by the mother mite (Donzé and Guerin 1994).

The deutonymphs molt directly to the adult phase, and mating takes place in the cell. The bee hatches and leaves the cell along with the original female mite(s) and the new adult females. These females may stay on the emerging bee, though many quickly switch to other bees. This is the phoretic phase. After a variable period of time, the females begin the cycle again and invade new brood cells. They can do so immediately, though when they feed on an adult bee during this interval, their reproductive rate is increased (Beetsma and Zonnevel 1992).

Only the adult females survive after the bee emerges. Some female deutonymphs, protonymphs, eggs (rarely), and males are left when the bee emerges. All of these apparently die, or are removed and killed by the nurse bees.

The effective reproduction rate is defined as the mean number of new adult female mites that an original female mite produces during the time she spends in the honey bee brood cells. Considering the time available, a single female mite should produce two new females in a worker brood cell and four in a drone cell. Nevertheless, the final rate is much lower. In worker cells, a single reproducing mite produces a mean of about 1.3 new females, and in drone cells, a mean of 2.6 (Schulz 1984). These data are for European bees in Germany. When more than one original mite enters the cell, the effective reproduction rate is reduced to close to unity in worker brood. Each invading female mite produces about one new adult female (Fuchs and Landenbach 1989). In drone cells, the rate also goes down, lowering to 1.5 with three original females and close to unity with more than six invading mites. As a consequence, as colony level infestations increase, the geometric rate of increase of *Varroa jacobsoni* gradually decreases compared with that found in low-level infestations, at least for European bees in Germany.

Often one finds that a bee emerges with one or several adult female mites that have no offspring. Accumulations of mite feces are found, demonstrating that the mites fed on the developing brood but were not able to reproduce (De Jong et al. 1982b). The percentage of such "infertile" mites varies considerably and probably is an important factor in the resistance of certain strains of bees (Ritter and De Jong 1984).

In cells with only a single mother female, which is the rule in small to moderate infestations, the young females must necessarily mate with their brothers, which would result in considerable inbreeding. When females are reared without males, they produce only male eggs.

Varroa jacobsoni seems well-adapted (perhaps pre-adapted) to reproducing on worker brood of the new host, *Apis mellifera*, as the bees succeed in developing to maturity and leaving the cells, even with a relatively high mite load, up to six and sometimes more mites. In this way the mites have an opportunity to reach maturity and leave the sealed cells. Only a few bees are killed or otherwise obviously damaged, except when the brood cells contain six or more female mites (De Jong et al. 1982b).

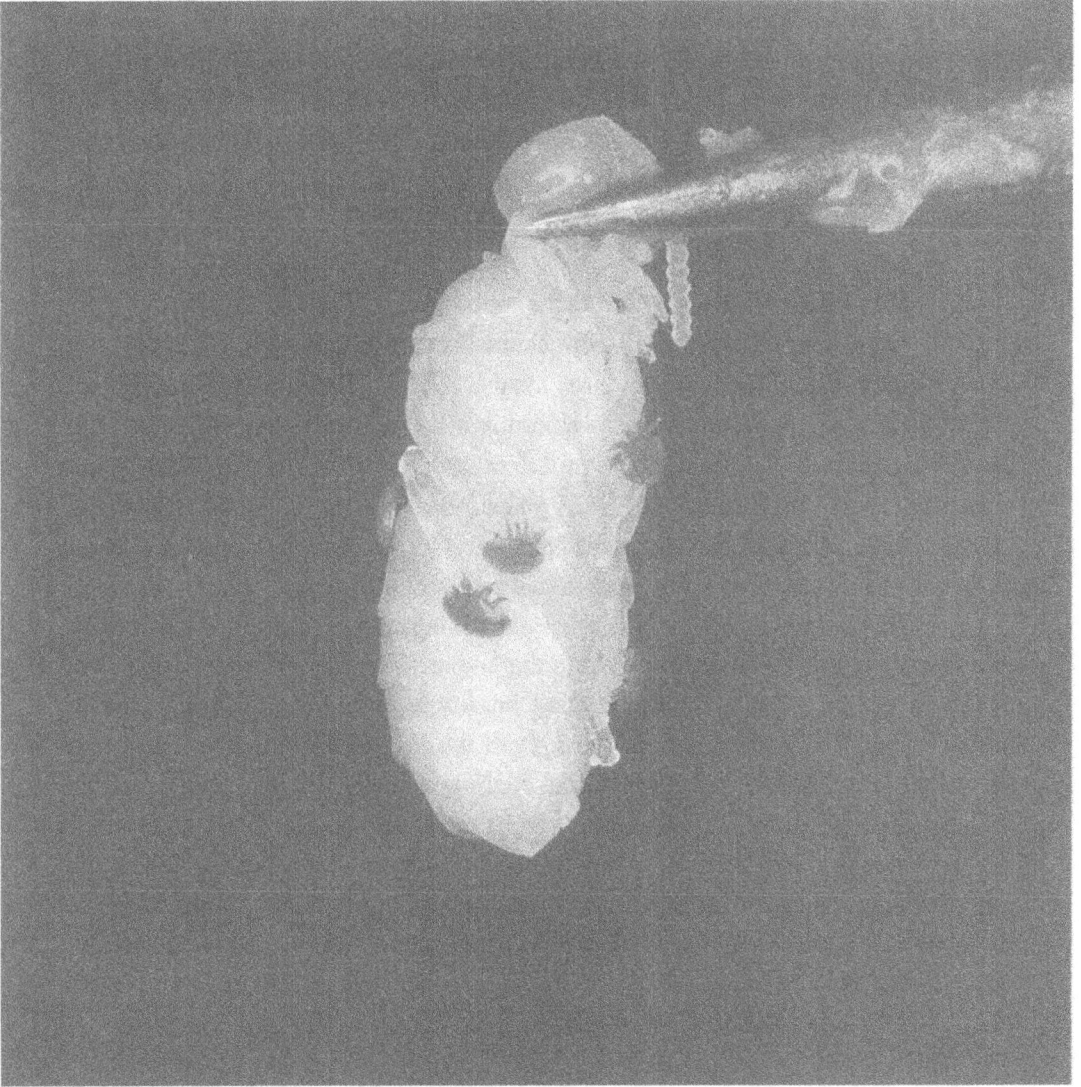

Figure 14.2. *Varroa* mites can be found, and are most easily seen on larvae that have been removed from capped cells. A more reliable test, however, is the insertion of an Apistan® strip into a colony for two days, with a sticky board on the bottomboard. (Hansen photo).

Female mites can go through more than one reproductive cycle. As many as seven cycles, producing up to 30 eggs, have been demonstrated by artificially transferring mites (de Ruijter 1987). Few data exist on the number of reproductive cycles in natural infestations. Fries and Rosenkranz (1993) found that 13 percent of introduced mother mites reproduced at least three times, which is considerably greater than earlier estimates (Fries et al. 1994).

Reproduction in the laboratory

Some researchers have developed means of reproducing mites in the laboratory so as to study the process more closely. Abbas (1990) placed young drone prepupae in plastic cells sealed with cotton wool. The mites were collected from foragers. Alternatively, the mites were fed on larval blood through a thin condom membrane. Issa and Gonçalves (1986) placed drone larvae from recently sealed brood cells in artificial beeswax cells each with a beeswax seal containing small holes for ventilation. The mites were taken from recently sealed brood cells.

Donzé and Guerin (1994) made a detailed study of reproductive behavior in transparent polystyrol cells in which the queen was induced to lay eggs. They observed that the mites generally defecate in a single communal location, normally on the cell wall. The adult females are agile in avoiding being caught by the movements of the larvae and create a space for feeding on the pupae by actively pushing the legs aside. Some eggs and protonymphs are killed by the bees' movements during pupation. The mother mite establishes a feeding site that is used by the nymphal stages and the male, which otherwise could not feed, demonstrating parental care. All of the mites confine their activity to two regions, the fecal accumulation (whitish in color and easily seen) and the feeding site. Matings are mostly confined to the fecal accumulation site.

Factors that trigger reproduction

For some time, it was presumed that female *V. jacobsoni* were stimulated to initiate egg production through the action of juvenile hormone (JH III) found in the host hemolymph. This was an attractive theory because these JH levels in larvae peaked at the time that egg production began, and *V. jacobsoni* was already known to be so very specialized that it incorporates many host proteins intact, without breaking them down (Tewarson and Engels 1982). JH levels in the brood were also thought to stimulate cell invasion behavior. Later research showed no such function, however (Rosenkranz et al. 1990, 1993a). JH levels are low in nurse bees, which is the stage preferred by *V. jacobsoni* (Steiner 1993). Varroa reproduces at different rates in different climates and in different kinds of bees. The rate of varroa reproduction in the tropics is normally lower than in temperate climates, and reproduction is reduced in Africanized bees and in *Apis mellifera lamarckii* compared to *Apis mellifera carnica*. However, no differences in JH hormone titers were found in the fifth-instar worker larvae of Africanized vs. European bees (Rosenkranz et al. 1990). Therefore, the precise triggering mechanisms that promote reproduction in *V. jacobsoni* remain unknown.

Apparently, the mites need to feed on bee blood shortly before cell invasion so as to begin reproduction (Rosenkranz and Stürmer 1992). Otherwise, no eggs are produced. Female mites must also feed on the hemolymph of a late fifth-instar bee larva. When they are placed on brood 24 hours after capping, when the larva is already spinning its cocoon, no eggs are produced (Steiner et al. 1994).

Invasion of worker brood cells

The initiation of reproduction by *V. jacobsoni* involves the choice of an appropriate brood cell by the mites. The honey bee carrying the mite has to come

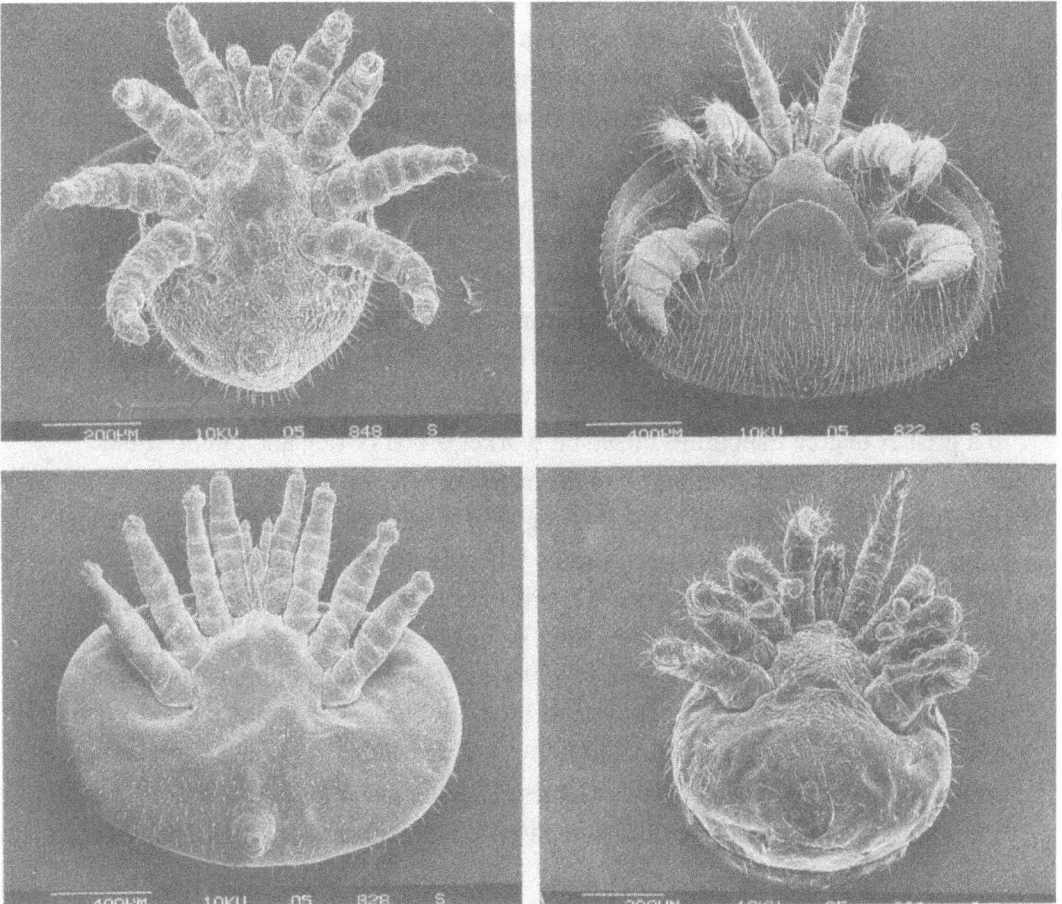

Figure 14.3. Electron scanning micrographs of *Varroa jacobsoni*, ventral views. *Top left:* protonymph. *Top right:* adult female. *Bottom left:* female deutonymph. *Bottom right:* adult male. (Steiner photos).

quite close to the cell opening before the mite will leave and invade the brood cell. Whereas it is clear that drone brood should be and is preferred, we also find that worker brood cells are not uniformly invaded by female *V. jacobsoni*, implying that factors exist that influence their choice. The number of female mites entering brood cells infested can be quantified as an "invasion rate." For instance, if we find 30 adult female mites in 100 sealed worker brood cells containing larvae or white pupae, the invasion rate is 30 mites/100 cells of brood. If this corresponds to 25 cells infested (meaning that some brood cells had more than one mite), then the infestation rate is 25 percent of the worker brood. Therefore, the invasion rate can be more than 100/100 cells, but the percentage of infested cells can never exceed 100.

Preference of Varroa jacobsoni for drone brood

Adult female *V. jacobsoni* preferentially invade drone brood cells. This makes sense for several reasons. In colonies of the original host, *Apis cerana*, nearly all reproduction is accomplished in drone brood cells, so the mites could be merely retaining an already inherent behavior on the new host (De Jong 1988). Also, the mites have a much greater reproductive success in drone brood cells. Fuchs and Langenbach (1989) reported an average of 1.4 new female mites per original mother mite in worker cells and 2.2 in drone cells. Generally about eight to 10 times more mites are found per cell in drone cells than in worker cells of European bees (Fuchs 1990, Fries et al. 1994, Boot 1995), though this ratio can vary considerably depending on the ratio between drone and worker brood cells, the season of the year, the race of the bees infested, and the type of climate. Possibly the origin of the mites also has an influence, as mites reared on drone brood prefer drone larvae to a greater degree than mites raised on worker brood (Otten and Fuchs 1988).

Drone brood is attractive to the mites for a period two to three times longer than worker brood (Ifantidis 1988, Wieting and Ferenz 1991, Boot et al. 1992), which helps to explain why it is more heavily infested. Drone cells also have a larger surface area (1.7 times larger than a worker cell), and drone larvae are inherently more attractive, as demonstrated by tests in the laboratory (Rosenkranz and Engels 1985, Otten and Fuchs 1988, Issa and Gonçalves 1982). Preference is increased when there are few drone cells, and it tends to decrease toward the end of the season in temperate climates (Fuchs 1990). Preference for drone cells varies considerably among colonies. In 12 feral colonies in Brazil, the ratios ranged from 2.3 to 46 times as many mites in drone cells, compared to worker brood cells (Gonçalves et al. 1982).

Table 14.2. Ratio of *Varroa jacobsoni* infestations in drone vs. worker brood.

Ratio	Bee Race	Location	Season	Reference
3.6	European	China	Summer	Fan and Li 1988
5.3	European	China	Spring	Fan and Li 1988
6.0	European	Germany		Ritter 1988
7.2	European	Japan		Takeuchi and Sakai 1986
7.2	European	Yugoslavia		Sulmanovic et al. 1982
8.6	European	Germany		Schulz 1984
9.8	European	Canada		Szabo 1994
11.6	European	Netherlands		Boot et al. 1995
12.1	European	Germany		Fuchs 1990
14.3	Africanized	Brazil		Gonçalves et al. 1982
5.5	Africanized	Brazil		Issa and Gonçalves 1984

Distribution

The original distribution of *V. jacobsoni* is associated with that of its natural host, the Asian bee *Apis cerana*. Repeated introductions of *A. mellifera* into Japan, China, Eastern U.S.S.R., the Philippines, and other regions of Asia allowed continued contact between the two bee species, and by about 1960, colonies of *A. mellifera* in those areas were also found to be infested. Further spread to areas outside the range of *A. cerana* occurred accidentally through the transport of infested *A. mellifera* to Eastern Europe from Eastern U.S.S.R., to South America from Japan, and to Africa from Eastern Europe, and of infested *A. cerana* from Southeast Asia to Western Europe (De Jong et al. 1982b).

Interestingly, the two bee species were in close association for almost 100 years in both Japan and Eastern U.S.S.R. before it became apparent that *A. mellifera* was infested (Ruttner 1983). One may hypothesize that some adaptation to the new host was made by the mite during that time.

Presently, *V. jacobsoni* is found throughout mainland Asia; in Indonesia, New Guinea, and the Philippines; in much of the Middle East; and throughout Europe. It has also become established in Northern Africa. All of North America and nearly all of South America are infested. Soon it will reach the remaining South American and Central American countries. The only major beekeeping areas believed to be free of *V. jacobsoni* as of 1997 are Australia and New Zealand.

Varroa was first reported in North America in September 1987. It apparently started in Florida as a result of illegal importations of queens from South America. By 1989 it was already found in 19 states and now is found throughout the United States and has already infested much of Canada. As was the case in Europe,

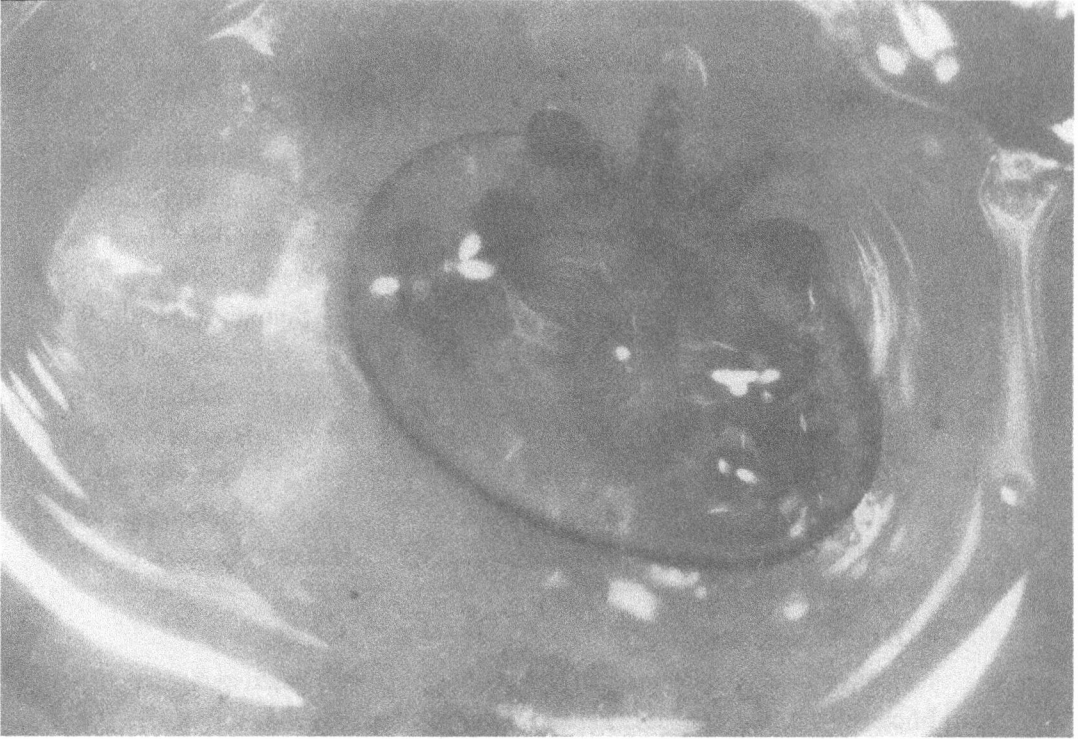

Figure 14.4. Adult female *Varroa jacobsoni* immersed in the food of a last-stage honey bee larva at the bottom of a brood cell. (De Jong photo).

all attempts at control by restrictions on bee colony movement were essentially ineffective. The restrictions themselves caused considerable economic loss. The discovery of *V. jacobsoni* in the United States provoked a ban on importation of queens and packages from the United States into Canada, which severely damaged the U.S. package bee industry. This probably delayed the entry of *V. jacobsoni* into that country, though within a few years Canada was also infested.

The mites are spread by intercolony drifting of infested adult worker and drone bees, by the movement of swarms, and by bees robbing weakened colonies. Large-scale increases in the range of infestation have been caused by migratory beekeeping. In Yugoslavia, *V. jacobsoni* spread from the Bulgarian border to the Mediterranean, a distance of 500 km in less than a year, through movement of hives by beekeepers (Ruttner 1983). In Brazil, infested colonies were transported over 2,500 km from the state of São Paulo north to the state of Piaui in a few days (De Jong and Gonçalves 1981).

Biotypes have been proposed for the *V. jacobsoni* from different regions of the world, as there is some variation in characters and in the damage that the mites cause to the bees (Delfinado-Baker 1988b). Biotype A would be the *V. jacobsoni* that attacks and is a deadly parasite of populations of European bees, attacking both worker and drone brood. Biotype B corresponds to the *V. jacobsoni* found in *A. cerana* colonies in Asia, in which essentially only the drone brood is affected, and colony mortality is not a problem. Biotype C corresponds to the *V. jacobsoni* populations of Africanized bees in South America and those found on African races in Africa, which attacks both worker and drone brood, but does not normally induce colony death. Despite the difference in impact on the bees, there is no evidence that such a division can be made. Furthermore, the mites found on Africanized bees in Brazil, and in Paraguay, where there is no evidence of *V. jacobsoni*-provoked colony mortality, have the same origin as and have populations contiguous with those found on European bees in Argentina, where annual treatments are essential to maintain the colonies alive (De Jong et al. 1984). A morphometric study of *V. jacobsoni* from 17 countries showed that the mites do not vary greatly. However, it was possible to determine that mites from the United States are more similar to those from Brazil than they are to the *V. jacobsoni* in Europe or Asia (Delfinado-Baker and Houck 1989). This suggests that the mites in the United States originated in South America.

To reduce the chance of new introductions, bee brood should not be imported from an infested area. Bees without brood, that is, package bees and queens in cages with accompanying bees, can be treated chemically (Henderson 1988), though with present techniques it is impossible to kill all the mites without destroying the bees.

Because *V. jacobsoni* does not attack honey bee queens, queens theoretically can be transported safely; it is necessary, however, to treat the bees in the queen cages with an acaricide and then to remove, kill, and replace the attendant bees inside a closed, bee-tight room. This is best done by transferring the queen to another queen cage and destroying all the original worker bees. Such a practice is feasible on a small scale under carefully controlled conditions. Unfortunately, whenever the bees are disturbed, the mites have a tendency to leave the cages, and there is a chance that other bees attracted to the queens will be accidentally infested. Ideally, the bees should be treated in the queen cages both at the place of origin before shipping and in a closed room at the destination.

Damage to honey bee colonies

Many reports exist of severe damage and loss of hundreds of thousands of *A. mellifera* colonies caused by infestation with *V. jacobsoni*, especially in temperate areas of Europe and Asia (Chun 1965, Ritter 1981, De Jong et al. 1982b, Choi 1986). Varroa is generally considered the most severe threat to world beekeeping.

Colonies severely infested with *V. jacobsoni* appear restless, the brood is neglected, often with clinical symptoms of European foulbrood, and eventually one finds that only a handful of bees remain along with the queen (Ritter et al. 1984). How *V. jacobsoni* infestations actually kill bee colonies is still not clear. Shimanuki et al. (1994) have suggested the terminology "parasitic mite syndrome" for colonies that have *V. jacobsoni* (and sometimes *Acarapis woodi*). The symptoms are a reduced adult bee population, evacuation of the hive by crawling bees, queen supersedure, spotty brood, sick brood with symptoms resembling European foulbrood and sacbrood disease. Young larvae are also affected; these turn a light brown color. Scales resulting from the dead brood are easy to remove, and the dead larvae and pupae do not have a "ropy" consistency of the type one sees with brood infected with foulbrood. Although no associated organism has been described, an acute bee paralysis-like virus is suspected.

When *V. jacobsoni* is first introduced to a new area, infestations of individual bee colonies increase slowly over a period of several years. When each colony has only a few to a few hundred mites, there is little sign of damage, and the problem often goes unnoticed. During that time, the mites are spreading to other colonies. Eventually, bees with damaged wings can be observed crawling from their hive entrances (Oku et al. 1983); when the mite population reaches 30 to 40 percent or more of the adult bee population, the number of adult bees declines rapidly, there is severe damage to the brood that superficially resembles the results of European foulbrood disease, and the colony dies, usually in the late summer or fall.

Infestation levels in worker brood increase dramatically in late summer and often reach an average of more than one mite per brood cell (De Jong et al. 1984, Rosenkranz and Engels 1985). Normally, all colonies that do not receive chemical treatment die within two to four years. This assumes a colony has not received additional mites from outside during this time. Continuous influxes of mites from neighboring colonies and apiaries can result in colony collapse in as little as a single season.

Damage to individual bees

The mites tear the integuments of the bees with alternate movements of their saw-toothed chelicerae and suck up the hemolymph with their mouthparts. The mite's chelicerae, labrum, and hypostomum, together with the hollow styli, form a preoral region (Cromroy and Kloft 1980).

Damage to individual bees that hatch from mite-infested brood cells includes a 6 to 25 percent weight loss of workers, depending on the degree of infestation (De Jong et al. 1982a), and reduction of 34 to 68 percent in mean adult life span (De Jong and De Jong 1983, Schneider 1986). Many of the infested bees that hatch in autumn do not survive the winter (Kovac and Crailsheim 1988). Feeding activity of the mites during the brood stage causes appreciable losses (15-50 percent) in hemolymph protein content and total hemolymph volume of the emerging bees (Engels and Schatton 1986).

Nevertheless, the damage to individual bees can be difficult for the bee-keeper to detect, except at heavy infestation levels. Daly et al. (1988) found virtually no morphological changes in adult bees infested during the brood phase with one to two mites, and very small reductions of 1 to 3 percent in wing length and vein measurements. Damage to the exoskeleton appears to be minor at all but the highest infestation levels. These observations were made on Africanized bees in Brazil.

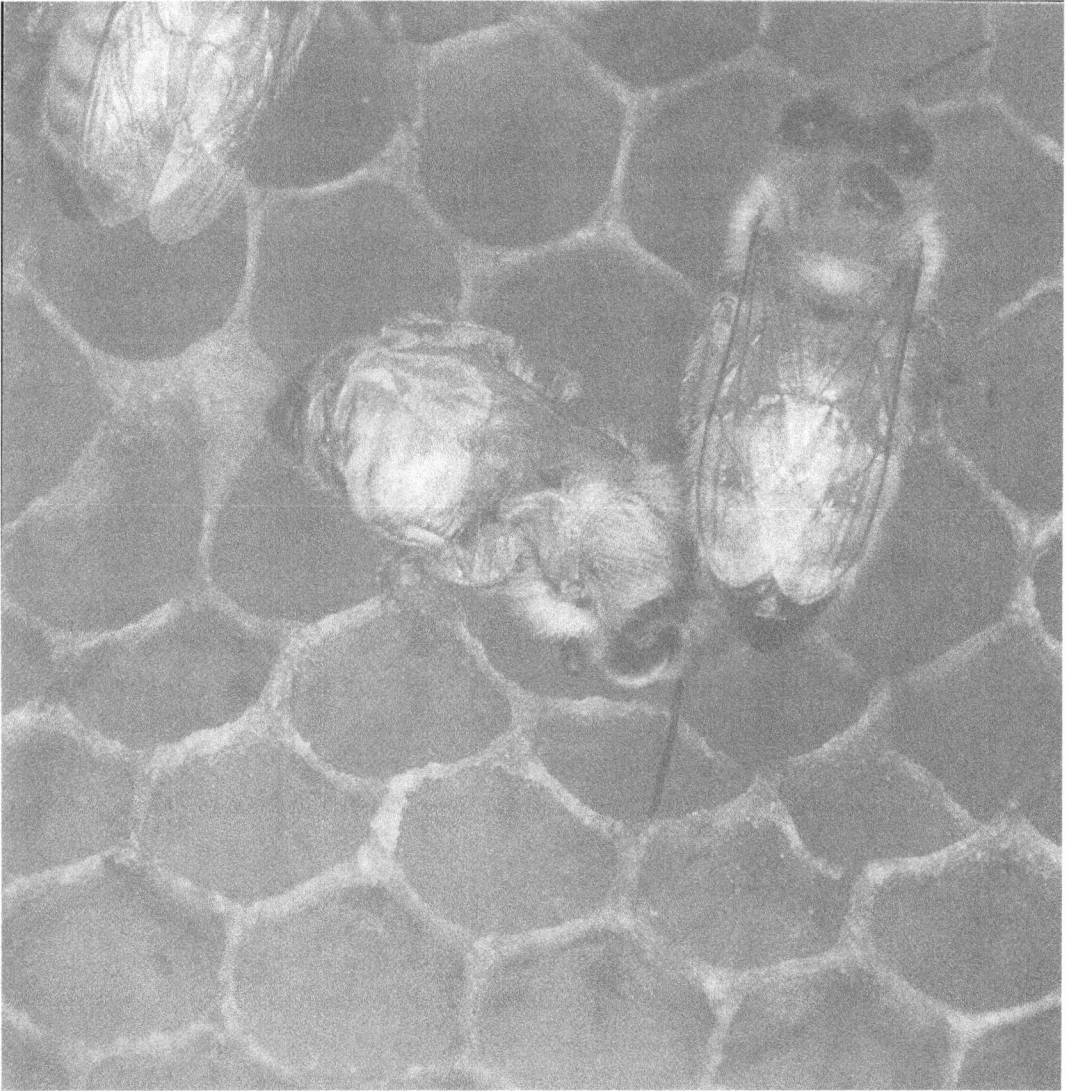

Figure 14.5. Worker bee (left) that shows typical damage from mite feeding during the brood stage, compared to normal bee (right). Damaged bees probably have deformed wing virus. (Hansen photo).

In Argentina, emerging European bees had major damage, especially wing defor-
mations, with infestations as low as one mite (Marcangeli et al. 1992a). About 25
percent of bees with one to three mites were severely damaged, and with four or
more mites the rate of damaged bees increased to 50 percent and more.

When five or more mites enter a single brood cell, there is a high probability that the bee developing in that cell, if it survives at all, will emerge with wing
damage. Even at a moderate to heavy level of colony infestation, however, only a
small percentage of parasitized brood cells have so many mites. Furthermore, although bees parasitized by even one or two mites during development emerge as
adults that are significantly smaller than average and have shortened life spans (De
Jong et al. 1982a, De Jong and De Jong 1983), these effects are not visible to the
beekeeper. Thus, when bees are seen crawling from the entrance with distorted and
shortened wings, the visible effects represent only a small fraction of the damage to
the colony, and the infestation is already extremely advanced.

Bees from dying *V. jacobsoni*-infested colonies have been found to be
infected with acute bee paralysis virus (Ball 1983, 1985, Ritter et al. 1984), which
apparently can be transmitted by *V. jacobsoni*. According to Ball (1986), "Acute
paralysis virus was the primary cause of both adult bee and brood mortality in
German honey bee colonies severely infested with *Varroa jacobsoni*." Apparently
the introduction of foreign proteins, such as digestive enzymes from *V. jacobsoni*,
into the bees' hemolymph can stimulate virus replication (Ball 1985, Ball and Allen
1988). Other bee pathogens that apparently are spread by *V. jacobsoni* include
Proteus vulgaris (Horn 1984) and (experimentally induced) *Hafnia alvei* (Strick and
Madel 1986). Many honey bee pathogens have been found on the surface of the
mites, but it is not clear if *V. jacobsoni* is important in disease transmission. Viable
spores of *Bacillus larvae* have been found on varroa in American foulbrood-
infested colonies in Argentina (Alippi 1992). Experiments made with *Serratia
marcescens*, a bacterium that has a characteristic red color, showed that *V. jacobsoni*
can act as a vector of bacteria. The mites ingested this bacteria from brood that was
artificially infected and could transmit it, in turn, to other healthy pupae (Glinski and
Jarosz 1992).

Because *V. jacobsoni* weakens the bees, it is more difficult for an infested
colony to maintain correct sanitary and environmental conditions within the hive;
the bees are therefore more susceptible to other disease agents. In addition, the
puncturing of an infested bee's integument facilitates invasion by pathogens, especially those carried by the mites. Nevertheless, it remains to be demonstrated
whether viral or other infections are the primary cause of honey bee mortality in *V.
jacobsoni*-infested colonies or merely secondary problems that occur because of
an accumulation of physiological damage to a critically high percentage of the
adult bees in the colony. (See also Chapter 2, Viruses.)

Drones that are infested during the brood stage have a reduced emergence weight, and the seminal vesicles and mucous glands are reduced in size (Ritter 1988).

Drone flight frequency is also decreased (Schneider 1986).

Critical levels of infestation

What is a critical infestation rate? Normally, in pest control programs, one determines a critical infestation rate, above which the control measures should be

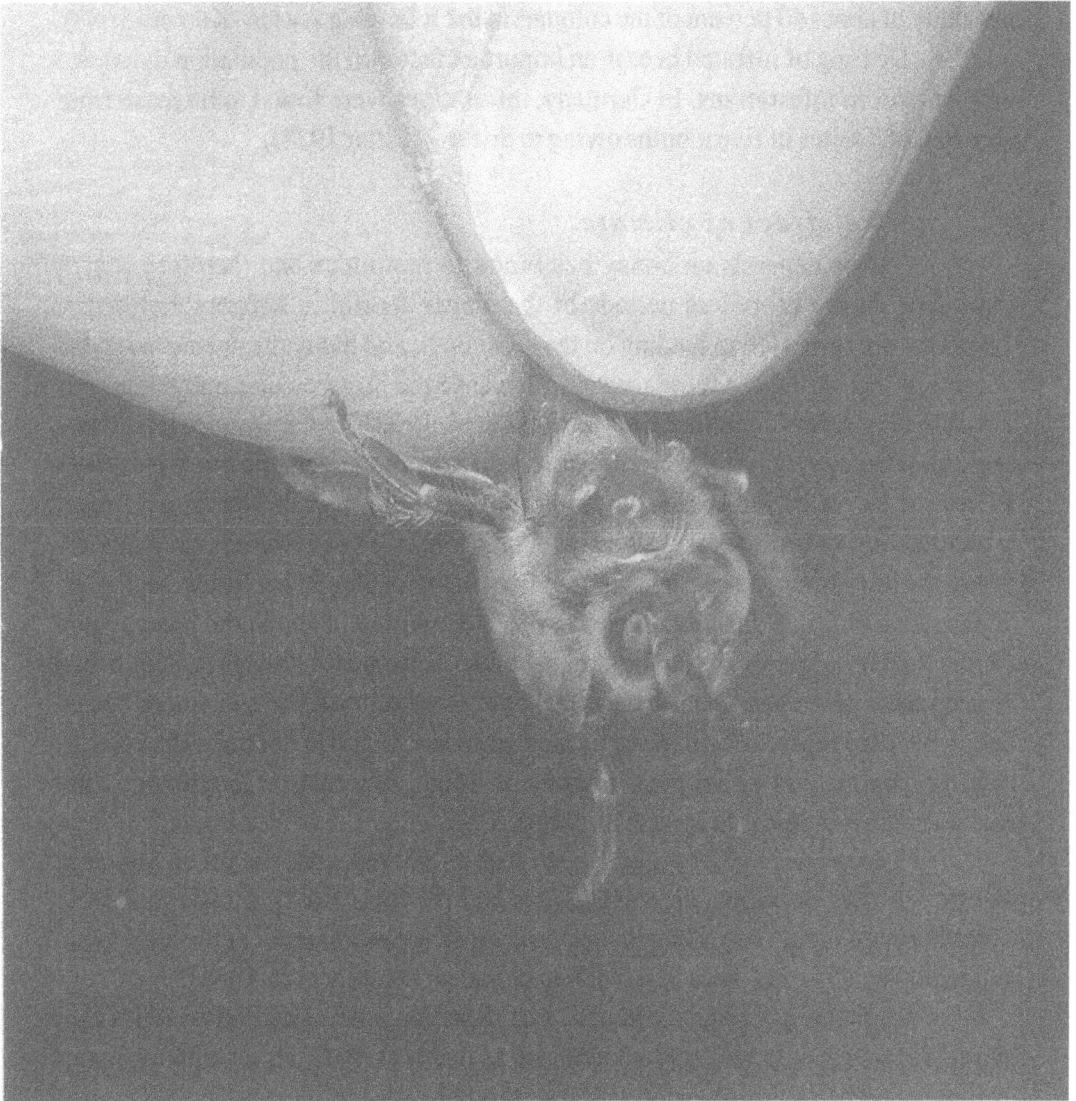

Figure 14.6. Adult mites can be found nearly anywhere on an adult bee, but most commonly on the abdomen, between segments, where they can reach the intersegmental membrane for feeding. (Hansen photo).

applied. Varroa infestations increase continually, so the critical value will depend on the time of year the sample is taken. For example, while 10 percent brood or adult infestation is tolerable in the fall in temperate climates, it is not during the summer, because the rate will increase to 30 percent or more by fall, resulting in considerable colony mortality. Unlike other diseases, in which one sees a gradual weakening of the colonies, *V. jacobsoni* can kill an apparently normal colony in a few months. The progression of the infestation is so fast that colonies can be strong, produce a large crop of honey in late summer, and die by late fall. In Iran, before chemical treatments were widely applied, infestation rates of about 10 mites /100 adult bees resulted in mortality of about 40 percent of the colonies in the following season (Komeili 1988).

Drifting of infested bees is an important factor in the population dynamics of *V. jacobsoni* infestations. In Germany, infestations were found to increase from zero to 6,000 mites in five months owing to drifting (Ritter 1988).

The effect of climate

Varroa depends on honey bee brood to reproduce and therefore cannot reproduce during broodless periods of the winter season in temperate climates. The mites are restricted to feeding on the adult bees, and many die during extended winters, either through natural causes or because the host bee dies and the mite is not able to transfer to another bee (Korpela et al. 1992). In the warmer (southern) regions of Europe, where the brood-rearing season is extended, the mites increased to critical levels of infestation faster (in fewer years) than in Northern Europe, where the active brood-rearing season is quite short. In the Rhine valley of Germany the brood-rearing season is longer than in the cooler Black Forest region, so *V. jacobsoni* infestations took longer to reach damaging levels in the latter region (Ritter 1988). Logically one would expect that in tropical climates, where brood rearing takes place year-round, the effect of *V. jacobsoni* will be even more devastating because reproduction never ceases; however, that has not been the case. In fact, in tropical and subtropical regions of South America, *V. jacobsoni* is not considered a problem (De Jong et al. 1984).

The percentage of female mites that do not reproduce is much higher in the tropics than in temperate areas (Ritter and De Jong 1984). Infestation levels remain low and relatively stable in tropical and subtropical areas but increase to the point of killing bee colonies in temperate climates (De Jong et al. 1984).

In a long-term study in Brazil, the infestation level of bee colonies in a cool climate at an altitude of 1,400 meters was as much as 10 times as high as that of colonies kept in a warmer region at 300 meters above sea level, even though the two locations were within 150 km of each other and the colonies were headed by sister queens (Moretto et al. 1991b).

The population dynamics of *V. jacobsoni* differs considerably in the various parts of the world. In Germany, the mite population increases ten-fold from spring to summer (Schulz et al. 1983), whereas in Brazil and parts of Uruguay the mite has been established more than 15 years and the populations are maintained at low levels without treatment (De Jong et al. 1984, Ruttner et al. 1984, Gonçalves 1987, De Jong 1990b). Average infestation rates in Brazil are around one to three mites per 100 adult bees, and there are no reports of colony mortality. Average infestation rates of *V. jacobsoni* on *A. mellifera* in South Vietnam are around 5 percent, without chemical treatments (Woyke 1987a).

One of the reasons for the much greater impact of *V. jacobsoni* on bee colonies in temperate regions is that the seasonal fluctuations and peaks in the infestation levels within the colonies are much greater in the cooler regions, in comparison with tropical climates. Hybrid Italian/Africanized bees kept in one of the coolest regions of Brazil, São Joaquim, Santa Catarina, had mean infestation levels that varied from about 6 to 7 percent (in adult bees) in late spring and early summer (1986-1988) to 25 to 33 percent in fall and winter, while in Rio Sul, which is nearby, was infested at about the same time, but is warmer due to a lower altitude, the mean infestations for the apiary varied from about 3 to 12 percent during the same period. São Joaquim is one of the few places in Brazil that receives snow. In Ribeirão Preto, São Paulo state, which is more tropical than Santa Catarina, the infestations varied even less, from about 2.5 to 5 percent (Moretto et al. 1991b). In the temperate climates of other regions such as Argentina and Europe, the final infestation levels and the seasonal fluctuations are much greater, to the point that the colonies are killed by the mites (De Jong et al. 1984).

Bees in temperate climates are especially vulnerable to *V. jacobsoni* because of the necessity to survive a winter season. Workers that emerge in late September (early fall in the Northern Hemisphere) should survive until April. About 70 percent do so in Austria, whereas among those that were infested during the brood stage, survival is reduced to around 10 percent (Kovac and Crailsheim 1988). When the brood areas drop in late summer, the relative infestation rate of the brood increases, just when winter bees are being reared (Ritter 1988).

Seasonal climatic differences appear to be important determinants of the severity of *V. jacobsoni* infestations. In spring and autumn, the brood infestation rate is higher than in summer (De Jong 1984, Kulincevic and Rinderer 1988, Moretto et al. 1991b). The number of mites entering from other colonies is low in spring and increases considerably in summer, staying at a high level until autumn (Sakofski et al. 1990). This can be measured by keeping a plastic *V. jacobsoni*-control strip (Bayvarol or Apistan) in the colony constantly and monitoring the number of mites that fall to the bottom.

Possible resistance factors

Beekeepers and researchers around the world are searching for European bees resistant to *V. jacobsoni*. Considerable variation in susceptibility has been found in European bees in several regions of the world (Boecking and Ritter 1994), with up to sevenfold differences in mite population growth (Büchler 1994). Nevertheless, it has not been possible to maintain these "resistant" bee lines without some kind of mite control.

Several breeding programs have started with the surviving colonies in regions where *V. jacobsoni* has caused heavy losses (Kulincevic and Rinderer 1988, Morse et al. 1991). Dead and mutilated mites are found at the bottom of such colonies, suggesting another parameter for selection. A long-term project initiated in the former Yugoslavia in collaboration with the USDA, made a bidirectional selection for resistant and susceptible lines (Kulincevic et al. 1992). A clearly genetic component was demonstrated. However, when these lines were introduced into the United States for testing, they did not maintain resistance levels to those mites found in the U.S. (Rinderer et al. 1993).

The relative innocuousness of *V. jacobsoni* infestations in much of South America is apparently attributable, at least in part, to resistance mechanisms in Africanized honey bees. Infestation levels are higher in European bees and first-generation European/Africanized hybrids than in "wild-type" Africanized colonies kept under the same conditions (Steiner et al. 1984, Message 1985, Engels et al. 1986, Mendoza et al. 1987, Moretto et al. 1991b).

Fertility of the mites

A key factor, which is strongly correlated with the infestation level in the colonies, is the relative fertility of the mites. During a normal infestation cycle, the female mite enters a brood cell, feeds on the hemolymph of the developing bee, and lays eggs that develop into nymphs and eventually new adult mites. However, many mites go through these cycles, feed on the bee pupae, but do not reproduce.

In some areas of the world, the percentage of fertile females and the infestation rates gradually decrease after the introduction of the mites into a region. This evidence suggests that there is an accommodation in the host-parasite relationship. This phenomenon has been reported from Uruguay (Ruttner et al. 1984), Brazil (Message 1985), and Yugoslavia (Sulimanovic et al. 1986).

The percentage of "infertile" mites can be determined by counting the number of mites that do not reproduce and dividing it by the total number of mites sampled in the brood cells. This is best done when the pupae are in a phase in which the eyes are colored and the body is beginning to darken. Wide variations in this mite infertility rate have been found. Only about 40 to 50 percent of the mites reproduce in bee colonies in Brazil, where *V. jacobsoni* infestations are so low that treatment is not necessary, whereas in Germany more than 80 percent do so (Ritter and De Jong 1984, Rosenkranz and Engels 1994). High rates of infertile female mites have also been found in bees in Tunisia (Ritter et al. 1990) and Uruguay (Ruttner et al. 1984). In Argentina, in a region with European bees, the percentage of brood cells containing non-reproducing females was 28 percent in spring and 44 percent in fall (Marcangeli et al. 1992b).

A significantly lower percentage of mites reproduce on Africanized bee brood than on brood of European (Italian or Carniolan) bees (Camazine 1986, Rosenkranz 1986, Moretto 1988) kept in the same regions of Brazil. This effect was observed even when the Africanized and European brood were present within the same colonies (Camazine 1986).

The colony rearing the brood also has an effect on the percentage of reproducing mites, independent of the origin of the brood and mites (Fuchs 1994), demonstrating that colony care factors are involved.

Duration of the capped-cell stage

Another resistance factor that influences mite reproduction is the bees' development time. Varroa have been experimentally introduced into colonies of the Cape honey bee (*Apis mellifera capensis*), which develops considerably faster than the European races (9.6 days for the sealed-cell stage of Cape bee workers vs. 12 days for European workers). Although the mites laid eggs in the worker cells of the Cape bees, most mite nymphs did not have time to develop to the adult stage (Moritz and Hänel 1984). The infestation levels were found to be low. Selection experiments with Cape honey bees demonstrated that the capping period is genetically controlled and can be reduced (Moritz 1985). Selection was more pronounced when drone brood development time was used as the critical parameter because of a greater phenotypic variation and because of male haploidy. The post-capping stage of queens also varies and is correlated with that of the workers, so some progress could be obtained by making the initial selections on developing queens (Le Conte et al. 1994).

The length of time of the capped-cell stage of worker brood has been found to be positively correlated with the *V. jacobsoni* infestation rate in the colonies (Büchler and Drescher 1990). An average 8.7 percent reduction of the mite infestation rate was calculated for each one-hour reduction of the capped period in a large-scale comparison of European hybrids. However, although the sealed-cell stage in Africanized worker bees is significantly shorter (about 20 hours less) than that of European workers, the number of mite offspring per reproducing female mite is only slightly different (Rosenkranz 1986), so although it probably contributes, sealed-cell stage duration is apparently not a key factor in the resistance of Africanized bees to *V. jacobsoni* (Rosenkranz and Engels 1994).

Hygienic behavior (removal of infested brood)

In honey bees, hygienic behavior is measured by determining the rate at which freeze-killed or perforation-damaged brood is removed. Apparently this ability is correlated with the ability to detect and remove diseased brood, helping to control disease by interrupting the normal disease cycle. Honey bees are able to detect and remove *V. jacobsoni*-infested brood, and bees that demonstrate hygienic behavior, by fast removal of freeze-killed pupae, also remove *V. jacobsoni* from infested cells at a greater rate than do less hygienic bees (Boecking and Drescher 1992, Spivak et al. 1994). This removal response is stronger for brood cells containing two mites than for cells with only one mite (Boecking and Drescher 1992). *Apis mellifera intermissa* in Tunisia has been found to be especially efficient at removing *V. jacobsoni* from artificially infested brood cells (Boecking and Ritter 1993) when compared to European races of bees. Workers of Africanized bee colonies are more efficient at removing brood infested with *V. jacobsoni* than are European bees kept in the same conditions (Guerra et al. 1994).

To determine the capacity of bees to selectively remove *V. jacobsoni*-infested brood, researchers introduce the mites into sealed brood cells, either by making a small opening in the capping and resealing the cell or by removing the bottoms of the brood cells in special plastic combs developed for rearing queens (Boecking et al. 1992). Making an opening in the cell capping can in itself induce removal of the brood by the bees, which makes it necessary to open and reseal control cells into which no mites are introduced. The artificial combs allow one to introduce *V. jacobsoni* without disturbing the cell capping (Spivak et al. 1994). The removal response of bees is significantly greater with plastic combs than it is for natural wax combs (von Posern 1988, Boecking and Drescher 1992), which has prompted some efforts to exploit such combs for *V. jacobsoni* control. Wieting and Ferenz (1991) suggest that bees kept on plastic brood combs have fewer *V. jacobsoni* because the period during which the worker larvae are vulnerable to invasion is shortened from the normal 14 hours to six hours. Apparently *V. jacobsoni* invades the cell when the larva completely fills the bottom, which occurs slightly later in the plastic cells owing to a larger diameter at the bottom.

Grooming behavior

Bees groom themselves and each other when they are irritated by the mites (Boecking et al. 1993). This behavior is evident in *A. cerana*, but it also occurs in *A. mellifera* and has been demonstrated in cage tests. Africanized bees groom and rid themselves of *V. jacobsoni* at a considerably greater rate than European honey bees (Moretto et al. 1993, 1995).

Attractiveness of the brood

The relative attractiveness of the brood to the mites varies between colonies and has been found to be correlated with infestation rates (Büchler 1990, 1994). Brood from more highly infested colonies is more attractive, in comparison with brood from colonies that are less infested. Wide variations have been determined between local populations of bees in Europe.

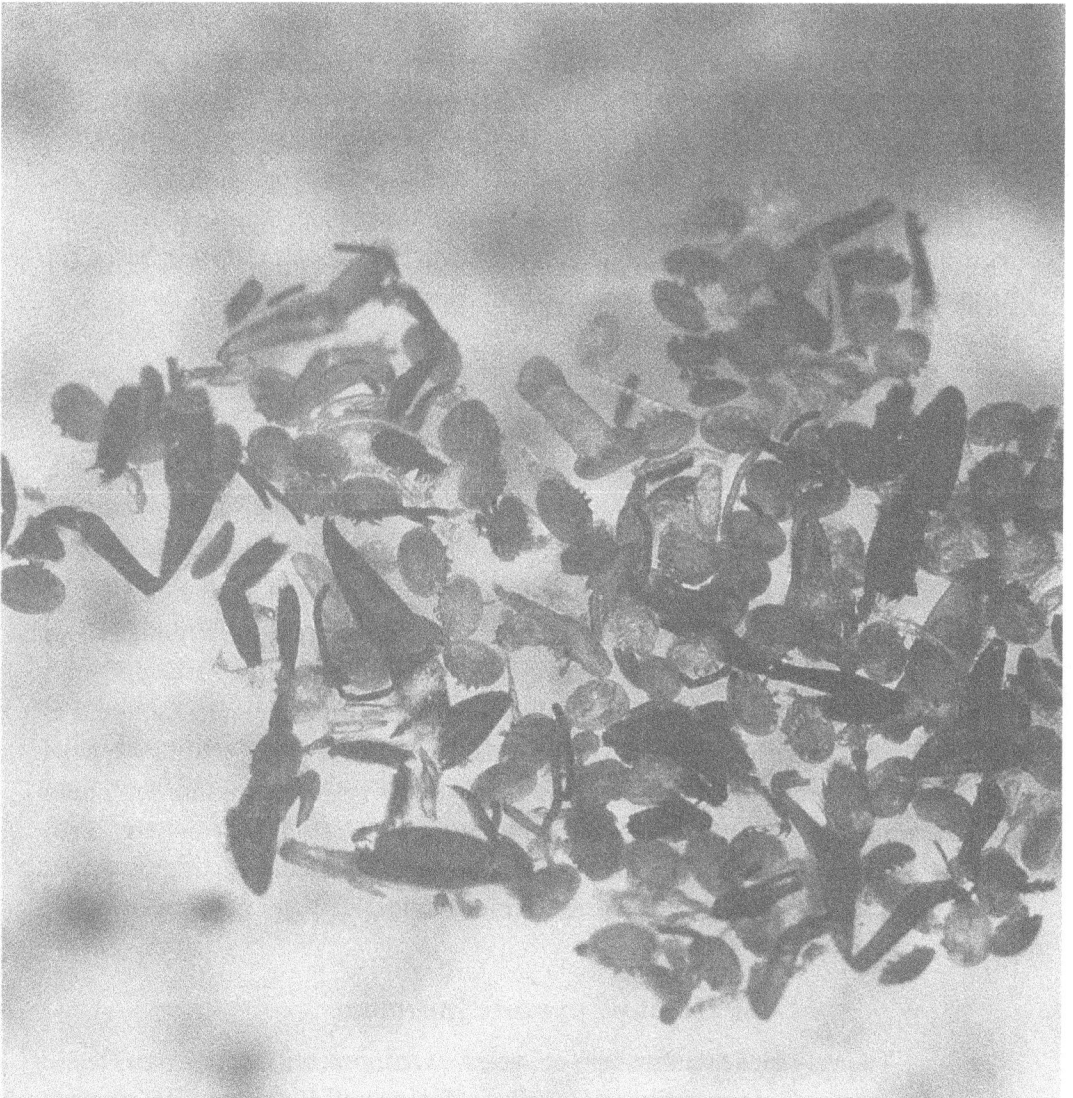

Figure 14.7. One result of grooming behavior is finding mites on the floorboard of a colony, often with mandible-induced wounds – severed legs, dented shields. (Hansen photo).

Various factors have been found to influence the invasion rate of *V. jacobsoni* (De Jong 1984). When queen cells are being built, the infestation rate in the worker brood increases severalfold, decreasing the relative infestation of the adult bees and increasing that of the brood (De Jong 1981).

Normally drone brood is much more preferred by female mites for reproduction in comparison with worker brood. However, even within the same colony we can find considerable variation in the infestation rate of worker brood. Cells that have protruding capping caused by irregularities in the comb are more heavily infested than are other neighboring cells (De Jong and Morse 1988). This phenomenon also occurs when cell depth is artificially altered by dropping melted wax into the cell (Ruijter and Calis 1988). Cell diameter can also have an effect. Drone larvae reared in drone-sized cells are infested at a greater rate than drone larvae in worker-sized cells (Issa et al. 1993). In a similar way, worker larvae reared in (larger diameter) European-size brood cells are infested at a greater rate than are larvae from the same queen reared in Africanized-size brood comb (Message and Gonçalves 1995).

The distance between the bee larva and the cell opening apparently affects the rate of invasion by the female mites (Goetz and Koeniger 1993). Invasion begins at between 7.5 and 7.0 mm and increases exponentially with decreasing distances until the cells are sealed.

Interval between reproductions

When an infested bee emerges from its brood cell, the accompanying female mites leave the cell and quickly transfer to adult bees, or they stay on the original host on which they hide and feed. The time interval until the next brood cell invasion varies and is one of the determining factors for the rate of increase of the *V. jacobsoni* population. When this interval is short, the rate of reproduction is effectively increased because more reproductive cycles are accomplished per unit time. Some mites can invade brood cells immediately. Boot et al. (1993) found in three different measurements, that 50 percent of the mites reentered brood cells within 2.0, 4.3, or 8.3 days. A longer time spent on the adult bees before beginning a new reproduction is correlated with a lower infestation rate in the colonies (Otten 1991). This time interval can be determined indirectly by calculating the proportion of mites on adult bees versus those in the brood cells (De Jong 1983).

The effect of juvenile hormone

The very close relationship between *V. jacobsoni* and the host bees led to a theory that reproduction in this mite is affected by juvenile hormone (JH) levels in the host. Maximum JH levels in bee brood coincide with the initiation of egg laying by the mites, and researchers looked to JH as a possible explanation for the differences in resistance of the various types of bees. A comparison between *A. m. carnica* (susceptible to *V. jacobsoni*), *A. m. lamarckii*, and Africanized bees (both of the latter have resistance characteristics), however, showed no differences in JH titers between the three types of bees (Rosenkranz et al. 1990).

Juvenile hormone titer and reproduction of *Varroa jacobsoni* in capped brood stages of *Apis cerana indica* in comparison to *Apis mellifera ligustica* have been studied to determine the role of JH in controlling mite reproduction. The JH III hemolymph titer of late larval and early pupal states did not differ significantly in the two species of bees. In drone brood, slightly higher JH III concentrations were found (Rosenkranz et al. 1993a). These results do not agree with the hypothesis that reproduction of the mite is regulated by host-derived JH. The fertility of mites in worker brood in Brazil is reduced but is not associated with a low JH III titer in worker larvae (Rosenkranz et al. 1990).

Mites mutilated by the bees

Mites found at the bottom (on the bottomboard) of infested colonies are often damaged, and such damage has been attributed to biting behavior of bees that detect and remove mites from other adults or from brood cells (Morse et al. 1991, Moretto et al. 1991a). Although some of this damage could be a result of natural drying out and breakage of dead mites or manipulation by the bees of mites already dead, damage such as bite marks on the dorsal shields of live mites found on the bottom and legs that have been cut demonstrates that the bees were actually responsible for the damage (Ruttner and Hanel 1992, Corrêa-Marques et al. 1994). The percentage of damaged mites was found to be inversely correlated with *V. jacobsoni* infestation levels in *Apis mellifera* colonies in Austria (Moosbeckhofer 1992) and Germany (Büchler 1993). The percentage of damaged mites varies with season, ranging from about 10 percent in March to about 60 percent in early summer (Büchler 1993). Some *V. jacobsoni* resistance selection programs have been based on this percentage of damaged mites (Wallner 1990a).

Attractants and repellents

Varroa mites discriminate among the various types of bee brood. The mites prefer drone brood over worker brood and worker brood over queen cells. They also prefer nurse bees over foragers and adult drones over adult worker bees. It has been suggested that host seeking in *V. jacobsoni* is mediated by kairomones (interspecific odor signals that benefit the receiver). Le Conte et al. (1989) found that the mites can discriminate between the odors of drone versus worker larvae, and they suggested the involvement of simple aliphatic esters based on studies with larval extracts. They concluded that methyl palmitate, found at higher levels in drone than in worker larvae, was a key attractant for the mites. Boot (1995) tested this substance on larvae, however, and found that it did not increase cell infestations.

Both alarm pheromone, which would characterize a bee that had stung or had been stung, and Nasonov pheromone, which would be present in greater quantities in foragers verses nurse bees, tend to repel *V. jacobsoni* (Hoppe and Ritter 1988, Kraus 1990). This means that the mites are sensitive to very specific bee odors that help them to find the most suitable adult bee hosts. Possibly, with more research, attractants, repellents, or both could be used to control *V. jacobsoni* behavior.

Kraus et al. (1994) tested various natural oils and determined that oil of clove and oil of cinnamon were attractive to *V. jacobsoni*, whereas oil of citronella was repellent. When oil of clove was incorporated into beeswax foundation at 0.1 percent, there was an increase in the infestation of the brood in such combs. Oil of marjoram foundation had the opposite effect.

Resistant bees

The natural resistance of Africanized honey bees in South America is evident from the low levels of infestation. In 1986, eight years after the mites were first discovered in Brazil, the average infestation of adult bees was found to be about two to three percent (Gonçalves 1987, Shimanuki et al. 1991). There is considerable evidence of an adaptive process in the resistance of Africanized bees in Brazil. The mean infestation rate in São Joaquim in spring (when infestations are lowest) was about seven mites per 100 bees in 1986, whereas in 1993 it was 2.03 per 100 bees. The same tendency was found in other regions of Santa Catarina state (Moretto et al. 1995b). An experiment made to select bees for the ability to remove *V. jacobsoni* from their bodies demonstrated that this characteristic is inheritable (Moretto et al. 1995b). However, Africanized bees are adapted to tropical and subtropical climates. That and their highly defensive behavior make them unsuitable candidates for introduction into many parts of the world where varroa is a problem.

Some of the search for resistant bee lines has been directed toward Africa in an effort to find African bees that can resist *V. jacobsoni*, but without the strong defensive characteristics and the inability to winter of *Apis mellifera scutellata* (the African subspecies introduced into Brazil in 1956). *Apis mellifera monticola* was introduced into Sweden for such an attempt. Hybrids with *A. m. ligustica* had lowered infestation rates as a result of a reduced percentage of fertile female mites and a shorter brood development period (Thrybom and Fries 1991). This apparent resistance appears to be a preadaptation because *A. m. monticola* has no contact with *V. jacobsoni* in the highlands of Kenya, where it originates.

Detection and evaluation of infestations

A beekeeper who wishes to determine whether *V. jacobsoni* exists in an apiary has several options. A quick technique is to uncap large areas of drone brood in the pupal stage with a long, sharp, serrated knife such as a bread knife and then strike the frame on a hive cover to dislodge the pupae. Alternatively, the brood can be pulled out with a capping scratcher (Szabo 1989). The mites can be seen readily on the white bodies of the drone brood. A rapid, relatively thorough survey can be made by examining several colonies of the apiary.

In the United States, the ether roll method is popular for detection. About 200 to 400 bees are placed in a glass jar, and these are sprayed with a jet of ether from an aerosol can (of the type used for helping to start cold car engines in winter). The jar is rolled and shaken for about 10 seconds, after which mites sticking to the jar's sides can be seen (Ellis et al. 1988).

For long-term monitoring, it is useful to place a sheet of light-colored material, such as plastic or stiff paper, on the bottomboard of the hive under a 2- to 4-mm mesh screen; dead mites fall through the screen and can be readily seen when the sheet is pulled out (Ritter 1981). The hive debris should be examined every three to seven days. Application of a fumigant with acaricidal properties increases the chances of finding mites (Ellis et al. 1988). Technicians in Israel (Slabezki et al. 1991) put paper strips with a few drops of Mavrik (agricultural formulation containing fluvalinate) inside the hive and burned them to provoke mite fall.

A comparison of several sampling methods showed that fumigant smoke with fluvalinate was the most efficient, with fluvalinate smoke > amitraz smoke > tobacco smoke > ether-roll (Witherell and Bruce 1990). In another comparison for first detection of *V. jacobsoni* in colonies previously free of this mite, inspection of hive debris was found be more efficient than sampling brood, which in turn was more efficient than sampling live adult bees (Fries et al. 1991).

Quantitative determinations

Adult bees can best be sampled by brushing 200 to 400 workers from brood combs into about 200 ml of a 0.5 to 1 percent detergent solution or 25 to 95 percent alcohol. The sample should be agitated vigorously for several minutes, then strained through a screen to remove the bees. The liquid that passes through the screen can be further strained through a white cloth on which the mites can easily be seen (De Jong et al. 1982c). A quantitative estimate of infestation rate can be obtained by counting the mites and bees. This is usually given as the number of mites per 100 adult bees.

The brood infestation rate is measured by uncapping sealed worker and/ or drone cells individually with a pair of forceps. A good, focused light source,

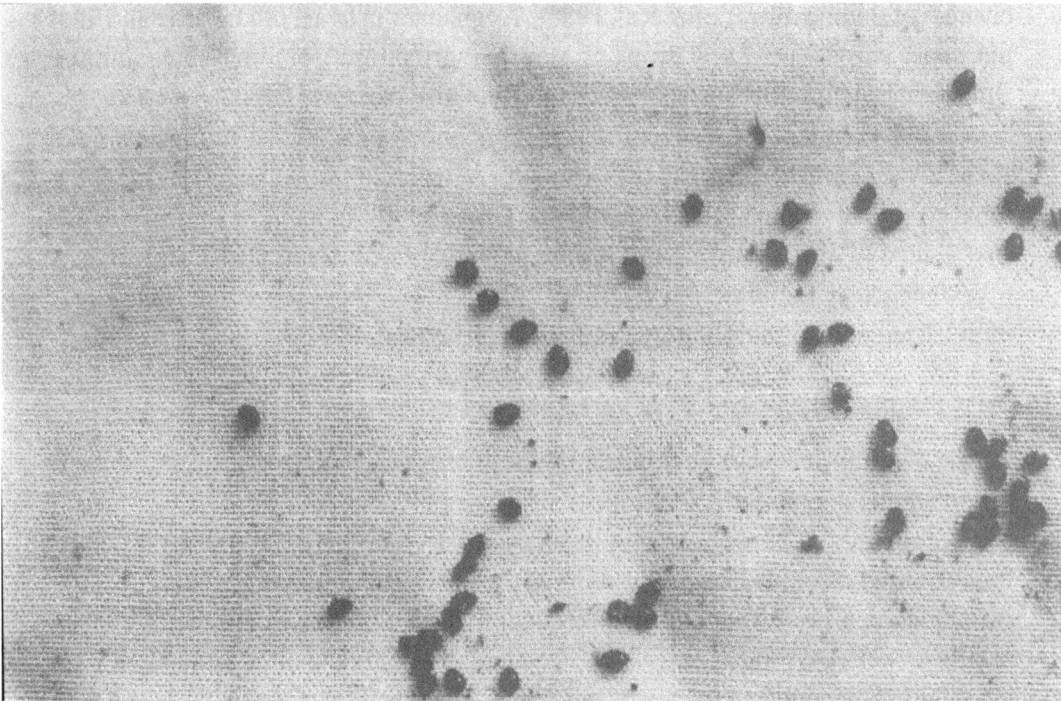

Figure 14.8a, b. After agitating 200-400 bees in a detergent or alcohol solution, they are placed on a mesh screen and the liquid, containing any mites, passes through. This is further screened to catch and observe mites, which can be easily seen and counted. (Morse photo).

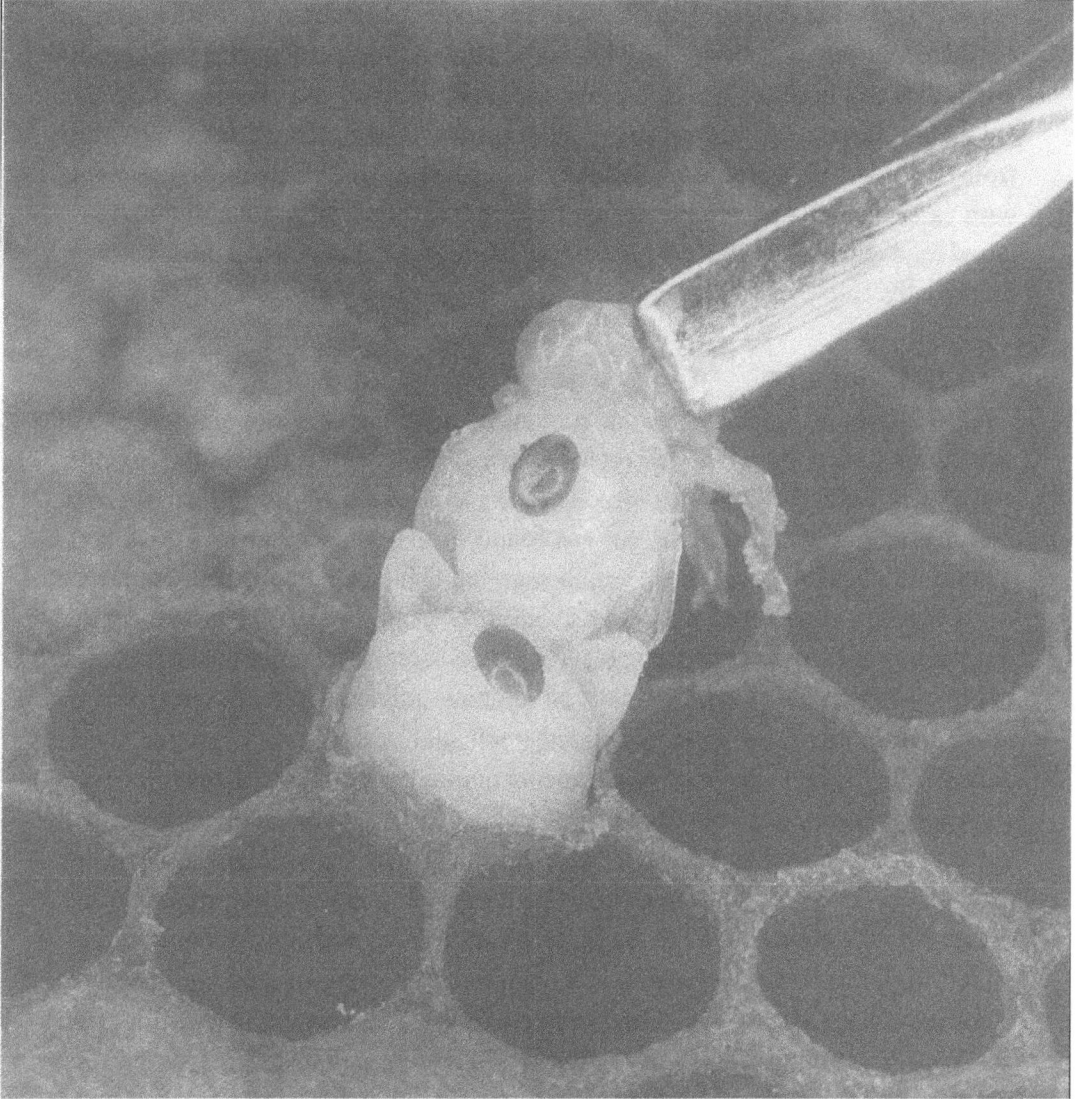

Figure 14.9. Examining individual cells requires planning. A good light source, forceps, and a knowledge of the age of the larvae to be examined is required. Plan to examine at least 100 cells. Drone brood is usually more heavily infested, but worker brood gives a better evaluation. (Hansen photo).

such as a fiber-optic lamp, an otoscope, or a small, focused flashlight is required for the mites to be seen on the bee larvae or pupae and on the cell walls. Approximately 100 cells or more should be examined in each colony to provide a reasonable estimate. The drone brood is nearly always more heavily infested, but an examination of worker brood provides a much better evaluation of the impact on the colony.

Some planning is necessary so as to maximize the usefulness of the brood infestation data. To obtain a count of adult mites that have entered brood cells, examine only capped worker brood less than 18 days old (i.e., pupae whose bodies are still light). The original female mites are darker in color than their offspring. To evaluate mite reproduction, check 18-day-old brood (bodies of worker pupae starting to darken), because mite offspring are easier to count and classify when they are further developed. For comparative purposes, it is important not to mix data from bee brood of diverse ages. Recently sealed larvae can be examined if removed with care, though they rupture easily, making it difficult to find the mites. If any larval food remains, it should be examined for mites still immersed in the worker (or drone) jelly.

Control

Once *V. jacobsoni* enters a country, all attempts at stopping its spread and eliminating it from infested regions have so far been unsuccessful (Koeniger and Fuchs 1989). Quarantines to control *V. jacobsoni* are generally short-lived, inefficient, and unenforceable, as was found in the United States (Anonymous 1988a). Eradication procedures, which were implemented in numerous countries, including the United States, the former U.S.S.R., Czechoslovakia, Chile, and Mexico, did little or nothing to slow the spread of *V. jacobsoni*. The major consequence in many cases was additional mortality of colonies and economic loss provoked by moving restrictions, owing to misguided eradication and control programs. This is partially the fault of the way governments normally operate. It is easy to obtain funding for emergency "elimination" of an exotic pest and much more difficult to get funds for control or research once it is clear that the problem has become established. In the United States in 1979, a suspected introduction based on two preserved mites found in a reused sample vial from an acarology laboratory provoked importation restrictions by all countries importing queens and packages from the United States. This "false" infestation was resolved only by killing and completely sampling all colonies within a 5 km radius and by extensive sampling outside this region (De Jong et al. 1981) at a considerable cost of resources and manpower.

Unlike many other diseases, keeping the colonies strong does not help to stop *V. jacobsoni*. All infested colonies die unless they are treated (except in some tropical and subtropical regions). Beekeepers who choose not to treat their colonies, or who do not believe that *V. jacobsoni* will be a problem, have lost entire apiaries. Normally no colonies are spared.

All treatments need to be done on an apiary or a regional level to avoid reinfestation. It is not feasible to treat some colonies and leave the rest because reinfestation will bring the infestation levels up to threatening levels in a very short time. Treated colonies become as infested as untreated controls in the same apiary in a relatively short time (Korpela et al. 1992). Thus good *V. jacobsoni* control necessitates a coordinated effort to treat all nearby apiaries at the same time.

Biological control

Some attempts have been made to control *V. jacobsoni* infestations by restricting brood rearing and by removing and destroying newly sealed combs (Ruttner et al. 1980), or by using drone brood as a trap to capture and then destroy a major percentage of the mites (Mel'nik and Muravskaya 1981, Schulz et al. 1983). This technique has been used successfully in tests in Sweden (Fries and Hansen 1993) and in Germany. However, the method is costly in labor and added work for the bees, and often so many mites succeed in reproducing in the worker cells that the colonies eventually become severely infested despite the control effort (Rosenkranz and Engels 1985, Marletto et al. 1990).

Heat treatments

One can selectively kill *V. jacobsoni* by heating the host adult bees with hot air at 42°-48°C (108°-118°F) (Khrust 1978, Komissar 1978). However, the adult bees must be removed from the brood and must be mixed constantly during treatment, and many bees die or are damaged unless the temperature is well-controlled. Heat treatments have not proved practical under field conditions. More recently, use of hot air (54°C for 15 minutes) treatment alone on honey bee colonies was found to be ineffective, but when it was combined with 5 ml of wintergreen oil, large numbers of mites were killed, with minimal bee mortality (Hoppe and Ritter 1989). Considerable bee mortality can result from heat treatment of the whole colony, but capped brood can be heated separately (Engels and Rosenkranz 1992). Above 38°C, the mites begin to be affected, and at 44°-45°C, brood is injured.

Chemical controls

Unfortunately, throughout most of the world, it has become necessary to apply chemical treatments (acaricides which selectively kill *V. jacobsoni*) at least annually so as to keep colonies of *A. mellifera* alive. The importance of beekeeping and the large number of countries involved have resulted in the development and use of many alternatives. Some of the principal chemicals have various trade names, depending on the country where they are used. Each country has its own regulations so that some of the substances are not approved in parts of Europe but are in use elsewhere. A good acaricide is one that is highly toxic for the mites, has a low toxicity for bees, and is innocuous for humans. Price and ease of application are also important. Some of the formulations developed for use in European apiculture, where high per colony costs and the necessity for multiple visits to the apiary are less significant considerations, are not suitable for commercial apiaries in countries such as Australia and the United States, where one beekeeper must treat a thousand or thousands of colonies.

The currently available control chemicals can be divided into four major categories: aerosols, fumigants, systemics, and materials that kill the mites by contact. These are modes of application and are listed in more or less historical order, though recently, problems with contamination of hive products and resistance to some of the control substances have provoked a return to some of the earlier controls.

Aerosols are sprayed directly into the hive entrance or over the brood frames. They are normally mixed with water. Amitraz was widely used in the past in this way. The mites are killed by contact or by breathing in the substance. Aerosols require a spraying machine and are time-consuming to apply. They are normally not very expensive, but multiple treatments at intervals of several days to one week are needed to achieve effective control. Generally speaking, aerosols have been the least effective mode of application. Acaricide applied in this way easily contaminates the various parts of the hive, and great care has to be taken to avoid applying aerosols at times of the year when bees are producing honey.

Fumigants are either evaporated in the hive (e.g., formic acid) or the material is burned directly (tobacco) or impregnated into a $NaNO_3$-soaked paper strip and then burned or smoldered (fluvalinate, amitraz). Fumigants are generally inexpensive and cause a quick knockdown of the mites, but they are normally not very efficient because the mites in the brood cells are relatively protected, making multiple treatments necessary. Generally, three treatments are made at seven-day intervals. The low cost of the chemicals is more than offset by the necessity for and cost of multiple visits. The mites are killed when they breathe the acaricide in the air. Generally speaking, fumigants do not contaminate the hive products as much as the other control methods because the effect is very short-term.

Systemics must be applied in a carrier. They are mixed with water (Perizin) or syrup (Apitol) and poured on the bees in the spaces between the brood frames. Bees that are wet with the acaricide mixture contact other bees and eventually spread the substance throughout the colony. Acaricides mixed with syrup are consumed by the bees and through the mutual exchange of food (trophyllaxis) the chemicals are distributed among the bees. Systemics are normally somewhat more expensive than fumigants, and they must be carefully dosed so that each bee gets enough, but not so much as to provoke toxicity. Ideally, they should be used during a broodless period, such as autumn in temperate climates, to be maximally effective and to reduce the risk of contamination of honey and other hive products. Application is slow and laborious and must be repeated several times. Dosages are more critical than with other means. Acaricides that are mixed with sugar syrup could potentially contaminate the honey in the hive, so they can only be applied well before the honey flow, or after harvest (when hives also normally have little or no brood). Nevertheless, especially in temperate climates, the mite populations are at critically high levels in autumn. Often the colonies are already dying, and treatments are no longer effective. Systemic chemicals are consumed by the bees and enter into the blood (hemolymph). The mites die when they ingest the contaminated bee hemolymph; the effectiveness of these chemicals depends on trophyllactic food exchange and distribution within the colony, which is not a completely efficient way to distribute effective dosages (van Buren et al. 1993). Coumaphos (the active ingredient of Perizin) has been found in beeswax from treated colonies (van Buren et al. 1992).

Contact control chemicals are currently the most widely accepted treatment method. They are incorporated into a special porous plastic carrier (a plastic strip) and are normally hung between frames of brood. The active ingredient is released slowly to the surface, and bees that walk over the strip become contaminated. These in turn contaminate other bees as they go through the hive. The main advantage is that only one application is necessary. The strips stay in the hive for six to eight weeks and are then removed. Contamination of colony products is generally minimal, and the strips are relatively safe for the beekeeper to handle because there is only a limited contact with the active ingredient. However, these products, as well as all acaricide formulations, should be handled with gloves. While the strips are in the hive, the release rate is relatively constant, permitting a good degree of control with minimal toxicity for the bees. Unfortunately, these slow-release strips are expensive. A typical cost per colony for one treatment is $4(U.S.) for the product alone (Clark 1994). Sometimes more than one treatment is necessary. In addition, the problem of contamination of hive products, although minimized when the strips are used according to recommendations, is not eliminated.

Currently, the most commonly used chemical controls are pyrethroids embedded in a plastic strip. These include fluvalinate (tau-fluvalinate) sold as Apistan® strips (Sandoz) and flumethrin, sold as Bayvarol strips (Bayer). Beekeepers place two Apistan® or four Bayvarol strips in each colony once a year, or if reinfestation is a problem, twice a year. These same chemicals have been used extensively in a clandestine form. Beekeepers have found that the same active ingredient is available for spraying crops. Mavrik and Klartan are two such acaricides, normally used to control mites of agricultural importance (Slabezki et al. 1991). The beekeepers make dilutions in water to approximate the plastic strip active ingredient dosages, soaking dry wooden strips, about 3 by 15 cm, 2 to 4 mm thick for one to several days. The strips are then left to drip dry and are inserted into the colonies. Such strips cost only a few cents each but are illegal throughout much of the world. Despite the much greater risk of contamination of honey and wax with these homemade wooden strips, in some countries nearly all beekeepers use them. Such homemade remedies have numerous disadvantages and one major advantage. The plastic strips cost around $4(U.S.) per colony per treatment (Clark 1994), whereas homemade substitutes cost only a few cents. In many countries, the commercial formulation plastic strips are not available because competition with the unofficial substitute prevents them from selling well enough to justify the costs of registering and selling them.

Fluvalinate (Apistan®)

In recent years, fluvalinate has been the preferred control for *V. jacobsoni* throughout much of the world because of the ease of use and the relatively low risk of contaminating the hive products with acaricide. Fluvalinate kills the mites by contact. Its commercial form, approved in many countries for use in honey bee colonies is called Apistan®. Apistan® strips are made of a porous plastic, and they

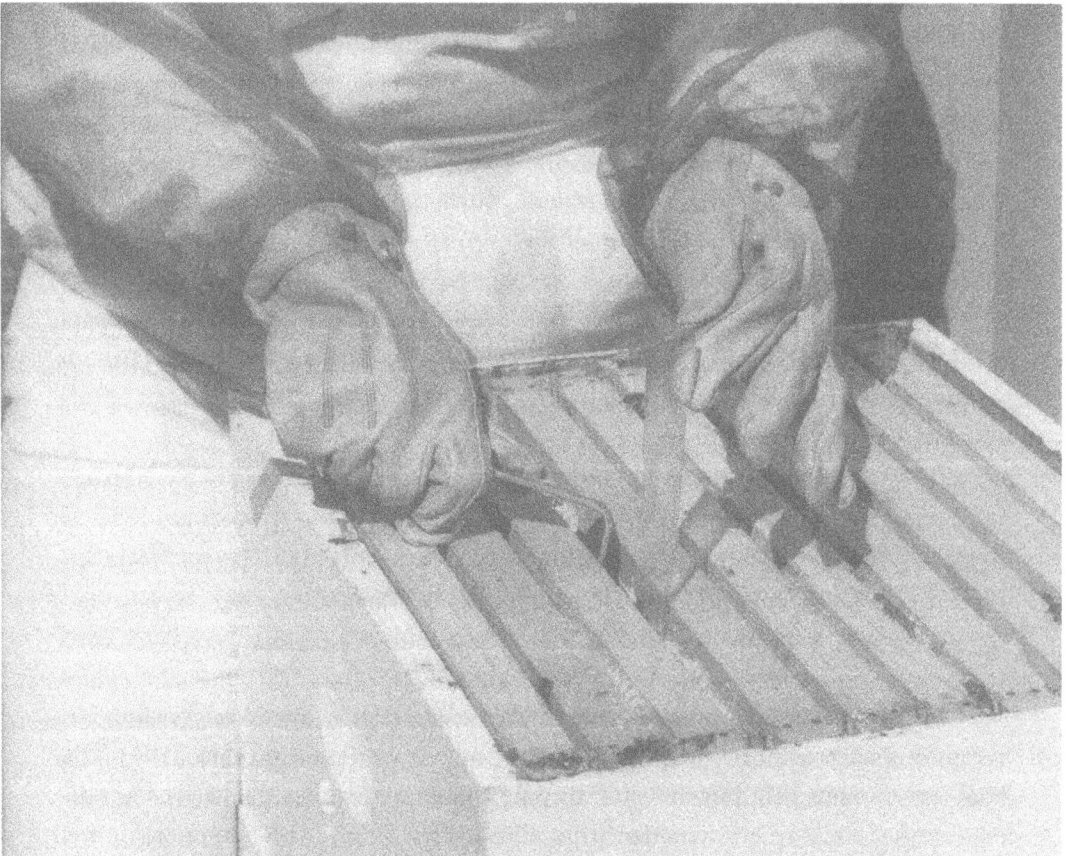

Figure 14.10. An Apistan® strip is placed between brood frames. (Sandoz Agra photo).

contain 10 percent (w/w%) of the active ingredient (tau fluvalinate). Two strips are normally placed in each colony and are hung vertically between brood frames, so that the bees will contact the strip surface. The bees become contaminated with the acaricide and in turn contaminate other bees. Mites on such contaminated bees are killed. Apistan® strips are left in the colonies six to eight weeks, so that the mites emerging from at least two full cycles of brood are controlled. The manufacturer recommends one treatment per year, though in some regions more than one application may be necessary (Azuma 1992).

The effectiveness of Apistan® can vary, but in general it is efficient. In Germany about 95 to 99.9 percent of the mites were found to be killed in 1989 (Moosbeckhofer and Kohlich 1989). Nearly 100 percent control was also found in the United States (Witherell and Herbert 1988).

Treatment of package bees and queen cages

Apistan® has been used effectively to treat bees in packages (Henderson 1988, Herbert et al. 1989). Nearly all mites are killed; nevertheless, it does not eliminate 100 percent of the mites, and therefore, package bees produced in *V. jacobsoni*-infested regions cannot be safely declared mite-free and would not be suitable for sale to regions of the world where *V. jacobsoni* has not yet arrived.

For the treatment of bees in queen cages, there is a product called Queen Tabs (Sandoz), which contains 1 percent fluvalinate. Queen mortality and introduction success is not affected by such a treatment (Pettis et al. 1991).

Contamination of honey and wax

Fluvalinate has been detected in both wax and honey, but residues in honey are normally low. Slabezki et al. (1991) found 0.06 ppm in honey after six to eight months of treatment with wooden fluvalinate-treated strips. Beeswax absorbs much more of this acaricide. They found 0.54 ppm and 0.83 ppm in the wax after six weeks and six to eight months, respectively. Contamination was reduced, but effective control was still maintained by placing the wooden strips in the entrance.

Although bee products can be contaminated by the use of Apistan®, the danger of contamination is considerably greater with homemade fluvalinate strips because the dosage is not well-controlled, and beekeepers sometimes introduce them while still wet with solution, directly contaminating the wax.

Apistan® strips also contaminate the combs. Residues of 900 to 1,200 ppm in beeswax after four weeks of treatment, and 24,000 ppm were found when the strips were left in the hives for 13 months (Liu 1992a). Mites continue to die at a high rate for three months or more after the strips are removed. Residues are also found in honey.

Figure 14.11. Apistan® Queen Tab being placed in a queen shipping cage. (Sandoz Agra photo).

Resistance to Apistan® (fluvalinate)

In Northern Italy, there have been numerous reports of resistance of *V. jacobsoni* to this product (Loglio 1993, Milani 1995, Boecking and Ritter 1994). An average of between four and 90 percent of the mites were killed by normal Apistan® treatments, with considerable mortality of colonies, even after treatment. One would expect resistance to develop eventually because of dependence on and repeated use of one product, although misuse has probably accelerated this process. Bee-keepers often neglect to remove the Apistan® or wooden fluvalinate strips after the recommended six weeks, often leaving them in year-round. That would mean that the mites would be constantly exposed to a gradually lower and lower dosage, liberated by the plastic or wooden strip. This creates a perfect environment for selection for resistance. Some beekeepers reuse the Apistan® strips. Though these used strips still effectively kill mites after the recommended six weeks (Moosbeckhofer and Kohlich 1989), less than the optimal amount of active ingredient is available, again creating an environment favorable to the development of resistance. Bioassays have shown that mites from areas where treatment with fluvalinate no longer controls infestations were 25 to 50 times more resistant to this chemical (Milani 1995). A cross-resistance was also found to flumethrin (the active ingredient of Bayvarol). Both of these chemicals are synthetic pyrethroids.

Application by fumigation or as an aerosol

Fluvalinate is also effective as a fumigant, though several treatments are necessary. Kulincevic et al. (1991) used a dose of 3.5 mg (active ingredient) in a paper strip impregnated with $NaNO_3$, three times at one-week intervals. The strip is ignited and allowed to smolder within the hive. Senegacnik (1990) applied 4.8 mg per strip, five times at four-day intervals.

Control of *V. jacobsoni* in Yugoslavia was accomplished with an aqueous emulsion of Klartan (Senegacnik 1990), applied as an aerosol. Klartan contains 240 g fluvalinate per liter. An emulsion was made with one part Klartan to 1,000 parts water. The hives were treated by applying 15 ml four times, at seven-day intervals, in late September and early October.

Problems with contamination

Wooden strips soaked with fluvalinate or flumethrin are at the same time a viable solution for the control of *V. jacobsoni* and a serious problem for beekeeping. To make them, beekeepers need to work with concentrated solutions, which can contaminate the handler. The original formulations, which are sold for application on crops, are sometimes mixed with insecticides, which on a number of occasions has resulted in extensive bee colony mortality. The beekeeper needs to make precise dilutions, often without proper orientation, and can easily overdose or make a treatment too weak to be effective. To determine a uniform dosage for treatments, the absorption capacity of the wooden substrate would have to be determined, a step rarely taken. Often these strips are inserted into the colonies while still wet, which greatly increases the risk of contamination of the wax and honey. Because the wood strips are inexpensive and are the same color as the wooden frames, they are even less likely to be removed after the treatment period than are the commercial plastic strips. This maintains the acaricide in the colony practically indefinitely, which certainly accelerates the development of resistance by the mites.

The commercial formulations (plastic strips) of the contact acaricides, when correctly used, are unlikely to seriously contaminate honey. But there is still a detectable residual contamination, which is higher than acceptable according to official rules in some countries. Even after the Apistan or Bayvarol strips are removed, mites still continue to die, demonstrating that the active ingredient remains in the wax combs for months or more (Büchler and Maul 1991, Moosbeckhofer 1991). This material can then diffuse into the honey, contaminating it (Wallner 1992a). Normally beeswax is contaminated by fluvalinate treatments (Lodesani et al. 1992). Combs from treated colonies also will kill mites when placed into untreated colonies (Moosbeckhofer and Kohlich 1990). This type of contamination can help provoke resistance by *V. jacobsoni* to the acaricide.

The more lucrative honey markets, such as Japan and Germany, do not accept contaminated honeys, and have rejected shipments and barred importations due to chemical contamination, making "organic" controls and other alternative strategies mandatory.

Further control possibilities
Organic acids

Various organic acids have become popular alternatives for the control of *V. jacobsoni* because they are natural substances and are not normally a contamination problem. They are also generally less expensive than the alternative control chemicals.

Formic acid was widely used in Germany and other parts of Europe in the late 1970s. Later, other chemicals became more popular owing to greater effectiveness and ease of application. At that time, beekeepers kept a small bottle of this acid in the hive. A wick slowly liberated the material and controlled the mites. Sometimes the bottles would fall over and kill the colony, and the beekeepers were often burned when this highly corrosive acid splashed on their hands. Formic acid does, however, have some important advantages. It is relatively inexpensive; a colony-treatment dose normally costs 5 to 10 cents. It is also a natural substance, found in small quantities in honey. Even when these levels increase temporarily, because of the treatment, formic acid is volatile and quickly evaporates. It does not have the disadvantage of a zero tolerance, which is often the case with artificial acaricides. For these reasons, formic acid is now regaining popularity. Apparently the mites are affected by formic acid more than the bees because of their small size, which makes it difficult for them to counteract the effects of the acid (Lupo and Gerling 1990).

Beekeepers apply formic acid by pouring it onto an acid board (an absorbent board inside a cover normally used for repellents) or onto paper toweling placed over the top bars (Fries 1991). The acid evaporates and the fumes flow down into the hive. The bees become nervous, and at high concentrations some queens and brood are killed. Beekeepers should use rubber or good quality plastic gloves, goggles to protect the eyes, and a respirator or other protective mask to avoid inhaling the fumes. Lower concentrations, around 65 percent, are a bit less dangerous, but still should be handled with extreme care. Hive inserts made with absorbent material, already containing the proper dose of formic acid, are much safer to handle and have been developed for beekeeper use. The beekeeper perforates a protective cover and inserts the formic acid "plate" into the hive. The plates are made of a kind of cardboard, about 30 cm by 20 cm by 0.15 cm thick, and contain about 20 ml of 65 percent formic acid (Hoppe et al. 1989). Though this means of application is more expensive than others, it is also much safer for the beekeeper, and there is less risk of contaminating honey or damaging the bees. Formic acid, in a gel solution, shows good control, safety to the beekeeper, and only moderate expense. it has been tested in the United States, but at press had not yet been approved. Approval, however, is expected.

Formic acid has the advantage of killing mites in the brood cells, though dosage must be carefully regulated to avoid killing brood, especially the smaller larvae (Fries 1991).

Lactic acid

Lactic acid has been used successfully in Italy along with trap combs to control *V. jacobsoni*. It can be applied with a five-liter knapsack sprayer at the rate of 8 ml of 15 percent lactic acid per comb side, giving a mean efficiency of 97.5 percent (Kraus 1991, 1992). Some mortality of bees' eggs has been found.

Thymol

Powdered thymol, alone or mixed with powdered sugar, has been used to control *V. jacobsoni* in Italy (Chiesa 1991). Half a gram of thymol was sprinkled over each comb filled with bees, four times at two-day intervals. Some bee mortality occurs, but mite populations are reduced by more than 98 percent. These treatments are done when there is little or no brood. In Israel 10 percent thymol was found to be effective for controlling *V. jacobsoni* (Gal et al. 1992). Thymol has a strong odor.

Mixtures of etheric oils

A recently developed mixture of etheric oils, consisting of 74 percent thymol, 16 percent eucalyptus oil, four percent menthol, and four percent camphor in an absorbent material, gives about 90 percent efficiency. Water suspensions with clove oil and wintergreen oil kill 99 percent of mites, but the effective dose is close to the dose for bee mortality (Hoppe and Ritter 1989).

Varroa jacobsoni on its original host, *Apis cerana*

Colonies of *A. cerana*, the natural host of *V. jacobsoni*, are normally little affected by the mites (Kitaoka 1983). There has been much speculation about resistance mechanisms in this Asian bee, and about how they might be 'bred into' *A. mellifera.*

Koeniger et al. (1981) found that *V. jacobsoni* on *A. cerana* in Sri Lanka reproduced only in drone cells, though the mites occasionally entered worker cells. However, some reproduction in worker cells occurs in South Korea (De Jong 1988), although infestations in *A. cerana* colonies are generally low, and no treatment is necessary. In India, natural reproduction is restricted to drone brood (5 to 34 percent infestation) and occurs only in the spring. The maximum infestation in worker brood is less than 1 percent with drone brood present, but can increase up to 6 to 8 percent when no drone brood and little worker brood are present. All of these worker brood infestations were apparently unsuccessful reproduction attempts; no eggs or nymphs were detected (Tewarson et al. 1992).

Most mite reproduction occurred in the spring.

The low tendency of *V. jacobsoni* to invade and reproduce in worker brood cells of *A. cerana* eliminates the principal cause of damage found in *A. mellifera* colonies. In the latter, such a large percentage of the worker brood is killed or debilitated that eventually the colony is weakened and killed, owing to the lack of replacement bees. Moreover, the mites are restricted in their reproduction because fewer drone cells are present in *A. cerana* colonies, and drones are produced during only part of the year.

Aside from the low tendency of *V. jacobsoni* to attack worker brood in this bee species, a behavioral factor is apparently involved in keeping mite populations low. When individual mites are introduced into colonies, the *A. cerana* workers quickly detect, bite, and kill the mites. If the mite succeeds in climbing onto an adult bee, the bee becomes agitated and performs grooming solicitation movements, and other bees quickly groom her and remove the mite. The reaction to the mites is so strong that it seems remarkable that any survive in the colonies. However, the high grooming rates originally reported by Peng et al. (1987a) were probably caused by odors from the *A. mellifera* colonies from which they were collected (Rosenkranz et al. 1993b). When mites are taken from the same *A. cerana* colony, they still tend to be attacked, but to a lesser degree (Tewarson et al. 1992). *A. cerana* can also detect and remove *V. jacobsoni* from infested brood cells (Rath and Drescher 1990) and will even remove *A. mellifera* brood infested with *V. jacobsoni* if it is introduced into their colonies (Peng et al. 1989b).

On *A. cerana, V. jacobsoni* is normally confined to the ventral part of the first abdominal segments, and it inserts itself between the sternites, where it is protected and can feed. This is in contrast to *V. jacobsoni* on *A. mellifera,* where the mites are often found on the surface of the thorax and abdomen (Boecking et al. 1993).

However, serious damage by *V. jacobsoni* to *A. cerana* has been reported from India, (Phadke et al. 1966, Pandey 1967). In heavy infestations, honey production is reduced, and up to four mites are found in some worker cells; colony mortality may be caused by secondary infections of the weakened colonies with sacbrood virus (Tewarson and Singh 1986).

The cocoon spun by the larvae of *A. cerana* drone is different from that of the worker. The cell capping is conical and has a small apical orifice, about 0.3 mm in diameter through which *V. jacobsoni* cannot pass. It is densely knit and difficult to remove. Apparently this cell capping helps protect the developing drone from being removed by the worker bees, but it also effectively protects the mites. Though *A. cerana* quickly removes a large percentage of mites that enter into worker brood cells, it does not do so in the case of drone brood, unless the cocoon cap is (artificially) removed and replaced with wax (Rath 1992, Rath and Drescher 1990).

The drones line the inner part of the wax cell cap with their spun cocoon, cutting it open when they are ready to emerge. This apparently protects the drones themselves from being cannibalized by the bees, and consequently protects the mites. When several mites infest a drone brood cell, the developing drone becomes so damaged that it dies, and as a consequence, the mites cannot leave the cell, and they die, too.

Varroa jacobsoni on *Apis koschenikovi*

V. jacobsoni has also been found on a newly described species of *Apis*, *Apis koschenikovi*, in Sumatra (Otis 1991). This bee is similar to *A. cerana*, but is slightly larger. As with *A. cerana*, *V. jacobsoni* mainly attacks the drone brood.

Varroa underwoodi

Another mite species, *Varroa underwoodi*, has been found at low levels in drone cells of *A. cerana* in Nepal (Delfinado-Baker and Aggarwal 1987) and South Korea (De Jong 1990b). It closely resembles *V. jacobsoni* in appearance, except that it has much longer setae on the lateral margin of the dorsal shield, and it is much smaller; the adult female is about 760 μm long by 1,160 μm wide. The male is about 675 μm long by 600 μm wide. Female *V. jacobsoni*, by comparison, have a mean length and width of about 1,160 μm by 1,696 μm. The similarity of these species is evident from an apparent character displacement in a region of Nepal where they are found together (in sympatry). The *V. jacobsoni* collected from the same region where *V. underwoodi* was found were the largest found in a morphometric study of mites from 17 countries (Delfinado-Baker and Houck 1989). They averaged 1,126 μm in length, even though they were collected from *A. cerana*, which normally has smaller *V. jacobsoni* than those found on *A. mellifera*.

Euvarroa sinhai

Euvarroa sinhai is a parasitic brood mite of the Asian bee, *A. florea*. The adult female *E. sinhai* is somewhat smaller than the *V. jacobsoni* female, about 1,040 µm long by 1,000 µm wide, and is brown and more or less pear-shaped (Delfinado and Baker 1974). Akratanakul (1975) studied the biology of this mite in Thailand and found it to be similar to that of *V. jacobsoni*, except that the mites enter only into drone brood cells, where all reproduction occurs. As in *V. jacobsoni*, the adult male is smaller than the female, is pale and has chelicerae modified for sperm transfer. The adult females are phoretic on both workers and drones (Akratanakul and Burgett 1976). According to Kapil and Aggarwal (1986), reproduction is cyclic, with peaks occurring in spring and autumn, in India. *E. sinhai* has also been found in Sri Lanka (Koeniger et al. 1983). *A. florea* does not exhibit grooming or other defensive behavior toward *E. sinhai* (Koeniger et al. 1993).

E. sinhai has been found in *A. florea* colonies in Iran. Infestation levels are low, less than 1 percent on adult bees. The adults survive for long periods (four to 10 months) between cycles of drone production (Mossadegh 1990).

Experimental infestations of *A. mellifera* and *A. cerana* demonstrate that these hosts are potential candidates for parasitism by *E. sinhai* (Mossadegh 1990, Koeniger et al. 1993), though the latter bee species actively attacks the mites.

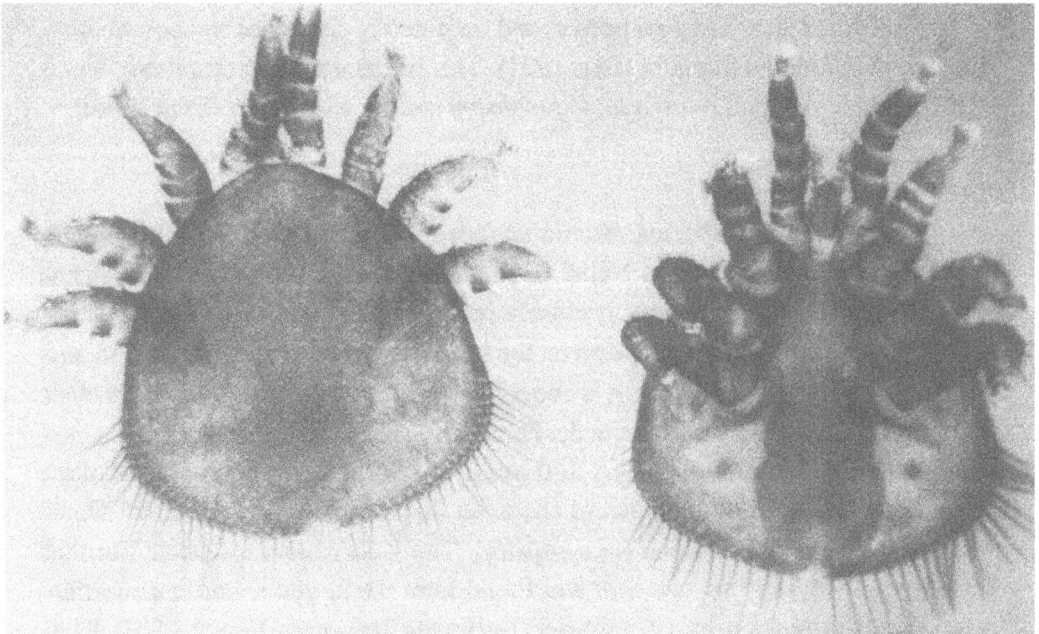

Figure 14.12. *Euvarroa sinhai* collected from *Apis florea* in Sri Lanka. Left, dorsal view; right, ventral view. (Koeniger photo).

Euvarroa wongsiri on *Apis andreniformis*

Another species from this genus, *Euvarroa wongsiri,* was recently described from specimens collected from drone brood of *Apis andreniformis* in Sumatra (Otis 1991).

Tropilaelaps clareae

Tropilaelaps clareae is a medium-sized (adult female: 1,030 μm long by 550 μm wide), elongated, active, light, reddish-brown mite native to Asia and parasitic on *A. dorsata.* The males are almost as large as the females, and they are less sclerotized. It was first described from specimens collected in *A. mellifera* colonies in the Philippines (Delfinado and Baker 1961) and has created problems for *A. mellifera*-based beekeeping in tropical Asia. The dorsum of adult mites is hard and the venter soft. The adult females swell considerably when they feed on bee brood. Normally they are about 300 μm thick, increasing to 500 to 600 μm (Woyke 1987c). Mating behavior is similar to that known for *V. jacobsoni* (Rath et al. 1991). Males introduce sperm between the third and fourth leg of the female with their hollow-tipped chelicerae.

This mite does not stay on the adult bees as much as does *V. jacobsoni* but often moves freely on the combs. *T. clareae* has relatively primitive, unspecialized chelicerae and is therefore incapable of piercing the membranes of adult bees to obtain food (Griffiths 1988). Thus, this species is phoretic on adult bees but depends on the brood for feeding.

A. dorsata is the native host. Infestations on this bee are relatively low. In northern Thailand, among 17 nests, 13 were found to be infested, with brood infestation rates of 0.2 to 6.8 percent (Burgett et al. 1990). Many of the mites found in colony debris under the comb are damaged, indicating that *A. dorsata* can kill these mites (Rath 1991). The adult mites are found principally in the petiolar region, probably to escape grooming (Buchler 1992a).

The damage to *A. mellifera* is quite severe (Burgett and Akratanakul 1985). Infested bee pupae are found to have darkly colored spots, mainly on extremities. Adult bees from infested colonies are often deformed. Injuries appear at lower degrees of infestation than with *V. jacobsoni* (Ritter 1988). Because of rapid reproduction, *T. clareae* is a much more serious problem than *V. jacobsoni* in regions where both mites occur (Woyke 1989). Rath (1991) found 13 times more *T. clareae* than *V. jacobsoni* in *A. mellifera* colonies in Thailand. These mites are often found in the debris of *A. mellifera* colonies, but unlike in *A. dorsata* colonies, few show bee-inflicted damage (Rath and Delfinado-Baker 1991).

This mite is widely distributed throughout Southeast Asia. It is generally restricted to the range of *A. dorsata*, but it is also found in Afghanistan (Woyke

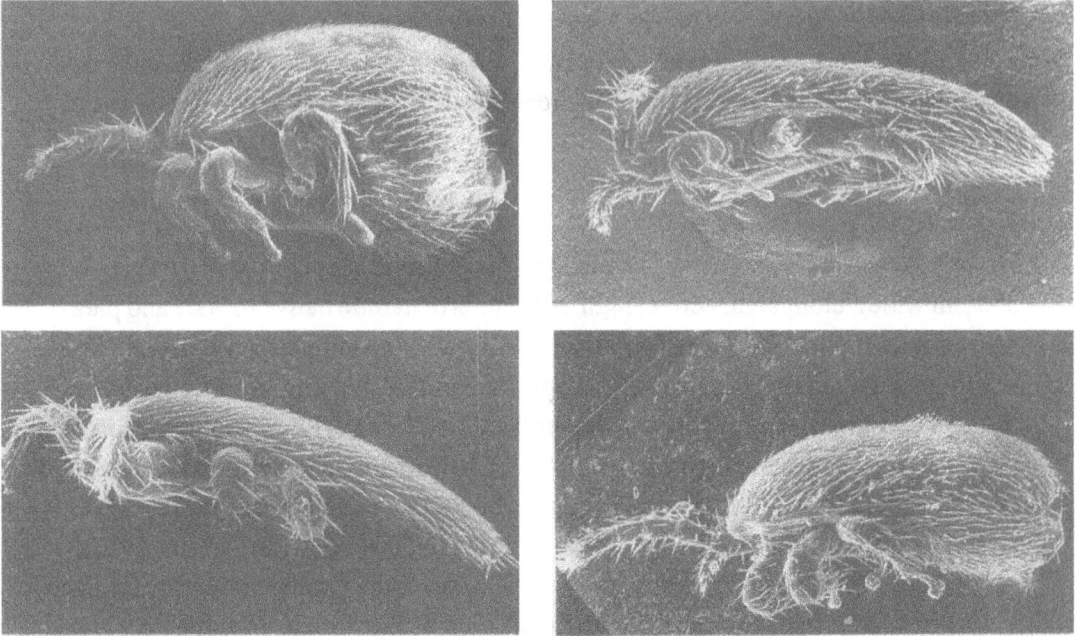

Figure 14.13. Females of *Tropilaelaps clareae*. Upper left: young female moving freely on comb. Lower left: mated female on brood, 0-4 hours after cell sealing. Upper right: gravid female, 20-24 hours after cell sealing. Lower right: gravid female, 48-52 hours after cell sealing. (Woyke photos, From J. Apic Res. 28: 196-200, 1989).

1984b) and in areas of China (Fan and Li 1988), where there are no *A. dorsata*. It has also been introduced into Kenya (Kumar et al. 1993).

T. clareae has been found in brood combs of *A. cerana* in Burma and Pakistan (Delfinado-Baker 1982). Aggarwal (1988) reported *T. clareae* at low infestation levels in colonies of *A. florea* in India, though there is no evidence of successful reproduction by this mite on this bee species. P. Akratanakul (personal communication 1977) found no *T. clareae* on *A. florea* in Thailand.

The life cycle of *T. clareae* is similar to that of *V. jacobsoni*. Its larva is mobile, however, and feeds, in contrast to the *V. jacobsoni* larva, which develops entirely within the cell (Kitprasert 1984). The first egg is laid about 48 hours after the cell is sealed (Woyke 1989). The preference for drone brood cells is less marked than for *V. jacobsoni*, being only about 1.5 to 3.0 times, when compared to worker brood (Woyke 1987a, Ritter 1988). Both males and females can be found outside brood cells. Females can reproduce more than once (Woyke 1994a).

The lengths of the life stages of *T. clareae* in *A. mellifera* brood are: egg, 0.3 to 0.4 days; larva, 0.3 to 0.6 days; protonymph, 1.7 to 2.0 days, and deutonymph 3.0 to 3.8 days, for a total of six days. Generally, all of the mites finish their development before the bee emerges, which helps explain their more efficient reproduction when compared to *V. jacobsoni* (Woyke 1987c).

Adults are phoretic on adult bees, but without access to bee brood, they can survive only a short time. It appears that *T. clareae* cannot feed on adult bees (Kitprasert 1984, Woyke 1984). At any one time, a great majority of these mites are found in brood cells. Woyke (1987b) found that only 3 to 4 percent of adult mites were on the adult bees, about one-tenth the percentage common for *V. jacobsoni*.

In areas of the world where there is a seasonal break in brood rearing, these mites apparently cannot survive. Different from *V. jacobsoni*, *T. clareae* often enters open brood cells to feed on the older larvae. They can survive at least four weeks when worker larvae at least four days old are present; younger larvae are not adequate (Woyke 1994).

In infested colonies, adult *T. clareae* can often be observed running over the brood combs, and the mites are easily dislodged and observed if the combs are shaken or hit sharply over a piece of paper (Laigo and Morse 1968). In a colony with a heavy infestation, many bee larvae are deformed, there is an irregular brood pattern, and emerging adult bees have damaged wings, shrunken abdomens, or both.

On *A. dorsata*, its original host, *T. clareae* can attain high infestation

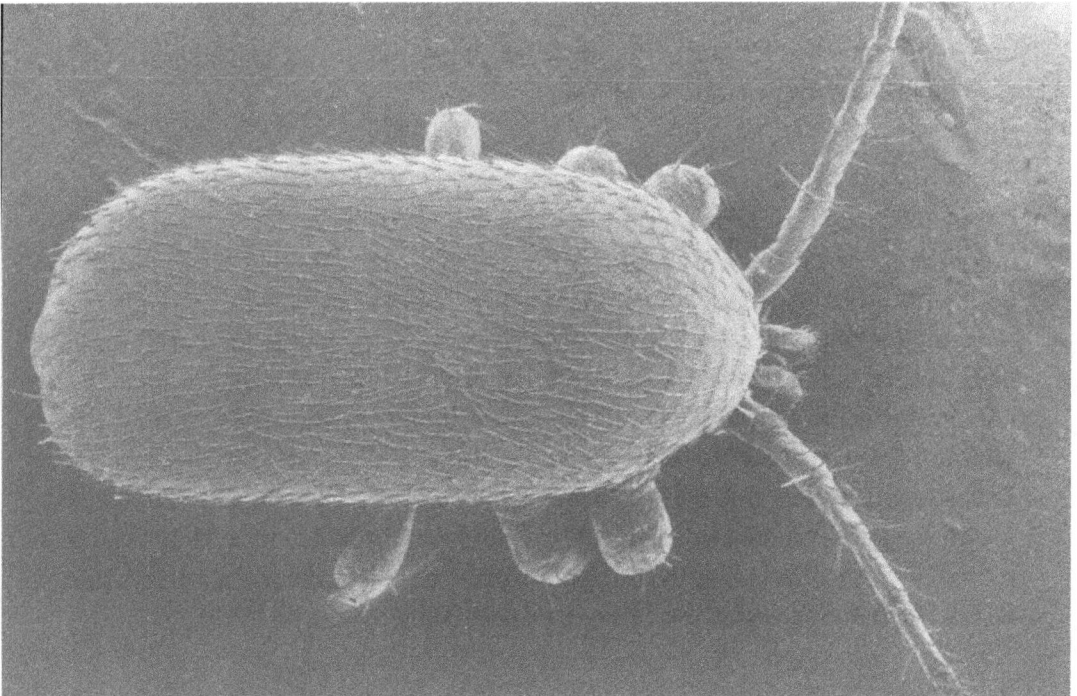

Figure 14.14. Dorsal view *T. clareae*.

levels, damaging 30 percent or more of the brood (Laigo and Morse 1968, Woyke 1985). But damage is much more severe in *A. mellifera* colonies. For example in Afghanistan, 90 percent of *A. mellifera* colonies were destroyed by these mites (Woyke 1984).

T. clareae can be cultured in the laboratory on *A. mellifera* brood. Swollen adult females taken from fifth-instar larvae are transferred into plastic cavities, in an ELISA microtiter rack, together with larvae taken from sealed brood cells. It is possible to obtain a complete reproductive cycle under these conditions (Rath 1991).

When *T. clareae* and *V. jacobsoni* are found together, the former predominates (Woyke 1987a), though apparently *T. clareae* does not prevent the reproduction of *V. jacobsoni* in brood cells infested by both. However, the reproductive rate of *V. jacobsoni* is reduced, just as it is when there is multiple infestation with *V. jacobsoni* (Rath et al. 1995). Woyke (1987a) found worker brood infestation rates of 5 percent for *V. jacobsoni* and 46 percent for *T. clareae* in *A. mellifera* colonies in South Vietnam, whereas on adult bees the corresponding rates were 6.3 and 3.4 percent. In Thailand, brood infestation rates were found to be 3 percent for *V. jacobsoni* and about 15 percent for *T. clareae* (Rath et al. 1995).

Of the two principal mites parasitic on honey bee brood, *V. jacobsoni* is generally a serious pest in temperate regions but not in the tropics, whereas in tropical and subtropical areas *T. clareae* is much more virulent, though it is easily controlled in temperate areas. For the future, the greatest risk is the spread of *T. clareae* into the tropics outside of Asia.

Biological control

Successful treatment of infested colonies can be achieved by creating a broodless environment (Woyke 1985). Recommendations for controlling this mite without the use of chemical controls include: removal of all brood combs (the remaining mites will die before new brood is available), caging the queen for 21 days, cutting out the remaining sealed brood, and caging the queen for nine days and uncapping and shaking out all the sealed brood (Woyke 1984). Control by selective removal of sealed drone brood probably would be effective because this mite strongly prefers this type of brood (Woyke 1987a). To minimize production losses, the queens should be caged toward the end of the honey flow. Brood can be placed in another hive, near the original, to make a nucleus colony. Two days after all the bees have emerged, the colonies can be reunited (Woyke 1993).

Chemical control

Most chemicals that control *V. jacobsoni* are also adequate for *T. clareae*, but the mode of application is important. Whereas fumigation works to some extent with *V. jacobsoni*, it is not adequate for *T. clareae* because of the short time that this mite spends on adult bees. At any one time a large percentage of the mites are protected inside brood cells (Woyke 1987b). However, they are more sensitive to formic acid treatment than is *V. jacobsoni*.

Tropilaelaps koenigerum

Tropilaelaps koenigerum is found on *A. dorsata* in Sri Lanka (Delfinado-Baker and Baker 1982). It is smaller than *T. clareae* (adult female: about 700 μm long by 450 m wide), oval, and light brown. The adult males are considerably smaller than the females, in contrast to *T. clareae*. *T. koenigerum* was collected on pupae in queen and worker cells, and on adult bees, in the brood area of *A. dorsata* comb. It is also found on *Apis laboriosa*.

CHAPTER FIFTEEN

———— ✵ ————

Amphibians and Reptiles
Keith S. Delaplane and Richard Nowogrodzki

Amphibians
Reptiles
Control

CHAPTER FIFTEEN
Amphibians and Reptiles

Amphibians are cold-blooded vertebrate animals with naked skin, aquatic (gill-breathing) larvae, and more or less terrestrial (lung-breathing) adults. They require moist habitats, and most species are nocturnal. Amphibians occur globally, except in polar areas and on most oceanic islands. Most species occur in areas of the tropics that have abundant rainfall (Frost 1985, Blair et al. 1968). In this chapter, we discuss three families of significance to beekeeping—the toads (family Bufonidae), tree frogs (family Hylidae), and frogs (family Ranidae).

Reptiles are cold-blooded, vertebrate, land-dwelling animals. Their skin is typically dry and protected by horny scales. Most can move rapidly to capture prey. Like amphibians, reptiles are best suited to the tropics, but they also extend into temperate regions (Blair et al. 1968). Lizards (order Squamata, suborder Sauria) are the only reptiles known to eat bees.

Many amphibians and reptiles eat insects, but for most of these predators, bees are relatively unimportant. For smaller predators the bee sting is a powerful deterrent, and larger predators generally prefer prey bigger than bees. Consequently, amphibians and reptiles are considered minor pests of beekeeping. The only exceptions are some toads, frogs, and lizards that raid apiaries in tropical and subtropical areas of the world.

Amphibians

Toads (family Bufonidae) in the genus *Bufo* are the most serious amphibian pests of beekeeping. Typically, toad predators sit near the entrances of bee hives at dusk or at night and eat adult bees. On hot nights when bees are clustered at entrances, toads can eat large numbers of bees quickly. If bees are not clustered outside, toads may jump and knock their heads against a hive and eat the bees that come out to investigate (Hilse 1986). This feeding activity can cause a steady and serious drain on the strength of colonies.

Toad species known to eat bees include *B. boreas, B. bufo, B. marinus, B. rangeri, B. regularis, B. terrestris, B. viridus*, and *B. woodhousei*. Of these, the cane or marine toad, *B. marinus*, is the most serious pest. *B. marinus* is a large toad native to the American tropics. Its natural range extends from Northern South America to Southern Texas. It was introduced to many places in the world to control insect pests of agriculture. Today it exists in the Antilles, Australia, Fiji, Hawaii, New Guinea, Philippines, Ryukyu Island, Taiwan, many Pacific islands (Frost 1985), Bermuda (Grant 1948), and Florida (Blair et al. 1968; DeJong, personal communica-

Figure 15.1. The toad *Bufo marinus*, a predator of honey bees in the tropics and subtropics of the Western Hemisphere and also Australia and Hawaii. (©Camazine photo).

tion 1987). *B. marinus* is a beekeeping pest in Fiji where it occurs throughout the island and catches bees as they leave the colony (Anderson 1990). In Hawaii, *B. marinus* is abundant in wet areas of the islands where it eats large numbers of honey bees at hive entrances (Messing 1991). *B. marinus* was imported in 1935 from Hawaii into Queensland, Australia, where it has become a "serious, persistent and voracious feeder" on honey bees. In Queensland, toads move at night from wetlands into apiaries located on drier ground where they eat bees (Gold 1969).

There are other records of *Bufo* predation on bees. Toads are pests in the Dominican Republic, where beekeepers take control measures against them (Morse 1975a). In one California apiary, large numbers of toads (probably *B. boreas*) were active at night and were the suspected cause of weakness in small queen mating colonies. Guts of toads collected at this apiary contained bees, and one night, 16 toads were found along a row of 12 beehives (Eckert 1934). *B. bufo*, a European species that has spread through Asia to the Pacific Ocean, eats bees at hive entrances in Great Britain (Herrod-Hempsall 1937) and has consumed them in the laboratory (Cott 1936, Lescure 1966). Bee predation is known by *B. viridis* in North Africa (Lescure 1966) and by *B. rangeri* and *B. regularis* in South Africa (Johannsmeier 1978). Toads eat bees at hive entrances in Thailand (Wongsiri et al.

1986). In Arizona, the Colorado River toad eats large numbers of bees in summer (Shimanuki et al. 1992).

Authors have measured with numerous parameters the rate of toad predation on bees. In Queensland, Australia, one *B. marinus* may eat from 100 to 500 bees per night (Gold 1969), whereas in Great Britain, *B. bufo* eats about eight bees per night (Herrod-Hempsall 1937). One large female *B. bufo* ate 32 bees in one hour and 95 bees in nine evenings (Lescure 1966). Guts of *B. boreas* collected at an apiary contained an average of 23.3 bees each (Eckert 1934), and guts of *B. marinus* contained 50 to 60 bees each (Grant 1948). Hilse (1986) examined stomach contents of 215 *B. marinus* collected in his 36-hive apiary over eight nights and estimated that toads ate over 30,000 bees per week—roughly the population of one bee colony.

Although toads can be locally serious, worldwide they are a relatively minor problem. In Oman, they are considered a minor pest (Hussein 1989). In North America, O'Brien and Marsh (1990) surveyed apiarists in 48 states and nine provinces to assess vertebrate pests of beekeeping. Only respondents from Arizona and Hawaii listed toads as a pest, although one Arizona respondent ranked toads as the second most damaging vertebrate pest in that state.

Frogs (family Ranidae, genus *Rana*) and tree frogs (family Hylidae, genus *Hyla*) are the only other notable amphibian predators of bees. Records of bee predation exist for *R. catesbeiana* in Missouri, U.S.A. (Korschgen and Moyle 1955), *R. esculenta* in Poland (Kawiak 1955), *R. temporaria* in England (Herrod-Hempsall 1937), and an unidentified *Rana* species and *Hyla* species in Thailand (Wongsiri et al. 1986). Huheey (1980) induced the North American tree frog, *H. cinerea*, to eat bees in the laboratory.

Bees are an insignificant fraction of the total diet of most amphibians. During a 12-week census in Arkansas, Brown (1974) examined the stomach contents of 1,301 toads and frogs and found only one honey bee in the stomach of *B. woodhousei* (Brown, personal communication 1975).

Bee stings can be a powerful deterrent to amphibians. Young or inexperienced toads readily attack bees, but most learn to avoid them after one or more sting experiences (Cott 1936, Brower et al. 1960, Brower and Brower 1962, 1965; Huheey 1980). Cott (1936) described a toad reacting to a bee sting with a "sudden violent jump, jerky vertical bowing motion of the head, repeated withdrawal of the eyes into the head, exaggerated swallowing movements often accompanied by mouth-wiping reactions with a forelimb, gaping mouth, and protruding tongue"; toads subsequently retreated when bees were introduced into their cages, and this learned response was repeated when bees were reintroduced after a two-week period free of bees.

Sting deterrence seems to depend in part on the relative size of the predator. In the laboratory, a small tree frog (*H. cinerea*, maximum body length of 5.4 cm [Blair et al. 1968]) died from a bee sting to the heart, whereas medium-sized toads (*B. terrestris*) avoided bees after being stung, and one large toad continued eating bees even though it was repeatedly stung and obviously in pain (Huheey 1980). In the laboratory, *B. terrestris* is more deterred by the bumble bee *Bombus americanorum* than by the much smaller honey bee (Brower and Brower, 1965). Toads, presumably large or otherwise tolerant of bees, often feast on bees with seeming disregard of stings. In a South African apiary, toads kept eating bees even though stings were embedded in their mouths, throats, and esophagi (Johannsmeier 1978). Cott (1936) and Brower and Brower (1962) recorded certain toad individuals that ate honey bees even after days of repeated stings.

Reptiles

Lizards, the only reptiles known to eat bees, are a large group with varied life histories including arboreal, burrowing, and terrestrial species. They are most numerous in warm temperate or tropical and subtropical regions of the world. Unlike the cosmopolitan toad *Bufo marinus*, no one species of lizard is a widely recognized pest of beekeeping. Lizards are minor pests of beekeeping, with some local, mostly tropical, exceptions. Much of the literature is anecdotal.

Seeley et al. (1982) observed feeding behavior of agamid lizards (*Calotes mystaceus* and other species) on *Apis cerana* colonies in Thailand. The lizards "dart to the nest entrance, snatch up a bee from the outer portion of the entrance tunnel, and leap away" and "probably exert a small but steady drain on colonies." Also in Thailand, the lizard *C. emma* and the gecko *Hemidactylus frenatus* feed on *A. mellifera* at hive entrances (Wongsiri et al. 1986).

In southern coastal North Carolina, U.S.A., the six-lined race-runner (*Cnemidophorus sexlineatus sexlineatus*) eats honey bees at hive entrances (J.T. Ambrose, personal communication 1987). Grout (1946) listed lizards as "minor enemies" of beekeeping in California. In a survey of apiculturists from North American states and provinces, lizards were listed as "always minor" pests of beekeeping by one respondent in Arizona (O'Brien and Marsh 1990).

In Sudan, the lizard *Mabuya quinquetaeniata* captures bees while they visit watering containers placed in the apiary by beekeepers (J.T. Ambrose, personal communication 1987). Lizards are predators of honey bees in Zimbabwe (Papadopoulo 1964), India (Gandheker 1959), and Oman (Dutton et al. 1981, Hussein 1989). In a two-year (1985-1986) survey of beekeeping pests of Oman, lizards were a problem in all months except July, August, and September (monsoon season); lizards comprised 7.2 percent of the recorded pests of honey bees in 1985 and 9

Figure 15.2. *Mabuya quinquetaeniata*, a bee-eating lizard from Sudan. (Ambrose photo).

percent in 1986 (Hussein 1989).

Insects are the dietary mainstay for many lizards. Some specialize in certain prey taxa such as ants (as with *Anolis distichus*, Schoener 1968), whereas others are more catholic, eating a variety of prey and plant material (as with *Leiocephalus* spp., Schoener et al. 1982). In Brazil, bees in the families Apidae, Halictidae, and Megachilidae were recovered from the guts of *Tropidurus* spp. (Vitt 1993), and apids were recovered from *Polychrus acutirostris* (Vitt and Lacher 1981). In Kansas, bees in the families Andrenidae and Halictidae were recovered from the guts of *Holbrookia maculata maculata*, and halictids were recovered from *Sceloporus undulatus thayerii* (Burt 1928). In these studies, bees made up a moderate or very small fraction of total gut contents. This may not be true of all lizards, however. With *Tropidurus melanopleurus* in Bolivia, Hymenoptera other than ants made up 5 to 30 percent, of consumed prey (Perez-Mellado and De la Riva 1993).

Bee defensive behavior no doubt deters some feeding by lizards. In Thailand, *A. cerana* defended their nest from agamid lizards by "hurling themselves" at the predator and "probably attempting to sting it" (Seeley et al. 1982). In the laboratory, *Sceloporus undulatus thayerii* refused to eat honey bees, and one medium-sized individual died from a sting (Burt 1928).

Control

Beekeepers can protect hives from toads by placing hives on stands 60 cm (2 feet) off the ground (higher than the toad's 45 cm jumping range) or by fencing the apiary with chicken wire (Roff and Brimblecombe 1953, 1963, Gold 1969, Morse 1975a, Roff 1975, Anderson 1990, Messing 1991). O'Brien and Marsh (1990) listed night hunting as another option for controlling toads. Roff (1966) recommended placing colonies close together in a circle, entrances inward, so that toads cannot reach hive entrances.

CHAPTER SIXTEEN

Birds

John T. Ambrose

Major predators
  Meropidae (Bee-eaters)
    Distribution
    Behavior
    Control
  Indicatoridae (Honeyguides)
    Distribution
    Behavior
    Control
Minor predators
  Apodidae (Swifts)
    Distribution
    Behavior
    Control
  Laniidae (Shrikes)
    Distribution
    Behavior
    Control
  Paridae (Titmice)
    Distribution
    Behavior
    Control
  Tyrannidae (Tyrant-flycatchers)
    Distribution
    Behavior
    Control
  Picidae (Woodpeckers)
    Distribution
    Behavior
    Control
    A Beneficial Woodpecker
  Galbulidae (Jacamars)
    Distribution
    Behavior
    Control
Occasional predators

CHAPTER SIXTEEN
Birds

Many birds include insects as a minor or major part of their diets, and it should not be surprising that birds are sometimes a serious problem for beekeepers and honey bees. The history of birds as predators of bees dates back at least to Aristotle in the Old World, and many examples have been added since then. There are only a few references to birds as predators of bees in the New World, but honey bees were brought to the Americas only a few hundred years ago, and their native predators were left behind in the Old World.

Aristotle identified the tit (titmouse), family Paridae, and the bee-eaters, family Meropidae, as enemies of bees (Fry 1984). Later, Virgil enlarged the list to include the green woodpecker (*Picus viridis*), the red-backed shrike (*Lanius collurio*), and the blue-titmouse (*Parus caeruleus*) (Fry 1984). Since then, several bird species and even entire families have been classed as major or minor honey bee predators. However, the standard American and English beekeeping references for the past 100 years give the impression that birds are bee pests of very little significance unless they are feeding on queen bees in a queen-rearing operation (Root 1880, Root 1974, Gochnauer et al. 1975); such references are generally limited to the eastern kingbird (*Tyrannus tyrannus*).

Birds generally are not significant pests to beekeeping in most of the temperate world, where the majority of the standard beekeeping texts are produced. Honey bees are not native to the Americas, and when they were imported from Europe, Asia, and Africa, their major bird predators were left behind. However, the bird families Meropidae (bee-eaters) and Indicatoridae (honeyguides) evolved in much the same areas as did the honey bees. Today, all species of bee-eaters include honey bees in their diets, and the various honeyguide species include either bees or wax.

In addition to the bee-eaters and the honeyguides, several other (primarily Old World) bird families contain species that prey on bees. These families include the shrikes (Laniidae), swifts (Apodidae), titmice (Paridae), tyrant-flycatchers (Tyrannidae), and, somewhat surprisingly, the woodpeckers (Picidae). The jacamars (family Galbulidae), a New World group, have also been described as bee predators in the South American tropics.

In this chapter, the birds that prey on bees are divided by family into major predators, minor predators, and occasional predators. The first group is limited to the Meropidae and the Indicatoridae, in which a substantial number of the species rely on honey bees or beeswax for a major portion of their diet. The second group contains families with only a small number of species that are serious pests to beekeepers, plus families in which several species prey on honey bees but as only a small part of their total diet. The designation minor refers to the degree of bee predation of the family as a whole; for example, spine-tailed swifts are major pests of beekeeping in the Philippines, but their family (Apodidae) is listed with the minor predators. The final category lists birds that feed on bees occasionally but not to any serious degree.

Major predators
Meropidae (Bee-eaters)
Distribution

The Meropidae (bee-eaters) is a distinctive coraciform family of Old World birds comprising three genera with 24 species. All species feed on insects, but only those in the genus *Merops* have been recorded as causing problems for beekeepers. The group is widely distributed (except in the New World), and many of the species are migratory; some are transcontinental migrants, such as the large green bee-eater (*Merops philippinus*), which leaves India to overwinter in the East Indies,

Figure 16.1. Two bee-eaters (*Merops apiaster*) in Kenya. (Demong photo).

Figure 16.2. *Merops apiaster* at their nesting holes in the bank of a streambed in Kenya. (Demong photo).

and the rainbow bird (*Merops ornatus*) from New Guinea, which flies to Australia for its breeding season. Within this family, only the European bee-eater (*Merops apiaster*) breeds in the temperate zone rather than in tropical areas (Van Tyne and Berger 1976, Harrison 1979, Fry 1984, Perrins 1990).

Behavior

Bee-eaters feed largely on venomous Hymenoptera. All species eat honey bees; the percentage of bees in the diet varies from species to species (Fry 1984) and area to area, such as in Europe, where the European bee-eater feeds primarily on bumble bees instead of honey bees (Perrins 1990). Although bee-eaters often feed in flocks and can consume large numbers of bees in an apiary, meropid attacks on honey bees have been reported as causing minor or only occasionally serious problems in many areas: Russia (Atakishiev 1970), Australia (Roff and Brimblecombe 1953, Anonymous 1986), India (Gandheker 1959), Nepal (*Merops orientalis* preying on *Apis dorsata* at cooler temperatures; B. A. Underwood, personal communication 1989), Iran (Esmali 1974), and South Africa (Botha 1970). A notable exception is the report of Jenn (1973) that bee-eaters seriously disrupted beekeeping operations in Algeria and forced many beekeepers to shift their queen-breeding operations to the months of October or November, after the migratory departure of *M. apiaster.*

General beekeeping conditions, the normal range of the predator and the prey, and other factors can cause a predator to become a serious problem in one geographic area rather than another. For example, I observed the carmine bee-eater (*Merops nubicus*) having a fairly serious effect on hives of *Apis mellifera* in Khartoum, Sudan. The birds would flock to the university apiary and perch on nearby tree branches, from which they would swoop down and capture flying bees. Environmental conditions in that desert area precluded the development of large bee populations in the apiary, so the bird attacks were serious. Reports indicate that the gregarious carmine bee-eater may nest in colonies of up to 10,000 adult pairs (Perrins 1990).

Bee-eaters have developed an interesting technique for dealing with stinging prey. The following description of how bee-eaters "devenomize" their prey is from Fry (1984: 17):

> Except for Carmines, which sometimes evidently de-sting bees without breaking flight, bee-eaters always return with a captured bee to their perch. There they deftly juggle it until it is held crosswise at the tip of the (bird's) mandibles near the end of the abdomen. Bouts of hitting the head end of the insect to one side of the perch alternate with rapid rubs of the tail with its sting to the other, like somebody vigorously using an india-rubber. I call it bee-rubbing, and it is peculiar to the treatment of stinging Hymenoptera.

Despite the bee-rubbing, the bee-eater may still occasionally be stung, and the question has been raised whether the birds are immune to the insects' venom. Fry's (1984) work with a captive-reared bee-eater (*Merops bullocki*) and her observations on nestlings suggest that the birds are at least partially immune to the effects of bee venom. Her work with the captive-reared bird also indicated that the bee-rubbing behavior occurred spontaneously and that the bird's skill in avoiding stings increased with experience even though it was not exposed to other birds. Thus, both instinct and learning play important roles in the birds' behavioral ability to avoid insect stings.

Control

The need for control varies considerably from area to area, and consideration should be given to the birds' effect on the environment as a whole and not just their tendency to eat honey bees. In some areas, bee-eaters are definitely considered beneficial because they prey on pest insects (Atakishiev 1970).

Specific recommendations for control of the bee-eaters include scaring, trapping, or shooting the birds, or moving the apiary. Singh (1962) noted that *Merops superciliosus* and *M. orientalis* are problems for beekeepers in some parts of India, but that they are also instrumental in keeping insect pest populations in check. He recommends scaring the birds away from an apiary by the noise from shooting if they become a nuisance. Roff and Brimblecombe (1953) found that the rainbow bird (*M. ornatus*) can be a problem for Australian beekeepers at certain times of the year and recommended moving the apiaries to new locations. Tutkun (1982) noted that the European bee-eater became a problem in some parts of Turkey and described a trap made of metal pipes for controlling the birds.

Other writers have advised destroying the bee-eaters in certain situations. Botha (1970) recommended shooting the European bee-eater in South Africa when it becomes a nuisance in the apiary. Atakishiev (1970) concurred that European bee-eaters can be major pests and suggested destroying them if necessary. His recommendations include several poisons, shooting, and even the use of falcons. Atakishiev claims, however, that other species of bee-eaters are beneficial and should not be destroyed even if they eat a few bees. Jenn (1973) also recommended shooting the European bee-eater when it became a problem in Algerian apiaries. Beekeepers reportedly kill several thousand European bee-eaters each year in Africa (primarily) and Europe (Perrins 1990).

Several techniques have also been tried, usually unsuccessfully, to scare the bee-eaters from an apiary. One of the most dramatic was cited by Jenn (1973). Beekeepers in Algeria strung the bodies of dead bee-eaters around the hives to scare away attacking European bee-eaters, but this was ineffective. Fry (1984) suggested that the most rational approach to bee-eater attacks on apiaries was simply to evade the problem either by confining the bees to the hive for a period of time or by moving the apiary. Mishra and Kaushik (1993) summarized methods used to discourage the green bee-eater (*Merops orientalis*) in Indian apiaries, but the techniques are generally ineffective.

The question of controlling bee-eaters cannot be addressed properly without considering their beneficial effects. Consumption of numerous pest insects by bee-eaters, such as locusts by the carmine bee-eater (Fry 1969a, b), is of obvious benefit. Although the destruction of locusts may not offset a beekeeper's concern for the loss of bees, Fry (1983, 1984) showed that various *Merops* species also eat large numbers of predators and pests of honey bees, including wasps, hornets, flies, and other bees. She found that the typical bee-eater consumed six bee predators for every 100 honey bees it ate and that the European bee-eater in particular consumed as many as one bee predator for every four honey bees.

Indicatoridae (Honeyguides)
Distribution

The family Indicatoridae includes 11 species of small birds grouped in four genera. They are closely related to the toucans (family Ramphastidae). Nine species are found in Africa south of the Sahara, excluding Madagascar, and the two remaining species live in Asia, one in the Himalayas and the other throughout Burma, Thailand, Sumatra, Malaysia, and Borneo. The birds are non-migratory except for local movements (Friedmann 1955, Van Tyne and Berger 1976, Harrison 1979).

Behavior

Very distinctive characteristics set the Indicatoridae apart from other birds. They are completely brood parasitic (i.e., they lay eggs in the nests of birds of other species, which then raise the chicks that hatch, apparently unable to detect that the chicks are not their own), they are cerophagic or wax-eating (from honey bees or in the form of scale insects), and they form a symbiotic relationship with certain mammals in which the mammalian symbiont is led to honey bee nests. Of all the bird families known to practice brood parasitism, only the Indicatoridae are completely brood parasitic (Friedmann 1955, Harrison 1979). For example, *Indicator* lays its eggs in the nests of hole-nesting birds whose eggs resemble its own. The newly emerged honeyguide will kill most or all of the host's offspring by pecking at them with the sharp hooks on the tip of its bill. These hooks later fall off, and the parasitic chick is raised by the unsuspecting host birds. One of the host groups of the honeyguide is the bee-eater family, whose members are themselves honey bee predators (Harrison 1979).

Honeyguides are the only bird family to eat wax. All honeyguide species feed on beeswax, except for two species in the genus *Prodotiscus* that eat wax in the form of scale insects (family Coccidae) (Friedmann 1955, Harrison 1979). Some of the birds are so fond of beeswax that two species, *Indicator indicator* and *Indicator minor*, are reported to have entered churches and pecked at the beeswax candles (Perrins 1990). The honeyguides of the genus *Indicator* eat a variety of insects but rely heavily on honey bees for food. Friedmann (1955) found that honeyguides prefer beeswax to honey or bee larvae. He was able to keep two lesser honeyguide (*Indicator minor*) individuals alive for 29 and 31 days, respectively, on a diet of beeswax alone.

The "guiding" behavior exhibited by at least some honeyguide species is perhaps their most fascinating characteristic. The birds have developed symbiotic relationships with certain mammals (humans; ratels or honey badgers, *Mellivora capensis* [see Chapter 17]; and possibly some species of baboons, genets, and mongooses). They lead a mammal to a honey bee nest, wait while the nest is plundered, and then eat the scattered remnants of the comb. The birds' guiding behavior ("indicating" bee nest locations) is the basis for both the common name and the scientific name of the family. The following is Friedmann's (1955:32) account of *Indicator indicator:*

> When the bird is ready to begin guiding, it either comes to a person and starts a repetitive series of churring notes or it stays where it is and begins calling these notes and waits for the human to approach it more closely. These churring notes are very similar to the sound made by shaking a partly full, small matchbox rapidly lengthwise. If the bird comes to the person to start leading him, it flies rapidly about within 15 to 50 feet from him, calling constantly, and fanning its tail, displaying the white outer rectrices. If it waits for a potential follower to approach it for the trip to begin, it usually perches on a fairly conspicuous branch, churring rapidly, fanning its tail, and slightly arching and ruffling its wings so that at times its yellow "shoulder" bands are visible. As the person comes to within 15 to 50 feet from it, the bird flies off with an initial conspicuous downward dip, with its lateral rectrices widely spread, and then goes off to another tree, not necessarily in sight of the follower, in fact more often out of sight than not. Then it waits there, churring loudly until the follower again nears it, when the action is repeated. This goes on until the vicinity of the bees' nest is reached. Here the bird often (usually, in my experience) suddenly ceases calling and perches unobtrusively in a nearby tree or shrub and there waits for the follower to open the hive, and it usually remains there until the person has departed with his loot of honeycomb, when it comes down to the plundered bees' nest and begins to feed on the bits of comb left strewn about. The time during which the bird may wait quietly may vary from a few minutes to well over an hour and a half.

Of the 11 species of honeyguides, the only two for which the guiding behavior has been recorded are *I. indicator* (the black-throated honeyguide) and *I. variegatus* (the scaly-throated honeyguide). This behavior is apparently not necessary for the birds to obtain beeswax. In fact, Friedmann (1955) noted that the guiding behavior seemed to be declining in areas of Africa where traditional honey hunting was declining, and there is some controversy as to whether the bird actually acts as a guide (Bambara 1987). I have heard African honey hunters swear by the birds' guiding ability, however, and Isack and Reyer (1989) present strong scientific evidence to support these claims.

In parts of Asia, a complex relationship exists between honey bees and the orange-rumped honeyguide, *Indicator xanthonotus.* In a study conducted in Nepal, Cronin and Sherman (1977) found that these birds did not seem to exhibit guiding behavior but did incorporate honey bee nests into their reproductive behavior. The authors reported that the birds eat beeswax and that the males actively defend territories that contain honey bee nests, keeping out other males but allowing females in, particularly during the mating season. The guarding male will allow a female to eat some beeswax from the protected nests, and then mating generally takes place on or near a bee nest. Males without territories seemed to be unsuccessful at mating. Cronin and Sherman listed the bees involved as being *Apis dorsata*, but more recent research indicates that the bees are probably *Apis laboriosa*, the Himalayan honey bee (Underwood 1992). The bees nested in multiple-colony aggregations, with each colony building a single, large, exposed comb. The bees were migratory and abandoned the comb nests each fall; however, the male orange-rumped honeyguides appeared to defend the bee nests even when the bees were gone, during the birds' breeding and non-breeding periods. Male birds allowed females with which they had previously mated to gain access to the bee nests for feeding, even during the non-breeding season (Cronin and Sherman 1977, Underwood 1992). Cronin and Sherman suggest that the protected bee nests can be considered "externalized secondary sexual characteristics" of the male birds.

Control

Control of honeyguides is not a serious problem. Friedmann (1955) notes that the guiding behavior is not likely to be a problem for beekeepers in areas with relatively high human populations, where the numbers of honey badgers and baboons, which might seriously damage colonies, are reduced. In addition, honeyguides that consume beeswax without the aid of a mammalian symbiont seem to limit their activities to exposed combs of beeswax in feral bee colonies and do not molest box hives.

Minor predators

Apodidae (Swifts)

Distribution

The family Apodidae has a worldwide range with the exception of some oceanic islands, such as New Zealand, and areas that have a scarcity of flying insects because of extreme cold or high winds, such as Northern North America, Northern Asia, and Southern South America. The Apodidae may be conservatively divided into nine genera and 75 to 80 species of strong-flying gregarious birds, with many migratory species (Van Tyne and Berger 1976, Harrison 1979).

Behavior

Most swifts feed in flocks and catch their prey on the wing, their main food source being "aerial plankton" (primarily insects) (Harrison 1979). It should thus not be surprising that swifts are serious pests of beekeeping in some parts of the world, particularly the Philippines (Morse and Laigo 1969b) and Southern Africa (the alpine swift, *Apus melba*; Fry 1984).

The spine-tailed swift (*Chaetura dubia*) is probably the bird best known as a bee predator, and its islandwide distribution in the Philippines has made it a major obstacle to beekeeping there (Morse and Laigo 1969b). Morse and Laigo reported that the spine-tailed swifts are found in Philippine apiaries throughout the year but are most prevalent on cool, cloudy, and windy days – bee-feeding behavior that is typical of other flock-feeding birds such as Meropidae. The swifts were reported to be in the apiaries from 8 a.m. to 3 p.m., and the largest congregations, flocks of up to 300 birds, were present from 11 a.m. to noon.

The stomach contents of 18 dissected spine-tailed swifts showed that the birds were feeding primarily on honey bees (Morse and Laigo 1969b). Three different species of honey bees were being eaten: primarily *Apis mellifera* but also *A. cerana* and, in small numbers, *A. dorsata*. The stomachs contained, on average, 88 honey bees each, with maximums of 245 (234 *A. cerana*, six *A. mellifera*, five *A. dorsata*) and 164 (162 *A. mellifera*, two *A. cerana*). Most of the consumed bees were workers, and the birds' mouths were filled with detached bee stings, suggesting that neither the sting itself nor the venom deters these bee predators.

Control

Because the spine-tailed swift tends to feed in large flocks, control can be difficult. Morse and Laigo (1969b) observed that the people who lived in their study area used sections of fish nets attached to long bamboo poles to capture the swifts when the birds were attacking or near the apiary; however, this method was successful only on a short-term basis. For longer-term protection against the bee-eating swifts, Morse and Laigo suggested relocating the apiaries to areas where the birds were not prevalent.

As with most insect-eating birds, swifts probably also consume a large number of harmful or pest insects. In most areas, control is neither necessary nor beneficial.

Laniidae (Shrikes)
Distribution

Shrikes are primarily an Old World family of birds, with two species found in North America. The Old World distribution includes Europe, Africa, and Asia south to Timor and New Guinea. This is a large family of birds, with 72 species ranging in size from medium (16 cm) to fairly large (37 cm); species in the northern ranges tend to be migratory (Van Tyne and Berger 1976, Harrison 1979).

Behavior

Shrikes are predators of small reptiles, small mammals, and insects, including honey bees. They tend to hunt from exposed perches from which they swoop down upon their prey. Virgil (Fraser 1931) listed the red-backed shrike (*Lanius collurio*) as a bee pest. More recent reports limit the shrikes as bee pests to the former U.S.S.R. and Ethiopia.

Atakishiev (1970) ranks the Caucasian shrike (*Lanius cristatus kobylini*) and the gray shrike (*L. excubitor*) as minor pests in the apiary. However, he does describe the black-headed shrike (*L. minor*) as an important pest that feeds on bees in the former U.S.S.R. from April through August. Rea (1974) observed *L. elegans* capturing bees in front of hives in Ethiopia but did not evaluate its importance as a bee predator.

The shrikes of the genus *Lanius* have been observed impaling their insect prey (probably including honey bees) on thorns or the barbs of fences; this behavior may serve to kill live prey (such as a venomous insect) or to store food. This presumed storing habit by the red-backed shrike (storing of frogs, bats, and other birds) resulted in the development of its alternative English common name, *butcher bird* (Harrison 1979, Perrins 1990). In addition, some species of shrikes also exhibit the devenomizing techniques used by bee-eaters (Fry 1983; see Meropidae, above).

Control

Atakishiev (1970) suggested that the black-headed shrike may be a serious pest in the apiary and that shooting them is probably the best solution. He believed that neither the Caucasian shrike nor the gray shrike kills enough bees to warrant its destruction.

Paridae (Titmice)

Distribution

Tits can be found throughout the wooded areas of Europe, Asia, Africa, India, and the Americas south to Guatemala. The titmice or chickadees consist of three genera and 46 species of active, gregarious wood or scrubland birds ranging in length from 11 to 20 cm. Most species are at least irregularly migratory, and movements involving millions of blue tits (*Parus caeruleus*) and great tits (*P. major*) have been recorded in Europe (Van Tyne and Berger 1976, Harrison 1979, Perrins 1990).

Behavior

Virgil (Fraser 1931) identified both the great tit and the blue tit as bee enemies. Other pest species include the south-Caspian great tit (*Parus major karelini*) and the Caucasian long-tailed tit (*Aegithalos caudatus*) in the former U.S.S.R. (Atakishiev 1970). Cheshire (1888) and Manley (1948) identified both the blue tit and the great tit as bee pests in Great Britain, and Royds (1918) identified them in Southern Europe.

The tits are primarily insect eaters, but during the cold months they can switch to a diet of seeds. These birds have adapted well to man-made situations and are very common at bird-feeding stations (Harrison 1979). Both Cheshire (1888) and Atakishiev (1970) noted that the tits cause the greatest damage to beekeeping during the winter, when the birds "lure" the bees from the hives and then devour them. Manley (1948) observed that tits also collect large numbers of bees during the summer but that they are mostly dead bees that the adult tits feed to their offspring.

Control

None of the references recommends destroying tits, even though some species may become serious pests, particularly during the winter (Cheshire 1888, Atakishiev 1970). The consensus seems to be that the benefits of the birds' feeding habits (eating insects other than honey bees) generally outweigh their potential pest nature. Manley (1948) and Atakishiev (1970) both recommend frightening tits away from the apiary but do not mention any particular methods.

Tyrannidae (Tyrant-flycatchers)

Distribution

The Tyrannidae, a large and diverse New World family of birds, occurs throughout the Americas and the West Indies, and a few species have been recorded as storm-blown vagrants in Europe. One of the largest families of perching birds, it contains approximately 115 genera divided into about 360 species, with individuals ranging from 8 to 23 cm in length. Most of the species are thought to be migratory (Van Tyne and Berger 1976, Harrison 1979, Perrins 1990).

Behavior

Two species of Tyrannidae have developed reputations as bee predators: the eastern kingbird, *Tyrannus tyrannus* (Beal 1915, Grant 1945, Coleman 1986), and the western or Arkansas kingbird, *T. verticalis* (Beal 1915, Grant 1945). Because the eastern kingbird is also known as the bee martin and the bee-bird in some areas, it is likely to be thought of as a beekeeping pest. Most apiculturists in the United

States agree that the eastern kingbird is not a serious predator of worker honey bees, but its consumption of queens may be a major problem in areas where queen-rearing operations are conducted (Root 1880, Dadant 1927, Root 1974, Gochnauer et al. 1975, Coleman 1986).

The eastern kingbird sits on an exposed perch and then swoops down on its victim (Root 1974) (as do the bee-eating shrikes, *Lanius* spp.). Information on the bee-feeding habits of this species was obtained when the stomach contents of 665 eastern kingbirds, collected throughout the United States, were examined. Only 22 bird stomachs contained honey bees, and 51 of the 61 bees found were drones. In a similar study on the western kingbird, 62 examined bird stomachs contained only 30 honey bees (the number of bird stomachs that contained bees was not given), and 29 of the bees were drones (Beal 1915). These data indicate that king-birds selectively eat drone honey bees and thus avoid the stings of worker bees. Such selection could be based on size of prey, which may indicate that kingbirds could be a serious threat to honey bee queens on mating flights.

Whereas the western and eastern kingbirds do not seem to be major preda-tors of worker honey bees, other kingbirds may be a problem. Grant (1945) reported that the gray kingbird (*Tyrannus domincensis*) will eat worker honey bees; his report was based on analyses of an unspecified number of bird stomachs. After examining the remains of bees found near feeding birds, he concluded that the western kingbird ate only the bee's abdomen. Root (1966) reported that the king-bird does not swallow the bee's body, but this was disputed by Selwyn (1970), a Canadian beekeeper who dissected one unnamed species of kingbird in Quebec and found whole worker bees in the bird. Because honey bees are not native to the Americas, it is conceivable that different species of Tyrannidae have developed various techniques for adding this immigrant insect species to their diets.

Control

Under most circumstances, the destruction of honey bees by kingbirds is probably far outweighed by the bird's predation on harmful or pest insects. The only exception to this generalization occurs when the birds begin to prey on virgin bee queens in a queen mating yard. Only two methods of control are found in the literature, and neither seems to be completely reliable; one is to shoot the birds (Root 1974), and the other is to locate the queen mating yard in an area with moder-ate tree cover so that the virgin queens can fly among the branches and be afforded some protection against bird attacks (Coleman 1986).

Picidae (Woodpeckers)

Distribution

The family Picidae includes 208 species divided into three subfamilies, and the subfamily Picinae (true woodpeckers) consists of 33 genera with 170 species. This large family of birds has a worldwide distribution with the exception of Madagascar, Australia, New Zealand, New Guinea, the extreme North (Greenland), and most oceanic islands. The birds are generally non-migratory and are well-known for their ability to dig insect prey out of wood or from under the bark of trees (Van Tyne and Berger 1976, Harrison 1979, Perrins 1990).

Behavior

Two species, the green woodpecker (*Picus viridis*) and the great spotted or great variegated woodpecker (*Dryobates major*), have been identified as being at least minor pests of honey bees in Russia (Atakishiev 1970), Great Britain (Manley 1948), and France (Guilloux 1969). The record of woodpeckers as pests of beekeeping dates back at least to Virgil's observations on the green woodpecker (Fraser 1931). *D. major* is generally considered a nuisance species that destroys a few adult bees from time to time (Manley 1948, Atakishiev 1970). Other woodpeckers, such as the red-headed woodpecker (*Melanerpes erythrocephalus*) and the Lewis woodpecker (*Asyndesmus lewis*), have also been identified as occasional eaters of honey bees in the United States (Grant 1945).

The green woodpecker (*P. viridis*) can be a serious pest (Manley 1948, Guilloux 1969, Perrins 1990). It has a long protrusible tongue that can actually be extended 5 cm past the tip of the bird's bill; the tongue also has small barbs and sticky saliva to increase the efficiency of capturing wood-inhabiting insects (Harrison 1979). The green woodpecker's tongue, combined with its ability to peck through beehives, makes it a formidable enemy of honey bees. Manley (1948) reports that the bird is most troublesome during winter, when its regular food sources are unavailable, and that it will generally peck holes near the hive entrance and through the recessed handholds of box hives. Morse and Hooper (1985) confirm that even in an area where bees have been kept for a considerable time without any problems, in a cold winter the green woodpecker may begin attacking hives for food and making holes up to 5 cm in diameter. Such attacks damage both the hive and the interior combs, and the damage may attract other pests such as rats and mice. In addition, many bees are lost. Because honey bee populations are at their lowest levels in winter, repeated woodpecker attacks can eventually destroy a bee colony.

Control

Control recommendations for woodpeckers vary from none to shooting the pest birds. Atakishiev (1970) lists both the green woodpecker and the great variegated woodpecker as beneficial species and recommends that control be limited to scaring the birds from the apiary when necessary. Guilloux (1969) recommends wrapping the hives with wire netting to prevent damage by the green woodpecker. Manley (1948) suggests covering only the hive entrance and any recessed areas, such as handholds, with wire mesh or shooting the birds if necessary.

A beneficial woodpecker

There is evidence that at least one species of the Picidae family may actually benefit honey bees. Williams (1990) reported that the yellow-bellied sapsucker (*Sphyrapicus varius*) may provide some foraging for honey bees during winter. In his study in Virginia (U.S.A.) he observed honey bees foraging at holes drilled by the yellow-bellied sapsuckers in sugar maple trees (*Acer saccharum*) during February where the bees were collecting maple sap.

Galbulidae (Jacamars)

Distribution

The Galbulidae, a small family found throughout the New World tropics, ranges from northeastern Argentina to Panama. The jacamars are classified in five genera divided into 17 species, and individuals range in length from 18 to 28 cm. They are thought to be non-migratory (Van Tyne and Berger 1976, Harrison 1979, Perrins 1990).

Behavior

The jacamars are generally classed with the Piciformes birds, which include the honeyguides and woodpeckers as well as toucans, barbets, and puffbirds, but they share many characteristics with the Meropidae, or bee-eaters (Harrison 1979, Fry 1984). In fact, Fry (1970, 1984) describes the jacamars as 'ecological analogues' of the bee-eaters. Similarities between the two families include coloration patterns, range of body sizes, and diet.

In one analysis of feeding habits, Fry (1984) found that the rufous-tailed jacamar, *Balbula ruficauda* (classified as *Galabula ruficauda* by Perrins 1990), and the red-throated bee-eater, *Merops bullocki*, showed definite similarities in eating habits. The jacamar's diet was 86 percent Hymenoptera, and the bee-eater's diet was 79 percent Hymenoptera, with most of the hymenopteran superfamilies equally represented in the diets of the two birds. The jacamar ate more social wasps and the bee-eater ate more flying ants, but this difference may have been caused by variations in local availability of prey species. There is no evidence, however, that jacamars devenomize their prey as do the bee-eaters (Fry 1984).

There is little direct evidence that jacamars are major predators on honey bees, but their seeming similarities to the bee-eaters certainly suggest this possibility. As more areas of tropical South America are opened to managed beekeeping, the data on honey bee eating by jacamars may become more significant.

Control

At present no control measures appear to be needed.

Occasional predators

Many birds are insectivorous or rely on insects for at least part of their diet, so it should not be surprising that they eat bees. Honey bees, as social insects, frequently occur in large numbers and remain in areas where other, non-perennial insects are sometimes not available to hungry birds. A list of all the birds known to feed on bees would be unduly long. Even in the United States, where honey bees are not native, Grant (1945) listed over 40 bird species known to have eaten honey bees, including mockingbirds (*Mimus polyglottus*), robins (*Turdus migratorius*), and ring-necked pheasants (*Phasianus colchicus*). However, none of the birds on that list appear to be serious predators of honey bees, with the possible exception of the eastern kingbird (*Tyrannus tyrannus*).

In Great Britain, several bird families have been implicated as bee predators: the Old World flycatchers (Muscicapidae: Cheshire 1888, Wedmore 1932, Manley 1948), swallows (Hirundinidae: Wedmore 1932, Manley 1948), and even ducks (Anatidae: Wedmore 1932). None of those families is regarded as a hindrance to beekeeping in Great Britain. Sparrows (*Passer* spp., family Passeridae: Wedmore 1932, Manley 1948) may present a problem, however, because of their habit of flying to apiaries in large flocks and eating the field bees. No reliable methods are listed for sparrow control.

Atakishiev (1970) lists the European pern or honey buzzard (*Pernis apivorous*) as an occasional bee predator in Azerbaijan (*apivorus* means bee-eating). He describes it as a narrowly specialized entomophagous bird whose diet consists primarily of the larvae of social Hymenoptera such as wasps and bumble bees. Fry (1984) states that the bird feeds on pupae, wax, and honey at bees' nests and that its head is closely covered with small, scaly feathers that probably protect it against stings. Schmidt et al. (1985) reported that the honey buzzard is also a predator on *Apis dorsata* in Sabah (North Borneo). Atakishiev (1970) notes that honey buzzards do kill some bees but that they compensate for it by destroying numerous wasp nests; he recommends frightening the birds away from the apiary if they become pests.

Several members of the drongo family, Dicruridae, have been listed as occasional bee predators in various parts of the world. B. A. Underwood (personal communication 1989) observed *Dicrurus leucophaeus* and *D. aeneus* preying on *Apis laboriosa* Smith in Nepal. In winter, the birds picked off individual bees that left their (combless) winter clusters when the birds approached; in some instances, a colony of bees broke up its cluster and moved to a new location, apparently in response to heavy bird predation. In early spring, just after colonies of bees had built their comb nests, the birds were seen preying on individual bees in flight. In winter and spring, Underwood saw flocks of both drongo species feeding on bees at the same time, along with the king crows (*Dicrurus macrocercus* and *D. aster*) as occasional predators of honey bees in apiaries in India, particularly on cloudy days. He did not attach much importance to these attacks, however, and did not suggest destroying the birds.

In the United States, members of the wren family, Troglodytidae, have been reported as bee predators, but Root (1880) noted that the birds are actually beneficial and spend most of their time gathering dead bees and wax moth larvae in the apiary. The purple martin (*Progne subis*, family Hirundinidae) is particularly well known in the United States, and sometimes beekeepers become concerned (Perrins 1990). The bird undoubtedly eats some bees, but that loss is probably more than offset by predation on mosquitoes and other pest insects. Hinkle (1975) reported that western tanagers (*Piranga ludoviciana*) visited his California apiary over a period of several years and destroyed one-fifth of his field bees. He suggests placing wire screening over hive entrances and shooting as possible control measures.

On a more positive note, two Canadian beekeepers (Ferland 1986, Neilson 1986) report that chickadees (*Parus atricapillus*, family Paridae) performed a valuable housecleaning function for their bee colonies. The birds overwintered near their apiaries and collected dead bees from bottomboards, preventing blockage of hive entrances. In addition, Neilson (1986) allowed the chickadees to eat the dead bees from the frames of hives that did not successfully overwinter, thereby cleaning the frames for reuse by the beekeeper.

CHAPTER SEVENTEEN

———— ⚬⚬⚬ ————

Mammals

W. Michael Hood and Dewey M. Caron

CHAPTER SEVENTEEN
Mammals

Introduction

Mammals can be a serious threat to beekeeping in many parts of the world. For centuries, humans have contributed to damage from vandalism, pesticide poisoning, and diseases spread, but only in the last few years have beekeepers themselves had a major negative impact on beekeeping, particularly through mite introductions and spread in the Americas. A large mammal, the bear, is a primary pest that often leaves behind destruction and sometimes causes great losses to the beekeeper. Among smaller mammals, shrews and mice enter colonies to feed and construct nests within the hive to protect themselves against the bees. Skunks and badgers pay nocturnal visits and disturb bee colonies to the extent that the bees may be prone to sting and difficult to manage. Other mammals reviewed here can be considered occasional pests for which practical and inexpensive control measures are available. The interactions between mammals and bees are well understood in some, but not all, cases.

Primates

Humans are increasingly becoming a serious mammalian pest of the honey bee. In a recent survey of Canadian and United States bee specialists and inspectors, 30 of the 53 respondents listed humans as one of the top three damaging mammals to honey bee colonies through vandalism; 14 listed humans as the number one mammalian pest (Hood 1996). In an earlier survey conducted by O'Brien and Marsh (1990), eight state and provincial apiary inspectors recorded humans as one of the most damaging species to honey bee colonies. In this earlier survey, respondents estimated human damages at $6,900 annually.

Several other researchers, both in the United States and abroad, share this view (Flottum 1991a, Botha 1970, Couston 1972). Root (1966) listed humans as the most serious bee enemy, particularly with regard to actions that further the spread of foulbrood. Jaycox (1969) indicated that humans are serious pests in out-apiaries, which, at least in the United States, are becoming more numerous and widely scattered.

Humans shoot holes into hives with guns. Taber (1993) reported an incident in which the front handhold of one of his hives was used as a bull's eye for target practice. He found 30, 22-caliber slugs on the bottomboard. Humans may damage hives with rocks, steal honey, leave covers off, and tip hives over. The age-old tipping-and-running vandal attack, especially by adolescents, is a problem in

Figure 17.1. Although tough, beehive woodenware is no match for a bullet. When used for target practice the damage can be extensive. (Caron photo).

some areas.

The most disturbing stories of humans as beehive pests are those of competing beekeepers poisoning colonies or rampaging through an apiary at night in battering-ram pickups, destroying everything in sight (Flottum 1991a). Humans are also responsible for poisoning thousands of bee colonies annually through widespread pesticide use. Many coastal South Carolina beekeepers claimed they suffered great colony losses when areawide mosquito abatement programs followed Hurricane Hugo in September 1989. Most pesticide poisonings are unintentional, but competing beekeepers and angry neighbors have purposely poisoned bee colonies on occasion.

Loss of entire bee colonies to bee rustlers is an example of human depredation. Honey crops may be taken, but more often entire colonies are stolen. In the September 1975 issue of the *New Jersey Beekeepers News*, for example, four out of 10 articles were on beehive thefts. Rustlers stole over 2,000 hives in Florida, New Jersey, and California according to these reports. Incidents of brood theft have been reported in which entire apiaries were stripped of brood frames during the spring buildup period, and all frames were neatly replaced by old, empty frames.

This human thievery will likely continue, mainly because of shortages of bee colonies for pollination in many areas.

Many bee colonies have been lost in the last several years in the United States as a result of the introduction of two mite species, tracheal and varroa mites. Humans were responsible for the importation of both mites and their widespread distribution through colony, package, and queen movement. Beekeepers are faced now with the enormous challenge of controlling these mites without further injury to the beekeeping industry. Illegal use of unregistered pesticides by beekeepers for mite control is one serious potential problem as reported by a concerned state apiarist (Hood 1996). The headline, "Feds Test for Toxic Honey," in the August 10, 1990, edition of the *Boston Herald* remains firm in the memories of many persons in the beekeeping industry (Gamber 1990).

Humans were responsible for the introduction of the African honey bee into Brazil in 1956. Although some Brazilians have been pleased with the Africanized bees, other countries where the bees have spread report "consternation, concern, and even fear" (Morse 1988). This introduction by humans has caused much unwarranted concern by the general public, even those who may live hundreds of miles from an Africanized bee colony.

As honey increases in value and as bee foraging areas continue to decrease because of human population pressures, colonies are inevitably located in and around populated areas. Vandalism and colony theft are likely to increase. Care in selecting apiary locations and removing colonies from locations where trouble occurs are the best solutions. Out-apiaries should be selected with care and should be visited regularly. Beekeepers should brand the exterior of colonies with a quickly identifiable mark and post their name in each apiary. Some beekeepers brand each frame with their last name. O'Brien and Marsh (1990) noted that respondents to their survey recommended placing colonies in exposed areas where vandals are more likely to be observed. Taber (1993), on the other hand, stated that it is still best to hide colonies from "prying eyes" in rural areas and, if possible, to place them near the house of someone who is sympathetic to your cause. Some beekeepers camouflage their colonies with paint to blend in with the environment to prevent casual observers and hunters from spotting the apiary. Even with these precautions, vandalism and colony movement necessitated by human activities will undoubtedly result in the regrettable loss of valuable bee forage and crop pollination.

Non-human primates

Three other primates, monkeys, baboons, and chimpanzees, have been referenced as pests of honey bees. Das Gupta (1945) and Crane (1980) noted that black-faced monkeys of the genus *Langur* are a pest to Indian beekeepers.

Subhapradha (1957, 1961) stated that monkeys open hives and remove and destroy frames. Bee stings do not seem to drive the monkeys away. Punjabi (1961) indicated that monkey (and bear) damage in both the plain and hill sections of India is "noteworthy."

Subhapradha (1957) diagramed a hive stand she designed to keep monkeys and ants (another serious pest in her area) away from the bees; it was made with solid pipe and had a motor oil barrier for ant protection. Each hive was securely fastened to the hive stand by steel strapping over the top of the hive. Monkeys cannot remove covers or enter the hives when they are tightly fastened in this manner. Although such straps may impede colony manipulation, the extra precautions are worthwhile in areas where monkeys are a problem.

Botha (1970) stated that baboons usually will not tamper with well-constructed hives because of the stings they receive. They do, however, raid and break up boxes set out for trapping bee swarms. Once they start such activities, they may continue and raid larger, established colonies. Baboons will tip colonies over repeatedly until the combs and frames come apart. They frequently carry the frames away from the bee yard to consume the honey. Friedmann (1955) and Crane (1975a) relate tales of baboons opening wild bee nests after having been led to the nesting sites by a honeyguide (see Chapter 16). According to Friedmann (1955), baboons (but never monkeys) occasionally follow a honeyguide.

Smith (1960) listed baboons and chimpanzees as pests of bees in the tropics. Deschodt (1969) described the activities of a group of 20 chimpanzees raiding a wild bee nest in a granite wall crevasse. One large male climbed to the nest site, removed pieces of comb, and threw them to the others on the ground. The stings did not drive the chimpanzees away, although there was much yelling and jumping about. Only the young chimps did not participate; they snuggled their faces against the chests of their mothers to avoid the stings.

Merfield and Miller (1956) watched chimpanzees use sticks as tools to obtain honey. The chimpanzees poked a long stick into a hollow in a tree that contained a wild bee nest. They then withdrew the stick and ate the honey on the stick. In studies of the tool-using behavior of chimpanzees at Mt. Assirik in the Parc National du Niokolo-Koba, Senegal, West Africa, Bermejo et al. (1989) reported finding four sticks in "wild beehives" that were used by chimpanzees to remove honey.

One other primate group may raid bee nests. Sabater Pi (1960) observed a gorilla as it chewed wax from a ground bee nest. Schaller (1963) was told by natives that gorillas will attack wild bee nests. He never observed them doing so, however, even in instances when bee nests and gorillas were in close proximity.

Carnivores

Several carnivore species are described as pests of honey bee colonies. Most of these mammals are not nearly as numerous as they once were. Some have pelts valuable to humans, and others are hunted and eliminated either because they occasionally attack domestic animals or simply because of their reputations as hunters and predators.

Das Gupta (1945) believed that a jackal (*Canis* sp.) was responsible for hives being thrown to the ground and their combs scattered about during the night. Problems were encountered during the honey flow when the smell of ripening honey was strong. Johnsen (1955) included a fox, *Vulpes vulpes*, and a cat, *Felis domestica*, in a listing of occasional bee pests. Zappi-Recordati (1971) listed a fox (no species given) as a pest attracted to honey. No other details were provided, and we have found no further references to bee pests in the cat or dog families.

The raccoon (*Procyon lotor*) is often labeled by beekeepers as a pest of the bee colony. Raccoons are omnivorous and frequently get into mischief around humans and their habitations. Maher (1968) identified the raccoon when responding to a letter inquiring about a pest that scratched at the entrance and chewed

Figure 17.2. Raccoons can cause significant damage to a colony. This one was caught in a live trap in a bee yard, using simple dry cat food for bait. (Flottum photo).

substances in front of colonies.

O'Brien and Marsh (1990), in their survey of state and provincial apiary inspectors, reported that raccoons were listed 11 times as causing some degree of damage. Only two respondents listed the raccoon in the top three categories of mammals responsible for damages to honey bees. A survey response from Minnesota noted that raccoon damage can sometimes be confused with bear damage. Although the raccoon is capable of tipping over a colony and scattering its contents in search of food, it is not considered a serious bee pest.

Mustelidae

Several members of the weasel family, Mustelidae, are occasional honey bee pests. Probably other members of this family not reported here should be included. Beljavsky (1927) listed the sable (*Mustela martes*) and the stone marten (*Mustela foina*) as occasional pests of beehives in the former U.S.S.R. Both have a great fondness for honey and may enter a colony repeatedly to feast on the honey stores. Dzierzon (1878) and Delibes (1978) mentioned the stone marten's honey-eating habit.

Smith (1960) and Beljavsky (1927) described the Himalayan yellow-throated marten (*Martes* [= *Charronia*] *flavigula*) as a pest of honey bee colonies. In parts of Asia, this pest is commonly called the honey dog. Yellow-throated martens range from 45 to 60 cm (18 to 24 inches) in length and have a tail about three-fourths of the body length. The fur is variegated; deep brown, black, and yellow are common colors. The yellow throat is margined by dark bands. This species inhabits the temperate zone of the mountains of Asia. The animals forage in a restless manner, either by day or by night, on the ground or in trees. They eat a wide variety of foods, including small mammals, bird eggs, carrion, insects, fruits, and nectar from flowers. One flower frequently visited is the large scarlet bloom of the silk cotton tree, and martens apparently assist in pollinating this plant (Prater 1965).

Wegner (1949) described the damage done by a *Martes* species to 20 *Apis cerana* hives. This marten may capture flying bees and consume them. Singh (1962) stated that the animal (he referred to it as a pine marten) inserts its tail into a hive entrance and eats the bees that become entangled in the tail hairs.

Other martens (genus *Mustela*), commonly called weasels, polecats, or ferrets, may eat bee and wasp larvae and adults. Ordetx and Espina Perez (1966) and Zappi-Recordati (1971) identified the polecat (probably *Mustela putorius*) as a honey-seeking pest. Beekeepers who cut drone, burr, or brace comb from frames and leave it on the ground around colonies may encourage such feeding. Egger (1973) believed weasels were beneficial in killing mice and shrews that could potentially inhabit bee colonies and that they should be protected around the bee yard.

Beljavsky (1927) and Johnsen (1955) listed another weasel family member, the badger (*Meles meles*), as a bee pest. Badgers were once found in temperate Europe, but now they exist mainly in Asia. Their bodies are dark gray; the head is whitish with a black stripe on either side extending through the eye and ear. They have short legs, a small tail, and a stocky body. Badgers are omnivorous and include bee and wasp larvae in their diet. Beljavsky (1927) stated that the usefulness of these creatures far outweighed the small amount of harm they might cause to honey bees. The American badger (*Taxidea taxus*) apparently does not feed on bees; it consumes rodents and other animals.

Another badger, the ratel or honey badger (*Mellivora capensis*), is a well-known honey bee pest that is native to tropical Asia and Africa. It was once common in India and much of Africa, but according to Keats (1970), the honey badger is now rarely found outside of game preserves in South Africa. These creatures are short-legged and have short tails and powerful front claws. They are whitish above and black below. Ratels are nocturnal, resting during the day in burrows that they dig with their front claws. They live on small animals and some vegetable matter; in addition to bees, they sometimes attack poultry.

Smith (1953) and Botha (1970) called the ratel a fierce animal. Beekeepers' encounters with it in the bee yard can be dangerous because the animal will attack with its powerful front claws. The ratel can push heavy hives from their stands and smash lumber up to two cm (one inch) thick. It usually carries combs away from the hive before consuming them, probably to escape the wrath of stinging bees. Ravages of the ratel may be followed by appearances of the giant white-tailed mongoose (*Ichneumia*) and the civet cat (*Civettictis*), both of which eat whatever the ratel leaves behind (Smith 1960). The ratel is often associated with the bird *Indicator indicator*, the honey guide, (see Chapter 16).

Smith (1960) reported that beekeepers in Africa hang their hives in trees to protect against pests such as the honey badger (and baboons). May (1969) strongly recommended placing hives on stands high enough and strong enough to deter honey badgers. Other beekeepers surround their apiaries with a fence of wire netting buried 30 cm (one foot) deep and angled outward. Some African beekeepers keep bee colonies in the European style by placing them in strongly built bee houses to avoid damage from badgers and other pests. The Peace Corps beekeeping training manual (Gentry 1982) recommends trapping or hunting badgers when necessary.

Skunks

Perhaps the best known of the Mustelidae bee pests are the skunks. Several species in two genera, *Mephitis and Spilogale*, are of concern to beekeepers. They are natives of the Americas and are especially common in North America. They forage at night in a slow, deliberate fashion. Skunks emit an offensive odor that causes enemies to cease an attack and flee.

Eckert and Shaw (1960:395) listed the skunk as "probably the greatest destroyer of bees among the smaller mammals." Apparently, skunks acquire a strong appetite for honey bees and return nightly to feast at an apiary. According to Morse (1975b), beginning beekeepers do not keep honey bee colonies very long before they encounter skunks.

O'Brien and Marsh (1990) reported that skunks were listed most often (48 times) by survey respondents as one of three vertebrate species causing most losses. Twenty-five respondents gave a total estimate of $423,050 as the annual damages to their colonies caused by skunks. Most of this estimated damage occurred in California ($175,000) and Texas ($100,000). A similar survey conducted by Hood (1996) confirmed that skunks are the most frequent cause of honey bee losses; 41 out of 53 respondents listed skunks as one of the three mammals causing the most damage. One Canadian respondent noted that bear damages are normally

Figure 17.3. Skunks nearly always forage at night. A mother skunk will train her offspring to forage at an apiary, and a family can seriously deplete the entire apiary's population. Look for scratch marks on and just above the landing board, bare ground in front of the hive, and fecal material. (Fell photo).

spectacular, but if the losses are fairly calculated, mice and skunks are probably responsible for higher annual losses.

Storer and Vansell (1935) gave a thorough description of striped skunks eating bees. The first indications of skunk visitation are scratches on the earth in front of hives and on the front of hive boxes. The skunk typically scratches at the entrance of a bee colony and eats adult bees that respond to the disturbance. Skunks may feed for hours in front of bee colonies. They gradually destroy the vegetation in front of the entrance, sometimes making holes as deep as 15 cm (six inches). All traces of dead bees vanish from in front of the hives. Paint is scraped from the hive front, and dirt is frequently evident on the front and entrance board of the colony. If one colony becomes too aggressive, the skunk switches to another to feed. Bees in hives being visited by skunks frequently become extremely aggressive and difficult for the beekeeper to manipulate without incurring numerous stings.

Goodrich (1968) reported skunks to be most active at daybreak or just before dark. He also indicated that skunks love honey and wax in addition to adult bees. Pellet (1916) felt that skunks were fond of honey, and he fed it to captive skunks. He said that they seldom succeeded in feeding on the honey in a colony.

Skunks may seriously weaken bee colonies by depleting the colony population, posing the most serious problems in the fall. They usually leave bee colonies alone during May and June, when beetles form a large portion of their diet and in summer, when they prefer crickets (Thorpe 1967).

In addition to eating bees, skunks dig under or around bee colonies. Townsend et al. (1965) reported that skunks do much damage to colonies packed in tar paper during the winter. They tear at the tar paper and expose the colony to adverse weather conditions. Probably the greatest problem posed by skunks, beyond the destruction of bees, is that the colonies they attack become very mean and vicious.

Jaycox (1969) reported that skunks are becoming more numerous, in part because fewer people are trapping them for their pelts. Root (1966), who also indicated that skunks are increasing in numbers, attributed their increase to the legal protection given them. According to a survey by Hood (1996), skunks are given legal protection in 14 out of 52 states and provinces. Thirty-seven respondents from states or provinces where skunk damage occurs reported damage levels as stable. Only three respondents (California, Nova Scotia, and Alberta) reported an increase in skunk damage, four (Florida, Iowa, Massachusetts, and West Virginia) reported a decrease, and six reported that skunk damage was nonexistent in their state or province.

Skunks also feed on bumble bees and other ground-nesting insects. Plath (1923) reported the destruction of bumble bee nests by a skunk. Skunks foraging in the wild on yellowjacket ground nests should be classified as beneficial.

Vansell and Storer (1930) reported that striped skunks (*Mephitis occidentalis*) were a serious problem at the University of California apiary in 1929; they captured several skunks with a steel trap. The palates, tongues, and throats of the captured skunks contained many stings because when attacked, bees try to sting skunks and protect their colony. Stings were not found embedded in the remainder of the digestive tract, and most of the bees taken from the stomachs of the skunks still had their stings. Later, Storer and Vansell (1935) examined some skunks that had been caught in the bee yard. These skunks also had numerous bee stings in their tongues, palates, and gums. Two skunks had their stomachs crammed with bees. In one skunk, bee fragments, including parts of at least 145 different individual bees, accounted for 17 percent of the stomach contents. One skunk dropping contained the remains of 108 bees. Thirty-three stings were found on the head of a skunk that was trapped beside a bee colony (Eckert 1932). Other stings were found elsewhere on the body, but all could have been received after the skunk was trapped. In addition to the external stings, 16 stings were found in the mouth,

Figure 17.4. One method of control is using a live trap. A trapped skunk 'usually' will not cause odor problems if the cage is covered. Once trapped, the skunk can be dispatched, or relocated, depending on local laws. (Flottum photo).

five on the tongue, seven in the throat, and seven in the walls of the stomach. The esophagus and intestine were free of stings.

Control

Killing skunks with strychnine is described by many writers as a quick and effective means of control (e.g., Eckert and Shaw 1960, Townsend et al. 1965, Goodrich 1968, Johansen and Mayer 1976). This poison, however, is now illegal for skunk control in the United States (Salmon 1980). No toxic materials are currently registered in the United States for skunk control (O'Brien and Marsh 1990).

Thorpe (1967) was able to control skunks with chemical repellents. Rags soaked in carbolic acid and benzaldehyde repelled skunks until the chemicals became diluted with rainwater. Placing paradichlorobenzene (PDB) crystals in jars with perforated lids around the bee yard was successful in eliminating skunk depredation in Michigan. Sprinkling the skunks' paths with powdered lye (Haydak 1948) or fertilizer (O'Brien and Marsh 1990) drove them away. Rowe (1966) reported that skunks could be eliminated by scattering a few moth balls around the apiary. The moth balls should be placed one meter or more from a hive.

Storer and Vansell (1935) suggested fencing apiaries to keep skunks away. The authors recommended a fence 91-122 cm (three to four feet) high with a mesh not larger than 5 cm (two inches). The netting should be buried 15 to 30 cm (six to 12 inches) deep. A barbed wire run at ground level will discourage digging by the skunks. The top level of the fence should have a sheet of tin or galvanized iron cap 36 cm wide, and the gate must be close-fitting and meet a bottom sill. Goodrich (1968) cautioned against the use of a fence unless it is sturdy and well-constructed. Pike (1947) and Eckert (1954) recommended using chicken wire laid flat on the ground around the outside of the fence as an added precaution against a skunk digging into a fenced bee yard.

Morse (1972) and Ambrose (1975) recommended elevating colonies to help prevent visits by skunks. Unfortunately, elevated hive stands can make beekeeping manipulations more difficult to carry out.

Various traps can be used to capture skunks that raid bee colonies. Storer and Vansell (1935) trapped skunks with an unbaited, exposed steel trap placed in the areas where skunks were scratching in front of the hive. Trapped skunks were picked up by a long board attached to a chain on the trap and then drowned in a garbage can or large pail of water.

Jaycox (1975) recommended trapping skunks with a wire cage trap. The cage can be baited with a chicken head, a dead mouse, or a piece of comb with brood and covered with burlap for ease in handling. Trapped skunks can be released away from the apiary or drowned while still in the cage. If the cage is covered

Figure 17.5. A board with protruding nails will inhibit a skunk from scratching, thus reducing predation. (File photo).

with paper or a plastic bag, the skunk can be killed by running vehicle exhaust into the enclosure.

Morse (1972) and Gillespie and Dick (1966) cautioned against the use of traps in skunk control because of the unpleasant odor released by a trapped skunk. Anderson (1941) indicated that box traps keep the offensive odor from being released.

Arnold and Bland (1954) and Goodrich (1968) recommend shooting is best practiced very early in the morning or just before dark, when skunks actively feed in the apiary.

Gochnauer et al. (1975) recommended combining skunk and mouse control. Screening placed over the colony entrance will control mice, and if the screen is allowed to extend outward from the entrance and is folded upward, it can also provide a measure of skunk control. The extended screening impedes the skunk's attempts to scratch on the hive, and it exposes the vulnerable underside of the skunk to stings by the bees. Other beekeepers place a large mesh wire such as chicken wire in front of colonies to keep skunks away from the entrance.

Webster (1947) suggested another possible control measure. A board 15 cm high or a row of bricks is placed several centimeters in front of the hive entrance. Skunks will dig under such obstructions but apparently will not harm the bees.

Figure 17.6. Another method of skunk control is to block the entrance with a wire cage. Skunks trying to scratch the front of a colony expose their undersides, causing discomfort and reducing their ability to capture bees at the entrance. (Root photo).

Campbell (1936) trapped a mother skunk and three young ones by placing two strips of Tanglefoot in front of his hives. He reported a minimum of odor and no disturbance to the bees. Respondents in a survey by O'Brien and Marsh (1990) listed trapping as the preferred method of skunk control, followed by exclusion, poisoning, elevating colonies, and shooting, in that order.

There are many suggested means for the control of skunks pestering bee colonies. Because it is not difficult to detect skunk activity, control is not usually a problem.

Bears

Probably the best known of all mammalian honey bee pests are the bears. Bears, which are members of the family Ursidae, typically, have large heads, short tails, and short, rounded ears. Their feet are five-toed and feature powerful non-retractable claws. The fur is coarse and shaggy. Bears are excellent climbers and can run as fast as 48 km (30 miles) per hour. Most bears can swim readily and well.

They are omnivorous and have molars for crushing instead of shearing. Bears have a fondness for sweets and are one of the few wild animals that are susceptible to tooth decay. They have a keen sense of smell and acute hearing.

Bears are solitary, but the young stay with their mothers for about two years. In colder climates, bears sleep in protected locations during the winter. Currently eight genera and 10 species are recognized in the bear family, of which five species are known as bee pests.

The black bear (*Euarctos americanus*) once ranged over most of North America from Mexico to Alaska but is now limited to wilderness areas in that range. The black bear can be black, cinnamon, or various shades of brown and has been assigned many "species" names. Black bears weigh from 90 to 225 kg (200 to 500 pounds). The average length is 152 cm (60 inches), and the average height is 89 cm (35 inches) at the shoulder. The black bear is shy, but it can be a problem in parks and campgrounds where it forages for food and garbage. Black bears can acquire a taste for livestock and honey, and they will return repeatedly to an area (or apiary) to forage.

The Eurasian brown bear (*Ursus arctos*) was once widely distributed in the temperate region of the Northern Hemisphere from Spain to Russia and Japan. It is now common only in parts of Russia. Its color may range from yellowish to brown and even black. In Russia, members of this species can weigh over 315 kg (700 pounds; the Siberian brown bear), but individuals in Europe and Japan (the Japanese brown bear) are always smaller.

Three additional bear species of more limited distribution are listed by Toumanoff (1939) and Walker et al. (1975) as pests of honey bees. The sloth or honey bear (*Melursus ursinus*) is a small bear, weighing, on average, about 103 kg (225 pounds) and reaching only about 153 cm (60 inches) in length. It has long, coarse fur and is generally black except for a patch of white or lighter fur on the chest. It inhabits forested areas of India and Sri Lanka.

Sloth bears search for tree hives of *Apis cerana* or for the exposed comb nests of *A. dorsata*. The bears attempt to knock down *A. dorsata* combs, especially after dark, to disorganize the guards. When stealing from *A. cerana*, the bears receive numerous stings and can be heard "bawling with the pain of the stings"; nevertheless, they persist (Perry 1970).

The Asiatic black bear (*Selenarctos thibetanus*) is black with white patches on the chin and chest. Members of this species are about the same size as the sloth bear. They occur from Eastern Asia through the Himalayan Mountains to China, Japan, and Siberia.

The Malayan sun bear (*Helarctos malayanus*) is the smallest bear species. These bears average slightly over 100 cm (39 inches) in length and generally weigh under 65 kg (143 pounds). They are black and have an orange or light-colored breast, muzzle, and feet. They occur in Southeast Asia but are seldom seen because they are nocturnal.

Maehr and Brady (1982) reported bear damage of $104,868 in Florida in a survey conducted in 1981. Peak bear damage occurred between April and June in Northern Florida. Robinson (1965) estimated that bears destroy thousands of dollars worth of bee colonies and honey annually in Florida. Thousands of acres of good bee pasture remain unused for beekeeping in Florida state and national parks because of actual or potential losses to bears; production of gallberry and palmetto honey is especially limited. He reported that bear damage is on the increase.

Garshelis (1989) noted the single greatest annual monetary loss ($10,000) attributed to bears in Minnesota was beehive depredation which occurs from May to August.

Jadczak (1994) noted that approximately 85 percent of Maine is considered "bear range," which supports a population of about 21,000 animals. Wildlife officials estimate bear density at one bear per square mile (2.6 km²) in Northern Maine and one bear per two square miles (5.2 km²) in the South. Jadczak stated that electric fences are necessary in Maine, especially where colonies are used for blueberry pollination and raspberry honey production.

Goltz (1990) stated that Northern California apiaries are especially vulnerable to bear depredation during "the long, warm autumn days when bears range widely, and sometimes fearlessly, in search of food." One California beekeeper claimed to have average annual losses of $6,000 as a result of bear damage from 1974 to 1988 (Hartshorn 1988).

Bears are also a problem in India (Singh 1962) and Russia (Galton 1971). In her history of beekeeping in Russia, Galton listed the bear as the great enemy of the early beekeepers there. Toumanoff (1939) listed species of bears that create problems in Central Asia; Okada (1959) named two species of problem bears in Japan.

Nelson (1974) provided the following estimates (in Canadian dollars) of bear damage in three Canadian provinces: Manitoba, 1968, $10,500; Saskatchewan, 1969, $10,600; Alberta, 1971, $63,000, 1972, $120,000, and 1973, $200,000. Losses were probably higher than reported. A second estimate of bear damage in Manitoba in 1968 was $8,896 (McRory 1969). In 1974, beekeepers' losses of $196,491 were reported in British Columbia (Warren 1975).

The beekeepers of the Peace River district in Alberta, Canada, have experienced considerable bear problems (Nelson 1974). In 1973, 400 bears were trapped and removed from the area, but losses were still estimated at $200,000. Losses without the bear removal program could have been nearly $500,000, according to Nelson (1974).

In Manitoba, a beekeeper reported 1992 bear damage losses of $50,000 in his operation (Anonymous 1993). The beekeeper noted that he had killed an estimated 220 bears to protect his apiaries from bears over the past 20 years.

Lord and Ambrose (1981a) surveyed all continental U.S. states and Canadian provinces on bear damage to beehives. Thirty-nine of the 62 states and provinces replied, reporting 1978 losses ranging from U.S. $150 (Maryland) to $52,059 (British Columbia), with an average of $16,209. Hood (1996) reported survey results of 1993 bear damage estimates in Canadian provinces as follows: Alberta ($10,000), British Columbia ($7,500), Manitoba ($60,000), New Brunswick ($20,000), Nova Scotia ($20,000), Ontario ($70,000), Quebec ($10,000), and Saskatchewan ($10,000).

Bears were rated the most serious vertebrate pest of honey bee colonies according to a survey by O'Brien and Marsh (1990). Nineteen Canadian and U.S. respondents listed bears in the "always serious" category. The total estimate of annual damages attributed to bears was $623,000, as provided by 22 respondents. Although another 14 respondents listed bears as pests, they did not give damage estimates. The two highest estimates of bear damage were from Florida ($275,000) and Ontario ($75,000).

Twenty-two out of 53 Canadian and U.S. respondents listed bears as the number one mammal pest to honey bee colonies in a survey conducted by Hood (1996). Twenty-eight respondents estimated 1993 bear damages at $638,500, with high estimates reported from California ($200,000), Ontario ($70,000), Wisconsin ($68,000), and Manitoba ($60,000). These figures exclude potential losses in beekeeper income from future honey production or pollination fees.

The majority of respondents (25 out of 41) residing in states or provinces where bear losses occur noted that bear damage has increased over the past 10 years (Hood 1996). Only four states (Colorado, Maine, Michigan, and Utah) reported a decrease in bear damage over the past 10 years, and one of these (Maine) noted that the decrease was a result of successful use of electric fences in his state. According to respondents, bear depredation prohibits or limits bee-

Figure 17.7. This bear, from Wyoming, like all bears, is solitary. Its range can include several apiaries. (File photo).

keeping in some areas in 23 (56 percent of respondents) states or provinces where damage occurs.

Toumanoff (1939) stated that in the central region of the former U.S.S.R., 700 to 800 colonies were destroyed annually by bears. Genov and Wanev (1992) cited 1,015 incidents of brown bear damage to domestic animals and bee colonies in Bulgaria from 1984 to 1988; 637 colonies were destroyed. Rajchev (1988) reported that bears destroyed up to 12 percent of honey bee colonies located in the Central Balkan Range of Bulgaria. Bears also create problems for beekeepers in Siberia. According to Toumanoff (1939), young bears are more of a problem than older ones.

Bears can cause more problems for beekeepers under unusual natural conditions such as wildfires, drought, or failure of a berry or acorn crop. Once individual bears acquire a taste for brood and honey, it is difficult to keep them away from apiaries, and even electric fences may not deter them. Typically, a bear will destroy one to three colonies nightly until the entire apiary is demolished. They may carry hives or hive parts several meters before smashing them and removing the honey and brood (Anderson 1941). According to Goltz (1990), bears can be very skillful and remove frames from the hive without damaging the super or hive body, but then attack frames and comb without mercy. Morse (1972) believed that bears are more interested in eating brood than honey. Webster (1947) and Dacy (1939) felt that bears particularly favor honey but will eat bees as well.

Bear protection

Lord and Ambrose (1981a) reported bear control programs in 29 of the 39 states and provinces in their survey. Bears were a protected species in 36 states and provinces according to respondents of a survey by Hood (1996). Six respondents (Alabama, Alaska, Michigan, New Brunswick, New Hampshire, and Saskatchewan) indicated no protection was given bears in their state or province.

After losing 250 bee colonies to bears in 1992, a Manitoba beekeeper claims the situation has worsened to where "conservation and government agencies must decide whether this region is for farming or a wildlife preservation district" (Anonymous 1993a). He stated that the pendulum has swung too far in favor of bear protection and conservation in his area. He noted that a good example of increased bear protection occurred in 1992 when the bear-hunting season was shortened from three months to two months.

A majority of respondents (17 out of 31 who answered this question) reported that present efforts by state or provincial government agencies were helping the beekeeping industry in its effort to resolve bear problems (Hood 1996). Eight respondents noted that government agencies had a negative impact on their beekeeping industry relative to bear problems. Thirty-five respondents said that their states or provinces have bear removal or relocation programs. In North Carolina, the Wildlife Resources Commission claims that proper management of the black bear in their state has become "a great challenge as human development continues to diminish and to fragment existing bear habitat" (Anonymous 1993a). The report noted that simply catching and relocating every bear that someone sees is not an option. The state has simply run out of adequate remote areas where bears will not become a nuisance, and both bear hunters and citizens concerned about bears object to a relocation program. The Massachusetts respondent stated that although his state has a relocation program, the program is ineffective because the state is so small that relocated bears return to their original territory within two weeks.

Compensation for bear damage

Compensation by government agencies that seek to protect the bear population is desirable. The amount of money awarded, however, usually does not compensate the beekeeper fully for destroyed bee colonies, nor does it address the issue of lost profits from future honey crops or colony rentals. Currently, 11 states and provinces (Colorado, Idaho, Manitoba, New Hampshire, Ontario, Pennsylvania, Saskatchewan, Vermont, West Virginia, Wisconsin, and Wyoming) compensate beekeepers for colonies damaged by bears (Hood 1996). These programs vary in the amount of compensation, and there are some restrictions. The respondent from Manitoba noted that his province pays beekeepers for 75 percent of the value of lost equipment, bees, and honey, but not for lost potential production. According to Maule (1972) and Weaver (1979), Pennsylvania will compensate only for bear damage that occurs within 900 feet (274 meters) of the beekeeper's residence and that is reported within 70 days.

Ten states and provinces (British Columbia, Colorado, Maine, Maryland, Minnesota, Montana, Pennsylvania, Saskatchewan, Tennessee, and Wisconsin) have government agencies that provide financial support to beekeepers for constructing electric or solar-powered fences (Hood 1996). Saskatchewan is currently reviving its compensation program in the form of a subsidy to the beekeeper for fencing materials. The beekeeper is subsidized for 60 percent of expenses for materials up to a maximum of $1,000 in purchases. The rebate program for the province is limited to $15,000 annually for the beekeeping industry. Maine has a cooperative program conducted by the USDA and the state for lending electric fencing on a temporary basis to farmers and beekeepers experiencing bear damage (Jadczak 1994).

Control

Unfortunately, many good honey bee foraging areas overlap prime bear habitat in the United States and Canada. The expected future urban development in both countries will result in increased bear depredation problems for beekeepers in many states and provinces. Beekeepers who wish to continue beekeeping activities in these high-risk areas will have to commit time and energy to minimizing bear depredation.

Twenty-nine survey respondents listed electric fences as the primary method of protection used by beekeepers in their state or province to reduce bear damage (Hood 1996). A combination of electric and solar-powered fences was listed by nine respondents, and one respondent (Saskatchewan) listed solar powered fences only as the primary method of protection. Two respondents (Alaska and Idaho) listed platforms as a primary method of protection, and one respondent (Mississippi) reported trapping/removal and moving colonies to other locations.

Avoidance was listed by two survey respondents (Nevada and Louisiana) as a primary method of protection from bears (Hood 1996). Jadczak (1994) noted that a beekeeper may avoid bear damage by locating colonies away from prime bear travel routes such as knolls, tree lines, and stream banks because bears move along land formations that limit their visibility. However, he added that electric fences are highly recommended in areas of high bear density.

Bruce (1990) recommended that beekeepers explore an area to discover possible bear signs before placing colonies. But he added that an area may be safe even though bear signs are present if the beekeeper notes closely where the bear travels. Bruce claimed to have harvested a good crop of honey from an apiary located 160 yards from a well-traveled bear run. Although he lived in bear country in the Northwestern United States, Bruce did not have any bear damage in his 20 years of beekeeping. He attributed this perhaps to luck but admitted to having a good understanding of bear habits from his many years of bear hunting. He noted that bears "usually" have a circuit of about 80 km (50 miles) that follows one food source to another. For instance, the bear will travel from one campsite trash can to another unless chased by dogs. The bears have regular routes such as "hardwood stands where they eat acorns, salmon runs where dead fish can be easily found after egg laying, riffles where they can pick up fish working through the shallow water, and berry patches for dessert." A preventive measure listed by Bruce is placing every piece of scrap frame or comb, especially drone brood, into a bucket and removing it from the area. He noted that because bears have very poor eyesight, beekeepers should paint their hives to blend in with the local vegetation. This camouflaging technique not only gives the bee yard a low profile to prevent a bear from investigating a sudden intrusion into his territory, but it also minimizes damages caused by human pilferers and hunters. Bruce noted that he has seen many white hives destroyed and entire apiaries left in ruins sometimes just outside the legal distance of his colonies. In many instances, the beekeeper placed the apiary on the bears' runway.

Shooting bears was once commonly practiced by beekeepers in the United States. However, as bear populations continue to decline in many areas, the wisdom of shooting or other lethal control measures is increasingly questioned. Atkinson (1974a), for example, reported wide denunciation of a program that involved trapping and shooting bears around apiaries in the Peace River district of Alberta, Canada. California and other states and provinces may issue a permit to beekeepers to shoot a bear if the beekeeper satisfactorily demonstrates actual or threatened damage from their bear (Cummings and Wade 1975), but six U.S. states prohibit the killing of bears even in defense of property (Lord and Ambrose 1981a). In Florida, a beekeeper was fined for poisoning a bear that had been causing damage (Green 1986).

Currently 16 states or provinces issue permits to beekeepers to kill bears (Hood 1996). Twenty-five survey respondents reported that their state or province does not issue permits to beekeepers to kill bears. Although beekeepers in Tennessee are not issued permits to kill problem bears, the beekeeper may kill a bear if it is damaging property.

Several people have sought to develop a practical and inexpensive bearproof fence. Storer et al. (1938) published one of the earliest studies on the use of electric fences to protect bee colonies from bears. The results of tests of several designs indicated that a substantial fence is necessary. Three or four barbed wire strands connected to a good portable electrical source with a large ground were deemed necessary. The basic design developed by these authors, shown in Figure 17.8, has been repeatedly cited by others.

Robinson (1961, 1963) tried several electric fence designs before he was able to find an arrangement that satisfactorily kept bears from molesting out-apiaries. Robinson (1965) listed three criteria for electric fences. They must be (1) easy to construct and relatively inexpensive, (2) dependable, and (3) designed to operate in a wilderness area. He recommended a structure 122 cm (four feet) high of ordinary woven field fencing. Two electrically charged barbed wire strands, one 20 cm (eight inches) from the ground and the other 20 cm (eight inches) from the top of the woven wire, are also needed. It is necessary to use heavy insulators, which are screwed into posts to hold the charged wires. A wire mat should extend 46 cm (18 inches) out from the fence. The mat should be fastened to the bottom strand of the woven fence by pegs. This mat helps prevent bears from digging under the woven fence and serves as an effective ground for the electrically charged wires.

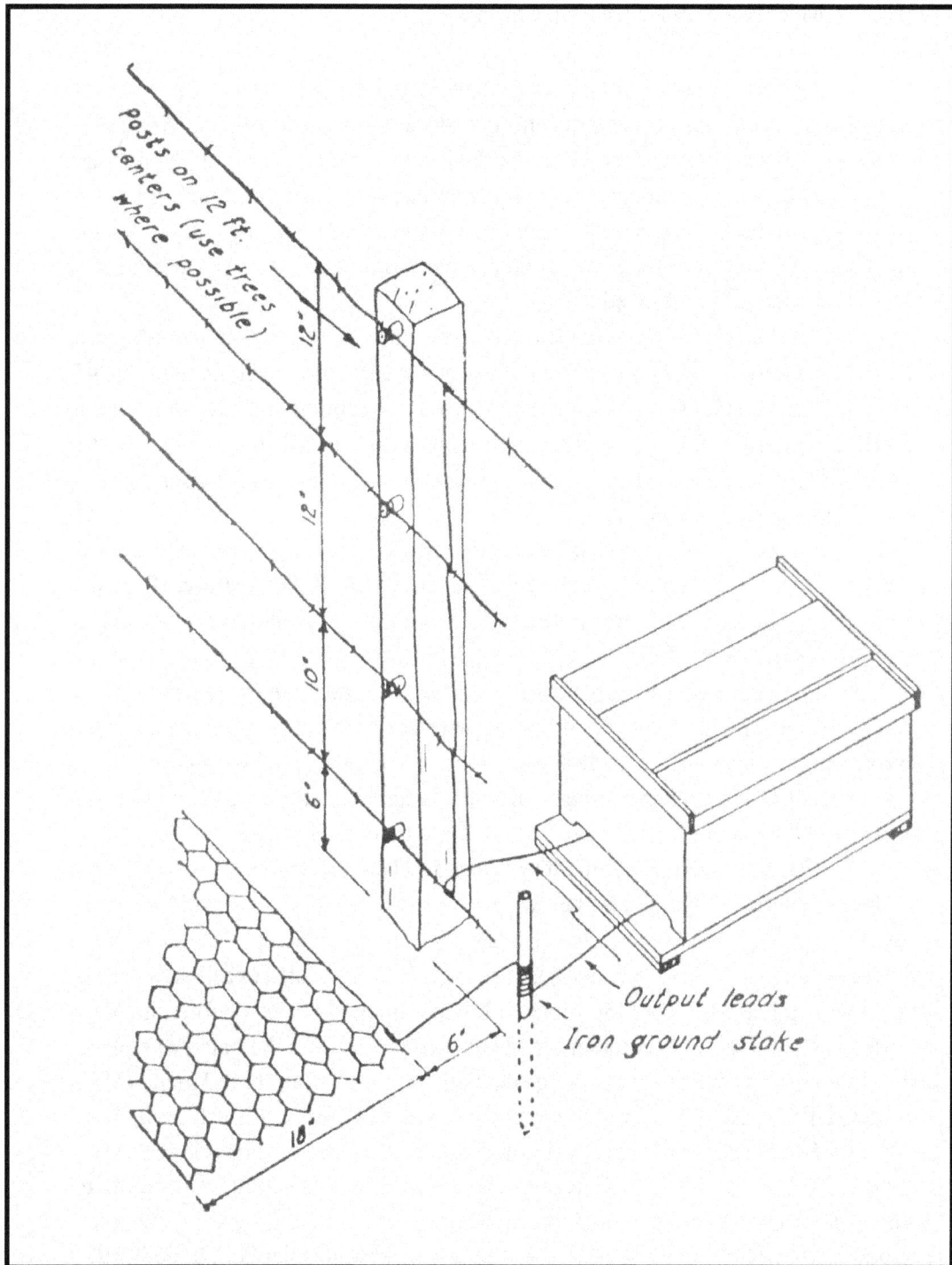

Figure 17.8. Typical setup of a fixed-post electric fence system. The bottom wire should be no more than 6 inches from the ground, and the height should be a minimum of 3.5 feet. Five or six feet is desirable, along with a lockable gate. The battery should be inside the fence, but accessible through the wires. A mat of chicken wire is sometimes used, as a ground, and to keep bears from digging beneath the fence. Weeds must be kept from touching the wires. Electric fencing material is readily available at farm supply stores. (Diagram modified from Storer, et al, 1938).

Nelson (1974) described a fence that can be assembled in two hours. It consists of metal bars driven into the ground and covered with plastic piping. Three strands of electrically charged barbed wire are strung on the plastic; care must be taken to avoid puncturing the plastic or letting the wire touch the metal posts. A ground wire can be run at the bottom and attached to each metal post, or the plastic piping can be cut in the middle to carry the center wire as a ground. The four corners should be braced.

Another bear fence consisting of four electric barbed wire strands is described by Nelson (1974). The strands, four-point barbed wire, are located 15, 41, 71, and 102 cm (six, 16, 28, and 40 inches) from the ground, and are connected in parallel. Chicken wire, 61 cm (two feet) wide, is put on the ground outside the electric fence starting about 15 cm (six inches) from the posts. Posts are placed about four meters (12 feet) apart.

Nelson (1974) attributed another fence design to the Manitoba Department of Mines and Natural Resources, Canada. This design features T-shaped metal posts placed about four meters apart. Sixteen-gauge barbed wire is placed on snap-on insulators at 15, 51, 91, and 122 cm (six, 20, 36, and 48 inches) from the ground. The wire is placed outside the posts except at the corners, where it is held on the inside. The four strands of barbed wire are connected at opposite corners and electrically charged. A negatively charged mat of chicken wire is placed around the enclosure. The power source (a six-volt battery) is connected to the barbed wires, and the negative lead is connected to the mat and a ground rod.

The Manitoba Department of Mines and Natural Resources tested a fence with five strands of charged barbed wire at 15 cm (5.9 inches) intervals. The total fence height was 76–81 cm (30-32 inches). A wire mat was not used but may be desirable in dry areas (Nelson 1974). Johansen (1975) presented excellent details on constructing bearproof electric fences. He recommended three or four strands of wire and the use of a 12-volt battery. He also suggested that the battery be placed in a hive body with a beehive above it to help protect it from theft. Knepp (1987) reported detailed instructions for an electric fence enclosure and construction costs.

Lord (1980) described a simple and inexpensive fence developed for migratory beekeepers in North Carolina. The fence has three strands: two hot wires insulated at each metal post and a middle wire firmly connected to the post as a ground. Each corner is braced outside. The total cost was under U.S. $100 in 1980; the fence can be assembled by two people in one hour and can be disassembled easily for relocation.

A bear fence should have a reliable battery to hold the electric charge and good electrical connections to avoid shorting from rust or corrosion. Overhanging trees and limbs should be removed, and weed control should be practiced outside and inside the fence. Currently, the herbicide paraquat applied semiannually is recommended for weed control. In 1965, a fence designed to enclose 55 colonies was built at a cost of U.S. $209 for supplies (Robinson 1965). Another article (Anonymous 1968) gave a figure of approximately $69 as the cost of a 15-m-square enclosure. This included four strands of barbed wire, electrification, and a chicken wire mat.

Butler (1992) reported on a cooperative program conducted jointly by USDA, APHIS, ADC, and the Maine Department of Inland Fisheries and Wildlife to develop fences to protect crops from animal destruction; this included fences to protect beehives from bear damage during pollination of blueberries and other fruit crops. The program was designed to accommodate as many growers and beekeepers as possible with flexinet fencing and an energizer on a first-come, first-served basis. The equipment is loaned on a temporary basis, and evaluation forms are completed by each participant. Not all systems were equally effective in different areas of the state. Butler stated that the program has been conducted annually and costs an average $2,000 per year (Personal communication 1994).

Jadczak (1994) described a bear fence system that has been used success-

Figure 17.9. A flex net system used in Maine, as described in the text. (Jadczak photo).

fully in Maine. The fence is constructed of three or four strands of polywire or flexinet with a variety of energizers available. One of the more popular energizers is powered by six "D cell" flashlight batteries, which will operate a fence for four to six weeks. Solar panels that are "bulletproof" are also available with this system. The key to these new energizers is that they carry a very short but extremely powerful energy pulse. Fence testers (voltmeters) are available with this system and are highly recommended to provide the beekeeper a reliable system check. The polywire fences normally have wire strands at 25-30 cm (10-12 inches) intervals. The four-strand fence has a wire at 13, 38, 63, and 89 cm (five, 15, 25, and 35 inches) above ground level. This polywire arrangement and the flexinet type fence also provides skunk protection.

Jadczak recommends proper grounding to ensure adequate apiary protection by placing chicken wire around the fence perimeter with a steel ground rod attached to the chicken wire. The cost of one of these energizers and complete materials to protect a typical apiary is approximately $190. Maine beekeepers commonly attach fresh bacon strips or a punctured can of sardines to the hot wires to give the bear an effective shock on the nose or tongue.

In an effort to save the dwindling Florida bear population, beekeepers

Figure 17.10. A predator platform used in North Carolina. (Hood photo).

were asked to protect their bee colonies by the use of elevated platforms or electric fences. According to the State Game and Freshwater Fish Commission, bears attracted to bee colonies were being killed illegally (Anonymous 1971). Platforms are expensive and inconvenient for examination and manipulation of colonies. In the Florida panhandle, where spring floods are a problem, platforms for both bear and flood protection are feasible. Robinson (1963) described Florida platforms; Goolsbey (1974) described a platform of metal piping used by a beekeeper in Idaho. Johansen (1975) discussed platforms and showed one that can be towed from site to site.

Some beekeepers report becoming frustrated with attempting to use electric fences to protect their apiaries from bear damage; they have begun to use predator platforms. Flanigan (1989) noted his displeasure with electric fences, including weak batteries, wire shorting by weed growth, and lack of fresh bacon strips on the wire rendering them ineffective. A properly built beehive platform is foolproof, maintenance-free, permanent, and protects against other hive predators. Flanigan described the platform concept by visualizing the "design of a simple milking stool on which the short, straight legs are attached to the underside of the seat at a location well in from the edge." The key to the design is the 76 cm (2 1/2 feet) overhang to prevent bears from climbing to the top of the platform. Bears easily climb the support poles but cannot reach over the overhang. Flanigan described a 3.7 m (12 feet) high structure platform that has a surface area of 3 by 5.5 m (10 by 18 feet) that will support several tons of weight. He noted that the cost of the structure, $93 (1989), was less than a typical electric fence which cost $136.

Johansen (1975) described cages of stout wire (hogwire) that a beekeeper used successfully on Vancouver Island, Canada. A 15-cm mesh wire was attached to the pallet and enclosed the colonies. Bears could jostle the colonies, but stings drove them off before they could get any honey. This protection is expensive but effective.

Fraser (1937) described the techniques once used by beekeepers in the Ural Mountains of the former U.S.S.R. to protect log gums (old-fashioned hives made of logs) from the numerous bears of the area. The beekeepers suspended their hives from trees from which the branches had been trimmed. A set weapon was rigged to shoot an arrow into the marauding bear. It must be emphasized that such weapons are illegal today in many countries. In any case, no weapon should be set to kill a bear because of the danger to humans or other innocent animals.

Fraser (1937) also described a trap he attributed to beekeepers in the former U.S.S.R. They constructed a platform and tied it to a tree trunk next to a suspended hive. The bear, upon climbing to reach the hive, finds the platform a convenient place to rest. Then, as it attempts to reach the hive, it trips a cord that releases the platform, which swings away from the tree trunk. If the bear remains on the platform, it can be shot by the beekeeper; if it jumps to the ground, it will be impaled on stakes placed in the ground.

Various other traps or snares that are more humane can be used to capture bears. Thurber (1975) described a method of snaring bears with a dead log fall. He illustrated that trapping was the least effective bear control measure.

Rao (1968) recommended the suspension of bee colonies to protect against bears in India. He said that colonies could be tied by ropes to branches and then suspended a meter or so below the branch, two to three meters above the ground. Inspection of suspended colonies could be a problem, as could bee orientation to colonies that twist and change direction.

Shiner (1983) described a bear control technique that was used succesfully in Northern Pennsylvania. Recycled soda bottles partially filled with human urine were placed in the ground at approximately 3 m (10 feet) intervals around the apiary. The bottles were placed at an angle to minimize rain water dilution.

Nelson (1974) attempted, without success, to find other bear control methods. A radio playing in an apiary may serve to frighten bears. Attempts to use chemical repellents, noisemakers, and ultrasonics have not offered much help. Aversive conditioning has also been attempted without much success. Goolsbey (1969) reported that tying voodoo dolls to trees, sprinkling pepper on covers, covering the ground with sulfur, and other "treatments" were equally unsuccessful. Goolsbey, however, re-

Figure 17.11. One way to recharge batteries in the field is to use a solar charger. The battery is 'charged' during the day, and remains charged overnight. Solar panels often serve as hunters' targets, though, and are easily removed. (Flottum photo).

ported one method of aversive conditioning that was completely successful in bear control. Strips of bacon rind are tied to an electrically charged wire. The bacon rind is highly attractive to a bear, and when a bear touches the bacon with its nose, it receives a shock that, according to Goolsbey (1969), discourages the bear from returning. Graff (1961) also endorsed the bacon-rind method.

Robinson (1963) tried strapping individual colonies together to make it difficult for bears to gain entry. He thought that the extra time and effort required by the bears to gain entry would result in more stings. However, bears were able to find a board (often the bottomboard or cover) that could be broken, and they eventually gained entry to and destroyed colonies that were strapped together.

Gallant (1994) described his system of bear control for his home yard in Laytonville, California. The system is a combination of a fluorescent light and his watchdog, a Norwegian elkhound. Gallant prefers a special five-watt fluorescent, "no flicker" light that is powered by any waterproof 110-volt source. The system produces 40 watts of light for pennies a day. He has the light installed inside a window of his honey house, which is located about 15 m (50 feet) from his six hives. The light and dog system has proved effective in controlling raccoons and skunks as well. When animals approach the bee yard, the dog chases them up a tree and alerts the beekeeper. Although he lives in an area where bears are a problem, Gallant

Figure 17.12. Using two batteries for a bear fence increases its potency, and, hopefully, its protection. (Jadczak photo).

claims his system has worked for five years without failure.

Bears are a fact of life in many prime honey bee foraging areas and sometimes where bees are used for pollination services. The beekeeper should ask neighbors or local residents about possible bear activities before placing bees in unfamiliar territory. The only two reliable means of protecting bee colonies from bears are a well-maintained electric/solar-powered fence or a properly constructed beehive platform. If the beekeeper plans to place colonies in prime bear country, the necessary protection should be arranged before apiary establishment. Protecting bee colonies is more difficult once a bear gets a taste for honey bees and honey. As Bonney (1994) noted, "Speaking from experience, I can say there is nothing more disheartening than to come upon a bee yard after a bear has stopped in for a snack."

Rodents

Several rodents are pests of the honey bee. Rodents are small to medium-sized and represent the largest mammalian group. All have teeth and jaws uniquely adapted for gnawing. Rodents are frequently abundant in an area. Two rodents, the Norway rat and the house mouse, have, like the honey bee, been introduced from Europe by humans traveling to other continents.

Mice

Mice are the leading rodent pests of honey bee colonies (CAPA 1991). O'Brien and Marsh (1990), in their survey of state and provincial apiary inspectors, reported that the house mouse was frequently rated as a "sometime serious" or "always minor" pest. Respondents listed the house mouse as one of the three most serious pests 40 out of a possible 57 times. Fifteen respondents estimated annual mice damage at $100,450. In a more recent survey, seven out of 53 state and provincial bee specialists and inspectors listed mice as the number one mammalian pest (Hood 1996). Overall, the more recent survey indicated mice following bears, humans, and skunks in priority as the most damaging mammal.

Roff and Brimblecombe (1963) listed mice as a problem for Australian beekeepers; they are described as a pest in South Africa (Anonymous 1908), Ireland (Doupe 1921), Great Britain (Manley 1948), Canada (Townsend et al. 1965), Hawaii (Eckert and Bess 1952), the continental United States (Anderson 1941, Caron 1974, Gochnauer et al. 1975), India (Singh 1962), Japan (Okada 1959), the former U.S.S.R. (Beljavsky 1927), and Europe and Asia (Borchert 1974, Dyulgerov 1979).

Several species are listed as pests of honey bees. The common house or domestic mouse (*Mus musculus*) and the wood mouse (*Apodemus sylvaticus*) are the most common species mentioned as entering beehives and destroying stored

Figure 17.13. The common house mouse (*Mus musculus*) is an introduced species, and is one of the most common mice causing problems for beekeepers.

bee equipment. Borchert (1974) and Toumanoff (1939) also listed several other species in the genera *Mus, Micromys,* and *Clethrionomys* as problems for beekeepers. One or more species of *Peromyscus* may also be included. Zappi-Recordati (1971) and Aran (1969) identified door mice as pests; these could be members in any one of an additional six genera.

Bonney (1993) reported deer mice and white-footed mice as being the more common rodent pests in the Northeastern United States and throughout much of the rest of the country. O'Brien and Marsh (1990) noted that survey respondents in New York and Massachusetts estimated annual deer mice damage at $6,000 and $1,500, respectively.

Mice feed on pollen, honey, and bees. Their attacks can lead to destruction or to serious weakening of an entire colony. Because mice chew combs and frames to provide space to build their nests, nest construction usually results in the destruction of hive equipment. Mice can successfully build a nest even in a strong bee colony, and they survive the winter without being stung by the bees. The mice move in and out of the colony entrance when the bees are inactive; the nest protects the mice while they are inside the bee hive (Bonney 1993).

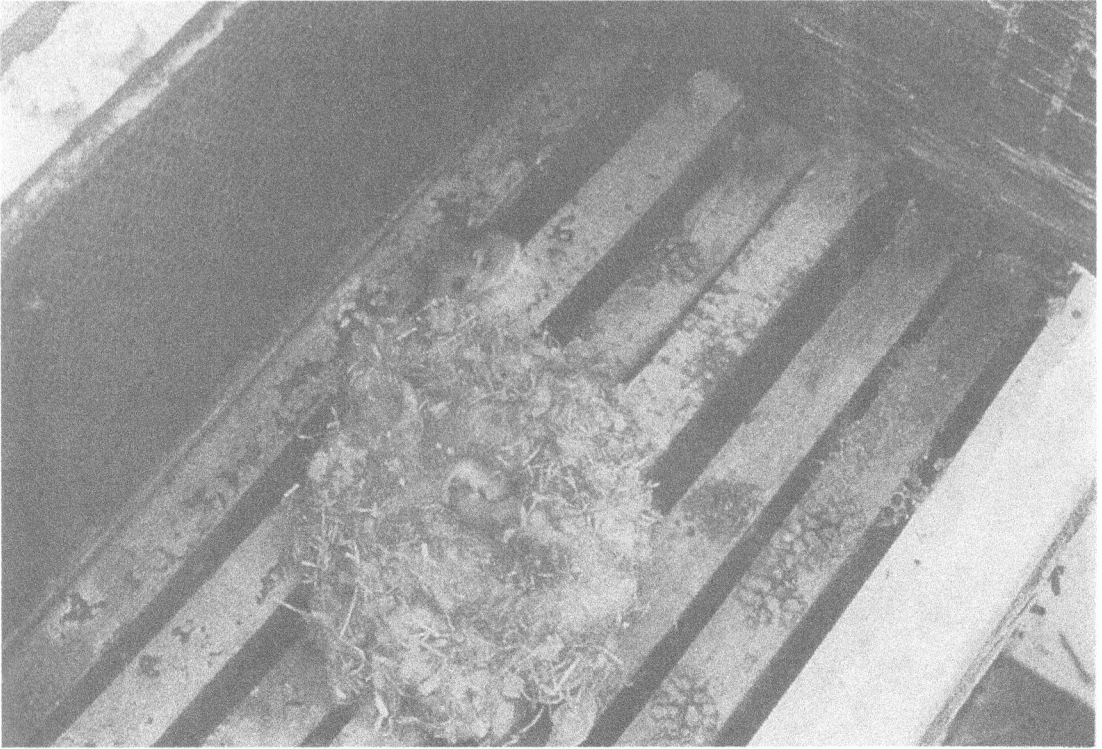

Figure 17.14. A deer mouse nest in an overwintered bee colony.

Other harmful effects of mice include the smell and odor created by their urine and droppings. Langstroth (1860) felt that bees would abandon a hive in the spring because of the smell from a mouse. Mice in a bee colony can also disturb and disrupt cluster behavior. Disturbance by mice may increase the incidence of dysentery in overwintering honey bee colonies (Egger 1973).

Mice are primarily a problem during the autumn after the evenings become cold. Pellett (1916) found mice both in colonies wintered outdoors and in those wintered in cellars. Colonies packed for winter in most packing materials have problems with mice (Borchert 1974). The type of hive apparently does not matter to the mouse, provided the opening is large enough to permit entry. Colonies in fields or at the edges of forests are especially vulnerable. During the 1935-36 winter season in New Jersey, beekeepers surveyed to determine the cause of winter losses reported the loss of 1,902 colonies, 41 of which were thought to have been killed by mice (Pitt 1937). Hood (1996) reported problems with mice moving into pulp pot bait hives in South Carolina with the onset of cooler weather in late season.

Control
Mouse control is not too difficult. The most widely practiced method is

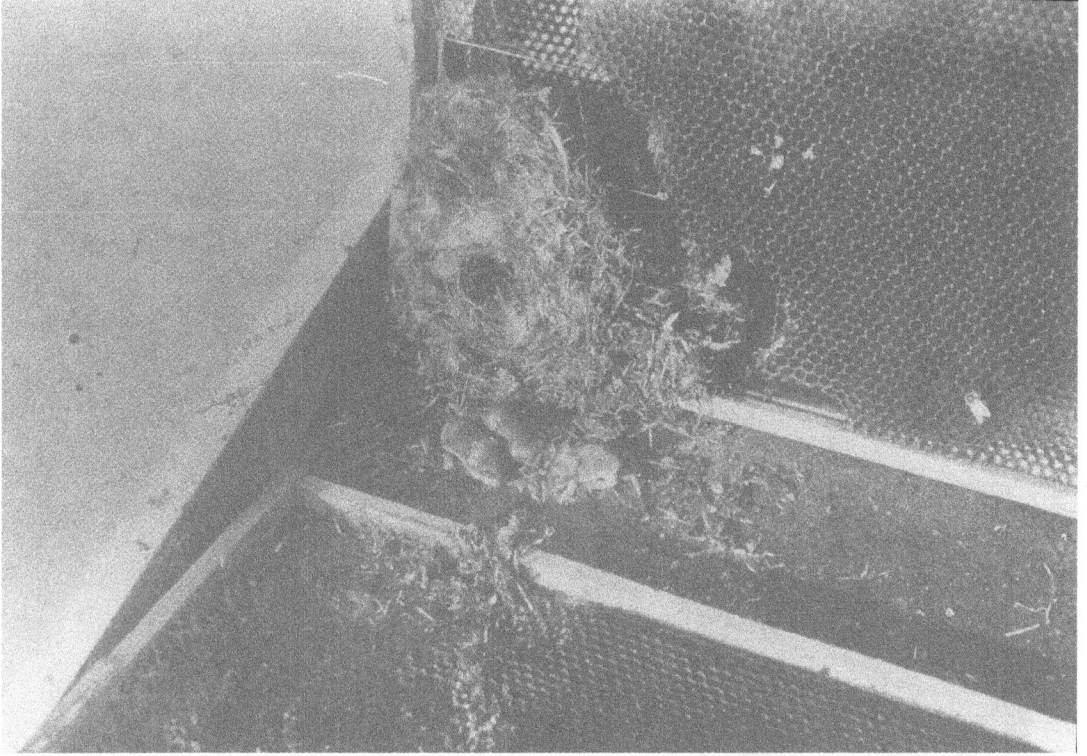

Figure 17.15. Mice nesting in hives during the winter can cause serious damage. Typically, nest construction damages several combs, frames, and maybe the entrance reducer. (Caron photo).

reduction of the size of the colony entrance before mice start to move from the field into the colony. Almost anything that mice cannot chew or gnaw can be used to reduce the entrance (G. Morse 1979). Although the hive entrance should be made small enough to keep mice and shrews out, it must not be made too small for bees to exit for their winter cleansing flights. O'Brien and Marsh (1990) noted that small mice can get through a 1 cm (3/8 inch) high by 10.5 cm (three inches) wide opening of a regular entrance reducer. An entrance reducer that leaves a .65 cm (1/4 inch) high clearance is more effective (Anonymous 1987c). Howes (1979) reports that using several 1 cm (3/8 inch) diameter holes as openings instead of one large opening provides better mouse control.

An alternative to entrance reduction is to place a barrier at or near the entrance to keep out rodents and small mammals. Roff and Brimblecombe (1963) suggested placing a queen excluder between the bottomboard and the brood chamber or tacking zinc queen excluder material over the colony entrance. Manley (1948) recommended the use of entrance screens. Entrance screens and queen excluders may, however, reduce ventilation in the hive and interfere with normal housecleaning activities, leading to a buildup of debris within the hive. Such methods should

Figure 17.16. A simple mouse guard has a mesh small enough to prevent mice entering, but large enough to let bees pass freely. Stapled to the bottomboard, this can be left on year round.

be used with care.

Bonney (1993) recommended a manufactured or a homemade metal guard for mouse exclusion. The metal guard can be made from sheet metal cut to the shape of an entrance reducer, or a guard can be made from a piece of .65 cm (1/4 inch) wire mesh bent to shape. Bonney advised checking the lower part of the exterior of the hive body to discover any weak areas or cracks that a mouse might gnaw and gain entrance. Staple a thin piece of metal such as a tin or aluminum can lid over weak areas.

Townsend et al. (1965) recommended the use of commercial rodent poison to control mice. Atkinson (1974b), however, experienced problems with certain formulations of the commercial bait. The ground meal formulation disappeared but did not eliminate the mice.

If a poison of any kind is used, care should be taken to determine that only the offending rodents are killed and that other creatures are not harmed. Baits work in some instances, but not in others. Generally, they do not remain effective outdoors for any length of time.

Root (1966) suggested using traps to eliminate mice and other rodents.

Traps may be useful in buildings such as honey houses or in areas where supers and bee equipment are stored. Borchert (1974) suggested baited traps with wheat flour or ripe fruit such as plums. Paillot et al. (1949) suggested the use of traps as well as entrance reducers and poison baits. They also recommended sprinkling wood ashes or pine needles around bee colonies. These materials are apparently somewhat repellent to mice. The removal of grass and weeds from the areas around hives may also help to keep mice out of colonies (Mace 1934).

If mice enter a colony before measures are taken to prohibit entry, their nest should be removed. Mice will usually attempt to escape when the nest is disturbed. If a mouse nest is discovered and removed during the winter, the winter cluster should be disturbed as little as possible. Damaged combs should be replaced if possible, and the hive should be protected against re-entry. Any remaining mouse scent is attractive to other mice.

Mice are certainly capable of creating a mess and even causing the loss of an entire colony during the winter. Mouse control is not very difficult, and beekeepers need not suffer the depredations of mice, given the numerous and inexpensive control measures available.

Rats

Pellett (1916) stated that rats are destructive to stored equipment in honey houses. Manley (1948) indicated that rats will pull out loose entrance blocks in colonies but that they are seldom a serious problem in Great Britain. Singh (1962) listed rats as an occasional problem in India. Borchert (1974) listed the house rat (*Rattus rattus*) and the Norway or migratory rat (*R. norvegicus*) as being especially troublesome for beekeepers.

Control of rats is similar to mouse control. Placing traps in honey houses or buildings where bee equipment is stored and reducing the size of colony entrances are the most useful control procedures.

Sugden and Collins (1990) described a serious problem with rats causing bait hive losses in their trapping program for the Africanized honey bee in Southern Texas. In one month in 1989, an estimated 10 percent of all pulp pot bait hives along the Texas-Mexico border were damaged or destroyed by rats. The rats entered the single entrance but then gnawed additional entrances through the trap lid. Rat excrement inside the trap, along with nesting materials, made the traps ineffective. "Burglar bars" were developed and placed over the port entrances to exclude rats. The burglar bars were made of 10 cm (4 inch) semicircles of 1.3 cm (1/2 inch) hardware cloth fastened to the hive interior over the entry port with staples and adhesive. Spaces were cut through the wire to allow insertion of the pheromone clip. Tests confirmed burglar bars did not alter the traps effectiveness in attracting bee swarms.

Squirrels

Squirrels, such as the gray squirrel, *Sciurus carolinensus,* may gain access to stored bee equipment and do some damage, especially during the winter. Smith (1960) listed squirrels as occasional pests in the tropics. Parks (1930) listed them as a pest in North America, and Johnsen (1955) stated that *Sciurus vulgaris* is a pest in Europe.

Squirrels chew and destroy combs in an apparent effort to obtain honey and pollen. To protect against this destruction, supers and bee equipment should be stored in rodentproof areas. Alternatively, each stack of equipment should rest on a closed bottomboard or pestproof piece of wood and should be covered with a queen excluder or snug top cover. Poison baits are available for squirrels and can be used when necessary. Trapping is also possible.

Grey squirrels and flying squirrels (*Petaurus norsolcensis*) are a common cause of pulp bait hive trap destruction in South Carolina. The squirrels moved into the traps in early fall at the onset of cooler weather. They made the traps ineffective by placing bedding materials in the traps and gnawing extra entry holes. The squirrels were very persistent and gained entry at the edges of protective barriers.

Large livestock

Large livestock such as horses, sheep, or cattle may damage bee colonies by toppling hives. Some animals will scratch to remove flies or pests or become skittish around buzzing insects and as a result may inadvertently upset bee hives. To protect against these possibilities, colonies should be maintained outside areas where large livestock are kept, or the apiary site should be fenced (Stanger et al. 1971). Colonies should not be located so that the bee flight path is in an area where large livestock are present.

O'Brien and Marsh (1990) reported that Canadian and U.S. survey respondents listed horses as sometimes serious (2), always minor (4), sometimes minor (22), or no problem (21). Cattle were listed as sometimes serious (6), always minor (6), sometimes minor (24), or no problem (14). Sheep were listed as sometimes serious (1), always minor (3), sometimes minor (6), or no problem (37). The Arkansas respondent listed cattle as the most damaging pest, with estimated annual damage of $4,000. Fencing was the preferred livestock control technique followed by avoiding and relocating apiaries. Grouping colonies closely together to avoid hives being knocked over was another suggestion when other techniques were not possible.

Cattle were listed by three respondents (Iowa, Missouri, and Wyoming)

Figure 17.17. Large livestock can inadvertently damage colonies in their area. Fencing is the only way to eliminate this problem. (Caron photo).

as the third most damaging mammal in a Canadian and U.S. survey conducted in 1993 (Hood 1996). A South Carolina beekeeper tells of an experience with a prized bull many years ago. The beekeeper placed several colonies inside a fenced pasture not knowing that a bull occupied the same pasture. After unloading the colonies, the beekeeper noticed that the colonies were extremely low on food stores, so he decided to feed. He placed a large cut-in-half iron tank on the ground and filled it with approximately a barrel and a half of thick sugar water. The bull discovered the sugar water that night and took a liking to the sweet syrup. He drank about half the tank of syrup and became seriously ill. The farmer discovered his swollen bull the next morning and summoned a veterinarian to the site. The bull lived, but the beekeeper paid dearly for a full day of veterinary care.

Marsupials

The opossum (*Didelphis marsupialis*), the single marsupial species reported as a pest of bees, ranges from Canada to South America. It is grayish white to black, and individuals can grow to nearly one meter (3 feet) in length, tail included. The muzzle is characteristically wedge-shaped and features large, promi-

Figure 17.18. The opossum (*Didelphis marsupialis*) can be an occasional pest. Control is similar to that of skunks. (Teasley photo).

nent, dark, beadlike eyes and thin, naked ears. The tail is long, naked, and prehensile. Opossums are omnivorous; eggs and fruits are usually included in their diet. They generally forage at night, moving with a slow, ambling gait from a sheltered den located in the ground or in a hollow tree.

Because opossums eat insects, honey bees would seem to be a possible food, but we found few references to their feeding on bees. Carroll (1972) shot two opossums in an apiary in Texas on successive evenings. He estimated that the opossums had consumed a total of one kg (two pounds) of bees. Crushed bees and stings were found in the mouths of the dead opossums. The editor of the beekeeper newspaper where this report appeared said that he has also, on occasion, seen bee yards bothered by opossums. White (1968) reported the capture of two opossums in traps set for skunk; he said that opossums work at the colony entrance in a manner similar to that of skunks.

Opossums were listed as one of the leading mammalian pests by two U.S. and Canadian survey respondents (Maryland and Pennsylvania) (Hood 1996). O'Brien and Marsh (1990) reported that nine survey respondents listed opossums as "causing some degree of damage to honey bee colonies." Although annual colony damages were estimated at only $100, the Delaware respondent reported that the opossum was the second leading vertebrate pest of honey bee colonies surpassed only by humans.

The opossum is apparently only a very occasional bee pest. With its nocturnal feeding habits, an opossum feeding in the apiary probably goes undetected in most instances but may weaken bee colonies and thereby decrease the chances of those colonies wintering successfully. Eliminating opossums by shooting individuals as they forage in the apiary or by trapping them seems feasible. Elevating bee colonies on hive stands one meter or more above the ground might also prove effective.

Insectivores

The insectivores are among the smallest and most active of mammals. As the name implies, these animals are chiefly insect eaters, but some have a more diversified diet. Most are nocturnal and possess anatomical adaptations for burrowing or for living in trees.

Hedgehogs

The common hedgehog (*Erinaceous europaeus*) is found over all of Eurasia but not in North America. It has spines on the upper part of its body, coarse hair underneath, and a short tail. It has a long snout and small eyes and averages about 25 cm (10 inches) in length. The claws are long but weak. When startled, an individual rolls into a ball. Its sharp, unbarbed spines project in all directions and create an effective defensive barrier. The hedgehog's diet consists primarily of insects but may include small mammals, bird eggs, and even plant materials and fruit. The hedgehog is nocturnal.

Toumanoff (1939), Johnsen (1955), and Zappi-Recordati (1971) list the hedgehog as a bee pest. Hedgehogs are especially attracted to hives that have bees hanging on the entrances or bottomboards in the evening. They also like foraging bees laden with nectar. Hedgehogs may blow or create a disturbance at the hive entrance and kill and eat the bees that emerge. Beljavsky (1927) reported that hedgehogs eat drone larvae thrown from the beehive by the beekeeper during inspections.

Borchert (1974) also listed the hedgehog as an occasional predator of honey bees. He indicated that hedgehogs usually take bees from the ground in front of bee colonies. Borchert recommended driving hedgehogs away from the apiary or elevating hives on a stand above the reach of these creatures.

Where hedgehogs are a problem, elevating hives seems to be an efficient means of preventing their depredation. Proper colony management to provide extra cluster space to keep bees within the hive would also seem advisable.

Shrews

Shrews, sometimes called shrew mice, are small mammals belonging to the family Soricidae. They have short fur and are gray to brown with long, pointed snouts that project beyond the lower lip. The tail is usually shorter than the body. Glands located on the sides of the body make shrews unpalatable to many predators. Shrews are nocturnal feeders and consume insects, snails, and worms. Most shrews must eat nearly continuously to support their high metabolism.

Parks (1930) was one of the first to identify the shrew as a honey bee pest in North America. He captured and identified the least shrew (*Cryptotis parva*) as a pest and surmised that either it or two other shrew species were responsible for damage to combs and frames, especially during the winter. He felt that most shrew activities went undetected because of the small size of the animals and their nocturnal habits. Mice are usually blamed for damage actually done by shrews, according to Parks (1930).

Paillot et al. (1949) stated that shrews eat many bees, especially those that are already dead. They also indicated that shrews do not usually attack combs to eat the wax and honey, as mice do. Seifert (1968) felt that shrews, especially the pygmy shrew, cause considerable damage. He indicated that shrews feed abundantly on honey bees from the winter cluster.

Toumanoff (1939) reported that shrews are dangerous to bees. He indicated that they may disturb wintering bee clusters in addition to making holes in equipment and nesting inside the hive. Toumanoff identified three shrew species as bee pests: the house shrew (*Crocidura aranea*), the pygmy shrew (*Sorex pygmaeus*), and the common shrew (*S. vulgaris*). Tomsik et al. (1955) also listed shrews in the genera *Sorex* and *Crocidura* as bee pests.

Pech (1966) felt that shrews cause limited damage, although the disturbance they create may contribute to elevated levels of dysentery and nosema disease. Pech noted that hives bothered by shrews had dead bees on their bottomboards and that many of the corpses had their wings removed and their thoraxes opened and emptied.

Borchert (1974), on the other hand, believed that the members of the family Soricidae cause little harm to bees. He labeled the wood shrew mouse (*Sorex aranus*) as the most troublesome species. In Germany, shrews are protected animals, and killing them is not permitted. Borchert (1974), Seifert (1968), and Paillot et al. (1949) labeled the shrew as a friend of the beekeeper, despite its use of the bee hive as a nest and its occasional diet of honey bees.

The measures favored for shrew control are the same as those used to keep mice out of colonies. Parks (1930) and Toumanoff (1939) recommended reducing the size of the hive entrance. An alternative control measure is to construct a slick, inclined surface to prevent shrews from climbing to the hive entrance (Borchert 1974). Pech (1966) recommended the use of spring-type mouse traps with sausage or oats as bait.

Moles

Moles are small to moderate-sized, burrowing insectivores. They are roughly cylindrical, with short legs and tail. The front legs and claws are conspicuously enlarged for burrowing. The dense fur is usually brown but may be darker.

Moles are not generally reported as honey bee pests. One of us (DMC) discovered a starnosed mole (*Condylura cristata*) in a weak but active bee colony in Maryland. Where moles are a problem, commercial traps may be used to capture individuals and reduce the population. Reducing the size of entrances would effectively keep moles out of bee colonies.

Edentates

The order Edentata, which occurs primarily in South and Central America, includes armadillos, anteaters, and sloths; no species are major bee predators. One species, the nine-branded armadillo (*Dasypus novemcinctus*), is found in the United States in several Southern and Southwestern states. The armadillo is classified as omnivorous; termites and other insects, carrion, and roots are commonly included in its diet. According to M. Levin (personal communication 1976), the armadillo digs into ground nests of honey bees in Brazil. Another edentate, the true anteater of South America, is well-adapted for digging into ant and termite nests. There is no record, however, of any anteater attacks on ground or aerial bee nests.

Table 1. The mammalian pest species that 53 state and provincial apiary inspectors and specialists listed as the three most damaging to honey bee colonies (Hood 1996).

Pest Species	Number of Responses			
	Most losses	2nd Most losses	3rd Most losses	Total
Bears	22	7	6	35
Humans	14	11	5	30
Skunks	10	14	17	41
Mice	7	17	8	32
Cattle	0	0	3	3
Opossums	0	0	2	2
Raccoons	0	0	2	2
Shrews	0	0	1	1

CHAPTER EIGHTEEN

---∞∞∞---

Variations, Abnormalities, and Noninfectious Diseases
Nicholas W. Calderone and Kenneth W. Tucker

CHAPTER EIGHTEEN
Variations, Abnormalities, and Noninfectious Diseases

Honey bee queens, workers, and drones can all manifest unusual conditions that are not the result of pests, parasites, or pathogens. These conditions, which include physical malformations, genetic and physiological malfunction, and normal phenotypic variation, can be seen in both immature and mature stages of development. These conditions may be the result of environmental factors such as inadequate nutrition or social dysfunction; or they may be a consequence of gene mutation, recombination, chromosomal rearrangements or atypical combinations of the products of meiosis. Assignment of causes should be considered provisional, however, until all factors impinging on the phenotype, including biochemical and biophysical ones, are understood (Martignoni 1964, O'Brien 1967).

Our knowledge of noninfectious diseases and abnormalities in bees is generally superficial, often obtained as a byproduct of studies with other goals. For instance, even though gynandromorphs (composite individuals composed of male and female structures) are the most widely studied type of abnormal bees, the literature is still replete with fundamental questions on this subject. Other conditions such as parasitic mite syndrome (see Parasitic mite syndrome below) have only recently been described. The following provides an account of variation in development and form in *Apis mellifera*, arranged generally in the sequence of the bee's life continuum from egg through adult.

Sterile eggs

Infrequently, a newly mated queen begins apparently normal egg laying, but most of the eggs do not hatch (Hitchcock 1956, Fyg 1959). These sterile eggs eventually shrivel if they are not removed by worker bees of the colony. Sometimes the queen may lay a second egg in a cell that already contains a shriveled egg. Nurse bees may deposit brood food in a few of the cells that contain unhatched eggs. Sterile eggs laid in drone comb survive no better than those laid in worker comb. The stage at which embryonic development stops is variable (Hitchcock 1956), ranging from first cleavage to near completion of larval development (Fyg 1972).

A few eggs of such progenies do hatch and develop into apparently normal workers and drones (Hitchcock 1956), and one queen was recorded to have been reared from such a larva (H. H. Laidlaw and K. W. Tucker, personal observation 1964). The proportion of survivors is quite low. Hitchcock's (1956) estimate of about 0.1 percent is probably an overestimate because it did not take into account the possible sequence of eggs that died and were replaced during the two weeks preceding his observation. A better estimate might be about 0.03 percent. The per-colony incidence of sterility is usually low. It may not occur at all among thousands of daughters of some queen mothers and may occur in only a few per thousand daughters of others. In one instance, it occurred in 35 of 200 queens shipped by a queen producer (Hitchcock 1956).

The cause of egg inviability in bees is not known, but it probably includes a genetic component. This hypothesis is supported by evidence of clustering of occurrences within individual queen progenies (Hitchcock 1956) or in queen prog-enies of related mother queens (Roberts, cited in Hitchcock 1956; K. W. Tucker, personal observation, 1969). However, several cases of egg sterility cited by Hitchcock (1956) from the reports of others may well have had different causes than the one case, involving five queens, that he observed.

The most likely cause of egg sterility in which survivors are infrequent is that the mothers of the progenies are triploid, at least in their germ tracts. In such a case, most of the eggs would be aneuploid and inviable, and the rare survivors would be euploid or nearly so. Chaud-Netto (1980a) produced triploid honey bee queens, the progeny of normal queens and diploid drones, and induced them to lay without mating; he reported that 44 percent of their eggs failed to hatch. Borstel and Rekemeyer (1959) observed similar sterility in progenies of known triploid female braconid wasps, *Bracon hebetor* (= *Habrobracon juglandis*).

For egg sterility with a higher proportion of survivors, the mother of the progeny could be heterozygous for an inversion or translocation. This condition is well-known in other insects (Borstel and Rekemeyer 1959). The egg sterility re-ported by Isaacson (cited in Hitchcock 1956), with up to 5 percent surviving worker bees, might be consistent with this hypothesis. Some of the reports cited by Hitchcock (1956) involved queens that produced normal eggs, then later produced sterile eggs. The explanation of this condition, if not supersedure by an abnormal daughter queen, is probably other than those above.

Scattered brood

Departures from the compact and solid brood pattern generally associ-ated with efficient brood production and populous colonies are frequently encoun-tered in honey bees. Such a condition is referred to as scattered brood, shot brood, pepper brood, or spotty brood. The amount of brood lost can vary greatly among colonies, from less than 10 percent to 50 percent or more. Scattered brood can result from non-pathogenic or pathogenic conditions or from a combination of these. Therefore, a quick assignment of its cause may be impossible.

Scattered brood slows the rate of growth of the worker population, and colonies with moderate to severe brood loss do not attain high populations (Woyke 1976a). The effect of low but constant brood loss has not been measured precisely, but such colonies are probably slow to develop a large worker population. A reduced rate of growth of the worker population is a liability where time is limited before the main honey flow but may be an asset where only fall flows exist. Known causes of scattered brood are discussed below.

Diploid drones

One cause of non-viability of fertilized eggs has been traced to a major locus often called the sex locus (Mackensen 1955, Woyke 1962a). The non-viable individuals have been identified as diploid drones (Woyke 1963a). Under colony conditions, diploid drone larvae are eaten by worker bees within six hours of hatching from the egg (Woyke 1962a, 1963b), apparently because they differ chemically from other larvae (Woyke 1967, Dietz and Lovins 1975). Diploid drones can be successfully reared under special laboratory conditions. These individuals become sexually mature drones that produce diploid, rather than haploid, sperm (Woyke and Skowronek 1974) that may be used in the instrumental insemination of queens to produce triploid workers (Chaud-Netto 1980b; but see Woyke 1986).

The non-viability of diploid drones can be traced to the honey bee's sex determining mechanism, which Mackensen (1951) found to be similar to that of the braconid *Bracon hebetor* in which diploid males are partially viable (Whiting 1943). Somewhere on one of the honey bee's 16 chromosomes is the X locus. Estimates of the number of sex alleles (alternative forms of the gene at the X locus) include 11 (Mackensen 1955), 12 (Laidlaw et al. 1956, Kerr 1967), and 19 (Adams et al. 1977), but these have been challenged by Ekbohm and Ebbersten (1979). Individuals heterozygous at this locus are female, whereas a hemizygous (haploid) or homozygous individual is male. Individuals homozygous at this locus are diploid males. The heterozygotes (workers and queens) and hemizygotes (haploid drones) survive, but the homozygous (diploid male) larvae are eaten by adult worker bees under colony conditions.

The percentage of diploid brood that is male can vary between zero and 50 percent. Queens inseminated by a single drone (presumably achieved only by instrumental insemination in this polyandrous species) produce either no diploid drones, which occurs when the sire drone and queen have no sex alleles in common, or 50 percent diploid drone brood, which occurs when either of the queen's two sex alleles is the same as the sire drone's allele. When a queen has been inseminated by more than one drone, as usually occurs in natural mating, any level of brood loss throughout the zero to 50 percent range is possible, depending on the proportion of sperm that has sex alleles in common with the queen. The average

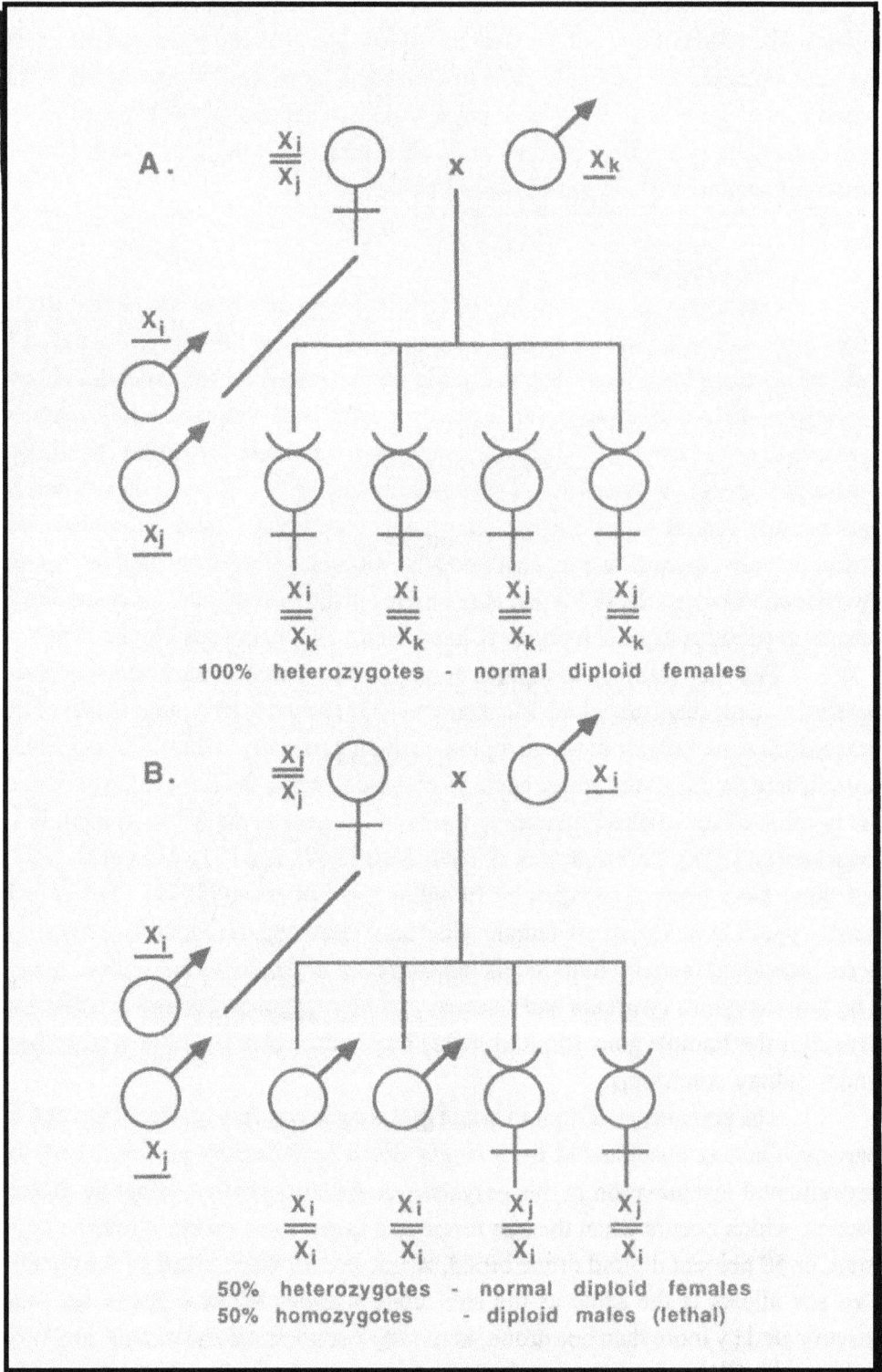

Figure 18.1. Diagram detailing results of the presence of diploid drones. (Calderone & Tucker).

expectation for a large sample of progenies from a randomly mating bee population with 12 sex alleles in equal proportion is an 8 percent loss for unrelated matings (Woyke 1976a); the expected loss is higher in matings between relatives.

The severity of this problem has been documented for the closed and somewhat inbred bee population on Kangaroo Island, South Australia (Woyke 1976a). Scattered brood was found in many colonies, none of which became very populous. A sample of 34 colonies showed an average brood loss of 24 percent (range 2 to 47 percent), with an estimate of only six sex alleles present in that population.

The sex-determining mechanism in the honey bee is of considerable significance for the design of breeding programs because the production of diploid drones imposes severe limitations on colony viability. Page and Marks (1982) and Page and Laidlaw (1982a, b) established the relationship between brood viability (measured as percent diploid drones) and population size (number of colonies), number of sex alleles present, number of matings by queens, and age of the population (number of generations) in a random-mating, closed breeding population. They found that the mean brood viability is dependent only on the number of sex alleles present in the population. The variance in brood viability is dependent on both the number of sex alleles in the population and the number of matings.

Genetic subviability

Another condition that can cause scattered brood is semilethality or genetic subviability. Losses caused by genetic subviability vary from one progeny to another, from low to as high as 18 percent for both workers and drones (Mackensen 1958). Because this condition has been observed only incidentally to genetic studies, comprehensive data are lacking. It seems that genetic subviability can explain scattered drone brood and must also be considered when investigating scattered worker brood.

The superficial appearance of the dead brood is known only for sealed brood, although death at all stages is possible. During the last nine to 14 days before eclosion, drone brood may die as larvae, prepupae, pupae, or imagoes that fail to chew open their cell cappings. Often, more dead prepupae are found than other stages. Dead larvae and prepupae appear gray or black and are saclike, with watery contents (K. W. Tucker, personal observation 1965). Fyg (1959) described a noninfectious phenocopy of sacbrood that may be part of this syndrome.

To explain genetic subviability, Mackensen (1958) suggested that there are two separate semilethal genes, one linked to the chromosomal location known as the brick (*bk*) locus, the other to the snow (*s*) locus. An alternative hypothesis is that six or more polygenes for subviability act jointly to produce one lethal event.

Under minimal developmental stress, more subviable genes are required for death; under greater developmental stress, fewer subviable genes are required (as with certain wing mutants; see Laidlaw et al. 1965, Tucker 1980). It is possible that an unknown infection is one of the stresses involved in genetic subviability.

Queen's egg-laying behavior

Scattered worker brood is often attributed to scattered egg laying by the queen, but this connection has not yet been documented. Drone brood in worker comb is often scattered, which may be in part a consequence of irregular egg-laying behavior. Harbo (1976) found irregular placement of eggs in worker comb to be more frequent by unmated queens and queens inseminated with dead sperm (33-94 percent) than by queens inseminated with live sperm (0 to 7 percent). Of the irregularly placed eggs, some were sideways on the cell bases and others were attached to the sides of the cells. Harbo did not follow further development, but perhaps scattered brood could ensue if some of the irregularly placed eggs were removed or eaten by worker bees.

Drone brood in worker cells

A queen may lay an increasing number of drone eggs in worker cells, especially as she ages, although this is not an essential requirement. Inadequate mating is also a cause of this type of queen failure. Egg laying by such queens, called drone-layers, may continue to deteriorate over time. This leads to a decline in the worker population and to the eventual loss of the colony if the queen is not replaced. There is usually some mortality of drone brood in worker combs, even though the drone brood from the same queen or from laying workers is reasonably viable in drone comb. In worker cells, well-fed drone larvae develop normally until the cells are sealed, but some larvae die soon after sealing. If the dead larvae are removed by the bees, a scattered brood pattern results. The dead larvae do not superficially resemble larvae that have died of common brood diseases, but they cannot be distinguished from larvae that have died of genetic subviability. They appear gray or black and are saclike, with watery contents. The cause of death of drone larvae in worker comb is not known, but it may be confinement within cells that are too small.

Cannibalism

Another cause of scattered brood is cannibalism by worker bees. Remaining portions of partially cannibalized pupae generally appear well-formed, with the body being eaten progressively from the anterior end. This is a major cause of spotty brood in the plains variety of A. cerana (Woyke 1976b) and is an occasional cause in *A. mellifera* (Woyke 1977). Cannibalism seems to be a response to a lack of pollen (Newton and Michl 1974). In *A. cerana*, solid brood patterns exist only during good pollen flows (Woyke 1976b).

Previous disease

Scattered brood is often a consequence of a brood disease from which the colony has recovered. The scattered appearance of the brood results from diseased brood having been removed from the comb (Morse 1976). In such cases, however, there is usually some residual evidence of disease.

Pupal anomalies

Fyg (1959) described two pupal anomalies that caused bees to die as pupae. In one of these, called white-headed bees (not to be confused with white-eyed bees, an eye-color mutation discussed below), the pupa's entire head and appendages remain white after the rest of the cuticle has darkened. Fyg (1959) believed the white heads were caused by a lack of oxygen resulting from a blockage of the prothoracic spiracles. In the other pupal anomaly, called Muttenz anomaly, the abdomen appears to be greatly shortened because it is telescoped anteriorly, the head appears to be enlarged, and the digestive system is displaced anteriorly. Laidlaw and Eckert (1962) mention "humpbacked" queen pupae with enlarged thoraxes and compressed heads.

Chilled brood

Brood will die if it becomes too cold. Chilling may occur during the spring buildup if the amount of brood is greater than the adult worker bees can cover and keep warm when tightly clustered during a cool night. It can also happen during cold weather if the adult worker bee population of a colony has been suddenly reduced through insecticide poisoning, for example, or by the partial or complete desertion of a newly made division of a larger colony.

Chilled brood can occur during the colder months in temperate and subtropical climates and may also be possible at very high altitudes in the tropics. A certain length of time at temperatures below a critical temperature is necessary; however, neither the time nor the critical temperature has been determined. Brood can survive for a considerable length of time at temperatures below 35°C (95°F), and this ability varies with the age of the brood (Jay 1963).

Chilled brood is usually found at the lateral and lower peripheries of the cluster. Sometimes all the brood in an outer comb dies, or a crescent of dead brood may be found below healthy brood in a comb in the cluster's center. The appearance of chilled brood is variable (Borchert 1966). It is often yellowish white and tinged with black on segmental margins, or it can be brownish or black. Chilled brood can be dry and crumbly, greasy, or watery, but is never ropy, as with American foulbrood. The odor is usually weak, but sometimes disagreeable and sour. Sometimes the cappings of sealed cells of dead brood are perforated. Microscopic observations of smears of dead, chilled larvae usually reveal no microorganisms, but occasionally some bacteria are found; these may invade the larvae after death (Borchert 1966).

Overheated brood

Death of brood from overheating is likely if a colony's ability to cool itself is impaired. Such a condition could occur if many of the hive bees as well as foragers were killed by pesticides (Atkins 1975). Brood death caused by overheating was reported by Root (1966) for colonies that had been confined on hot days. Death of brood in incubators is well known. The minimum lethal temperature for brood is reported to be 37°C (99°F) (Himmer 1927, cited in Betts 1928d).

Half-moon disorder

New Zealand beekeepers report a condition known as half-moon disorder (Reid 1984), so called because larvae are found dead in the early coiled stage (c-stage), usually dried out and discolored. No causative pathogens have been isolated from these larvae (Vandenberg and Shimanuki 1990b), and the condition is believed to be brought about by stress, possibly in conjunction with nutritional and genetic factors. It is not generally a serious problem according to Reid (1984), and it is treated by requeening.

Parasitic mite syndrome

Reports from Europe (Ritter 1988) and the United States (Shimanuki et al. 1994) describe a deterioration of brood associated with high levels of infestation by the parasitic mite, *Varroa jacobsoni*. Affected brood is characterized by symptoms similar to those of European foulbrood, American foulbrood, and sacbrood. Unlike these diseases, however, affected larvae, prepupae, and pupae can be of any age. The honey bee pathogens associated with these diseases have not been isolated from affected brood. This condition, termed parasitic mite syndrome (Shimanuki et al. 1994), is relatively refractory to treatment with oxytetracycline but is highly responsive to treatment with fluvalinate (Calderone, personal observation in apiaries during 1993, 1994 at Bee Research Lab, Beltsville, Maryland). This suggests that *V. jacobsoni*, which is highly susceptible to fluvalinate, is involved in some way. The mite could act as a vector of various pathogens, or it could result in a reduction in the level of brood care to a point where immature bees die or are rendered more susceptible to invasion by pathogens.

Other causes of brood mortality

Brood may die of environmental hazards other than temperature. Pollen and nectar contain a variety of compounds, depending on the species, that may be harmful to the developing bee. Davis (1942) reported that locoweed can be toxic, resulting in the death of individuals from the late pupal stage to just before top emergence. Vansell et al. (1940) describe several stages of poisoning caused by California buckeye (*Aeschulus californica*). Larvae generally die soon after hatching. In severe cases, queens may produce eggs that do not hatch or from which larvae die soon after egg hatching. Affected queens may lose the ability to lay eggs altogether, or they may lay only drone eggs. Typically, one finds large numbers of light-colored and deformed, immature bees near the entrance. In the southern United States, the summer titi (also known as leatherwood), *Cyrilla racemiflora*, causes a condition called "purple brood," so called because the dead brood turn blue or purple. Burnside and Vansell (1936) describe symptoms and causes of many instances of plant poisoning.

Consumption of pesticide-contaminated pollen by nurse bees can also

Figure 18.2. Parasitic mite syndrome. (Calderone photo).

poison the brood they feed, or the brood may die of starvation, chilling, or over-heating if many nurse bees die. Brood mortality from either pesticides or plant toxins can persist if the colony's pollen stores are contaminated. Starvation is possible if there are too few hive bees to feed the larvae. Normally, in a colony with insufficient food reserves, larvae are removed (see Cannibalism, above) by nurse bees. Brood can be killed by jarring the combs (K. W. Tucker, personal observation 1971), probably during the prepupal and pupal stages when it is vulnerable (Jay 1963).

Bald-headed brood

Infrequently, brood cells contain apparently normal pupae but are un-capped or partially capped. Milne (1942) describes brood with a similar appearance among samples sent to Rothamsted Experiment Station for diagnosis. He referred to this condition as "bald-headed" brood. No cause was identified, but there appeared to be an association with fecal material from the lesser wax moth, *Achroia grisella*. The greater wax moth, *Galleria mellonella*, may also produce brood with this appearance (Whitcomb 1936). Pupae in affected cells generally complete development. Other causes of bald-headed brood may also exist.

Visible Mutations

Departures from the usual appearance of bees occur occasionally. These abnormal bees can be consequences of cytological accidents, unfavorable environment during development, or mutant genes. Well over 40 mutant phenotypes have been found in honey bees (Rothenbuhler et al. 1968, Tucker 1986). These include more than 20 for eye color, as well as some for eye structure, wing characteristics, sting structure, body color, and body hair. Most mutants are discovered in drones because they are haploid and therefore readily show phenotypes of recessive mutations. Most of the known honey bee mutations occur spontaneously, although Soares (1981b) reported the induction of the eye mutant chartreuse-limao (ch^l) using gamma radiation from ^{60}Co. Many mutations were first found by commercial beekeepers.

The mutant eye colors range from almost white through various shades of yellow, red, and brown; each color is characteristic of a particular mutation. The colors result from incomplete formation of eye pigments (see Dustmann 1975a, Cruz-Landim et al. 1979). Dustmann (1969, 1973, 1975b) has assigned most of the mutations to specific reactions in the biosynthetic pathways of these pigments. Among drones, most eye-color mutants cannot fly well enough to mate naturally (Cornuet 1978). The genes for all known eye-color mutants are inherited as Mendelian recessives (Tucker 1980).

Of the eye structure mutants, eyeless (*e*) has neither ommatidia nor testes, reduced facet (*rf*) has a variable reduction in ommatidia, and cyclops (no symbol) has only one eye over the top of the head (Laidlaw and Tucker 1965; see also Fyg 1959). Usually, eyeless and reduced facet occur only in drones, and cyclops occurs in both workers and drones. Reduced facet appears almost exclusively along with brick- or garnet (*g*)-colored eyes. The gene for eyelessness is inherited from one heterozygous queen to the next as a Mendelian recessive. The causes of reduced facet and cyclops are probably genetic because these phenotypes cluster within individual progenies, but their mode of inheritance is obscure.

The wing mutants include phenotypes with modified structure and one phenotype with modified comportment (Rothenbuhler et al. 1953, Laidlaw et al. 1965, Witherell 1973, Tucker 1986). The phenotypes of short (*sh*) and diminutive (*di*) are short and narrow wings, respectively; whereas truncate (*tr*) wings are quite short and squared at the tip, and wrinkled (*wr*) bees have a wrinkled area in the distal portion of the forewing. Wrinkled and short include anomalous wing venation. The wings of the droopy (*D*) phenotype droop to the sides. Of these wing mutant phenotypes, only diminutive and wrinkled can fly. The gene for each of these is inherited as a single Mendelian trait; droopy is dominant and probably lethal in drones, and the others are recessive.

The body color mutant cordovan (*cd*) dilutes the color of normally black portions of the cuticle to a cordovan brown (Mackensen 1951). The gene for cordovan body color is inherited as a single Mendelian recessive. Other genes for body color are sometimes classed as mutants when used as genetic markers in bee populations where their frequency is low, such as a gene for black body color in golden bees (Laidlaw and el-Banby 1962). It may also be possible that the cordovan gene is fixed in homozygous condition in isolated bee populations in desert oases in northern Africa. Other body color mutants are listed in Tucker (1986).

Soares (1980, 1981a) has recorded the phenotype open sting or split sting (*sps*). Bees with this condition have a modified sting structure that makes them incapable of stinging effectively. In the phenotype of the body hair mutant hairless (*h*), body hair is produced normally but is sloughed off after eclosion (Mackensen 1958). The gene for hairlessness is inherited as a Mendelian recessive.

Varying degrees of subviability accompany several mutant gene phenotypes (see, for example, Kuz'mina et al. 1979a, b). Most strongly affected are truncate, short, and eyeless. To a lesser extent, other mutant genes and even wild types are subviable. Subviability for the same gene varies among progenies and at different times within progenies (Laidlaw et al. 1965), which may indicate stresses during development imposed by variable genetic load and environment.

Contributions to research

Abnormalities in honey bees ought not to be viewed as mere curiosities but as valuable research tools to help us understand normal, as well as abnormal, development and behavior. Mutant markers have been used to study a variety of aspects of honey bee biology.

Cordovan has been used in studies of kin recognition, including studies on agonistic behavior (Getz and Smith 1983, Evers and Seeley 1986), swarming (Getz et al. 1982), and brood care (Noonan and Kolmes 1989). Frumhoff (1991), however, presented data suggesting that cordovan may not be a neutral marker with respect to nestmate interactions. Cordovan (Visscher 1989) and cordovan in conjunction with eye mutants snow (s) and tan (s^t) (Page and Erickson 1988) have been used to study worker reproduction (laying workers). Cordovan has also been used to study genetic effects on division of labor (Frumhoff and Schneider 1987, Frumhoff and Baker 1988, Oldroyd et al. 1991a, b, Oldroyd et al. 1992). Oldroyd et al. (1991a) used cordovan to study genetic components of the dance language. Kirchner and Sommer (1992) studied communication during the dance language using the wing mutant diminutive.

The mating range of the honey bee queen was examined using cordovan by Peer and Farrar (1956), who found that queens and drones can mate when their colonies are up to six miles apart. Roberts (1944) demonstrated multiple mating based on variation in abdominal color patterns of worker progeny produced by queens selected for yellow body color that were allowed to mate naturally in an area with yellow and black drones. Subsequently, polyandry has been studied using cordovan (Taber 1954, Peer 1956, Taber and Wendel 1958). Sperm utilization patterns have been studied using cordovan (Taber 1955a), as well as cordovan in conjunction with snow and tan (Laidlaw and Page 1984). Sperm homogenization has been examined by Harbo (1990) using the mutant chartreuse-Benson (ch^B) and by Moritz (1983) using the mutants cordovan, diminutive, and pearl (pe). The survival of honey bee spermatozoa in liquid nitrogen was investigated by Harbo (1977) using the eye-color mutants snow and tan.

Mutants have also been used to study neurophysiology, especially as it affects behavior and vision. Gribakin (1990) used the eye mutant snow as a model for the study of optical properties of visual pigments. The photoreceptor optics (Gribakin 1988) and the electrolyte contents of eye and hemolymph of the mutants snow, snow-laranja (s^{la}), ivory umber (i^u), and chartreuse-1 (ch^1) have also been studied (Gribakin et al. 1987). Lopatina et al. (1984) found that workers with the snow-laranja mutation exhibited depressed functional activity of the nervous system compared to wild type workers. This depression was subsequently attributed to the effects of increased levels of serotonin in the mutant workers (Lopatina and

Dolotovskaya 1985).

Gynandromorphs

Gynandropmorphic bees are sex mosaics whose bodies are partly male and partly female in structure (Rothenbuhler 1958b, Fyg 1959). Sometimes, especially in older references, they have been called hermaphrodites, but this is a misnomer because a hermaphrodite produces both sperm and eggs, but gynandromorphs do not.

The distribution of male and female tissues in various parts of the gynandromorph's body can take several patterns. Female and male tissue may be divided by body region, such as a worker head with a drone thorax and abdomen, or bilaterally, or there may even be patches of male and female tissues. The most likely arrangement is male tissues anteriorly and female posteriorly, and tissues near one another are most likely to be the same sex (Drescher 1975, Milne 1976, 1977). Most often the female parts are diploid worker and the male parts haploid drone.

The female and male tissues of gynandromorphs can have several different origins, which are detectable in genetically marked stock. The most frequently found so far have been biparental female tissues and paternal male tissues. Gynandromorphs with female and male tissues of this type are zygogenetic-androgenetic, or Z-A (Rothenbuhler et al. 1952, Drescher and Rothenbuhler 1963). In these, the female tissues arise from the usual zygote, and the male tissues arise from one or more accessory sperm, which in non-gynandromorphic bees are not involved in development. Another, far less frequent origin is biparental (zygotic) female tissues and maternal male tissues (Mackensen 1951, Woyke 1962b, Drescher and Rothenbuhler 1963). In this type, it is not certain whether fertilization takes place after cleavage or the combination results from a single fertilization of a binucleate egg. Gynandromorphs are known also from unmated queens and seem to originate from unusual binucleate eggs in which haploid cleavage precedes union of descendants of the two-egg nuclei to give rise to female tissue (Tucker 1958). In these, all tissues are maternal, and the male tissues are mosaics of the two haploid products of meiosis. Gynandromorphs other than the Z-A type have been too infrequent to be studied adequately.

The tendency to produce Z-A gynandromorphs is inherited (Rothenbuhler 1955). The frequency of gynandromorphs was increased in a gynandromorph-producing stock by selection and inbreeding but was lost in four generations of outcrossing to drones of non-gynandromorph-producing stock. The reciprocal crosses took three generations to establish a low level of gynandromorph production in a non-gynandromorph-producing stock. Thus, the inherited factors are at least partly chromosomal; whether there is also cytoplasmic inheritance is not clear.

Gynandromorphs are rare. If they appear at all, there are usually only a few in a given progeny. But occasionally Z-A gynandromorphs appear in greater numbers in some stocks (Rothenbuhler et al. 1949), and with subsequent selection and inbreeding can be increased to as high as 40 percent of the progeny from fertilized eggs (Rothenbuhler 1958).

The frequency of Z-A gynandromorphs can be conditioned by environmental factors. Within gynandromorph-producing stocks, gynandromorphs were most frequent during periods of accelerating oviposition (Rothenbuhler, cited in Drescher and Rothenbuhler 1963). Furthermore, gynandromorphs were induced by chilling very young eggs (20-30 minutes old) of some stocks that were otherwise non-gynandromorph-producing, although this treatment was only successful in some stocks (Drescher and Rothenbuhler 1963). Heat treatment of eggs has also been reported to induce gynandromorphs (Hachinohe, cited in Drescher and Rothenbuhler 1963).

How Z-A gynandromorph-producing eggs differ from normal eggs is not known. Drescher and Rothenbuhler (1963) suggest that some chemical that inhibits development of accessory sperm nuclei in normal eggs is somehow inactivated in gynandromorph-producing eggs. Whatever the system, it is affected by the queen's genes (and not the zygote's genes) and the environment. Tucker and Laidlaw (1966) suggested that an isogenic sperm line could be produced using gynandromorph-producing stock with androgenic male tissue. Kubasek et al. (1980) proposed a mating scheme for isogenic sperm line maintenance in the honey bee.

Other mosaics

Several other mosaics have been found in genetically marked stocks. Clearly related to Z-A gynandromorphs in origin are mosaic drones, whose zygogenetic component is diploid drone, rather than female, along with androgenetic haploid drone tissue (Rothenbuhler 1957, Drescher and Rothenbuhler 1964, Harbo 1980). For several years, these mosaic drones were the only evidence of the existence of diploid drone tissue, until Woyke (1962a) demonstrated the viability of diploid drones. Laidlaw and Tucker (1964) found gynandromorphs with paternate (androgenetic) female tissue, indicating development from union between accessory sperm. Two other types of mosaics seem to originate from binucleate eggs: mosaic drones similar to the mosaic drone tissue of gynogenetic gynandromorphs from unfertilized eggs (Tucker 1958, Woyke 1962b), and mosaic females presumably from doubly fertilized binucleate eggs (Taber 1955b, Woyke 1962b). Woyke (1962b) found a three-way mosaic worker of parthenogenetic female and two different biparental origins, so all four haploid products of meiosis, two of them fertilized, could have been involved.

Parthenogenetic females

Although parthenogenetic females are apparently normal bees, they originate in an unusual way from unfertilized eggs that usually produce drones. Bees of a colony usually rear parthenogenetic females as workers, but it is possible to rear queens from parthenogenetic female larvae (Mackensen 1943). Parthenogenetic females have been found in various stocks, usually at low frequency (Mackensen 1943, Tucker 1958, Woyke 1962b). Their incidence is higher in the Cape honey bee of South Africa (*A. mellifera capensis*) (Jack 1916). They can be found in progenies of either laying workers or unmated queens.

Parthenogenetically produced females apparently result from automixis, parthenogenesis in which two of the four haploid nuclei resulting from meiosis fuse to form a single diploid nucleus; this nucleus then develops normally. Tucker (1958) inferred this mechanism from patterns of segregation observed among workers parthenogenetically produced by heterozygous unmated queens. Verma and Ruttner's (1983) direct microscopic observations of eggs of laying workers of the Cape honey bee confirmed this conclusion and clarified the underlying cytological mechanism.

The frequency of parthenogenetic females is highest during accelerating oviposition, either during initial laying or following an interruption in laying (Tucker 1958). The tendency to produce parthenogenetic females can be increased by selection (Tryasko 1969).

Dwarf bees

Very small but apparently well-formed worker bees are found from time to time. They can occur among normal-sized workers reared by old bees (K. W. Tucker, personal observation 1970). Fyg (1959) reported them from colonies with nosema disease. Dwarf workers probably develop from larvae that are poorly fed during late larval life, since it is possible to produce them by starving older worker larvae (Jay 1963). Dwarf queens also can result from underfeeding (Fyg 1959, Jay 1963). Daly and Morse (1991) produced dwarf worker honey bees in colonies that were given only the larger drone-size cells. They suggested that in the absence of worker-size cells, the nurse bees behaved abnormally, and as a result, the developing larvae were starved.

Dwarf drones develop from drone brood in worker cells. Whether they are partially starved by nurse bees or stop eating as cramped larvae has apparently not been studied. However, dwarf drones can be produced by starving older larvae (Jay 1963).

Crippled bees

Crippled bees are found occasionally in colonies known to be of normal stock. The symptoms resemble those that can be caused by the parasitic mite *V. jacobsoni* Oudemans (see Chapter 14). Most often the wings are crippled so that the bees cannot fly. The wings may be missing, they may be present but not expanded, or they may be partially expanded and crumpled. All three castes can have crippled wings, but those most often seen are drones. Queens with crumpled wings usually emerge from queen cells reared near the periphery of the cluster during cool weather. Crippling of this sort is probably caused by subnormal temperatures during the pupal stage (Fyg 1959, Jay 1963), although jarring of queen cells during sensitive periods of pupal development can also produce this condition (Calderone, personal observation).

Crippling of the antennae, mouthparts, and legs of workers and drones caused by abnormal pupation was described by Fyg (1959). Such cripples can be produced in the laboratory with normal stock, but also occur spontaneously in some colonies (Schneider and Bruger, cited in Fyg 1959). This anomaly seems to be inherited, as it persisted for three related generations.

Internal anomalies of queen bees

Fyg (1964) described several abnormal conditions of the queen's internal structure: ovaries that did not develop; one or both of the lateral oviducts missing (mentioned also in Laidlaw and Eckert 1962); two spermathecae instead of the usual one; and accessory ovarioles not included in the ovaries but located elsewhere in the abdomen. In the considerable information on the histology of sick queens (Fyg 1964, Vecchi 1976), conditions are described whose fundamental functional lesion is not influenced by an infection. With a few exceptions, however, most of these maladies still require experiments to elucidate their etiology.

Catalepsy in queen bees

Queens rarely faint, but they have been observed to do so by beekeepers who have handled large numbers of queens. This condition has been called catalepsy (Brunnich 1922), epilepsis (Laidlaw and Eckert 1962), fainting (Miles 1922), or shock (Latham 1922). When it happens, it occurs just after a queen is picked off the comb by her wings. According to Latham (1922), the queen hooks her abdomen forward, stiffens momentarily, becomes motionless for a few minutes, and then gradually revives and returns to normal activity. Not every queen who hooks her abdomen is so affected, but Latham believed it most likely to occur in large queens with enlarged abdomens that are laying heavily. In Miles' (1922) experience, catalepsy happened only to young queens; moreover, most of the cases described by Brunnich (1922) were also young queens. In some cases, the queen does not revive, and death results (Miles 1922). The cause of catalepsy is unknown, but may result from a temporary nervous disorder.

Overheated bees

Overheating can develop in worker bees when they are confined during hot weather and have no access to water. Even before the bees are irreversibly damaged, they may be stressed sufficiently to affect their subsequent performance (Weaver 1969). Before the bees die of overheating, they crawl rapidly and flutter their wings, which become cloudy (K. W. Tucker, personal observations, 1951, 1969, 1983). If released, overheated bees disperse by crawling in a disorderly manner, and they regain no vestige of organized activity. Bees that have died of overheating are often wet (Grimsley and Sadler 1936), perhaps from regurgitating fluids in vain attempts to cool themselves, as bees do under hot conditions (Esch 1976). The temperature in an overheated cluster probably exceeds 38°C (100°F) and may be as high as 50°C (122°F), as can be inferred from data on upper lethal temperatures (Free and Spenser-Booth 1962, Lensky 1964).

Overheating bees is a hazard to be expected during hot weather when starting queen cells in a swarm box (Newswander 1977) and when shipping package bees (Grimsley and Sadler 1936). The hazard is obviated if the bees can be kept cool; Weaver (1969) suggests 18°C (64°F) as optimal for package bees.

Starvation of adults

Bees cannot survive without honey. Colonies with inadequate stores die, and the deceased workers can usually be found, each with nearly its entire body head first in a cell. Often hundreds of such workers are found in contiguous cells. Starvation is frequently found toward the end of winter but can occur at any time of the year.

Laying workers

Although typically not productive in queenright colonies, worker bees can lay unfertilized eggs when their ovaries are functionally developed. Ovary development in workers is suppressed by the queen's pheromones, but released when this influence is removed (Butler 1957; see review in Free 1987). Therefore, laying workers develop most frequently under queenless conditions, but the rate at which they occur varies with different stock. The Cape honey bee of South Africa develops laying workers after four to eight days of queenlessness, whereas other honey bee species take much longer (Ruttner et al. 1976, Anderson 1977, Crewe

1984). In our experience with the mixture of European races present in the United States, laying workers develop after about two weeks of queenlessness and broodlessness. In some stocks, however, workers begin laying before all the brood emerges, and in other stocks, sometimes no workers will lay.

Occasionally, worker bees will lay at least a few eggs in queenright colonies (Page and Erickson 1988, Visscher 1989). Estimates of the proportion of workers with some development of the ovaries have ranged from 1 percent to 30 percent (Verheijen-Voogd 1959, Jay 1968, Kropácová and Haslbachová 1969, 1970, 1971), and partial ovarian development was found to be common in queenright bees of a swarm or colonies with poor queens (Koptev 1957). Laying workers are probably responsible for the few cells of drone brood sometimes found above a queen excluder when the queen is confined below. In some colonies, when supersedure of a poor queen is prevented, workers can produce considerable drone brood in addition to that of the queens. If a queen is successfully introduced into a laying worker

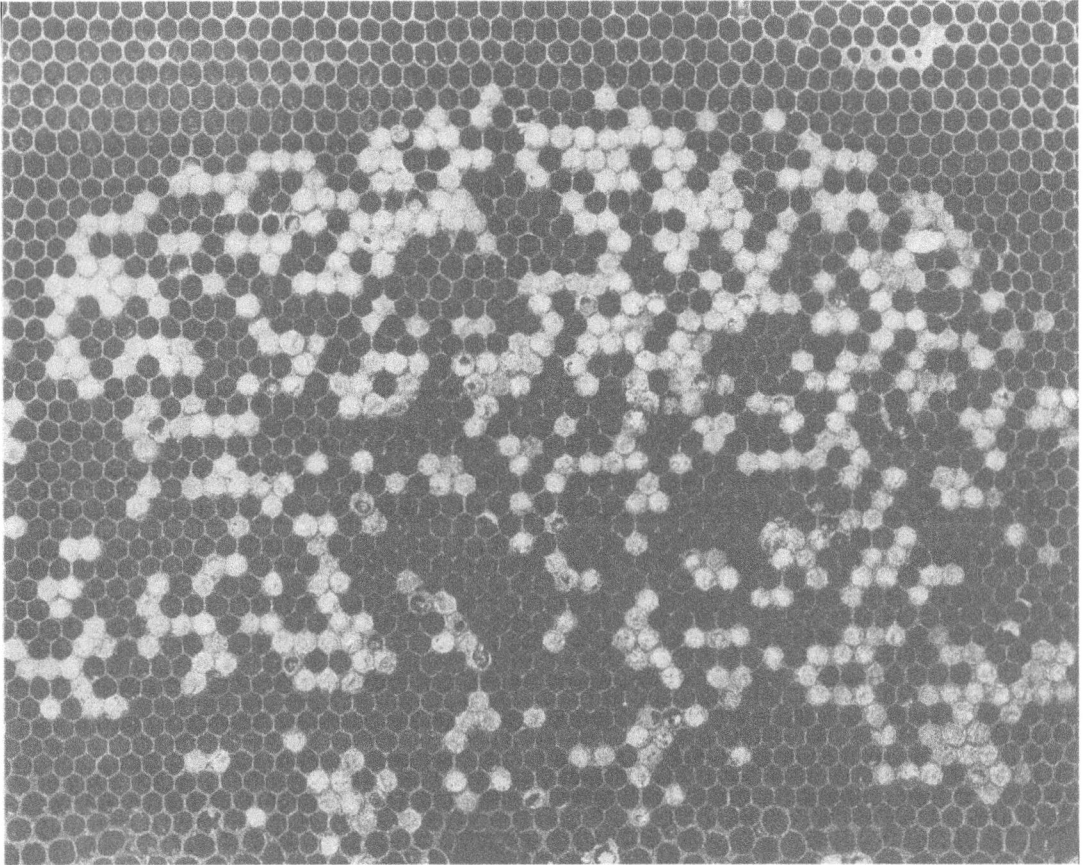

Figure 18.3. Drone brood in worker cells, and a scattered pattern are indicative of laying workers. (File photo).

colony, the workers may continue to lay at a gradually diminishing rate along with the queen before they stop completely. Each of these possible instances of coexistence of laying workers and a queen has been observed in cases in which the progeny of the workers and the queen were distinguishable by body color or by mutant eye-color genes (Page and Erickson 1988, K. W. Tucker, personal observation 1953, 1965, 1975). Moreover, these observations are consistent with the experimental results of Jay and Nelson (1973).

Laying workers are usually revealed by the presence of eggs and brood in a queenless colony and by the brood's appearance. The pattern of egg laying is usually diagnostic: There are many eggs in a cell (rather than only one, as occurs when the queen lays), and some of the eggs are attached to the walls rather than to the base of the cell. Some of the eggs develop into larvae, but many eggs and young larvae must be eaten to leave only one larva in each cell. The brood is virtually all drone brood, which is usually but not always scattered. Atypically, the egg-laying pattern of some laying workers is so queenlike that much time is lost by the beekeeper in a futile search for a queen. Page and Erickson (1988) report that the eggs laid by workers in a colony during the first few days of worker egg-laying activity are placed normally and preferentially in drone cells. This indicates a caste-specific adaptation for oviposition behavior because drones reared in drone cells are larger and presumably more competitive than those laid in worker cells.

The usual fate of laying worker colonies when left alone is that the adult workers eventually die and the colony dwindles. Three to four weeks following the onset of laying by workers, there are considerable numbers of small drones, sometimes more drones than remaining worker bees. In times of dearth, laying worker colonies may be robbed of their honey, and the remaining bees may starve. Laying worker colonies of the Cape honey bee dwindle more slowly, however (Anderson 1963), because parthenogenetic workers are frequent enough in that race to replace the adult population to some extent.

Laying worker colonies often have the opportunity to raise a new queen but fail to do so. Contrary to popular belief, one or a few female larvae are available in many laying worker colonies. Although parthenogenetic females are infrequent in terms of numbers of larvae, they may not be rare. Tucker (1958) found six of eight laying worker colonies producing one or more worker bees each. The nurse bees may not recognize the sex of the unusual female larvae in these cases, but it is more probable that the process of queen rearing is interfered with in some other way. In light of the finding that laying workers produce queen substance (Ruttner et al. 1976), it is likely that the production of queen cells could be inhibited in laying worker colonies. Even in laying worker colonies of the Cape honey bee, where parthenogenetic females are common, queen rearing is less frequent than might be expected (Anderson 1963).

Treatments

Some of the conditions described are undesirable and can threaten the life or productivity of a colony. Others are simply transient or sources of interest and curiosity. Beekeepers may wish to eliminate conditions deemed undesirable. In such cases, the treatment will depend on the specific condition involved.

Broadly speaking, treatments for these conditions fall under the heading of sound management practices. Beekeepers should always provide the best possible environment for their bees. This begins with the selection of the apiary site. Desirable traits for an apiary can be found in any general text on apiculture. Management for overwintering is also essential to the maintenance of healthy colonies. Only healthy, strong, disease-free colonies with adequate reserves of both pollen and honey should be overwintered. Hives should be prepared in an appropriate manner for winter. Depending on your location, this may include upper entrances, entrance reducers, mouse guards, and insulation. Protection from prevailing winds is also important. Treatment for parasitic mites should be used regularly, before the colony shows signs of deterioration. Proper spring management, especially for disease control, swarm control, supering, and inspection for the presence of adequate stores, is also essential. In general, a seasonal program of colony management, with well-defined goals at each stage, should be adhered to rigorously.

A young, vigorous queen, as evidenced by large areas of similarly aged brood with a solid pattern, is the foundation for a healthy colony. A regular requeening program can prevent many problems from arising. Regular requeening will also decrease the incidence of drone layers in an operation.

Sterile eggs and most causes of scattered brood can be eliminated by requeening. Cannibalism generally indicates inadequate stores and should be treated either by supplementary feeding or relocating colonies to apiaries with adequate nectar and pollen sources. Chilled brood can be minimized by maintaining colonies in hives with no more room than is actually needed. Adding supers before they are required, especially in the spring, may hamper the colony's ability to thermoregulate the nest adequately. Overheated brood and bees can usually be treated by supplying colonies with clean water close to the nest. Poisoning from toxic nectar and pollen is usually transitory and may affect only a small proportion of colonies in an apiary. In areas in which toxic nectar sources predominate, however, the only treatment is to relocate the colonies while the harmful sources are in bloom.

Mutations and other genetic anomalies are not generally deleterious to a colony. In fact, they are usually a source of some interest to both scientist and beekeeper. Should the need arise to eliminate such a condition, requeening is the only solution.

Perhaps no other condition is as frustrating as that of laying workers. The practical beekeeper's concern is usually how to return a laying worker colony to a productive queenright condition. Laying worker colonies usually will not accept a laying queen. Chances of acceptance are improved if their brood is removed and replaced with emerging worker brood. Örösi-Pál (1929) reported that queen acceptance is returned to a satisfactory level if all the combs are removed and the bees of the laying worker colony are confined without food for two days. During this time the ovaries of the worker bees will return to a non-functional condition. Trying to save a colony with laying workers can require considerable time. In large operations, it is probably best to eliminate these colonies rather than attempt to save them.

CHAPTER NINETEEN

Poisoning by Plants
John A. Skinner

CHAPTER NINETEEN
Poisoning by Plants

A poison is a chemical that has a harmful effect, including illness and death, on humans or animals (DuBois and Geiling 1959, Ottoboni 1991). Technically, a poison causes harm in very small quantities (Ottoboni 1991). The term toxin is currently used to describe a chemical that causes a harmful effect without respect to the quantity needed to generate a response. The science of toxicology is historically defined as the study of poisons and their actions (DuBois and Geiling 1959), but this definition has been recently expanded to include information from most branches of biology, chemistry, mathematics, and physics (Gallo and Doull 1991).

A poisonous plant is one that produces chemicals that interfere directly or indirectly with the metabolic processes of living organisms. One important concept that remains relevant today is found in the early writings of the Swiss physician, Paracelsus, who wrote in 1567, "All substances are poisons; there is none which is not a poison. The right dose differentiates a poison from a remedy." The dose is the amount or quantity of chemical present. It is critical to understand the concept of dose because a low dose of a chemical toxin from a poisonous plant may be a valuable medicine, whereas a higher dose of the same chemical can be deadly. A substance that is toxic to one organism may not be toxic to another. If honey produced from the nectar of one plant is toxic to humans, it cannot be assumed that the nectar or honey from this plant is also toxic to bees. The converse is also possible.

How many plants are known to be poisonous to humans or animals? Approximately 20,000 species of seed plants are native or naturalized in the United States, and a similar number of cultivated ornamental plants occur there (Fuller and McClintock 1986). Poisonous plants are considered rare; 700 to 1,000 species are estimated to show toxic properties, and only a few of these are considered very toxic (Kingsbury 1964). Most reviews include only plants known to be toxic to humans, usually after they eat a portion of the plant. Some plants cause disorders and illness in livestock after the animal has ingested the plant (MacLean and Davidson 1970). The degree of toxicity and severity of poisoning varies with the portion of the plant consumed. When toxins are found in the leaves or other plant parts, we cannot assume the toxin is also present in the nectar or pollen.

This review is intended as a reference for beekeepers, horticulturists, scientists, students, and gardeners who have an interest in plants poisonous to bees. Information regarding plants that produce nectar or honey toxic to humans is also included. Geographically, emphasis is given to plants of North American origin or those naturalized in this region; other locations are included when data are available. Ornamental cultivars of many plants are found worldwide wherever they will grow. Taxonomically, 27 families of seed-bearing plants (Spermophyta) are examined, including 25 families of flowering plants (angiosperms) and two families of conifers (gymnosperms) following the classification of Benson (1979).

The information was obtained by searching literature, emphasizing as many scientific accounts as possible as well as some observations from popular sources, and personal communications from colleagues. Accounts that were considered purely opinions were disregarded for lack of validation.

Reference books about plants poisonous to humans and livestock are available and somewhat useful (Kingsbury 1964, 1965, Lampe 1985, Fuller and McClintock 1986, Gibbons et al. 1990); unfortunately, limited information is provided concerning bees and whether nectar or honey from the plant is toxic to bees or people. Illustrated field guides and atlases of poisonous plants are available for several regions of the world to identify plants known or suspected to be sources of toxins. These areas include Australia (Everist 1981), Britain (North 1967, Cooper and Johnson 1984), Europe (Frohne and Pfander 1984, Woodward 1985), East Africa (Vercourt 1969), India (Chopra 1977), New Zealand (Connor 1977), Venezuela (Blohm 1962), and North America (Muenscher 1960, Schmutz and Hamilton 1979, Stephens 1980, Lampe 1985, Westbrooks 1986, and Foster and Caras 1994). Unfortunately, many accounts and experiments, although relevant, were conducted before toxins had been identified and before procedures and methods of analytical chemistry had been improved.

Concerning plants poisonous to bees, the picture is rarely complete. Full knowledge would include the identity of the toxin, how it is acquired, and at what concentration it becomes toxic. We should also know: (a) whether the toxin is obtained from nectar, pollen, honey, honeydew, or plant sap; (b) all symptoms associated with toxic response; (c) whether the response is confined to foraging adults or if all bees, including nurse bees and immatures, are affected; and (d) if fresh or stored honey, made from the suspect plant, is toxic. In many situations, the researcher(s) seemed close to finding a key factor or toxin, and the work was abruptly discontinued. This review is incomplete because the information used to compile it is incomplete. I encourage new approaches using modern toxicological methods and creative science to stimulate more research to improve our knowledge about poisonous plants. We desperately need research-based information to determine whether suspected cases of poisoning are real or imagined.

Anacardiaceae

The genus *Toxicodendron* in the family Anacardiaceae includes poison oak (*T. toxicarium*), poison ivy (*T. radicans*), and poison sumac (*T. vernix*), all of which produce an oily sap containing urushiol, a non-volatile phenol that causes skin irritation to 75 percent of persons coming into contact with the plant (Gibbons et al. 1990). The nectar and honey from these species have not been implicated as toxic, and Lovell (1935) quotes Vansell as describing poison oak as "an important source of excellent honey." Persons consuming portions of the plant have suffered stomach upset and even death. The toxin apparently does not have the same effect on animals because livestock and other animals appear to be immune to it (Gibbons et al. 1990).

Apocynaceae
All plant parts of oleander (*Nerium oleander*) contain the alkaloids nerrin and oleandrin, which are poisonous to humans and livestock. Honey made from oleander nectar is bitter and toxic (Gibbons et al. 1990) and poisonous (James 1973). California has many roadside plantings of oleander because the plant is drought tolerant and produces many showy blossoms. The author has never observed bees visiting the flower, and when flowers were examined, no nectar was found.

Asteraceae
Honey made from tansy ragwort (*Senecio jacobaea*) contains six pyrrolizidine alkaloids (PAs) that are also found in the plant tissues. These PAs have been found to be toxic to livestock that ingest them. Levels of susceptibility to a given dose of toxin varied among the animals: Cattle and horses are much more susceptible than were sheep, goats, or rabbits (Cheeke 1989). Liver diseases in humans have been attributed to ingestion of PAs in teas made from *Senecio* spp. (Huxtable 1989). Bees collect nectar and consume honey with no apparent adverse effect. Honey produced from ragwort is reported to be bitter and not palatable unless blended with another honey. The level of contamination ranges from 0.3 to 3.9 mg/kg.

Boraginaceae
In Australia, honey made from nectar gathered by bees foraging on *Echium plantagineum* (Paterson's curse, salvation Jane) has been found to contain several pyrrolizidine alkaloids (PAs), mainly echimidine (Culvenor et al. 1981), that are responsible for livestock poisoning (Cheeke 1989). *Echium* is considered an excellent source of nectar for bees at a time when little other forage is available. A man named Paterson introduced the plant to his garden in 1880. Thereafter, beekeepers praised the plant in honor of Paterson's wife, Jane, calling it "Salvation Jane." Cattle producers, however, call the plant "Paterson's curse" because the plant poisons live-

stock, due to the PAs. Here is an example of how toxicity can vary among animals. Honey bees do not appear to be susceptible to poisoning, either by nectar or honey. The lethal dosage of PAs varied – small herbivores (sheep, rabbits, gerbils) were tolerant to PAs, whereas larger grazers (horses, cattle) were more susceptible (Cheeke 1989). PAs may provide a defense mechanism for the plant to prevent or reduce the effects of large grazers and enhance its survival.

Convolvulaceae

The dodder (*Cuscuta* spp.), a leafless parasitic vine, bears small flowers whose nectar is collected by honey bees. Beekeepers from Zavala County, Texas, reported that bees visiting dodder died shortly after foraging on the flowers. In some apiaries, colony losses reached 50 percent, when drought conditions had removed all other forage (M. 1930). In California, the author and others (Thorp, personal communication 1995) have observed honey bees and halictine bees visiting dodder with no immediate ill effect. Conditions in California may have been different from those present in the Texas case mentioned above or possibly another species of dodder was involved.

Coriariaceae

More than 200,000 ha (1,000 square miles) in northern New Zealand have been closed to beekeeping because honey from the tutu (*Coriaria arborea*) growing in this area has poisoned humans that consumed it. The sap from the tutu contains the picrotoxin, tutin. Toxicological studies can be challenging. If a toxin is present in the liquid sap of a plant, one cannot assume that the nectar from the plant is also toxic. Discovering that tutu was a "source" of the toxin and describing the ecological and chemical factors of toxicity involved cooperative studies starting in 1945, resulting in an eight-part series published later (Harris and Filmer 1947, Palmer-Jones 1947a,b, Paterson 1947, Sutherland and Palmer-Jones 1947a,b, and Palmer-Jones 1949). Two recent reviews that describe the history of events and identification of toxins in this system differ substantially (Love 1990, Sutherland 1992). The more recent review indicates that at least three chemical fractions are toxic, but only one has been identified, and the name for this material is in doubt (Sutherland 1992). Material presented in these reviews will be of special interest to readers who enjoy analytical chemistry.

The mystery began in the 1890s with reports from medical personnel of periodic human poisoning from an unknown source of honey in the Bay of Plenty district in New Zealand. Reports were made at irregular intervals (Palmer-Smith 1947a, 1965). Honey from swarms, extracted honey, section honey, and honey mead were all implicated as dangerous (Palmer-Jones 1965). Persons consuming as little as one-half teaspoon of liquid honey were severely affected (Palmer-Jones 1947a). Patients suffered from giddiness, confusion, abdominal and head pain, vomiting, convulsions, coma, and loss of memory.

In the Bay of Plenty district, rangiora (*Brachyglottis repanda*), buttercup (*Ranunculus rivularis*), and tutu were suspected of being the source of the toxin. Rangiora bloomed from August to October, whereas the toxic honey was apparently produced in February and March. Beekeepers claimed that honey produced in swamp areas where buttercup predominated was innocuous, whereas that produced from upland sites was more dangerous. In 1927, an experienced doctor had a patient who appeared to suffer from honey poisoning, and tutu was the only plant found within several miles of the swarm believed responsible for the honey. Because the tutu was not known to produce nectar, beekeepers surmised that the poison came from pollen that was found in the honey. An agricultural chemist failed to isolate tutin from the honey, and tutu pollen caused no symptoms when fed to rats. The connection of tutu to the poisonous honey appeared doubtful.

Beekeepers were advised not to place hives in the suspect area. After a 12-year period of no reports involving poisonous honey, colonies were moved near to the area, and initially no problems were noted. But an outbreak of honey poisoning occurred in 1945, stimulating definitive studies to determine the source of honey. Twenty-seven people had symptoms similar to those reported earlier from honey poisoning. Enquiries by C. R. Paterson, apiary instructor from the Department of Agriculture, indicated the honey had come from an apiary in the Pongakawa Valley (Paterson 1947). The toxic honey had come from the third extraction made by the beekeeper at the end of February. It (200 lbs) was collected and used for experimental study. The apiary was relocated. Botanical examination of toxic honey to identify pollen revealed that this honey contained pollen from numerous plants, including a small amount from tutu (Harris and Filmer 1947).

Toxicity tests were conducted with toxic honey by administering a dose of honey to pigs, sheep, rabbits, mice, rats, bees, and guinea pigs (Palmer-Jones 1947a). Rats showed mild susceptibility, whereas guinea pigs were highly susceptible. Bees and the other animals listed above showed no susceptibility to the honey. Guinea pigs were chosen as the test subject on which to study symptoms and compare effects of other honey samples.

In September 1945, two test hives were established on the original apiary site that had produced toxic honey (Paterson 1947). Samples of honey from comb and pollen (via pollen traps) were collected every two weeks and sent to the laboratory to be tested on guinea pigs. By the end of December, the flowering period of tutu was over and berries were forming. The test hives had accumulated a large amount of honey believed to come from teatree (*Leptospermum scoparium*) and rewarewa (*Knightia excelsa*). All samples of honey collected through December were non-toxic to guinea pigs. While Paterson (1947) was picking berries from tutu for examination in January 1946, he made an interesting discovery that provided the key to solving the mystery. He found a sticky substance on many of the lower leaves of the plants. This substance, honeydew, was being secreted by the passion vine hopper (*Scolypopa australis*), a sap-feeding insect (actually a fulgorid) that was abundant on the plants. Honey bees were known to collect honeydew from plants, especially when nectar was scarce. Initially, samples of honey collected in January were non-toxic; in February, however, they were found to be toxic. This time period coincided with dry weather conditions that limited nectar in other plants. Vine hopper activity and production of honeydew were at their highest level during this time period.

As Palmer-Jones (1965) pointed out, several factors were required to produce toxic honey, including large numbers of tutu plants, heavy infestation of the passion vine hopper, large excretions of honeydew, and hot, dry weather that reduced nectar production from other sources and stimulated honey bees to collect honeydew. Danger from toxic honeydew can be eliminated or reduced by managing the passion vine hopper, reducing the number of tutu plants in the area, providing other plants that produce a safe and more attractive source of nectar, or by harvesting honey before honeydew production. Species of *Coriaria* are found worldwide and tutu plants have been imported into the United States (Barker 1990).

Corynocarpaceae

The New Zealand Laurel or karaka (*Corynocarpus laevigatus*), a small tree native to New Zealand and grown as an ornamental in California, produces nectar containing the toxic glycoside karakin, which kills bees (Palmer-Jones and Line 1962, Palmer-Jones 1968, 1970, Fuller and McClintock 1986). The karaka produces masses of flowers, whose copious nectar is very attractive to honey bees. It blooms from August to December, spring in New Zealand, when other food sources are often scarce.

Losses of adult bees in the spring had occurred for many years in the Hawke's Bay district with no apparent cause (Palmer-Jones and Line 1962). A local beekeeper observed that karaka trees in bloom were observed attracting large numbers of bees. When colonies were moved away from this area during the bloom period, no more losses occurred. Investigations of 20 colonies placed near a blooming grove of karaka (Palmer-Jones and Line 1962, Palmer-Jones 1970) revealed that foraging workers were affected with as many as 4,000 dead or dying bees from a single colony. Affected bees appeared weak and could not fly; they died gradually. Queens and brood appeared unaffected. Field bees collected on karaka flowers and placed in cages with sugar syrup suffered 25 percent mortality compared to no mortality for control bees collected from other nectar sources. Hive workers that were fed karaka nectar dissolved in water in November and December showed signs of weakness and died outside, being unable to enter the hive. Another hive fed in May as above was unaffected. Karaka nectar apparently became non-toxic after 26 weeks in storage. Barker (1990) explains that the toxin, karakin, would be degraded by B-glucosidase to 3-nitropropanol, which would be toxic in a one percent syrup; however, this material is also volatile and may be lost in storage. Bees collected little pollen from karaka, and no effect of pollen was tested.

Guinea pigs that were dosed by fresh karaka nectar applied with a stomach tube showed symptoms, including temporary paralysis, tremors, and convulsions after handling. Guinea pigs dosed in the same manner with karaka nectar that had been stored for nine weeks showed no symptoms, suggesting that the toxin was no longer present in a reactive form. Honey from karaka has not been shown to be poisonous to humans. Beekeepers are advised to move their apiaries away from areas that have blooming karaka.

Cyrillaceae

The plant *Cyrilla racemiflora* is most often called southern leatherwood

Figure 19.1. Summer titi. (Cutts photo).

or summer titi. Other common names include titi, black titi, white titi, red titi, iron-wood, swamp cyrilla, leatherwood, and huckleberry. When common names are used without reference to the scientific name, confusion can result because a re-lated plant, *Cliftonia monophylla*, is also called titi. It is similar in appearance and occurs in the same locations as *C. racemiflora*. In this review, we refer to *C. racemiflora* as summer titi. Summer titi usually grows as a bushy shrub (Figure 19.1) but may also take the form of a tree reaching heights of 25 feet. It occurs in swampy areas from southern Virginia to Florida and west to Louisiana and Texas. The plant blooms from May through July. Its abundant white flowers are fragrant and attractive to bees.

Losses of half the bee colonies in northwestern Florida in 1932 prompted studies to determine the cause (Burnside 1933a, Burnside and Foster 1933). Burnside (1935a) observed summer titi to be the cause of these poisonings, and he named the malady purple brood because affected larvae appeared blue or purple. He reported that purple brood comes on suddenly, in late May or early June, and all the colonies in the affected area develop symptoms at the same time. Larvae, pupae, and pre-emergent adults may be affected, and the brood may continue to die for a month before the disorder subsides. Colonies may be slightly affected or seriously weak-ened. The weakened colony may recover, but honey production will be consider-ably reduced. Purple brood could not be induced by transferring affected brood, combs, or stored pollen, but brood died when honey (in comb or extracted) was transferred to healthy colonies (Burnside and Foster 1933).

Purple brood has been found throughout the Southeast, linked to the presence of summer titi, and it is still considered a serious problem periodically in Florida (Sanford 1994), Georgia (Delaplane, personal communication 1994), Missis-sippi (Fulton, personal communication 1994), and South Carolina (Hood, personal communication 1994). The prevalence of purple brood seems related to the local abundance of summer titi and the lack of other food sources (Sanford 1994). Sanford (1994) recommends that in areas containing abundant summer titi, beekeepers move their colonies out of the area or feed them with sugar syrup to dilute the plant's effect.

Ericaceae

Rhododendron species, including *Azalea*, are widely cultivated through-out the world as ornamental shrubs because of their attractive bloom and foliage (Figure 19.2). The clustered flowers produce nectar and pollen and are most often visited by bumble bees and pollen bees. Few references are available that indicate pollen or nectar is directly toxic to bees. Allen (1940) observed that bees carrying *Rhododendron* pollen were affected, and Brown (1939) observed dead bees among those collecting propolis from buds before opening.

Carey et al. (1959) conducted laboratory toxicity tests by feeding caged honey bees nectar from 17 species and 14 hybrids of *Rhododendron*. The results indicated that *R. thompsonii* and some of its hybrids are especially toxic, though the toxicity of some hybrids varied unpredictably. One hybrid derived from four parent species, three of which had toxic nectar, was non-toxic itself. *Rhododendron* species known to poison bees include *R. arboreum*, *R. luteum* (=*Azalea pontica*), *R. occidentale*, *R. prattii*, *R. ponticum*, and *R. thomsonii* (Plugge 1891, Poltew 1956, Carey et al. 1959, Piskovoi and Kolonshchakov 1961, Poltev et al. 1976, Barker 1990).

It is difficult to understand how honey bees can produce honey from nectar that is toxic to them. The effect on the bee is probably related to dose at an individual as well as a colony level. The bee must consume a minimum amount of the toxin before it is affected. If the bee is visiting other non-toxic plants before returning to the colony, the toxin from the poisonous nectar may be diluted. Another factor may be the time it takes the toxin to act. The bee may eventually succumb, but not before it brings the toxic nectar back to the colony. At the colony, a toxic nectar could be mixed with non-toxic nectar, diluting the toxin in the honey. For a water-soluble toxin, as the nectar is processed and water removed to make honey, the concentration of the toxin may increase to a level that causes a response.

Historically, plants from the heath family are among the earliest cited sources of toxic honey. Honey from *Rhododendron ponticum*, growing near Trebizonde in Asia Minor, was believed to have poisoned Greek soldiers of Xenophon in 401 B.C. (the retreat of the ten thousand) (Plugge 1891, Carey et al. 1959, Patwardhan and White 1973, Cooper and Johnson 1984). Some writers suggest that only unripe honey, before capping by bees, is toxic (Howes 1948, Atkins 1992).

The toxin from *Rhododendron* is a resinous substance, a gray anotoxin called andromedotoxin; it is present in all parts of the plant, including the pollen (Hardin and Arena 1969, Gibbons et al. 1990). "Even honey made from the nectar is bitter and toxic. When sufficient quantities of the leaves, pollen, or honey are ingested, humans begin watering in the mouth, eyes, and nose. Within six hours, nausea and vomiting begin. Sweating, abdominal pain, headache, weakness, tingling of the skin, slow pulse, incoordination, low blood pressure, and respiratory difficulty may also occur. Symptoms often culminate in convulsions, progressive paralysis, coma, and death, the latter due to respiratory failure" (Gibbons et al. 1990). Additional toxicology studies have examined occurrence and analysis of toxic diterpinoids in honey, including two other unnamed grayanotoxins (Scott et al. 1971).

Figure 19.2. Typical *Rhododendron* flower grown as an ornamental. (Skinner photo).

Two other members of the family, the widely cultivated fetterbush (*Pieris floribunda*) and Japanese andromeda (*P. japonica*) also contain andromedotoxin, which has been cited as producing nectar and honey toxic to humans (Lampe 1985).

The mountain laurel (*Kalmia latifolia*) is a native evergreen shrub growing in moist woodlands and hillsides of upper elevations in the Appalachian Mountains (U.S.). When in bloom, its numerous clusters of predominant white flowers make an impressive visual display (Fig. 19.3). As with *Rhododendron*, all parts of the *Kalmia* plant contain an andromedotoxin that poisons humans and has been found in honey (Hardin and Arena 1969, James 1973). Persons eating a spoonful of this honey may feel numbness and lose consciousness for several hours (Atkins 1992). A toxic sample of honey from Virginia (Fell, personal communication 1994), presumed to be from *K. latifolia*, was found to contain significant levels of grayanotoxins I and III; there were, however, no concurrent reports of bee losses associated with this discovery. Nectar or pollen of *Kalmia* has not been reported to be toxic to honey bees, which is surprising, because it often blooms alone in dense stands.

Wild rosemary (*Ledum palustre*) produces a honey containing ericolin, a glucoside that is toxic to humans. Species of *Andromeda*, and indeed all members

Figure 19.3. *Kalmia latifolia* blossom. (Skinner photo).

of the Ericaceae, are often listed as capable of producing toxic honey because all are suspected to contain andromedotoxins (Palmer-Jones 1965). More research is needed to prove that the toxin is present and that the honey produced is toxic.

Euphorbiaceae

The spurge family contains the genus *Euphorbia*, which has about 680 species throughout the world, including the well-known poinsettia, snow-on-the-mountain, and spurge. The white, milky sap contains many potentially toxic materials, but none have been positively identified as toxic to bees. Honey produced from snow-on-the-mountain (*E. marginata*) has been reported to be toxic to humans (Kingsbury 1964, Pellett 1978), but some beekeepers from Texas reported that during a drought in 1921, the plant produced a surplus of dark amber honey; another beekeeper ridiculed the notion that the honey was poisonous (Pellett 1978). Deodikar et al. (1958) reported poisoning of bees foraging on *E. geniculata*. Pollen grains of this species were found in the gut and pollen loads of poisoned bees. The bees were paralyzed, had curved abdomens and outstretched wings, and could not fly. Feeding pollen substitutes to colonies reduced bee visits to the poisonous plants. This implies that the bees were visiting primarily for pollen. More research is needed to determine the identity of the toxin.

Fagaceae

Salin (1973) reported that oak (*Quercus* spp.) was responsible for bee poisoning in the U.S.S.R. Compared to other years when bees collected pollen only from oak, in 1972 they also collected nectar. Bee colonies switched their visitation from orchard and berry crops to oak. Bee losses were observed in colonies near the end of oak bloom. Flight from colonies was reduced, queens reduced laying, and larvae died, causing spotty brood patterns. Colonies that were fed with sugar syrup during and after oak bloom appeared to be in better shape (Salin 1973). The combination of colder temperature and rain reduced foraging and was suspected to increase toxic effects by inducing bees to consume honey stores containing large amounts of oak honey. Salin (1973) also speculated, without data, that the toxicity could have been related to the presence of large amounts of lime in the soil. Barker (1990) suggests that fungal poisoning associated with honeydew may have been responsible, but Salin (1973) states that secretion of honeydew was not observed. Weather factors that induced bees to collect oak nectar and consume oak honey seem to be the most likely causes. No other reports have been discovered that attribute bee poisoning to oak.

Hippocastanaceae

The California buckeye (*Aesculus californica*), which occupies more than 15 million acres in California, has been a constant source of concern for California beekeepers by causing large losses of honey bee colonies throughout its range (Vansell et al. 1940). *A. californica* varies widely in form: It is a short, shrubby plant in the north and a medium-sized tree (40 feet) in the south. The white or pink blossoms produce a showy display on their terminal panicles (Figures 19.4 & 19.5). The pollen is bright red. The severity of losses reported by beekeepers (Vansell 1926a, Burnside and Vansell 1936) prompted additional study to map the precise distribution of buckeye (Vansell and Watkins 1936, Vansell et al. 1940). The toxicology of the plant, however, is virtually unknown, as Barker (1990) explained. Vansell (1926b) suspected that an unstable glucoside was the toxin. He explained that California Indians used an extract from buckeye for "stupefying fish." A coumarin derivative, aesculin, has been isolated and listed as the toxin, although no data were presented (Fuller and McClintock 1986).

Vansell (1926b) describes the symptoms of buckeye poisoning as follows: "The effects on bees of buckeye honey, pollen, nectar, and sap are sometimes very severe. Not only the field bees, but the adult queen and drones are affected as well as the larvae and emerging young adults. In severe cases, the whole colony dies with the hive full of honey. The majority of the larvae being fed are killed outright and are, in the main, devoured by the adults. Those young which are not killed will pupate and emerge if they are not so badly deformed that they are unable to do so. Adults with but four normal legs are very common."

Another report (Vansell et al. 1940) described several stages that observers might use to diagnose the severity of poisoning. In mild cases, the colony simply fails to become populous and there is a "spotty" brood pattern with mixed empty, uncapped, and capped cells. This spotty condition in the brood nest is the first symptom of buckeye poisoning. If many buckeye products are received in the food, then only eggs, day-old larvae, and capped brood are present. The absence of uncapped brood is a typical symptom of a "buckeyed" colony. With continued poisoning, only eggs remain. "The cells do not contain the dead brood, as with some diseases or other kinds of poisoning." In most cases of buckeye poisoning, piles of light-colored, immature, and deformed bees accumulate at the hive entrance. Fertilization also seems affected because queens from "buckeyed" colonies that had been laying drones only started laying fertile worker eggs when moved to uncontaminated food. Vansell (1926b) also reported that sap collected from buckeye twigs, leaves, and buds may have contributed to poisoning. The sap exuded from feeding punctures caused by the bug, *Urbisea solani*.

The intensity of the effects of buckeye poisoning is directly related to the availability of other food sources to bees during the four- to six-week period buckeye is in bloom. In years of sufficient rainfall, bees feed on other food sources and the negative effect of buckeye is reduced.

Vansell (1926b) recommended that beekeepers move colonies during buckeye bloom or relocate colonies in areas that have large numbers of trees. Feeding sugar syrup during buckeye bloom was also suggested to dilute toxins coming from buckeye.

A relative of the buckeye, the horse chestnut (*A. hippocastanum*) is native to southeastern Europe, but has been hybridized and cultivated throughout the world. After reports of poisoning from California buckeye reached Europe, researchers there discovered bees showing the same symptoms (Maurizio 1945a, Velthoen 1947). Pollen from *A. hippocastanum* or *A. parvia* (dwarf buckeye) was found in the gut of affected bees. Poisoning was attributed to the presence of these trees because bees foraging on them returned to the colonies carrying the characteristic bright red pollen. Maurizio (1945a) maintains that colonies showing symptoms of poisoning were not affected by mites or nosema. Feeding bees in cages with extracts of pollen and aqueous extracts of blossoms of *A. hippocastanum* and *A. parvia* caused noticeable symptoms after three days and greatly shortened the life span of these bees, compared to bees fed with sugar syrup.

Figure 19.4 & 19.5. California buckeye, *Aesculus californica*. Top, close-up of flower. (Mussen photo). Bottom, buckeye bush. (Thorp photo).

Saponin, a toxin, has been found in the sap, nectar, and pollen of *A. hippocastanum* (Johnsen 1952b, Schulz-Langner 1966). Barker (1990) suggests that all *Aesculus* species contain toxic saponins, but they are rarely collected at lethal doses (Maurizio 1945a, 1968). The saponin content of several species of *Aesculus* tested by Schulz-Langner (1967) varied widely on a yearly basis and among individual trees. She could not prove conclusively that mass death of bees was caused by horse chestnut. Saponins reduced bee longevity in laboratory tests, but she believed that in the field, bees selected other nectar sources that did not contain these toxins. We do not know whether bees have the ability to detect saponins in the nectar. The "dose effect" again seems to repeat itself. Why would bees visit a toxic source of food? The obvious answer may almost be too simple to suggest: The bees visit the toxic source because nothing else is available.

Lamiaceae

Stachys arvensis (field nettle betony, staggerweed), a member of the mint family, is widely distributed in the northeastern United States, but no problems are reported as a result of its presence. This plant has poisoned sheep in Australia (Barker 1990), although livestock generally avoid consuming it unless there is a shortage of food (Goodacre 1940). In Australia, nectar from staggerweed was suspected to cause dwindling by reducing the population of bees in the colony, especially nurse bees (Goodacre 1940). When the plants were plowed under, all symptoms disappeared. The identity of the toxin is unknown.

Leguminoseae

Astragalus species (locoweeds, tragacanth) can occasionally accumulate toxic levels of selenium (Kingsbury 1964); but selenium is not known to be poisonous to bees (Barker 1990). In Utah, U.S., plants not containing selenium were believed responsible for infrequent, but serious losses to bees (Sturtevant et al. 1941). Sudden losses of adult bees were thought to be related to blooming of locoweeds (Vansell 1935, Davis 1942, Knowlton 1944). Young foragers were found dead or dying near blooming *Astragalus lentiginosus* (specklepod loco) (Vansell and Watkins 1934). The toxic nectar also weakened other foragers but was "slow acting," allowing these bees to bring some of the material back to the colony. Delayed negative effects on brood have been attributed to "loco" honey (Beckwith 1898, Riggs 1924, Knowlton 1944) and to pollen of *A. lentiginosus* stored in combs (Davis 1942). Knowlton (1944) reports that when a beekeeper moved honey suspected to be "loco" from one apiary where deaths had occurred to another apiary, additional deaths occurred. When the suspect "loco" honey was removed, the bees recovered.

Honey bees have been killed while foraging on timber milk vetch (*Astragalus miser* v. *serotinus*) in British Columbia, Canada (Majak et al. 1980). The nectar contained six percent miserotoxin. Details on the chemistry of the toxic compounds and precursors are available (Majak and Pass 1989). Caged bees died after being fed a one percent dosage in sucrose syrup. Signs of poisoning included loss of coordination, weakness, and inability to fly.

The yellow kowhai (*Sophora microphylla*), a tree distributed throughout New Zealand, produces large amounts of nectar when it flowers from July to October. Honey bees and birds are common nectar foragers. On some occasions, the nectar from kowhai has a narcotic effect on bees that may be followed by death when combined with low temperatures, which are common during blooming in the spring (Palmer-Jones 1970, Clinch et al. 1972). Only adult bees have been affected. About 75 percent of bees fed with nectar became unconscious within 30 minutes of consumption; most bees recovered after four hours in an incubator kept at 30°C. Extracts of yellow kowhai seeds, wood, and leaves, when added to syrup and fed to bees, caused narcosis and mortality similar to the effects of feeding nectar. Clinch et al. (1972) assumed these effects were caused by toxic alkaloids known to be present in plant parts. No effect of toxicity was observed for birds during the same time period. Considerable variability was noted in degree of mortality in treated bees and in amount of total dissolved solids in nectar samples with respect to date of sample, location, and even among individual trees at a given location. Some trees produced nectar that was not toxic at all. Sugar concentration and degree of toxicity did not appear related. Barker (1990) indicated that nectar samples supplied by Clinch contained no toxic sugars. This toxicity from specific trees is certainly a mystery that needs further investigation. Bees were not observed to collect pollen from yellow kowhai, and no attempt was made to determine toxicity of pollen.

Suhayda (1968) reported that both nectar and pollen from *Sophora japonica* (Japanese pagoda tree or Japanese acacia) were toxic to bees. Clinch et al. (1972) infer that nectar from other species of *Sophora* may be toxic in the same manner as above for *S. microphylla* because the presence of alkaloids is a characteristic of this genus. Further study is needed to determine if other species have toxic nectar and if the toxic factor is an alkaloid before any assumption can be made concerning toxicity of an entire taxon. Beekeepers should carefully observe their bees for any adverse symptoms when colonies are placed near these plants and, if possible, move them away from plants in bloom.

Liliaceae

Poltow (1956) listed pollen of *Allium cepa* (common onion) as toxic to bees in the U.S.S.R., but more information is needed to verify this contention.

Figure 19.6. *Gelsemium sempervirens*, (yellow jessamine vine) in flower. (Hood photo).

Honey bees have been used to pollinate onion for seed production throughout the United States and no poisoning has been reported. Bees visit onion flowers to obtain nectar and pollen. The presence of potassium in onion nectar has been suggested to reduce the attractiveness of onion to honey bees (Waller et al. 1972), but when other sources are not present, the bees collect nectar and pollen with no apparent ill effect (McGregor 1976).

Tulipa spp. (tulip) are common garden flowers. Body fluids of bees that died after visiting tulips contained either mannose or galactose (Geissler and Steche 1962), sugars that are toxic to honey bees (Barker and Lehner 1974). Because galactose is the predominant sugar in the stigmatic exudates of tulip (Barker and Lehner 1976), Barker (1990) suggested that tulip nectar may have been responsible for the bee deaths.

Veratrum spp. (corn lily, false hellebore, Indian poke) are sources of hellebore, which has insecticidal properties (Barker 1990); toxic alkaloids including veratrin (Harden and Arena 1969); and glycoalkaloids (Crosby 1971) that have been shown to be active teratogenically (Frohne and Pfander 1984). *Veratrum lobelianum, V. dahuricrum*, and *V. nigrum* have poisoned bees in the U.S.S.R. (Perepelova 1927, Zakatimova 1948, Poltew 1956).

Vansell (1935) and Vansell and Watkins (1933) reported that bees were poisoned by the corn lily (*V. californicum*) in El Dorado County, California, which they visited because their normal forage was not available. In this case, adult bee populations were killed off in a three-day period, and brood in a nearby colony died from starvation rather than from poisoning. Dead and dying adult bees were found on the flowers and at the hive (51 bees found per minute at one hive). This evidence suggested that nectar from corn lily was the source of toxin (Vansell and Watkins 1933, Burnside 1935b). In the U.S.S.R., the pollen of white hellebore (*V. album*) killed caged bees when it was fed in sugar syrup; young bees were more susceptible than old bees.

Zygadensis venenosus (death camas) produces several toxic alkaloids similar to those found in *Veratrum,* including the steroidal alkaloid, zygadine. Children have been poisoned by eating the flowers (Hardin and Arena 1969). Perrins (1933) reported that death camus killed adult bees in Utah. Bees foraging in cages on death camus were observed paralyzed on the floor of the cage nine hours later (Hitchcock 1959). Hitchcock demonstrated poisoning of bees fed with centrifuged death camus nectar or with sugar syrup to which pollen had been added. He added it was fortunate that the plants produced very small quantities of nectar and bloomed for only ten days.

Loganiaceae

Gelsemium sempervirens, commonly known as yellow jessamine or Carolina jessamine, is a woody evergreen vine with showy, yellow, fragrant, trumpet-shaped flowers (Fig. 19.6). The 1-inch-long flowers appear in late winter, spring, and sometimes in fall. The native distribution includes the Coastal Plain and lower hills from Virginia to Texas (Kingsbury 1965, Gibbons et al. 1990). The plant is also commonly planted as an ornamental farther north. The toxic alkaloids gelsemine, gelseminine, and gelsemicine are found in all parts of the plant with greatest concentrations occurring in the roots and nectar of flowers (Hardin and Arena 1969).

Humans have been poisoned and died from sucking the nectar from flowers or from eating honey made from this plant (Howes 1949, Kingsbury 1965, James 1973, Patwardhan and White 1973).

Bees foraging on the flowers of yellow jessamine appeared intoxicated, became paralyzed, and died (Brown 1879). Symptoms of poisoning disappeared after yellow jessamine stopped flowering. Huggins (1942) suggested that the toxin was in the pollen because young nurse bees were more affected than were field bees. Brood does not seem to be affected (Burnside 1933a). Burnside and Vansell (1936) reported that the dead young bees contained a light-colored solution, presumed to be honey because the fluid contained a high concentration of reducing sugars. Because only a few of these bees had consumed yellow jessamine pollen, the toxin was suspected to be in the honey. Oertel (1961) caged 150 bees and allowed them to feed on fresh flowers plus sugar syrup, sugar syrup made with water in which 50 blossoms had been soaked, syrup with finely ground blossoms, and syrup containing pollen from 30 blossoms. Only a few bees died in any of the treatments, and he concluded that yellow jessamine was not poisonous to the bees in this study. His conclusion may seem appropriate for this study under these "conditions," but the data do not justify a general conclusion that yellow jessamine is not poisonous to bees. The amount of toxin present in the treatments was unknown, and the total volume of syrup used was unspecified; therefore, the effects of the toxin may have been substantially diluted by syrup to a non-toxic level. The overall evidence indicates that yellow jessamine does sometimes affect bees; more conclusive research is needed, however, to determine whether the toxic effect is caused by pollen, nectar, or both. Yellow jessamine has been reported to cause periodic poisoning of bees in Georgia (Delaplane, personal communication 1994) and in Mississippi (Fulton, personal communication 1994).

Yellow jessamine has been cited as causing recent serious losses to bee colonies in Florida (Sanford 1994). In this case, reports of severe bee losses in southern Florida in spring 1994 resulted in an investigation by Lawrence Cutts and his staff of bee inspectors at the Division of Plant Industry. Varroa did not appear to be the problem, and tracheal mite levels were low. Veteran bee inspector Tom Dowda suggested analyzing gut contents; bees' intestines were found to be loaded with pollen from yellow jessamine. Investigation by the Beltsville Bee Laboratory confirmed the pollen identification and revealed no diseases. The losses appeared to be related to extensive bloom of yellow jessamine, whereby a gap in the bloom pattern of maple and willow allowed a period of time when yellow jessamine was the only plant in bloom. Sanford (1994) suggests several questions that need to be answered: Is yellow jessamine nectar or pollen toxic to bees? Will bees work this flower if other bloom is available? If yellow jessamine is toxic, what percentage is necessary to be lethal? To answer these questions would require a substantial research effort. Yellow jessamine pollen, nectar, and honey should be analyzed chemically to determine the identity and concentration of toxin(s). Toxicology studies are needed such as feeding colonies varying doses of pollen, nectar, and honey and monitoring effects, including death rate of bees over time. Affected bees should be analyzed for presence of toxin and its concentration. Foraging rates of bees should be compared when several food sources are blooming at the same time as yellow jessamine to determine if bees will selectively visit other plants more frequently.

Oleaceae

Syringa amurensis, a cultivated lilac from northern China, was cited as causing spring and summer losses of bees in the Primorye Territory of the U.S.S.R. after the bees had visited its poisonous blossoms (Gorbacheva 1953). Losses were unrelated to the presence of honeydew and were highest when pollen was scarce. The reference cited above is the only report of lilacs (*Syringa* spp.) causing poisoning of bees. Other references indicate lilac secretes nectar that is sought after by bees but only when the nectar in the floral tube pools in sufficient quantity to allow the relatively short tongue of the honey bee to reach it (Howes 1948).

Papaveraceae

The poppy family contains the well-known opium poppy (*Papaver somniferum*), which has been used as a medicinal plant for more than 2,000 years. It is cultivated for production of opium, alkaloids, seeds, or oil. The medicinally valuable alkaloids morphine, codeine, and noscapine are present only in this species (Frohne and Pfander 1984). All poppies contain alkaloids that could be suspected as potential poisons, yet poppies are reported to be a source of abundant pollen for honey bees (Howes 1948, Pellet 1978). The same authors mention anecdotal reports of bees appearing narcotized after foraging on this plant. Howes (1948) describes a German observer who indicated that 90 percent of his bees returning to the hive with poppy pollen had difficulty in finding the entrance of their hive. Pettis (personal communication 1994) reported that a beekeeper noticed bees performing irregular acrobatics after visiting red poppies.

Pinaceae

Schneider (1951) described a case of *Waldtrachtkrankheit* (Forest-sickness) occurring in Switzerland where bees were paralyzed after collecting aphid honeydew. Two species of aphids were found feeding on silver fir (*Abies alba*). Numerous dead bees were found at hive entrances, and live bees appeared paralyzed. Colonies quickly recovered when fed sugar syrup or when moved to another location without fir trees.

Proteaceae

Barker (1990) reports that macadamia (*Macadamia integrifolia*) killed bees when they were caged with trees in bloom. Cyanide was detected in cylinders enclosing flowers and was assumed to have killed the caged bees. An anonymous report cited by McGregor (1976) also mentioned that macadamia flowers produce a gas that is highly toxic to bees and may have a somewhat repelling effect on them. McGregor's (1976) summary of macadamia pollination suggests honey bees should be used to provide adequate pollination. Honey bees collect pollen from macadamia (Urata 1954, Gary et al. 1972), and nectar is secreted at the base of the flower in such small quantities that the amount present may not be sufficient to attract insects (McGregor 1976). As Barker suggests (1990), it is uncertain whether honey bees are affected by the "gas" in normal field conditions. The confinement of bees in cages may have increased the effect of the "gas."

Ranunculaceae

Aconitum spp. (monkshood, wolfsbane) contain alkaloids, including aconitine (Crosby 1971), which is very toxic to mammals in small amounts (Frohne and Pfander 1984). Dengg (1926) reported that claims have been made in alpine areas where monkshood is abundant and visited industriously by bees. The honey shows toxic properties, probably less for bees themselves than for humans. Bumble bees have shown no symptoms of poisoning. Dengg (1926) also reports a case where three men mowing hay became aware of a sizable bumble bee nest provided with plenty of honey. Two of the men were overjoyed with this find and proceeded to consume about three tablespoonfuls when they both felt, hardly one-half hour later, a biting on the tongue and a pain in the left wrist. The pain became worse each minute and passed to the feet and finally to the abdomen and the chest. The men became faint, convulsions occurred, and one man died, presumably because he was not able to vomit the material he had ingested. Monkshood flowers blooming nearby were assumed to be the source of the problem.

Pollen from monkshood was believed responsible for poisoning honey bee colonies in Siberia (Koptev 1948). When pollen was fed to colonies, adult workers became paralyzed, and drones and queens were also affected.

The author has never observed honey bees visiting monkshood. The monkshood flower is considered adapted for bumble bees because entry into the flower requires considerable force from the visitor (Proctor and Yeo 1972). The elongate nectaries make the nectar difficult to reach unless the bee has a long tongue such as that of a bumble bee.

All plant parts of *Ranunculus* species (buttercup, crowsfoot) (Figure 19.7) contain a highly toxic volatile oil, protoanemonin. This toxin is found only in the growing plant and is synthesized by the glycoside, ranunculin (Barker 1990). Protoanemonin degrades into a non-toxic anemonine when the plant dies. Serious cases of poisoning caused by *Ranunculus* pollen have been reported by Morganthaler and Maurizio (1941), occurring in Switzerland and elsewhere in Europe. Maurizio (1944, 1945b) referred to it as the "Bettlach May-Sickness" because it occurred in the district between Biel and Solothurn and especially near Bettlach. She attributed the occurrence of the disease to cold weather, which retarded devel-

Figure 19.7. *Ranunculus* (buttercups) are used as ornamentals, and often escape. (Stone photo).

opment of the usual pollen sources, inducing the bees to collect more than normal amounts of buttercup pollen. Maurizio (1941, 1945b) used the term anemol to describe the toxic extract she removed from buttercup pollen. Toxins have been found in *Ranunculus acer, R. acontifolius, R. acris, R. alpestris, R. aurecomus, R. breyninus* (= *R. nemorosus*), *R. lanugginosus, R. polyanthemos, R. puberulus, R. repens,* and *R. steveni* (Maurizio 1941, 1945b, Muller 1948).

Maurizio (1944) wrote that buttercup poisoning affects all young bees. Bees taking care of the brood appeared at the hive entrance and began to tremble, unable to fly, moving excitedly around on the landing board and the soil, losing control of their legs, and rotating violently on their backs. Paralysis and death occurred shortly thereafter.

It seems unusual that cases of buttercup poisoning have not been reported elsewhere. Howes (1948) speculated that the May-sickness observed in Switzerland did not occur in England where buttercups are common because other pollen plants were available during the time buttercups bloom.

Sapindaceae

There are 15 species of cultivated soapberry (*Sapindus*) (Barker 1990).

The berries are used to make "soap." Toxic saponins with insecticidal properties have been found in fruits of *Sapindus marginatus*, the Florida soapberry (Crosby 1971). This small tree bears clusters of small white flowers (Preston 1966), whose nectar has been reported to kill thousands of bees in the United States (Maurizio 1945b). Although soapberry is cited as a honey plant (Pellet 1978, Lovell 1926), no losses to honey bees have been reported in recent years.

Scrophulariaceae

Digitalis purpurea (common foxglove) was introduced from Europe into North America where it has been naturalized, and is widely planted as an ornamental because of its attractive floral display (Fig. 19.8). The plant contains several cardiac glycosides, including digitonin, that have been used medicinally (Hardin and Arena 1969). Children have been poisoned after sucking on the flowers. Foxglove pollen has been reported to poison bees (Muck 1936), and an extract of digitonin from seeds was highly toxic when fed to bees in sugar syrup (Maurizio 1945b).

Solanaceae

Datura stramonium (jimsonweed, thorn apple) is native to North America.

Figure 19.8. *Digitalis purpurea* (common foxglove). (Kingsbury photo).

All parts of the plant, including the pollen, contain toxic alkaloids such as atropine and scopolamine (Gibbons et al. 1990, Westbrooks 1986). Honey made from *Datura* contains the toxic alkaloid hyoscine, atropine, or both (Crane 1975a). Though toxic to vertebrates, these alkaloids are not very toxic to insects (Barker 1990).

Hyoscyamus niger (black henbane) was introduced to North America from Europe and has been naturalized. This plant contains the same alkaloids mentioned above for *D. stramonium*. Shaginyan (1956) reported in Armenia that 24 colonies died and 50 colonies were weakened in an apiary of 150 colonies because of black henbane poisoning. Adult bees and brood were affected, and the bees were very vicious, attacking people and animals nearby. When the apiary was moved, the problem disappeared. The toxin responsible for the poisoning was not identified.

Taxaceae

Leaves and berries of the widely cultivated yew (*Taxus* spp.) contain the toxic alkaloid taxine. In antiquity, extracts of leaves were used for murder and suicide (Frohne and Pfander 1984), and Virgil described honey poisonous to humans from bees foraging on Corsican yew trees (Pryce-Jones 1944). Howes (1948) reports that *T. baccata*, growing in England, does not produce nectar and that bees will sometimes collect pollen from male flowers when other sources of pollen are in low supply. Jordan (1961) detected large amounts of yew pollen in the guts of poisoned bees.

Theaceae

Camellia reticulata (netvein camellia) nectar killed thousands of larvae in several colonies in Taiwan; when mixed with other nectar, however, no harm resulted to larvae or adult bees (Fan-Tsung 1952). The trisaccharide sugar raffinose has been found in nectar in concentrations greater than can safely be fed in sugar syrup to honey bees (Barker and Lehner 1974). Nectar of plants in this family containing raffinose included *Camellia japonica*, 5 percent; *C. sasanqua*, 11 percent; and *Thea sinensis* (tea), 14 percent (Echigo et al. 1972).

Tiliaceae

In general, species of *Tilia* (basswood, linden, lime tree) are considered excellent honey plants. Pellett (1978) considers the American basswood (*T. americana*) "one of the best known sources of honey in the eastern states." The tree produces numerous yellow-white, fragrant flower clusters providing nectar and pollen. The author has observed numerous bumble bees and honey bees foraging for nectar on basswood in Tennessee with no apparent ill effect.

Anderson (1976), studying the pollination biology of four species of *Tilia* in Nebraska and Connecticut, collected 66 species of insect visitors from 29 families, of which bees and flies were the most common diurnal visitors and moths were most common at night. He found that maximum nectar production coincided with stigma receptivity and that odor at this time was strong and persistent. Bagging experiments indicated extremely limited self-fertility, with the possible exception of *T. cordata*. The protandrous flowers are male first, with the stamens maturing before the stigma is receptive. This dichogamy promotes outcrossing. The outcrossing requirement, production of abundant rewards of nectar and pollen, and providing additional visual and odor cues, indicate the plant is adapted for insect pollination as Anderson concluded.

In certain locations and situations, however, these trees are reported to be toxic to bees and other insects. Native and introduced species grown in Europe and the United States are reported to cause poisoning. These include big leaf linden (*T. platyphyllos*) little leaf linden (*T. cordata*), and white or silver linden (*T. tomentosa*)(= *T. argentea, T. peteolaris*). When these trees bloomed in summer months in Europe, especially in Switzerland and Germany, large numbers of bees visiting them for nectar and pollen were reported to have *Waldtrachkrankheit* (forest blossom disease) (Bratschi 1942, Maurizio 1943, 1945b, 1953). Affected honey bees, bumble bees, and solitary bees were found under the trees, unable to fly, appearing paralyzed, and dragging their abdomens on the ground. Examinations of the digestive systems of affected honey bees revealed *Tilia* pollen in the intestines and in the swollen honey stomachs. When bee colonies near blooming linden were examined, many dead, paralyzed, and crawling bees were found on landing boards and in front of the colonies. One report (Maurizio 1943) indicated that 50 percent of the foraging workers from a colony had been lost. Howes (1948) observed a "stupefying effect" on foragers, which when picked up, seemed to recover and fly off, but, if they remained on the ground, were subject to predation by birds. Honeydew from trees infested with aphids has also killed bees in late summer, long after the trees had bloomed (Butler 1943, Maurizio 1943). When caged bees were fed nectar extracts from the three species of *Tilia* mentioned above, symptoms of paralysis and death started one or two days after initiation of feeding and continued for up to 12 days until all bees had succumbed. Melezitose was suspected by Maurizio to be the toxic factor in honeydew poisoning, but Barker (1990) reports this substance not to be toxic to bees. The toxin in nectar is suspected to be water soluble (Louveaux and Lavie 1949), and melezitose is not soluble in water. The toxic factor is most likely from sugars discovered in the nectar (Geissler and Steche 1962), including galactose, melibiose, and mannose, which are toxic to honey bees. Several theories have been proposed concerning the mechanism by which mannose poisons bees. One theory (Sols et al. 1960) indicates that mannose and a metabolite mannose-6-

phosphate (also toxic to bees) interferes with metabolism of non-toxic sugars, such that the latter are not metabolized and the blood sugar level drops (Crane 1977, 1978a). Another theory (Van Handel 1971) indicates that honey bees can metabolize mannose, but so slowly that they become paralyzed. The mechanism of toxicity seems to be unclear and warrants more study. We know that the toxicity of certain sugars depends on their concentration in the diet (Geissler and Steche 1962, Barker and Lehner 1974). The poisoning problem with linden has been noticed only in certain years in certain locations (Maurizio 1943, 1945b), especially under drought conditions. As in many poisoning situations, it appears that a lack of other forage caused by drought induces bees to forage exclusively on linden, and this constancy causes the problem. In summary, the same plant can be a good honey source or be poisonous, depending on local weather conditions.

Conclusions

Plants can produce chemicals that are toxic to honey bees. The toxin may be present in plant sap, pollen, nectar, or honeydew. Some plants, such as the linden, that are normally considered excellent sources of nectar, have been shown to poison bees under certain specific stress conditions. These specific stress conditions seem to occur repeatedly in most cases of poisoning, and they affect the dose of poison the bees receive. When environmental conditions, especially soil moisture, reduce other sources of nectar, the bees are forced to forage from the toxic source because it is the only food available. Under normal moisture, other available sources dilute the amount of toxic substance to a level below the threshold of a toxic response.

Plants are known to produce toxins, termed secondary plant substances, that are metabolites of the plant. These metabolites may be waste products or they may function to defend against other plants, microorganisms, and herbivorous insects or mammals. Pollination by bees benefits the plant by starting the process of fertilization that permits seed production, allows outcrossing, and permits the process of selection to occur. Why would a plant produce a product toxic to honey bees, a potential pollinator that could improve the survival of the plant? We really do not know the answer to this question although several theories or hypotheses can be suggested. In some locations, such as North America, the honey bee is an introduced species. The honey bee was not present in North America at the time the plant adapted a chemical defense system in its food to prevent non-pollinating visitors from robbing the food. Possibly the honey bee is not currently adapted to forage on certain plants native to North America. In this scenario, the food source may be toxic to the honey bee, whereas other native bees may be able to detoxify the chemical and use the food safely. We know that insects maintain the genetic diversity to change and become resistant to chemicals, either man-made ones such as pesticides or chemicals produced by plants themselves. It would seem doubtful that selective pressure exerted by the herbivore would outweigh the advantage to the plant of producing seed; however, the plant may be protecting its limited reward of nectar by producing toxins that discourage nectar thieves (Rhoades and Bergdahl 1981). Another suggestion is that the entire process is random and that we are trying to define a process where none exists.

How can a beekeeper determine if losses of bees are caused by a poisonous plant? First, eliminate all other potential causes of loss, including diseases and pests. Second, find out if any plant in the flight range of the colony is documented to be poisonous to bees and if the bees are foraging on it. In addition, determine if weather conditions have reduced bloom of other plants and induced bees to forage on the toxic source. What symptoms at the hive suggest poisoning? Samples of bees, brood, and comb can be sent to a laboratory for chemical analysis. Make sure to package the sample according to recommended procedure in your county, province, state, or country. Contact the agricultural extension agent in your area for assistance. Diagnosing plant poisoning requires systematic study of every possible factor and may require cooperation of many individuals. Maintain accurate records of when the effect was noted and include a detailed description of the situation.

What can the beekeeper do if a poisonous plant is suspected to be causing damage? Reduce the amount of toxin the bees are being exposed to. If possible, move the colonies out of flight range from the toxic source of food while the plant is blooming or producing honeydew. Feeding with sugar syrup usually dilutes incoming toxin and may supply enough sugar to reduce foraging on the toxic source. Feeding with a pollen supplement may also be helpful if the source of toxin is pollen.

Poisonous plants can harm bees and should not be ignored as an additional stress to weaken a colony. More information is sorely needed to improve the current listing of plants that are toxic to bees. It will allow us to protect our bees and make our food supply more abundant and safe.

Acknowledgments

Special thanks are extended to Roy Barker, author of this chapter in the Second Edition, for generously providing many references, including translations of German and Russian literature, original letters from G. H. Vansell and others, and his personal notes, which contained many insights. Thanks are also extended to many colleagues who provided information about poisonous plants in their states.

CHAPTER TWENTY

Antibiotic Systems in Nectar, Honey, and Pollen
D. Michael Burgett

Honey and nectar
    Acidity
    Osmotic pressure
    Glucose oxidase
        Origin
        Presence in other hymenopterans
        Function
    Other antibiotic factors
Honey flora
    Yeasts
    Bacteria
Pollen
Conclusion
Author's note

CHAPTER TWENTY

Antibiotic Systems in Nectar, Honey, and Pollen

The food reserves in a honey bee colony — honey, nectar, and pollen — represent potentially luxuriant media for exploitation by microorganisms. As Rosebury (1969) pointed out, "Any naturally fabricated organic compound in the free environment is rapidly found by a living thing that can make use of it, and is accordingly destroyed for its food energy or incorporated into the substance of the living thing." The caloric stores of a honey bee colony would seem ripe for such "incorporation."

During the course of evolution, many selective pressures have been exerted on the honey bee. One result has been the development of several defensive systems that preclude the growth and development of microbial organisms in stored honey, nectar, and pollen. Because of these antibiotic systems, honey especially has been universally recognized as a pure food product (White 1975b), that is, a food that is free of microbial contaminates and therefore safe for human consumption with little or no processing. In addition to using honey as a basic sweetener, humans have for millennia recognized its antibiotic properties and have incorporated its use in medicine. Molan (1992a,b) recently reviewed the antibacterial activity of honey with special emphasis on its use as a medicament.

In the following discussion of the antimicrobial properties of colony energy stores, honey and nectar are considered as a single entity because their caloric value is equal on a dry-weight basis. Pollen is viewed as a second (proteinaceous) component of hive stores.

Honey and nectar

Three known systems are responsible for the majority of antibiotic activity found in honey and nectar. These involve acidity, osmotic pressure, and what has been called inhibine. Inhibine is the enzymatic production and accumulation of hydrogen peroxide in diluted honey or nectar (White et al. 1963). Acting either singly or in combination, these systems preclude the establishment or growth of most invading microorganisms.

Acidity

In a classic study of honey, White et al. (1962a) reported an average pH of 3.9 for the 504 samples of American honey they examined. Honey pH generally ranges from 3.2 to 4.5 (White 1975b). This degree of acidity alone will prevent the growth of many bacterial pathogens (Frazier 1967, White 1975b, Molan 1992a).

Molan (1992a) points out that honey acidity alone, though frequently not a primary antibiotic mechanism in honey, cannot be ruled out as a contributing factor to honey's freedom from microbial contamination.

The known acid constituents of honey are summarized in the table below. Of the many acids found in honey, gluconic acid predominates (Thompson et al. 1954, Stinson et al. 1960). It is produced from the action of glucose oxidase on the glucose substrate in honey (White 1975b). Ruiz-Argüeso and Rodriguez-Navarro (1973) suggested that some of the gluconic acid present in honey is produced by bacteria of the genus *Gluconobacter*. These bacteria may be present in ripening nectar but are absent in honey.

Table 20.1 The Acids of Honey

ACID	REFERENCE
acetic	7
butric	7
citric	1,5,7
formic	9
gluconic	7
glucose-6-phosphate	8
glycollic	4
lactic	7
maleic	1
malic	1,3,5,7
oxalic	6
pyroglutamic	7
pyruvic	4
succinic	5
tartaric	2
α-ketoglutaric	4
α- or ß-glycerophosphate	8
2- or 3-phosphoglyceric	8

After White 1975a.
1. Goldschmidt and Burkert 1955
2. Heiduschka and Kaufmann 1913
3. Hilger 1904
4. Maeda et al. 1962
5. Nelson and Mottern 1931
6. Philipsborn 1952
7. Stinson et al. 1960
8. Subers et al. 1966
9. Vogel 1882, cited in Farnsteiner 1908

Osmotic pressure

Honey is a supersaturated solution of carbohydrates; as such it is a hyperosmotic medium with very little "free water" available to support the growth of most microorganisms. When vegetative forms of most prokaryotes (such as bacteria) enter such a medium, the osmotic pressure differential is so great that the organisms lose water by osmosis. Sustained dehydration will eventually kill the organisms.

While the hyperosmosity of honey is universally accepted as conferring antibiotic properties, few data exist on the osmotic pressure of honey or nectar. Work by Sidie and Burgett (personal observation 1975) has shown the osmotic pressure of fully ripened honey to be greater than 2,000 milliosmols/kg.

Glucose oxidase

Sackett (1919) investigated the possibility that honey could serve as a carrier for a variety of disease organisms, including the typhoid pathogen. Neither ripened nor diluted honey would support the growth of any of the organisms he tested. Sackett suggested that the osmotic pressure of fully ripened honey would prevent the growth of bacteria. He expressed surprise, however, at the lack of microbial growth in diluted honey and was unable to explain the phenomenon. He concluded that "the possibility of honey acting as a carrier of typhoid fever, dysentery, and other diarrheal affections is very slight" (Sackett 1919). This pioneer work was the first of many investigations that have shown that honey acts against the growth of a wide range of bacteria and other microorganisms (Molan 1992a). Nester et al. (1978) lists additional human pathogens that have been screened for sensitivity to the antibiotic nature of honey.

Dold et al. (1937) suggested the term *"inhibine"* for the observed antibacterial activity found in diluted honey. Their work on honey inhibine was confirmed by Prica (1937), Plachy (1944), and Warnecke and Duisberg (1958). Europeans, especially, have attached great significance to inhibine in honey. The absence of inhibine has been used as an indication that honey has been overheated ("pasteurized"), a process abhorred by some consumers, especially those interested in eating only natural foods.

The demonstration of an antibiotic factor in dilute honey led to many queries into its specific makeup. Most investigators concurred with Dold and his co-workers that inhibine was heat- and light-labile (Duisberg and Warnecke 1959, Sipos et al. 1960, Stomfay-Stitz and Kiminos 1960, Khristov and Mladenov 1961, Linder 1962, White and Subers 1964a,b).

Dold and Witzenhausen (1955) described an assay to standardize inhibine assessment, based on the microbiological methodology of Dold et al. (1937). The assay involves the inoculation of the bacterium *Staphylococcus aureus* Rosenbach onto agar plates containing honey at known concentrations ranging from zero to 25 percent. After incubation at 37°C (98.6°F) for 24 hours, the plates are inspected for bacterial growth, which is rapid in the absence of honey. If complete inhibition of bacterial growth is observed when the honey concentration is only 5 percent, the honey is assigned an inhibine value of 5.0. If it takes a 10 percent honey concentration to prevent growth of the bacteria an inhibine value of 4.0 is given. If bacterial growth is not inhibited even at 25 percent honey (the highest concentration), the inhibine value is zero. Intermediate values are assigned accordingly (see Adcock 1962 for more details).

White et al. (1962b, 1963) were the first to state that the liberation and accumulation of hydrogen peroxide (H_2O_2) in honey by way of the enzyme glucose oxidase accounts for all the known properties of honey inhibine. Previous work into the enzymatic production of gluconic acid in honey (Stinson et al. 1960) had led to the discovery of the honey glucose oxidase system.

Hydrogen peroxide has been recognized as an effective antibiotic for many years. The notatin factor was recognized as a major antibacterial component of some of the early penicillin drugs. Coulthard et al. (1945) showed notatin to result from the production of H_2O_2 by way of a glucose oxidase system.

Gauhe (1941), by demonstrating the presence of glucose oxidase in the hypopharyngeal glands of the honey bee, was the first to discover a glucose oxidase system in the animal kingdom. She did not, however, connect its presence there with the inhibine phenomenon in honey. Glucose oxidase had previously been reported only in plants or plantlike material – mold (Muller 1928), true fungi (Coulthard et al. 1945), bacteria (Dowling and Levine 1956), red algae (Bean and Hassid 1956), and citrus fruits (Bean et al. 1961).

The substrate specificity of the various glucose oxidases has been well-studied. Dixon and Webb (1958), Schepartz and Subers (1964), and Schepartz (1965a, 1966a) found that glucose oxidases from several biological sources displayed a high substrate specificity for b-D-glucose. Dixon and Webb (1958) commented that any alteration of the b-D-glucose molecule would reduce the activity of the enzyme enormously.

The following characteristics have been reported for the honey glucose oxidase system: optimal pH, 6.0 to 7.0 (Schepartz and Subers 1964); optimum temperature for peroxide evolution, 40°C (104°F) (Schepartz and Subers 1964); complete inactivation by temperatures over 60°C (140°F) (Schepartz and Subers 1964); optimal substrate concentration, 2.7 molar glucose (Schepartz 1965b); and molecular weight, two fractions corresponding to approximately 120,000 and greater than 200,000 (White and Kushnir 1967). Schepartz and Subers (1964) indicated a molecular weight of 150,000 to 152,000 for a glucose oxidase derived from *Penicillium*.

Work by White and Subers (1964a) showed that inhibine — measured as antibacterial activity and as peroxide accumulation — is heat-sensitive. In the range 55° to 70°C (131° - 158°F), the half-life of the peroxide accumulation is a logarithmic function of temperature. In assays of 29 honey samples from a variety of floral sources, White and Subers found wide variation in sensitivity to heat. They felt that such variation would negate the value of inhibine as a useful parameter of heat exposure.

White and Subers (1964b) and Dustmann (1972) studied the effect of visible radiation on peroxide production. Both papers reported variation in inhibine photodeterioration, again corresponding to the floral source of the honey. Radiation, particularly in the 425 to 525 nm range, was most effective in photo-oxidizing the enzyme system. Sunlight and radiation from fluorescent lamps were more effective than light from incandescent lamps in reducing peroxide levels. White and Subers (1964b) recommended that an inhibine value not be used as a measure of honey quality because of the wide variation in susceptibility of honeys of different floral sources to light and heat. White (1975b) reiterated the uselessness of levels of glucose oxidase activity and other honey enzymes as valid measures of honey quality.

The presence of the enzyme catalase in honey was demonstrated by Schepartz (1966b) and Schepartz and Subers (1966). They suggested that catalase brings about the enzymatic degradation of peroxide, and in this manner may serve as a throttle mechanism for the peroxide equilibrium in a dilute honey.

White and Subers (1963) reported on a standardized assay for peroxide in dilute honey. They were able to show a linear relationship between the inhibine value (a microbiological assay) and the logarithm of peroxide accumulation (mg H_2O_2 per gram of honey per hour).

In summary, the peroxide level attained in a diluted honey or nectar depends on the previous exposure to heat and light; the rate of peroxide evolution, a function of enzyme activity; and the rate of destruction of evolved peroxide by honey components.

Origin

There has been no definitive statement on the origin of glucose oxidase in honey. All experimental evidence, however, indicates that the hypopharyngeal glands of worker honey bees are the source. Gauhe (1941) found a glucose oxidase present in these glands. Brangi and Pavan (1954), not knowing inhibine to be of enzymatic origin, suggested that the antibacterial properties of honey originated in the floral source. Maurizio (1975) stated that the hypopharyngeal glands are the source of the enzymes diastase, invertase, and glucose oxidase.

The best-studied function of the hypopharyngeal glands is the production of royal jelly (Ribbands 1953). These glands are maximally developed in young nurse bees that are actively feeding larvae. After nurse bees abandon the temporally determined task of larval feeding, their hypopharyngeal glands atrophy. Ribbands (1953) and Cruz Landim (1969) suggested that the hypopharyngeal glands continue their enzyme secretory function even after the glands cease production of royal jelly. Nye et al. (1973) concluded that the gluconic acid found in royal jelly is produced by a glucose oxidase secreted by the hypopharyngeal glands of nurse bees.

Presence in other hymenopterans

Many species of eusocial insects collect saccharin exudations (nectar and honeydew) and elaborate these into honey (Burgett and Young 1974, Crane 1975b, 1990). Gonnet et al. (1964) were the first to report the presence of inhibine in honey other than that produced by *Apis mellifera*. They used the microbioassay of Dold and Witzenhausen (1955) to assign inhibine values to honey from five species of stingless bees (tribe Meliponini). They found high levels of antibiotic activity in these honeys, even after some of the honeys were heavily diluted. Their paper made no mention of inhibine activity being a function of peroxide accumulation resulting from a glucose oxidase system.

Burgett (1974) reported that the inhibine system is present in the honeys of many honey-storing, eusocial insects. The species examined included four species of honey bees (*Apis*), two species of stingless bees (*Trigona*), two species of bumble bees (*Bombus*), a honey wasp (*Protonectarina*), and a honey ant (*Myrmecocystus*). These species represent the three major superfamilies (Apoidea, Vespoidea, and Formicoidea) in the order Hymenoptera. Burgett hypothesized that glucose oxidase is produced by most, if not all, honey-storing eusocial species.

Function

Nearly all reports of glucose oxidase activity have come from investigations involving the dilution of honey. Little or no peroxide is present in fully ripened *Apis mellifera* honey (less than 18.6 percent moisture) (White et al. 1963).

Nectar, when collected and brought into the hive by forager bees, is a dilute solution of carbohydrates. At this time, it is most susceptible to microbial contamination. In converting the dilute nectar to honey, the bees, and presumably the other hymenopterans known to possess an inhibine system, add glucose oxidase to the ripening nectar. I have found high levels of peroxide in ripening nectar removed from the combs of *A. mellifera* (personal observation 1973). The peroxide accumulated during the ripening process protects the nectar until enough moisture is removed to promote the antibiotic function of increased osmotic pressure.

The speed at which nectar is accumulated in the colony can affect the amount of glucose oxidase added by the bees. During periods of heavy nectar intake, the amount of time spent processing nectar will be reduced, and therefore, the amount of enzyme added is probably also reduced.

Humans have used honey for centuries as a medicine or a medicinal adjuvant. Only during this century have we begun to understand that the ancients were justified in prescribing honey as a medicinal aid (Molan 1992a,b). In speaking of the ancient Egyptian physicians' use of honey, Majno (1975:118) wrote that they "happened to choose an ingredient that was practically harmless to tissues, aseptic, antiseptic, and antibiotic."

Other antibiotic factors

Several reports state the existence of antibiotic activities in various honeys that are independent of glucose oxidase activity, osmolarity, and acidity. These accounts are well-reviewed by Molan (1992a). He comments that "the finding of antibacterial activity in honey that is stable to heating has been an indication . . . of the existence of non-peroxide antibacterial factors." It has been shown that honeys from specific plant species can be associated with non-peroxide germicidal effects. The New Zealand honey mauka (*Leptospermum scoparium*) is reported to possess such non-peroxide mechanisms (Molan and Russell 1988), as is honey from viper's bugloss (*Echium vulgare*) (Allen et al. 1991).

Sources of such antibiotic activity are claimed for flavonoids (Bogdanov 1984, 1989) and the bacteriolytic enzyme lysozyme (Mohrig and Messner 1968), although Molan (1992a) casts serious doubts on the efficacy of such materials as antibacterial factors in honey. Antibacterial phytochemicals have also been considered as a potential source for some of the reported non-peroxide bacteriocidal activities and are reviewed by Molan (1992a). He points out that "There is clearly much variation in the findings of non-peroxide antibacterial factors in honey, and in the quantitative importance of these factors in the antibacterial activity of honey."

Honey flora

It is the rare defense system that is unassailable. Several microbial taxa can tolerate the high osmotic pressures found in honey. Of these, however, only the yeasts are able to carry on all life functions.

Yeasts

White (1975b) summarized the information on most of the described yeasts found in honey and nectar. He did not use the most current yeast taxonomy, however, and he failed to consider the work of Lodder (1970), which incorporates the nomenclatural scheme most accepted by yeast taxonomists. This table details the revised nomenclature of the yeasts presently known to be associated with honey. The older nomenclature is also presented.

Osmophilic yeasts have been found occurring naturally in nectar (Lochhead and Heron 1929), on the bodies of bees (Klöcker 1907), in apiary soil (Lochhead and Farrell 1930), and in the honey house and honey-extracting equip-

Table 20.2 Revised nomenclature of yeasts identified in honey

Species[1]	Synonym[2]	Reference
Candida lodderiae	*Torulopsis lodderiae*	Kumbhojkar 1981
Candida onishii	*Torulopsis onishii*	Kumbhojkar 1981
Hansenula anomala		
Nematospora ashbya gossyppi		Aoyagi and Oryu 1968
Saccharomyces bailii	*Saccharomyces acidifaciens*	Feiler 1969
Saccharomyces bailii variety *osmophilus*		Tysset & de Rautlin de la Roy 1974
Saccharomyces bisporus variety *mellis*	*Zygosaccharomyces mellis*	Fabian and Quinet 1928
	Zygosaccharomyces richteri	Lochhead & Heron 1929
	Zygosaccharomyces perspicillatus	Sacchetti 1932
	Zygosaccharomyces nectarophilus	Lochhead 1942
	Saccharomyces mellis	Rodriguez-Navarro 1968
Saccharomyces heterogenicus		Feiler 1969
Saccharomyces lochheadii		Kumbhojkar 1978
Saccharomyces rosei	*Saccharomyces torulosus*	Aoyagi and Oryu 1968
Saccharomyces rouxii	*Zygosarccharomyces mellisacidi*	Richter 1912
	Zygosaccharomyces priorianus	Fabian and Quinet 1928
	Zygosaccharomyces barkeri	Lochhead and Heron 1929
	Zygosaccharomyces nussbaumeri	Lochhead and Heron 1929
	Zygosaccharomyces japonicus	Lochhead and Heron 1929
	Zygosaccharomyces ravennatis	Sacchetti 1932
	Zygosaccharomyces rugosus	Lochhead 1942
	Saccharomyces rouxii polymorphus	Feiler 1969
Schizosaccharomyces octosporus		Lochhead and Farrell 1931
Schizosaccharomyces slooffiae		Kumbhojkar 1972
Schwanniomyces occidentalis		Aoyagi and Oryu 1968
Torula melis[3]		Fabian and Quinet 1928

After Lodder 1970.
[1]Species are arranged in alphabetical order.
[2]Synonyms of a species are arranged chronologically.
[3]Validity of species in question.

ment (Lochhead and Heron 1929). Their ubiquitous nature makes it very difficult, if not impossible, to remove them from honey mechanically.

Bacteria

It is a commonly held misconception that honey does not possess a bacterial flora. Honey, as a hyperosmotic medium, does not support the vegetative life stage of any bacterial species. In honey, however, some bacterial species can maintain themselves in the spore stage. Experienced beekeepers have long known that honey can serve as a vector for the causal organism of American foulbrood disease (Sturtevant 1932). More recently, it has been shown that honey can serve as a carrier of spores of the bacterial organism responsible for human botulism (Arnon 1980; see Author's note, below).

The causal agent for American foulbrood disease is *Bacillus larvae* White (see Chapter 3); that of botulism is *Clostridium botulinum*. Both species are in the bacterial family Bacillaceae, which includes the endospore-forming rods and cocci (Bergey 1994). Endospores (spores) serve as an intermediate stage in the normal development of several bacterial genera. Spores are characteristically resistant to the effects of heat, desiccation, chemical disinfection, and radiation (Wistreich and Lechtman 1980). Because bacterial spores require very little free water, they can survive, but not germinate, in honey. Their removal from or inactivation in honey would be mechanically and/or economically impractical with current honey-processing methods.

Pollen

Pollen is the major source of protein and fat in the honey bee diet. The chemical composition of pollen varies greatly, depending on the floral species. Echlin (1974) provided the following ranges for the major chemical constituents of pollen: 3 to 15 percent water, 10 to 35 percent protein, 3 to 8 percent starch, and 5 to 20 percent fat. Stanley and Linskins (1974) gave ranges of the following order: up to 50 percent water, 6 to 36 percent protein, 1 to 35 percent starch, 1 to 20 percent fat. Carbohydrates can account for up to 50 percent of the dry mass of pollen (Stanley and Linskins 1974). Honey bees collect and store pollen during periods of pollen abundance. Morse (1972) stated that pollen reserves of a colony can range from one to 15 pounds (0.4-7 kg); it is important to bees that this highly nutritious food reserve be protected.

There are several mechanisms for in-hive pollen preservation. Because bees moisten pollen with honey stomach contents during the collection process, it has been assumed that the hydrogen peroxide present in the nectar used to moisten the pollen functions as an antibacterial system (Morse 1972). This may be a valid assumption, although no experimental evidence exists to support it.

A second protective system involves bees "topping off" pollen storage cells with a layer of honey (Morse 1972). Cells in which pollen is stored are usually filled to about three-quarters of their capacity. When pollen is not being consumed rapidly, bees will add a layer of honey on top of the pollen. The protection here is twofold: First, a physical barrier has been erected that will exclude microbial contamination, and second, the honey barrier itself possesses the antibacterial systems previously discussed.

Stored pollen is neither chemically nor nutritionally equal to freshly collected pollen. Stored pollen undergoes chemical changes that make it less amenable to bacterial despoilers. Pain and Maugenet (1966) studied the microbial changes that take place in bee-stored pollen. They reported the presence of lactic acid-producing bacteria of the genus *Lactobacillus*. Lactic acid promotes the development of a natural ensilage. The same authors also found *Pseudomonas* bacteria and yeast of the genus *Saccharomyces*. *Pseudomonas* is thought to contribute to the anaerobiosis required by *Lactobacillus*. The yeast may aid in the mechanical and biochemical breakdown of pollen.

Conclusion

Honey bees depend on plant nectars for their carbohydrates and on pollen for their protein and fat. The antimicrobial systems used by honey bees to protect their calorically rich food stores are adaptations that have contributed significantly to their evolutionary success. Other honey-storing eusocial insects have independently evolved similar systems of food protection.

Author's note

In 1978, honey was implicated as a spore vector for *Clostridium botulinum*, the causal agent of botulism (Anonymous 1978). This article was primarily concerned with reporting the identification of a newly recognized human syndrome named "infant botulism." Before 1978, intoxication by *C. botulinum* toxins was believed to be a phenomenon of adult humans only. *C. botulinum* spores are nearly ubiquitous and are known to be widespread in soil and as surface contaminants on a multitude of agricultural products. Unfortunately, honey is not immune to inadvertent contamination by *C. botulinum* spores. A 1979 survey found 10 to 15 percent of the honey samples examined to be contaminated by botulinal spores (Brown 1979).

In a review article concerning the relationship between infant botulism and honey, Lawrence (1986:485) pointed out the present consensus in all segments of the beekeeping industry in the United States: "Honey has by no means been proven to cause infant botulism, but it has been shown conclusively to be a risk factor associated with the condition. This is a fine distinction but an important one to make. Since honey is not necessary for the nutrition of infants, and since it is an avoidable source of botulinal spores, it is a risk that need not be taken."

CHAPTER TWENTY-ONE

———— ⨋ ————

The Genetic Basis of Disease Resistance
Robert E. Page Jr. and Ernesto Guzmán-Novoa

CHAPTER TWENTY-ONE
The Genetic Basis of Disease Resistance

Introduction

A casual examination of a honey bee colony will reveal that individual workers vary in size, in color, and in the tasks that they perform. If you examine several colonies, you will find that they differ from each other in the characteristics of their workers; their organization; and the quantities of brood, honey, and pollen. Much of this variability is genetic in origin (Page and Robinson 1991, Page et al. 1993). You probably will also notice that some colonies have more evidence of disease, such as sacbrood, chalkbrood, American foulbrood, and infestation with varroa. Microscopic examination will reveal even more variability in colony loads of parasites and pathogens. Much of this variability is probably the result of accidental exposure of colonies to parasites and pathogens or the result of differences in other environmental components that weaken or strengthen a colony's defenses against these plagues. Some observed variability may, however, also be a consequence of differences in the genetic composition of colonies. The genetic differences between individual workers and between colonies constitute the raw material for natural or artificial selective breeding of disease-resistant strains of honey bees.

Beekeepers are confident that genetic differences are at least partially responsible for observed differences in disease loads of colonies. This confidence is demonstrated by the common practice of requeening diseased hives. The expectation is that the new family derived from the new queen will have fewer disease problems than the one currently inhabiting the hive. This practice of culling out diseased stock and replacing it by trial and error is unlikely to result in the development of a uniformly resistant strain. Controlled breeding must be practiced to produce uniformly disease-resistant stock.

Principles of breeding

Expectations

Expectations for breeding programs should be realistic. The elimination of disease problems through selective breeding is not a realistic objective. Instead, a reduction in incidence or a reduction in the need to treat hives chemically may be attainable; the complete elimination of the need to treat chemically is not. Improvements in apicultural practices will often have greater effects than selective breeding for economically important traits. For example, good hygienic practices by beekeepers will reduce the spread of American foulbrood (AFB) and probably have a greater effect on the incidence of disease than will selection for resistance.

Selection is an ongoing process that is necessary to produce and maintain resistant stocks. It is unrealistic to expect to maintain stocks that have selected characteristics without practicing continuous selection. Selective progress will begin to deteriorate as soon as selection on the population is relaxed.

Assay

Before selecting for any trait, it is necessary to have a reliable assay. An assay is simply a method to categorize or quantify the variability you observe for the characteristic you wish to improve. The most important consideration for a good assay is to control the environment in which the assay is performed. That usually requires having the best apicultural practices possible before you begin selecting. Assays should have a high degree of repeatability under a given set of environmental conditions. Examples of assays are: (1) the time it takes a colony to uncap and remove a specific number of cells of diseased or freeze-killed larvae and pupae; this test is used for evaluating resistance to AFB and chalkbrood disease (Rothenbuhler 1964a, b, Taber 1982, Gilliam et al. 1983, Spivak and Gilliam 1993); and (2) the number of adult tracheal mites found in the tracheae of young workers five days after they are introduced into colonies infested with tracheal mites (Page and Gary 1990).

Assays can be conducted in field hives, nucleus colonies, or in the laboratory. Environmental conditions are more easily standardized in laboratory tests, but results from selection using laboratory tests or nucleus colonies should be verified under commercial field conditions. You cannot directly select for resistance to diseases and parasites you do not have, nor can you do so while chemically treating for the disease. In the latter case, the disease or parasite can no longer act as a selective agent. You may, however, select directly for known mechanisms such as hygienic behavior. The results from the selection must be tested with diseased or parasitized colonies. Mechanisms selected for resistance under constrained conditions may have little or no effect on full-sized hives in commercial apiaries.

Fundamentals

Selective breeding requires four simple steps: (1) Colonies or individual workers, drones, or queens are evaluated in some base population from which the breeding stock is selected; (2) queens of colonies having the desired characteristics are selected as breeders for the next generation; (3) matings are controlled between queens and drones raised from the selected breeders; (4) progeny of the new queens are evaluated, either as individuals or as a colony. If a sufficient proportion of the observed trait variation in the parental population was caused by genetic differences among them, the progeny will resemble their parents and selective improvement will be obtained. Hence, selective breeding is a process of measurement and trial and error.

Standard stocks

Numerous unsubstantiated claims of disease-resistant stocks are found in the beekeeping trade journals. But it is necessary to evaluate and document the performance of stock that has been produced through selection. To do so, a benchmark is needed against which selective progress can be compared so as to separate the effects of the selection program from those of a changing evaluation environment. The benchmark for comparison should be another population of bees that has not been subjected to the same selection.

Three methods are commonly used: (1) comparison with a "standard" stock that remains genetically unchanged throughout the time selection is occurring. This is very difficult with honey bees because it requires the maintenance of inbred lines that are genetically constant through time, or the long-term survival of a set of queens whose progeny can be used for comparison with subsequent generations of the selected population; (2) two-way selection. This requires maintaining two selected populations, one selected for higher values of the selected trait, the other for lower values. Its advantage is that it allows quick determination of the genetic basis of observed colony differences within a breeding population. Its disadvantages are that it requires maintaining commercially worthless low-trait colonies as well as additional evaluation of the high-trait population against other unselected stocks. Without the additional evaluation, it is impossible to determine if only the low-trait colonies changed with selection, only the high, or if both changed; (3) comparison with a "random" stock that represents the unselected gene pool from which the original breeder queens were selected to initiate the breeding program. The advantage of this method is that it is not necessary to maintain one extra line (standard or susceptible); the main disadvantage is that variation within the randomly selected stock may yield different results in different tests, confounding the evaluation of the selected strain.

Sources of bees

Before beginning a breeding program, resistant stock must be located. There are four potential sources: local commercial bees, commercial bees from disease-afflicted geographical locations, different races of bees, and feral honey bees.

Local commercial bees

Locally available commercial stocks probably will have genetic variability for resistance to a given disease if it exists anywhere. Strains of bees resistant to AFB disease (Rothenbuhler 1975a), tracheal mites (Page and Gary 1990), bee paralysis (Kulincevic and Rothenbuhler 1975), and varroa (Kulincevic et al. 1992) were all derived from existing commercial stocks.

Commercial bees from afflicted areas

Commercial bees from areas that have an historical association with the specific disease may constitute the foundation population for breeding. These bees may already have been subjected to artificial or natural selection and already demonstrate some genetically based resistance.

Races of bees

Specific races of bees may be imported and tested for resistance. Some races may have a longer association with the particular pathogen or parasite and already be naturally resistant. Such is the claim for the Carniolan "Yugo" bees imported to the United States because of their resistance to varroa and tracheal mites (Rinderer et al. 1993). Or, as with Africanized honey bees in the New World, imported bees may be "preadapted" and naturally resistant (Steiner et al. 1984, Message 1985, Engels et al. 1986, Mendoza et al. 1987, Moretto 1988, and see below). Alternatively, imported bees may be no more resistant than local commercial bees, but when they are crossed with local commercial stocks, their progeny may demonstrate hybrid vigor, resulting in reduced disease symptoms or incidence. This apparently occurred for tracheal mites when British bees were imported to the United States and crossed with local commercial bees (Gary et al. 1990).

Feral bees

Feral bees may be used as a reservoir for genetic diversity and may potentially have naturally selected disease resistance. Most studies of feral populations have found them to be relatively free of disease (Bailey 1958, Taber 1979, Gilliam and Taber 1991, Ratnieks et al. 1991). Natural selection may also produce disease-resistant populations of feral honey bees. Although no cases have been documented, circumstantial evidence exists. For instance, *Apis cerana* has had a long-term exposure to varroa in Asia and is relatively resistant to severe damage by this parasite. American foulbrood disease devastated the beekeeping industry of Hawaii in the 1930s, resulting in its near total collapse (Keck 1949, Eckert 1950). By the late 1940s, however, AFB was a less severe problem, and larvae from Hawaii demonstrated resistance to infection from AFB spores relative to tested susceptible strains (Thompson and Rothenbuhler 1957). Little is known about the genetic composition of feral bees, however. Nielsen et al. (1994) found that the feral bees of California demonstrated local genetic differentiation for the enzyme malate dehydrogenase, suggesting that they have undergone local adaptation and are thus probably genetically differentiated from commercial honey bees. Their study supported an earlier study by Daly et al. (1991), who found that California feral bees differed in size throughout their geographical range and differed from commercial bees in the sizes of various body parts.

Kraus and Page (1995) estimated that more than 85 percent of all feral colonies in some areas of California were killed by varroa between 1990 and 1994. They also estimated that the introduction of varroa into California reduced the life expectancy of a feral colony from about 3.5 years to less than one year. Therefore, it is possible that strong natural selection in feral colonies, along with their at least partial genetic isolation from commercial bees, has resulted in some degree of disease resistance in the population. Unfortunately, there are no studies of naturally nesting feral bees to determine levels of resistance to diseases.

Mechanisms of resistance

Genetic variability has been demonstrated for three general classes of disease resistance in honey bees: physiological, behavioral, and anatomical. Mechanisms of more than one type may simultaneously operate against the same pathogen or parasite. In addition, the same mechanism may operate against more than one pathogen, parasite, or both.

Physiological mechanisms

Physiological resistance is what first comes to mind when we think of resistance to diseases. Somehow, the larva or adult bee produces a physiological product that impedes the growth, development, or reproduction of the pathogen or parasite. Genetic variability for this kind of resistance has been demonstrated for AFB disease in larvae (Rothenbuhler and Thompson 1956, Bambrick and Rothenbuhler 1964, Lewis and Rothenbuhler 1961, Bambrick 1964, Hoage and Rothenbuhler 1966) and is implicated in the development of strains resistant to the viral agent of bee paralysis (Kulincevic and Rothenbuhler 1975, Rinderer and Rothenbuhler 1975). Rate of larval development is another possible physiological mechanism of larval resistance to AFB. Sutter et al. (1968) showed that larvae of a resistant strain were larger than those of the susceptible strains during early larval development, perhaps making them less susceptible to infection by Bacillus larvae spores.

Honey bee larvae show a second kind of physiological resistance to varroa infestation. Moritz and co-workers have demonstrated genetic variability for the amount of time that drone and worker brood are capped (Moritz 1985, Moritz and Mautz 1990, Moritz and Jordan 1992). Colonies with workers that develop more quickly into adults (have shorter post-capping stages) are relatively resistant to varroa compared with colonies with longer post-capping stages. This resistance is a consequence of a restricted development time for mites to mature to adults in cells with shorter post-capping periods (Moritz and Hanel 1984; see below).

Yet another kind of physiological resistance to American foulbrood disease is manifested through the feeding activities of nurse bees. Thompson and Rothenbuhler (1957) showed that nurse bees derived from stocks imported from Hawaii, and believed to be resistant to AFB, conferred resistance to the larvae they fed when colonies were inoculated with *B. larvae* spores in sugar syrup. Rose and Briggs (1969) cultured *B. larvae* spores in brood food derived from the AFB-resistant Brown and the susceptible Van Scoy lines and found that brood food derived from nurse bees of the Brown line inhibited the germination of *B. larvae* spores, relative to brood food from the susceptible Van Scoy line.

Behavioral mechanisms

The best-known example of disease resistance in honey bees is the hygienic behavior that was originally described by O. W. Park (1937). Apparently, there are two independent behavioral activities, uncapping cells and removing diseased larvae from uncapped cells (Woodrow and Holst 1942). These two activities were shown by Rothenbuhler (1964a,b) to be under the control of two independent genetic mechanisms. He hypothesized that one gene with two alleles controlled each activity; Moritz (1988) however, reanalyzed Rothenbuhler's data and proposed that at least three genes, and probably more, are involved in hygienic behavior. Hygienic behavior is also an effective behavioral mechanism against chalkbrood disease and varroa mite infestations (Gilliam et al. 1983, Boecking and Drescher 1991, Büchler 1992a,b, Spivak and Gilliam 1993).

Recently, there has been considerable interest in the effects of grooming behavior of bees on colony levels of varroa infestation. This interest is a consequence of the finding of Peng et al. (1987a,b) that *A. cerana* workers groom mites from their bodies and kill the mites by puncturing them with their mandibles. We discuss this mechanism in more detail below.

Anatomical mechanisms

Sturtevant and Revell (1953) reported that strains of honey bees differed in their abilities to reduce the numbers of spores of *B. larvae* in stored "honey" produced from contaminated sugar syrup. They suggested that the bees filtered the spores through the proventricular valve, the "honey stopper". Plurad and Hartman (1965) confirmed their results using the AFB-resistant Brown and suscep-tible Van Scoy lines of Rothenbuhler. Workers of the resistant Brown line were more efficient than Van Scoy line workers in removing spores from sugar syrup contain-ing suspensions of *Bacillus thuringiensis*, *Bacillus cereus*, and *B. larvae*. These studies demonstrate genetic variability for removal of suspended spores but do not confirm the mechanism. Plurad and Hartman (1965) suggest, as did Sturtevant and Revell (1953), that it could be the honey stopper, but they also propose that it could be the result of other bacteriostatic or bactericidal mechanisms.

Selection programs

There are remarkably few documented examples of controlled breeding for resistance to honey bee diseases and parasites. This is unexpected when consider-ing the perceived economic importance of honey bees and the dramatic results obtained in the few cases where selective breeding was practiced. In this section, we outline the breeding methods and results of some of the better-documented selection programs. In the following section, we present in detail what is known about potential mechanisms of resistance to *Varroa jacobsoni*, the most serious problem for beekeeping worldwide.

Resistance to American foulbrood: Park et al.

The first successful breeding program for resistance to AFB was imple-mented in September 1934 by O. W. Park, F. B. Paddock, and F. C. Pellett (Park 1937, see also Rothenbuhler 1958a). Over the next 15 years, they successfully selected a resistant stock and documented their progress (see Figure 1).

Source of bees

Park et al. initiated their breeding program by requesting beekeepers to supply them with colonies that were presumably resistant to AFB. Altogether Park et al. received 45 colonies from throughout the United States, representing Carniolan, Italian, and Caucasian races. Colonies were consolidated in a single apiary near Atlantic, Iowa, and were tested for resistance.

Assay

Each colony received one rectangle of brood comb, approximately 5 x 6.2 cm (ca. 200 cells) containing about 75 to 100 AFB scales. Comb sections were inserted into brood combs in the center of the brood nest. Colonies were of equiva-lent strength, occupying at least eight modified Dadant brood combs. Colonies were examined frequently for evidence of AFB disease.

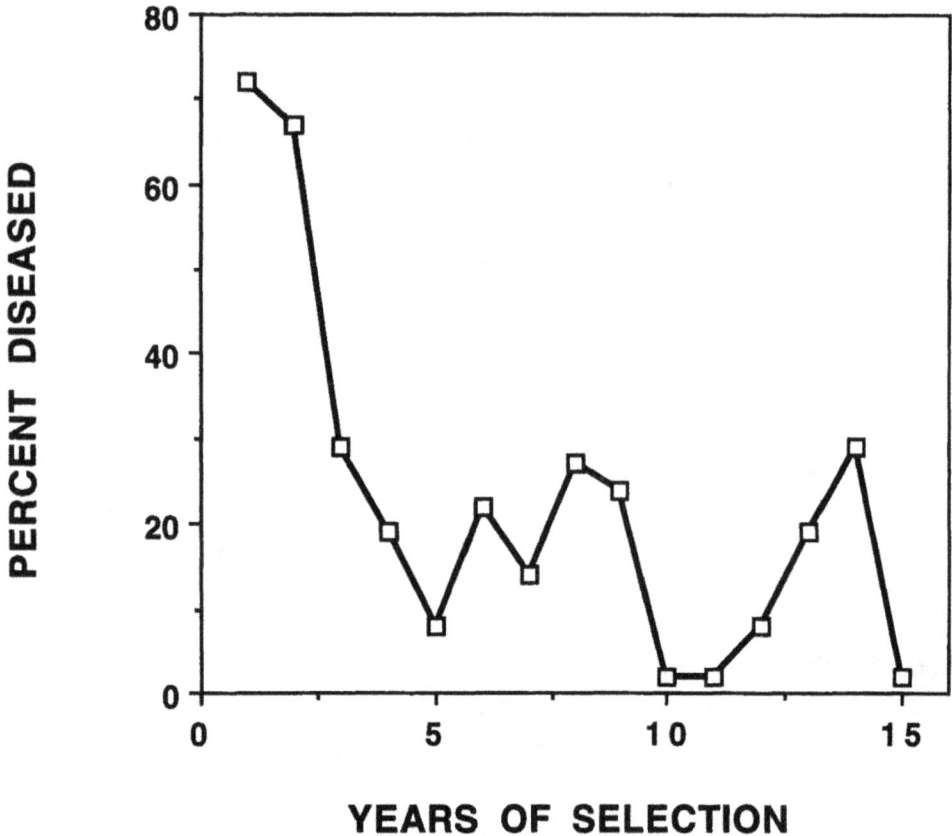

Figure 1. The percentage of AFB-inoculated colonies that became diseased during 14 years of selective breeding. Data are from the selection program of O.W. Park, covering the years 1935-1949, as presented by Rothenbuhler (1958).

Mating system

Isolated mating yards were established in a large citrus grove near Weslaco, Texas, and in the sandhills southeast of San Antonio, Texas. Queens and drones of resistant stocks mated naturally from these apiaries, but it is unknown how much outcrossing actually occurred because of incomplete isolation.

Mechanisms of resistance

Park (1937) demonstrated that one mechanism of resistance was behavioral. He found that some colonies quickly tore down and removed comb inserts containing scales of AFB. Colonies that removed the diseased combs had a lower incidence of disease. He also demonstrated that they were responding to the presence of the disease, not just the foreign comb. Woodrow (1942) concluded that there was no evidence for physiological larval resistance to AFB. Woodrow and Holst (1942) suggested that the mechanism of resistance was the ability of the resistant stocks to recognize and remove diseased larvae before they became infectious. Park (1937), however, suspected that physiological mechanisms besides housecleaning behavior were involved.

Resistance to American foulbrood: Rothenbuhler et al.

Rothenbuhler (1958), after reviewing the work of Park (see Thompson and Rothenbuhler 1957), also concluded that hygienic behavior alone was not sufficient to explain all resistance to AFB. As a consequence, he initiated a breeding program in 1954 specifically to study the potential genetic mechanisms of resistance.

Source of bees

In 1954, Rothenbuhler obtained 10 queens from Edward G. Brown that were believed to be resistant to AFB. Brown had operated a wax rendering plant for many years. During this time, he had been selecting for disease resistance in his apiary by exposing diseased combs, brought to him for rendering, to robbing bees. Queens from this apiary constituted the foundation for the resistant "Brown" line.

Rothenbuhler also started a susceptible line. He believed that the best way to study resistance was to compare putative resistant stocks to those that are susceptible. Therefore, in 1950, he obtained a single, mated queen from Homer Van Scoy of New York state from which he initiated a disease-susceptible line, the "Van Scoy" line.

Assay

The assay used by Rothenbuhler to breed for resistance and susceptibility was the same as that of Park (1937), discussed above.

Mating system

Brown raised new queens from his contaminated apiary and allowed them to mate naturally, resulting in a kind of "natural" selection for disease resistance. Rothenbuhler used instrumental insemination to maintain inbred lines of his resistant and susceptible stocks.

Mechanisms of resistance

Using these resistant and susceptible strains, Rothenbuhler and his associates demonstrated one behavioral, three physiological, and a single, putative anatomical mechanism (see discussion of mechanisms above). The results of studies of AFB resistance demonstrate the diversity of resistance mechanisms that can occur simultaneously when colony selection is used. Colony selection focuses on the occurrence of the disease, not the specific mechanisms responsible for resistance. Single-factor resistance may occur if only a single mechanism is selected, such as hygienic behavior. Multi-factor resistance is probably more effective, however.

Resistance to tracheal mites

Many circumstantial claims have been made of the development of genetically determined resistance of honey bees to infestation with *Acarapis woodi*. Most of these claims are anecdotal. Many believe that resistance to tracheal mites developed in Great Britain as a consequence of their devastation of the bee industry between 1900 and 1920. Evidence for this resistance is the decline in the incidence of tracheal mite-infested bees in colonies in England and Wales after 1927 (Figure 2). Bailey and Ball (1991), however, argue that the incidence of tracheal mite infestation declined with the numbers of bee colonies over this time. That is the opposite of what is expected with an increase in resistance; beekeeping should have recovered, not continued to decline. They attribute the decline in tracheal mite incidence to epidemiological factors associated with lower densities of honey bee colonies; having fewer colonies reduces the rate of spread of tracheal mites between colonies.

R. A. Morse (1989) imported 26 queens to the United States that had been collected from diverse sources throughout England, Wales, and Scotland. They were imported specifically to compare the relative resistance of British bees to those from North American stocks. Morse's study failed to demonstrate superior resistance of the British bees but did demonstrate that the British X North American hybrids were more resistant than the North American stocks, suggesting hybrid vigor for resistance.

Rinderer et al. (1993) imported Carniolan bees (*A. m. carnica*) from Yugoslavia that were presumed to be somewhat resistant to varroa. They tested them against U.S. commercial stocks in Florida and reported that the Yugoslavian bees and their hybrids were relatively more resistant to tracheal mites, but not to varroa, than the U.S. commercial stocks tested.

Studies have revealed significant differences in resistance among North American commercial stocks. Gary and Page (1987) compared young workers with respect to their attractiveness to adult female mites (see below for more details). They found consistent differences among colony sources for the number of female mites that invaded the prothoracic tracheae. Milne et al. (1991) compared colonies derived from queens of a single Texas queen producer with those of a single California queen producer. They found that the Texas stock was relatively more resistant to tracheal mites than the stock from California. The Texas stock was derived

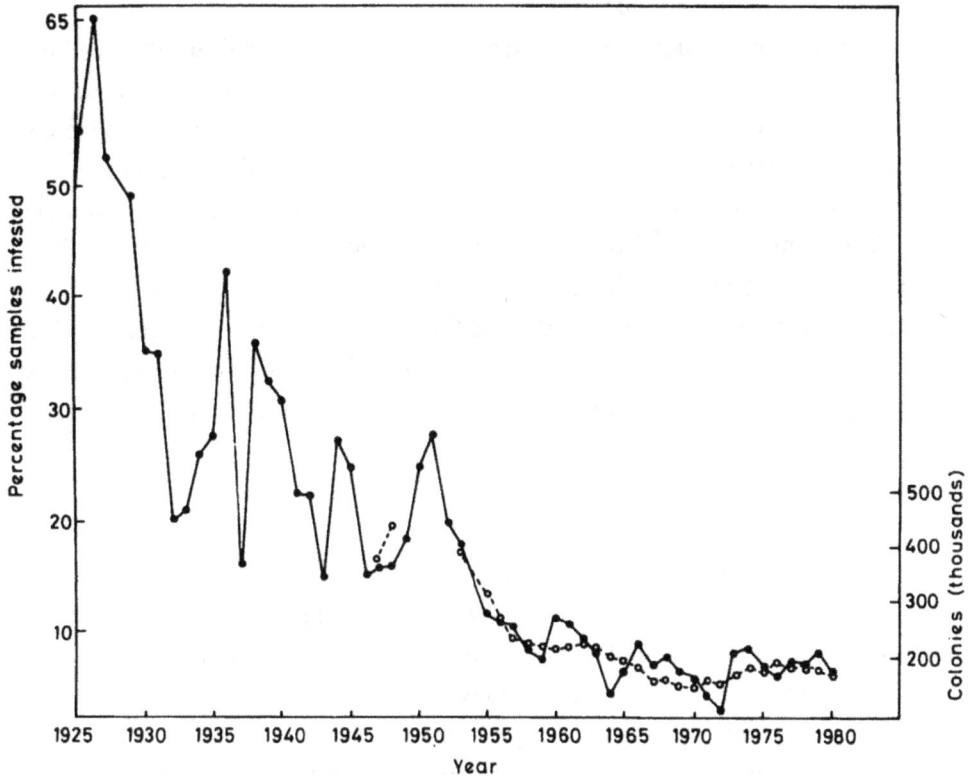

Figure 2. The percentage of samples of bees infested with tracheal mites (closed circles) in England and Wales and the number of commercial hives (open circles). From Bailey and Perry (1982), used with permission.

from the "Buckfast" strain developed at Buckfast Abbey, England. Because they compared bees from only two genetic sources, however, there is no way of knowing if the Buckfast strain is in general superior to North American stocks or if the single California source was relatively susceptible.

    The only controlled, documented, tracheal mite-resistance selection study was conducted by R. E. Page and N. E. Gary (Gary and Page 1987, Page and Gary 1990). They conducted two-way selection for the attractiveness of young workers to infestation by adult female mites. After a single generation of selection, workers from the susceptible strain were 40 percent more likely to become infested than workers of equal age from the resistant strain. After the second generation of selection, the workers from the susceptible strain were 2.4 times more likely to become infested. Page and Gary also tested their generation two high- and low-strain bees against local commercial bees. The commercial bees were intermediate between the selected strains for levels of tracheal mite infestation.

Source of bees

The breeding population used by Page and Gary (1990) was derived from a test of phenotypic variability conducted on 22 colonies derived from queens of a closed honey bee population developed and maintained in Madison, Wisconsin (Severson et al. 1986). An additional queen from a commercial source was included in their study, for a total of 23. The closed population was originally established from 25 queens obtained from queen producers located throughout the United States and was believed to be genetically heterogeneous. Colonies were tested near Lake Placid, Florida, and the queens that produced worker progeny with the highest and lowest likelihood of becoming infested in laboratory cages were selected as queen and drone mothers for the susceptible and resistant lines, respectively.

Assay

Page and Gary used two types of assays for their studies. The first was a laboratory assay in which they caged combs of emerging brood and placed them in an incubator. Workers from all colonies to be tested emerged during the same 24-hour period and were individually tagged with plastic numbered tags for identification, then placed into screen-wire cages. Each of several test cages contained 10 workers from each source colony, the "target" bees, plus an equivalent number of workers taken from the top bars of the brood nest of a highly infested colony, the "host" bees. Target bees remained in cages in the incubator, in contact with host bees for seven days, after which time all target bees were sacrificed, dissected, and the number of mites counted in each prothoracic tracheal trunk.

Page and Gary concurrently tested target bees in highly infested host colonies. Ten workers from each source colony were tagged in the same way as for the cages, then placed in colonies for five to seven days, after which they were removed and dissected. All breeder queens from each generation were selected on the basis of the laboratory test. The field colony assays were used to confirm that selection based on the laboratory assay was resulting in changes in field colonies.

Mating system

Two-way selection was performed. Relatively resistant and susceptible breeding populations were established from the original foundation population of 23 that was tested for phenotypic variability (Gary and Page 1987). Queens from the foundation population were ranked from one to 23 on the basis of the average number of adult mites found in the tracheae of their progeny in the cage studies. Each generation – drones and several virgin queens – were raised from surviving queens with high (most mites per worker) and low ranks (fewest mites per worker) to produce the next generation of susceptible and resistant workers, respectively. For generation one, four resistant and three susceptible queens were selected as breeders; for generation two, five resistant and four susceptible queens were used. Each virgin queen was instrumentally inseminated with the semen from a single drone that came from a corresponding, unrelated high- or low-ranking queen.

Mechanism of resistance
The mechanism of resistance is unknown.

Resistance to varroa
Kulincevic and Rinderer (1986) initiated a two-way breeding program with *A. m. carnica* colonies in Yugoslavia. The parental generation of the susceptible line averaged 14.8 percent of brood cells infested, whereas the parental colonies of the resistant line averaged 5.6 percent. After four generations of bidirectional selective breeding, resistant colonies had about half the number of infested cells as the susceptible colonies (Kulincevic et al. 1992, Figure 4). Moritz (1994), however, points out that the results of Kulincevic et al. (1992) are not clear because environmental effects and genotype X environment interactions did not allow them to separate genetic from environmental effects. The second generation of both lines suffered severe winter mortality, and only two colonies from the resistant line and one from the susceptible line survived. These colonies were then used to produce the third generation; the parents of the fourth generation were chosen for general vigor rather than infestation level. The genetic "bottleneck" caused by this manipulation and by the low number of colonies used from the second generation may have reduced the genetic variability for further separation of the selected lines. Moreover, because of an increase in the number of mites in both lines, they had to treat their experimental colonies with miticides, which meant that the third and fourth generations could not be compared directly to each other or to the first generation. Regardless of the criticisms of the Yugoslavian bee breeding program, the study did demonstrate consistent differences in varroa susceptibility between strains over several seasons, suggesting the feasibility of breeding for resistance.

Rinderer et al. (1993) imported queens from the resistant strain into the United States and tested them against U.S. stocks. The Yugoslavian bees were no more resistant than any of the U.S. stocks against which they were compared.

Source of bees
The breeding population used by Kulincevic et al. (1992) was derived from *A. m. carnica* colonies that had survived a varroa infestation in Yugoslavia.

484   Robert E. Page Jr. and Ernesto Guzmán-Novoa

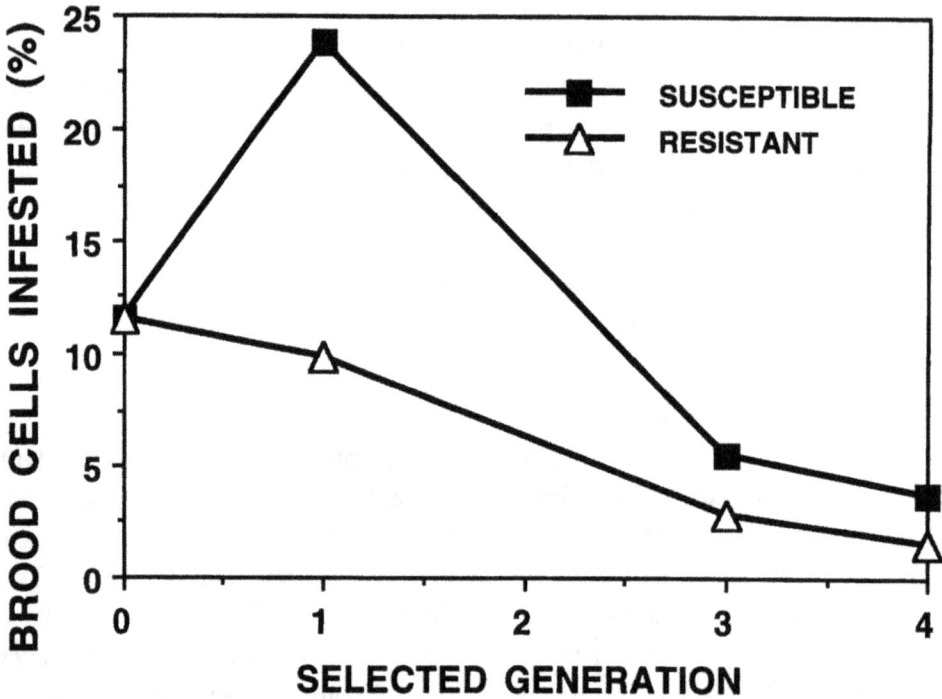

Figure 3. Percentage of inspected, capped brood cells found to be infested with varroa in colonies of selected resistant and susceptible strains. Authors did not show data for the second generation due to high winter colony mortality. Data are from Kulincevic et al. (1992).

Assay

The evaluation assay was conducted to estimate the percentage of capped worker brood cells infested by reproducing mites (Kulincevic and Rinderer 1988, Kulincevic et al. 1988, 1991). In each colony, the researchers opened and examined samples of 100 capped worker brood cells to determine the number of female, male, and immature mites. The two queens of the colonies having the highest infestation level were chosen to develop a susceptible line, whereas the two queens from the colonies with the lowest infestation level were chosen to develop a resistant line.

Mating system

Queens were reared from the selected queen mothers and were permitted to open mate in the same apiary location.

Mechanism of resistance

The mechanisms of resistance were not identified.

Potential mechanisms of resistance to varroa

Colony infestation with varroa is the most serious problem for beekeeping worldwide. As a consequence, it is remarkable that there is only a single example of successful artificial selection for resistance. Much is known, however, about the biology of varroa and how it interacts with its natural host, *A. cerana,* and its new, accidental host, *A. mellifera.* This understanding suggests potential mechanisms of resistance, some of which have been demonstrated to vary genetically. Because of the importance of varroa, this section contains more details than those above. Here we focus on what are considered to be resistance mechanisms with potential for selection.

Natural variation in resistance to varroa

The western honey bee, *A. mellifera,* is not the natural host of varroa. Its natural host, *A. cerana,* demonstrates several resistance mechanisms that apparently have evolved in response to a long-term association with varroa (see below). Most temperate colonies of *A. mellifera* succumb to varroa, however, and seem to have little resistance. Nevertheless, some colonies survive even in locations that have been devastated by varroa, suggesting that those colonies may have some degree of resistance to the parasite (Engels et al. 1986, Kulincevic and Rinderer 1988, Morse et al. 1991). An alternative explanation is that the mites in the surviving colonies are better adapted to their host and are thus less virulent (Ruttner et al. 1984). Moosbeckhofer et al. (1988) demonstrated large differences in varroa infestation among colonies in Germany. Kulincevic and Rinderer (1988) reported differences in infestation levels among colonies of Yugoslavian Carniolan bees that were used to initiate their varroa resistance program. Genetic differences for varroa resistance were also found among lines of Italian, Carniolan, and Caucasian bees in Israel (Ron and Rosenthal 1989).

Although varroa has been a problem for honey bees in most countries, it does not appear to be a serious pest in the South American tropics. The tropical climate and tropically adapted bees, such as Africanized honey bees, both play a role in maintaining reduced levels of infestation (DeJong et al. 1984, Ruttner et al. 1984, Engels et al. 1986, Gonçalves 1987, Mendoza et al. 1987, Moretto et al. 1991). Moretto et al. (1991) demonstrated that infestation levels were higher in colonies located in cooler climates of Brazil, and in European, rather than Africanized, bees.

Studies also suggest that temperature and climate strongly influence levels of mite reproduction (Camazine 1986, Rosenkranz 1986, Moritz and Mautz 1990). Ritter and DeJong (1984) found that 40 to 75 percent of female mites do not reproduce in tropical climates of South America. In Germany and Turkey, only about 20 percent of female mites fail to reproduce. This explains why varroa appear to be less pathogenic in the South American tropics. Other factors may also contribute to this apparently reduced pathogenicity (see below).

Mechanisms of resistance

Several mechanisms of natural resistance against varroa have been reported for *A. cerana*. These same mechanisms have been reported to occur at much lower levels in populations of *A. mellifera* but offer hope for selection programs designed to enhance them.

Grooming behavior

A. cerana has at least four behavioral mechanisms that may contribute to its relative resistance to varroa (Peng et al. 1987a, Büchler et al. 1992). Workers infested with varroa groom themselves extensively to remove the mites. If they cannot remove them, they perform a grooming dance and nearby workers use their mandibles to remove the mites. After removing a mite from a nestmate, a grooming worker frequently punctures the mite with her mandibles and removes it from the nest. When mites are particularly difficult to dislodge, the bee attracts a large number of workers that then engage in group grooming of the infested individual.

All of these grooming behavioral responses to varroa have been directly or indirectly observed in *A. mellifera* but at much lower frequencies than occur in *A. cerana* (Peng et al. 1987a, Morse et al. 1991, Rath 1991, Boecking and Drescher 1991, Boecking 1992, Ruttner and Hänel 1992, Moretto et al. 1993). Peng et al. (1987a) found that only 16.6 percent of the European honey bee workers infested with varroa attempted to remove them from their bodies, and only 0.3 percent of them were successful.

Dead mites retrieved from many colonies have apparently been killed by mandibular punctures (Wallner 1990, Boecking and Drescher 1991, Morse et al. 1991, Rath 1991, Boecking 1992a, Moretto et al. 1993). Ruttner and Hänel (1992) found that 30 to 50 percent of the dead mites collected from some colonies had damaged legs. They compared the structure of the mandibles of *A. mellifera* and *A. cerana* and concluded that there were no substantial differences between the two species of bees. They thus assumed that observed damage to mites was caused by the workers' mandibles. Ruttner and Hänel (1992) also showed that colonies with mite damage had slower rates of mite population growth. Moosbeckhofer (1992) reported a significant negative correlation between the number of damaged mites and the total infestation level of 111 colonies, suggesting that grooming behavior decreased the mite infestation level.

Africanized honey bees of South America seem better able to defend themselves from varroa than are Italian honey bees. Moretto et al. (1993) reported that in Brazil, Africanized workers were seven times more efficient than Italian workers in eliminating mites from their bodies; 38.5 percent of the Africanized workers infested were able to discard the mites from their bodies, as compared with only 5.75 percent of the Italian workers. The heritability of grooming behavior in Africanized bees was estimated to be 0.71 (Moretto et al. 1993). This high heritability index suggests that selection for this trait is possible.

Hygienic behavior

A. cerana workers are able to detect capped brood cells that are infested with mites. They open the cells, remove the mites, and kill them (Peng et al. 1987b, Rath and Drescher 1990). Boecking and Drescher (1991) found that workers of *A. m. carnica* also detected and removed varroa mites from infested brood cells in a way similar to *A. cerana*. If an infested cell is uncapped, the immature mites die; the adult females must search for another cell or be killed by the bees. This should slow the population growth of varroa in the colony. Boecking and Drescher (1991) found that 10 days after infestation, up to 100 percent of the mites were removed in some of the colonies that they observed. Büchler (1992a) found a negative correlation among four lines of *A. m. carnica* between level of hygienic behavior activity and susceptibility to mites.

Duration of the post-capping stage

The post-capping stage is one of the best-studied mechanisms of resistance of *A. mellifera* against varroa. The amount of time that a bee remains within a capped cell could influence the reproductive rate of varroa. A female mite enters a cell shortly before it is capped, then waits about 60 hours before laying her first egg, which requires about 6.2 to 7.5 days to develop into an adult (Ritter 1981, De Jong 1984). Eggs are laid singly at approximately 30-hour intervals (Infantidis 1983, De Jong 1984, Rehm and Ritter 1989). Therefore, eggs laid with fewer than seven capped-cell days left are not likely to survive. Thus, if the post-capping stage could be shortened, it might be possible to reduce the reproductive output of females.

The duration of the postcapping stage varies with bee race. European honey bees require about 12 days from sealing of the cell to adult emergence, whereas the Cape bee *A. m. capensis* needs only 9.6 days (Moritz 1985). Moritz and Mautz (1990) demonstrated reduced population increase of varroa on *A. m. capensis* hosts compared to *A. m. carnica*. Africanized bees have a post-capping stage of about 11 days (De Jong 1984). This shorter post-capping time may at least partially explain their relative resistance to varroa.

The duration of the post-capping stage is less variable among races of European honey bees (Büchler and Drescher 1990). Schousboe (1986) found differences of up to 1.15 days among colonies of *A. m. ligustica*, whereas Büchler and Drescher (1990) reported differences of up to nine hours among different strains of *A. m. carnica, A. m. mellifera*, and the Buckfast strain. Individual colonies varied by up to 19 hours. Three studies actually estimated the heritability for the post-capping stage in various races of honey bees to range from 0.23 to 0.80 (Moritz 1985, Büchler and Drescher 1990, Harbo 1992). If this is true, then a breeding program that selects for this characteristic may be successful.

Le Conte and Cornuet (1989) confirmed significant differences in the duration of the post-capping stage for three European races and their hybrids. Büchler and Drescher (1990) estimated a reduction of 8.7 percent in the final mite infestation for each hour reduced from the capped period. They also found a positive correlation between the length of the capped stage and the varroa infestation level (r = 0.48). Harbo (1992), working with 26 different colonies, found an average of 285.4 hours for the capped period. He predicted that selective breeding from 10 percent of the population should reduce the average capped period of workers by five hours in a single generation.

Some authors, however, do not believe that a shorter post-capping stage has a significant effect on reducing varroa population growth. Regardless of the difference in post-capping stage between European and Africanized bees (about 20 hours), Rosenkranz (1986) estimated that the number of offspring per female mite would be nearly the same. The difference between Africanized and European honey bees would need to be larger for this characteristic to be as important in conferring resistance as hypothesized by Moretto (1993). Other mechanisms of resistance such as hygienic and grooming behavior are greater in Africanized bees than in European bees (Cosenza and Silva 1972, Lengler 1977, Message 1979, Moretto 1993).

Moritz and Jordan (1992) suggested breeding for varroa-resistant honey bees by selecting drones – for instrumental insemination of queens – that have a reduced post-capping stage. They concluded that selecting drones should allow rapid genetic progress, not only because they show a large phenotypic variance for the duration of the post-capping stage, but also because of male haploidy. In haploid individuals, genes exist in single copies, and dominance effects cannot interfere with the genetic evaluation of the program. The problem with selecting bees with a shorter post-capping stage is that such a breeding program may also lead to selection of mites with a shorter development time.

Attractiveness of brood and adult bees

Varroa females favor oviposition in drone rather than worker cells (De Jong 1984). If this bias is based on the relative attractiveness of drone and worker brood, then perhaps the reproductive capacity of varroa is reduced by selecting for less attractive worker brood. Büchler (1989) combined in individual frames sections of young brood from different genetic origins. These mixed combs were placed in colonies that were highly infested with varroa. Büchler (1989) found a significantly lower infestation rate in *A. m. mellifera* than in brood from the Buckfast strain. His results were supported with a positive correlation between his comb assay and varroa population growth in field colonies (Büchler 1990).

In a similar experiment, Guzmán-Novoa et al. (1994) found that European brood were twice as susceptible to varroa infestation than as were brood from

Figure 4. Percentage of inspected, capped brood cells infested with immigrant female varroa. F_1 workers were of two types: those with Africanized queen mothers (Am) and those with European mothers (Em). European and hybrid brood were nearly twice as attractive as Aficanized brood. Data are from Guzmán-Novoa et al. (1994).

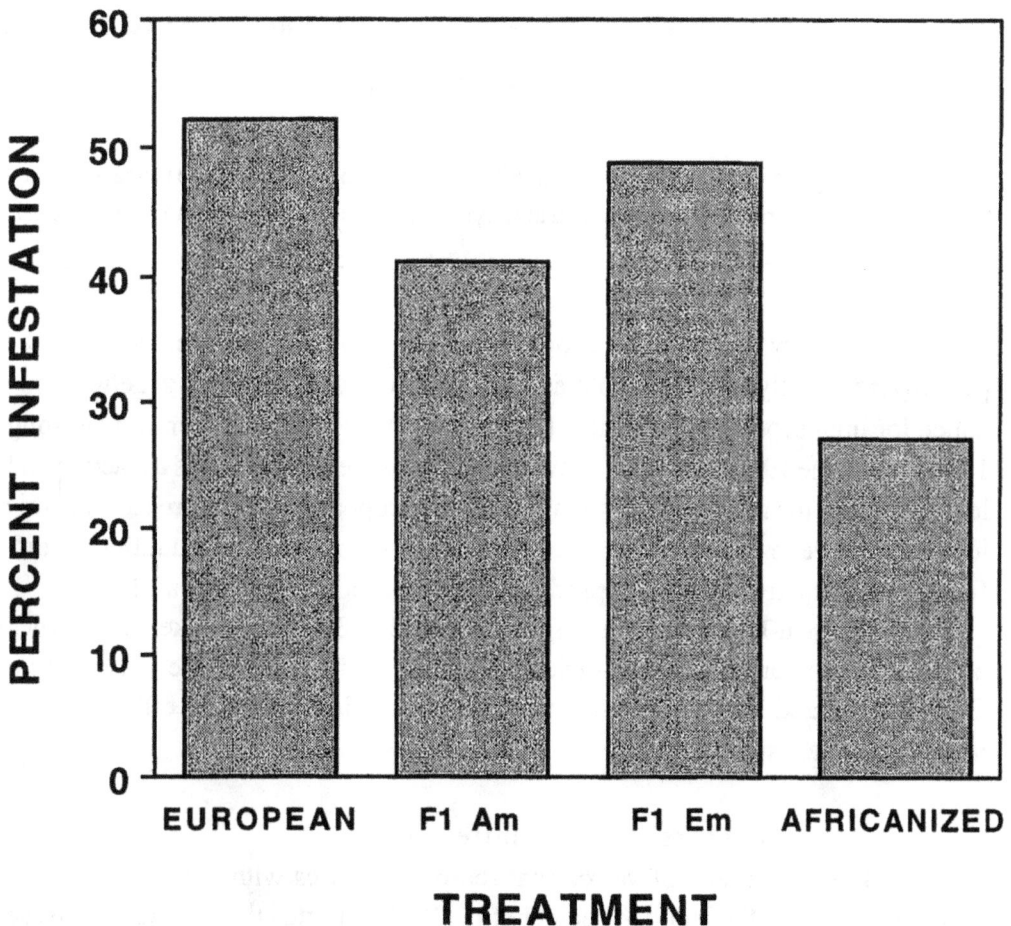

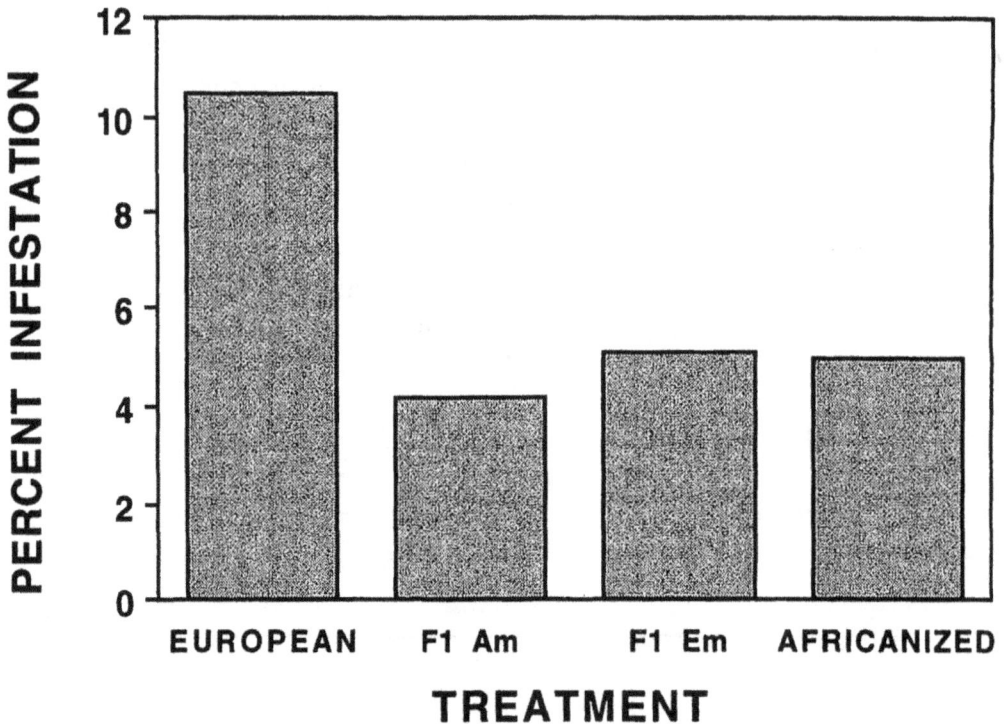

Figure 5. Percentage of adult bees infested with varroa. European bees were twice as attractive to adult female mites as Africanized and hybrid bees. Data from Guzmán-Novoa et al. (1994).

Africanized honey bee colonies. Brood from hybrid colonies (Africanized X European) were equally as susceptible as European brood, suggesting genetic dominance for high brood attractiveness (Figure 4). Other studies (Ritter and De Jong 1984, Camazine 1986) have also shown that Africanized brood is less attractive and less favorable for reproduction than brood of European bees. The reason for this lower attractiveness is not clear, but it may involve qualitative or quantitative differences in attractive chemicals produced by the brood (Trouiller et al. 1994).

Guzmán-Novoa et al. (1994) found that adult European workers were twice as attractive to varroa as were Africanized bees. Hybrid bees were not different from Africanized bees, suggesting a possible genetic dominance effect for low adult attractiveness (Figure 5).

Factors limiting mite fertility

The percentage of varroa that reproduce varies with host species, host race, and the sex of the brood. Koeniger et al. (1981) reported that varroa females do not reproduce when they infest worker brood of *A. cerana*. Likewise, varying proportions of female mites do not reproduce on brood of different races of *A. mellifera.* Values for percentage reproduction of varroa on European bees are 47 to 84 percent on worker brood and 90 to 95 percent on drone brood; for Africanized bees the values are 43 to 49 percent on worker brood, and 86 to 87 percent on drone brood (Ritter and De Jong 1984, Camazine 1986, 1988, Kulincevic et al. 1988, Fries et al. 1994). If variation for this characteristic results at least in part from genetic differences, then selective breeding for this trait might lead to the development of stocks resistant to varroa.

Relative effect of resistance mechanisms

Fries et al. (1994) developed a model of varroa population dynamics to assess the relative effect of different mechanisms and factors known to affect reproduction. According to their analysis, parameters directly affecting the number of mite offspring and the number of reproductive cycles per female were the factors that most influenced varroa's population dynamics. Their analysis predicted that selection programs based on mite infertility and a shorter post-capping period would yield bees significantly more resistant to varroa.

Perspectives and conclusions

The worldwide eradication of any honey bee disease is unrealistic. Thus, a reduction in economic damage in the beekeeping industry must depend on methods that maintain pathogens and parasites at reduced levels. The widespread use of chemical treatments has serious potential costs and risks resulting from the development of chemically resistant strains of disease agents and chemical contamination of hive products. Thus, the development of non-chemical control methods should receive more attention. The evidence presented above suggests that there is sufficient genetic variability for resistance to diseases to make selective breeding a viable component of commercial honey bee management.

Selective breeding programs will not succeed, however, without economic incentives. Current prices paid for queens produced in the United States will not support the added expense of industry-driven breeding programs. Institutional breeding programs at state-supported universities and the United States Department of Agriculture have never succeeded, partly because of the failure of the bee industry to adopt the stocks they produced. If the beekeeping industry is not willing to pay higher prices for selected stocks or is not willing to support and accept stocks produced by institutional stock-improvement programs, then the only alternative for the future is the continued use of dangerous and expensive chemicals.

Chapter Twenty-Two

——— ∞∞∞ ———

Summary of Control Methods

Hachiro Shimanuki and David Knox

Nonspecific disease cleanup
> *Procedures*
> *Burning*
> *Lye bath*
> *Ethylene oxide and radiation*

Feeding drugs
> *Bulk feeding*
> *Dusting*
> *Antibiotic extender patties*

Prevention and control of American and European foulbrood disease
> *Oxytetracycline*
> *Shaking bees*
> *Other treatments for American foulbrood disease*

Control of nosema disease
> *Chemotherapy*
> *Heat treatment*
> *Acetic acid fumigation*

Control of wax moths
> *Paradichlorobenzene (PDB)*
> *Carbon dioxide*
> *Heat treatment*
> *Cold treatment*
> *Aluminum phosphide*
> *Bacillus thuringiensis*
> *General precautions in wax moth control*

Control of *Varroa jacobsoni*

Control of other parasitic brood mites

Control of *Acarapis woodi*

Control of insect and mammalian pests

Disease diagnosis service
> *How to address and package samples*
> *Samples of brood*
> *Samples of adult honey bees*

CHAPTER TWENTY-TWO
Summary of Control Methods

The control of bee diseases is a rapidly changing field, especially regarding the use of chemicals. For various reasons, we have no firm recommendations for safe and effective methods for controlling some of the problems mentioned in this book. In certain cases, not enough research has been done to develop such methods. In others, no universal recommendation can be made because climate can often profoundly influence the effectiveness of a control method. Furthermore, each country, and often state or provincial governments within a country, has its own laws and regulations that may affect control methods and procedures. In many areas, there is a strong move to use nonchemical control whenever feasible. Even where this book makes clear recommendations or suggestions, it is important to consult with local control officials before using any chemical or other control method.

Nonspecific disease cleanup
Procedures
In general, the spread of American foulbrood disease and several other bee diseases can be prevented by destroying or thoroughly disinfecting all combs and other hive parts in infected colonies. Often, the thicker wooden equipment can be disinfected and reused. Burning contaminated bee equipment has been the most widely practiced bee disease control method and remains the principal method used by most apiary inspection services. Lye baths are often used to clean contaminated frames (after the wax comb has been cut away) and sometimes other hive parts. This procedure enables the beekeeper to salvage wooden equipment that might otherwise have to be destroyed. Much care and caution are required for the successful use of lye baths; they are not automatically approved by all inspection services. Ethylene oxide fumigation is another effective method for disinfecting equipment and controlling the spread of disease, but ethylene oxide is dangerous and its use is generally impractical.

Burning
A pit at least 45 cm (18 inches) deep and wide enough to hold all the material to be burned should be dug in a place not likely to be disturbed. After all

the bees have been killed,[1] the hive is placed on pieces of burlap or cardboard; this makes it easier to recover and destroy bits of comb, honey, or dead bees that would otherwise fall to the ground.

A brisk, hot fire is necessary to burn all of the brood and honey in an infected colony. The fire should be burning strongly before these materials are added to it. Several large tree limbs or timbers placed across the fire pit will support and hold the hive parts above the fire so that they can be completely burned. After all the material has turned to ashes, the pit should be covered with dirt.

All of the brood and honey must be burned completely. Bottomboards, hive bodies, inner covers, or outer covers can be salvaged by first scraping and then scorching them. It is considered best to salvage only wood that is at least 2 cm (about 3/4 in.) thick. Careful scraping is necessary so that the insides of all cracks and crevices may be exposed and well-scorched. All scrapings should be collected and burned. Scorching may be done by holding the items over a hot fire, by covering them with kerosene and then burning off the kerosene, or by using a blowtorch. Bees will accept and use wooden hive parts that have been charred.

Lye bath

A boiling lye[2] solution (454 grams of lye to 38 liters of water, or one pound of lye to 10 gallons of water) is useful for the removal of possibly contaminated wax and propolis from hive equipment. The frames are most commonly treated with a lye bath, but bottomboards or supers may also be treated. In the case of frames, the combs should first be cut from the wood; after careful rendering by the beekeeper or a rendering plant, this wax may be used in the same way as wax from healthy bees. Comb foundation made from properly rendered beeswax taken from colonies infected with American foulbrood disease or other diseases has never been found to be a source of reinfection. In modern rendering plants, beeswax is subjected to high temperatures and is filtered before being made into new foundation. While cutting and treating the frames, however, care must be taken not to leave the infected comb exposed in an area where it can be visited by robbing bees.

Before treatment in a lye bath, all equipment should be scraped so that the lye solution is not wasted on dissolving large pieces of wax and propolis that can be removed easily by hand. Scraping also allows the lye solution to penetrate cracks and crevices so as to remove any material that might harbor disease spores. The bee equipment should be completely immersed for five to 20 minutes in the boiling lye solution; the time varies with the amount of material to be removed.

---

[1]Resmethrin, a synthetic pyrethroid, has been approved for killing bees in diseased colonies in the United States (see Appendix 4).
[2]Lye as sold in the United States is 100 percent sodium hydroxide.

Wooden parts should not be left for more than 20 minutes because the wood fibers may be weakened by being kept in the bath for too long; bees will chew wooden parts that have been treated excessively with lye. It is helpful to dip the wood in clean boiling water immediately after removal from the boiling lye solution. It is usually necessary to re-nail equipment after it has been through a lye bath because the joints become weakened.

Precaution: Read the precautions on the label of the container before using lye. Lye is caustic and can cause severe burns. Never put lye into boiling water; the lye must be added before the water is heated. Do not use aluminum utensils for lye solutions; iron or enamel utensils are preferred.

Ethylene oxide and radiation

Ethylene oxide fumigation is effective against numerous honey bee pathogens and pests, including *Bacillus larvae,* the cause of American foulbrood (Cantwell et al. 1975). Ethylene oxide vapor is highly toxic to humans, however, and its use is restricted to people who have received appropriate training. Furthermore, it can only be applied in special high-pressure chambers. In the United States, the use of ethylene oxide on beekeeping equipment is still permitted in some states; however, that situation is being reviewed by the EPA, and may be revoked. A very few states have modified surplus pressure chambers for use on beekeeping equipment. Other states use or allow the use of the Miskoe portable fumigator, a small, pressurized unit designed especially for fumigating beekeeping equipment (Ellison 1977).

The concentrations of ethylene oxide used range from 450 to 700 mg per liter, with an exposure time of eight to 24 hours at $38° + 10°C$ ($100° + 18°F$) and a relative humidity of 80 percent or more. Because ethylene oxide does not penetrate honey, all honey must be extracted from the combs before they are fumigated to ensure that no bacterial spores survive the treatment.

Gamma radiation from a cobalt-60 source (Hornitzky and Wills 1983) and high-velocity electron beam radiation (Shimanuki et al. 1984) are also effective against *Bacillus larvae* and numerous other bee pathogens and pests. Ethylene oxide and radiation have two main advantages over other control methods: they are effective against many different organisms, and their use does not destroy beekeeping equipment. Ethylene oxide chambers and irradiation facilities have a limited availability, however, and in many cases it is not feasible to transport equipment to such facilities.

Feeding drugs

The discovery of sulfa drugs and antibiotic substances has brought about a profound change in bee-disease control strategies. Before the introduction of drugs, control programs consisted of responses to the outbreak of a disease. Suddenly, the feeding of drugs as a preventive measure was possible.

Drugs have been used successfully by many beekeepers. Although their use is widespread, they are not recommended uniformly by all inspection services. The specific drugs prescribed for treatment of individual diseases and the precautions to be followed are discussed elsewhere in this chapter.

Three methods that have been approved by the U.S. Food and Drug Administration for the feeding of chemotherapeutic agents to honey bees are bulk feeding, dusting, and the use of extender patties. Some drugs, such as bicyclohexylammonium fumagillin (available as Fumidil-B), are effective only when fed in a syrup (bulk-fed). Oxytetracycline HCl (available as Terramycin) is generally more effective when introduced by dusting or in extender patties.

Delivering the correct dose of a drug may be critical to a disease-control program. How much of a product to feed depends on how much of the active ingredient is present per pound of packaged product. In recent years, new formulations of Terramycin have been made available in the United States, and at the same time, other formulations are no longer on the market. Terramycin Soluble Powder (TSP; used for cattle and other livestock as well as bees and is sold by beekeeping supply dealers and agricultural supply stores) has 25 g (0.88 oz) of oxytetracycline HCl in each pound (0.45 kg) of the packaged mixture. TM-50D has 50 g of oxytetracycline HCl per pound, and TM-100D has 100 g of oxytetracycline HCl per pound. For an excellent summary of mixing and use of Terramycin, see Delaplane and Lozano (1994).

Bulk feeding

Bulk feeding is the feeding of medicated sugar syrup to a colony. Syrup consisting of sucrose (beet or cane sugar) and water is naturally attractive to bees, and they will readily collect it from a feeder. Both sodium sulfathiazole (generally called sulfathiazole) and Fumidil-B have been fed effectively in this manner. Oxytetracycline HCl can also be introduced by bulk feeding, but the other delivery techniques are usually preferable because oxytetracycline HCl breaks down faster in solution (Gilliam and Argauer 1975a,b). In a subsequent publication (Argauer and Herbert 1992), it was demonstrated that pollen can be contaminated with oxytetracycline HCl from medicated syrup . Because oxytetracycline is more stable in pollen, its efficacy may be prolonged long after it dissipates from the sugar syrup.

With any bulk feeding, attention to beekeeping conditions is important. If the syrup is fed at a time when there is not much food in the hive, it will be soon sumed rapidly; the medication will be ingested and will take effect immediately, as desired. But if the syrup is fed when food is available in the field, it is likely to be removed from the feeder slowly and stored, and there is little or no control of its effectiveness. There is also danger that the medication may remain in the hive and eventually contaminate harvested honey. It is mainly for these reasons that some beekeepers and apiculturists refuse to recommend the feeding of medicated syrup.

Bulk feeding is carried out in the same manner as the feeding of unmedicated sugar syrup. It is most popular to use a friction-top can, a pail with a screened opening in the lid, or large glass jar with many small holes in the lid, which is inverted over the frames, preferably immediately above the brood nest. Division-board feeders are also widely used.

Entrance feeders (Boardman feeders) are not very effective, especially in temperate areas, and their use is not advised. Feeding medicated sugar syrup in a large open container, such as a barrel or tub, which is available to many colonies at the same time, is also not advised because it is difficult to control the dosage level and because colonies with small populations would probably fail to collect a sufficient quantity of the syrup.

Precaution: See Dusting.

Dusting

Certain medications may be mixed with fine powdered (confectioners') sugar and sprinkled (dusted) over the top bars of the frames in a hive to be treated. It is important that the dust not fall between frames with open brood where it may mix with the food placed in the cells by nurse bees; such direct feeding of a medicated dust to honey bee larvae usually results in their death. The dust is usually placed in a circle on the top bars so that it circumscribes the brood nest.

This method of introducing medication gives quick results and does not require special feeders. The major disadvantage of dusting is that it is difficult to control the dosage of the drugs. Dusts are usually consumed by the bees in three to four days; in some cases, this makes weekly applications necessary. The quantities to be used are listed below.

Precaution: The feeding of medicated sugar syrup or medicated dust should cease at least 45 days before the beginning of the honey flow, and which time all honey should be removed from the hive. A 45-day period is considered sufficient to ensure that all traces of the drug are consumed and no contaminated honey is present in the hive.

Antibiotic extender patties

Extender patties have the consistency of thick paste and may be conveniently handled and shaped. They are composed of solid vegetable oil or petroleum jelly, sugar, and the drug to be administered. The oil or jelly deters rapid feeding by honey bees, so the mixture is consumed over an extended period, usually six to eight weeks. The patties are placed on the top bars of the brood nest frames. In the original paper on their use, it was stated that "adult bees apparently removed the patty, not because they enjoyed eating it, but rather because it was in their way and annoyed them" (Wilson et al. 1970). It appears that bees do not store the patty material because it is not normal for them to store fat or fat products. Wilson (1974) observed that the honey from colonies treated with extender patties in April (in Wyoming) had no drug residue when harvested that summer.

Oxytetracycline and sulfathiazole, both of which have been used for prevention and control of American foulbrood disease, were tested in extender patties; only oxytetracycline is effective in this form (Wilson et al. 1971). Gilliam and Argauer (1975a,b) showed that oxytetracycline remains stable in extender patties for many weeks.

Several formulas have been tested, but the one considered most effective is Number 3 in the Wilson et al. (1970) paper: 1/3 pound (150 g) vegetable fat, 2/3 pound (300 g) granulated sugar, and 2 tablespoons (12.5 g) TM-25 for each patty. In later tests, however, it was found that a patty half as large, with ingredients in the same proportions, was satisfactory, and that such a patty would last for six weeks (W. T. Wilson, personal communication 1977). The vegetable oil originally used and still recommended is hydrogenated and is sold under the trade name Crisco®.

Precaution: Antibiotic extender patties have been approved by the U.S. Food and Drug Administration for use in the United States. They can now be purchased as a premix.

Prevention and control of American and European foulbrood disease

The major course of treatment used for American foulbrood disease is the burning of all frames, combs, honey, brood, and bees from any colony found to be infected, and the scorching of bottomboards, top boards, and supers as described above. For European foulbrood disease, the main approach is to requeen the colony and strengthen it in various ways: feeding honey or sugar syrup; supplying clean, fresh water; adding frames of healthy, capped brood from other colonies; and making sure the hive is in a dry, sunny spot and is supported above the ground.

For either disease, drugs can be used in prevention and treatment, provided they are administered carefully. Feeding antibiotics avoids destroying hive equipment and bees, and can be used routinely as a prophylactic measure. It is possible, however, that the drugs may be introduced into honey meant for human consumption.

Oxytetracycline

Both American foulbrood disease and European foulbrood disease will yield to treatment by oxytetracycline HCl (Terramycin). Terramycin can be applied by bulk feeding or dusting, as described below, or in extender patties, as described above. The extender patties have not proven effective in the control of European foulbrood disease.

Commercial beekeepers, however, have found that a fall feeding of oxytetracycline is effective. In areas where treatment is routine, both fall and spring feedings of oxytetracycline are advised. Sulfathiazole is not approved for control of bee diseases in the United States, but it is available for use in other countries.

Specific recommendations follow. Except for bulk feeding of sodium sulfathiazole, all proportions given here are on a weight:weight basis.

Oxytetracycline (see Delaplane and Lozano 1994)
     TSP; fall or spring feeding
     For bulk feeding: mix 1 part TSP with 600 parts 1:1 sugar syrup (equal parts sugar and water)
     Each feeding: 1.9 liters (0.5 gallon) of mixture
     Feeding frequency: 3 times, at 4- to 5-day intervals
     For dusting: mix 2 parts TSP with 15 parts powdered sugar
     Each feeding: 28 g (1 ounce) of mixture
     Feeding frequency: 3 or more times, at 4- to 5-day intervals
Sodium sulfathiazole; fall feeding only
     For bulk feeding: mix 1 g (0.035 ounces) sodium sulfathiazole with 3.8 liters (1 gallon) of 2:1 sugar syrup (2 parts sugar to 1 part water, by weight or volume)
     Each feeding: 3.8 liters (1 gallon) of mixture
     Feeding frequency: 3 times, at 4- to 5-day intervals
     For dusting: mix 1 part sodium sulfathiazole with 32 parts powdered sugar
     Each feeding: 30 to 60 ml (2 to 4 tablespoons) of mixture
     Feeding frequency: 3 times, at 4- to 5-day intervals

Precaution: No medication should be fed to the colonies when there is danger of contaminating the honey crop. All drug feeding should be terminated at least 45 days before the start of the surplus honey flow.

When colonies are found to be infected with American foulbrood disease, especially where a drug treatment is routine, all bees, combs, and frames should be burned and all wood more than 2 cm (about 3/4-inch) thick should be scorched. No strain of *Bacillus larvae* (the cause of American foulbrood disease) resistant to sulfathiazole or oxytetracycline has yet been found; with continued use of these drugs, however, such a strain can be expected to develop in the future.

Shaking bees

A little-used method of treating colonies infected with American foulbrood disease, which has the advantage of allowing one to save the bees, is called shaking. The method involves shaking the bees from their combs; installing them in an empty, clean hive on new foundation and in clean equipment; burning the infected combs; and scorching the heavy wooden parts from the infected hive. This method is dangerous, however, and is costly. It must also be done at a time of year when the bees will be able to draw the new foundation and store sufficient honey for the winter or dearth period.

The dangers involved in shaking are many. Once exposed, however briefly, infected equipment may be robbed by bees from clean colonies. Clothing, hive tools, and smokers become contaminated with honey, wax, and refuse that may contain *Bacillus larvae* spores. Honey, burr comb, pollen, and even dead brood may be shaken from frames with infected brood and fall on the ground where robber bees may pick it up. In the process of shaking, it is necessary, as is normal, to smoke the colony. This causes the bees to engorge on honey contaminated with American foulbrood spores that they can carry to a new hive. Lastly, shaking requires skilled personnel as well as time and money invested in the shaking itself and in the preparation of new or clean equipment.

Done properly, shaking can be undertaken so that the reinfection rate is less than 5 percent. The following precautions are necessary: Colonies to be shaken are moved to an area remote from other colonies. The ground where the shaking is done is covered with plastic, cardboard, or some other material, which must be collected and burned upon completion of the shaking. Shaking is usually done in the early evening, when there is less chance that the shaken bees will drift or that robbing bees will be in the vicinity. A funnel is needed for the shaking process. The bees are shaken into packages or cardboard boxes provided with ventilation screens, in which they are held for at least four hours. During this time, the bees clean themselves and each other, so that any drops of honey on the bees or in the box are consumed. It is generally agreed that *B. larvae* spores that pass beyond the honey stomach will be voided with feces, probably far enough away from the hive that they will not reinfect any bees. The bees are not fed during the time they are held in captivity. After the period of confinement, the bees are put onto frames of foundation so that they have no place to store any residual honey they may be carrying. This further guarantees that the new equipment will not become contaminated with spores. Lastly, all equipment and clothing that may have become contaminated must be thoroughly cleaned. Hive tools, funnels, shoes, and so forth, must be scrubbed clean of honey, wax, and propolis using a chlorinated cleanser.

The shaking method was practiced widely before the advent of drugs for American foulbrood control. It is now considered good practice to treat the newly established shaken colonies with drugs; the use of drugs, however, in no way negates the need for care in shaking.

The attraction of the shaking method is obvious, but before undertaking it, the beekeeper should carefully investigate the costs involved. If the bees remain contaminated with *B. larvae* spores, the new frames and foundation will also be lost.

Other treatments for American foulbrood disease

Several treatments have been attempted for American foulbrood disease; most were devised to save the contaminated combs and frames. These included soaking solutions of alcohol-formalin, water-formalin, water-formalin-soap, iodine, and acetone-formalin. A commercial alcohol-formalin mixture was available during the 1920s. Most of these sterilizing solutions were tested by Sturtevant (1926), who found that all were ineffective when the combs contain honey or sealed brood.

Chlorine gas and formaldehyde gas have been tested for treatment of American foulbrood–contaminated combs. Formaldehyde gas is not an effective sterilant for combs (White 1906), and chlorine weakens frames and metal parts. It will contaminate honey, making it unsuitable for consumption by bees or humans (Hutson 1927).

Control of nosema disease

Fumagillin (Fumidil-B) fed to bees kills the protozoan *Nosema apis* (see Chapter 4), but viable *N. apis* spores will persist in beekeeping equipment. These spores can be killed if all hive parts are exposed to heat or fumigated with ethylene oxide or acetic acid; acetic acid is not registered for this use in the United States, but is used in some other countries. In practice, many commercial beekeepers in the United States feed fumagillin, but few, if any, treat their equipment. Irradiation, discussed above, has been used as a treatment, and anecdotal evidences suggests equipment so treated is free of spores. This bears further investigation.

Chemotherapy

Fumagillin is the only drug that has been approved by the U.S. Food and Drug Administration for the control of nosema disease. Fumagillin should be fed only in sugar syrup (bulk feeding); dusting and antibiotic extender patties are not effective. Dosages and administration are given below.

For wintering colonies

Prepare 4.5 g (1 level teaspoon) of Fumidil-B in 3.8 liters (1 gallon) of 2:1 sugar syrup (2 parts sugar to 1 part water, by weight or volume)

Each feeding: 3.8 liters (1 gallon)

Feeding frequency: two times in the fall (September-November in North America), early enough so that the syrup will be consumed by the time brood rearing stops, which is October-November in the northern United States; or two times in the winter (January-March in North America), for colonies normally fed sugar syrup in winter

For packages installed in spring

Prepare 4.5 g (1 level teaspoon) of Fumidil-B in 3.8 liters (1 gallon) of 1:1 sugar syrup (equal parts sugar and water, by weight or volume)

Each feeding: 3.8 liters (1 gallon)

Feeding frequency: one feeding at time of installation is usually sufficient; however, if the bees are unable to fly because of unfavorable weather, repeat the treatment two weeks after package installation

Heat treatment

Heat treatment of combs and other hive equipment is one method of de-contaminating a potential source of nosema infection. The treatment involves heat-ing all hive equipment for 24 hours at 49°C (120°F) in a room where the temperature is uniform and thermostatically controlled. Hot spots in the room should be avoided because higher temperatures could melt combs or cause them to sag.

Precaution: Combs should contain little or no honey or pollen. They should be heated in their normal, upright position (not on ends or sides). Boxes and combs must have spaces between them for air circulation. The temperature should not exceed 49°C (120°F), even in small areas within the room. In large rooms, the air should be circulated. The combs should be cooled to the outdoor temperature before they are removed from the room.

Acetic acid fumigation

Like heat treatment, acetic acid fumigation renders spores of *Nosema apis* nonviable. Acetic acid is not approved for this use in the United States. All bees must be removed from the equipment to be treated. The hive bodies with the frames and combs they contain are placed on a floorboard outdoors or in an open shed. A pad of cotton or other absorbent material soaked in 118 ml (1/4 pint) of 80 percent acetic acid is placed on the top bars of the frames in each hive. The entrance is blocked and the stack is covered. Cracks in or between the hive bodies should be sealed with tape. The stack should be left undisturbed for one week. After fumigation, the equipment should be aired for about 48 hours before use.

Control of wax moths

Wax moths can attack combs both within and outside a bee colony. For combs in hives occupied by bees, the most effective control method is to maintain a strong colony. Frequently, the same colonies that can successfully resist wax moth intrusions when healthy become prone to invasion by the moths if weakened by disease or other causes.

Beekeepers are also concerned with the protection of stored empty combs and of the honey in the combs (comb or chunk honey) that is intended for human consumption. U.S. federal regulations allow the use of paradichlorobenzene (PDB) for the protection of stored extracting combs, but this material is not approved for the protection of comb honey. Treatment with heat or cold and fumigation with carbon dioxide do not fall within federal regulations and may be freely practiced by beekeepers; certain precautions must be taken, however (see below).

Paradichlorobenzene (PDB)

Paradichlorobenzene (PDB) is readily available in the form of crystals, which are fairly inexpensive and convenient to use. (Only one product is available that is labeled for this use.) PDB is effective against wax moths at temperatures higher than about 21°C (70°F). This material has a characteristic odor, and its fumes can be absorbed by honey, which is thereby rendered unfit for human consumption. PDB should be used to treat only empty stored combs or combs with honey to be used exclusively as bee feed. All combs that have been treated with PDB should be aired for a few hours before being placed on colonies.

When the crystals are exposed to air, the PDB is emitted in gaseous form. Because this gas is heavier than air, the bottoms of the stacks of hive bodies (no more than five full-depth or 10 half-depth supers) should be sealed. No special chambers are required, but all cracks between the supers should be sealed with tape.

Approximately 85 g (3 oz or 4 tablespoons) of the crystals are placed on a piece of paper or cardboard on the top bars of the frames on the top of the stack. The stack of supers should be covered. Stacks of supers should be examined for the presence of crystals at two- to three-week intervals. Replenishing the supply of crystals is necessary for the PDB to be effective. PDB does not kill the egg stage of the wax moth.

Precaution: Comb honey intended for human consumption must not be fumigated with PDB.

Carbon dioxide

An airtight room or fumigation chamber is required to fumigate combs with carbon dioxide. A temperature of approximately 38°C (100°F), a relative humidity of 50 percent, and a carbon dioxide concentration of 98 percent for four hours are the optimal conditions for killing all stages of the wax moth. In practice, somewhat different conditions, which are easier to establish and maintain, are used. Those used in one large-scale commercial operation for control of wax moths in comb honey are described in Chapter 7. Carbon dioxide is not toxic to humans in the normal sense, but a person cannot survive in an atmosphere of carbon dioxide, and it is thus necessary to air the storage room before people enter it. Carbon dioxide is heavier than air and will flow rapidly from a room when the door is opened.

This treatment can be used for honey in combs intended for human consumption. There are no state or federal regulations governing or controlling the use of carbon dioxide to control wax moths, and there is no reason to believe that any legislation will be enacted.

Precaution: Persons should not enter a chamber where fumigation is being conducted or while it is filled with carbon dioxide. An excess of carbon dioxide can cause narcosis and death.

Heat treatment

Heat can be used to kill all stages of the wax moth. At 46°C (115°F), an exposure time of 80 minutes is sufficient; but when the temperature is increased to 49°C (120°F), only 40 minutes of exposure is required. The timing should commence when combs reach the proper temperature.

Precaution: Do not expose combs to temperatures greater than 49°C (120°F), as combs can be damaged by the high temperature. Heat only combs with little or no honey, and do not remove the combs until they are cooled. Combs should be treated in an upright position and not on ends or sides. To avoid hot spots, use a fan to circulate the hot air. Beeswax melts at about 64°C (148°F). At temperatures between 49° and 64°C, combs will sag and the cells can become misshapen.

Cold treatment

Low temperatures can be used to kill wax moths on combs that contain honey and pollen. The only limitation of this method is that a large freezing chamber is expensive.

At -7°C (20°F), 4 1/2 hours are required to kill all stages of the wax moth; at -12°C (10°F), three hours; and at -15°C (5°F), two hours. Honey in combs intended for human consumption can be subjected to such temperatures with no ill effects.

At least one beekeeper in Florida, U.S.A., protects several thousand supers of stored combs from wax moths by keeping them in a refrigerated building. The building is well-insulated, and standard refrigeration equipment is used. The storage room is maintained at about 48°F (9°C), a temperature low enough to inhibit wax moth reproduction. Several in-room thermocouples that can be checked without opening the room allow the temperature to be monitored frequently from the outside. Electrical costs are moderate (W. Perry, personal communication 1989). This is practiced in many parts of the country, with various refrigeration devices used. This does not prevent wax moth development, but slows it to a unnoticeable level.

Precaution: Combs must be handled carefully at temperatures below freezing because the wax becomes brittle. When timing cold treatments, be sure to adjust for the size of the load; the times listed above are for exposures starting when the combs reach the target temperature.

Aluminum phosphide

Aluminum phosphide is registered for the control of wax moths in the United States. It is on a list of "restricted use" chemicals and should be used only by licensed pesticide applicators. The material is available under various trade names either as tablets or as pellets (see Appendix 4).

Bacillus thuringiensis

Certan®, a biological insecticide that contains the bacterium *Bacillus thuringiensis* Berliner is not being manufactured at the present time. It was developed to protect honeycombs in storage. The formulated material (one part) is mixed with 19 parts water and sprayed on each comb. This strain of *B. thuringiensis* was selected specifically for its pathogenicity to wax moth larvae. The entire surface of the comb should be covered with the spray. The degree of protection may be lost once the combs are placed in an active hive.

General precautions in wax moth control

All treated combs should be stored in rooms that prevent the entry of adult wax moths. Untreated combs should never be mixed with treated combs while in storage. Stored combs should be inspected monthly (more frequently in tropical areas) for signs of new wax moth activity.

Control of *Varroa jacobsoni*

Around the world, a wide variety of chemicals have been used to control *Varroa jacobsoni* Oudemans. The list of acaricides in use has changed considerably in recent years and will continue to change as new products are developed, new regulations and restrictions are imposed, and strains of mites develop that are resistant to some of the chemicals. For all control measures, best results are obtained when a colony is treated during a brood-free period and all colonies in an area are treated simultaneously to reduce the chance of reinfestation.

In the United States, as of this writing, the only approved control chemical is manufactured by the Sandoz Agro/Novartis Corporation, under the trade name Apistan®. Apistan® consists of a plastic strip (for hives or shipping package bees) or a plastic tab (for queen-shipping cages) impregnated with fluvalinate, a pyrethroid (synthetic pyrethrin). Current regulations require treatment as follows:

1. To treat a colony, remove all honey supers and hang one Apistan strip for each five combs of bees in each brood chamber. Suspend the strips with the pre-cut tab so that they do not touch any comb. Hang one strip between frames 3 and 4, the other between frames 7 and 8 (in a 10-frame hive body) if the super is full. If not, or if the population is off-center, place the strips symmetrically in the super. Leave the strips in the hive for 42 to 56 days.

2. To treat a package of bees, nail or staple one Apistan package-bee strip inside the package so that it hangs freely among the bees for at least five days.

3. To treat a queen cage, place one Apistan queen tab in the cage so that the queen and attendant workers are exposed for at least three days.

Control of other parasitic brood mites

Tropilaelaps clareae apparently cannot persist in regions where honey bee colonies have a seasonal break in brood rearing. In other areas, artificially creating a broodless period has eliminated the mites from *Apis mellifera* colonies (Woyke 1985b).

For the other known parasitic brood mites – *Varroa underwoodi*, *Euvarroa sinhai*, and *Tropilaelaps koenigerum* – control measures have not been developed and, in fact, may not be necessary.

Control of *Acarapis woodi*

Menthol crystals have been shown to be effective in reducing the infestation levels of *Acarapis woodi* to below the economic threshold. Experiments indicate that 50 g (1.8 oz) of menthol should be placed in screened bags on top of the winter cluster or, in the absence of a winter cluster, on the bottomboard of the hive. For menthol to be effective, minimum daytime temperatures should be 21°C (70°F). Beehives should be treated after the surplus honey crop has been removed. All treatments must cease one month before the surplus nectar flow.

    Precaution: In the United States, menthol has been registered for the control of *A. woodi* in most states, but restrictions may still vary by state.

    At publishing, formic acid had not yet been approved for use in the United States, but is in use in Canada and other countries. Although formic acid is caustic, several methods and devices have evolved to reduce exposure to the applicator. These include slow-release bags, gels and wick-type devices.

    In early 1997, the USDA, spurred to action by the U.S. beekeeping industry granted funds to test a formic acid gel material to control mites in a beehive. Previous efficacy data, residue data, and other required studies had already been completed. This final project, it is believed, should lead to registration for use by U.S. beekeepers.

    In late 1996, several studies were undertaken to determine the usefulness of a limited number of essential oils, that is, oils derived from some plants, for controlling parasitic mites in beehives. Initially promising, these products still require significant testing for efficacy, residue, and other benefits or problems.

Control of insect and mammalian pests

    Although there are many insect and mammalian pests of honey bees, none (other than wax moths) is widespread and a menace or problem everywhere bees are kept. Pests, predators, and problems vary from area to area. Some pests may cause beekeepers to suspend operations for short periods of time; queen rearing, for example, may be impossible for one to three weeks in some areas while dragonflies are active. Other pests, such as bears, may make it impossible to keep bees in certain areas.

    Large mammalian pests are often protected by state, provincial, or other governments; a small number of such governmental units may compensate beekeepers for losses sustained by protected animals. It is best to contact a local beekeeper or government officer for details and applicable regulations.

    See Tables 22.1 and 22.2 for a summary of control measures for these pests.

Table 22.1. Summary of control measures for insect pests

Order	Common name	Method of control
Diptera	flies	None usually necessary
Hymenoptera	ants, wasps, bees	None usually necessary; see Chapter 9 for problem areas
Odonata	dragonflies	Move colonies; relocate queen mating locations; discontinue queen rearing during period of peak activity
Orthoptera	cockroaches mantids	Maintain strong, active colonies None usually necessary
Isoptera	termites	Treat wooden parts with a wood preservative; use treated hivestands
Psocoptera	booklice or barklice	None necessary
Hemiptera	bugs	Maintain strong, active colonies; reduce size of colony entrance; move colony
Neuroptera	lacewings, etc.	None necessary
Strepsiptera	twisted-winged	None necessary parasites or stylops
Coleoptera	beetles	For large species, reduce size of colony entrance; fumigate stored equipment; none necessary for most species

Table 22.2. Summary of control measures for mammalian pests		
Mammal group	**Common name**	**Method of control**
Marsupials	opossums	Trap or shoot; elevate colonies
Insectivores	hedgehogs	Elevate colonies
	shrews	Reduce size of colony entrance
	moles	None needed
Endentates	armadillos	None needed
Rodents	mice	Reduce size of colony entrance; trap or set out poison baits
	rats	Trap
	squirrels	Trap
Carnivores	dogs, foxes, cats	None needed
	raccoons	Elevate colonies, trap
	martens, sables, weasels	Use baited traps or move colonies
	ratels	Use protective fence and/or elevate colonies
	skunks	Trap; set out poison baits; use protective fence; elevate colonies
	bears	Use protective fence; elevate colonies; move colonies
Ungulates	horses, cattle, sheep	Place colonies outside livestock areas; fence apiary
Primates	humans, monkeys, baboons, chimpanzees	Move colonies; securely fasten hives on elevated stands

Disease diagnosis service

Samples of dead bees and brood that are suspected of having been killed by an infectious disease in the United States will be examined at no charge at the Bee Research Laboratory of the U.S. Department of Agriculture. Because an accurate diagnosis depends on how well the sample has been preserved, it is important to follow the directions listed below.

How to address and package samples

1. All samples should be sent to the Bee Research Laboratory, BARC-East, Building 476, Beltsville, MD 20705.

2. Your name and address and a short description of the problem should accompany the sample. Your name and address should be plainly written on the top of the box.

3. Indicate on the outside of the box or envelope that the package contains honey bee disease samples.

4. Mail the sample in a wooden or heavy cardboard box. The sample can be loosely wrapped in a paper bag, paper towel, newspaper, or the like. Avoid such wrappings as plastic bags, aluminum foil, waxed paper, tin, and glass; these materials allow mold to grow on samples.

Samples of brood

1. The sample of comb should be about four inches square (10 x 10 cm).

2. The sample should contain as much of the dead or discolored brood as possible.

3. No honey should be present in the sample.

4. If a comb cannot be sent, the probe used to examine cells may contain enough material for tests. It can be wrapped in paper and sent to the laboratory in an envelope. However, this method is unsatisfactory for verification of virus-induced diseases.

Samples of adult honey bees

1. Send at least 100 bees in a sample.

2. Select, if possible, bees that appear moribund or have just died. Bees that have been dead for some time are decayed and are unsatisfactory for diagnosis.

3. Bees being submitted for mite identification should be placed in 70 percent ethyl or methyl alcohol as soon as possible after collection. These samples should be in leak proof containers.

4. Bees being submitted for virus examination should have died recently. The bees should be sent dry and not in alcohol.

APPENDIX ONE

—⊷⊷⊷—

Species of *Apis*
Roger A. Morse

The *A. cerana* group
The *A. dorsata* group
The *A. florea* group
Summary and author's opinion

APPENDIX ONE

Species of *Apis*

Roger A. Morse

Since 1862, most authorities have considered the genus *Apis* to consist of four species of honey bees: *cerana, dorsata, florea,* and *mellifera* (Alexander 1992). *A. mellifera* is a native of Europe and Africa and has been spread deliberately to North and South America, Australia, New Zealand, and several Asian countries. It is the honey bee of commerce in most parts of the world. The other three species are Asian in origin, but *A. florea,* has spread outside its native habitat to the Near East and North Africa. Its spread was probably not natural but rather a byproduct of the international shipment of boxed products in which a natural nest was hidden.

The reason for accepting these four groupings of species is that each is distinct morphologically. Equally important, the genus has two distinct nest types. Several obvious bonds make this genus distinct: For example, all honey bees control their brood nest temperature, separate their brood and food, store their food above their brood, and live exclusively on pollen and honey.

Maa (1953) wrote a long monograph on *Apis* dividing the genus into a large number of species based on minute *characters*. His paper is still widely cited because, although he no doubt "oversplit," he made careful notes about how each of his "species" differed from the next. As we study Asian honey bees, we find that several of Maa's species are valid. But he did not criticize the four main groupings of *Apis*, with no truly close relatives, listed above. Although there are well over 20,000 bee species, the stingless bees and bumble bees are most closely related to *Apis* and are distantly related to honey bees.

Iran is the only country in the world where all four species groups are found naturally, although the European honey bee, *Apis mellifera,* is never contiguous with the other three; it is restricted to western Iran and the others to the eastern portions, except for the recent movement of *A. florea.* The Iranian mountains form a mighty barrier separating the Asian honey bee species from those in Europe.

Our primary concern in this book is the European-African *A. mellifera.* Because Asian honey bees are the source of one major disease, and little is known about the effects of Western bee disease on Eastern bees, Asian bees have become of greater interest to those who study diseases.

The *A. cerana* group

Apis cerana, also called the Indian honey bee, is the most widespread of the Asian species. It is the second best known honey bee species because it has been used commercially in several Asian countries. *Apis cerana* is found in all Asian countries and lives as far west as Iran. It nests in protected cavities that enable it to survive outside warm areas. The bee adapts readily to a small version of the Langstroth hive and may be managed as far as swarm control, queen rearing, and honey production are concerned. It is the native host of the dreaded mite, *Varroa jacobsoni.*

Recently, several researchers have agreed that the larger, reddish bee, *Apis koschenikovi,* from Sumatra is not merely an aberrant variant of *Apis cerana* but a valid, closely related species. What little is known about this species is discussed by Otis (1991), whose paper includes a picture of the bee with a varroa mite on its thorax. We presume this *Varroa jacobsoni* is not a variant.

Apis nuluensis is yet another high altitude bee belonging to the *A. cerana* group about which too little is known to write intelligently. It is probably not found outside of Sabah, Borneo, Malaysia. Apidologie 27(5), 1996, contains several papers pertinent to new finds in this area.

The *A. dorsata* group

The giant bee, *A. dorsata,* has fascinated researchers and beekeepers because it is about twice the size of *A. mellifera,* suggesting great potential as a honey gatherer and wax producer. In fact, it stores little honey, and nests containing as much as 10 kilos are rare (Morse and Laigo 1969a). The U. S. Department of Agriculture attempted to bring this species into the United States around the turn of the last century.

Based on its markedly different biology, I am willing to accept *A. laboriosa* as a distinct species, but several researchers have argued against that view because the genitalia of the two appear to be the same.

The *A. florea* group

A. florea are the smallest of the honey bees and are often called dwarf bees. There is now good evidence that this group contains two species of which *A. florea* is the most common. *A. florea* is the one species of honey bee that has been reported in a new area in recent years. Its original distribution was Asia west to Iran, but it has now spread, probably accidentally, into Oman and more recently the Sudan (Tomasko 1989). Tomasko reports that when she left the Sudan in 1988, "well over a hundred" nests had been reported in the vicinity of the capital city of Khartoum, indicating that the species is firmly established there. The first nest had been found in the Sudan in 1985. Glaiim (1992) reports that *A. florea* is established in Iraq. Cool weather appears to be the only barrier to prevent its further spread into other parts of Europe and Africa.

The second species in this group is *A. andreniformis*. This bee was described as a separate species in the last century but was not accepted as a species in its own right until 1986. The logic behind this decision and further information may be found in Otis (1991).

A major difference between the *dorsata-florea* and the *mellifera-cerana* groups is that both of the former nest under branches and on the walls of cliffs where they are little protected from the elements and are thus restricted to the warm parts of the world.

Summary and author's opinion

All those who work with bees agree that there are four distinct species groups of honey bees in the genus *Apis*, all of which are more closely related to each other than to any other bees. And, within the *A. mellifera* group, because all races can mate with one another, it is clear that there is only one species ranging from Norway in the north of Europe to South Africa in the south of that continent. Among these, however, there are at least 26 distinct races. Each of the Asian species groups has already been split into at least two species and, especially because of new techniques in systematics, further splitting will occur.

APPENDIX TWO

———— ∞∞∞ ————

Subspecies of *Apis mellifera*
Walter S. Sheppard

Introduction
How do subspecies differ?
On the origin of subspecies
The list of subspecies of *A. mellifera*
Apis mellifera adamii Ruttner
Apis mellifera adansonii Latreille
Apis mellifera anatoliaca Maa
Apis mellifera capensis Escholtz
Apis mellifera carnica Pollman
Apis mellifera caucasica Gorbachev
Apis mellifera cecropia Kiesenwetter
Apis mellifera cypria Pollman
Apis mellifera iberica Goetze
Apis mellifera intermissa von Buttel-Reepen
Apis mellifera lamarckii Cockerell
Apis mellifera ligustica Spinola
Apis mellifera litorea Smith
Apis mellifera macedonica Ruttner
Apis mellifera major Ruttner
Apis mellifera meda Skorikov
Apis mellifera mellifera L.
Apis mellifera monticola Smith
Apis mellifera sahariensis Baldensperger
Apis mellifera scutellata Lepeletier
Apis mellifera sicula Montagano
Apis mellifera syriaca von Buttel-Reepen
Apis mellifera unicolor Latreille
Apis mellifera yemenitica Ruttner

APPENDIX TWO
Subspecies of *Apis mellifera*

Introduction

Worldwide, the western honey bee, *Apis mellifera,* is undoubtedly the most economically important honey bee species. In recent years, it has displaced other *Apis* species as the predominant honey bee of apiculture in countries such as Thailand, China, and Japan. Although it now can be found throughout most of the Old and New Worlds, its original distribution included only Europe, Africa, and the Middle East. The beginning of worldwide colonization by *A. mellifera* can be traced to its transport to the New World by European immigrants. The distribution of European and African honey bees to new regions over the past few hundred years presents intriguing opportunities to observe unfolding population genetics and evolutionary phenomena such as the spread of Africanized honey bees in the Americas and the interspecific host shift of the parasitic mite *Varroa jacobsoni* from *A. cerana* to *A. mellifera*. However, to understand the extent of such micro-evolutionary events and postulate mechanisms for them, an appreciation of the diversity of honey bees within their regions of endemism is required.

Across the vast geographic range of *Apis mellifera,* restrictions to "gene flow" (interbreeding) have occurred, either because of geological barriers such as mountains and seas or climate-influenced phenomena such as deserts, glaciers, and shifts in vegetation. The result of restricted gene flow over thousands of years was an accumulation of genetic differences among isolated populations and ulti-mately the development of subspecies or geographic races. Ruttner (1975) defines geographic races of bees as distinct "units" that have acquired genetic differences as a result of natural selection in their homeland.

One of the earliest observers of racial variation within the honey bee was Aristotle, who noted behavioral differences between color varieties of bees in ancient Greece. The development of modern morphological analysis in this century led to the current taxonomic classification of the honey bee into about two dozen subspecies. Therefore, although they are all members of the species *Apis mellifera* and can interbreed freely, enough measurable differences exist that specialists con-sider it appropriate to classify honey bees into various groups at a level "below" that of species. A list of the subspecies of *A. mellifera* appears at the end of this text. Note that subspecies are given formal trinomial status, that is, a third name follows the standard binomial of *Apis mellifera*. So in the case of the well-known Italian honey bee, the scientific name is *Apis mellifera ligustica,* with the names referring to the genus, species, and subspecies, respectively.

How do subspecies differ?

Without the yardstick of "reproductive isolation" that biologists use to define species, the classification of honey bees at the "sub" species level depends on current knowledge of the amount and distribution of morphological or physical variation present in populations throughout the range of the species. Taxonomists have named subspecies based on perceived natural groupings within the species as suggested by this "phenotypic" variation (see Ruttner 1988 and references therein). The logical underpinning is that phenotypic variation reflects an underlying genetic variation. The distribution of genetic variation among populations provides a measure of gene flow and, thus, can be used to compare overall genetic differentiation within the species and among subspecies.

In addition to variations in morphology, which includes such characters as color, wing venation, body measurements, hair length, etc., behavioral characters also vary among honey bee subspecies. Although not a formal part of the taxonomy, these behavioral differences strengthen the hypothesis that a degree of genetic differentiation exists between races. They include overwintering behaviors (such as the extent of propolization of the nest and propensity to eliminate drones in the fall), degree of defensiveness, swarming behavior, and tendency for seasonal absconding. Some of the behaviors seem to be adaptations for particular ecological conditions and are common within broad racial groupings. Therefore, subspecies endemic to tropical Africa share the characteristic of seasonal absconding during periods of nectar dearth (Fletcher 1978, Schneider and McNally 1992), whereas subspecies of the colder temperate areas rarely exhibit this trait. Analysis of the genetic basis of behavioral variation among honey bee subspecies is an area of active research that in the future should provide insight into the diversity of *Apis mellifera*. In the section below, which provides brief descriptions of the subspecies, it is evident that our knowledge of some subspecies is limited.

Over the past 20 years, new approaches have been increasingly used to investigate genetic diversity in the honey bee. These include study of protein variation, which, like morphology, is used to infer genetic variation and, more recently, study of variation in the genetic material, or DNA itself. These new approaches have added to our understanding of diversity and relationships in the endemic subspecies and of population genetics and gene flow in the honey bees introduced onto other continents, including Africanized honey bees to the Americas.

Early studies of enzyme (protein) variation revealed differences in the frequency of certain variants (alleles) among some honey bee subspecies (Badino et al. 1983, 1984, 1988, Sheppard and McPheron 1986). These differences were used initially to characterize several Old World honey bee subspecies and possibly to

522   Appendix 2

"identify" Africanized honey bees (Sylvester 1982, Cornuet et al. 1986, Del Lama et al. 1990, Comparini and Biasolo 1991). It became evident, however, that various groups of subspecies shared many of the alleles and, while providing markers useful for the study of racial hybridization (Manino and Marletto 1984, Comparini and Biasolo 1991, Meixner et al. 1993), the reported enzyme variation clearly was too low to provide much improvement in subspecific taxonomy. In the populations introduced to areas outside the Old World, the level of enzyme polymorphism was even lower, because of "founder effects," although this limited allelic variation proved useful to study certain population interactions (Cornuet 1979, Sheppard 1988, Lobo et al. 1989, Oldroyd et al. 1992, Schiff et al. 1994, Sheppard et al. 1991a, Smith et al. 1989). "Founder effects" refers to the loss of genetic variation that accompanies the introduction and establishment of a species into a new habitat.

DNA studies have included analysis of variation in nuclear and mito-chondrial DNA sequences as a means to compare genetic changes within and among populations and subspecies. The use of such "molecular" methods pre-dominates much of the ongoing research on honey bee evolution and population genetics. Early findings of subspecific DNA variation were used to test hypoth-eses about the Africanization process in the Americas (Hall and Muralidharan 1989, Rinderer et al. 1991, Sheppard et al. 1991a,b, Smith et al. 1989). DNA data, together with those derived from analysis of enzyme polymorphism and morphol-ogy, were used to infer the degree of genetic introgression between the introduced African and European subspecies in South and Central America. Similar methods have been applied to study population genetic processes in the introduced honey bee populations of Australia and North America (Oldroyd et al. 1992, Schiff et al. 1994, Schiff and Sheppard 1995).

DNA analysis of Old World subspecies revealed significantly more varia-tion than had been detected by analysis of enzyme polymorphism and was used to study natural introgression between subspecies (Smith et al. 1991, Garnery et al. 1992, Meixner et al. 1993) and to test hypotheses of genetic relationships between the subspecific groupings themselves (Garnery et al. 1991, Arias and Sheppard *in press*). Some interesting findings are beginning to emerge from these studies and are mentioned in the subspecies descriptions below. At present, however, the re-sults must be interpreted with caution, because baseline molecular data from popu-lations from throughout the geographic range of *Apis mellifera,* comparable to those collected for morphology, are lacking.

On the origin of subspecies

Subspecies are generally believed to result from isolation of populations that eventually resulted in the accumulation of genetic differences. For many of the

European subspecies, this isolation apparently occurred during the Pleistocene glaciation of Europe, when populations were relegated to existence in various refugia (Ruttner 1975). The last major glacial advance occurred almost 20,000 years ago and, following the retreat of the European ice sheets, the newly formed subspecies expanded their range northward. In some cases, geographic barriers such as mountain ranges served to maintain the relative isolation of the subspecies. We now know, however, that significant hybridization occurs between some subspecies. Examples include *A. m. mellifera* and *A. m. carnica* across the steppes of Russia (Ruttner 1975), *A. m. mellifera* and *A. m. ligustica* in northwestern Italy (Badino et al. 1984, Manino and Marletto 1984), *A. m. ligustica* and *A. m. carnica* in northeastern Italy (Comparini and Biasilo 1991, Nazzi 1992, Meixner et al. 1993) and *A. m. mellifera* and *A. m. iberica* in northeastern Spain (Smith et al. 1991).

The formation of the African subspecies is somewhat less clear, although the source of isolation of the honey bee populations may have been shifting patterns of desertification and vegetation that characterized the continent's recent geological and ecological history (Potts and Behrensmeyer 1992). Here, too, we find subspecies that, although adapted to different ecological zones, exhibit interbreeding in areas where they commingle. Examples include the mountain bee (*A. m. monticola*) and the savanna bee (*A. m. scutellata*) in East Africa (Meixner et al. 1994) and *A. m. capensis* and *A. m. scutellata* in South Africa (Crewe et al. 1994).

Whatever the status of endemic or Old World subspecies relative to their isolation one from another, the situation in the New World (the Americas and Australia) is completely different. In these regions, a subset of the subspecies was introduced over several centuries, and the introduction assured the possibility for mixing different genetic stocks (Sheppard 1989). The 1950s introduction of the sub-Saharan African subspecies, *A. m. scutellata,* to Brazil and movement of Africanized honey bees northward into the United States are recent historical events demonstrating the adaptability of the species to varied ecological situations. Population genetic studies of the Africanization process have sometimes disagreed regarding the extent of gene flow between temperate- and tropical-adapted honey bee lineages (Hall and Muralidharan 1989, Lobo et al. 1989, Smith et al. 1989, Del Lama et al. 1990, Rinderer et al. 1991, Sheppard et al. 1991a). But there is clear agreement that distinct genetic differences exist among honey bee subspecies and are reflected in variations in defensive behavior, disease and pest resistance (tolerance), overwintering ability, and honey production, among others. Beekeepers, queen breeders/producers, and apicultural scientists need to identify and use these differences in selection and management programs to attain their shared goal of well-adapted and agronomically optimized honey bees.

The list of subspecies of *A. mellifera*

The list is presented alphabetically. The subspecies known to have been introduced to the United States are so identified. Brief descriptions of geographic range and notable characteristics were compiled from the literature, especially the work of Ruttner (1975, 1988, 1990, 1992) and Adam (1983, 1987). For several subspecies, detailed studies are lacking and are reflected in the sketchy information provided below. The citation giving the original description and the author of the subspecies name is listed at the end of each entry.

Apis mellifera adamii Ruttner

One of several named subspecies endemic to islands in the Mediterranean Sea, *A. m. adamii* is found on the island of Crete. Behavioral characteristics on Crete include year-round brood rearing with a strong peak in the spring, construction of up to 100 queen cells in colonies preparing to swarm, and variable defensive behavior.

Morphologically, the subspecies are variable in color, ranging from yellow to dark, with medium-length hair on the abdomen, broad tomenta, and low cubital index (ratio of the length of two wing veins of the cubital cell). Despite its geographic location, *A. m. adamii* is morphometrically closer to the subspecies of the eastern Mediterranean and western Turkey than to those of Greece.

Original description: Ruttner, F. 1980. *Apis mellifera adamii* (n. ssp.), die Kretische Biene. Apidologie 11: 385-400.

Apis mellifera adansonii Latreille

This subspecies occurs in western Africa and has been assigned to samples taken to Zaire from Senegal, Mali, and Niger (Ruttner 1988). Ambiguity exists as to the extent of its range and the demarcation between *A. m. adansonii* and *A. m. scutellata* toward the east and south. Behavioral characteristics include those typical for many sub-Saharan tropical subspecies: rapid colony growth, high rate of swarming, propensity to abscond during nectar dearth, and high levels of defensiveness. Morphologically, it is a small yellow bee and appears most similar to *A. m. scutellata* and *A. m. litorea*.

Original description: Latreille, P. A. 1804. Notice des espècies d'abeilles vivant en grande societé, et formant de cellules hexagonales, ou des abeilles proprement dites. Annales du Museum D'Histoire Natururelle. Paris. 5: 161-178.

Apis mellifera anatoliaca Maa

This subspecies occurs throughout Turkey, with the exception of the northeast and southeast corners. In the transition to these regions, genetic mixing occurs with *A. m. caucasica* and *A. m. meda*, respectively (Kandemir and Kence 1995). In western Turkey, the subspecies shows a morphological affinity with *A. m. adamii* of Crete. *A. m. anatoliaca* is slightly larger in size than the similar appearing *A. m. ligustica*, with a low cubital index. Although considered a "yellow bee," it ranges more accurately from orange through brown. Behavioral and apicultural characteristics associated with *A. m. anatoliaca* include brood rearing closely tied to availability of food resources, excellent overwintering ability even in cold conditions, and a propensity to swarm and use propolis. The defensive nature of the subspecies is noteworthy ("extremely bad-tempered and aggressive," [Adam 1987]).

Original description: Maa, T. C. 1953. An inquiry into the systematics of the Tribus Apidini or honey bees (Hymenoptera). Treubia 21: 525-640.

Apis mellifera capensis Escholtz

A subspecies existing in the extreme southwestern corner of the African continent, this race is most notable for its high level of expression of thelytoky, or the genetic trait whereby unfertilized eggs develop into diploid adults. Thus, it was in this subspecies that the ability of eggs laid by workers to develop into female offspring was first noted (Jack 1916). Recent work suggests that the range of *A. m. capensis* may be larger than previously thought and further defines a zone of hybridization with the neighboring subspecies, *A. m. scutellata* (Crewe et al. 1994).

Behavioral characteristics associated with the subspecies include gentleness on the combs, overwintering ability in the cool, humid conditions of the Cape Town region, the ability to raise queens in queenless swarms, and the ability of *A. m. capensis* workers to act as social parasites in *A. m. scutellata* colonies, by entering their colonies and taking over reproductive function from the *A. m. scutellata* queen. Morphologically it is a small bee of dark body color, with a slender abdomen and relatively long tongue.

Original description: Escholtz, J. F. 1822. Entomographien 1. Reimer, Berlin.

Apis mellifera carnica Pollman

The original homeland of the Carniolan honey bee included Austria, Hungary, Rumania, Bulgaria, and the former Yugoslavia. Areas of hybridization with *A. m. mellifera* and *A. m. ligustica* are noted under those subspecies. Behavioral characteristics often associated with *A. m. carnica* include rapid spring buildup, gentleness on the combs, "sensible" adaptation of brood production to the available resources, and excellent overwintering ability in cold climates (Ruttner 1988). Morphologically, the Carniolan honey bees are large, dark bees with short brownish-gray hair covering the abdomen, dense tomenta, and a characteristically high cubital index.

A. m. carnica is an important subspecies for U. S. beekeepers. It was first introduced into North America in 1877. Although it never attained the popularity of A. m. ligustica (probably owing to its resemblance to the less desired A. m. mellifera), "Carniolan" strains are commercially available from queen producers.

Original description: Pollman, A. 1889. Wert der verschiedenen Bienenrassen und deren Varietäten. 2nd ed. Voigh, Berlin Leipzig. Ruttner (1988) notes that the first edition with a description of A. m. carnica was published in 1879 but is difficult to find.

Apis mellifera caucasica Gorbachev

This subspecies is known to apiculturists as the Caucasian or gray Caucasian honey bee. Its natural range is limited to the area of the Caucasus Mountains, including their southern valleys and the "Little Caucasus" (Ruttner 1988). Behavioral and apicultural characteristics associated with this subspecies include gentleness on the combs (considered by some to be the most gentle race), a slow spring buildup to a medium population size, low tendency to swarm, high use of propolis, and susceptibility to nosema disease. Morphologically, it is similar to A. m. carnica, with general dark pigmentation, broad gray tomenta, and short, lead-gray cover hairs. The cubital index is lower than in Carniolans. Caucasian drones are noted for the black color of their thorax hairs, reportedly a unique feature (Ruttner 1988).

This subspecies was introduced into the United States in the 1880s with some fanfare, owing to its gentle disposition. The bees were never highly popular in this country, although some U. S. queen producers currently sell queens of presumed "Caucasian" ancestry.

Original description: Gorbachev, A. N. 1916. The gray mountain Caucasian bee (Apis mellifera caucasica) and its place among other bees. Tiflis (Russian with English summary).

Apis mellifera cecropia Kiesenwetter

This subspecies occurs across southern Greece and the Peloponnes. Morphologically, it is larger than A. m. macedonica and has the highest cubital index of any A. mellifera subspecies.

Original description: Kiesenwetter, E. A. H. 1860. Über die Bienen des Hymettos. Berlin Entomologische Nachrichten 1860: 315-317.

Apis mellifera cypria Pollman

This subspecies is native to the island of Cyprus in the eastern Mediterranean Sea. Noted behavioral and apicultural traits include good overwintering ability, continuous high levels of brood production from spring until fall, production of numerous queen cells in preparation for swarming, and a high level of defensiveness. The subspecies is morphologically related to the nearby *A. m. anatoliaca, A. m. syriaca,* and *A. m. meda.* Differences with neighboring *A. m. anatoliaca* include smaller size, relatively long legs and tongue, and high cubital index. The color is a rather distinctive "shining carrot-orange" (Ruttner 1988).

This race was introduced into the United States in the early 1880s with many importations coming from a queen rearing operation set up for the purpose of exportation (Benton 1880, Pellett 1938). The defensive nature and perceived poor wintering abilities of the bees caused U. S. beekeepers to lose interest in this subspecies after only a few years.

Original description: Pollman, A. 1889. Wert der verschiedenen Bienenrassen und deren Varietäten. 2nd ed. Voigh, Berlin Leipzig.

Apis mellifera iberica Goetze

This is the name given to the honey bees of the Iberian Peninsula. Behavioral traits noted for this subspecies include moderately high defensiveness, good overwintering ability, use of propolis, and a somewhat high rate of reproductive swarming. In a morphological comparison to *A. m. intermissa,* Ruttner (1988) reported *A. m. iberica* to be larger and to have narrower wings, a broader abdomen, and a lower cubital index. It is a dark bee, similar in color to neighboring *A. m. mellifera* and *A. m. intermissa.*

Original description: Goetze, G. 1964. Die Honigbiene in natürlicher and küsntlicher Zuchtauslese. Parey, Hamburg.

Apis mellifera intermissa von Buttel-Reepen

This North African subspecies occurs from the Atlas Mountains to the Atlantic or Mediterranean coasts and extends over a distance of 2,500 km from east to west (Ruttner 1988). Behavioral traits associated with *A. m. intermissa* include a "bimodal" cycle of brood rearing (with high levels of brood produced in the spring and fall), well-developed defensive behavior, production of a large number of queen cells during reproductive swarming, and the persistence of numerous virgin queens in the colony until one is mated and begins to lay eggs. Morphologically, this subspecies has black pigmentation, little hair cover, and moderate body size.

One of the African subspecies introduced into the United States during the last century, *A. m. intermissa* was known to apiculturists of the late 19th century as the "Punic Bee." The bees were tried briefly by beekeepers but, following an article dismissing them as "spiteful stingers" (Benton, in Shepherd 1892), they soon disappeared from the marketplace.

Original description: Buttel-Reepen, H. 1906. Apistica. Beiträge zur Systematik, Biologie, sowie zur geschichtlichen und geographischen Verbreitung der Honigbiene (*Apis mellifera* L.), ihrer Varietäten und der übrigen Apis-Arten. Veröff Zool. Mus. Berlin. 118-201.

Apis mellifera lamarckii Cockerell

Commonly known as the Egyptian honey bee, this subspecies is restricted to the Nile River Valley. It was used for thousands of years in migratory operations that transported it down the Nile River on large barges. Behavioral characteristics associated with the Egyptian honey bee include a high level of brood rearing throughout the year, moderate tendency to swarm, production of numerous queen cells during swarming, and moderate to high level of defensiveness. Morphologically, *A. m. lamarckii* is the smallest subspecies of honey bee, and it does not adapt well to standard-sized Langstroth equipment (S. Kamel, personal communication). It is described by Ruttner as a "small, very slender, short-tongued, short-winged and short-legged bee" (1988). The subspecies is noted for its striking yellowish or copper body coloration and white or silvery hair on the thorax.

This subspecies was initially imported into Germany in 1863 and into the United States in 1866 (Gerstäcker 1866, Parmly 1867). Although not considered successful as a honey bee of modern U. S. apiculture, genetic markers peculiar to this subspecies have been found in low frequency, but widely dispersed, in the feral honey bee population of the United States (Schiff et al. 1994).

Original description: Cockerell, T. D. A. 1906. New Rocky Mountain bees, and other notes. Canadian Entomologist 38: 160-166.

Apis mellifera ligustica Spinola

The Italian honey bee ranges throughout the Ligurian Peninsula, south of the Alps. Areas in northwestern and northeastern Italy show populations introgressing with nearby subspecies, *A. m. mellifera* and *A. m. carnica,* respectively. Behavioral characteristics often associated with *A. m. ligustica* include a rapid spring buildup to large colony size, docility on the combs, tendency to rear brood late in the fall, and relatively good resistance to diseases (Adam 1987). Morphologically, *A. m. ligustica* is well-known for the presence of yellow pigmentation, although this trait is variable, and some populations have distinctly brownish coloration. It is similar in morphology to *A. m. carnica,* but has a slightly smaller abdomen, shorter hair, and differences in wing venation.

This race figured prominently in U. S. beekeeping after its importation in 1859 (Sheppard 1989). It was highly popular because of its gentle behavior and golden color. By the late 1920s it was noted that U. S. strains of Italian honey bees were more uniformly yellow than the subspecies in Italy, probably as a result of selection by beekeepers (Alpatov 1929). Even today, U. S. beekeepers prefer this race of honey bee to all others, as measured by the predominance of "Italians" among varieties offered for sale by queen producers.

Original description: Spinola, M. 1806. Insectorum Liguriae Species Novae 1., Genuae: 159.

Apis mellifera litorea Smith

This subspecies is named from samples collected along the eastern coast of Africa and ranges from Kenya to Mozambique. Occurring in the moist coastal lowlands, it occupies a different ecological zone than that of the neighboring savanna honey bee, *A. m. scutellata*. Little is known of its biology, although Ruttner mentions that absconding is common during periods of nectar dearth (1988). When collecting samples of *A. m. litorea* from top bar hives in Kenya, they were found to be quite defensive, even with the judicious use of smoke (W. S. S. personal observation). Morphometrically, it is smaller than *A. m. scutellata* and has a long proboscis relative to its body size.

Original description: Smith, F. G. 1961. Races of honey bees in East Africa. Bee World 42: 255-260.

Apis mellifera macedonica Ruttner

This recently named subspecies occurs in parts of Ukraine, the former Yugoslavia, Bulgaria, Romania, and northern Greece. Behavioral and apicultural attributes include low defensive tendency, little inclination to swarm, reduction in brood rearing in the fall, good overwintering abilities, and high use of propolis. Morphologically, *A. m. macedonica* differs from the nearby *A. m. carnica* by having a smaller body, longer legs, longer proboscis, higher cubital index, and more yellow pigmentation.

Original description: Ruttner, F. 1988. Biogeography and Taxonomy of Honey bees. Springer-Verlag, Berlin.

Apis mellifera major Ruttner

This subspecies is restricted to the Rif Mountains of North Africa in the area close to the Mediterranean Sea (Cornuet et al. 1988, Ruttner 1992). Behaviorally, it is reported to be similar to the nearby *A. m. intermissa* (Ruttner 1988). Morphologically, the bees are very distinct, being much larger for several characters than the other subspecies of North Africa.

Original description: Ruttner, F. 1975. Ein metatarsaler Haftapprarat bei den Drohnen der Gattung Apis (Hymenoptera, Apidae). Entomologica Germanica 2: 22-29.

Apis mellifera meda Skorikov

The distribution of this subspecies includes Iran, northern Iraq, and southeastern Turkey. Behavioral traits associated with *A. m. meda* include good overwintering ability in cold regions and a strong tendency to swarm. Defensiveness is reported to vary across its range. It is morphologically similar to *A. m. ligustica,* although it can be differentiated by discriminant analysis.

Original description: Skorikov, A. S. 1929. Eine neue Basis für eine Revision der Gattung Apis L. Rep. Appl. Entomol. 4: 249-264 (Russian with German summary).

Apis mellifera mellifera L.

This bee is commonly known as the Dark Bee of northern Europe or the German Black Bee. Its original distribution ranged from the British Isles and Scandinavia eastward to the Ural Mountains of Russia and south to the Alps and Pyrenees. Ruttner (1975) reported the occurrence of gradual introgression with *A. m. carnica* across the grassland steppes of southern Russia. Zones of hybridization also occur with *A. m ligustica* in northwestern Italy and *A. m. iberica* in northeastern Spain.

Behavioral characteristics often associated with *A. m. mellifera* include a slow spring buildup of brood, an early cessation of brood rearing in the fall, "nervousness" or propensity to run on the combs, and an excellent overwintering ability in cold climates. Ruttner (1990) ascribes this subspecies' reputation of being prone to sting to hybridization rather than to a trait of the pure subspecies. Morphologically, it is a large honey bee with relatively long abdominal hairs and dark pigmentation.

This subspecies is important in the history of U. S. beekeeping because it was the first to be imported to North America, having become established in Virginia by 1620 (see references in Sheppard 1989). The Dark Bee spread rapidly throughout the continent ahead of European colonists and was known to native Americans as the "white man's fly" (Jefferson 1788). After the arrival of Italian honey bees in the United States in the mid-1800s, *A. m. mellifera* fell out of favor with beekeepers.

Original description: Linnaeus, C. 1758. Systema Naturae. 10th ed. Holmiae Laur Salvii.

Apis mellifera monticola Smith

This subspecies is unique in having a clearly disjunct distribution in isolated, forested, mountain areas of East Africa such as Mt. Kilimanjaro, Mt. Kenya, and Mt. Meru. It may also occur in the highlands of Ethiopia. There is some introgression with *A. m. scutellata,* especially in the lower elevations (Meixner et al. 1994), but *A. m. monticola* appears to persist as a relic from a previously widespread distribution during a moister period in East Africa's history (Ruttner 1988). Behavioral characteristics reported for this subspecies include a lack of defensiveness, a tendency to reduce brood rearing during shortages of forage, and an ability to fly at lower temperatures than *A. m. scutellata.* Morphologically, *A. m. monticola* is a medium-sized bee (but larger than most sub-Saharan races) that can vary in color.

Original description: Smith, F. G. 1961. Races of honey bees in East Africa. Bee World 42:255-260.

Apis mellifera sahariensis Baldensperger

This subspecies occurs in isolated oases within the overall range of *A. m. intermissa* in North Africa. Like *A. m. monticola, A. m. sahariensis* may represent relic populations of a subspecies that was once much more widespread. Behaviorally, *A. m. sahariensis* is different from *A. m. intermissa* because it is reported to be much less defensive, to have a moderate tendency to swarm, to produce fewer queen cells, and rapidly to eliminate excess virgin queens (Adam 1983, Ruttner 1988). Morphologically it is a small, yellow- (or tan) colored bee.

Original description: Baldensperger, Ph. J. 1922. Sur l'apiculture en orient. Proceedings of the International Congress of Apiculture 6: 59-64.

Apis mellifera scutellata Lepeletier

This subspecies occurs across an extensive range in the savannas of East and South Africa from Ethiopia to the Cape. Behavioral traits associated with this subspecies include those typical of sub-Saharan honey bees (see *A. m. adansonii* above). Morphologically, it is somewhat larger than the lowland *A. m. litorea* and is variable in color with both black and yellow pigmentation.

This subspecies was introduced into Brazil in the mid-1950s, and descendants of the importation became known as "Africanized" honey bees. They are most newsworthy for their defensive behavior and their ability to exist in previously marginal beekeeping areas of South America. Africanized honey bees were detected in the United States in 1990 and have subsequently been detected across the southwestern tier of states (Texas, New Mexico, Arizona, and California).

Original description: Lepeletier, A. 1836. Histoire naturelle des insectes.

Hymenopteres 1: 400-407. Roret, Paris.

Apis mellifera sicula Montagano

This subspecies refers to the bees endemic to the island of Sicily. Behavioral and apicultural traits associated with the subspecies include low defensive tendencies, almost continuous brood rearing, use of propolis, strong swarming tendencies, high number of queen cells reared during reproductive swarming, and temporary coexistence between virgin daughters and the queen mother before swarm departure. Morphologically, *A. m. sicula* is a dark bee, in contrast to the nearby yellow to brown *A. m. ligustica*. Moreover, the Sicilian honey bee has shorter forewings, shorter proboscis, longer hair, and a lower cubital index.

Original description: Montagano, J. 1911. Relation sur l'*Apis sicula*. Proceedings of the International Beekeeping Congress 5: 26-29.

Apis mellifera syriaca von Buttel-Reepen

This subspecies refers to the honey bees endemic to the region east of the Mediterranean Sea, including Israel, Jordan, Lebanon, and Syria. Behavioral traits associated with *A. m. syriaca* include extreme defensiveness, sensitivity to cold, production of high numbers of queen cells during swarming, and "survival of dozens of young queens in the colony until the first of them is mated and laying eggs" (Ruttner 1988). Morphologically, this is one of the smallest honey bee subspecies. It has short hair covering the abdomen, a low cubital index, and bright yellow pigmentation.

This subspecies, like *A. m. cypria*, was imported into the United States in the 1880s and tried briefly by beekeepers. Although Frank Benton was responsible for many of these early introductions, he later denounced them as "inferior in temper and wintering qualities to the races already here [in America]" (Benton 1899).

Original description: Buttel-Reepen, H. 1906. Apistica. Beiträge zur Systematik, Biologie, sowie zur geschichtlichen und geographischen Verbreitung der Honigbiene (*Apis mellifera* L.), ihrer Varietäten und der übrigen Apis-Arten. Veröff Zool. Mus. Berlin 118-120.

Apis mellifera unicolor Latreille

This is the name given to the honey bees endemic to the island of Madagascar. According to Douhet (1965), behavioral traits vary between coastal lowland and central highland forms, with the highland bees being better suited to apiculture. Morphologically, the subspecies is characterized by a deep black pigmentation, short tongue, and relatively long wings.

Original description: Latreille, P. A. 1804. Notice des espècies d'abeilles vivant en grande societé, et formant de cellules hexagonales, ou des abeilles proprement dites. Annales du Museum D'Histoire Naturelle Paris 5:161-178.

Apis mellifera yemenitica **Ruttner**

This subspecies, adapted to some of the most extreme desert conditions of the species' range, can be found in Yemen, Oman, Somalia, Saudi Arabia, and Chad. Behaviorally, it is considered only mildly defensive, with a rapid buildup and frequent reproductive swarming under good forage conditions. Morphologically, *A. m. yemenitica* is a very small bee with some characteristics overlapping those of the allopatric congener, *A. cerana*.

Original description: Ruttner, F. 1975. Ein metatarsaler Haftapprarat bei den Drohnen der Gattung Apis (Hymenoptera, Apidae). Entomologica Germanica 2: 22-29.

Appendix Three

Synonymy in Bee Diseases
Hachiro Shimanuki

Bee diseases have been described in fascinating ways as long as bees have been kept, but the terminology used has sometimes been so confusing that earlier descriptions cannot be related to the diseases known today. Indeed, the only data given may be descriptions of bee behavior or speculations concerning cause and effect. Thus, this appendix can only reflect opinions of the author about synonymy in bee diseases.

The term "foulbrood" was used originally to describe a disease of brood, but the first descriptions (before the early 1900s) were so vague that the disease described could have been either American or European foulbrood disease. Phillips and White (1912) reported that Molitor-Mühlfeld described two types of foulbrood in 1868, the "infectious" (virulent) kind and the "mild" (nonvirulent) kind. Molitor-Mühlfeld believed that the mild foulbrood was caused by neglected brood that had become chilled and that virulent foulbrood was the result of an attack by a small parasitic ichneumonid wasp. But the description he gave for the infectious form is quite applicable to American foulbrood disease. Among the symptoms given were punctured cappings, ropiness, stickiness, and an ill-smelling mass that the bees could not remove.

Phillips and White (1912) reported that Dzierzon, in 1882, had also described two types of foulbrood, a mild form and a virulent form. Both forms were said to be contagious, but the mild form was curable. The descriptions given by Dzierzon for the mild form and the malignant form of foulbrood fit European and American foulbrood disease, respectively, as we know them today. Dzierzon, however, assumed that the two diseases were caused by the same organism.

Cheshire and Cheyne (1885) agreed with Dzierzon that there were two forms of larval disease of honey bees. They isolated *Bacillus alvei* from one of them and believed that it caused the mild form of the disease. Howard (1900) studied New York disease (or black brood) and named the causal organism *Bacillus milii* (New York disease is now thought to be European foulbrood). Today *B. alvei* is known not to cause European foulbrood disease but so frequently accompanies it that many scientists use it as a criterion for diagnosis of the disease. Phillips, in the preface to White (1906), designated European foulbrood as "that disease in which *Bacillus alvei* is present." As a result, foulbrood induced by *Bacillus larvae* was henceforth called American foulbrood disease even though White had earlier referred to *B. larvae* as *Bacillus "X"* and the disease as "X brood."

Today, the disease we call American foulbrood is called that in many languages. In others, the name is the equivalent of "malignant" or "virulent" foulbrood. European foulbrood is also called benign, mild, sour, or acid foulbrood. "Parafoulbrood" is sometimes cited as a separate disease that is supposedly caused by *Bacillus para-alvei*. *B. para-alvei*, however, is not recognized as being different from *B. alvei;* parafoulbrood is generally accepted as European foulbrood disease.

Septicemia is an example of a poorly named disease of honey bees. The term literally means septic blood or blood that contains a poison as a result of invasion of the bloodstream by any virulent microorganism. Septicemia in bees, however, refers only to the disease that results from the infection of bees by *Pseudomonas apiseptica* Burnside.

Pickled brood and sacbrood are names for the same disease; pickled brood is the earlier name. The disease was first attributed to a fungus, though it is now known to be caused by a virus. White (1913) suggested the name sacbrood instead and defined it as a brood disease where no microorganisms are present. The virus could not be seen with a light microscope.

Amoeba disease or ameba disease is also called malpighamoebiosis and amoebiasis apis, especially in European literature. This disease of the adult bee is caused by *Malpighamoeba mellificae;* the earlier name for the organism was *Vahlkampfia mellificae* (Steinhaus 1949a).

Nosema disease is most frequently referred to as nosema and most correctly, though rarely, as nosemosis. The primary name of the disease is derived from the name of the pathogen, *Nosema apis* Zander.

Chalkbrood is also called hard brood, lime brood, or pericystis mycosis, from the former name of the fungal pathogen, *Pericystis apis* (now *Ascosphaera apis* Olive and Spiltoir). Infrequently, chalkbrood is referred to as ascosphaeriosis. Stonebrood is another disease of fungal origin. This disease can be caused by more than one *Aspergillus* species, however, so the disease should be called aspergillus mycosis or aspergillosis. Because stonebrood is sometimes called hard brood, it is clearly necessary to be as precise as possible.

A disease that remains poorly characterized is Isle of Wight disease. Beekeepers and scientists are still speculating about the possible cause(s) of this disease, which "was said to have reached epidemic proportions in the British Isles on at least three occasions between 1905 and 1919" (Bailey 1963a). Rennie (1921) believed that he had found the causative agent when he discovered *Acarapis woodi*, the tracheal mite, in bees from affected colonies. (Many authors have since used the term acarine disease for problems attributed to tracheal mites, although this term could apply to disease caused by any mite species.) Bailey (1963a, 1964) challenged Rennie's claim that the mites were the primary cause of Isle of Wight disease. He argued that a variety of diseases, weather conditions, and poor beekeeping practices could have caused the reported losses, and that the number of colonies that died has been greatly exaggerated. Bailey (1964) concluded that "we have no evidence that any [single] parasite we know today was the cause of the wholesale losses of bees." The factor he believed caused the Isle of Wight disease "myth" was the lack of knowledge among beekeepers of the parasites and noninfectious diseases of honey bees. Adam (1968), however, still believed that the Isle of Wight disease was caused by *Acarapis woodi*. In my opinion, it is unlikely that any one pathogen or organism was responsible. Indeed, the length of time during which bee populations were reduced seems to rule out the one-pathogen theory. I agree with Bailey that several diseases and several factors probably led to the loss of bees. But the primary cause of colony losses seems to be *A. woodi*. Consequently, the use of Isle of Wight disease as a synonym of *A. woodi* seems justifiable. Usually, beekeepers and scientists tend to observe the loss of bees and colonies and attribute it to a single cause. Many diseases, communicable and noncommunicable, can cause the death of bee colonies; although the diseases may be different, the symptoms may be similar. For instance, the presence of bees crawling on the ground near the hive entrance can be a result of paralysis, tracheal mites, nosema disease, chemical poisoning, lack of energy to fly, or chilling.

Nutritional diseases of honey bees have also confused beekeepers and scientists. For example, much has been written about May disease, a disease caused by pollens of *Ranunculus* species (buttercups). This one disease has many names: *Laufkrankheit* (running-about illness), *Sandläuferei* (running-in-the-sand sickness), *Tollkrankheit* (frenzy sickness), *Flügellähme* (wing paralysis), *Zitterkrankheit* (trembling sickness), *Flugunfähigkeit* (flight incapacity), *Paralyse* (paralysis), *Kreiselkrankheit* (reeling sickness), and Bettlach disease. Clearly, each name describes only one of the many symptoms that can be observed. In my opinion, the alimentary poisonings by toxic pollens should be so classified. Then the term "May disease" and all its synonyms should be replaced by the more descriptive subcategory toxic pollen (*Ranunculus* species).

I propose that the hairless-black syndrome and bee paralysis be named chronic bee paralysis disease. In both cases, an etiologic agent, a virus, has been isolated, and they appear to be similar. Unfortunately, this is not the case with *mal des forêts* or *mal de las selvas* (forest disease), *Waldtrachtkrankheit* (forest foraging disease), black sickness, black disease, bee paralysis (Russian), and forest sickness. It is impossible to prove that these diseases are of viral or nutritional origin. If the diseases are nutritional, the floral sources, and not merely the symptoms, should be cited.

On the other hand, dysentery should not be called a disease. It is a symptom that can result from several causes including fermented honey, nosema disease, and sometimes honeydew.

"Disappearing disease" is a classic misnomer. In the first place, the bees disappear, not the disease. In addition, in my opinion, the term is used as an umbrella for what may well be many maladies. "Disappearing disease," like the Isle of Wight disease, may be more than one disease. Certainly, attempts to find one common cause for an effect that occurs under a variety of conditions have thus far been unsuccessful. We do know that beekeepers observe an unexplained loss of adult bees in either spring or fall. No laboratory test is yet available to confirm the presence of any disease in the colony – communicable or noncommunicable.

Foote (1966) stated that autumn collapse, described only from California, U.S.A., was also called "disappearing disease." From conversations with beekeepers, state officials, and researchers who have done some preliminary laboratory studies, I am convinced that autumn collapse is primarily a nutritional problem, that is, a lack of some needed component in the diet, which inhibits brood rearing. This explanation does not rule out all other causes, because the nutrition of honey bees can be modified by genetics, climate, and floral sources. In addition, the gross symptoms of disappearing disease and autumn collapse resemble what was previously called "fall dwindling" or "spring dwindling." In other words, we again have the confusion that results when beekeepers and scientists try to explain all unexplained bee losses because no pathogen is present. I am convinced that the terms "disappearing disease" and "autumn collapse" should be dropped. I suggest that we use "spring," "fall," or "winter dwindling syndrome." If proof is found that these are indeed one disease, a new name should then be coined.

APPENDIX FOUR

———— ⌘ ————

Laboratory Diagnosis of Honey Bee Diseases
D. A. Knox and Hachiro Shimanuki

Bacterial diseases
  American foulbrood disease
     Detection of *Bacillus larvae* spores in hive products
  European foulbrood disease
  Powdery scale disease
  Septicemia
  Spiroplasma
Fungal diseases
  Chalkbrood
  Stonebrood
Viral diseases
  Sacbrood
  Chronic bee paralysis
  Filamentous virus (rickettsial disease)
Protozoan diseases
  Nosema disease
  Amoeba disease
Parasitic honey bee mites
  Honey bee tracheal mite (Acarapis woodi)
  Varroa jacobsoni
Directions for sending diseased brood and adult honey bees for diagnosis
  Samples of brood
  Samples of adult honey bees
  How to address samples

Apiary inspectors and beekeepers must be able to recognize bee diseases and parasites. They should also be acquainted with sampling techniques and routine laboratory procedures for occasions when laboratory diagnosis or verifications are necessary.

Bacterial diseases

Most bacterial diseases of the honey bee are diagnosed by observing the associated microorganisms with a compound microscope. The modified hanging drop technique (Michael 1957) is very useful for differentiating some bacterial diseases of the brood. Residue from a suspected cell is first mixed with water. Then a drop of this suspension is placed on a cover glass. The suspension used should always be dilute and only slightly turbid. The resultant smear is dried and fixed under a heat lamp, or the smear can be air-dried and heat-fixed by passing it rapidly through a Bunsen burner flame two or three times. The fixed smear is stained with carbol fuchsin [Solution A: 0.3 g basic fuchsin (90% dye content), 10 ml ethyl alcohol (95%); Solution B: 5 g phenol, 95 ml distilled water; combine solutions A and B] or a suitable spore stain for 10 seconds. Enough stain should be placed on the cover glass to blanket the entire smear. The excess stain is then washed off with water. While the smear is still wet, the cover glass is inverted with the smear side down and placed onto a standard microscope slide previously coated with a very thin layer of immersion oil. The slide is gently blotted dry and examined with a microscope using the oil immersion objective. Results: Organisms that are not fixed to the cover glass are caught in areas where pockets of water have formed in the oil, and the organisms usually exhibit Brownian movement.

Alternatively, using a simple stain, bacteria can be differentiated solely by morphology. A drop of the suspension is placed directly on a microscope slide. The smear is heat-fixed and stained as described in the previous section. Carbol fuchsin, methylene blue, and safranin are examples of stains that can be used. The stained smear is air-dried or gently blotted. A drop of immersion oil is placed directly on the smear. No cover glass is necessary. The slide is examined using the oil immersion objective. Results: Organisms are uniformly stained and easily distinguished.

The Gram stain is a standard microbiological method that can be substituted for the simple stain. Briefly, the procedure is as follows: A fixed smear is stained with crystal violet, immersed in iodine solution, decolorized in ethyl alcohol, and counterstained with safranin. Results: Gram-positive organisms are blue; Gram-negative organisms are red.

American foulbrood disease

Bacillus larvae is the bacterium that causes American foulbrood (AFB). *B. larvae* is a slender rod with slightly rounded ends and a tendency to grow in chains. The spore is oval and approximately twice as long as wide, about 0.6 by 1.3 mm. When stained with carbol fuchsin, the spore walls appear reddish-purple and clear in the center. The spores may form clusters and appear to be stacked. If the larva has been infected for less than 10 days, the vegetative cells are present and some newly formed spores may be seen.

The modified hanging drop technique can be very useful for differentiating American foulbrood from other brood diseases. In areas of the smear where pockets of water are formed in the oil, the spores of *B. larvae* exhibit Brownian movement. This is a valuable diagnostic technique because the spores formed by the other *Bacillus* species associated with the known bee diseases usually remain fixed. Brownian movement can be affected by slide preparation; in addition, debris and other bacteria can exhibit this motion. Therefore, Brownian movement must not be used as the sole criterion for diagnosis but must be considered together with the characteristic morphology of the spores and the gross larval symptoms. If microscopic examination is not conclusive, cultural tests can be made using the same suspension.

Routine culture media such as nutrient broth will not support the growth of *B. larvae*. Good vegetative growth occurs on Difco brain heart infusion fortified with 0.1 mg thiamine HCl per liter of medium (BHIT) and adjusted to pH 6.6 with hydrochloric acid. The medium can be liquid, semisolid (0.3% agar), or solid (2% agar). To culture *B. larvae,* prepare spore suspensions by mixing diseased material with sterile water in screw-cap tubes. The suspension is heat-shocked at 80°C for 10 minutes (effective time) to kill non-spore-forming bacteria. A sterile glass rod (bent) or sterile cotton swab is used to spread a portion of the suspension (approximately 0.2 ml) evenly over the surface of a BHIT agar plate, which is then incubated for 72 hours at 34°C. Individual colonies are small (1 to 2 mm) and opaque; if large numbers of viable *B. larvae* spores are inoculated, however, a solid layer of growth may cover the plate.

Detection of *Bacillus larvae* spores in hive products

Honey: Occasionally it may be necessary to examine honey for the presence of *B. larvae*. Because of the high concentration of carbohydrate and other natural bacteriostatic substance(s) in honey, the examination of honey requires special considerations. The classical method (Sturtevant 1932, 1936) is to dilute the honey 1:9 with water, centrifuge the mixture to concentrate the spores in the sediment, and then examine the sediment microscopically for the presence of spores. Cultural techniques are required, however, to demonstrate conclusively the presence of *B. larvae* spores in honey.

Direct inoculation (Hansen 1984a,b): Aliquots (5 g) of the honey to be examined are placed in 50-ml sterile beakers, which are then set in a water bath for five minutes (effective time) at 88°C to 92°C. After heating the honey, using an inoculating loop, three agar plates containing appropriate media are consecutively streaked. This is equal to a sample of approximately 0.08 g honey. The plates are incubated at 37°C for 72 hours and examined for colonies of *B. larvae*. The method can detect *B. larvae* when more than 2,000 spores are present in 1 g honey. The direct inoculation of honey onto agar plates is quick and easy, but it is limited by the amount of honey that can be sampled and the likelihood of contaminants.

Dialysis (Shimanuki and Knox 1988): Honey to be examined is heated to 45°C to permit easier handling and to decrease viscosity for more uniform distribution of any spores that may be present. The honey (25 ml) is placed in a 50-ml beaker and diluted with 10 ml of sterile water. The diluted honey is then transferred into a 1.75-inch (44-mm) dialysis tube. The open end is tied after the tube is filled, and the tube is submerged in a water bath with three to four water changes over 18 hours. Following dialysis, the contents of the tube are centrifuged at about 2,000 g for 20 minutes. The supernatant is carefully removed with a pipette, the remaining residue is resuspended in 9 ml of water in a screw-cap vial and heat shocked at 80°C (effective time) for 10 minutes to kill non-spore-forming bacteria. Next, 0.5 ml of the suspension is spread onto a plate of BHIT agar. The plate is incubated at 37°C for 72 hours and examined for colonies of *B. larvae*. Using the dialysis technique, the presence of *B. larvae* can be determined when more than 80 spores are present in one g honey.

Pollen: *Bacillus larvae* spores can also be recovered from bee-collected pollen pellets by physically removing bits of AFB scale. A series of sieves of different sizes is helpful. If scales are not detected, one may pass a water-pollen suspension through No. 2 filter paper, centrifuge the filtrate, and culture the pellet as described above (Gochnauer and Corner 1974).

Beeswax: Spores morphologically similar to those of *B. larvae* can be recovered by melting beeswax in boiling water, removing the beeswax cake after cooling, and centrifuging the water at 2,000 X G for 20 minutes. The sediment is then examined microscopically for the presence of spores. Positive identification of the spores is not possible, however, because the recovery technique renders the spores nonviable.

European foulbrood disease

Melissococcus pluton is the bacterium that causes European foulbrood (EFB). This bacterium is generally observed early in the infection cycle before the appearance of the varied microflora associated with this disease. The *M. pluton* cell is short, non-spore-forming, and lancet-shaped. The cell measures 0.5–0.7 by 1.0 mm and occurs singly, in pairs, or in chains.

It is difficult to isolate *M. pluton* in vitro because of its growth requirements and competition from other bacteria. In addition, identification of *M. pluton* is difficult because of its pleomorphic nature, especially when grown on artificial media. *Melissococcus pluton* will grow on a medium developed by Bailey (1959): 1 g yeast extract (Difco), 1 g glucose, 1.35 g potassium dihydrogen phosphate (KH_2PO_4), 1 g soluble starch, 2 g agar, and distilled water to make 100 ml; the pH is adjusted to 6.6 with potassium hydroxide (KOH), and the mixture is autoclaved at 10 lb/inch^2 (116°C) for 20 minutes. A water suspension prepared from larvae (apparently healthy, infected, or dead), can be streaked on freshly prepared medium and incubated anaerobically at 34°C. The "Gas Pak" (BBL) Anaerobic System (disposable hydrogen and carbon dioxide generator) is used to obtain anaerobic conditions. Small white colonies of *M. pluton* should appear after four days.

Some organisms do not cause European foulbrood, but they influence the odor and consistency of the dead brood and can be helpful in diagnosis. These secondary invaders include the following:

Bacillus alvei: The bacterium *B. alvei* is frequently present in cases of European foulbrood disease (EFB). It is a rod 0.5–0.8 mm wide by 2.0–5.0 mm long. Spores measure 0.8mm by 1.8–2.2 mm. Like *B. larvae*, the spores may be clumped and appear stacked. The sporangium may be attached to the spore. On nutrient agar, *B. alvei* spreads vigorously, may show "motile colonies," and produces an unpleasant odor.

Bacillus laterosporus: The vegetative form of *B. laterosporus* measures 0.5–0.8 mm by 2.0–5.0 mm, and the spores 1.0–1.3mm by 1.2–1.5 mm. An important diagnostic feature is the production of a canoe-shaped parasporal body that stains very heavily along one side and the two ends and remains firmly adherent to the spore after lysis of the sporangium. The clear portion with the finely outlined wall is the spore. *Bacillus laterosporus* grows on nutrient agar, becoming dull and opaque, and spreads actively if the agar surface is moist.

Enterococcus faecalis: Ovoid cells (elongated in the direction of chain) are 0.5 to 1.0 mm in diameter and are usually in pairs or short chains. This organism resembles *Melissococcus pluton* and may exhibit Brownian movement when the modified hanging drop technique is used. Growth occurs on nutrient agar usually within one day. Colonies are generally smaller than 2 mm; they are smooth and convex, with a well-defined border. When magnified, the colonies appear light brown and granular.

Bacillus apiarius: The bacterium *B. apiarius* is rarely encountered and may or may not be associated with EFB. Rods are 0.6 to 0.8 mm in diameter and often less at the poles. Special diagnostic features include the ridged, thick, rectangular spore coat and the stainable remnants of the sporangium, which remain attached for a considerable time. Growth can occur on nutrient agar and Sabouraud dextrose medium.

Powdery scale disease

Bacillus pulvifaciens is the bacterium that causes powdery scale disease. This disease is seldom reported, perhaps because the average beekeeper is unable to identify it. A useful diagnostic characteristic is the scale that results from the dead larva. The scale is light brown to yellow and extends from the base to the top of the cell. The scale appears to be rigid, but crumbles into a powdery dust when handled. *Bacillus pulvifaciens* vegetative cells measure 0.3–0.6 mm by 1.5–3.0 mm. The spores are 1.0 mm by 1.3–1.5 mm. The bacterium can be isolated on nutrient agar, but growth is heavier on glucose agar. When first isolated, the organism produces a reddish-brown pigment that can be lost by subculturing. *Bacillus pulvifaciens* closely resembles *B. larvae,* but the spores do not exhibit Brownian movement in the modified hanging drop technique. *B. pulvifaciens* can also be differentiated from *B. larvae* by its ability to grow at 20°C and on nutrient agar.

Septicemia

Pseudomonas aeruginosa is the bacterium that causes septicemia in adult honey bees. This disease results in the destruction of connective tissues of the thorax, legs, wings, and antennae. Consequently, the affected bees fall apart when handled. Dead or dying bees may also have a putrid odor. Vegetative cells measure 0.5–0.8 mm by 1.5–3.0 mm. They are Gram-negative and occur singly, in pairs, or in short chains. A bacterial smear can be prepared by removing a wing from the thorax and dipping the wing base in a drop of water on a microscope slide. The smear, after heat fixing, can be stained with the Gram stain procedure. To culture this organism, streak the base of a wing across Difco *Pseudomonas* isolation agar or *Pseudomonas* Agar F. The optimal temperature for growth is 37°C. *Pseudomonas aeruginosa* in culture is characterized by the excretion of diffusible yellow-green pigments that fluoresce in ultraviolet light (wavelength below 260 nm).

Septicemia disease can also be diagnosed by reproducing the disease symptoms in healthy caged bees. This is accomplished by preparing a water extract (macerate the equivalent of one suspect bee per ml of water) and inoculating healthy bees in the thorax with a syringe and needle or dipping bees in the water extract. Bees with septicemia die within 24 hours and exhibit the typical odor and the "break apart" symptom after approximately 48 hours.

Spiroplasma

Spiroplasma is a helical, motile, prokaryote with no cell walls that is found in the hemolymph of infected adult honey bees. The organism is a tiny, coiled, and sometimes branched filament 0.7 to 1.2 mm in diameter. Its length increases with age and ranges from 2 to >10 mm (Clark 1977b, 1978a). Spiroplasma can be seen in the hemolymph using a darkfield condenser on the microscope. Hemolymph can be taken from adult bees by puncturing the intersegmental membrane directly behind the first coxa with a fine capillary tube made from the tip of a Pasteur pipette. This organism can be cultured in mycoplasma broth medium.

Fungal diseases
Chalkbrood

Ascosphaera apis, the fungus that causes chalkbrood disease, is a heterothallic organism that develops a characteristic spore cyst when opposite thallic strains (+ and -) fuse. Chalkbrood disease can be easily identified by its gross symptoms. *A. apis* grows luxuriantly on potato dextrose agar fortified with 4 g yeast extract per liter. Growth and sporulation also occur on malt agar, but less profusely and with no aerial hyphae; this facilitates subculturing and microscopic examination. Cultures have a characteristic fruity odor similar to that of fermenting peaches. The optimal temperature for growth is 30°C. *A. apis* can be easily isolated from newly infected larvae or fresh mummies. These can be placed directly on the medium and incubated. New mycelial growth is usually visible within 24 hours. If only one thallus (+ or -) is isolated, a fluffy, cottonlike growth will eventually cover the plate. When both the + and - thalli are present, black spore cysts form throughout the culture. *Ascosphaera apis* can also be isolated from old mummies by placing them on water agar (agar with no added nutrients), incubating them, and transferring the new mycelial growth to a nutrient medium. *A. apis* may sometimes fail to grow or may be overgrown by other fungi, especially from contaminants on old mummies.

Stonebrood

Stonebrood is usually caused by *Aspergillus flavus, A. fumigatus,* and *A. niger* or other *Aspergillus* species. An infected larva is hard and is difficult to remove from the cell and crush. The fungus can erupt from the integument forming a false skin covered with powdery fungal spores. The spores of *A. flavus* are yellow green, *A. fumigatus* spores are gray green, and those of *A. niger* are black. These spores can become so numerous that they fill the cells and cover the comb. Stonebrood can usually be diagnosed from gross symptoms, but positive identification of the fungus requires its cultivation in the laboratory and subsequent examination of its conidial heads. *Aspergillus* spp. can be grown on potato dextrose or Sabouraud dextrose agars.

Viral diseases

Sacbrood

Sacbrood disease is the only common brood disease that is caused by a virus. Because sacbrood-diseased larvae are relatively free from bacteria, laboratory verification is usually based on gross symptoms (White 1917) and the absence of bacteria. Positive diagnosis requires the use of a specific antiserum.

Chronic bee paralysis

Ideally, the diagnosis of paralysis disease is made using serological techniques. Because this is beyond the capability of most laboratories, diagnosis is usually made by observing symptoms in individual adult bees; bees appear hairless and greasy. Paralysis disease can also be diagnosed by reproducing the disease symptoms in healthy bees by spraying, feeding, or inoculating a water extract made from the suspect bees.

Filamentous virus (rickettsial disease)

Filamentous virus, previously thought to be a rickettsia, can be diagnosed by examining the hemolymph of infected adult bees using phase-contrast or darkfield microscopy. The hemolymph of honey bees infected with this virus is milky white and contains many spherical to rod-shaped viral particles of a size close to the limit of resolution for light microscopy. The viral particles consist of a folded nucleocapsid within a viral envelope and are 0.4 by 0.1 mm (Clark 1978b).

Protozoan diseases

Procedures useful for examining adult bees for protozoa include wet mounts and the removal of digestive tracts. To prepare a wet mount, macerate a portion of the sample in water. Place a drop of the suspension on a microscope slide, and carefully drop the cover glass on it to minimize air pockets. No stain is required. The wet mount is usually examined using the dry objectives of a microscope. Results: Organisms refract light and are therefore visible on the slide. A phase-contrast microscope may be helpful, especially if an oil immersion objective is required.

Digestive tracts can easily be obtained by first removing the head of the bee to free the digestive tract and then grasping as much of the stinger as possible with a pair of fine tweezers. Finally, gently remove the entire digestive tract using a steady action. Freshly killed honey bees are required for this procedure.

Nosema disease

Nosema apis spores are large, oval bodies, 4–6 mm long by 2–4 mm wide. The spores develop exclusively within the epithelial cells of the ventriculus of the adult honey bee. Positive diagnosis can be made only by microscopic examination of suspect bees or their fecal material for the presence of *Nosema* spores. Samples of bees to be examined can be dried or preserved in alcohol. For quick examinations, the abdomens from 10 or more bees are removed, placed in a dish with 1.0 ml water per bee abdomen, and ground with a pestle or the rounded end of a clean test tube. A cleaner preparation can be obtained by grinding only the digestive tracts. A wet mount is prepared from the resulting suspension and examined under the high dry objective of a compound microscope.

Nosema can be detected without sacrificing workers or queens by examining their fecal material. A colony can be sampled by collecting feces of worker bees on glass plates placed near the hive entrance, scraping off the deposit, mixing it with water, and preparing a wet mount from the resulting suspension (Wilson and Ellis 1966). Queens can be held in small dishes or in glass tubes and allowed to walk freely. They usually defecate within one hour. Queen feces appear as drops of clear, colorless liquid, which are then transferred to a microscope slide with a pipette or capillary tube. A cover glass is placed over the feces before examination with a high dry objective (L'Arrivee and Hrytsak 1964).

Amoeba disease

Because *Malpighamoeba mellificae* cysts are found in the Malpighian tubules of adult bees, diagnosis is made by the removal and microscopic examination of the tubules for their presence. The Malpighian tubules are long, threadlike projections originating at the junction of the midgut and the hindgut. The tubules can be teased away from the digestive tract with a pair of fine forceps and then placed in a drop of water on a microscope slide. A cover glass is positioned over the tubules while applying uniform pressure to obtain a flat surface for microscopic examination. Cysts measure 5–8 mm in diameter and can be seen through the tubule wall using a high dry objective and then changing to the oil immersion objective for more detail.

Parasitic honey bee mites

Honey bee tracheal mite (*Acarapis woodi*)

Positive diagnosis can be made only by microscopic examination of the tracheae for mites; *Acarapis woodi* is the only mite found in adult honey bee tracheae. The trachea must be examined carefully; it may not always be discolored when mites are present, and a cloudy or discolored trachea does not always contain mites. The trachea of a severely infested bee has brown or black blotches with crustlike lesions and is obstructed by many mites in different stages of development.

Methods for dissecting honey bees to obtain tracheae are listed below. Each of these methods has advantages and disadvantages.

1. Grasp the bee with forceps or between your thumb and forefinger and remove the head and first pair of legs. Then with a scalpel, razor blade, or fine pair of scissors, cut a thin transverse section from the anterior face of the thorax in such a way as to obtain a disk. Place the disk on a microscope slide and add a few drops of lactic acid. This makes the material more transparent and also helps to separate the muscle. With the aid of a dissecting microscope (10–20X), carefully separate the muscles to expose the tracheae. Gently remove tracheae that appear to contain mites and place them in a drop of lactic acid on a glass slide. Place a cover glass over the tracheae and examine using a compound microscope (40–100X). This procedure is useful for the rapid examination of a few bees.

2. Prepare transverse-section disks from the thoraces of 50 honey bees as described in method 1, place them in 5–10% potassium hydroxide (KOH), and incubate at 37°C for 16 to 24 hours. The KOH dissolves the muscle and fat tissue, leaving the tracheae exposed. Then examine the disk-tracheae suspension under a dissecting microscope (10–20X). Remove suspicious-appearing trachea from the disks and examine the tracheae microscopically (40–100X). This procedure is recommended when large numbers of bees are to be examined.

3. Differentiation of live mites from dead mites (Eischen et al. 1986) is the method of choice for evaluating chemicals used to control tracheal mites. Anesthetize live bees with carbon dioxide, and remove the abdomens with a scalpel to prevent being stung during examination. Remove the head and first pair of legs of each bee by holding the bee on its back and gently pushing this section off with a downward and forward motion. Place the bee under a dissecting microscope, and remove the first ring of the thorax with fine forceps. This exposes the tracheal

attachment to the thoracic wall. Remove tracheae that appear abnormal with the forceps and transfer to a microscope slide containing a thin film of glycerol. Then dissect the tracheae using a pair of fine needle probes. Mites are considered dead if they do not move; dead mites also often appear discolored and desiccated. Living mites have a translucent gray or pearl color and move within a few seconds after dissection.

Varroa jacobsoni

The adult female mite is oval and flat, about 1.1 mm long and 1.5 mm wide, and pale to reddish brown; it can easily be seen with the unaided eye. The mites attach to the adult bee between the abdominal segments or between body regions and may therefore be difficult to detect. *Varroa jacobsoni* can be dislodged by shaking the bees in 70% alcohol. The mites are recovered by passing the bees and alcohol through a wire screen (8- to 12-mesh) to remove the bees and then sieving the alcohol through a 50-mesh screen or cotton cloth. The screen or cloth is then examined for mites. On brood, *Varroa* can be easily seen against the white surface of worker or drone pupae after they are removed from their cells.

Note that the bee-louse, *Braula coeca,* resembles *Varroa jacobsoni* in size and color. *Braula,* however, being an insect, has six legs that extend to the side. *Varroa,* an arachnid, has eight legs that extend forward.

Directions for sending diseased brood and adult honey bees for diagnosis

The Bee Research Laboratory has seen an increase in the number of brood samples that we in the laboratory have chosen to call the "parasitic mite syndrome." It is likely that this syndrome is the result of a secondary infection in colonies infested with *Varroa jacobsoni.* The microbial flora associated with this syndrome may require biochemical tests. Accordingly, now more than ever, an accurate diagnosis depends on the sample.

Samples of brood

• The sample of comb should be at least 2 x 2 inches and contain as much of the dead or discolored brood as possible. *No honey should be present in the sample.*

• The comb can be loosely wrapped in a paper bag, paper towel, or newspaper and sent in a heavy cardboard box. Avoid wrappings such as *plastic bags, aluminum foil, waxed paper, tin, or glass because they facilitate decomposition.*

• If a comb cannot be sent, the probe used to examine a diseased larva in the cell may contain enough material for tests. The probe can be wrapped in paper and sent to the laboratory in an envelope.

Samples of adult honey bees

•Send at least 100 bees in a sample.

•If possible, select bees that are dying or that died recently. Decayed bees are not satisfactory for examination.

•Bees should be placed in 70% ethyl or methyl alcohol as soon as possible after collection and carefully packed in leak-proof containers.

•If alcohol is unavailable or if you suspect a virus disease, bees can be loosely wrapped in a paper bag, paper towel, or newspaper and sent in a mailing tube or heavy cardboard box. Avoid *plastic bags, aluminum foil, waxed paper, tin, and glass.*

How to address samples

•Samples are accepted from anyone and in most cases are processed within one working day.

•Send all samples to Bee Disease Diagnosis, Bee Research Laboratory, Beltsville Agricultural Research Center-East, Building 476, Beltsville, MD 20705.

•A short description of the problem, along with your name and address, should be attached to the outside of the package.

APPENDIX FIVE

Honey Bees and Pesticides
Roger A. Morse

History
The DDT era
The carbamate era
The beekeeper indemnity payment program
The Environmental Protection Agency
The current situation
Symptoms of pesticide losses
Control

Pesticides are chemical substances, sometimes manufactured, that are used to kill pestiferous plants and animals. Although they are usually made in a laboratory, several are patterned after natural chemicals that plants use for protection against animals, especially insects, that might eat them. Some chemicals, such as rotenone, pyrethrums, arsenic, and nicotine, that are used as pesticides are extracted directly from plants. The ideal pesticide acts rapidly and kills only the target pest; in reality, however, nontarget species are sometimes affected. Thus, honey bees often suffer from the misuse of pesticides. Pesticides are not of primary concern in this book. But, they are mentioned here because they can kill honey bees. This appendix summarizes the history of honey bee-pesticide losses in the United States and suggests ways to distinguish honey bee losses caused by disease from those caused by pesticides. The honey bee-pesticide problem is less serious than it was two to five decades ago because of new legislation, a greater public awareness, and the fact that agriculture is becoming more concentrated. A list of the common insecticides used in the United States and their relative toxicity to honey bees has been prepared (Atkins et al. 1981).

History

The first recorded death of honey bees in the United States because of a pesticide took place in 1881 when a farmer noted that his bees were being killed after visiting a flowering pear tree that he had sprayed with Paris green (an insecticide containing arsenic). Shaw (1941) reviewed this and the pertinent literature and found that arsenicals, used both as herbicides and insecticides, had been causing serious honey bee losses. Dusts were more toxic to honey bees than were liquid formulations. A few people surveyed by Shaw reported problems with rotenone. Shaw's paper included the results of a survey, to which 62 people responded, that indicated that poisoning of honey bees was increasing. The arsenicals continued to be a problem for bees and beekeepers through the 1940s until they were replaced by DDT and related compounds.

The DDT era

The insecticide DDT was designed in the late 1930s and used widely to protect U.S. troops by killing malaria-carrying mosquitoes, chiggers, and other disease-carrying and noxious insects during World War II. Following the war, DDT was tested and used on a variety of agricultural crops. There is no record of a colony of honey bees being killed by a single application of DDT despite its widespread use. DDT is a selective insecticide and is much less toxic to some groups of insects and other animals than are other chemicals. Several related chlorinated hydrocarbons were equally innocuous. During the DDT era, the use of arsenic came to a halt, and honey bee deaths caused by arsenic-containing substances ceased (Anderson and Atkins 1968, Johansen 1977).

The carbamate era

The picture changed in 1959 when the carbamate insecticide Sevin was introduced. Morse (1961) observed the effects of Sevin applied for gypsy moth control to over 73,000 acres of woodland, over several days, where 21 colonies of honey bees were distributed in five apiaries within the spray plot. Although no colony was killed, those within the plot lost an average of 20,000 bees each, whereas check colonies three and a half miles away lost only about 3,000 bees during the 47-day spray and post-spray period. Higher than normal losses of honey bees continued within the spray area for up to three weeks following the applications as bees in the hives consumed the contaminated pollen that had been collected. Worker honey bees collecting pollen contaminated with Sevin are not affected immediately, and they live long enough to carry contaminated pollen loads back to the hive, where the pollen is stored. In the experiment described above, the insecticide was applied in late May and early June in New York state; by fall, the colonies, which were returned to the Ithaca, New York, area that usually has a fall goldenrod honey flow only, had recovered enough to produce as much honey as the check colonies. Summer or fall applications of the same insecticide to a flowering crop or weeds might well result in colony deaths because the bees would not have time to rebuild their populations. In my experience, colonies are rarely killed as a result of a single application of a pesticide in the spring of the year that contaminates the flowers from which they collect pollen. However, multiple applications of an insecticide in the spring often lead to the death of colonies because of the greater accumulation of the chemical.

The beekeeper indemnity payment program

The beekeeper indemnity payment program was authorized in section 804 of the U.S. Agricultural Act of 1970. Under this legislation, which was retroactive to 1967, the federal government assumed responsibility for honey bee losses as a result of pesticide use. The act was terminated with a notice in the September 9, 1980, issue of the Federal Register (page 59299). During the time it was in effect, nearly $40 million was paid to beekeepers for losses they suffered from pesticides.

Most beekeepers did not lament the program's termination. A typical philosophy expressed by over 100 people interviewed by USDA staff investigators was that "The person who used the poison should be the one to pay for my damages." Otherwise, "there's little or no motivation"... "to remedy the poisoning of honey bees" (Morse 1981b).

The Environmental Protection Agency

The Environmental Protection Agency (EPA) was created by the federal government in 1970. The greatest boon to the beekeeping industry brought about by this agency is that warning labels on containers of the bee-toxic pesticides clearly state that they should not be applied to plants in flower or in situations where the material may drift onto flowering plants. This indicates that applicators are responsible for honey bee-pesticide losses should they occur. For example, a current Penncap-M label states, in a specially boxed guideline: "This product is highly toxic to bees exposed to direct treatment or residues on crops or blooming weeds. Do not apply this product or allow it to drift to blooming crops or weeds if bees are visiting the areas to be treated."

In its infancy, the EPA used the USDA's Pesticide Regulation Division's Notice 68-19, dated November 29, 1968, as a guideline in coping with honey bee-pesticide loss questions and problems. The title of this document was "Notice to Manufacturers, Formulators, Distributors, and Registrants of Economic Poisons." It makes clear that there was an intent to address the problem. More important, it illustrates the thinking that led to the formation of the indemnity program and perhaps the EPA itself. The regulation imposed no restriction on the application of a pesticide and was not enforceable. Soon after the collapse of the indemnity program, the EPA wrote a stronger label statement that included the words "do not apply." The new statement also considers the factors of residual toxicity and drift. The intent was to provide enforceable statements and protection across a variety of crops, weather conditions, and geographical locations. The final rule, Data Requirements for Pesticide Registration, was published in the Federal Register on October 24, 1984, pages 42856-42905 (Volume 49, Number 207).

The current situation

In 1993, a survey of 27 apiculturists across the United States indicated that many fewer honey bee-pesticide losses occur today than several decades ago (Morse 1993). Furthermore, most respondents agreed that existing EPA honey bee warning labels on packaged insecticides serve a useful role, although several felt they could be more explicit. The majority believed that additional legislation to protect honey bees is not needed at this time. Several respondents stated that grower and beekeeper education continues to be important. A few locations in intense agricultural areas, especially in Texas, Arizona, California, and Washington state, continue to have honey bee-pesticide problems. Misuse and failure to follow label recommendations continue to pose small difficulties, but the situation is greatly improved.

Symptoms of pesticide losses

Almost all honey bee-pesticide losses come about as a result of pollen being contaminated with a chemical and the bees collecting it and carrying it back to the hive, where they and the house bees feed on it. Very few bees are killed because of nectar contamination, however. Generally speaking, flowers display their pollen and hide their nectaries and nectar. Biologically, the reasons for this are simple. Flowers want bees to cover their bodies with pollen so that it may be transferred from their male to female parts, and so it is displayed so it can be seen. The color of pollen is usually attractive to bees, and pollens with less attractive colors appear to contain chemicals that are given off as attractive odors. Although nectar is of great importance to bees, plants usually force bees to search for it. In the process of searching, the male and female parts of flowers are mauled, and as a result, there is pollen transfer.

It is important to emphasize that applying pesticides to the outside of a hive, such as with a ground sprayer or an airplane, has little and usually no effect on the bees in a hive. If the nozzle of a sprayer is directed into a hive entrance, the whole colony may be killed. If there is strong flight during an airspray, the bees flying through the spray as it falls to the ground may be killed. However, the number of bees affected by an airspray is usually low unless the brood nest is contaminated. If the queen and the eggs, larvae, and pupae, as well as the young nurse bees, are not contaminated, the colony will usually recover.

There is no one symptom of a honey bee-pesticide loss that may be used consistently for diagnosis, but several are indicative of the problem. One extreme is the appearance within a short period of time, often less than 24 hours, of a pile of many thousands of dead bees in front of a hive. When a large number of dead bees are seen in front of a hive, the cause is almost always pesticide-contaminated pollen; the two most serious insecticides causing problems of this nature today are Sevin and Penncap-M (Morse 1993). Some formulations of Sevin are less toxic to honey bees than others. The time of year is important. In no bee disease situation do we find a large pile of dead bees in front of a hive at a time when a number of flowers are in bloom and there is active honey bee foraging. The other extreme is the sudden disappearance of a colony's field force. We see fewer instances today where a field force may disappear as a result of a pesticide killing bees foraging in the field because the chemical causing the greatest problem in this regard was parathion, which is approved for use on a smaller number of crops than formerly and is used much less widely today. When a pesticide loss occurs as a result of a chemical being applied and drifting onto a flowering plant(s) within a mile of an apiary, all of the colonies are usually equally affected since most bees forage within this distance of their hive. However, bees from colonies in an apiary do not forage equally in all directions. I have seen instances where an insecticide was applied to a field more than a mile from an apiary and only some of the colonies, those with bees foraging in the treated area, were affected. Having only some of the colonies in a apiary affected makes diagnosis more difficult.

Four to eight weeks after a honey bee-pesticide loss, it is common for European foulbrood, sacbrood and chilled (dead) brood to be found in affected colonies. These diseases are most likely to show themselves when colonies are under stress. In the case of a honey bee-pesticide loss, the stress of the lack of a field force to collect fresh pollen and the loss of these same bees that may assist in protecting brood by keeping the hive sufficiently warm on cool or cold nights or sufficiently cool during hot days may cause problems.

Control

There is no good method to prevent honey bees from foraging on pesti-cide-contaminated flowers. As agriculture becomes more intense worldwide, the areas where pesticides are used also become more concentrated, leaving large areas where few, if any, pesticides are used. For this reason, some beekeepers have moved their colonies away from high pesticide use areas to those of low pesticide use.

Covering colonies or using other methods to restrict honey bee flight from colonies so as to reduce foraging is costly and even dangerous because colonies may have difficulty controlling their hive temperature when confined. When thinking about methods of controlling honey bee flight, remember that bees are not domesticated animals and they respond, as they always have, to natural conditions.

APPENDIX SIX

———⬦⬦⬦———

The Federal Honey Bee Act and Amendments
Hachiro Shimanuki and Phil Lima

The first written record of a honey bee importation into North America was made in the 1600s by the Jamestown Co. The greatest number of importations and diversity most likely occurred in the latter half of the 1800s. In 1921, a report was published implicating the mite, *Acarapis [Tarsonemus] woodi* in causing Isle of Wight disease. As a result of this report, in 1921, the USDA conducted a limited survey of 202 colonies from 38 states, Canada, and the District of Columbia and found no *A. woodi* (Phillips 1922). A second survey, conducted in 1922, also found no *A. woodi* in the samples from 36 states and Canada (Phillips 1923).

The apiculture section of the Association of Economic Entomologists (merged with the Entomological Society of America in 1952) recommended to the Secretary of Agriculture that "if possible, this scourge be kept out of the country." At the time this recommendation was made, only the result of the 1921 survey was available, but the group felt justified based on the potential seriousness of this problem.

The Secretary of Agriculture recommended to the Postmaster General that the "postal regulations be amended as to prohibit the receipt" of queens and accompanying worker bees through the mails from all foreign countries except Canada. The revised postal regulation was announced on March 21, 1922. A bill to regulate foreign importation of adult honey bees was subsequently introduced into both houses of Congress in April 1922. On August 31, 1922, the President signed into law what is today known as "The Honeybee Act" (See Addendum. Note that the Honeybee Act did not become effective until May 15, 1923, when "the rules and regulations" were approved by the Secretary of Agriculture.)

The Honeybee Act of 1922 established the authority to "prohibit, destroy, or export" all adult honey bee imports into the United States. The act stated that only the United States Department of Agriculture (USDA) can import adult honey bees and only for "experimental or scientific purposes." Provisions were made in the act's regulations to permit persons other than employees of the Department of Agriculture to import adult honey bees for experimental or scientific purposes. The Honeybee Act empowered the Secretary of Agriculture and the Secretary of the Treasury to prescribe rules and regulations for the import of honey bees from countries where no diseases dangerous to adult honey bees exist. The act also established the penalty for violation of the act – upon conviction, "a fine not exceeding $500 or imprisonment not exceeding one year, or both."

Three important amendments to the Honeybee Act of 1922 were enacted in 1947, 1962, and 1976. The 1947 amendment revoked a regulation which allowed importations by organizations and individuals other than the USDA. The 1962 amendment (See Addendum) expanded the import prohibition to all species of *Apis*. This was precipitated after *A. woodi* was found in *Apis cerana*. This amendment also permitted the Agricultural Research Service (ARS) to import honey bees from any country "for experimental and scientific purposes" and established the permit system for importation of honey bees. A new section to the Honeybee Act authorized the administrator of the ARS "to prescribe rules and regulations to prevent the introduction and interstate spread of harmful diseases, parasites, and germ plasm dangerous to honeybees."

The most comprehensive change to the act resulted from the amendment passed in 1976 (See Addendum). The importation of honey bees and honey bee semen was now permitted only from countries free of "undesirable species or sub-species" as determined by the Secretary of Agriculture. The amendment included authority for the Secretary of Agriculture "either independently or in cooperation with States or political subdivisions . . . farmers' associations, and similar organizations and individuals . . . to carry out operations or measures in the United States to eradicate, suppress, control, and to prevent or retard the spread of undesirable species and subspecies of honeybees."

The final rules and regulations for implementation of the 1976 amendment were published in two separate issues of the Federal Register in June 1985. Canada was exempt from all regulations and could ship honey bees and honey bee semen into the United States without restriction, and another category of countries was established by the USDA, Animal and Plant Health Inspection Service (APHIS).[1] Under the new regulations, APHIS could issue a permit for the USDA now to import honey bees and honey bee semen from Australia, Bermuda, France, Great Britain, New Zealand, and Sweden for experimental and scientific purposes. The amendment also increased the penalty for violation of the Honeybee Act from $500 to $1,000; the term of imprisonment remained one year.

In a separate regulation (CFR 319.76, June 10, 1985) to protect the honey bee industry further, additional restrictions were placed on the importation of any live and dead bees in the Superfamily Apoidea, in addition to honey bees, which were not covered in the Honeybee Act and amendments. Used bee boards, hives, nests and nesting material, used beekeeping equipment, beeswax unless it has been liquefied, pollen for bee feed, and honey for bee feed were placed on a restricted list. These articles may be imported into this country only after APHIS and the state agricultural official of destination approve the request by issuing a permit to the importer.

---

[1]APHIS was a division of ARS until 1971 when it became a separate entity.

Impact of the Honeybee Act on the honey bee industry

Queen importations

There has been much discussion among scientists and beekeepers on the impact of the Honeybee Act on beekeeping. Many have felt that the regulations are too restrictive. The USDA has imported queens using the permit system on four separate occasions, once each from Sweden and Yugoslavia and twice from England. The importations from Yugoslavia and England were all made in the past five years. In addition, the USDA has made three importations of queens and package bees from New Zealand. These importations were made for scientific purposes and under rules established by the receiving state and APHIS.

Germ plasm diversity

North America has a diverse bee germ plasm. Schiff and Sheppard (1993), ARS Bee Research Laboratory in Beltsville, Maryland, studied the mitochondrial DNA from 194 feral colonies collected in 1991 to 1992 in New Mexico, Texas, and Louisiana, and found that 20 percent of the colony parentage could be traced to importations made from Europe as early as the 17th century. One colony had a mitochondrial DNA pattern that suggested its origin from queen imports from Africa made in the 19th century. Perhaps scientists need to use the feral population for genetic solutions to present-day industry problems.

Importation of used bee equipment

The intent of the regulation is not to impede or restrict legitimate importations for such uses as museum displays. In fact, APHIS has several regulatory treatments to sanitize used bee equipment for individuals and museums. The costs of such treatments must be paid by the importer.

Honey bee products

It is unlikely that honey is ever imported strictly for bee feed. In the past, however, pollen has been imported for the formulation of pollen supplements. In fact, there is an often-quoted but unsubstantiated belief that chalkbrood disease was introduced into this country via pollen. To date, no permit request has been received by APHIS to import pollen or honey for bee feed. Pollen, however, is being imported by botanists and health food entrepreneurs.

Importation of pests and pathogens of honey bees

Even though *Acarapis woodi*, *Varroa jacobsoni* and *Ascosphaera apis* are now considered established in the United States, they are still on the restricted list and cannot be imported without approval by APHIS. The rationale for this action is the concern that there may be strains of the pests and pathogens that are even more devastating than those present in the United States.

Future of the Honeybee Act

Despite the presence of Africanized honey bees, *Varroa jacobsoni,* and *Acarapis woodi* within our borders, the restrictions of the Honeybee Act remain the same. Other mites such as *Euvarroa sinhai* and *Tropilaelaps clareae* are parasitic on *Apis mellifera.* We must maintain our vigilance not only against these known mites but for other unknown diseases, pests, and undesirable germ plasm. The Honeybee Act is only effective when beekeepers and scientists comply with its provisions.

Addendum

The Honeybee Act

The Act of August 31, 1922 (Public No. 293-627th Congress), entitled "An Act to regulate foreign commerce in the importation into the United States of the adult honeybee (*Apis mellifica*)," provides as follows:

Sec. 1. That, in order to prevent the introduction and spread of diseases dangerous to the adult honeybee, the importation into the United States of the honeybee (*Apis mellifica*) in its adult stage is hereby prohibited, and all adult honeybees offered for import into the United States shall be destroyed if not immediately exported: *Provided,* That such adult honeybees may be imported into the United States for experimental or scientific purposes by the United States Department of Agriculture: *And provided further,* That such adult honeybees may be imported into the United States from countries in which the Secretary of Agriculture shall determine that no diseases dangerous to adult honeybees exist, under rules and regulations prescribed by the Secretary of the Treasury and the Secretary of Agriculture.

Sec. 2. That any person who shall violate any of the provisions of this act shall be deemed guilty of a misdemeanor and shall, upon conviction thereof, be punished by a fine not exceeding $500 or by imprisonment not exceeding one year, or both such fine and imprisonment, in the discretion of the court.

Amendment (1962)

To further amend the Act of August 31, 1922, to prevent the introduction and interstate spread of harmful diseases, parasites, and germ plasm dangerous to the honey bee cultures of the United States, and for other purposes. *Be it enacted by the Senate and House of Representatives of the United States of America in*

Congress assembled, That section 1 of the Act of August 31, 1922 (42 Stat. 833; 76 Stat. 169; 7 U.S.C. 281), is amended to read as follows:

Sec. 1. In order to prevent the introduction and interstate spread of harmful diseases, parasites, and germ plasm dangerous to honeybees, the importation into the United States of all honeybees, including all life stages of bees and semen, of the genus *Apis* is prohibited and all such material offered for import or intercepted entering the United States shall be destroyed or immediately exported: *Provided,* That honeybees may be imported into the United States by the United States Department of Agriculture for experimental or scientific purposes: *Provided further,* That honeybees may be imported into the U.S. from countries in which the Secretary of Agriculture determines that no diseases, parasites, or germ plasm harmful to honeybees exist and that adequate precautions have been taken by such countries to prevent the importation of honeybees from countries where such diseases, parasites, or germ plasm exist.

Sec. 2. Section 2 of such Act is redesignated as 3 and is amended by inserting the words, "or any regulation issued thereunder" after the word "act."

Sec. 3. The Act of August 31, 1922, is further amended by adding a new sec. 2 to read as follows: "Sec. 2. The Secretary of Agriculture is authorized to prescribe rules and regulations to prevent the introduction and interstate spread of harmful diseases, parasites, and germ plasm dangerous to honeybees."

Amendment (1976)

(a) In order to prevent the introduction and spread of diseases and parasites harmful to honeybees, and the introduction of genetically undesirable germ plasm of honeybees, the importation into the United States of all honeybees is prohibited, except that honeybees may be imported into the United States –

(1) by the United States Department of Agriculture for experimental or scientific purposes, or

(2) from countries determined by the Secretary of Agriculture –

(A) to be free of diseases or parasites harmful to honeybees, and undesirable species or subspecies of honeybees; and

(B) to have in operation precautions adequate to prevent the importation of honeybees from other countries where harmful diseases or parasites, or undesirable species or subspecies, of honeybees exist.

(b) Honeybee semen may be imported into the United States only from countries determined by the Secretary to be free of undesirable species or subspecies of honeybees, and which have in operation precautions adequate to prevent the importation of such undesirable honeybees and their semen.

(c) Honeybees and honeybee semen imported pursuant to subsections

(a) and (b) of this section shall be imported under such rules and regulations as the Secretary of Agriculture and the Secretary of the Treasury shall prescribe.

(d) Except with respect to honeybees and honeybee semen imported pursuant to subsections (a) and (b) of this section, all honeybees or honeybee semen offered for import or intercepted entering the United States shall be destroyed or immediately exported.

(e) As used in this Act, the term "honeybee" means all life stages and the germ plasm of honeybees of the genus *Apis*, except honeybee semen. Any person who violates any provision of this Act or any regulation issued under it is guilty of an offense against the United States and shall, upon conviction, be fined not more than $1,000, or imprisoned for not more than one year, or both. The Secretary of Agriculture, either independently or in cooperation with States or political subdivisions thereof, farmers' associations, and similar organizations and individuals, is authorized to carry out operations or measures in the United States to eradicate, suppress, control, and prevent or retard the spread of undesirable species and subspecies of honeybees.

The Secretary of Agriculture is authorized to cooperate with the governments of Canada, Mexico, Guatemala, Belize, Honduras, El Salvador, Nicaragua, Costa Rica, Panama, and Colombia, or the local authorities thereof, in carrying out necessary research, surveys, and control operations in those countries in connection with the eradication, suppression, control, and prevention or retardation of the spread of undesirable species and subspecies of honeybees, including but not limited to *Apis mellifera adansonii,* commonly known as African or Brazilian honeybee[2]. The measure and character of cooperation carried out under this Act on the part of such countries, including the expenditure or use of funds appropriated pursuant to this Act, shall be such as may be prescribed by the Secretary of Agriculture. Arrangements for the cooperation authorized by this Act shall be made through and in consultation with the Secretary of State. In performing the operations or measures authorized in this Act, the cooperating foreign country, State, or local agency shall be responsible for the authority to carry out such operations or measures on all lands and properties within the foreign country or State, other than those owned or controlled by the Federal Government of the United States, and for such other facilities and means as in the discretion of the Secretary of Agriculture are necessary.

Funds appropriated to carry out the provisions of this Act may also be used for printing and binding without regard to section 501 of title 44, United States

---

[2]The honey bee queens imported into Brazil are now believed to have been *Apis mellifera scutellata,* and the common name of the hybrids as they exist in the new world is the Africanized bees.

Code, for employment, by contract or otherwise, of civilian nationals of Canada, Mexico, Guatemala, Belize, Honduras, El Salvador, Nicaragua, Costa Rica, Panama, and Colombia for services abroad, and for the construction and operation of research laboratories, quarantine stations, and other buildings and facilities.

There are hereby authorized to be appropriated such sums as may be necessary to carry out the provisions of this Act.

APPENDIX SEVEN

State Bee Disease Laws
David A. Knox and Hachiro Shimanuki

Introduction
Authors' opinions
Acknowledgments

The first apiary inspection law in the United States was established in San Bernardino County, California, in 1877. By 1883, a statewide law had been passed by the California legislature. In 1906, 12 states had laws relating to foulbrood disease. At present, almost all states have laws regulating honey bees and bee-keeping.

The state laws and regulations relating to honey bees and beekeeping are designed primarily to control bee diseases. They attempt to regulate movement and entry of bees and beekeeping equipment, issuance of permits and certificates, apiary location control and quarantine, inspection, and methods of treating diseased colonies. Although the destruction of American foulbrood-diseased colonies is included in most state laws, many states also allow the use of drugs[1] for treatment of this disease.

As noted in Tables A7.1 and A7.2, state bee laws and regulations are not uniform, but there is considerable agreement on specific points of law. For instance, most states require registration of apiaries, certificates of inspection for interstate movement of bees and equipment, and right of entry of the inspector.

Authors' opinions

The key enforcer of state bee laws and regulations is the apiary inspector. The inspector's jurisdiction may include the entire state, a county or counties, or a community. His or her efforts are directed primarily toward locating American foulbrood-diseased colonies and eliminating the source of contamination. Recently, the role of inspectors has been expanded to the detection of parasitic mites and Africanized honey bees in the United States. In some states, the apiary inspector may have extension responsibilities, especially in assisting beekeepers and growers in pollination services; still other states may employ only part-time inspectors. The national trend has been reduction and in some cases even elimination of bee inspection.

The effectiveness of bee laws and regulations depends on the compliance of beekeepers. In the final analysis, responsibility for disease control remains with beekeepers, who should routinely examine colonies for disease as a part of their management program and take the necessary steps when disease is found.

More uniformity in state laws regulating beekeeping is desirable, especially because package bees and queens, as well as honey bee colonies for honey production and pollination, are transported over great distances and in many cases only for short durations. Uniform state laws would provide for more expedient

---

[1]Currently only oxytetracycline HCl (Terramycin) is registered for this use in the United States.

movement of honey bee colonies and could lead to a more productive beekeeping industry.

Acknowledgments

The authors acknowledge the help of state apiarists and state entomologists for providing information and comments in the preparation of this summary.

SUMMARY OF STATE HONEY BEE LAWS
Table A7.1 - Apiary Inspection Requirements

State	Registration of Colonies	Identification of Apiary	Related Fees	Apiary Locations Controlled	Inspector Right of Entry	Inspection of Apiary	American Foulbrood Quarantine
Alabama	X	X	X	X			X
Alaska	X			X	X		X
Arizona (1)							
Arkansas	X	X		X	X	X	X
California	X	X	X		X		X
Colorado					X		
Connecticut	X				X		X
Delaware	X				X	X	X
Florida	(2)	X	X		X	X	X
Georgia	X		X		X	X	X
Hawaii						(3)	
Idaho	X	X		X	X	X	X
Illinois	X	X			X	X	
Indiana	X				X	X	
Iowa					X	(3)	X
Kansas	(4)		X		X	X	X
Kentucky					X		
Louisiana	X	X	X		X		X
Maine	X		X		X	X	X
Maryland	X	X			X		
Massachusetts					X	X	
Michigan							
Minnesota	X		X		X		
Mississippi	(5)		X		X	X	X
Missouri					X		X
Montana	X	X	X	X	X	X	X
Nebraska	X	X	X		X	X	X
Nevada	X	X	X		X		X
New Hampshire	X	X			X	X	X
New Jersey	X	X			X	X	X

State	Registration of Colonies	Identification of Apiary	Related Fees	Apiary Locations Controlled	Inspector Right of Entry	Inspection of Apiary	American Foulbrood Quarantine
New Mexico	X		X		X	X	X
New York					X	X	X
North Carolina					X	X	X
North Dakota	X	X	X				
Ohio	X	X	X		(6)		
Oklahoma	X				X		
Oregon	(7)		X		X		
Pennsylvania	X		X		X	X	X
Rhode Island	X	X			X	X	X
South Carolina(4)				(4)	X		X
South Dakota	X	X	X	X	X	X	X
Tennessee	X				X	X	X
Texas	X	X	X			X	X
Utah	X	X	X		X	X	X
Vermont	X			X	X	X	
Virginia (8)					X	X	X
Washington	X	X	X		X		
West Virginia	X	X			X	X	X
Wisconsin					X		
Wyoming	X	X	X	X	X	X	X

(1) All apiary laws were repealed as of July 1994.
(2) Registration of beekeeper required; each hive must have a registration number.
(3) Only if presence of deleterious pathogens, parasites, or pests suspected.
(4) Requirement pertains only to colonies moved into the state on entry permits.
(5) Registration required only for beekeepers who own 50 or more colonies.
(6) Beekeeper may refuse entry.
(7) Registration required if beekeeper owns more than five colonies.
(8) Inspection certificate must be attached to each colony.

Table A7.2 - Requirements for the Movement of Honey Bee Colonies and Equipment

State	Inspection Certificate	Interstate Entry Permit	Arrival Notification	Intrastate Mite Certification	Inspection Certificate	Moving Permit
Alabama	X				X	X
Alaska (1,2)	X		X			
Arizona						
Arkansas	X	X		(5)	X	X
California	X		X	(3,4)	(3)	
Colorado	X	X	X	(5)		
Connecticut	X			(5)	X	X
Delaware	X	X		(4)	X	
Florida	X			(4)	X	X
Georgia	X	X		(5)	X	
Hawaii (2,6)						
Idaho	X	X(7)				
Illinois	X	X	X	(4)	X	X
Indiana	X	X		(5)		
Iowa	X	X(7)		(5)		
Kansas (2)	X	X(7)		(4)		
Kentucky						
Louisiana (2)	X	X				
Maine	X	X(7)		(4)		
Maryland	X	X		(4)		
Massachusetts	X	X		(5)	X	X
Michigan						
Minnesota	X	X		(4)		
Mississippi	X	X		(5)	X	X
Missouri	X	X(7)		(5)		
Montana	X	X(7)	X			
Nebraska	X	X	X	(4)		
Nevada	X	X(7)		(4)		X
New Hampshire	X			(4)		
New Jersey	X		X	(4)	X	
New Mexico	X		X	(5)	X	
New York	X				X	
North Carolina (1)	X			(5)		
North Dakota		X	X			
Ohio	(8)	X	X	(5)		
Oklahoma	X			(5)		
Oregon						
Pennsylvania	X	X	X	(4)		
Rhode Island	X			(5)		X
South Carolina	X	X		(4)		

State	Interstate Inspection Certificate	Entry Permit	Arrival Notification	Intrastate Mite Certification	Inspection Certificate	Moving Permit
South Dakota	X	X		(4)		
Tennessee (2)	X	X	X	(5)	X	X
Texas	X	X	X	(4)	X	X
Utah	X	X	X			
Vermont	X	X	X	(4)		
Virginia	X	X		(5)		
Washington	X	X				
West Virginia	X	X	X	(5)	X	X
Wisconsin (2)	X			(4)		
Wyoming	X	X(7)	X		X	X

(1) Entry of bees on combs prohibited.
(2) Special requirements for used bee equipment.
(3) Certification requirement is a county option.
(4) *Varroa jacobsoni*
(5) *Varroa jacobsoni* and *Acarapis woodi*
(6) Importation of live or dead honey bees prohibited.
(7) Entry fee is required. In Iowa, fee does not apply to residents.
(8) If no certificate, colonies may enter provided inspection conducted upon entry; inspection fee charged.

APPENDIX EIGHT

Canadian Laws
Don Nelson

Canadian laws
Import requirements
Provincial regulations

The laws covering the importation of honey bees into Canada and movement of bees and equipment within Canada were last summarized in 1978 for the first edition of *Honey Bee Pests, Predators, and Diseases*. Since then, many changes have occurred in the Canadian bee industry and in certain regulations of both the federal and provincial governments.

During the last 25 years, the Canadian honey bee industry has been working toward a greater degree of self-sufficiency, by increasing emphasis on wintering colonies, both indoors and out; by producing nucs, packages, and queens in areas such as southern British Columbia and southern Ontario; and by managing of colonies on the prairies to produce nucs in June with new queens for winter replacements and colony increase.

In the late 1970s and 1980s, Canadian beekeepers and apicultural specialists began to express concern about the potential introduction of "exotic" mites into Canada. In response to these concerns, research projects were initiated on wintering techniques and breeding programs to develop winter-hardy bees, queen rearing courses were offered, and technical assistance was available to beekeepers for wintering of honey bees.

When honey bee tracheal mites (HBTM) were found in the United States, Canada implemented a honey bee tracheal mite certification program that listed criteria to be met by states wishing to ship bees, queens, or both, to Canada. Although this did not prevent the inevitable importation of HBTM, it certainly delayed high levels of mites being imported via package bees.

In addition, during this period improved methods of disease control and measures for the control of tracheal and varroa mites were being planned. In many ways, the time lag and planning paid off for the Canadian beekeeping industry because approximately 80 percent of Canadian colonies were being wintered by the time the border was closed in 1987 to U.S. imports. Canada now has the following products scheduled for use against bee mites: Apistan, menthol, and formic acid.

Canadian laws

The threat of exotic mites led Canadian apiculturists, in consultation with the Canadian Honey Council (Canada's national beekeeping organization), and the federal government to review provisions under the Animal Health Act. After lengthy discussion, it was decided in the early 1980s, to name *Varroa jacobsoni* as a reportable pest in the act, but the tracheal mite (HBTM) was not included. In addition, a Varroa Action Plan was formulated to dictate how varroa would be handled once it was found in Canada.

Because varroa was now under federal legislation, the Canadian federal government was suddenly confronted with the task of formulating a response strategy that would address beekeeping interests throughout Canada.

Extensive consultation between industry groups and governments took place throughout the winter of 1987 and 1988. In 1987-88, no chemical control product was registered in Canada to control varroa.

The Animal Health Act was last amended in 1990. Section 16 of the regulations was revoked and replaced by the following: "10(1) Subject to this part, no person shall import an animal into Canada from a country other than the United States unless the person does so in accordance with a permit issued by the minister." In 1987, the Honeybee Prohibition Regulations were enacted for a two-year term and are reviewed biannually. These were last amended in 1993 and now read as follows: "No person shall import any bee of the genus *Apis,* commonly known as honeybees, into Canada, any part of Canada, or Canadian port, from the United States except from the state of Hawaii during the period beginning on the date of the coming of force of these regulations and ending on December 31, 1995." The regulations concerning bee equipment and wax are as follows: "No person shall import into Canada (a) used beehives or used beehive equipment or (b) beeswax, unless an inspector is satisfied that the beeswax has been liquefied."

It should be remembered that the closure of the Canadian-U.S. border was enacted because of varroa, not HBTM. The vast majority of beekeepers and beekeeping operations in Canada have remained free of varroa. Although the pest will continue to spread in the years to come, varroa's principal distribution is still concentrated in border areas. After six years of border closure, the policy must be recognized as having succeeded.

Import requirements

The importation of bees or queens into Canada is permitted only in accordance with a permit issued by the Minister. The only areas for which permits will be issued at present are: Hawaii – queens only; New Zealand – queens and package bees; Australia – queens and package bees. The ban on imports of honey bees from the continental United States is reviewed biannually.

Although varroa is named in the federal Animal Health Act, the federal government is no longer involved in surveys or quarantines. These are now a provincial responsibility. Each province has determined how it will deal with the first find of varroa and how it will minimize subsequent spread of the mite.

Provincial regulations

Provincial bee acts include regulations for beekeeper registration, control of bee diseases, movement of bees, and honey house standards (Table A8.1). The extent of the regulations varies from province to province. Most provinces now have requirements that must be met to move bees into the province, and some do not allow for any importation except by federal permit (Table A8.2). These include Saskatchewan, Ontario, Nova Scotia, and Newfoundland. The regulations relating to disease control, particularly American foulbrood, are shown in Table A8.3. These tables are meant to summarize the regulations, so any specific information should be obtained from the province or territory in question.

The Canadian Honey Council (CHC) meets annually and consults with the Canadian Association of Professional Apiculturists (CAPA) to formulate action plans to deal with disease and pest problems. A disease publication, *Honey Bee Diseases and Pests,* produced by CAPA, has been distributed to registered beekeepers in most provinces and has had a beneficial impact on beekeepers' ability to diagnose common bee diseases. This has allowed provincial inspectors to spend needed time on tracheal and varroa mite detection and control. The CHC and CAPA also meet jointly with the Research Branch of Agriculture and Agri-Food Canada approximately every five years to set research priorities.

Table A8.1. Summary of provincial apiary and honey house inspection regulations

Province	Registration of apiary required	Identification of apiary required	Inspection of apiary required	Inspection of honey house required	Inspector has right of entry	Inspection certificate required
Alberta	X		X 1	X 2	X	X
British Columbia	X		X	X	X	X
Manitoba		X	X		X	X
Newfoundland					X	
New Brunswick*	X	X	X		X	
Nova Scotia	X	X	X		X	X 4
Ontario	X	X	X 1	X 2	X	
Prince Edward Island	X		X			X 4
Quebec			X	X 3	X	X
Saskatchewan	X		X	X	X	X

* Currently under revision
1 Must be inspected when selling equipment
2 Federal inspection laws apply for packer-grader license
3 Regulations for farm products and food processors apply
4 Inspection certificate required for selling used equipment or moving bees to a new permanent location
Note: No honey bee regulations are currently in force in the Yukon or the Northwest Territories.

Table A8.2. Summary of provincial regulations relative to bee importation and interprovincial movement of bee colonies and equipment

Province	Entry			Certificate	
	prohibited	permit required	required with application	must accompany load	time limit for inspection (days)
Alberta		X	X	X	30
British Columbia		X	X	X	30
Manitoba		X	X	X	30
Newfoundland	X	X 1			
New Brunswick		X	X	X	30
Nova Scotia	X	X 1			
Ontario	X	X1,2	X		
Prince Edward Island	X				
Quebec		X	X	X	30
Saskatchewan	X	X 3			

1 Approval by federal permit from Australia and New Zealand and queens from Hawaii
2 Approval by permit from Vancouver Island, B.C., Nova Scotia, New Brunswick, and Prince Edward Island
3 Approval by federal permit from New Zealand, Australia, queens from Hawaii and the quarantine zone of Vancouver Island, B.C.

Table A8.3. Summary of provincial regulations relative to American foulbrood (AFB) control

	AFB diseased colonies or equipment					Drug treatment permitted for prevention
	Owner must be notified	must be quarantined	must not be exposed	must not be sold or transferred1	must be destroyed2	
Alberta	X		X	X	X	X
British Columbia	X	X	X	X	X	X
Manitoba	X	X	X	X	X	X
Newfoundland	X				X	X
New Brunswick	X	X	X	X	X	X
Nova Scotia	X	X	X	X	X	X
Ontario	X		X	X	X	X
Prince Edward Island	X	X	X	X	X	X
Quebec	X	X	X	X	X	X
Saskatchewan	X	X	X	X	X	X

1 Must be inspected before being sold; this is to protect the new owner.
2 To destroy in most provinces means destruction of diseased combs; most provinces have authority to destroy colonies, but is only used in severe cases of AFB infection.

Note: Only oxytetracycline hydrochloride (Terramycin) is registered for use in Canada.

APPENDIX NINE

Pesticides Registered by the U.S. Environmental Protection Agency

Al Vaughan and Hachiro Shimanuki

The pesticides listed in the table are registered for use on honey bees and beekeeping equipment in the United States (as of February 1996). In addition, the products Fumidil B, for control of *Nosema apis* Zander (active ingredient: fumagillin), and Terramycin, for control of bacteria (active ingredient: oxytetracycline HCI), are registered with the U.S. Food and Drug Administration. Bee repellents such as Benzaldehyde and Bee-Go (butyric anhydride) that are used to drive bees off combs for harvesting honey no longer require EPA registration. Because of frequent changes in the registration status of products and because pesticide use is also subject to state and other local restrictions, consult with local authorities for current information about pesticides. Also, for some pesticides, only use by a licensed applicator is permitted.

Pesticides registered by the U.S. Environmental Protection Agency for use on honey bees and beekeeping equipment in the United States

Active Ingredient	Trade Name of Product[1]	EPA Reg. Number
For control of tracheal mites:		
Menthol	Mite-A-Thol	61671-1
For control of varroa mites:		
Fluvalinate	Zoecon RF-318 Apistan Strip	2724-406
	Zoecon RF-348 Apistan Queen Tab	2724-429
	Zoecon RF-349 Apistan Strip	2724-430
For treatment of diseased beehives:		
Ethylene oxide[2]	Carboxide Sterilant-Fumigant Gas	ME91000500
		CA94001900
		NC95000200
For control of wax moths:		
Bacillus thuringiensis	Certan	55947-116
Paradichlorobenzene (PDB)	Para-Moth	61671-2
Aluminum phosphide	L-Fume Pellets	30574-1
	L-Fume Tablets	30574-4
	Fumiphos Pellets	43568-1
	Fumiphos Bags	43568-2
	Fumiphos Tablets	43568-3

Active Ingredient	Trade Name of Product[1]	EPA Reg. Number
	Quik-Fume Tablets	59209-1
	Quik-Fume Pellets	59209-2
	Quik-Fume Bags	59209-3
	Degesch Phostoxin Tablets-R	40285-1
	Degesch Phostoxin Pellets	40285-3
Magnesium phosphide	Magnaphos Bags[3]	59209-6

For killing colonies (diseased or nuisance):

Pyrethrins	Chemsico Aerosol Insecticide	9688-111
	No-Chlor Wasp and Hornet Killer	9852-70
	Speer Py-Perm Aqueous Killer #6	11715-258
Aluminum phosphide	L-Fume Tablets	30574-4
	Fumiphos Bags	3568-2
	Quik-Fume Pellets	59209-2
Diazinon	Science 5% Diazinon Dust	769-922
Potassium salts of fatty acids	M-pede insecticide	CA94001900

For killing colonies (Africanized bees):

Potassium salts of fatty acids	M-pede insecticide	CA94001900
Aluminum phosphide	Fumitoxin Tablets	5857-1
	Fumitoxin Pellets	5857-2
	L-Fume Pellets	30574-1
	L-Fume Tablets	30574A
	Fumiphos Pellets	43568-1
	Fumiphos Bags	43568-2
	Fumiphos Tablets	43568-3
	Quik-Fume Tablets	59209-1
	Quik-Fume Pellets	59209-2
	Quik-Fume Bags	59209-3

Wood protection treatment - beehives:

Zinc naphthenate	2% Zinc Nap-All	963-16
	2% Zinc Hydro-Nap	963-18
	Protecto-Zin	54734-2
Copper naphthenate	M-Gard S510HF	963-24

1 The list of trade names may be incomplete.
2 Although ethylene oxide has been registered in only a few states, its use is permitted in many other states.
3 Also labeled for the control of dermestid beetles.

APPENDIX TEN

International Trade Agreements And Honey Bee Diseases

Andrew Matheson

New multilateral rules governing agricultural trade will have a significant effect on international trade in bees and bee products. National laws affecting imports must be objective and justifiable, and adhere to principles set out in international agreements. This section sets out the background to these changes, what the changes entail, and likely consequences for apicultural trade.

Recent changes in international trade rules

Uruguay Round of GATT

Since 1947, the world trading environment has been regulated through the General Agreement on Tariffs and Trade, commonly known as GATT. This agreement was an attempt to avoid a repeat of the disastrous economic policies of the 1930s, and the resulting worldwide depression.

GATT was changed through a series of 'rounds' of negotiations, the most recent of which (the Uruguay Round) was both long – seven years – and very radical. The multilateral agreements arising from this round provide a whole new thrust toward freeing up international trade from tariffs and other barriers, and are much broader in scope than those formed in earlier rounds.

The Uruguay Round was the first GATT round to consider agriculture comprehensively, as previously, domestic agricultural policies were largely exempt from international trade rules because food security was tied up with national security and thus was non-negotiable. Many nations heavily subsidized farmers and protected agriculture.

Creation of the World Trade Organization

A new organization, the World Trade Organization (WTO), has been set up to encompass the revised General Agreement on Tariffs and Trade (GATT 1994) and all the 21 multilateral agreements established under the Uruguay Round. The WTO has over 120 members, including most of the world's major trading nations (see Table A10.1). Many nonmember countries are moving toward joining the WTO and are trying to adopt SPS principles as part of that process.

The WTO is based in Geneva, Switzerland, and has a number of functions. It administers and implements the multilateral and plurilateral trade agreements, which together come under the WTO, and acts as a forum for multilateral trade negotiations. It oversees national trade policies and can require member countries to remove import prohibitions and reduce import tariffs. The WTO is also a forum for resolving trade disputes, though while it has a formal disputes settlement procedure and can impose 'penalties,' its aim is to prevent disputes and resolve them before they are submitted to the settlement procedure.

The SPS agreement

As trade barriers are reduced as a result of the Uruguay Round agreements, pressure might come on some governments to use sanitary or phytosanitary measures (health-protection requirements) to restrict trade and protect domestic markets. Any sanitary or phytosanitary measure which is not actually needed to protect health can be an effective form of protectionism, and one which is difficult to challenge because of the technical complexity of such issues.

The Uruguay Round incorporated the new 'Agreement on the application of sanitary and phytosanitary measures,' commonly known as the SPS agreement (World Trade Organization, 1995), which built on previous GATT rules to restrict the inappropriate use of sanitary and phytosanitary measures to protect trade.

The SPS agreement is about how to apply sanitary (human and animal health) measures and phytosanitary (plant health) measures, establishing principles which bind WTO member countries when they set health-protection measures for the importation of plants, animals, and their products. It covers all SPS measures which affect or may affect international trade.

Some regional trade agreements incorporate very similar principles; examples include the North American Free Trade Agreement (NAFTA) and the Closer Economic Relationship (CER) agreement between New Zealand and Australia.

The main principles of the SPS agreement are outlined below.

Necessity of SPS measures

Members have the right to protect the life and health of their animal, plant, and human populations, provided the measures taken aren't inconsistent with the SPS agreement.

National sovereignty is preserved, but balanced against the commitments made when members agreed to the results of the Uruguay Round. In fact, the most fundamental challenge of making the SPS agreement work is to reach a balance between the sovereign rights of individual countries to set their own health-protection measures and the aim of facilitating trade.

The agreement defines necessity by reference to science rather than politics, as sanitary measures must be based on scientific principles and kept in place only while justified by scientific evidence.

Consistency

Sanitary measures must be applied consistently, and there are three main areas where WTO members have to be careful to avoid what the SPS agreement calls arbitrary or unjustifiable discrimination.

First, discriminating against foreign suppliers is not allowed. This 'national treatment' principle means that imports may not be treated differently from local trade in the same commodity; for instance, a country can't require imported bees to be free of a disease if there's no similar requirement for locally traded bees.

The second principle is called 'most-favoured nation'; members may not discriminate between members where identical or similar health conditions prevail, setting tough health standards for imports from one while being more liberal with imports from the other.

The third area of consistency deals with how a country is consistent in determining what level of risk it will accept for different areas of trade; different species or products, different countries, and even different sectors (for example, the risks accepted for poultry meat imports compared with the level of risk of introducing diseases on beef imports). This aspect of consistency is still poorly defined, and the subject of debate at the WTO's SPS committee.

Assessment of risk

Any WTO member must ensure that its sanitary measures are based on an assessment of risk. Risk analysis is a fast-evolving science which helps regulators assemble data in a thorough and consistent way, so their decisions can be made on a sound technical basis. The process also becomes more transparent, so anyone affected by a decision can see the assumptions and decisions made in developing sanitary measures.

Determining the appropriate level of protection

Once the risk analysis has been performed, regulators must decide the level of protection that is appropriate to their circumstances. Sanitary measures should be suitable to the need, which is dictated by the probability of a pest or disease being introduced or becoming established and the consequences of that happening.

Given a choice of sanitary measures which deliver the appropriate level of protection, members must choose the one which will be least distorting on trade.

Equivalence

The agreement also forces a move away from importing countries insisting that particular sanitary measures be applied to animals or animal products. In the post-Uruguay Round environment, it's results that are important, not processes. Different health measures used by an exporting country must be accepted by an importing country if it can be objectively shown that they achieve the importing country's appropriate level of protection.

For example, an importing country couldn't insist that beeswax be heated at a certain temperature for a certain time if the exporting country could show that another treatment (say irradiation) delivered the desired level of protection (e.g., reducing to the same level the risk of introducing *Bacillus larvae*).

Equivalence will mostly be applied on a bilateral basis, and the agreement envisages members setting up bilateral agreements based on this concept. A number of member countries are developing wide-ranging veterinary agreements based on recognition of equivalence over several product or animal sectors, including honey bees and bee products.

Harmonization

Harmonizing sanitary and phytosanitary measures is an important objective of the agreement, and members are obliged to base their sanitary measures on international standards, recommendations, and guidelines where they exist. For animal health, the international standards are those developed by the OIE, the *Office International des Epizooties* or the World Organization for Animal Health.

The OIE now has a much more important role than it did in the past. Since 1924, this intergovernmental organization has worked to share information on animal diseases, coordinate research, and harmonize regulations on international trade. Now its recommendations have a new status, and it is vitally important for countries to work to make these scientifically valid and up-to-date. All sanitary measures based on OIE standards are deemed to be acceptable under the SPS agreement.

WTO members may use higher sanitary standards than those developed by the OIE, either if there is scientific justification or if they can demonstrate a need, based on an analysis of risk, for a higher level of protection than the OIE standard would give. Any higher sanitary standard must still not be inconsistent with the other provisions of the SPS agreement.

Regional conditions

Sanitary measures should take account of demonstrable regional variations in health status in the exporting and importing regions. It's no longer appropriate to think of a whole country as being 'infected' with a disease, if there are real differences in the presence or incidence of that disease within the country.

To support a claim that a region is free of a disease or has a low incidence of a disease, an exporting country must provide objective evidence on issues such as effective surveillance, import control measures, and geographical or ecological factors maintaining the disease status.

Transparency

Probably the most immediate change in the way countries operate in the SPS environment is an opening up of information channels about the sanitary measures they use.

Members are obliged to notify other members of proposed sanitary regulations and allow time prior to implementation for comment (except for emergencies such as outbreaks of serious diseases). Other members are entitled to comment, and have their submissions discussed.

WTO members must also set up single enquiry points, so that any other member may ask about a wide range of sanitary measures including SPS regulations, internal procedures such as manuals used by inspectors, and even the risk analysis procedures used to develop import health standards.

Other provisions

The SPS agreement also covers other areas, including the following:
- Members should contribute to relevant international organizations, which for animal health means the OIE, and to the development and review of standards.
- Members are encouraged to provide technical assistance to other members.
- There are special provisions for developing countries, delaying full implementation for five years in the case of least-developed countries and two years for other developing countries (from the agreement's entry into force on January 1, 1995).
- The SPS committee, which meets in Geneva, is charged with implementing the agreement, and has a mandate to provide a forum for discussion on issues such as equivalence and harmonization.

Likely effects of the SPS agreement on trade in bees and bee products

The SPS agreement provides a framework for increasing trade in bees and bee products while allowing nations to protect bee health where necessary. Changes likely to result from this agreement include reductions in trade barriers and increased market opportunities.

Unjustified requirements

Importing countries are still requiring exporting countries to issue certification that is not justified. Live bee exports must often be tested for, and found free of, diseases which are present in the importing country but not under official control. Sometimes honey must be certified as originating in apiaries which are free of parasites which cannot be transmitted in honey (such as tracheal mites and varroa).

Unsustainable regulations

WTO members must bring their national legislation into line with international commitments contained in the SPS agreement. For instance, it is not fulfilling these commitments to ban importation of honey bees from all countries where diseases or parasites harmful to honey bees are present, as this blanket ban does not allow the importing country to fairly assess risk and determine an appropriate level of protection to be achieved through applying sanitary measures.

Sanitary measures based on such legislation are now open to challenge by other WTO members, if they are thought to unjustifiably discriminate between members or against imports.

Trade opportunities

The SPS agreement will make it easier for exporters of bees and bee products to sell their produce on world markets because WTO members have undertaken to scientifically justify the sanitary measures they impose. New trade opportunities will arise from:

- New markets. Previously closed markets will be opened up as trade policies are brought into line with SPS principles.
- Lower compliance costs. The cost of meeting unnecessarily rigid sanitary requirements can marginalize an otherwise viable export operation. As unjustified requirements are removed, these costs will be reduced.
- Certainty. Exporters will be able to plan ahead with more confidence, as WTO member countries are no longer able to impose arbitrary restrictions on an export industry.

Implementation

World adoption of SPS principles hasn't happened overnight, but gradually this agreement will lead to trade in animals and animal products being based on sound science. It has already begun to influence the behavior of regulators around the world, shifting the burden of proof onto those setting sanitary measures for imported bees and bee products.

The SPS agreement provides for trade to go ahead unless there are valid health reasons for it to be restricted, rather than for trade to be permitted only when this suits the importing country.

References

World Trade Organization (1995) *The results of the Uruguay Round of multilateral trade negotiations; the legal texts.* World Trade Organization; Geneva, Switzerland; pp 69-84 (available on the World Wide Web at http://www.wto.org/wto/agric/spsagr.htm).

Table A10.1. WTO membership as of January 1997

Members

The following 129 governments have accepted the Marrakesh Agreement establishing the World Trade Organization.

Angola, Antigua and Barbuda, Argentina, Australia, Austria, Bahrain, Bangladesh, Barbados, Belgium, Belize, Benin, Bolivia, Botswana, Brazil, Brunei Darussalam, Bulgaria, Burkina Faso, Burundi, Cameroon, Canada, Central African Republic, Chad, Chile, Colombia, Costa Rica, Côte d'Ivoire, Cuba, Cyprus, Czech Republic, Denmark, Djibouti, Dominica, Dominican Republic, Ecuador, Egypt, El Salvador, European Community, Fiji, Finland, France, Gabon, Gambia, Germany, Ghana, Greece, Grenada, Guatemala, Guinea Bissau, Guinea, Guyana, Haiti, Honduras, Hong Kong, Hungary, Iceland, India, Indonesia, Ireland, Israel, Italy, Jamaica, Japan, Kenya, Korea, Kuwait, Lesotho, Liechtenstein, Luxembourg, Macau, Madagascar, Malawi, Malaysia, Maldives, Mali, Malta, Mauritania, Mauritius, Mexico, Morocco, Mozambique, Myanmar, Namibia, Netherlands (Kingdom in Europe and Netherlands Antilles), New Zealand, Nicaragua, Niger, Nigeria, Norway, Pakistan, Papua New Guinea, Paraguay, Peru, Philippines, Poland, Portugal, Qatar, Romania, Rwanda, Saint Lucia, Saint Kitts and Nevis, Saint Vincent and the Grenadines, Senegal, Sierra Leone, Singapore, Slovak Republic, Slovenia, Solomon Islands, South Africa, Spain, Sri Lanka, Suriname, Swaziland, Sweden, Switzerland, Tanzania, Thailand, Togo, Trinidad and Tobago, Tunisia, Turkey, Uganda, United Arab Emirates, United Kingdom, United States, Uruguay, Venezuela, Zaire, Zambia, Zimbabwe.

Accessions

The following governments have requested to join the WTO, and their applications are currently being considered by accession working parties.

Albania, Algeria, Armenia, Belarus, Cambodia, People's Republic of China, Chinese Taipei, Congo, Croatia, Estonia, Georgia, Jordan, Kazakhstan, Kirgyz Republic, Laos, Latvia, Lithuania, Former Yugoslav Republic of Macedonia, Moldova, Mongolia, Nepal, Panama, Russian Federation, Saudi Arabia, Seychelles, Sudan, Sultanate of Oman, Tonga, Ukraine, Uzbekistan, Vanuatu, Vietnam.

APPENDIX ELEVEN

──◦◦◦◦◦◦──

Country Records For Honey Bee Diseases, Parasites And Pests

Andrew Matheson

World bee health

      This appendix is intended to serve as an accessible and reliable reference to the honey bee health status of almost every country in the world. From this, readers can infer something about the world distribution of the parasites, pests, and diseases surveyed, which is useful both when attempting to obtain supplies of healthy honey bees and when studying beekeeping in different areas. This review may also be useful for readers wanting to know more about bee health in their own countries, especially when published references are not readily available.

      The material summarizes work previously published by the International Bee Research Association (IBRA) in its journal *Bee World* (Matheson 1993a, 1995, 1996) and is available as a separate publication (Matheson 1993b). Records of sacbrood have been checked against the paper on virus distribution by Allen and Ball (1996) and updated where necessary. Further information on disease incidence will continue to be published by IBRA in *Bee World*.

Methodology

Species included in this review

      In general, only records from *Apis mellifera* and *A. cerana* are covered in this survey, as little management of other species takes place, and knowledge of their pathology is limited. (Any exceptions are identified in the tables.) For the sake of continuity, the same 10 diseases and parasites covered in earlier IBRA reviews are considered (see Table A11.1), even though objections can be made against the inclusion of some.

      For instance, braula (*Braula* spp.) is not a parasite and is regarded by many as having almost no economic significance. However, it is still an issue for some certifying authorities, and earlier reviews and maps of its distribution contained a significant number of inaccuracies.

      Most diagnoses of sacbrood are made on gross symptoms only, and a survey involving laboratory analysis will often produce startling evidence of the almost ubiquitous nature of sacbrood virus. Allen and Ball (1996) distinguish records relying on field symptoms only from those based on laboratory diagnosis. The disease is still significant for beekeepers, at least in part because it can confuse field diagnosis of more serious brood diseases.

      As with sacbrood virus, *Nosema apis* may well be fairly universal, found in most places where honey bees exist naturally or have been taken. However, while information on nosema is still patchy, it seems prudent to record what data are available.

      Because of debate on the importance of Kashmir bee virus, records on the occurrence of KBV have also been included.

Descriptions of bee health status

This report employs more stringent criteria for recording the status of a disease, parasite or pest than those used in earlier reviews, and I have endeavored not to present less reliable information as being authoritative. Many of the references used in the reports on which this appendix is based are less than perfect; to restrict information to only that based on completely reliable surveys would mean discarding much that is still of value, so I have used five statuses which are given in Table A11.2.

Bee health status is given on a country level. No information is given on the distribution of parasites or diseases within a country, unless relevant information exists for separate parts of it (e.g., Alaska, the United States; Northern Ireland, the United Kingdom), and especially where some control exists over movement of bees from the other part of that country. The data given relate to species established in a particular area, so records of parasites on bees imported under quarantine conditions have not been included.

Territorial descriptions

The map of the world has changed dramatically in recent years, especially in Europe and Asia. Areas such as the former Czechoslovakia and Yugoslavia have been retained for this review, as it is unlikely that their constituent republics will have significantly different bee health conditions given the free movement of bees and bee products that took place over several decades when these now independent states comprised single nations. However, records for new countries have been added so that information on their health status can be recorded as it becomes available.

Countries for which no information has been collected do not appear in the tables. Names given to countries are the commonly used English-language forms. The description of a territory makes no judgment about the political status of an area.

References

The reviews on which this material is based (Matheson 1993a, 1995, 1996) contain a full reference for each record in the tables, so that readers can make their own assessments of the validity of the health status reported. In choosing references, scientific papers or rigorous surveys were preferred, especially to justify negative reports. In order to keep the number of references manageable, recent authoritative reviews were used where available, but in most cases, the primary sources used for these reviews were checked and erroneous records discarded. Records from the OIE (*Office International des Epizooties*) have been used in places, but reliable bee science or beekeeping papers have been preferred where possible.

All references used are held in the Eva Crane IBRA Library, and full details of how to use this library service are available from the International Bee Research Association.

Trends in apparent distribution of reported species

The records in this appendix do not give a complete picture of the distribution of the diseases and parasites under discussion. Information is incomplete for many areas, and the data given often reflect the distribution (and enthusiasm) of research workers, the occurrence of bee disease surveys, and the extent to which field symptoms are followed up by laboratory work.

However, the data collected in this review allow some comment on the distribution of diseases and parasites and recent trends.

Brood diseases

American foulbrood (AFB) is widespread in North America, Europe, and Asia, but has not been reported from sub-saharan Africa nor from every country in South America. European foulbrood (EFB) has been reported from most of North America, South America, Europe, and Asia, and a few countries in Africa. Its status in Southeast Asia is largely unknown.

EFB is absent from New Zealand and, like AFB, from much of the South Pacific.

Little change from earlier reviews is apparent in the reported distribution of sacbrood and chalkbrood, though the work of Allen and Ball (1996) adds important reliability to some of the earlier records. Thai sacbrood virus has been reported from *Apis cerana* in Bangladesh, Bhutan, Burma, China, India, Nepal, Pakistan, Thailand, and Vietnam. Sacbrood virus has been reported from *Apis mellifera* in the following east Asian countries: China, Pakistan, Philippines, South Korea, and Thailand.

Adult diseases

Nosema apis has a very wide distribution. Reliable records do not exist for many countries on the status of amoeba (*Malpighamoeba mellificae*).

Parasitic mites

Tropilaelaps

Tropilaelaps clareae probably has a universal distribution within the natural range of its presumed host species *Apis dorsata*. Following the discovery of the mite in Papua New Guinea its spread through that country is being moni-tored, and with the finding of *A. cerana* on Australian islands in the Torres Strait, its chances of spreading to tropical parts of Australia are heightened. *T. clareae* has already been reported from tropical Africa where, if it becomes established, has the potential to cause immense harm to beekeeping.

Varroa

Varroa jacobsoni has been detected in more countries in South America and North America, and has been reported from countries closer to Central America (Belize in the north and Colombia in the south).

Varroa is now universal in Europe (except for Ireland) and also universal in Africa north of the Sahara. The status of this mite in sub-saharan Africa remains uncertain, despite the OIE records for it in Cape Verde, Niger and Senegal.

Varroa's spread through Papua New Guinea is, like that of *T. clareae*, a concern for Australia's beekeeping industry, as varroa has now been found on islands in the Torres Strait between those two countries. *Varroa underwoodi*, previously recorded in Nepal and Korea, has also been found in Papua New Guinea.

Tracheal mite

This species is now found in both Canada and the continental United States, and has been recorded in several Central American countries. Reports from Europe now indicate that only two Scandinavian countries are apparently free of the tracheal mite. Little is known of its distribution in sub-saharan Africa, but most reasonably thorough investigations have detected its presence there. Its distribu-tion appears patchy in the Middle East, but it has been found in most other Asian countries where it has been looked for. It is not found in Australia or New Zealand.

Other species

Significant corrections have been made to previously published records for braula.

Kashmir bee virus (KBV) has been isolated from *Apis cerana* in India and Papua New Guinea, and from *A. mellifera* in Australia, Canada, Fiji, New Zealand, Papua New Guinea, Solomon Islands, Spain, and the United States. It is quite likely, however, that further investigations using appropriate laboratory techniques will show that KBV has a much wider distribution.

Discussion

How accurate are disease status reports?

The reliability of any report on the presence or absence of a disease or parasite depends on the accuracy of the original investigation. Reports based on field symptoms are often significantly altered once a thorough investigation takes place.

Negative results (statements that a country is free of a particular disease or parasite) must be considered carefully. They are reliable only if a thorough survey is carried out, with sampling to a statistically significant level and laboratory analysis performed by competent personnel. Few such surveys have been completed, and importing countries often accept zoosanitary declarations based on inadequate information. This has led to widespread international movement of serious diseases and parasites.

The length of time for which a country's records remain accurate depends on the likelihood of infected bees or contaminated hive parts or hive products being imported. Experience shows that most countries with land borders are unlikely to be able to prevent diseases or parasites being introduced from neighboring countries, and that transfer of such species is restricted mainly by physical barriers (such as mountain ranges or deserts) which limit natural or assisted bee movements. Island nations are more likely to be able to prevent such introductions, though this is dependent on the existence of effective legislation and inspection systems.

Newly introduced parasites or diseases may remain undetected by beekeepers for some time. The chances of such infestations being discovered are greatly increased if there is a comprehensive colony inspection program or, more importantly, a program of active surveillance for exotic diseases.

Further information needed

An area of study requiring frequent updates is the mite fauna of Asian species of honey bees, especially as two of the serious mite parasites reviewed in this article originated in Asia. Related species may prove also to be serious parasites: *Varroa underwoodi* parasitizes *A. cerana*, and *Euvarroa sinhai* may parasitize *A. mellifera*. *Tropilaelaps koenigerum*, previously reported only from *A. dorsata* in Sri Lanka, has recently been found on both *A. mellifera* and *A. cerana* in northern India.

The distribution of parasites and diseases in subsaharan Africa, especially the status of the mites *Varroa jacobsoni* and *Tropilaelaps clareae*, needs investigation. It would be also be desirable to review the bee health status of the new republics of the former Soviet Union.

TABLE A11.1. Diseases, parasites, and pests surveyed for this review.

Common name	Causative organism
Brood diseases	
American foulbrood (AFB)	*Bacillus larvae*
European foulbrood (EFB)	*Melissococcus pluton*
Sacbrood	Sacbrood virus (Thai sacbrood virus in *Apis cerana*)
Chalkbrood	*Ascosphaera apis*
Adult diseases	
Nosema disease	*Nosema apis*
Amoeba disease	*Malpighamoeba mellificae*
Parasitic mites	
Tracheal mite	*Acarapis woodi*
Varroa	*Varroa jacobsoni*
Tropilaelaps	*Tropilaelaps clareae*
Other	
Braula (bee louse)	*Braula* spp.

Table A11.2. Descriptions of the status of parasites and diseases used in this report.

Description of status	Symbol used in Table A11.3	Criteria employed
Present	+	Reported in a peer-reviewed article, or an authoritative review or investigation; report by an apparently reliable witness of a disease with recognizable field symptoms.
Suspected present	+?	Anecdotal or hearsay report; field diagnosis of condition without obvious symptoms.
Suspected absent	-?	Limited investigation made with negative results; anecdotal reports about absence of disease with field symptoms.
Not present	-	Rigorous survey carried out; laboratory analyses made where appropriate.
No information		No information available; anecdotal reports of the absence of asymptomatic condition.

Table A11.3. Honey bee diseases, parasites, and pests in countries of the Americas.

	Brood Diseases				Adult Diseases		Parasitic Mites			Other
	AFB	EFB	Sacbrood	Chalkbrood	Nosema	Amoeba	Tracheal	Varroa	Tropilaelaps	Braula
North America										
Canada	+	+		+	+		+	+	-	+?
Mexico	+	+	+	+	+		+	+	-	+
USA (cont. states)	+	+	+	+	+		+	+	-	+
USA (Alaska)	+	+	+	+	+	+	+	-	-	
Central America										
Belize	+	+	+	+	+	+	+	+	+	
Costa Rica	+	+	+	+	+	+	+	+	-	
El Salvador	+	+	+	+	+	+	+	-	-	+
Guatemala	+		+	+	+		-	-	-	
Honduras	-?	+		+	+	-?	+	-?	-?	
Nicaragua	+	+	+	+	+		+	-	-	
Panama	+	+	+	+	+		+	-	-	+
Caribbean and other islands										
Antigua	-	-	-	-						
Bahamas	+	+	-							
Bermuda	+	+		-	+					
Cuba	+	+	-	+			+	-	-	
Dominica	-	-	-	-				-		
Dominican Rep.	-?	-?						-?		
Grenada	-	-		-			+	+		
Guadeloupe	-	+								
Haiti	+	+					+			

Jamaica	+	-	-	-			-		
Martinique	-	-	-	-					
Nevis (St Kitts-Nevis)	-	-	-	-					
Puerto Rico				+			+	+	
St Lucia	-	-	-	-					
St Vincent	+	-	-	-			+		
Trinidad and Tobago	-	+	+	-					+
South America									
Argentina	+	+	-	+	+	+	+	+	+
Bolivia					+			+	+
Brazil	-	+	+	-	+	+	+	+	+
Chile	-	+?	-	-	+	+	+	+	+
Colombia	+	+	+		+		+	+	+
Ecuador	+	+		+				-	
French Guiana	- ?	- ?	- ?	- ?					
Guyana	- ?	- ?	- ?	- ?					
Paraguay		+			+		+	+	+
Peru	+	+			-		-	+	+
Surinam	- ?	- ?	- ?	- ?				-	
Uruguay	+	+			+		+	+	+
Venezuela	-	+	+	+ *	+	+	+	+	+

*Species described as *Arrhenosphaera cranei*.

Table A11.4. Honey bee diseases, parasites, and pests in countries of Europe (including the former USSR).

	Brood Diseases				Adult Diseases		Parasitic Mites			Other
	AFB	EFB	Sacbrood	Chalkbrood	Nosema	Amoeba	Tracheal	Varroa	Tropilaelaps	Braula
Albania	+	+		+	+		+	+		+
Austria	+	+			+	+	+	+		+
Belarus	+	+	+							
Belgium	+	-?		+	+	+	+	+		+
Bulgaria	+	+			+		+	+		+
Corsica (France)						+	+			+
Croatia	+	+			+		+	+		
Cyprus	+	+	+		+		+	+		
Czechoslovakia (former)	+	+	+	+	+	+	+	+		+
Denmark	+	+	+	+	+	+	+	+		+
Estonia	+				+		+	+		
Finland	+	+	+	+	+	·	+	+		+
France	+	+	+	+	+	+	+	+		+
Germany	+	+	+	+	+	+	+	+		+
Guernsey (UK)	+	+			+			+		+
Greece	+	+	+	+	+	+	+	+		+
Hungary	+	+			+	+	+	+		+
Ireland	+	+			+	+	+	·		
Isle of Man (UK)										
Italy	+	+	+	+	+	+	+	+		+
Jersey (UK)								+		
Krygyzstan	+									

Country	1	2	3	4	5	6	7	8	9
Latvia	+	+	+	+	+	-	+	+	+
Lithuania	+	+	+		+		-?	+	-
Luxembourg	+				+		+	+	
Majorca (Spain)	+	+	+		+		+	+	
Malta	+	+	+		+		+	+	+?
Moldova	+	+	-?	+	+	-?	+	+	+
Netherlands	+	+	-?	+	+	+	+	+	+
N. Ireland (UK)	+	+	+	+	+		+	+	+
Norway	+	+	+	+	+	+	-	+	+
Poland	+	+	+	+	+	+	+	+	+
Portugal	+	+	+	+	+	+	+	+	+
Romania	+	+	+	+	+	+	+	+	+
Russia	+	+	+	+	+		+	+	
Sardinia (Italy)	+	+		+	+		+	+	
Slovenia	+	+			+	+	+	+	
Spain	+	+	+	+	+	-	+	+	+
Sweden	+	+	+	+	+	+	-	+	+
Switzerland	+	+	+	+	+	+	+	+	
Ukraine	+	+	+		+		+	+	
United Kingdom	+	+	+	+	+	+	+	+	+
U.S.S.R. (former)	+	+	+	+	+	+	+	+	+
Uzbekistan	+	+			+		+	+	
Yugoslavia (former)	+	+	+	+	+	+	+	+	+

Table A11.5. Honey bee diseases, parasites, and pests in countries of Africa.

	Brood Diseases				Adult Diseases		Parasitic Mites			Other
	AFB	EFB	Sacbrood	Chalkbrood	Nosema	Amoeba	Tracheal	Varroa	Tropilaelaps	Braula
North of the Sahara										
Algeria	+	+	+	+?	+	-	-	+		+
Canary Islands (Spain)	-	-		-	+		+	+		+
Egypt	-	-	+	+	+	+	+	+		+
Libya					+		+	+		
Morocco	+	+			-?	+	+	+		+
Tunisia	+	+		+	+			+		+
Sub-saharan										
Angola	-?	+?	+	+?	+		-			+
Botswana		+								+
Burundi	-	-?								+
Cape Verde								+		+
Cen. African Rep.	-?	+?			-?		-?			+
Congo										+
Ethiopia			+				+	-		+
Ghana	-	+			+		+	-	-	+
Guinea-Bissau		+			+					+
Ivory Coast	-?	-?	-?		-?		-?			+
Kenya	-?	-?		-?	+?		-	-	+	+
Madagascar	-?	+	-?		+?		-?	-		+
Malawi	-	+	-				+	-		+
Mauritius	-?	-?	-?	-?	+?		-?	-	-	+

Country								
Mozambique								+
Niger	-?		+		+	+	+	
Nigeria			+		+	-		-
Reunion	+?	+	+		+	+	-?	+
Rwanda	-?	-?	-?					+
Senegal	+	-?	-?		+	+	+	+
Seychelles	-?	-?	-?					-
South Africa	-	+	-	+*	+	-	+*	+
Sudan	-	-				-		
Tanzania	+	+			-	+	-	+
Uganda	-?	-?	-?					
Zaire	-?				+	-	-	+
Zambia	-	+					-	+
Zimbabwe				+		-	-	+

Country Records for Honey Bee Diseases, Parasites, and Pests

Table A11.6. Honey bee diseases, parasites, and pests in countries of Asia.

	Brood Diseases				Adult Diseases		Parasitic Mites			Other
	AFB	EFB	Sacbrood	Chalkbrood	Nosema	Amoeba	Tracheal	Varroa	Tropilaelaps	Braula
Middle East										
Iran	+	+	+	+	+	+	+	+	+	+
Iraq	-?	+	-?	-?	+			+		+
Israel	+	+			+			+		+
Jordan	+	+	+	+	+		+	+		+
Lebanon	+	+	+	+	+		-	+		+
Oman	+	-		+	+		-	-	-	
Saudi Arabia	+	+	+	+	+		+	+		+
Syria	+	+			+		+	+		+
Turkey	+?	+?		+	+?		+?	+		+
Un. Arab Emirates	-?	-?	-?	-?				+		+
Yemen	-?	-?	-?	-?						
Indian subcontinent										
Afghanistan	-?	-?	-?	-?						
Bangladesh	+	+	+		+		+	+		
Bhutan		+	+				+	+	+	
India	+	+	+	+	+	+	+	+	+	+
Nepal	-?	+*	+	-?	-?	-?	+	+	+**	+
Pakistan	-?	-?		-?	-?	-	+	+		
Sri Lanka	-?	-?	-?	-?				-		
East of Indian subcontinent										
Burma	+			+			+	+	+	+
Cambodia										

China	+	+	+	+	+	+	+	+	+
Hong Kong	+	+?	+	+		+	+	+?	
Indonesia	+	+				-?	+	+	
Japan	+	+	+	-	+?***		+	+	
Korea (N)	+								
Korea (S)	+	+	+		-		+	+	-
Laos									
Malaysia	+?	-?	-?				+	+	
Mongolia									
Philippines	+	-?	+	+			+	+	
Singapore									
Taiwan	+	+	+	+	+	+	+	+	
Thailand	+	+	+				+	+	
Vietnam	+	+	+	+			+	+	

* *M. pluton* found in *Apis laboriosa* and *A. cerana* in Nepal.

** *Tropilaelaps clareae* found on *A. mellifera*, *A. dorsata*, and *A. laboriosa*.

*** Records for *Acarapis woodi* in Japan are contradictory.

Table A11.7. Honey bee diseases, parasites, and pests in countries of Oceania (including Australia).

	Brood Diseases				Adult Diseases		Parasitic Mites			Other
	AFB	EFB	Sacbrood	Chalkbrood	Nosema	Amoeba	Tracheal	Varroa	Tropilaelaps	Braula
Australia (mainland)	+	+	+	+	+	+	-	-	-	-
Australia (WA)	+	-	+	-	+	-	-	-	-	-
Cook Islands	+	-	-	-	-	-	-	-	-	-
Fiji	+	-	+	-	+	+	-	-	-	-
Hawaii (USA)	+	+	+	+	+	+	-	-	-	-
New Caledonia (France)	+	+	+	+	+	-	-	-	-	-
New Zealand	+	-	+	+	+	+	-	-	-	-
Niue	+	+	+	-	+	-	-	-	-	-
Norfolk Island (Australia)	-	-	-	-	-	-	-	-	-	-
Palau	-?	-?	-?	-?	-	-	-	-?	-?	-
Papua New Guinea	+	-	+	-	+	-	-	+	+	+
Solomon Islands	-	+	+	-	+	-	-	-	-	+
Tasmania (Aus.)	+	+	+	-	+	-	-	-	-	-
Tonga	+	+	+	-	+	-	-	-	-	-
Torres Strait Islands (Aus.)	-	-?	-?	-?	-	-	-	+	-	-
Tuvalu	-	-	-	-	-	-	-?*	-?	-?*	-
Vanuatu	-?	-?	-?	-?	-	-	-	-?	-?	-
West Samoa	-	-	+	+	+	-	-	-	-	-

* Assumed free as bees have been imported only from New Zealand.

LITERATURE CITED

Abbas, N. D. 1990. In vitro-reproduktion der parasitischen bienenmilbe *Varroa jacobsoni*. [In vitro reproduction of the parasitic bee mite *Varroa jacobsoni*]. Ph.D. thesis, Naturwissenschaften, Eberhard-Karls-Universität, Tübingen, Germany. 90 pp.

Abeille de France et l'Apiculteur. 1987. Concerning Perizin [A propos du Perizin]. Abeille de France et l'Apiculteur (No. 714): 47-151.

Accorti, M. 1986. Food transfer in *Apis mellifera ligustica*: field observations using a systemic acaricide [La distribuzione del cibo in *Apis mellifera ligustica* Spin.: osservazioni in pieno campo tramite l'uso di un acaricida sistemico]. Apicoltura, Italy (No. 2): 121-127.

Accorti, M., and F. de Pace. 1983. *Varroa jacobsoni* Oud. nell'Italia centrale [*Varroa jacobsoni* Oud. in central Italy]. Apicoltore Moderno 74:3-6.

Adam, B. 1968. "Isle of Wight" or acarine disease: its historical and practical aspects. Bee World 49:6-18.

———. 1985. The acarine disease menace—short-term and long-term countermeasures. American Bee Journal 125:163-164.

———. 1987a. The honey-bee tracheal mite—fact and fiction. American Bee Journal 127:36-38.

———. 1987b. The tracheal mite—breeding for resistance. American Bee Journal 127: 290-291.

Adam, Brother. 1983. In search of the best strains of bees. Dadant and Sons, Hamilton, Illinois.

———. 1987. Breeding the honeybee. Northern Bee Books, Mytholmroyd/ Hebden Bridge, Yorkshire.

Adamovic, Z. R. 1949. La mouche rapace *Dasypogon teutonus* L. comme l'ennemie des abeilles [The predatory fly *Dasypogon teutonus* L. as an enemy of bees]. Arhiv Bioloskih Nauka 1:266-269.

———. 1950. Second contribution à l'étude des asilides comme les ennemis des abeilles [A second contribution to the study of asilids as bee enemies]. Arhiv Bioloskih Nauka 2:74-79.

———. 1963a. Ecology of some asilid species (Asilidae, Diptera) and their relation to honey bees (*Apis mellifica* L.) [in Serbian]. Museum d'Histoire Naturelle de Beograd, Editions hors série.

———. 1963b. The feeding habits of some asilid species (Asilidae, Diptera) in Yugoslavia. Archives of Biological Sciences 15(1-2):37-74.

———. 1971. Habitat distribution of the robber flies (Diptera, Asilidae) in the district of Ulcinj, Montenegro. Acta Entomologica Jugoslavica 7(2):25-35.

———. 1972. Bee-killing robber flies (Diptera, Asilidae) in the area of Ulcinj, Montenegro. Acta Entomologica Jugoslavica 8(1-2):33-37.

Adams, J., E. D. Rothman, W. E. Kerr, and Z. L. Paulino. 1977. Estimation of the number of sex alleles and queen matings from diploid male frequencies in a population of *Apis mellifera*. Genetics 86:583-596.

Adamson, A. M. 1943. Enemies and diseases of the honeybee in Trinidad. Trinidad and Tobago Agricultural Society Proceedings 43:37-53.

Adcock, D. 1962. The effect of catalase on the inhibine and peroxide values of various honeys. Journal of Apicultural Research 1:38-40.

Adlakha, R. L. 1975. Wasps: a notorious bee enemy in India. American Bee Journal 115:55.

———. 1976. Acarine disease of adult honey bees in India. American Bee Journal 116:324-344.

Adlakha, R. L., and O. P. Sharma. 1976. Stylops (Strepsiptera) parasites of honey bees in India. American Bee Journal 116:66.

Adlakha, R. L., and T. D. Verma. 1976. Scarab beetles and *Apis mellifera* colonies in India. American Bee Journal 116:53.

Adsay, M. F. 1950. Arilikta Denizli civcivlerinin rolu [Denizli chickens for controlling wasps in the apiary]. Aricilik Dergisi 6:124–125. *From* Apicultural Abstracts 43/53.

Aggarwal, K. 1988. Incidence of *Tropilaelaps clareae* on three *Apis* species in Hisar (India). *In* Africanized honey bees and bee mites. G. R. Needham, R. E. Page, Jr., M. Delfinado-Baker, and C. E. Bowman, editors. Ellis Horwood, Chichester, England.

Ahmad, R., N. Muzaffar, and Q. Ali. 1983. Biological control of the wax moths *Achroia grisella* (F.) and *Galleria mellonella* (L.) (Lep., Galleriidae) by augmentation of the parasite *Apanteles galleriae* Wlk. (Hym., Braconidae) in Pakistan. Apiacta (Bucharest) 18:15–20.

Ahmad, R., N. Muzaffar, and M. S. Munawar. 1986. Studies on the biological control of hornet predators of honey bees in Pakistan. Proceedings of the 30th International Congress of Apiculture, Nagoya, Japan; Apimondia. pp. 403-404.

Ahmadi, A. A. 1984. Incidence of honey bee *(Apis mellifera)* diseases and parasites in southern Iran. Bee World 65:134–135.

Akratanakul, P. 1975. Biology and systematics of bee mites of the family Varroidae (Acarina: Mesostigmata). M.S. thesis, Oregon State University, Corvallis.

———. 1977. The natural history of the dwarf honey bee, *Apis florea* F., in Thailand. Ph.D. dissertation, Cornell University, Ithaca, New York.

Akratanakul, P., and M. Burgett. 1975. *Varroa jacobsoni:* a prospective pest of honeybees in many parts of the world. Bee World 56:119–121.

Akratanakul, P., and M. Burgett. 1976. *Euvarroa sinhai* Delfinado and Baker (Acarina: Mesostigmata): a parasitic mite of *Apis florea.* Journal of Apicultural Research 15:11–13.

Akre, R. D., and D. F. Mayer. 1994. Bees and vespine wasps. Bee World 75:29–37.

Akre, R. D., W. B. Hill, J. F. McDonald, and W. B. Garnett. 1975. Foraging distances of *Vespula pensylvanica* workers (Hymenoptera: Vespidae). Journal of the Kansas Entomological Society 48:12–16.

Alber, M. A. 1953. Il calabrone, tigre dell'aria [The hornet, tiger of the air]. Apicoltore d'Italia 20:188–189, 192.

Alberts, D., D. Bray, J. Lewis, M. Raff, K. Roberts, and D. J. Watson. 1983. Molecular biology of the cell. Garland Publishing, New York.

Alcock, J. 1975. Territorial behavior of males of *Philanthus multimaculatus* with a review of territoriality in male sphecids. Animal Behavior 23: 889-895.

Aldrich, J. M. 1924. *Braula coeca* Nitzsch in Maryland apiaries. Journal Washington Academy of Science. 14(8): 181.

Aleksandrova, L. V. 1949. Growing the causative organism of European foulbrood *(B. pluton)* in pure culture [in Russian]. *In* Bolezni pchel [Bee diseases]. Gosudarstvennue Izdatel'stvo, Moscow.

Alekseenko, F. M., and A. J. Kolomiets. 1967. A study of the virus paralysis of bees in the Ukraine. Proceedings of the 21st International Apicultural Congress (Maryland) (abstract number 53:492).

Alexander, B. 1992. A cladistic analysis of the genus *Apis. In* Diversity in the genus *Apis.* D. R. Smith, editor. Westview Press, Boulder, Colorado.

Alexander, R. 1939. Bees and raspberries: scabious poisoning bees [letter to the editor]. Bee World 20:116.

Alfonsus, E. C. 1930. The mutillid wasp: an enemy of the honeybee in Europe. American Bee Journal 70:568–569.

Alford, D. V. 1975. Bumblebees. Davis-Poynter, London, England. 352 pp.

Al-Ghamdi, A. A. 1990. Survey of honeybee diseases, pests and predators in Saudi Arabia. Thesis, University of Wales, Cardiff, U.K. 171 pp.

Ali, A. D., M. A. Abdellatif, N. M. Bakry, and S. K. El-Sawaf. 1973. Studies on biological control of the greater wax moth, *Galleria mellonella.* 2. Impregnation of comb foundation with Thuricide-HP. Journal of Apicultural Research 12:125–130.

Alippi, A. M. 1992. Transporte de espora de *Bacillus larvae* por el ácaro *Varroa jacobsoni.* [Transportation of spores of *Bacillus larvae* by the mite *Varroa jacobsoni*]. Revista de la Facultad de Agronomía, La Plata 68:83–86.

Alizadeh, A., and M. S. Mossadegh. 1994. Stonebrood and some other fungi associated with *Apis florea* in Iran. Journal of Apicultural Research 33:213–218.

Allen, H. W., and M. H. Brunson. 1957. Control of Nosema disease of potato tuber worm, a host used in the mass production of *Macrocentrus ancylivorus*. Science 105:394.

Allen, K. L., P. C. Molan, and G. M. Reid. 1991. A survey of the antibacterial activity of some New Zealand honeys. Journal of Pharmacy and Pharmacology 43:817–822.

Allen, M. F., and B. V. Ball. 1995. Characterisation of Kashmir bee virus strains. Annals of Applied Biology (in press).

Allen, M. F., and B. V. Ball 1996. The incidence and world distribution of honey bee viruses. Bee World 77:141-162.

Allen, M. Y. 1940. Arena [letter to the editor]. Bee World 21:81–82.

Allsopp, M. H., and R. M. Crewe. 1993. The cape honey bee as a Trojan horse rather than the hordes of Genghis Khan. American Bee Journal 133:121–123.

Alonso, J. M., J. Rey, F. Puerta, J. Hermoso de Mendoza, M. Hermoso de Mendoza, and J. M. Flores. 1993. Enzymatic equipment of *Ascosphaera apis* and the development of infection by this fungus in *Apis mellifera*. Apidologie 24:383–390.

Alpatov, V. V. 1976. A fatal error in determining a race of honeybees. Priroda (Moscow) 1976(5):72–73.

Alpatov, W. W. 1929. Biometrical studies on variation and races of the honey bee (*Apis mellifera* L.). Quarterly Review of Biology 4:1–58.

Ambrose, D. P., and D. Livingstone. 1987. Biology of *Acanthaspis siva* Distant, a polymorphic assassin bug (Insecta, Heteroptera, Reduviidae). Mitt. Zool. Mus. In Berlin 63:321–330.

Ambrose, D. P., and D. Livingstone. 1990. Polymorphism in *Acanthaspis siva* Distant (Reduviidae: Heteroptera), a predator of the Indian honey bee *(Apis cerana)*. Journal of the Bombay Natural History Society 87(2):218–222.

Ambrose, J. T. 1975. A skunk in the bee yard. Gleanings in Bee Culture 103:385–386.

Amos, J. M. 1972. Beekeeping in Virginia. Virginia Extension Publication 372.

Amster, F. A. 1940. A tragedy from the sky. Gleanings in Bee Culture 68:562–563.

Anderson, D. L. 1984. A comparison of serological techniques for detecting and identifying honeybee viruses. Journal of Invertebrate Pathology 44:233–243.

———. 1990. Pests and pathogens of the honeybee (*Apis mellifera* L.) in Fiji. Journal of Apicultural Research 29:53–59.

———. 1994. Non-reproduction of *Varroa jacobsoni* in *Apis mellifera* colonies in Papua New Guinea and Indonesia. Apidologie 25:412–421.

Anderson, D. L., and A. J. Gibbs. 1988. Inapparent virus infections and their interactions in pupae of the honey bee (*Apis mellifera* Linnaeus) in Australia. Journal of General Virology 69:1617–1625.

Anderson, E. D. 1969. An appraisal of the beekeeping industry. U.S. Department of Agriculture ARS 42-150.

Anderson, E. J. 1941. Diseases and enemies of the honeybee. Pennsylvania Agricultural Extension Service Circular 156.

Anderson, G. J. 1976. The pollination biology of *Tilia*. American Journal of Botany 63(9):1203–1212.

Anderson, J. 1930. "Isle of Wight disease" in bees. I. and II. A check to the immunity hypothesis. Bee World 11:37–42, 50–53.

———. 1934. Brood diseases in Scotland. Rothamsted Conference 18:17–18.

———. 1938. Chalk brood. Scottish Beekeeper 14:106.

Anderson, L. D. and L. Atkins. 1968. Pesticide usage in relation to beekeeping. Annual Review of Entomology 13:213–238.

Anderson, R. H. 1963. The laying worker in the Cape honeybee, *Apis mellifera capensis*. Journal of Apicultural Research 2:85–92.

———. 1977. Some aspects of the biology of the Cape honey-bee. *In* African bees: taxonomy, biology and economic use. D. J. C. Fletcher, editor. Apimondia, Pretoria.

Anderson, R. H., B. Buys, and M. F. Johannsmeier. 1983. Union of South Africa, Department of Agriculture Technical Services Bulletin number 394.

———. 1983. Beekeeping in South Africa. Dept. Agriculture, Pretoria S.A. Bulletin #394.

Andrews, R. T. 1882. *Vespa crabro* L. preying on honey bees in England. Science Gossip 18:282.

Anonymous. 1908. Ants and mice and bees. Agricultural Journal of the Cape of Good Hope 32:282.

———. 1932. Injury to comb honey by *Vitula edmandsii* Pack. Journal of Economic Entomology 25:946.

———. 1959. Diseases of bees. Ministry of Agriculture, Fisheries and Food Bulletin number 100. Townsend and Sons, London.

———. 1968. Protection of bee hives. Canadian Beekeeping 1(5):5.

———. 1971. Black bear facing extinction. St. Petersburg (Florida) Times, Friday, June 30, 1971, pp. 1-B, 12-B.

———. 1972. Controlling the greater wax moth. . .a pest of honeycombs. U.S. Department of Agriculture Farmers' Bulletin number 2217.

———. 1975. Bee products: what we know about pollen. Bee World 56:155–158.

———. 1976a. Bears are in for a shock. Gleanings in Bee Culture 104:294.

———. 1976b. Bee products: harvesting pollen from hives. Bee World 57:20–25.

———. 1978. Honey exposure and infant botulism. Morbidity and Mortality Weekly Report 27:249–250, 255.

———. 1984a. *Acarapis woodi* mites found in three Texas locations. American Bee Journal 124:565–566.

———. 1984b. *Acarapis woodi* mite discovery brings federal quarantine and control program. American Bee Journal 124:633–634.

———. 1984c. *Acarapis woodi* found in Louisiana, other states negative. American Bee Journal 124:697.

———. 1984d. Hyattsville meeting recommends new mite guidelines. American Bee Journal 124:761.

———. 1984e. Acarine disease confirmed in South Dakota, New York, North Dakota and Florida. American Bee Journal 124:761–762 and 768–769.

———. 1984f. Illinois mite detection workshop attracts over 50 officials. American Bee Journal 125:6.

———. 1985. Nebraska tracheal mite survey summary. American Bee Journal 125:6.

———. 1986. Bee-eaters, a passing problem. Australasian Beekeeper 88:67.

———. 1987a. Varroa mites found in the United States. American Bee Journal 127:745–746.

———. 1987b. Colony, honey plant and market conditions during December. National Honey Market News, January 14, 1987.

———. 1987c. Winter/summer mouse guard and bottom board. Gleanings in Bee Culture 115:533.

———. 1988a. Apistan approved for varroa mite control. American Bee Journal 128:157.

———. 1988b. Quarantine killed. Gleanings in Bee Culture 116:361.

———. 1993a. Bear population increasing. Bee Culture 121:50.

———. 1993b. Guidelines for responding to bear problems. Published by N. C. Wildlife Resources Commission, Raleigh. 8 pp.

———. 1993c. USDA to release mite-resistant bees to selected breeders. American Bee Journal 133:115.

Aoyagi, S., and C. Oryu. 1968. The honey bee and the honey. III. Studies on yeasts in honey. Bulletin of the Faculty of Agriculture, Tamagawa University number 7/8:203–213.

Aran, S. 1969. Apicultura practica [Practical apiculture]. Graficas Yagues, Madrid.

Argauer, R. T., and E. W. Herbert, Jr. 1992. Stability of oxytetracycline residue in pollen pellets harvested from medicated research colonies of the honey bee. American Bee Journal 132:332–334.

Argo, V. N. 1926a. *Braula coeca* Nitzsch in Maryland. Journal of Economic Entomology 19(1): 130-181.

————. 1926b. *Braula coeca* or the bee louse. Gleanings in Bee Culture 54: 435-438.

Arias, M. C., and W. S. Sheppard. 1996 Molecular phylogenetics of honey bee subspecies (*Apis mellifera* L.) inferred from mitochondrial DNA sequences. Molecular Phylogenetics and Evolution 5:557-566, in press.

Arnold, H., U. Seitz, and G. Löhr. 1974. Die Hexokinase und die Mannosetoxizität der Biene. Hoppe-Zeyler's Zeitschrift für Physiologische Chemie 355(3):266-272.

Arnold, J. H., and S. E. Bland. 1954. Beekeeping in Saskatchewan. Saskatchewan Department of Agriculture Publication.

Arnon, S. S. 1980. Infant botulism. Annual Review of Medicine 31:541-560.

Atakishiev, T. A. 1970. Birds that prey on bees [in Russian]. Pchelovodstvo 1970(3): 32-33.

————. 1971a. Hive cohabitants in Azerbaidzhan [in Russian]. Uchenye Zapiski Kazanskogo Veterinarogo Instituta 109:261-265.

————. 1971b. The bee louse (*Braula coeca* Nitzsch) [in Russian]. Pchelovodstvo 1971(2):16-17.

Atkins, E. L. 1975. Injury to honey bees by poisoning. *In* The hive and the honey bee. Dadant & Sons, editors. Rev. ed. Dadant & Sons, Hamilton, Illinois.

————. 1992. Injury to honey bees by poisoning. *In* The hive and the honey bee. J. Graham, editor. Dadant & Sons , 1153-1208.

Atkins, L., D. Kellum, and K. W. Atkins. 1981. Reducing pesticide hazards to honey bees. University of California Division of Agricultural Sciences Leaflet 2883.

Atkinson, J. 1974a. Alberta beekeepers under bear attack. Canadian Bee Journal 85:28-29.

————. 1974b. Mouse control. Canadian Bee Journal 85:31.

Atwal, A. S. 1971. Acarine disease problem of the Indian honey bee, *Apis indica* F. American Bee Journal 111:134-135.

Atwal, A. S., and G. S. Dhaliwal. 1969. Robbing between *Apis indica* F. and *Apis mellifera* L. American Bee Journal 109:462-463.

Atwal, A. S., and O. P. Sharma. 1970. Acarine disease of adult honeybees: prevention and control. Indian Farming 20:39-40.

Atwal, A. S., and N. P. Goyal. 1971. Infestation of honey bee colonies with *Tropilaelaps* and its control. Journal of Apicultural Research 10:137-142.

Avery, S. W., and D. W. Anthony. 1983. Ultrastructural study of early development of *Nosema algerae* in *Anopheles albimanus*. Journal of Invertebrate Pathology 42:87-95.

Avitabile, A. 1978. Brood rearing in honeybee colonies from the autumn to early spring. Journal of Apicultural Research 17:69-73.

Azuma, H. 1992. The acaricide, Misubishi Apistan [in Japanese]. Honeybee Science 13:115-119.

Badino, G., G. Celebrano, and A. Manino. 1983. Population structure and Mdh-1 locus variation in *Apis mellifera ligustica*. Journal of Heredity 74:443-446.

Badino, G., C. Celebrano, and A. Manino. 1984. Population genetics of Italian honeybee (*Apis mellifera ligustica* Spin.) and its relationships with neighbouring subspecies. Bolletino del Museo Regionale di Scienze Naturali di Torino 2:571-584.

Badino, G., C. Celebrano, and A. Manino. 1988. Genetica di popolazione delle sottospecie mediterranee di *Apis mellifera* L. sulla base di varianti alloenzimatiche. (Population genetics of Mediteranean subspecies of *Apis mellifera* based on allozyme variante.) Apicoltore Moderno 79:233-239.

Bahr, L. 1916. Die Krankheiten der Honigbiene und ihrer Brut [Diseases of honey bees and their brood]. Deutsche Tierärztliche Wochenschrift 24:255-258, 264-266.

Bährmann, R. 1967. Über das Auftreten von Pylorusschorfen bei gesunden und nosemakranken Arbeiterinnen der Honigbiene unter verschiedenen experimentellen Bedingungen [The occurrence of pyloric scabs in healthy and nosema-infected worker honey bees under various experimental conditions]. Annales de l'Abeille 10:29-37.

Bailey, L. 1953. The transmission of nosema disease. Bee World 34:171-172.

————. 1954. The respiratory currents in the tracheal system of the adult honey-bee. Journal of Experimental Biology 31:589-593.

———. 1955a. Control of amoeba disease by the fumigation of combs and by fumagillin. Bee World 36:162–163.

———. 1955b. The epidemiology and control of Nosema disease of the honey-bee. Annals of Applied Biology 43:379–389.

———. 1955c. The infection of the ventriculus of the adult honeybee by *Nosema apis* (Zander). Parasitology 45:86–94.

———. 1955d. Results of field trials at Rothamsted of control methods for Nosema disease. Bee World 36:121–125.

———. 1956. Aetiology of European foul brood; a disease of the larval honeybee. Nature 178:1130.

———. 1957a. Comb fumigation for nosema disease. American Bee Journal 97:24–26.

———. 1957b. European foul brood: a disease of the larval honeybee (*Apis mellifera* L.) caused by a combination of *Streptococcus pluton* (*Bacillus pluton* White) and *Bacterium eurydice* White. Nature 180:1214–1215.

———. 1958. The epidemiology of the infestation of the honey bee, *Apis mellifera* L., by the mite *Acarapis woodi* Rennie and the mortality of infested bees. Parasitology 48:493–506.

———. 1959. An improved method for the isolation of *Streptococcus pluton* and observations on its distribution and ecology. Journal of Insect Pathology 1:80–85.

———. 1960. The epizoology of European foul brood of the larval honey bee, *Apis mellifera* Linnaeus. Journal of Insect Pathology 2:67–83.

———. 1961. The natural incidence of *Acarapis woodi* (Rennie) and the winter mortality of honeybee colonies. Bee World 42:96–100.

———. 1963a. Infectious diseases of the honey bee. Land Books, London.

———. 1963b. The pathogenicity for honey-bee larvae of microorganisms associated with European foulbrood. Journal of Insect Pathology 5:198–205.

———. 1964. The "Isle of Wight disease": the origin and significance of the myth. Bee World 45:18, 32–37.

———. 1965a. Paralysis of the honey bee, *Apis mellifera* Linnaeus. Journal of Invertebrate Pathology 7:132–140.

———. 1965b. Susceptibility of the honey bee, *Apis mellifera* Linnaeus, infested with *Acarapis woodi* (Rennie) to infection by airborne pathogens. Journal of Invertebrate Pathology 7:141–143.

———. 1966. Pathogens of the wax moths and other insects. Lecture leaflets. Central Association of Bee-keepers, Ilford, Essex, England.

———. 1967a. The effect of temperature on the pathogenicity of the fungus, *Ascosphaera apis,* for larvae of the honey bee, *Apis mellifera. In* Insect pathology and microbial control. P. A. van der Laan, editor. North- Holland Publishing, Amsterdam.

———. 1967b. The world distribution of viruses of the honey bee. Bulletin Apicole 10:121–124.

———. 1967c. The incidence of virus disease in the honey bee. Annals of Applied Biology 60:43–48.

———. 1967d. *Nosema apis* and dysentery of the honeybee. Journal of Apicultural Research 6:121–125.

———. 1968a. Honey bee pathology. Annual Review of Entomology 13:191–212.

———. 1968b. The measurement and interrelationships of infections with *Nosema apis* and *Malpighamoeba mellificae* of honey-bee populations. Journal of Invertebrate Pathology 12:175–179.

———. 1969a. The multiplication and spread of sacbrood virus of bees. Annals of Applied Biology 63:483–491.

———. 1969b. The signs of adult bee diseases. Bee World 50:66–68.

———. 1972a. *Nosema apis* in drone honeybees. Journal of Apicultural Research 11:171–174.

———. 1972b. The preservation of infective microsporidian spores. Journal of Invertebrate Pathology 20:252–254.

————. 1975. Recent research on honeybee viruses. Bee World 56:55–64.

————. 1976. Viruses attacking the honey bee. Advances in Virus Research 20:271–304.

————. 1981. Honey bee pathology. Academic Press, London.

————. 1982a. Viruses of honeybees. Bee World 63:165–173.

————. 1982b. A strain of sacbrood virus from *Apis cerana*. Journal of Invertebrate Pathology 39:264–265.

————. 1985. *Acarapis woodi*: a modern appraisal. Bee World 66:99–104.

————. 1986. The mite that roared. American Bee Journal 126:469.

————. 1989. Some notes on *Acarapis woodi* (Rennie). American Bee Journal 129:543–545.

Bailey, L., and B. V. Ball. 1978. *Apis* iridescent virus and "clustering disease" of *Apis cerana*. Journal of Invertebrate Pathology 31:368–371.

Bailey, L., and B. V. Ball. 1991. Honey bee pathology. Academic Press, London, 2d ed.

Bailey, L., B. V. Ball, J. M. Carpenter, and R. D. Woods. 1980a. Bee virus Y. Journal of General Virology 51:405–407.

Bailey, L., B. V. Ball, J. M. Carpenter, and R. D. Woods. 1980b. Small virus-like particles in honey bees associated with chronic paralysis and with a previously undescribed disease. Journal of General Virology 49:149–155.

Bailey, L., B. V. Ball, and J. N. Perry. 1981. The prevalence of viruses of honey bees in Britain. Annals of Applied Biology 97:109–118.

Bailey, L., B. V. Ball, and J. N. Perry. 1983a. Honey bee paralysis: its natural spread and its diminished incidence in England and Wales. Journal of Apicultural Research 22:191–195.

Bailey, L., B. V. Ball, and J. N. Perry. 1983b. Association of viruses with two protozoal pathogens of the honey bee. Annals of Applied Biology 103:13–20.

Bailey, L., J. M. Carpenter, D. A. Govier, and R. D. Woods. 1980b. Bee Vvirus Y. Journal of General Virology 51:405–407.

Bailey, L., J. M. Carpenter, and R. D. Woods. 1979. Egypt bee virus and Australian isolates of Kashmir bee virus. Journal of General Virology 43:641–647.

Bailey, L., J. M. Carpenter, and R. D. Woods. 1982. A strain of sacbrood virus from *Apis cerana*. Journal of Invertebrate Pathology 39:264–265.

Bailey, L., and M. D. Collins. 1982a. Taxonomic studies on *Streptococcus pluton*. Journal of Applied Bacteriology 53:209–213.

Bailey, L., and M. D. Collins. 1982b. Reclassification of *Streptococcus pluton* (White) in a new genus, *Melissococcus pluton* nom. rev.: comb. nov. Journal of Applied Bacteriology 53:215–217.

Bailey, L., and E. F. W. Fernando. 1972. Effects of sacbrood virus on adult honeybees. Annals of Applied Biology 72:27–35.

Bailey, L., and A. J. Gibbs. 1962. Cultural characters of *Streptococcus pluton* and its differentiation from associated enterococci. Journal of General Microbiology 28:385–391.

Bailey, L., A. J. Gibbs, and R. D. Woods. 1963. Two viruses from adult honey bees (*Apis mellifera* Linnaeus). Virology 21:390–395.

Bailey, L., A. J. Gibbs, and R. D. Woods. 1964. Sacbrood virus of the larval honey bee (*Apis mellifera* Linnaeus). Virology 23:425–429.

Bailey, L., and D. C. Lee. 1962. *Bacillus larvae*: its cultivation in vitro and its growth in vivo. Journal of General Microbiology 29:711–717.

Bailey, L., and R. G. Milne. 1978. Filamentous viruslike particles in honey bees in Britain. Journal of Invertebrate Pathology 32:390–391.

Bailey, L., and J. N. Perry. 1982. The diminished incidence of *Acarapis woodi* (Rennie) (Acari: Tarsonemidae) in honey bees, *Apis mellifera* L. (Hymenoptera: Apidae), in Britain. Bulletin of Entomological Research 72:655–662.

Bailey, L., and H. A. Scott. 1973. The pathogenicity of Nodamura virus for insects. Nature 241:545.

Bailey, L., and R. D. Woods. 1974. Three previously undescribed viruses from the honey bee. Journal of General Virology 25:175–186.

Bailey, L., and R. D. Woods. 1977. Two more small RNA viruses from honey bees and further observations on sacbrood and acute bee-paralysis viruses. Journal of General Virology 37:175–182.

Baker, E. W., and M. D. Delfinado. 1976. Notes on the bee mite *Neocypholaelaps indica* Evans, 1963. American Bee Journal 116:384–386.

Baker, E. W., and M. D. Delfinado. 1978. Notes on the driedfruit mite *Carpoglyphus lactis* (Acarina: Carpoglyphidae) infesting honeybee combs. Journal of Apicultural Research 17:52–54.

Baker, E. W., and M. Delfinado-Baker. 1983. New mites (Sennertia: Chaetodactylidae) phoretic on honey bees (*Apis mellifera* L.) in Guatemala. International Journal of Acarology 9:117–121.

Baker, G. M., and P. F. Torchio. 1968. New records of *Ascosphaera apis* from North America. Mycologia 60:189–190.

Ball, B. V. 1983. The association of *Varroa jacobsoni* with virus diseases of honey bees. *In* Varroa jacobsoni Oud. affecting honey bees: present status and needs. R. Cavalloro, editor. Balkema, Rotterdam.

———. 1985. Acute paralysis virus isolates from honeybee colonies infested with *Varroa jacobsoni*. Journal of Apicultural Research 24:115–119.

———. 1986. *Varroa jacobsoni*, a vector of honey bee viruses. Arbeitsgemeinschaft der Institute für Bienenforschung E.V. Abstracts of the Varroa Workshop in Feldafing/Starnberg, West Germany.

———. 1989. *Varroa jacobsoni* as a virus vector. *In* Present sense of varroatosis in Europe and progress in the varroa mite control. R. Cavalloro, editor. EEC, Luxembourg, 241–244.

Ball, B., and M. F. Allen. 1988. The prevalence of pathogens in honey bee (*Apis mellifera*) colonies infested with the parasitic mite *Varroa jacobsoni*. Annals of Applied Biology 113:237–244.

Ball, B. V., H. A. Overton, K. W. Buck, L. Bailey, and J. N. Perry. 1985. Relationships between the multiplication of chronic bee paralysis virus and its associate particle. Journal of General Virology 66:1423–1429.

Bambara, S. 1987. How do honey-guides guide? Gleanings in Bee Culture 115:97, 104.

Bamford, S., and L. A. F. Heath. 1989. The infection of *Apis mellifera* larvae by *Ascosphaera apis*. Journal of Apicultural Research 28:30–35.

Bamrick, J. F. 1964. Resistance to American foulbrood in honey bees. V. Comparative pathogenesis in resistant and susceptible larvae. Journal of Insect Pathology 6:284–304.

Bamrick, J. F., and W. C. Rothenbuhler. 1964. Resistance to American foulbrood in honey bees. IV. The relationship between larval age at inoculation and mortality in a resistant and in a susceptible line. Journal of Insect Pathology 3:381–391.

Banaszak, J. 1980. Badania nad fauna towarzyszaca w zasiedlonych ulach pszczelich [Investigations of the fauna associated with beehives]. Fragmenta Faunistica 25:127–177.

Barjac, H. de, and J. V. Thompson. 1970. A new serotype of *Bacillus thuringiensis: B. thuringiensis* var. *thompsoni* (serotype 11). Journal of Invertebrate Pathology 15:141–144.

Barker, R. J. 1977. Some carbohydrates found in pollen and pollen substitutes are toxic to honey bees. Journal of Nutrition 107:1859–1862.

Barker, R. J., and Y. Lehner. 1974. Acceptance and sustenance value of naturally occurring sugars fed to newly emerged adult workers of honey bees (*Apis mellifera* L.). Journal of Experimental Zoology 187:277–285.

Barker, R. J., and Y. Lehner. 1976. Galactase, a sugar toxic to honey bees, found in exudate of tulip flowers. Apidologie 7:109–111.

Barker, R. J. H. 1990. Poisoning by plants. *In* Honey bee pests, predators, and diseases. R. A. Morse and R. Nowogrodzki, editors. 2d ed. Cornell University Press, Ithaca, N.Y., 307–328.

Barthel, B. 1971. Der Kalkbrut auf der Spur [On the trail of chalkbrood]. Garten und Kleintierzucht, Ausgabe C, Imker 10(4):12–13.

Barton. 1866. [An account of poisonous plants.] American Bee Journal 2:14.

Batra, L. R., S. W. T. Batra, and G. E. Bohart. 1973. The mycoflora of domesticated and wild bees (Apoidea). Mycopathologia et Mycologia Applicata 49:13–44.

Basan, I. N. 1977. Equipment for treating Braula and Varroa infestations of honeybees. Veterinariya, Moscow, USSR (No. 4): 81.

Batra, S. W. T. 1980. Ecology, behavior, pheromones, parasites and management of the sympatric vernal bees Colletes inaequalis, C. thoracicus and C. validus. Journal of the Kansas Entomological Society 53:509–538.

Beal, F. E. 1915. Some common birds useful to the farmer. U.S. Department of Agriculture Farmers' Bulletin number 630.

Bean, R. C., and W. Z. Hassid. 1956. Carbohydrate oxidase from a red algae, Iridophycus flaccidum. Journal of Biological Chemistry 218:425–436.

Bean, R. C., G. G. Porter, and B. M. Steinberg. 1961. Carbohydrate metabolism of citrus fruits. Journal of Biological Chemistry 236:1235–1240.

Beck, S. D. 1960. Growth and development of the greater wax moth Galleria mellonella (L.) (Lepidoptera: Galleriidae). Transactions of the Wisconsin Academy of Sciences, Arts and Letters 49:137–148.

Becker, J. O., and F. J. Schwinn. 1993. Control of soil-borne pathogens with living bacteria and fungi: status and outlook. Pesticide Science 37:355–363.

Beckwith, I. W. 1898. Loco-plant and the bees. American Bee Journal 38:227.

Bedding, R. 1984. Large scale production, storage and transport of the insect-parasitic nematodes Neoaplectana spp. and Heterorhabditis spp. Annals of Applied Biology 104:117–120.

Beetsma, J., and K. Zonneveld. 1992. Observations on the initiation and stimulation of oviposition of the Varroa mite. Experimental and Applied Acarology 16:303–312.

Befus-Nogel, J., D. L. Nelson, and L. P. Lefkovitch. 1992. Observations on the effect of management procedures on chalkbrood levels in honey bee (Apis mellifera L.; Hymenoptera; Apidae) colonies. Bee Science 2:20–24.

Beljavsky, A. G. [also transliterated as Beliavskii]. 1927. The enemies of bees [in Russian]. Mysti, Leningrad.

———. 1929. The study of Braula coeca. Bee World 10(6): 84-85.

———. 1933. Blister beetles and their relation to the honey bee. Bee World 14:31–33.

———. 1935. The Mutilla europea L. as a bee enemy. Bee World 16:122.

———. 1936. Stylops melittae as a bee enemy. Bee World 17:32–33.

———. 1937. The hornet (Vespa crabro L.) as an enemy of bees. Bee World 18:75–77.

Belize Ministry of Agriculture. 1987. A Guide to Beekeeping. Beekeeping Section, Ministry of Agriculture, Belmopan, Belize; 82pp.

Benhamou, N., and G. B. Ouellette. 1985. Les anticorps monoclonaux: une technologie de pointe en phytopathologie [Monoclonal antibodies: a technology from the perspective of phytopathology]. Phytoprotection 66:5–15.

Benson, L. 1979. Plant classification. 2d. ed. D. C. Heath and Company, city.

Benton, F. 1880. Letter to the editor. American Bee Journal 16:51–52.

———. 1896. The honey bee: a manual of instruction in apiculture. USDA Bulletin 1-118.

Bergey, D. H. 1994. Bergey's manual of determinative bacteriology. 9th ed. J. G. Holt, editor. Williams and Wilkins, Baltimore, Md.

Bermejo, M., G. Illera, and J. Sabater-Pi. 1989. New observations of the tool-behavior of chimpanzees from Mt. Assirik (Senegal, West Africa). Primates 30(1):65–73.

Betts, A. D. 1912a. A bee-hive fungus, Pericystis alvei, gen. et sp. nov. Annals of Botany 26:795–799.

———. 1912b. The fungi of the beehive. Journal of Economic Biology 7:129–162.

———. 1919. Fungus diseases of bees. Bee World 1:132.

———. 1920. Nectar yeasts. Bee World 1:252.

———. 1928a. Hive yeasts. I. Bee World 9:108–110.

————. 1928b. Hive yeasts. II. Bee World 9:137–138.

————. 1928c. Hive yeasts. III. Bee World 9:154–155.

————. 1928d. Bee laboratory. Bee World 9:93.

————. 1929a. Hive yeasts. IV. Bee World 10:33–34.

————. 1929b. Press mirror. Bee World 10:142–143.

————. 1932. Chalk brood. Bee World 13:78–80.

————. 1936a. Press mirror. Bee World 17:28–30.

————. 1936b. Research notes. Bee World 17:33–35.

————. 1936c. The bee pirate (*Philanthus triangulum* F.). Bee World 17:64–66.

————. 1937. Research notes. Bee World 18:58–60.

————. 1939. Press mirror. Bee World 20:15–17.

————. 1951. The diseases of bees: their signs, causes and treatment. Hickmott, Camberley, England.

Bian, Z., H. M. Fales, M. S. Blum, Z. Bian, T. H. Jones, T. E. Rinderer, and D. F. Howard. 1984. Chemistry of cephalic secretion of fire bee *Trigona (Oxytrigona) tataira*. Journal of Chemical Ecology 10:451–461.

Bignell, D. E., and L. A. F. Heath. 1985. Electropositive redox state of the fifth-instar larval gut of *Apis mellifera*. Journal of Apicultural Research 24:211–213.

Bilsing, S. W. 1920. Quantitative studies in the food of spiders. Ohio Journal of Science 20:215–260.

Bissett, J. 1988. Contribution toward a monograph of the genus *Ascosphaera*. Canadian Journal of Botany 66:2541–2560.

Bitner, A. R., W. T. Wilson, and J. D. Hitchcock. 1972. Passage of *Bacillus larvae* spores from adult queen honey bees to attendant workers *(Apis mellifera)*. Annals of the Entomological Society of America 65:899–901.

Blair, W. F., A. P. Blair, P. Brodkorb, F. R. Cagle, and G. A. Moore. 1968. Vertebrates of the United States. 2d ed. McGraw-Hill, New York.

Blake, H. A. 1966. Beekeeping in Egypt (Gaza Strip). Gleanings in Bee Culture 94:168–170, 180.

Blake, J. O. 1946. Ambush bugs killing bees. Journal of Economic Entomology 39:295.

Blohm, H. 1962. Poisonous plants of Venezuela. Harvard University Press, Cambridge, Mass.

Blum, M. S. 1981. Chemical defenses of arthropods. Academic Press, New York.

Boch, R., D. A. Shearer, and A. Petrasovits. 1970. Efficacies of two alarm substances of the honey bee. Journal of Insect Physiology 16:17–24.

Boecking, O. 1992. Removal behavior of *Apis mellifera* towards sealed brood cells infested with *Varroa jacobsoni*: techniques, extent and efficiency. Apidologie 23:371–373.

————. 1992b. The removal response of *Apis mellifera* L. colonies to brood in wax and plastic cells after artificial and natural infestation with *Varroa jacobsoni* O. and to freeze killed brood. Experimental and Applied Acarology 16:321–329.

————. Reproductive success of *Varroa jacobsoni* in worked brood cells with regard to the duration of the post-capping stage. Journal of Apicultural Research (submitted).

Boecking, O., and W. Drescher. 1991. Response of *Apis mellifera* L. colonies to brood infested with *Varroa jacobsoni* O. Apidologie 22:237–241.

Boecking, O., and W. Drescher. 1992. The removal response of *Apis mellifera* L. colonies to brood in wax and plastic cells after artificial and natural infestation with *Varroa jacobsoni* O. and to freeze killed brood. Experimental and Applied Acarology 16:321–329.

Boecking, O., W. Rath, and W. Drescher. 1992. *Apis mellifera* removes *Varroa jacobsoni* from sealed brood cells in the tropics. American Bee Journal 132:732–734.

Boecking, O., W. Rath, and W. Drescher. 1993. Grooming and removal behavior—strategies of *Apis mellifera* and *Apis cerana* bees against *Varroa jacobsoni*. American Bee Journal 113:117–119.

Boecking, O., and W. Ritter. 1993. Grooming and removal behaviour of *Apis mellifera intermissa* in Tunisia against *Varroa jacobsoni*. Journal of Apicultural Research 32:127–134.

Boecking, O., and W. Ritter. 1994. Current status of behavioral tolerance of the honey bee *Apis mellifera* to the mite *Varroa jacobsoni*. American Bee Journal 134:689–694.

Bogdanov, S. 1984. Characterization of antibacterial substances in honey. Lebensmittlewissenschaft und Technologie 17:74–76.

———. 1989. Determination of pinocembrin in honey using HPLC. Journal of Apicultural Research 28:55–57.

Bohart, G. E. 1954. Honey bees attacked at their hive entrance by the wasp *Philanthus flavifrons* Cresson. Proceedings of the Entomological Society of Washington 56:26.

———. 1972. Management of wild bees for the pollination of crops. Annual Review of Entomology 17:287–312.

Boika, A. K. [also transliterated as Boyko]. 1939. Larvae of *Senotainia tricuspis* Meig. causing heavy losses of bees. C. R. Academy Science in USSR. 24(3): 304-306.

———. 1948. A new kind of myiasis in bumble bees [in Russian]. Doklady Akademii nauk SSR 61(2):423-424.

———. 1949. Apimyasis in bees [in Russian]. *In* Works of the 27th session of the Veterinary Section of the Lenin Academy of Agricultural Sciences, Moscow, pp. 115–135.

———. 1958. Senotainiosis [apimyiasis] of bees. Proceedings of the 17th International Bee-keeping Congress.

———. 1959. Methods of controlling senotainia infestation of bees [in Russian]. Pchelovodstvo 1959(3):40–45.

———. 1971. Senotainia parasitism of honeybees [La senotainiase des abeilles]. Bulletin Apicole de Documentation Scientifique et Technique et d'Information 14(1): 71-87.

Bond, V. 1947. How to keep out ants. American Bee Journal 87:544.

Bondarenko, O. I. 1966. On the frequency of detection of amoebae in honey bees. *In* Materials of the Annual Scientific Conference of the All-Union Institute of Experimental Veterinary Science, Moscow.

Bonney, R. 1993. Winter prep. Bee Culture 121:491–493.

———. 1994. Location! Location! Bee Culture 122:23–25.

Boot, W. J. 1995. Invasion of *Varroa* mites into honey bee brood cells. Ph.D. thesis, Wageningen University, The Netherlands. 119 pp.

Boot, W. J., N. M. Calis, and J. Beetsma. 1992. Differential periods of *Varroa* mite invasion into worker and drone brood cells of honey bees. Experimental and Applied Acarology 16:295–301.

Boot, W. J., J. N. M. Callis, and J. Beetsma. 1993. Invasion behavior of *Varroa* mites into honey bee brood cells: a matter of chance or choice? Journal of Apicultural Research 32:167–174.

Borchert, A. 1928. Beiträge zur Kenntnis der Bienen Parasiten *Nosema apis*. Archiv für Bienenkunde 9:115–178.

———. 1966. Die Krankheiten und Schädlinge der Honigbiene [The diseases and pests of the honey bee]. S. Hirzel, Leipzig.

———. 1974. Schädigungen der Bienenzucht durch Krankheiten, Vergiftungen und Schädlinge der Honigbiene [Damages in apiculture by diseases, poisoning, and pests of the honeybee]. S. Hirzel, Leipzig.

Borgmeier, T. 1924. Novos géneros e espécies de phorideos do Brasil [New genera and species of Brazilian phorids]. Boletim do Museu Nacional do Rio de Janeiro 1:167–202.

———. 1935. Sobre o cyclo evolutivo de *Chonocephalus* Wandolleck, e uma nova espécie de *Melaloncha* Brues, endoparasita de abelhas [On the evolution of *Chonocephalus*, and a new species of *Melaloncha* Brues, an internal parasite of bees]. Archivos do Instituto de Biologia Vegetal (Rio de Janeiro) 2:255–265.

Borneck, R. 1987. Varroa in 24 European countries. *In* Proceedings of the Workshop on Parasitic Bee Mites and Their Control, Pulawy, Poland, August–September 1987. Food and Agriculture Organization of the United Nations, AGS:BMC/87/16.

Bornus, L., J. Muszynska, and Z. Konopacka. 1977. Relation between worker bees condition and nosema disease [in Polish, English summary]. Pszczelnicze Zeszyty Naukowe 21:129–146.

Borror, D. J., D. M. DeLong, and C. A. Triplehorn. 1976. An introduction to the study of insects. 4th ed. Holt, Rinehart and Winston, New York.

Borror, D. J., C. A. Triplehorn, and N. F. Johnson. 1989. An introduction to the study of insects. 6th ed. Saunders College Publishing, Philadelphia.

Borstel, R. C. von, and M. L. Rekemeyer. 1959. Radiation induced and genetically contrived dominant lethality in *Habrobracon* and *Drosophila*. Genetics 44: 1053–1074.

Botha, J. J. C. 1970. About enemies of bees in South Africa. Gleanings in Bee Culture 98:100–103.

Böttcher, F. K., H. Hirschfelder, D. Mautz, and K. Weiss. 1973. Die Tätigkeit der Bayerischen Landesanstalt für Bienenzucht in Erlangen im Jahre 1972. Imkerfreund 28:74–84.

Böttcher, F. K., D. Mautz, and K. Weiss. 1974. Die Tätigkeit der Bayerischen Landesanstalt für Bienenzucht in Erlangen im Jahre 1973. Imkerfreund 29:70–80.

Böttcher, F. K., D. Mautz, and K. Weiss. 1975. Die Tätigkeit der Bayerischen Landesanstalt für Bienenzucht in Erlangen im Jahre 1974. Imkerfreund 30:75–85.

Bowman, C. E., and C. A. Ferguson. 1985. *Forcellinia galleriella*, a mite new to British beehives. Bee World 66:51–53.

Boyko, A. K. [also transliterated as Boiko]. 1939. Larvae of *Senotainia triguspis* Meig. [sic] causing heavy losses of bees. Doklady Akademiia Nauk S.S.S.R. (international edition) 24:304–306.

Bradbear, N. 1988. World distribution of major honeybee diseases and pests. Bee World 69:15–39.

Brangi, G. P., and M. Pavan. 1954. Sulle proprietà antibatteriche del miele, propoli, pappa reale e veleno di *Apis mellifera* L. [The bactericidal properties of honey, propolis, royal jelly, and venom of *Apis mellifera* L.] . Memorie della Società Entomologica Italiana 33:19–32.

Brar, H. S., G. S. Gatoria, J. S. Jhaji, and B. S. Chahal. 1985. Seasonal infestation of *Galleria mellonella* and population of *Vespa orientalis* in *Apis mellifera* apiaries in Punjab. Indian Journal of Ecology 12:109–112.

Brassler, K. von. 1929. *Ptinus raptor* Str. als Schädling im Bienenstock [*Ptinus raptor*, a pest of bee colonies]. Zeitschrift für Angewandte Entomologie 15:635–637.

Bratschi, H. 1942. Lindentracht und Bienentod [Linden flow and bee deaths]. Schweizerische Bienen-Zeitung 65:508.

Braun, W. 1957. Apimyiasis: mortandad de otõno. Apiculturo Americano 2:9–14.

Brcák, J., and O. Králík. 1965. On the structure of the virus causing sacbrood of the honey bee. Journal of Invertebrate Pathology 7:110–111.

Brcák, J., J. Svoboda, and O. Králík. 1963. Electron microscopic investigation of sacbrood of the honey bee. Journal of Insect Pathology 5:385–386.

Bregante, H. 1972. Mantidae preying upon honey bees (*Apis mellifera* L.) in Argentina. Proceedings of the 24th International Congress of Apiculture 24:372–374.

Bretschko, J. 1979. Nosemabefall des winterlichen Leichenfalles im Verlaufe der Winterruhe bei ungestörten und gestörten Völkern. Allgemeine Deutsche Imkerzeitung 13:341–343.

Bromley, S. W. 1930. Bee-killing robber flies. Journal of the New York Entomological Society 38:159–176.

———. 1936. The genus *Diogmites* in United States of America, with descriptions of new species (Diptera: Asilidae). Journal of the New York Entomological Society 44 (3): 225–237.

———. 1945. The robber flies and bee-killers of China (Asilidae). Lingnan Science Journal 21:87–105.

———. 1946. Beekilling Asilidae of southeastern U.S. (Diptera: Asilidae). proceeding Entomological Society of Washington; 48: 16-17.

———. 1948. Honey-bee predators. Journal of the New York Entomological Society 56:195–199.

Brower, J. V. Z., and L. P. Brower. 1962. Experimental studies of mimicry. 6. The reaction of toads *(Bufo terrestris)* to honeybees *(Apis mellifera)* and their dronefly mimics *(Eristalis vinetorum).* American Naturalist 96:297–308.

Brower, J. V. Z., and L. P. Brower. 1965. Experimental studies of mimicry. 8. Further investigations of honeybees *(Apis mellifera)* and their dronefly mimics (*Eristalis* spp.). American Naturalist 99:173–188.

Brower, L. P., J. V. Z. Brower, and P. W. Westcott. 1960. Experimental studies in mimicry. 5. The reactions of toads *(Bufo terrestris)* to bumblebees *(Bombus americanorum)* and their robberfly mimics *(Mallophora bomboides),* with a discussion of aggressive mimicry. American Naturalist 94:343–355.

Brown, J. P. H. 1879. Bee forage in the South. American Bee Journal 15:500–502.

Brown, K. C. 1939. Bees and rhododendrons. Bee World 20:80.

Brown, L. W. 1979. Commentary: infant botulism and the honey connection. Journal of Pediatrics 94:337–338.

Brown, R. 1988. Honey Bees. A guide to management. Crowood press, VT. 128pp.

Brown, R. L. 1974. Diets and habitat preferences of selected anurans in southeast Arkansas. American Midland Naturalist 91:468–473.

Bruce, J. 1990. Things I've learned about people, bees and bears. Gleanings in Bee Culture 118:448.

Bruce, W. A., D. L. Anderson, N. W. Calderone, and H. Shimanuki. 1995. A survey for Kashmir bee virus in honey bee colonies in the United States. American Bee Journal 135:352–355.

Bruce, W. A., and G. A. LeCato. 1980. *Pyemotes tritici:* a potential new agent for biological control of the red imported fire ant, *Solenopsis invicta* (Acari: Pyemotidae). International Journal of Acarology 6:271–274.

Brügger, A. 1935. Raubfliegen als Bienenfeinde [Robber flies as bee enemies]. Schweizerische Bienen-Zeitung 58:435–438.

———. 1946. The deathhead moth. Gleanings in Bee Culture 74:602–603, 651.

Brunnich, K. 1922. Catalepsy in queens. American Bee Journal 62:69–70.

Buchanan, E., and N. E. Gibbons, editors. 1974. Bergey's manual of determinative bacteriology. 8th ed. Williams & Wilkins, Baltimore, Maryland.

Bucher, G. E. 1958. General summary and review of utilization of disease to control insects. Proceedings of the 10th International Congress of Entomology 4:695–701.

Büchler, R. 1989. Attractivity and reproductive suitability for the Varroa-mite of bee brood from different origins. Proceedings of a meeting of the EC-experts group, Udine, Italy; 1988:139–145.

———. 1990. Possibilities for selecting increased *Varroa* tolerance in central European honey bees of different origins. Apidologie 21:365–367.

———. 1992a. Test auf Varroatoleranz im Rahmen von Leistungsprüfungen [Test for Varroa-tolerance performance]. Neue Bienen Zeitung 3:162–167.

———. 1992b. Zucht auf Varroatoleranz [Breeding for Varroa-tolerance]. Deutsches Imker-Journal 3(2):43–50.

———. 1993. Rate of damaged mites in natural mite fall with regard to seasonal effects and infestation development. Apidologie 24:492–493.

———. 1994. Varroa tolerance in honey bees—Occurrence, characters and breeding. Bee World 75:54–70.

Büchler, R., and W. Drescher. 1990. Variance and heritability of the capped developmental stage in European *Apis mellifera* L. and its correlation with increased *Varroa jacobsoni* Oud. infestation. Journal of Apicultural Research 29:172–176.

Büchler, R., W. Drescher, and I. Tornier. 1992. Grooming behavior of *Apis cerana, Apis mellifera* and *Apis dorsata* and its effects on the parasitic mites *Varroa jacobsoni* and *Tropilaelaps clareae.* Experimental and Applied Acarology 16:313–319.

Büchler, R., and V. Maul. 1991. Die Nachwirkung einer Bayvarolbehandlung auf später in die Beinenvölker eingebrachte Varroamilben [The active ingredient of Bayvarol continues to kill varroa in the colonies after it is removed]. Apidologie 22:389–396.

Büdel, A., and E. Herold, editors. 1960. Bienen und Bienenzucht [Bees and beekeeping]. Ehrenwirt, Munich.

Bulger, J. W. 1928. *Malpighamoeba* (Prell) in the adult honeybee found in the United States. Journal of Economic Entomology 21:376–379.

Bullington, S. W. 1978. Two new records for *Dasylechia atrox* (Williston) (Diptera: Asilidae), with review of all previous records. Entomological News 89:195–196.

Bunn, D. S. 1988. Observations on the foraging habits of the hornet *Vespa crabro* L. (Hymenoptera, Vespidae). Entomologist's Monthly Magazine 124:1492–1495.

Burges, H. D. 1978. Control of wax moths: physical, chemical and biological methods. Bee World 59:129–138.

Burgett, D. M. [also publishes under the name M. Burgett]. 1974. Glucose oxidase: a food protective mechanism in social Hymenoptera. Annals of the Entomological Society of America 67:545–546.

Burgett, D. M., P. A. Rossignol, and C. Kitprasert. 1990. A model of dispersion and regulation of brood mite *(Tropilaelaps clareae)* parasitism on the giant honeybee *(Apis dorsata)*. Canadian Journal of Zoology 68:1423–1427.

Burgett, D. M., and R. G. Young. 1974. Lipid storage by honey ant repletes. Annals of the Entomological Society of America 67:743–744.

Burgett, M., and P. Akratanakul. 1982. Predation on the western honey bee, *Apis mellifera* L., by the hornet *Vespa tropica* (L.). Psyche 89:347–350.

Burgett, M., and P. Akratanakul. 1985. *Tropilaelaps clareae*, the little known honey bee brood mite. American Bee Journal 125:112–114.

Burgett, M., P. Akratanakul, and R. A. Morse. 1983. *Tropilaelaps clareae:* a parasite of honeybees in south-east Asia. Bee World 64:25–28.

Burkholder, W. E., and M. Ma. 1985. Pheromones for monitoring and control of stored-product insects. Annual Review of Entomology 30:257–272.

Burnside, C. E. 1927. Saprophytic fungi associated with the honey bee. Papers of the Michigan Academy of Science, Arts and Letters 8:59–86.

———. 1928. A septicemic condition of adult bees. Journal of Economic Entomology 21:379–386.

———. 1930. Fungous diseases of the honeybee. U.S. Department of Agriculture Technical Bulletin 149.

———. 1933a. Brood poisoning of bees. American Bee Journal 73:460.

———. 1933b. Preliminary observations on "paralysis" of honeybees. Journal of Economic Entomology 26:162–168.

———. 1934a. Comparison of *Pericystis alvei* from Europe and the related species from America. Bee World 15:105–106.

———. 1934b. Studies on the bacteria associated with European foulbrood. Journal of Economic Entomology 27:656–668.

———. 1935a. A disease of young bees caused by a *Mucor.* American Bee Journal 75:75–76.

———. 1935b. Plant poisoning of bees and brood. Maryland Agricultural Society Report 1934:235–240.

———. 1935c. Purple brood of bees; plant poisoning by southern leatherwood. Gleanings in Bee Culture 63:717–718.

———. 1939. Fungus attacking queen larvae and sealed cells. News Letter, Bureau of Entomology and Plant Quarantine 6:28.

———. 1945. The cause of paralysis of honeybees. American Bee Journal 85:354–355, 363.

Burnside, C. E., and R. E. Foster. 1933. Plant poisoning of brood of bees; tracing down the cause of heavy loss of colonies in western Florida. Gleanings in Bee Culture 61:470–473.

Burnside, C. E., and I. L. Revell. 1948. Observations on nosema disease of honey bees. Journal of Economic Entomology 41:603–607.

Burnside, C. E., and G. H. Vansell. 1936. Plant poisoning of bees. U.S. Department of Agriculture, Bureau of Entomology and Plant Quarantine Mimeo E-398.

Burri, R. 1941. Zusammenfassung der Ergebnisse neuer Untersuchungen über den Erreger der Sauerbrut der Bienen [New investigations on the organism causing "sour brood" of honey bees]. Schweizerische BienenZeitung 64:496–498.

Burrill, A. C. 1926. Ants that infest beehives. American Bee Journal 66:29–31.

Burt, C. E. 1928. Insect food of Kansas lizards with notes on feeding habits. Journal of the Kansas Entomological Society 1:50–68.

Butler, C. G. 1943. Bee paralysis, May-sickness, etc. Bee World 24:3–7.

———. 1945. The incidence and distribution of some diseases of the adult honeybee (Apis mellifera L.) in England and Wales. Annals of Applied Biology 32:344–351.

———. 1957. The control of ovary development in worker honey bees (Apis mellifera). Experientia 13:256–257.

———. 1974. The world of the honeybee. Rev. ed. Collins, London.

Butler, E. 1992. Temporary electric fence again available for growers with deer, bear, and livestock damage. USDA, APHIS, ADC leaflet. 1 p.

Buttel-Reepen, H. von. 1920. Die neue(?) verheerende Milbenkrankheit der Biene [The new(?) destructive mite disease of honey bees]. Archiv für Bienenkunde 2:328–332.

Buys, B. 1975. A survey of honeybee pests in South Africa. Proceedings of the First Congress of the Entomological Society of South Africa, pp. 185–189.

———. 1977. A nosema disease affecting honeybee brood. In Apimondia symposium of bee biology and pathology, Merelbeke, Belgium 1976: Apimondia Publishing House, Bucharest: 73–76.

———. 1987. Competition for nectar between Argentine ants (Iridomyrmex humilis) and honeybees (Apis mellifera) on black ironbark (Eucalyptus sideroxylon). South African Journal of Zoology 22:173–174.

———. 1990. Relationships between Argentine ants and honeybees in South Africa. In Applied myrmecology. R. K. Vander Meer, K. Jaffe, and A. Cedeno, editors. Westview Press, Boulder, Colorado.

Buys, B. B. 1972. Nosema in brood. South African Bee Journal 44:2–4.

Buza, L. 1978. Control of varroa disease in Hungary. Apiacta (Bucharest) 13:176–177.

Buzzard, C. N. 1951. Ways of a hornet. Country Life, London 110:344–345.

Byers, C. F. 1930. A contribution to the knowledge of Florida Odonata. Biological Science Series, volume 1, number 1. University of Florida, Gainesville.

Calatayud, F., and C. Feuerriegel. 1987. Comparison of levels of infestation with Varroa and Braula [Comparacion entre niveles de infestacion de Varroa y Braula]. Vida Apicola (No. 26): 47-51.

Calderone, N. W., A. Bruce, G. Allen-Wardell, and H. Shimanuki. 1991. Evaluation of botanical compounds for control of the honey-bee tracheal mite, Acarapis woodi. American Bee Journal 131:589–591.

Calderone, N. W., and H. Shimanuki. 1992. Evaluation of sampling methods for determining infestation rates of the tracheal mite (Acarapis woodi R.) in colonies of the honey bee (Apis mellifera): spatial, temporal, and spatio-temporal effects. Experimental and Applied Acarology 15:285–298.

Calderone, N. W., and H. Shimanuki. 1995. Evaluation of four seed-derived oils for control for Acarapis woodi (Acari: Tarsonemidae) in colonies of the honey bee, Apis mellifera (Hymenoptera: Apidae). Journal of Economic Entomology 88:805–809.

Cale, G. H. 1973. Beginning with bees—IV. American Bee Journal 113:15.

Callan, E. M. 1954. Vespids attacking honey bees in Trinidad and British Guiana. Entomologist's Monthly Magazine 90:134.

Camazine, S. 1985. Tracheal flotation: a rapid method for the detection of honey bee acarine disease. American Bee Journal 125:104–105.

———. 1986. Differential reproduction of the mite Varroa jacobsoni (Mesostigmata: Varroidae) on Africanized and European honey bees (Hymenoptera: Apidae). Annals of the Entomological Society of America 79:801–803.

————. 1988. Factors affecting the severity of *Varroa jacobsoni* infestations on European and Africanized bees. *In* Africanized honey bees and bee mites. G. R. Needham, R. E. Page, Jr., M. Delfinado-Baker, and C. E. Bowman, editors. Ellis Horwood, Chichester, U.K.: 444–451.

Campbell, C. P. 1936. "Tanglefoot" catches skunks. Gleanings in Bee Culture 64:477.

Cantwell, G. E. 1970. Standard methods for counting nosema spores. American Bee Journal 110:222–223.

————. 1980. Control of the greater wax moth—an update. American Bee Journal 120:581–583.

Cantwell, G. E., E. G. Jay, G. C. Pearman, Jr., and J. V. Thompson. 1972a. Control of the greater wax moth, *Galleria mellonella* (L.), in comb honey with carbon dioxide. Part I. Laboratory studies. American Bee Journal 112:302–303.

Cantwell, G. E., and T. Lehnert. 1968. Mortality of *Nosema apis* and the greater wax moth, *Galleria mellonella* L., caused by heat treatment. American Bee Journal 108:56–57.

Cantwell, G. E., T. Lehnert, and J. Fowler. 1972b. Are biological insecticides harmful to the honey bee? American Bee Journal 112:255–258, 294–296.

Cantwell, G. E., T. Lehnert, and R. S. Travers. 1975. USDA research on ethylene oxide fumigation for control of diseases and pests of the honey bee. American Bee Journal 115:96–97.

Cantwell, G. E., and T. R. Shieh. 1981. Certan™—a new bacterial insecticide against the greater wax moth, *Galleria mellonella* L. American Bee Journal 121:424–426, 430–431.

Cantwell, G. E., and H. Shimanuki. 1969. Heat treatment as a means of eliminating nosema and increasing production. American Bee Journal 109:52–54.

Cantwell, G. E., and H. Shimanuki. 1970. The use of heat to control nosema and increase production for the commercial beekeeper. American Bee Journal 110:263.

Cantwell, G. E., and L. J. Smith. 1970. Control of the greater wax moth, *Galleria mellonella*, in honeycomb and comb honey. American Bee Journal 110:141.

Cantwell, G. E., L. J. Smith, and T. Lehnert. 1971. Effects of extreme cold and of microwaves on spores of *Nosema apis*. American Bee Journal 111:188.

CAPA. 1991. Honey bee diseases and pests. Canadian Association of Professional Apiculturists, Guelph, Canada. 16 pp.

Cardinale, S., A. Biasiolo, R. Piva, and R. Locci. 1994. Il complesso *Ascosphaera apis* (Maassen ex Claussen) Spiltoir et Olive (1955)—*Apis mellifera* L. [The *Ascosphaera apis* (Maassen ex Claussen) Spiltoir et Olive (1955)—*Apis mellifera* L. complex]. Micologia Italiana 23:61–69.

Carey, F. M., J. J. Lewis, J. L. MacGregor, and M. Martin-Smith. 1959. Pharmacological and chemical observations on some toxic nectars. Pharmaceutical Journal 183:241.

Caron, D. M. 1974. Mouse control. Gleanings in Bee Culture 102:285, 295.

————. 1992. Pseudoscorpions. Gleanings in Bee Culture 120(7):389.

————. 1993. What's bugging you. American Bee Journal 133:705–706.

————. 1994. Pseudoscorpions in Central America. Apicultura Moderna 5:10-12.

Caron, D. M., and P. W. Schaefer. 1986. Social wasps as bee pests. American Bee Journal 126:269–271.

Carroll, W. 1972. Bee-eating 'possums problem in Texas. Speedy Bee 1:3.

Chahal, B. S., H. S. Brar, G. S. Gatoria, and H. S. Jhajj. 1986. Aggressive behavior of *Apis florea* towards *Apis mellifera* in hive robbing and in foraging. Journal of Apicultural Research 25:134-138.

Chambers, S. R. 1982. Management, harvesting and processing of honey bee collected pollen in Western Australia. American Bee Journal 122:97–101.

Chaud-Netto, J. 1980a. Biological studies on *Pseudohypocera kerteszi* (Phoridae, Diptera). Experientia 36:61–62.

————. 1980b. Estudos biologicos com rainhas triploides de *Apis mellifera*. 1. Producao de ovos abortivos por rainhas virgens [Biological studies with triploid queen honey bees. 1. Production of abortive eggs by virgin queens]. Ciencia e Cultura 32:483–486.

————. 1980c. Estudos biologicos com abelhas triploides de *Apis mellifera* (Hymenoptera, Apidae) [Biological studies with triploid honey bees]. Ciencia e Cultura 32:611–615.

Cheeke, P. R. 1989. Pyrrolizidine alkaloid toxicity and metabolism in laboratory animals and livestock. *In* Toxicants of plant origin. P. R. Cheeke, editor. Volume 1, Alkaloids. CRC Press, Boca Raton, Florida

Cheshire, F. R. 1888. Bees and beekeeping. L. Upcott Gill, London.

Cheshire, F. R., and W. W. Cheyne. 1885. The pathogenic history and history under cultivation of a new bacillus *(B. alvei)*, the cause of a disease of the hive bee hitherto known as foul brood. Royal Microscopical Society Journal, series 2, 5:581–601.

Chiesa, F. 1991. Effective control of varroatosis using powdered thymol. Apidologie 22:135–145.

Chmielewski, W. 1971. Morfologia, biologia i ekologia *Carpoglyphus lactis* (L., 1758) (Glycyphagidae, Acarina) [Morphology, biology and ecology of *Carpoglyphus lactis*]. Prace Naukowe Instytutu Ochrony Róslin 13:63–166.

————. 1975. Roztocze (Acarina) wystepujace w zbieranym przez pszczoiy i przechowywanym pylku kwiatowym [Mites (Acarina) occurring in pollen collected and stored by honey bees]. Zeszyty Problemowe Postepow Nauk Rolniczych 171:237–244.

Choi, S. Y. 1974. New apiculture [in Korean]. Tip Hyun, Seoul.

————. 1986. Current status on the bionomics and control of bee mite *(Varroa jacobsoni* Oudemans) in Korea. Korean Journal of Apiculture 1(1):96–106.

Choi, S. Y., and K. S. Woo. 1974. Studies on the bionomics of the bee mite, *Varroa jacobsoni* Oudemans, and its chemical control (II) [in Korean]. Research Reports of the Office of Rural Development (Livestock) (Suwon) 16:69–76.

Chopra, R. N. 1977. Poisonous plants of India. Indian Council of Agricultural Research, New Delhi.

Chowdhury, S. 1953. A *Trichoderma* disease of honey bees. Current Science 22:348.

Christensen, M., and M. Gilliam. 1983. Notes on the *Ascosphaera* species inciting chalkbrood in honey bees. Apidologie 14:291–297.

Chun, L. 1965. Beekeeping in China. Gleanings in Bee Culture 93:38–39.

Ciencia y Abejas. 1976. *Mallophora ruficiauda* (parasitic fly of bees) [*Mallophora ruficauda* (moscardon cazador de abejas)]. Ciencia y Abejas 2(8): 47-48.

Claerr, G. 1977. La Varroase, fléau imminent pour l'apiculture européenne? [Varroasis, imminent plague of European beekeeping?]. Revue Française d'Apiculture 358:494–498.

Clapperton, B. K., P. A. Alspach, H. Moller, and A. G. Matheson. 1989. The impact of common and German wasps (Hymenoptera: Vespidae) on the New Zealand beekeeping industry. New Zealand Journal of Zoology 16:325–332.

Clark, K. 1994. Control of Varroa mites in British Columbia with either formic acid or Apistan. American Bee Journal 134:829.

Clark, K. J. 1990. 1990 field trials comparing vegetable oil and menthol as a control for tracheal mites. American Bee Journal 130:799–800.

————. 1992. Applications of formic acid liquid or gel for the control of honey bee tracheal mites. British Columbia Ministry of Agriculture, Fisheries and Food. Dawson Creek, B.C., Canada (unpublished report circulated).

Clark, K. J., and J. Gates. 1991. Tracheal mite control trials in British Columbia. American Bee Journal 131:773.

Clark, K. J., H. Huxter, N. J. Gates, and T. I. Szabo. 1990. Screening breeder honey bee stock for resistance to tracheal mites. American Bee Journal 130:800.

Clark, K. J., D. L. Nelson, and D. McKenna. 1989. Effect of menthol on queen rearing. American Bee Journal 129:811.

Clark, T. B. 1977a. Another virus in honey bees. American Bee Journal 117:340–341.

————. 1977b. *Spiroplasma* sp., a new pathogen in honey bees. Journal of Invertebrate Pathology 29:112–113.

———. 1978a. A filamentous virus of the honey bee. Journal of Invertebrate Pathology 32:332–340.

———. 1978b. Honey bee spiroplasmosis, a new problem for beekeepers. American Bee Journal 118:18–19, 23.

———. 1980. A second microsporidian in the honeybee. Journal of Invertebrate Pathology 35:290–294.

Clark, T. B., and R. F. Whitcomb. 1984. Pathogenicity of mollicutes for insects: possible use in biological control. Annales de Microbiologie (Paris) 135A:141–150.

Clark, W. H., M. H. Clark, and H. A. Rhodes. 1986. First record of *Acarapis woodi* Rennie in the honey bee from Baja California, Mexico. American Bee Journal 126:123–124.

Clauss, B. 1983. Bees and beekeeping in Botswana. Report, Ministry of Agriculture, Botswana.

———. 1985. The status of the banded bee pirate, *Palarus latifrons*, as a honeybee predator in southern Africa. Proceedings of the Third International Conference of Tropical Apiculture, Climates, Nairobi, pp. 157–159.

Claussen, P. 1921. Entwicklungsgeschichtliche Untersuchungen über den Erreger der als "Kalkbrut" bezeichneten Krankheit der Bienen [Historical development of the researches concerning the cause of the disease of bees known as "chalkbrood"]. Arbeiten aus der Biologischen Reichsanstalt für Land- und Forstwirtschaft 10:467–521.

Clinch, P. G. 1979. *Nosema apis* and mites in honeybee colonies in Papua New Guinea. Journal of Apicultural Research 18:298–301.

Clinch, P. G., T. Palmer-Jones, and I. W. Forster. 1972. Effect on honey bees of nectar from the yellow kowhai (*Sophora microphylla* Ait.). New Zealand Journal of Agricultural Research 15:194–201.

Clout, G. A. 1956. Chalk brood and hunchback flies. Bee Craft 38:135.

Cockerman, K. L., and E. Oertel. 1954. Control of ants in apiaries. Gleanings in Bee Culture 82:348–349.

Cockle, J. W. 1920. *Vitula serratilineella* Rag., a honey feeding larva. Proceedings of the British Columbia Entomological Society, Systematic Series 16:32–33.

Coleman, R. 1986. Birds take the buzz out of queen breeder profits. American Bee Journal 126:208–209.

Colin, M. E., J. Ducos de Lahitte, E. Larribau, and T. Boué. 1989. Activité des huiles essentielles de Labiées sur *Ascosphaera apis* et traitement d'un rucher [Activity of essential oils of Labiaceae on *Ascosphaera apis* and treatment of an apiary]. Apidologie 20:221–228.

Colwell, R. R. 1970. Polyphasic taxonomy of the genus *Vibrio*: numerical taxonomy of *Vibrio cholerae, Vibrio parahaemolyticus*, and related *Vibrio* species. Journal of Bacteriology 104:410–433.

Comparini, A., and A. Biasiolo. 1991. Genetic discrimination of Italian bee, *Apis mellifera ligustica* versus Carniolan bee, *Apis mellifera carnica* by allozyme variability analysis. Biochemical Systematics and Ecology 19:189–194.

Conner, H. E. 1977. The poisonous plants in New Zealand. E.C. Keating Government Printer.

Connor, L. J. 1974a. Chalk brood is now in Ohio. American Bee Journal 114:460.

———. 1974b. Honeybee diseases and other bee pests. Ohio State University Cooperative Extension Service Bulletin 582.

Cook, A. J. 1902. The bee-keepers guide. George W. York, Chicago.

———. 1908. Practical and potential pointers. American Bee Journal 48:176.

Cook, V. A., and C. E. Bowman. 1983. *Mellitiphis alvearius*, a little-known mite of the honeybee colony, found on New Zealand bees imported into England. Bee World 64:62–64.

Cooke, M. J. 1993. The cape bee problem in South Africa—an update. American Bee Journal 133:285–286.

Cooper, M. R., and A. W. Johnson. 1984. Poisonous plants in Britain and their effects on animals and man. H..M.S.O., London.

Cooreman, J. 1966. Présence de *Braula kohli* Schmitz, 1914, en Belgique. Bulletin et Annales de al Société Royal Entomoloque de Belgique. 102: 53.

Copello, A. 1922. Biología de *Mallophora ruficauda*, Wied. Physis. 6(21): 30-42.

———. 1925. El moscardon cazador de abejas [The fly preying on bees]. Rev. Apicultura 2(22): 6-9.

———. 1926. Biología de *Hermetia illucens* L. (la mosca de nuestros comenas) [Biology of *H. illucens*, the fly of argentina beehives. Rev. Soc. ent. argentina. 1(2)23-26.

———. 1927. Biología del moscardon cazador de abejas *Mallophora ruficauda* Wied [Biology of the fly of preying on bees]. Argentina Ministry of Agriculture Circular #699. 19pp.

Corréa-Marques, M. H., M. R. C. Issa, and D. De Jong. 1994. Estudo dos danos causados pelas abelhas africanizadas ao ácaro *Varroa jacobsoni* [A study of the damage to the mite *Varroa jacobsoni* caused by africanized bees]. Anais do IV Congresso Iberolatinoamericano de Apicultura y I Foro Expo-comercial Internacional de Apicultura. Río Cuarto, Córdoba, Argentina.

Corner, J. 1960. An incident of destruction of honeybee colonies in the interior of British Columbia by an ant, probably *Formica integra* Nylander. Proceedings of the Entomological Society of British Columbia 57:8.

Cornuet, J. M. 1978. Contrôle génétique des hybrides interraciaux d'*Apis mellifica* L. [Genetic control of interracial hybrids of *Apis mellifera* L.]. Apidologie 9:195–201.

———. 1979. The MDH system in the honeybees (*Apis mellifera* L.) of Guadaloupe. Journal of Heredity 70:223–224.

Cornuet, J. M. A. Daoudi, and C. Chevalet. 1986. Genetic pollution and number of matings in a black honey bee *(Apis mellifera mellifera)* population. Theoretical and Applied Genetics 73:223–227.

Cornuet, J. M., A. Daoudi, H. Mohssine, and J. Fresnaye. 1988. Etude biométrique de populations d'abeilles Marocaines. [Biometrical study of honey bee populations from Morocco.] Apidologie 19:355–366.

Cosenza, G. W., and T. Silva. 1972. Comparação entre a capacidade de limpieza de favos da abelha africana, da abelha caucasiana e de suas híbridas [Comparison of comb cleaning behavior between the African bee, the Caucasian bee and their hybrids]. Ciência e Cultura 24:1153–1158.

Cott, H. B. 1936. The effectiveness of protective adaptation in the hive-bee, illustrated by experiments on the feeding reactions, habit formation, and memory of the common toad *(Bufo bufo bufo)*. Proceedings of the Zoological Society of London 1936:111–133.

Cotton, W. C. 1842. My bee book. J. G. F. & J. Rivington, London.

Coulthard, C. E., R. Michaelis, W. F. Short, G. Sykes, G. E. H. Skrimshire, A. F. B. Standfast, J. H. Birkinshaw, and H. Raistrick. 1945. Notatin: an antibacterial glucose-aerode hydrogenase from *Penicillium notatum* Westling and *Penicillium resticulosum sp.* nov. Biochemical Journal 39:24–36.

Couston, R. 1972. Principles of practical beekeeping. Scottish and Universal Newspapers, Kilmarnock.

Cowan, T. W. 1881. Notes on a new bee-disease. British Bee Journal 9:33–34.

Cox, R. L., J. O. Moffett, and W. T. Wilson. 1989a. Techniques for increasing the evaporation rate of menthol when treating honey bee colonies for control of tracheal mites. American Bee Journal 129:129–131.

Cox, R. L., J. O. Moffett, W. T. Wilson, and M. Ellis. 1989. Effects of late spring and summer menthol treatment on colony strength, honey production and tracheal mite infestation levels. American Bee Journal 129:547–549.

Cox, R. L., W. T. Wilson, D. L. Maki, and A. Stoner. 1986. Chemical control of the honey bee tracheal mite. *Acarapis woodi*. American Bee Journal 126:828.

Crane, E. 1975a. History of honey. *In* Honey: a comprehensive survey. E. Crane, editor. Crane, Russak, New York.

———. 1975b. Honey from other bees. *In* Honey: a comprehensive survey. E. Crane, editor. Crane, Russak, New York.

———. 1977. Dead bees under lime trees. Bee World 58:129–130.

———. 1978a. Sugars poisonous to bees. Bee World 59:37–38.

———. 1978b. The varroa mite. Bee World 59:164–167.

———. 1979. Fresh news on the varroa mite. Bee World 60:8.

———. 1980. Conference on tropical apiculture. Bee World 61:108–117.

———. 1982. Report on apiculture in Mauritius. International Bee Research Association, unpublished.

———. 1990. Bees and beekeeping. Cornell University Press, Ithaca, N.Y.

Crewe, R. M. 1984. Differences in behaviour and morphology between *capensis* and *adansonii*. South African Bee Journal 56(1):16–20.

Crewe, R. M., H. R. Hepburn, and R. F. A. Moritz. 1994. Morphometric analysis of 2 southern African races of honey bee. Apidologie 25:61–70.

Cromroy, H. L., and W. J. Kloft. 1980. Problems raised by invasion of honey bees by acarine mites in the world. Apiacta (Bucharest) 15:61–65.

Cronin, E., and P. Sherman. 1977. A resource-based mating system: the orangerumped honeyguide. Living Bird 15:5–32.

Cros, A. 1932. Dégâts commis dans les ruches par les larves des méloés [Damage done in hives by the larvae of *Meloe*]. Congrés International d'Entomologie (Paris) 5:841–845.

Crosby, D. G. 1971. Minor insecticides of plant origin. *In* Naturally occurring insecticides. M. Jacobson and D. G. Crosby, editors. Marcel Dekker, New York.

Cross, E. A. 1965. The generic relationships of the family Pyemotidae (Acarina: Trombidiformes). University of Kansas Science Bulletin 45:29–275.

Cross, E. A., and J. C. Moser. 1975. A new, dimorphic species of *Pyemotes* and a key to previously-described forms (Acarina: Tarsonemoidea). Annals of the Entomological Society of America 68:723–732.

Cross, W. H., W. L. McGovern, and E. A. Cross. 1975. Insect hosts of the parasitic mites called *Pyemotes ventricosus* (Newport). Journal of the Georgia Entomological Society 10:1–8.

Crumb, S. E., P. M. Eide, and A. E. Bonn. 1941. The European earwig. U.S. Department of Agriculture Technical Bulletin number 766.

Cruz Landim, C. da. 1969. Ultrastructure of *Apis mellifera* hypopharyngeal gland. Proceedings of the Sixth Congress, International Union for the Study of Social Insects 6:121–130.

Cruz-Landim, C. da, J. Chaud-Netto, and L. S. Gonçalves. 1979. Morphological alterations in the compound eyes of eye-color mutants of *Apis mellifera* L. (Hymenoptera, Apidae). 1. Revista Brasileira de Genetica 2:223–231.

Culvenor, C. J., J. A. Edgar, and L. W. Smith. 1981. Pyrrolizidine alkaloids in honey from *Echium plantagineum* L. Journal of Agricultural and Food Chemistry 29:958.

Cummings, M. W., and D. A. Wade. 1975. Bears and bees. University of California Apiaries Newsletter, June 1975.

Cury, R. 1951. Micoses das abelhas e fungos das colméias [Mycoses of honey bees and fungi of beehives]. Biológico (Brazil) 17:214–220.

Cutbush, L. S. 1950. Poisoning of bees foraging on cut buckwheat *(Fagopyrum esculentum)*. Bee World 31:26.

Cymborowski, B., and M. Bogus. 1976. Juvenilizing effect of cooling on *Galleria mellonella*. Journal of Insect Physiology 22:669–672.

Dacy, G. H. 1939. Electrified fence. Gleanings in Bee Culture 67:619–621.

Dadant, C. P. 1911. Langstroth on the hive and the honey bee. Revised by C. P. Dadant. Twentieth-century ed. Dadant & Sons, Hamilton, Illinois.

———. 1927. Langstroth on the hive and the honey bee. Revised by C. and C. P. Dadant. 23d ed. American Bee Journal, Hamilton, Illinois.

———. 1933. Brood poisoning of bees [editorial]. American Bee Journal 73:431.

———. 1957. First lessons in beekeeping. Revised and rewritten by M. G. Dadant and J. C. Dadant. American Bee Journal, Hamilton, Illinois.

Dadant and Sons, editors. 1975. The hive and the honey bee. Rev. ed. Dadant & Sons, Hamilton, Illinois.

Dade, H. A. 1949. The laboratory diagnosis of honey-bee diseases. Monographs of the Quekett Microscopial Club (London) 4.

Dahm, K. H., D. Meyer, W. E. Finn, V. Reinhold, and H. Röller. 1971. The olfactory and auditory mediated sex attraction in *Achroia grisella* (Fabr.). Naturwissenschaften 58:265–266.

Dall, D. J. 1985. Inapparent infection of honey bee pupae by Kashmir and sacbrood bee viruses in Australia. Annals of Applied Biology 69:461–468.

Dallmann, H. 1966. Neue Wege bei der Bekämpfung der Kalkbrut in Bienenvölkern [New methods for the control of chalkbrood in bee colonies]. Garten und Kleintierzucht, Ausgabe C, Imker 5(9):10.

Dalton, J. 1931. How to control ants. American Bee Journal 71:159.

Daly, H. V., D. De Jong, and N. D. Stone. 1988. Effect of parasitism by *Varroa jacobsoni* on morphometrics of Africanized worker honey bees. Journal of Apicultural Research 27:126–130.

Daly, H. V., K. Hoelmer, and P. Gambino. 1991. Clinal geographic variation in feral honey bees in California, USA. Apidologie 22:591–609.

Daly, H. V., and R. A. Morse. 1991. Abnormal sizes of worker honey bees (*Apis mellifera* L.) reared from drone comb. Journal of the Kansas Entomological Society 64:193–196.

Daniels, G. 1978. A new species of *Dakinomyia* from Queensland (Diptera: Asilidae). Proceedings of the Linnean Society of New South Wales 103(4): 275–281.

Danielyan, S. G. 1963. Earwig: a spreader of infection [in Russian]. Pchelovodstvo 1963(2):29.

Danielyan, S. G., A. A. Markosyan, K. M. Nalbandyan and N. M. Akopyan. 1975. Gas smoke generator for treating bees. Veterinariya, Moscow, USSR. (No. 8): 67.

Danielyan, S. G., and K. M. Nalbandian. 1971. The Hungarian oil beetle *Meloe hungarus* as a dangerous parasite [in Russian]. Pchelovodstvo 1971(5):32.

Danka, R. G., Hellmich, R. L. and T. E. Rinderer. 1992. Nest usurpation, supersedure and colony failure contribute to Africanization of commercially managed European honey bees in Venezuela. Journal of Apicultural Research 31: 119-123.

Danka, R. G., and T. E. Rinderer. 1988. Social reproductive parasitism by Africanized honey bees. *In* Africanized honey bees and bee mites. G. L. Needham, R. E. Page, M. Delfinado-Baker, and C. E. Bowman, editors. Ellis Horwood, Chichester, England.

Danka, R. G., and J. Villa. 1994. Preliminary observations on the susceptibility of Africanized honey bees to American foulbrood. Journal of Apicultural Research 33:243–245.

Das Gupta, K. C. 1945. The romance of scientific beekeeping. Hem Prabha Devi Khadi Pratisthan, Calcutta.

Dave, M. L. 1943. Wasps. Indian Bee Journal 5:127, 131.

Davidson, E. W. 1970. Ultrastructure of peritrophic membrane development in larvae of the worker honey bee (*Apis mellifera*). Journal of Invertebrate Pathology 15:451–454.

Davis, A. C. 1943. Effect of California buckeye on ants. Journal of Economic Entomology 36:800.

Davis, A. R. 1942. Loco weed poisoning bees. Beekeepers' Item 26:112.

Davis, H. G., G. W. Eddy, T. P. McGovern, and M. Beroza. 1969. Heptyl butyrate, a new synthetic attractant for yellow jackets. Journal of Economic Entomology 62:1245.

Dawicke, B. L. 1991. A comparison of the migration of the honey bee tracheal mite (*Acarapis woodi* (Rennie)) in drone and worker honey bees (*Apis mellifera* L.). M. S. thesis, University of Guelph, Guelph, Ontario.

Deans, A. S. C. 1940. Chalk brood. Bee World 21:46.

DeBach, P. 1964. Biological control of insect pests and weeds. Chapman and Hall, London.

Deinzer, M. L., P. A. Thompson, D. M. Burgett, and D. L. Isaacson. 1977. Pyrrolizidine alkaloids: their occurrence in honey from tansy ragwort (*Senecio jacobaea* L.). Science 195:497–499.

De Azevedo, A. 1924. A *Braula coeca*, piolho das abelhas [*Braula coeca* the bee louse]. Correio-Agric. 2(11)333-4.

De Jong, D. 1976. Experimental enhancement of chalk brood infections. Bee World 57:114–115.

———. 1977. A study of chalk brood disease of honey bees. M.S. thesis, Cornell University, Ithaca, New York.

———. 1979. Social wasps, enemies of honey bees. American Bee Journal 119:505–507, 529.

———. 1981. Effect of queen cell construction on the rate of invasion of honey bee brood cells by *Varroa jacobsoni*. Journal of Apicultural Research 20:254–257.

———. 1983. Técnicas para estimação de taxa de aumento de população de *Varroa jacobsoni* [Techniques for estimating population increase in *Varroa jacobsoni*]. Resumos da 35a Reunião Anual de SBPC, Belém, Pará, Brasil. Ciência e Cultura 35(7):672.

———. 1984. Current knowledge and open questions concerning reproduction in the honey bee mite, *Varroa jacobsoni*. *In* Advances in invertebrate reproduction, 3. W. Engels, editor. Elsevier Press, Amsterdam.

———. 1988. *Varroa jacobsoni* does reproduce in worker cells of *Apis cerana* in South Korea. Apidologie 19:103–106.

———. 1990a. Insects: Hymenoptera (ants, wasps, and bees). *In* Honey bee pests, predators, and diseases. 2d ed. R. A. Morse and R. Nowogrodzki, editors. Cornell University Press, Ithaca, New York.

———. 1990b. Mites: varroa and other parasites of brood. *In* Honey bee pests, predators, and diseases. 2d ed. R. A. Morse and R. Nowogrodzki, editors. Cornell University Press, Ithaca, New York.

De Jong, D., D. M. Caron, H. Shimanuki, and I. B. Smith. 1981. Resolution of the *Varroa jacobsoni* problem in Maryland. Bulletin of the Entomological Society of America 27:267–270.

De Jong, D., and P. H. De Jong. 1983. Longevity of Africanized honey bees (Hymenoptera: Apidae) infested by *Varroa jacobsoni* (Parasitiformes: Varroidae). Journal of Economic Entomology 76:766–768.

De Jong, D., P. H. De Jong, and L. S. Gonçalves. 1982a. Weight loss and other damage to developing worker honey bees (*Apis mellifera*) due to infestation with *Varroa jacobsoni*. Journal of Apicultural Research 20:254–257.

De Jong, D., Morse, R.A. and G.C. Eickwort. 1982b. Mite pests of honey bees. Annual Review of Entomology 27:229-252.

De Jong, D., and L. S. Gonçalves. 1981. The varroa problem in Brazil. American Bee Journal 121:186–189.

De Jong, D., L. S. Gonçalves, and R. A. Morse. 1984. Dependence on climate of the virulence of *Varroa jacobsoni*. Bee World 65:117–121.

De Jong, D., and R. A. Morse. 1976. Chalk brood: a new disease of honey bees in the U.S. New York's Food and Life Sciences Quarterly 9(2):12–14.

De Jong, D., and R. A. Morse. 1988. Utilisation of raised brood cells of the honey bee, *Apis mellifera* (Hymenoptera: Apidae), by the bee mite *Varroa jacobsoni (Acarina: Varroidae)*. Entomologia Generalis *14:103–106*.

De Jong, D., R. A. Morse, and G. C. Eickwort. 1982b. Mite pests of honey bees. Annual Review of Entomology 27:229–252.

De Jong, D., D. A. Roma, and L. S. Gonçalves. 1982c. A comparative analysis of shaking solutions for the detection of *Varroa jacobsoni* on adult honey bees. Apidologie 13:297–306.

De Jong, D., J. Steiner, L. S. Gonçalves, and R. A. Morse. 1984. Brazilian Varroa research rates current treatments too expensive. American Bee Journal 124:111–112, 134.

Dekanadze, A. M. 1986. Disinfection of the bee bread and wax in the cases of sac brood disease. Apiacta (Bucharest) 21:47–49.

Delaplane, K. S. 1992a. Controlling tracheal mites (Acari: Tarsonemidae) in colonies of honey bees (Hymenoptera: Apidae) with vegetable oil and menthol. Journal of Economic Entomology 85:2118–2124.

————. 1992b. Survey of miticide use in Georgia honey bee hives. American Bee Journal 132:185–187.

Delaplane, K. S., and L. F. Lozano. 1994. Using Terramycin in honey bee colonies. American Bee Journal 134:259–261.

Delfinado, M. D. 1963. Mites of the honeybee in south-east Asia. Journal of Apicultural Research 2:113–114.

Delfinado, M. D., and E. W. Baker. 1961. *Tropilaelaps*, a new genus of mite from the Philippines *(Laelaptidae [s. lat.]: Acarina)*. Fieldiana, Zoology 44:53–56.

Delfinado, M. D., and E. W. Baker. 1974. Varroidae, a new family of mites on honey bees (Mesostigmata: Acarina). Journal of the Washington Academy of Sciences 64:4–10.

Delfinado, M. D., and E. W. Baker. 1982. A new species of *Tropilaelaps* parasitic on honey bees. American Bee Journal 122:416–417.

Delfinado-Baker, M. 1982. New records for *Tropilaelaps clareae* from colonies of *Apis cerana indica*. American Bee Journal 122:382.

————. 1984a. *Acarapis woodi* in the U.S. American Bee Journal 124:805–806.

————. 1984b. The nymphal stages and male of *Varroa jacobsoni* Oudemans a parasite of honey bees. International Journal of Acarology 10:75–80.

————. 1985. An acarologist's view: the spread of the tracheal mite of honey bees in the United States. American Bee Journal 125:689–690.

————. 1988a. The tracheal mite of honey bees: a crisis in beekeeping. *In* Africanized honey bees and bee mites. G. R. Needham, R. E. Page, Jr., M. Delfinado-Baker, and C. E. Bowman, editors. Ellis Horwood, Chichester, England.

————. 1988b. Variability and biotypes of *Varroa jacobsoni* Oudemans. American Bee Journal 128:567–568.

Delfinado-Baker, M., and K. Aggarwal. 1987. A new *Varroa* (Acari: Varroidae) from the nest of *Apis cerana* (Apidae). International Journal of Acarology 13:233–237.

Delfinado-Baker, M., and E. W. Baker. 1982a. A new record for *Aeroglyphus robustus* in beehive. American Bee Journal 122:110.

Delfinado-Baker, M., and E. W. Baker. 1982b. Notes on honey bee mites of the genus *Acarapis* Hirst (Acari: Tarsonemidae). International Journal of Acarology 8:211–226.

Delfinado-Baker, M., and E. W. Baker. 1982c. A new species of *Tropilaelaps* parasitic on honey bees. American Bee Journal 122:416–417.

Delfinado-Baker, M., and E. W. Baker. 1987. Notes on mites new to beehives in Puerto Rico and North America. American Bee Journal 127:365–366.

Delfinado-Baker, M., and M. A. Houck. 1989. Geographic variation in *Varroa jacobsoni* (Acari, Varroidae): application of multivariate morphometric techniques. Apidologie 20:345–358.

Delgado, L. M. and A. F. Afonson. 1986. On the asilids ("robber flies") of honeybees, in Portugal [Sobre os asilideos ("moscas corsarias") das abelhas, em Portugal]. Repositorio de Trabalhos do Laboratorio Nacional de Investigacao Veterinaria 18: 71-74.

Delibes, M. 1978. Feeding habits of the stone marten, *Martes foina* (Erxleben, 1977), in northern Burgos, Spain. Zeitschrift für Säugetierkunde 43:282–288.

Del Lama, M. A., J. A. Lobo, A. E. E. Soares, and S. N. Del Lama. 1990. Genetic differentiation estimated by isozymic analysis of Africanized honey bee populations from Brazil and from Central America. Apidologie 21:217–280.

Dengg, O. 1928. Die Bienen und die Giftpflanzen [Bees and poisonous plants]. Illustrierte Monatsblätter für Bienenzucht 28:16–19.

Deodikar, G. B., C. V. Thakar, R. P. Phadke, and P. N. Shah. 1958. Poisoning of honeybees foraging on *Euphorbia geniculata*. Bee World 39:118–120.

Deschodt, C. 1969. Let me tell you about the chimps and the bees. South African Bee Journal 41(1):13.

Deyrup, M. 1988. Review of adaptations of velvet ants (Hymenoptera: Mutillidae). The Great Lakes Entomologist 21:1–4.

626   Literature Cited

Dickens, J. C., F. A. Eischen, and A. Dietz. 1986. Olfactory perception of the sex attractant pheromone of the greater wax moth *Galleria mellonella* L. (Lepidoptera: Pyralidae), by the honey bee, *Apis mellifera* L. (Hymenoptera: Apidae). Journal of Entomological Science 21:349–354.

Dietz, A., and R. W. Lovins. 1975. Studies on the "cannibalism substance" of diploid drone honey bee larvae. Journal of the Georgia Entomological Society 10:314–315.

Dindo, M. L., and M. Monti. 1986. Effetti della densita di popolazione nei Lepidotteri: il caso di *Galleria mellonella* L. [Effect of population density in Lepidoptera: the case of *Galleria mellonella* L.] Bolletino dell'Istituto di Entomologia "Guido Grandi" della Universita degli Studi di Bologna 40:111–120.

Dingman, D. W., N. Bakheit, C. C. Field, and D. P. Stahly. 1984. Isolation of two bacteriophages from *Bacillus larvae*, PBL1 and PBL0.5, and partial characterization of PBL1. Journal of General Virology 65:1101–1105.

Dingman, D. W., and D. P. Stahly. 1983. Medium promoting sporulation of *Bacillus larvae* and metabolism of medium components. Applied and Environmental Microbiology 46:860–869.

Dixon, M., and E. C. Webb. 1958. Enzymes. Longmans, Green, London.

Dold, H., D. H. Du, and S. T. Dziao. 1937. Nachweis antibakterieller, hitzeund lichtempfindlicher Hemmungsstoffe (Inhibine) im Naturhonig (Blütenhonig) [Evidence of antibacterial inhibitors, sensitive to heat and light, in natural honey]. Zeitschrift für Hygiene und Infektionskrankheiten 120:155–167.

Dold, H., and R. Witzenhausen. 1955. Ein Verfahren zur Beurteilung der örtlichen inhibitorischen (Keimvermehrungshemmenden) Wirkung von Honigsorten verschiedener Herkunft [A method for judging the local bacterial suppressant effect of honeys of different origins]. Zeitschrift für Hygiene und Infektionskrankheiten 141:333–337.

Donaldson, J. M. I. 1989. *Oplostomus fuligineus* (Coleoptera: Scarabaeidae): life cycle and biology under laboratory conditions and its occurrence in bee hives. Coleopterists Bulletin 43:177–182.

Dönhoff and Leuckart. 1857. Über den Fadenpilz im Darm der Biene [About the fungi in the guts of bees]. Bienenzeitung 13:66–67, 72.

Donovan, B. J. 1992. Problems caused by immigrant German and common wasps in New Zealand, and attempts at biological control. Bee World 73:131–148.

Donovan, B. J., H. Moller, G. M. Plunkett, P. E. C. Read, and J. A. V. Tilley. 1989. Release and recovery of the introduced wasp parasitoid *Sphecophaga vesparum vesparum* (Curtis) (Hymenoptera: Ichneumonidae) in New Zealand. New Zealand Journal of Zoology 16:35–364.

Donzé, G., and P. M. Guerin. 1994. Behavioral attributes and parental care of *Varroa* mites parasitizing honeybee brood. Behavioral Ecology and Sociobiology 34:305–319.

Dörntlein, D., and G. Reng. 1970. Vergleichende Untersuchungen zur Frage der Infektiosität von *Nosema apis* Z. und *Nosema bombi* F&P. Dissertation, Technischen Hochschule, München, Germany.

Dougherty, E. M., G. E. Cantwell, and M. Kuchinski. 1982. Biological control of the greater wax moth (Lepidoptera: Pyralidae), utilizing *in vivo*– and *in vitro*–propagated baculovirus. Journal of Economic Entomology 75:675–679.

Douhet, M. 1965. Beekeeping in Madagascar. Proceedings of the International Beekeeping Congress 20:690–710.

Doull, K. M. 1961. A theory of the cause of development of epizootics of nosema disease of the honey bee. Journal of Insect Pathology 3:297–309.

Doull, K. M., and K. M. Cellier. 1961. A survey of the incidence of nosema disease (*Nosema apis* Zander) of the honeybee in South Australia. Journal of Insect Pathology 3:280–288.

Doull, K. M., and J. E. Eckert. 1962. A survey of the incidence of nosema disease in California. Journal of Economic Entomology 55:313–317.

Doupe, R. 1921. Irish bee notes. Bee World 2:81.

Dover, C. 1929. Wasps and bees in the Raffles Museum, Singapore. Bulletin of the Raffles Museum, Singapore, Straits Settlements 2:43–70.

Dowling, J. H., and H. B. Levine. 1956. Hexose oxidation by an enzyme system of *Malleomyces pseudomallei*. Journal of Bacteriology 72:555–560.

Drees, B. M. and S. B. Vinson. 1985. Fire ants and their control. Texas Agricultural Extension Service, Bulletin B-1536.

Dreher, K. 1938. Auftreten von Bienenkrankheiten in Niedersachsen und Braunschweig im Jahre 1937 [Occurrence of bee diseases in Lower Saxony and Braunschweig in the year 1937]. Niedersächsische Imker 73:282–284.

———. 1953. Zur Steinbrut (Aspergillusmykose) der Honigbiene [On stonebrood of the honey bee]. Zeitschrift für Bienenforschung 2:92–97.

Drescher, W. 1975. Die räumliche und anteilmässige Verteilung männlichen und weiblichen Gewebes in der äusseren Morphologie gynandromorpher Individuen bei *Apis mellifica* L. [Spatial and proportional division of male and female tissues in the external morphology of gynandromorphic individuals in *Apis mellifica* L.]. Insectes Sociaux 22:13–26.

Drescher, W., and W. C. Rothenbuhler. 1963. Gynandromorph production by egg chilling. Cytological mechanisms in honey bees. Journal of Heredity 54:194–201.

Drescher, W., and W. C. Rothenbuhler. 1964. Sex determination in the honey bee. Journal of Heredity 55:90–96.

Drum, N. H., and W. C. Rothenbuhler. 1963. Non-stinging aggressive responses of worker honey bees to hive mates, intruder bees and bees affected with chronic bee paralysis. Journal of Apicultural Research 22:256–260.

Dryenski, P. 1933. Die Krankheiten und Schädlinge der Honigbiene in Bulgarien in Jahre 1933 [Diseases and enemies of bees in our century in the year 1933]. Demy 8vo, 20pp.

Dubois, K. P., and E. M. K. Geiling. 1959. Textbook of toxicology. Oxford University Press, Oxford, England.

Dubois, L., and E. Collart. 1950. L'Apiculture au Congo Belge et au Ruanda-Urundi [Apiculture in the Belgian Congo and Ruanda-Urundi]. Ministère des Colonies. Direction de l'Agriculture, de l'Elevage et de la Colonisation, Brussels.

Du Buysson, H. 1900. Dégats du *Forficula auricularia* Linn. dans les ruches d'abeilles [Damage of *Forficula auricularia* Linn. to bee hives]. Bulletin de la Société Entomologique de France 1900:183.

Duisberg, H., and B. Warnecke. 1959. Erhitzungs- und Lichteinfluss auf Fermente und Inhibine des Honigs [Effect of heat and light on enzymes and inhibine in honey]. Zeitschrift für Lebensmitteluntersuchung und -forschung 111:111–119.

Dukov, D. 1964. Morphological and biological studies on *Braula orientalis* and its control [in Russian]. Zhivotnovodni Nauki 1:77–87.

Dunham, W. E. 1929. Breeding of the Indian meal moth in extracting honey combs. American Bee Journal 69:327–328.

Dunn, J. F. 1938. Getting rid of ants. Gleanings in Bee Culture 66:499.

Dustmann, J. H. 1969. Eine chemische Analyse der Augenfarbmutanten von *Apis mellifica* [A chemical analysis of the eye color mutants of Apis mellifica]. Journal of Insect Physiology 15:2225–2238.

———. 1972. Über den Einfluss des Lichtes auf den Peroxid Wert (Inhibin) des Honigs [Influence of light on the peroxide value (inhibine) of honey]. Zeitschrift für Lebensmitteluntersuchung und -forschung 148:263–268.

———. 1973. Gene-actions in mutants of the honey bee. Proceedings of the 24th International Apicultural Congress 24:321–323.

———. 1975a. Die Pigmentgranula im Komplexauge der Honigbiene *Apis mellifica* bei Wildtyp und verschiedenen Augenfarbmutanten [The pigment granules in the compound eye of wild type and various eye-color mutant honey bees, *Apis mellifera*]. Cytobiologie 11:133–152.

———. 1975b. Quantitative Untersuchungen zur Tryptophan-Ommochrom-Reaktionskette bei Wildtyp und Mutanten der Honigbiene *Apis mellifera* [Quantitative studies on the tryptophan-ommochrome reaction series in wild type and mutant honey bees, *Apis mellifera*]. Insect Biochemistry 5:429–435.

———. 1993. Natural defense mechanisms of a honey bee colony against diseases and parasites. American Bee Journal 133:431–434.

Dutky, S. R., and W. S. Hough. 1955. Note on a parasitic nematode from codling moth larvae, *Carpocapsa pomonella*. Proceedings of the Entomological Society of Washington 57:244.

Dutky, S. R., J. V. Thompson, and G. E. Cantwell. 1962. A technique for mass rearing the greater wax moth (Lepidoptera: Galleriidae). Proceedings of the Entomological Society of Washington 64:56–58.

Dutton, R. W., and J. B. Free. 1979. The present status of beekeeping in Oman. Bee World 60:175, 176–185.

Dutton, R., and J. Simpson. 1977. Producing honey with *Apis florea* in Oman. Bee World 58:71–76.

Dutton, R. W., F. Ruttner, A. Berkeley, and M. J. D. Manley. 1981. Observations on the morphology, relationships and ecology of *Apis mellifera* of Oman. Journal of Apicultural Research 20:201–214.

Dyce, E. J. 1953. Beekeeping in Costa Rica. American Bee Journal 93:296–298.

Dyulgerov, T. 1979. Rodent control in apiaries [in Bulgarian]. Veterinarna Sbirka 1979:22–24.

Dzierzon, J. 1878. Rationelle Bienenzucht [Rational beekeeping]. Verlag der Falschen Buchdrückerei, Brieg.

Echigo, T., T. Takenaka, and M. Ichimura. 1972. Studies on the inversion of sucrose in nectar during the process of honey formation. Bulletin of the Faculty of Agriculture, Tamagawa University 12:17–27.

Echlin, P. 1974. The biology of pollen. American Bee Journal 114:249, 274, 298–299, 314, 332–333, 345, 350, 352.

Eckert, J. E. 1932. Skunk receives many stings. Gleanings in Bee Culture 60:559–560.

———. 1934. The California toad in relation to the hive bee. Copeia 1934:92–93.

———. 1950. The development of resistance to American foulbrood by honey bees in Hawaii. Journal of Economic Entomology 43:562–564.

———. 1951. Beekeeping in Hawaii. Gleanings in Bee Culture 79:393–400, 468–472, 509.

———. 1954. A handbook on beekeeping in California. California Agricultural Extension Service Manual 15.

———. 1961a. Acarapis mites of the honey bee, *Apis mellifera* Linnaeus. Journal of Insect Pathology 3:409–425.

———. 1961b. Observations on the *Acarapis* mites of honey bees. American Bee Journal 101:183–188.

———. 1963. Injury to bees by poisoning. *In* The hive and the honeybee. R. A. Grout, editor. Rev. ed. Dadant & Sons, Hamilton, Illinois.

Eckert, J. E., and H. A. Bess. 1952. Fundamentals of beekeeping in Hawaii. Hawaii Cooperative Extension Bulletin 55.

Eckert, J. E., and F. R. Shaw. 1960. Beekeeping. Macmillan, New York.

Eckert, J. E., and M. C. West. 1948. Chlordane vs. bees and ants. American Bee Journal 88:584.

Edwards, R. 1980. Social wasps; their biology and control. Rentokil Limited, East Grinstead, England.

Egger, A. 1973. Die Dörrobstmotte [The dried fruit moth]. Imkerfreund 28:346–347, 350–351.

Egorova, A. I. 1971. Preservative microflora in stored pollen [in Russian]. Veterinariya 8:40–41.

Egorova, A. I., and I. P. Bab'eva. 1967. Yeast flora of the honey bee (*Apis mellifera* L.) [in Russian]. Akademiia Nauk SSSR Izvestiia Sibirskoe Otdelenie (Seriia Biologo-Meditsinskikh Nauk) 2:127–132.

Ehara, S. 1968. On two mites of economic importance in Japan (Arachnida: Acarina). Applied Entomology and Zoology 3:124–129.

Eickwort, G. C. 1988. The origins of mites associated with honey bees. *In* Africanized honey bees and bee mites. G. R. Needham, R. E. Page, Jr., M. Delfinado-Baker, and C. E. Bowman, editors. Ellis Horwood, Chichester, England.

Eischen, F. A. 1987. Overwintering performance of honey bee colonies heavily infested with *Acarapis woodi* (Rennie). Apidologie 18:293–304.

Eischen, F. A., D. Cardoso-Tamez, W. T. Wilson, and A. Dietz. 1989. Honey production of honey bee colonies infested with *Acarapis woodi* (Rennie). Apidology 20:1–8.

Eischen, F. A., and A. Dietz. 1990. Improved culture techniques for mass rearing *Galleria mellonella* (Lepidoptera: Pyralidae). Entomological News 10:123–128.

Eischen, F. A., A. Dietz, and J. H. Brower. 1984. Effect of aging on the mating competitiveness of irradiated male greater wax moths (Lepidoptera: Pyralidae). Journal of Economic Entomology 77:1534–1536.

Eischen, F. A., J. S. Pettis, and A. Dietz. 1986. Prevention of *Acarapis woodi* infestation in queen honey bees with amitraz. American Bee Journal 126:498–500.

Eischen, F. A., J. S. Pettis, and A. Dietz. 1987. A rapid method of evaluating compounds for the control of *Acarapis woodi* (Rennie). American Bee Journal 127:99–101.

Eischen, F. A., T. E. Rinderer, and A. Dietz. 1986. Nocturnal defensive responses of Africanized and European honey bees to the greater wax moth (*Galleria mellonella* L.). Animal Behaviour 34:1070–1077.

Eischen, F. A., W. T. Wilson, D. Hurley, and D. Cardoso-Tamez. 1988. Cultural practices that reduce populations of *Acarapis woodi* (Rennie). American Bee Journal 128:209–211.

Eischen, F. A., W. T. Wilson, J. S. Pettis, A. Suarez, D. Cardoso-Tamez, D. L. Maki, A. Dietz, J. Vargas, C. Garza de Estrada, and W. L. Rubink. 1990. The spread of *Acarapis woodi* (Acari: Tarsonemidae) in northeastern Mexico. Journal of the Kansas Entomological Society 63:375–384.

Ekbohm, G., and K. Ebbersten. 1979. On the estimation of the number of sex alleles in honeybees. Heredity 42:267–269.

Elbe, H., and W. Weide. 1961. Die Bekämpfung von *Pericystis apis* Maassen in Labor und Praxis [Control of *Pericystis apis* Maassen in the laboratory and in practice]. Wissenschaftliche Zeitschrift der Universität Halle, Mathematisch- Naturwissenschaftliche Reihe 10:83–86.

El-Borollosy, F. M., A. K. Wafa, and A. M. El-Hefny. 1972. Studies on the biology of *Philanthus triangulum* F. (Hymenoptera: Sphecidae). Bulletin de la Société Entomologique d'Egypte 56:287–295.

Eliseev, V. K. 1978. Control of varroa disease of honeybees with sulphur [in Russian]. Sbornik Nauchnykh Trudov, Moskovskaya Veterinarnaya Akademiya 99:128–129.

Ellis, M., R. Nelson, and C. Simonds. 1988. A comparison of the fluvalinate and ether roll methods of sampling for Varroa mites in honey bee colonies. American Bee Journal 128:262–263.

Ellis, M. D. 1994. Toxic effects of monoterpenoids on the honey bee, *Apis mellifera* L., and its tracheal mite parasite, *Acarapis woodi* (Rennie). Ph.D. dissertation, University of Nebraska, Lincoln.

Ellison, D. 1977. The Miskoe portable fumigator. Gleanings in Bee Culture 105:52–53.

El-Sawaf, S. K. 1950. The life-history of the greater wax moth (*Galleria mellonella* L.) in Egypt, with special reference to the morphology of the mature larva. Bulletin de la Société Fouad Premier d'Entomologie 34:247–297.

Elsen, P. 1972. *Afrocypholaelaps* gen. nov., un nouveau genre pour *Neocypholaelaps africana* Evans, 1963, et redescription de cette espèce (Acarina: Mesostigmata) [Afrocypholaelaps new genus, a new genus for *Neocypholaelaps africana* Evans, 1963, and redescription of that species]. Revue de Zoologie et de Botanique Africaines 86:158–162.

Engels, W., L. S. Gonçalves, J. Steiner, A. H. Buriolla, and M. R. C. Issa. 1986. *Varroa*-Befall von Carnica-Völkern in Tropenklima [Varroa infestation in Carniolan colonies in a tropical climate]. Apidologie 17:203–216.

Engels, W., and P. Rosenkranz. 1992. Hyperthermie-Erfahrungen bei der Varroatose-Kontrolle [Heat treatment for varroa control.] Apidologie 23:379–381.

Engels, W., and K. Schatton. 1986. Changes in hemolymph proteins and weight loss in worker bees due to *Varroa* parasitation. Arbeitsgemeinschaft der Institute für Bienenforschung E.V. Abstracts of the Varroa Workshop in Feldafing/ Starnberg, West Germany.

Eremie, N. G. 1932. Tobacco blossoms injure bees? American Bee Journal 72:372.

Erickson, E. H., and R. Medenwald. 1979. Parasitism of queen honeybee pupae by *Melittobia acasta*. Journal of Apicultural Research 18:73–76.

Erlandson, S. 1961. Bivargens, *Philanthus triangulum* F., nu Kanda utbrednig i de nordiska landerna [The present distribution of the bee wolf in Scandinavia]. Nordisk Bitidsskrift 13:32–36.

Esch, H. 1976. Foraging strategies in bees. American Bee Journal 116:568–569, 573.

Esmali, M. 1974. Bee-eaters: a problem for beekeepers in Iran. American Bee Journal 114:136–137.

Essig, E. O. 1940. Mediterranean flour moth breeding in comb of honeybee. Journal of Economic Entomology 33:949–950.

Evans, A. V., and A. Nel. 1989. Notes on *Macrocyphonistes kolbeanus* Ohaus and Rhizoplatys auriculatus (Burmeister) with comments on their melittophilous habits (Coleoptera: Melolonthidae: Dynastinae: Phileurini). Journal of the Entomological Society of Southern Africa 52:45–50.

Evans, H. E. 1955. *Philanthus sanbornii* Cresson as a predator on honey bees. Bulletin of the Brooklyn Entomological Society 50:47.

Evans, H. E., and C. S. Lin. 1959. Biological observations on digger wasps of the genus *Philanthus* (Hymenoptera: Sphecidae). Wasmann Journal of Biology 17:115–132.

Evans, H. E., and K. M. O'Neill. 1988. The natural history and behavior of North American beewolves. Cornell University Press, Ithaca, New York.

Everist, S. I. 1981. Poisonous plants of Australia. Angus and Robertson, Australia.

Evers, C. A. and T. D. Seeley. 1986. Kin discrimination and aggression in honey bee colonies with laying workers. Animal Behaviour 34:925–925.

Fabian, F. W., and R. I. Quinet. 1928. A study of the cause of honey fermentation. Michigan State College Agricultural Experiment Station Technical Bulletin number 92.

Fan, Z.-Y., and L.-S. Li. 1988. The distribution and damage of bee mites in China. *In* Africanized honey bees and bee mites. G. R. Needham, R. E. Page, Jr., M. Delfinado-Baker, and C. E. Bowman, editors. Ellis Horwood, Chichester, England.

Fang, Y., F. Feng, B. Dong, and Z. Guo. 1983. Detection of chronic bee paralysis virus by an enzyme-linked immunosorbent assay. Scientia Agricultura Sinica 1983(2):93–96.

Fantham, H. B., and A. Porter. 1912a. Microsporidiosis, a protozoal disease of bees due to *Nosema apis* and popularly known as Isle of Wight disease. Annals of Tropical Medicine and Parasitology 6:145–162.

Fantham, H. B., and A. Porter. 1912b. The morphology and life history of *Nosema apis* and the significance of its various stages in the so called "Isle of Wight" disease. Annals of Tropical Medicine and Parasitology 6:163–195.

Fantham, H. B., and A. Porter. 1912c. Note on certain protozoa found in bees. Journal of the Board of Agriculture and Fisheries, London 19:138.

Fantham, H. B., A. Porter, and L. R. Richardson. 1941. Some microsporidia found in certain fishes and insects in eastern Canada. Parasitology 33:186–208.

Fan-Tsung, D. 1952. Beekeeping in Formosa. Bee World 33:150–151.

Farnsteiner, K. 1908. Der Ameisensäuregehalt des Honigs [The formic acid content of honey]. Zeitschrift für Untersuchung der Nahrungs- und Genussmittel 15:598, 604.

Farrar, C. L. 1942. Nosema disease contributes to winter losses and queen supersedure. Gleanings in Bee Culture 70:660–661, 701.

———. 1947. Nosema losses in package bees as related to queen supersedure and honey yields. Journal of Economic Entomology 40:333–338.

——. 1968. Productive management of honey-bee colonies. Part VI. American Bee Journal 108:316–317.

Fattig, P. W. 1943. The Mutillidae or velvet ants of Georgia. Emory University Museum Bulletin 1, Emory University, Georgia.

——. 1945. The Asilidae, or robber flies, of Georgia. Emory University Museum Bulletin 3, Emory University, Georgia.

Faucon, J. P., J. C. Arvieu, and M. E. Colin. 1982. Possibilité d'utilisation du bromure de méthyle pour la désinfection du matériel apicole [Possibility of utilizing methyl bromide for disinfection of apicultural material]. Révue de Médecine Vétérinaire 133:207–210.

Faveaux, M. A. de. 1984. Les acariens et les insectes parasites et prédateurs des abeilles, *Apis mellifica intermissa* Buttel-Reepen, en Algèré [Parasitic mites and insect parasites and predators of honey bees *Apis mellifica intermissa*, in Algeria]. Bulletin Zool. Agri. Institut National Agronomique, Algiers 8: 13-21.

Federal Register, volume 50 (111), Monday, June 10, 1985/Rules and Regulations 24171–24174.

——, volume 50 (120), Friday, June 21, 1985/Rules and Regulations 25688–25691.

Feiler, E. 1969. Mikroorganismen in Bienenhonig und ihre Rolle bei der Schaumgärung [Microorganisms in honey and their role in fermentation]. Zeitschrift für die Gesamte Hygiene und Ihre Grenzgebiete 15:871–875.

Fekl, W. 1956. Die Bakterienflora der Tracheen und des Blutes einiger Insekten [The bacterial flora of the tracheae and blood of several insects]. Zeitschrift für Morphologie und Oekologie der Tiere 44:442–458.

Feldlaufer, M. F., D. A. Knox, W. R. Lusby, and H. Shimanuki. 1993a. Antimicrobial activity of fatty acids against *Bacillus larvae*, the causative agent of American foulbrood disease. Apidologie 24:95–99.

Feldlaufer, M. F., W. R. Lusby, D. A. Knox, and H. Shimanuki. 1993b. Isolation and identification of linoleic acid as an antimicrobial agent from the chalkbrood fungus, *Ascosphaera apis*. Apidologie 24:89–94.

Felizardo, J. G. 1925. Doenças e inimigos das abelhas [Diseases and pests of bees]. Egatea. 10(6):483-8.

Ferber, C. E. M., and H. E. Nursten. 1977. The aroma of beeswax. Journal of the Science of Food and Agriculture 28:511–518.

Ferland, M. 1986. Birds do the housecleaning for his colonies. American Bee Journal 126:350–351.

Fichter, B. L., L. A. Royce, D. M. Burgett, and G. W. Krantz. 1986. ELISA detection method for *Acarapis woodi*. Proceedings of the Honey Bee Tracheal Mite (*Acarapis woodi*, R.) Scientific Symposium, July 8–9, 1986, Saint Paul, Minnesota. American Association of Professional Apiculturists, Gainesville, Florida.

Fichter, B. L., and W. P. Stephen. 1987. Efficacy of selected fungicides against chalkbrood of the leafcutting bee. Journal of Apicultural Research 26:137–143.

Fielitz, H. 1925. Untersuchungen über die Pathogenität einiger im Bienenstock vorkommenden Schimmelpilze bei Bienen [Investigations on the pathogenicity to bees of some molds occurring in beehives]. Centralblatt für Bakteriologie, Parasitenkunde und Infektionskrankheiten 66:28–50.

Fingler, B. G., W. T. Nash, and T. I. Szabo. 1982. A comparison of two techniques for the measurement of nosema disease in honey bee colonies wintered in Alberta, Canada. American Bee Journal 122:369–371.

Finn, W. E., and T. L. Payne. 1977. Attraction of greater wax moth females to male-produced pheromones. Southwestern Entomologist 2:62–65.

Flanigan, T. C. 1989. Protecting bees and saving bear with predator platforms. American Bee Journal 129:721–722.

Flechtmann, C. H. W. 1977. A abelha "Africanizada" e a disseminação do *Acarapis woodi* (Rennie) no Brasil [The dissemination of Acarapis woodi in Africanized bees in Brazil]. Anais da Sociedade Entomologica do Brasil 6:130–131. From Apicultural Abstracts 947L/80.

———. 1980. Dois ácaros associados à abelha (*Apis mellifera L.*) no Perú [Two mites associated with the bee *Apis mellifera* L. in Peru]. Anais da Escola Superior de Agricultura "Luiz de Queiroz," Universidade de São Paulo 37:737–741.

Fletcher, D. J. C. 1976. New perspectives in the causes of absconding in the African bee (*Apis mellifera adansonii*). Part II. South African Bee Journal 48(1):6–9.

———. 1978. The African bee, *Apis mellifera adansonii*, in Africa. Annual Review of Entomology 23:151–171.

Flint, H. M., and J. R. Merkle. 1983. Mating behavior, sex pheromone responses, and radiation sterilization of the greater wax moth (Lepidoptera: Pyralidae). Journal of Economic Entomology 76:467–472.

Flottum, K. 1991a. Home sweet home. Gleanings in Bee Culture 119:334–338.

———. 1991b. Inner cover. Gleanings in Bee Culture 119:312, 351.

Foelix, R. F. 1982. Biology of spiders. Harvard University Press, Cambridge, Massachusetts.

Foote, H. L. 1966. The mystery of the disappearing bees. Gleanings in Bee Culture 94:152–153, 182.

Foote, L. 1971. California nosema survey, 1969–1970. American Bee Journal 111:17.

Foster, J. W., W. A. Hardwick, and B. Guirard. 1950. Antisporulation factors in complex organic media. I. Growth and sporulation of *Bacillus larvae*. Journal of Bacteriology 59:463–470.

Foster, S., and R. A. Caras. 1994. A field guide to venomous animals and poisonous plants. Houghton Mifflin, Boston.

Fowler, H. G. 1990. Carpenter ants (Camponotus spp.): pest status and human perception. *In* Applied myrmecology. R. K. Vander Meer, K. Jaffe, and A. Cedeno, editors. Westview Press, Boulder, Colorado.

Fraser, H. M. 1931. Beekeeping in antiquity. University of London Press, London.

———. 1937. Regurgitations. British Bee Journal 65:392–394.

Frazier, W. C. 1967. Food microbiology. McGraw-Hill, New York.

Free, J. B. 1970. The behaviour of wasps (*Vespula germanica* L. and *Vespula vulgaris* L.) when foraging. Insectes Sociaux 17:11–20.

———. 1987. Pheromones of social bees. Cornell University Press, Ithaca, New York.

Free, J. B., and C. G. Butler. 1959. Bumblebees. Collins, London.

Free, J. B., and Y. Spencer-Booth. 1962. The upper lethal temperatures of honeybees. Entomologia Experimentalis et Applicata 5:249–254.

Frey. 1914. Ein neuer Beinenfeind [A new enemy of bees]. Leipziger BienenZeitung 29:152.

Friedmann, H. 1955. The honey-guides. U.S. National Museum Bulletin 208. Smithsonian Institution, Washington, D.C.

Fries, I. 1988a. Comb replacement and nosema disease. Apidologie 19:343–354.

———. 1988b. Contribution to the study of nosema disease (*Nosema apis* Z.) in honey bee (*Apis mellifera* L.) colonies. Dissertation, Swedish University of Agricultural Sciences, Uppsala.

———. 1988c. Infectivity and multiplication of *Nosema apis* Z. in the ventriculus of the honey bee (*Apis mellifera* L.). Apidologie 19:319–328.

———. 1989. Observations on the development and transmission of *Nosema apis* Z. in the ventriculus of the honey bee. Journal of Apicultural Research 28:107–117.

———. 1991. Treatment of sealed honey bee brood with formic acid for control of *Varroa Jacobsoni*. American Bee Journal 131:313–314.

———. 1994. Amöbasjukan (*Malpighamoeba mellificae*) konstaterad i Sverige. Bitidningen 93:328–329.

Fries, I., A. Aarhus, H. Hansen, and S. Korpela. 1991. Comparison of diagnostic methods for detection of low infestation levels of *Varroa jacobsoni* in honey-bee (*Apis mellifera*) colonies. Experimental and Applied Acarology 10:279–287.

Fries, I., S. Camazine, and J. Sneyd. 1994. Population dynamics of *Varroa jacobsoni*: a model and a review. Bee World 75:5–28.

Fries, I., G. Ekbohm, and E. Villumstad. 1984. *Nosema apis*, sampling techniques and honey yield. Journal of Apicultural Research 23:102–105.

Fries, I., and F. Feng. 1995. Crossinfectivity of *Nosema apis* in *Apis mellifera* and *Apis cerana*. Proceedings of the 34th International Congress of Apiculture of Apimondia, Lausanne, September 15–19, 1995. Apimondia, Bucharest.

Fries, I., F. Feng, A. da Silva, S. B. Slemenda, and N. J. Pieniazek. 1995. *Nosema ceranae* sp. nov. (Microsporidia, Nosematidae), morphological and molecular characterization of a microsporidian parasite of the Asian honey bee *Apis cerana* (Hymenoptera, Apidae). European Journal of Protistology 32:356–365.

Fries, I., R. R. Granados, and R. A. Morse. 1992. Intracellular germination of spores of *Nosema apis* Z. Apidologie 23:61–71.

Fries, I., and H. Hansen. 1993. Biotechnical control of *Varroa* mites in cold climates. American Bee Journal 133:435–438.

Fries, I., and P. Rosenkranz. 1993. Number of reproductive cycles of the Varroa mite. Apidologie 24:485–486.

Frilli, F. 1983. *Varroa Jacobsoni*: the situation in Italy. *In* Proceedings of a meeting of the European Community Experts' Group, Wageningen, Netherlands. R. Cavalloro, editor. Balkema, Rotterdam.

Fritz, R., and D. H. Morse. 1985. Reproductive success and foraging of the crab spider *Misumena vatia*. Oecologia 65:194–200.

Frohne, D., and H. J. Pfander. 1984. A colour atlas of poisonous plants. Wolfe Publishing Limited, London.

Frost, D. R., editor. 1985. Amphibian species of the world. A taxonomic and geographical reference. Allen Press/Association of Systematics Collections, Lawrence, Kansas.

Frumhoff, P. C. 1991. The effects of the cordovan marker on apparent kin discrimination among nestmate honey bees. Animal Behaviour 42:854–856.

Frumhoff, P. C., and S. Schneider. 1987. The social consequences of honey bee polyandry: the effects of kinship on worker interactions within colonies. Animal Behaviour 35:255–262.

Frumhoff, P. C., and J. Baker. 1988. A genetic component to division of labour within honey bee colonies. Nature 333:358–361.

Fry, C. H. 1969a. The recognition and treatment of venomous and nonvenomous insects by small bee-eaters. Ibis 111:23–29.

———. 1969b. The evaluation and systematics of bee eaters (Meropidae). Ibis 111:555–592.

———. 1970. Convergence between jacamars and bee-eaters. Ibis 112:257–259.

———. 1983. Honeybee predation by bee-eaters, with economic considerations. Bee World 64:65–78.

———. 1984. The bee-eaters. T & AD Poyser Ltd., Calton, England.

Fuchs, S. 1990. Preference for drone brood cells by *Varroa Jacobsoni* Oud. in colonies of *Apis mellifera carnica*. Apidologie 21:193–199.

Fuchs, S. 1994. Non-reproducing *Varroa jacobsoni* Oud. In honey bee worker cells – status of mites or effect of brood cells? Experimental & Applied Acarology 18:309-317.

Fuchs, S., and K. Langenbach. 1989. Multiple infestation of *Apis mellifera* L. brood cells and reproduction in *Varroa jacobsoni* Oud. Apidologie 20:257–266.

Fuller, T. C., and E. McClintock. 1986. Poisonous plants of California. University of California Press, Berkeley.

Furgala, B. 1962a. Factors affecting queen losses in package bees. Gleanings in Bee Culture 90:294–295.

———. 1962b. The effect of the intensity of nosema inoculum on queen supersedure in the honey bee, *Apis mellifera* Linnaeus. Journal of Insect Pathology 4:429–432.

———. 1962c. Residual fumagillin activity in sugar syrup stored in wintering honeybee colonies. Journal of Apicultural Research 1:35–37.

Furgala, B., and R. Boch. 1970. The effect of Fumidil B, Nosemack, and Humatin on *Nosema apis*. Journal of Apicultural Research 9:79–85.

Furgala, B., S. Duff, S. Aboulfaraj, D. Ragsdale, and R. Hyser. 1989. Some effects of the honey bee tracheal mite (*Acarapis woodi* Rennie) on nonmigratory, wintering honey bee (*Apis mellifera* L.) colonies in east central Minnesota. American Bee Journal 129:195–197.

Furgala, B., and T. A. Gochnauer. 1969. Effect of treatment method with Fumidil B. American Bee Journal 109:380–381, 392.

Furgala, B., and R. A. Hyser. 1969. Minnesota nosema survey—distribution and levels of infection in wintering apiaries. American Bee Journal 109:460–461.

Furgala, B., R. A. Hyser, and E. C. Mussen. 1973. Enzootic levels of nosema disease in untreated and fumagillin treated apiaries in Minnesota. American Bee Journal 113:210–212.

Furuya, K., K. Takatori, O. Sonobe, and T. Mabuchi. 1981. Occurrence of chalk brood disease in honey bee larvae in Japan [in Japanese, with English summary]. Transactions of the Mycological Society of Japan 22:127–133.

Fyg, W. 1932. Beobachtungen über die Amöben-Infektion ("Cystenkrankheit") der Malpighischen Gefässe bei der Honigbiene [Observations on amoeba infection ("cyst disease") of the Malpighian tubules of the honey bee]. Schweizerische Bienen-Zeitung 55:562–572.

———. 1934. Beitrag zur Kenntnis der sog. "Eischwarzsucht" der Bienenkönigin (*Apis mellifera* L.)]. Contribution to the knowledge of the so-called "egg melanosis" of the queen bee (*Apis mellifera* L.)]. Landwirtschaftliches Jahrbuch der Schweiz 1934 48:65–94.

———. 1936. Beiträge zur Anatomie, Physiologie und Pathologie der Bienenkönigin (*Apis mellifica* L.). III. Eine Methode zur subkutanen Impfung von Bienen-königinnen als Hilfsmittel beim Studium der Melanose [Contributions to the anatomy, physiology and pathology of the queen bee (*Apis mellifera* L.) . III. A method of subcutaneous inoculation of queen bees as a means to study melanosis] . Landwirtschaftliches Jahrbuch der Schweiz 1936 50:867–880.

———. 1939. Die Bedeutung der Königinkrankheiten für die Bienenzucht [The importance of queen diseases for the beekeeper]. Schweizerische Bienen-Zeitung 62:547–552.

———. 1945. Der Einfluss der Nosema-Infektion auf die Eierstöcke der Bienenkönigen. Schweizerische Bienen-Zeitung 68:67–72.

———. 1954. Über das Vorkommen von Flagellaten im Rectum der Honigbiene (*Apis mellifica* L.) [The occurrence of flagellates in the rectum of the honey bee (*Apis mellifera* L.)]. Mitteilungen der Schweizerischen Entomologischen Gesellschaft 27:423–428.

———. 1959. Normal and abnormal development in the honeybee. Bee World 40:57–66, 85–96.

———. 1962. Beitrag zur Pathologie der Sackbrut [Contribution to the pathology of sacbrood]. Zeitschrift für Bienenforschung 6:93–103.

———. 1964. Anomalies and diseases of the queen honey bee. Annual Review of Entomology 9:207–224.

———. 1972. Über die Keimentwicklung in "tauben" (abortiven) Bieneneiern [On the embryonic development in sterile (abortive) bee eggs]. Apidologie 3:125–148.

Gal, H., Y. Slabezki, and Y. Lensky. 1992. A preliminary report on the effect of origanum oil and thymol applications in honey bee (*Apis mellifera* L.) colonies in a subtropical climate on population levels of *Varroa jacobsoni*. Bee Science 2:(4):175–180.

Gallant, D. 1994. About bears, and such. Bee Culture 122:328.

Gallo, M. and J. Doull. 1991. History and scope of toxicology. *In* Casarett and Doull's Toxicology; The basic science of poisons. Eds. M. O. Amdur, J. Doull, and C. D. Klaasen Pergamon Press, New York.

Galton, D. 1971. Survey of a thousand years of beekeeping in Russia. Bee Research Association, London.

Gamber, W. 1990. Fluvalinate scare should serve as a warning. American Bee Journal 130:629.

Gandheker, S. S. 1959. Some common bee enemies in India. Gleanings in Bee Culture 87:360–361, 363.

Garnery, L., J. M. Cornuet, and M. Solignac. 1992. Evolutionary history of the honey bee *Apis mellifera* inferred from mitochondrial DNA analysis. Molecular Ecology 1:145–154.

Garnery, L., D. Vautrin, J. M. Cornuet, and M. Solignac. 1991. Phylogenetic relationships in the genus *Apis* inferred from mitochondrial DNA sequences data. Apidologie 22:87–92.

Garshelis, D. L. 1989. Nuisance bear activity and management in Minnesota. *In* Bear-people conflicts: proceedings of a symposium on management strategies. M. Bromley, editor. Northwest Territories Department of Renewable Resources. Yellowknife, Northwest Territories.

Gary, N. E. 1966. Robbing behavior in the honey bee. American Bee Journal 106:446–448.

———. 1992. Activities and behavior of honey bees. *In* The hive and the honey bee. J. M. Graham, editor. Dadant & Sons, Hamilton, Illinois.

Gary, N. E., R. F. L. Mau, and W. C. Mitchell. 1972. A preliminary study of honey bee foraging range in macadamia (*Macadamia integrifolia*) Maiden and Betche. Hawaii Entomology Society Proceedings 21:205–212.

Gary, N. E., and R. E. Page, Jr. 1987. Phenotypic variation in susceptibility of honey bees, *Apis mellifera*, to infestation by tracheal mites *Acarapis woodi*. Experimental and Applied Acarology 3:291–305.

Gary, N. E., and R. E. Page, Jr. 1989. Tracheal mite (Acari: Tarsonemidae) infestation effects on foraging and survivorship of honey bees (Hymenoptera: Apidae). Journal of Economic Entomology 82:734–739.

Gary, N. E., R. E. Page, Jr., and K. Lorenzen. 1989. Effect of age of worker honey bees (*Apis mellifera*) on tracheal mite (*Acarapis woodi*) infestation. Experimental and Applied Acarology 7:153–160.

Gary, N. E., R. E. Page, Jr., R. A. Morse, C. E. Henderson, M. E. Nasr, and K. Lorenzen. 1990. Comparative resistance of honey bees (*Apis mellifera* L.) from Great Britain and United States to infestation by tracheal mites (*Acarapis woodi*). American Bee Journal 130:667–669.

Garza-Q., C., and J. H. Dustmann. 1993. The problem of acariosis in honey bees (*Apis mellifera* L.) in north-east Mexico; investigations into incidence, disease patterns, and possible therapy. Animal Research and Development 38:7–31. Institute for Scientific Cooperation, Germany.

Garza-Q., C., J. H. Dustmann, W. T. Wilson, and R. Rivera. 1990. Control of the honey bee tracheal mite (*Acarapis woodi*) with formic acid in Mexico. American Bee Journal 130:801.

Gauhe, A. 1941. Über ein glukoseoxydierendes Enzym in der Pharynxdruse der Honigbiene [On a glucose oxidizing enzyme in the pharyngeal gland of the honey bee]. Zeitschrift für Vergleichende Physiologie 28:211–253.

Geissler, G., and W. Steche. 1962. Natürliche Trachten als Ursache für Vergiftungserscheinungen bei Bienen und Hummeln [Signs of natural flows poisoning honey bees and bumble bees]. Zeitschrift für Bienenforschung 6:77–92.

Genov, P. W., and J. I. Wanev. 1992. Berichte über Angriffe des Braunbaren (*Ursus arctos* L.) auf Haustiere und Bienenvolker in Bulgarien [Reports of brown bear (*Ursus arctos*) attacks on domestic animals and beehives in Bulgaria]. Zeitschrift für Jaqdwissenschaft 38(1):1–8.

Gentry, C. 1982. Small scale beekeeping. Peace Corps Manual M-17. Peace Corps, Washington, D.C.

Gerstäcker, C. E. A. 1866. Quoted description. American Bee Journal 2:106.

Gertsch, W. J. 1949. American spiders. D. Van Nostrand, New York.

Getz, W. M., D. Druckner, and T. R. Parisian. 1982. Kin structure and the swarming behavior of the honey bee *Apis mellifera*. Behavioral Ecology and Sociobiology 10:265–270.

Getz, W. M., and K. B. Smith. 1983. Genetic kin recognition: honey bees discriminate between full and half sisters. Nature 302:147–148.

Giauffret, A., and Y. P. Taliercio. 1967. Les mycoses de l'abeille (*Apis mellifica* L.): étude de quelques antimycosiques [Fungal diseases of the honey bee (*Apis mellifera* L.): a study of some antimycotics]. Bulletin Apicole 10:163–174.

Giauffret, A., M. J. Tostain-Caucat, and Y. Taliercio. 1969. Possibilités de désinfection par l'oxyde d'éthylène en pathologie apicole [Possibilities of disinfection by ethylene oxide in bee pathology]. Bulletin Apicole 12:45–52.

Giauffret, A., C. Vago, M. Rousseau, and J. L. Duthoit. 1966. Recherche sur l'action d'un virus dans l'étiologie de la loque européenne de l'abeille *Apis mellifera* [Research on the action of a virus in the etiology of European foulbrood of the bee *Apis mellifera*]. Bulletin Apicole 9:123–124.

Giavarini, I. 1937. Sopra una gregarina trovata nell'intestino di api [A gregarine found in the intestine of the bee]. Revista di Apicoltura Anno VI-1937-XV; Fasc. I:1–7.

———. 1950. Sui flagellati dell'intestino dell'ape [Flagellates of the small intestine of the honey bee]. Bollettino di Zoologia Agraria e di Bachicoltura 17 (Supplement):603–608.

Gibbons, W., R. R. Haynes, and J. L. Thomas. 1990. Poisonous plants and venomous animals of Alabama and adjoining states. University of Alabama Press, Tuscaloosa, Alabama.

Giblin, R. M., and H. K. Kaya. 1984. Associations of halictid bees with the nematodes *Aduncospiculum halicti* and *Bursaphelenchus kevini*. Journal of the Kansas Entomological Society 57:92–99.

Gilbert, C. H. 1939. The Indian meal moth, a honeyhouse pest. Gleanings in Bee Culture 67:426.

Gil Collado, J. 1931. Notas sobre pupiparos de España y Marruecos del Museo de Madrid (Diptera, Pupiparae) [Notes on Pupiparae of Spain and Morocco in the Madrid Museum]. EOA 8(1): 29-41.

Gill, R. 1986. Acarine mite. California Plant Pest and Disease Report 5:228.

Gillespie, W. H., and O. M. Dick, Jr. 1966. Beekeeping in West Virginia. West Virginia Department of Agriculture Bulletin 33.

Gilliam, M. 1973. Are yeasts present in adult worker honey bees as a consequence of stress? Annals of the Entomological Society of America 66:1176.

———. 1978a. Chalkbrood—status today and hopes for control. American Bee Journal 118:468–471.

———. 1978b. Fungi. In Honey bee pests, predators, and diseases. R. A. Morse, editor. Cornell University Press, Ithaca, New York.

———. 1979. Microbiology of pollen and bee bread: the yeasts. Apidologie 10:43–53.

———. 1986a. Chalkbrood disease: presentations at the 30th International Apicultural Congress in Nagoya, Japan. American Bee Journal 126:493–496.

———. 1986b. Infectivity and survival of the chalkbrood pathogen, Ascosphaera apis, in colonies of honey bees, *Apis mellifera*. Apidologie 17:93–100.

———. 1989. Aspectos generales sobre el pollo escayolado y estrategias para su control [Current knowledge about chalkbrood disease and strategies for control]. Vida Apicola 36:18–19, 21–24.

———. 1990. Chalkbrood disease of honey bees, *Apis mellifera*, caused by the fungus, Ascosphaera apis: a review of past and current research. Fifth International Colloquium on Invertebrate Pathology and Microbial Control, Adelaide, Australia.

Gilliam, M., and R. J. Argauer. 1975a. How long is terramycin stable in diets fed to honey bee colonies for disease control. American Bee Journal 115:230, 234.

Gilliam, M., and R. J. Argauer. 1975b. Stability of oxytetracycline in diets fed to honeybee colonies for disease control. Journal of Invertebrate Pathology 26:383–386.

Gilliam, M., and B. J. Lorenz. 1993. Enzymatic activity of strains of *Ascosphaera apis*, an entomopathogenic fungus of the honey bee, *Apis mellifera*. Apidologie 24:19–23.

Gilliam, M., and B. J. Lorenz. 1994. Taxonomic implications of spore cyst size in *Ascosphaera apis*, an entomopathogenic fungus of the honey bee, *Apis mellifera*. Sixth International Colloquium of Invertebrate Pathology and Microbial Control, Montpellier, France.

Gilliam, M., and B. J. Lorenz. 1995. Problems in the identification of *Ascosphaera apis*. Thirty-fourth International Congress of Apiculture, Lausanne, Switzerland, in press.

Gilliam, M., B. J. Lorenz, and S. L. Buchman. 1994. *Ascosphaera apis*, the chalkbrood pathogen of the honey bee, *Apis mellifera*, from larvae of a carpenter bee, *Xylocopa californica arizonensis*. Journal of Invertebrate Pathology 63:307–309.

Gilliam, M., H. L. Morton, D. B. Prest, R. D. Martin, and L. J. Wickerham. 1977. The mycoflora of adult worker honey bees, *Apis mellifera*: effects of 2,4,5-T and caging of bee colonies. Journal of Invertebrate Pathology 30:50–54.

Gilliam, M., and D. B. Prest. 1977. The mycoflora of selected organs of queen honey bees, *Apis mellifera*. Journal of Invertebrate Pathology 29:235–237.

Gilliam, M., and H. Shimanuki. 1967. *In vitro* phagocytosis of *Nosema apis* spores by honey-bee hemocytes. Journal of Invertebrate Pathology 9:387–389.

Gilliam, M., and S. Taber III. 1973. Microorganisms and diseases encountered in continuous bee production. American Bee Journal 113:222–223.

Gilliam, M., and S. Taber III. 1991. Diseases, pests, and normal microflora of honeybees, *Apis mellifera*, from feral colonies. Journal of Invertebrate Pathology 58:286–289.

Gilliam, M., S. Taber III, B. J. Lorenz, and D. B. Prest. 1988. Factors affecting development of chalkbrood disease in colonies of honey bees, *Apis mellifera*, fed pollen contaminated with *Ascosphaera apis*. Journal of Invertebrate Pathology 52:314–325.

Gilliam, M., S. Taber III, and G. V. Richardson. 1983. Hygienic behavior of honey bees in relation to chalkbrood disease. Apidologie 14:29–39.

Gilliam, M., S. Taber III, and J. B. Rose. 1978. Chalkbrood disease of honey bees, *Apis mellifera* L.: a progress report. Apidologie 9:75–89.

Gilliam, M., L. J. Wickerham, H. L. Morton, and R. D. Martin. 1974. Yeasts isolated from honey bees, *Apis mellifera*, fed 2,4-D and antibiotics. Journal of Invertebrate Pathology 24:349–356.

Gilliam, M., and S. Taber III. 1991. Diseases, pests, and normal microflora of honeybees, *Apis mellifera*, from feral colonies. Journal of Invertebrate Pathology 58:286–289.

Giordani, G. 1952. Alcune osservazioni su lieviti rinvenuti nell'apparato digerente di api ammalate [Yeasts found in the digestive system of diseased bees]. Annali della Sperimentazione Agraria 7:633–646.

———. 1956. Contributo alla conoscenza della *"Senotainia tricuspis"* Meig., dittero sarcofagide, endoparassita dell'ape domestica [Contribution to the knowledge of *"Senotainia tricuspis"* Meig., (Diptera, Sarcophagidae), an internal parasite of the honey bee]. Bollettino dell'Istituto di Entomologia della Università degli studi di Bologna 21:61–84.

———. 1957. Il lievito nella alimentazione delle api mellifiche [Yeast in the nutrition of *Apis mellifera*]. Apicoltore d'Italia 24:125–160.

———. 1959. Amoeba disease of the honey bee, *Apis mellifera* Linnaeus, and an attempt at its chemical control. Journal of Insect Pathology 1:245–269.

———. 1967. Laboratory research on *Acarapis woodi* Rennie, the causative agent of acarine disease of the honey bee. Note 5. Journal of Apicultural Research 6:147–157.

Girardeau, J. H., Jr. 1972. Fumagillin use in honey bee queen mating hives to suppress *Nosema apis*. Environmental Entomology 1:519–520.

Glaiim, M. K. 1992. First definite record of *Apis florea* in Iraq. Beekeeping and Development 23:3 (see AA 1211/92).

Gleanings in Bee Culture Editorial Staff. 1974. Starting right with bees. A. I. Root, Medina, Ohio.

GliNski, Z. 1981. Studies on the effect of the fungus *Ascosphaera apis* on larvae of the honey-bee, *Apis mellifera* L. Polskie Archiwum Weterynaryjne 23:9–16.

———. 1986. Léceni zvápenatení vcelího plodu cholinovou solí N-glukosyl- polyfunginem (Ascocidin-Polfa) [Control of chalkbrood disease by choline salt of N-glukosylpolyfunginem (Ascocidin-Polfa)]. Veterinární Medicína (Praha) 31:442–448.

GliNski, Z., and M. Chmielewski. 1979. Antifungal activity of certain polyene antibiotics against Ascosphaera apis, the causative agent of chalk brood. Annales Universitatis Mariae Curie-Sklodowska, Section D, Medicina 34:1–7.

GliNski, Z., and M. Chmielewski. 1981. Badania nad wykrywaniem pozóstalosci soli cholinowej N-glukozylopolifunginy w miodach [Detection of residues of the choline salt of N-glucosylpolifungine in honey]. Medycyna Weterynaryjna 37:732–736. (English summary.)

GliNski, Z., and M. Chmielewski. 1989. *Ascosphaera apis*, virulence, biochemiké a sérologické typy [*Ascosphaera apis*: virulence, biochemical, and serological types]. Veterinární Medicína (Praha) 34:113–119.

GliNski, Z., and J. Jarosz. 1992. *Varroa jacobsoni* as a carrier of bacterial infections to a recipient bee host. Apidologie 23:25–31.

GliNski, Z., M. Kowalska, and T. Osipowski. 1981. Aktywnosc wybranych srodków dezynfekcyjnych w stosunku do *Ascosphaera apis* [Activity of some disinfectants against *Ascosphaera apis*]. Medycyna Weterynaryjna 37:277–280. (English summary)

GliNski, Z., and J. Rzedzicki. 1980a. Chalk brood disease of the honey bee (*Apis mellifica* L.). Toxicity of the polyfungine antibiotics and Amphotericine B to larval and adult honey bees. Polskie Archiwum Weterynaryjne 22:297–303.

GliNski, Z., and J. Rzedzicki. 1980b. Chalk brood disease of the honey bee (*Apis mellifica* L.). Stability of the polyfungine antibiotics and Amphotericine B in honey and sugar syrup. Polskie Archiwum Weterynaryjne 22:305–313.

GliNski, Z., and J. Rzedzicki. 1980c. Chalk brood disease of the honey bee (*Apis mellifica* L.). The "in vitro" studies of some antimycotics with particular reference to polyene antibiotics. Polskie Archiwum Weterynaryjne 22:315–322.

Gochnauer, T. A. 1951. Drugs fight foul brood diseases in bees. Minnesota Farm and Home Science 9:15.

———. 1963. Diseases and enemies of the honey bee. In The hive and the honey bee. R. A. Grout, editor. Rev. ed. Dadant & Sons, Hamilton, Illinois.

Gochnauer, T. A., and J. Corner. 1974. Detection and identification of *Bacillus larvae* in a commercial pollen sample. Journal of Apicultural Research 13:264–267.

Gochnauer, T. A., R. Boch, and V. J. Margetts. 1979. Inhibition of *Ascosphaera apis* by citral and geraniol. Journal of Invertebrate Pathology 34:57–61.

Gochnauer, T. A., and B. Furgala. 1969. Chemotherapy of nosema disease. Compatibility of fumagillin with other chemicals. American Bee Journal 109:309–311.

Gochnauer, T. A., B. Furgala, and H. Shimanuki. 1975. Diseases and enemies of the honey bee. *In* The hive and the honey bee. Dadant & Sons, editors. Rev. ed. Dadant & Sons, Hamilton, Illinois.

Gochnauer, T. A., S. J. Hughes, and J. Corner. 1972. Chalkbrood disease of honeybee larvae: a threat to Canadian beekeeping? Canada Agriculture 17:36–37.

Gochnauer, T. A., and V. J. Margetts. 1979. Properties of honeybee larvae killed by chalkbrood disease. Journal of Apicultural Research 18:212–216.

Gochnauer, T. A., and V. J. Margetts. 1980. Decontaminating effect of ethylene oxide on honeybee larvae previously killed by chalkbrood disease. Journal of Apicultural Research 19:261–264.

Goebel, R. L. 1981. Australian hover fly - pest of honey. American Bee Journal 121(8): 589–590.

Goerzen, D. W., and T. C. Watts. 1991. Efficacy of the fumigant paraformaldehyde for control of microflora associated with the alfalfa leafcutting bee, *Megachile rotundata* (Fabricius) (Hymenoptera: Megachilidae). Bee Science 1:212–218.

Goettel, M. S., G. M. Duke, G. B. Schaalje, and K. W. Richards. 1992. Effects of selected fungicides on in vitro spore germination and vegetative growth of *Ascosphaera aggregata*, the causative organism of chalkbrood in the alfalfa leafcutter bee, *Megachile rotundata*. Apidologie 23:299–310.

Goettel, M. S., K. W. Richards, and D. W. Goerzen. 1993a. Decontamination of Ascosphaera aggregata spores from alfalfa leafcutting bees (*Megachile rotundata*) nesting materials by fumigation with paraformaldehyde. Bee Science 3:22–25.

Goettel, M. S., K. W. Richards, and D. W. Goerzen. 1993b. Susceptibility to chalkbrood of alfalfa leafcutter bees, *Megachile rotundata*, reared on natural and artificial provisions. Journal of Invertebrate Pathology 61:58–61.

Goettel, M. S., K. W. Richards, and G. B. Schaalje. 1991. Bioassay of selected fungicides for control of chalkbrood in alfalfa leafcutter bees, *Megachile rotundata.* Apidologie 22:509–522.

Goettel, M. S., J. D. Vandenberg, G. M. Duke, and G. B. Schaalje. 1993b. Susceptibility to chalkbrood of alfalfa leafcutter bees, *Megachile rotundata,* reared on natural and artificial provisions. Journal of Invertebrate Pathology 61:58–61.

Goetz, B., and N. Koeniger. 1993. The distance between larva and cell opening triggers broodcell invasion by *Varroa jacobsoni.* Apidologie 24:67–72.

Gold, R. C. 1969. The giant cane toad and a "taste of honey." American Bee Journal 109:467.

Goldschmidt, S., and H. Burkert. 1955. Die Hydrolyse des cholinergischen Honigwirkstoffes und anderer Cholinester mittels Cholinesterasen und deren Hemmung im Honig [Hydrolysis of the cholinergic factor in honey]. HoppeSeyler's Zeitschrift für Physiologische Chemie 301:78–89.

Gonçalves, L. S. 1987. O combate à varroa em todo o mundo [The battle against varroa around the world]. Apicultura no Brasil 3(19):31–35.

Gonçalves, L. S., De De Jong, and R. H. Nogueira. 1982. Infestation of feral honey bee colonies in Brazil by *Varroa jacobsoni.* American Bee Journal 122:249–251.

Gonçalves, L. S., A. E. E. Soares, A. C. Stort, A. H. Buriolla, M. R. C. Issa, J. Steiner, and M. E. P. V. Veloci. 1981. Estudo sobre o ácaro parasita de abelhas *Varroa jacobsoni.* I. Grau de infestação em apiários do Estado de São Paulo (nota preliminar) [A study on the parasitic mite *Varroa jacobsoni.* 1. Extent of infestation in the state of São Paulo (preliminary note)]. Resumos do V Congresso Brasileiro de Apicultura e III Congresso Latino-Ibero-Americano de Apicultura: 91.

Gonnet, M., P. Lavie, and P. Nogueira-Neto. 1964. Etude de quelques caractéristiques des miels récoltés par certains Meliponines brésiliens [Study of some characteristics of honeys gathered by certain Brazilian *Melipona*]. Compte Rendu de l'Académie des Sciences (Paris) 258:3107–3109.

Gontarski, H. 1937/38. Beitrag zur Pathologie der Bienenkönigin. Eine neue Eierstockkrankheit der Bienenkönigin [An addition to the pathology of the queen bee: a new ovarian disease of the queen]. Deutscher Imkerführer 11:54–56.

———. 1949. Giftige Bienenpflanzen? [Plants poisonous to bees?]. Natur und Volk 79:180–186.

———. 1950. Ein Fall von parasitärer Melanose der Pharynxdrüsen bei pollensammelnden Bienen [A case of parasitic melanosis of the pharyngeal glands of pollen-gathering bees]. Zeitschrift für Bienenforschung 1:7–12.

Goodacre, W. A. 1923. A casual enemy of the bee. The dragon fly (*Hemianax papuensis*). Agricultural Gazette of New South Wales 34:373–374.

———. 1940. Dwindling troubles with bees. Australian Beekeeper 42:65–67.

Goodman, R. D., P. Williams, B. P. Oldroyd, and J. Hoffman. 1990. Studies on the use of phosphine gas for the control of greater wax moth (*Galleria mellonella*) in stored honey bee comb. American Bee Journal 130:473–477.

Goodrich, R. 1968. Skunks and bees. Gleanings in Bee Culture 96:201–204, 251.

Goolsbey, A. 1969. It's for the bears. Gleanings in Bee Culture 97:280–283.

———. 1974. Bear platform management. Gleanings in Bee Culture 102:185, 191.

Gorbacheva, A. 1953. Summer losses of bees in Primorsky region due to *Syringa amurensis* [in Russian]. Pchelovodstvo 1953(9):36–40.

Graaf, D. C., de. 1991. Tissue specificity of *Nosema apis.* Journal of Invertebrate Pathology 58:277–278.

Graaf, D. C., de, H. Raes, and F. J. Jacobs. 1994a. Spore dimorphism in *Nosema apis* (Microsporidia, Nosematidae) developmental cycle. Journal of Invertebrate Pathology 63:92–94.

Graaf, D. C., de, H. Raes, G. Sabbe, P. H. Rycke, de, and F. J. Jacobs. 1994b. Early development of *Nosema apis* (Microspora: Nosematidae) in the midgut epithelium of the honeybee (*Apis mellifera*). Journal of Invertebrate Pathology 63:74–81.

Graff, H. 1961. Bees and bears. Gleanings in Bee Culture 89:356.

Graham, J., editor. 1987a. Varroa mites found in the United States. American Bee Journal 127:745–746.

———, editor. 1987b. More states find varroa; no decision on national action. American Bee Journal 127:817–818.

Grant, C. 1945. Drone bees selected by birds. Condor 47:261–263.

———. 1948. Selection between armed and unarmed arthropods as food by various animals. Journal of Entomology and Zoology 40:66.

Grant, G. A., D. L. Nelson, P. E. Olsen, and W. A. Rice. 1993. The "ELISA" detection of tracheal mites in whole honey bee samples. American Bee Journal 133:652–655.

Grant, G. G. 1976. Courtship behavior of a phycitid moth, *Vitula edmandsae*. Annals of the Entomological Society of America 69:445–449.

Grassi, G. B., and D. C. Parona. 1881. Animali che devono essere conosciuti dagli apicoltori [Animals that should be known by beekeepers]. Apicoltore 10:38–39.

Green, M. 1986. Court fines beekeeper for killing bear. American Bee Journal 126:9, 58.

Greenfield, M. D., and J. A. Coffelt. 1983. Reproductive behaviour of the lesser waxmoth, *Achroia grisella* (Pyralidae: Galleriinae): signalling, pair formation, male interactions, and mate guarding. Behaviour 84:287–315.

Gribakin, F. G. 1988. Photoreceptor optics of the honeybee and its eye colour mutants: the effect of screening pigments on the long-save subsystem of colour vision. Journal of Comparative Physiology A 164:123–140.

———. 1990. "White eye" as a model for study of optical properties of visual pigments in insects. *In* Sensory systems and communications in arthropods. Advances in Life Sciences, Birkhauser, Verlag, Basel.

Gribakin, F. G., I. V. Burovina, Ye. G. Chesnokova, Yu. V. Natochin, Ye. I. Shakmatova, K. Yu. Ukhanov, and E. Woyke. 1987. Reduced magnesium content in non-pigmented eyes of the honey bee (*Apis mellifera* L.). Comparative Biochemistry and Physiology 86A:689–682.

Griffiths, D. A. 1988. Functional morphology of the mouthparts of *Varroa jacobsoni* and *Tropilaelaps clareae* as a basis for the interpretation of their life-styles. In Africanized honey bees and bee mites. G. R. Needham, R. E. Page, M. Delfinado-Baker, and C. E. Bowman, editors. Ellis Horwood Ltd., Chichester, England.

Grigortsovskaya, T. P., and N. I. Borodai. 1972. A study of the toxicity of cultures of fungi *Aspergillus niger, A. fumigatus, Trichoderma lignorum* on bees of Primorsk (USSR) [in Russian]. Mikologia i Fitopatologiya 6:345–346.

Grimaldi, D., and B. A. Underwood. 1986. *Megabraula*, a new genus for two new species of Braulidae (Diptera), and a discussion of braulid evolution. Systematic Entomology 11:427–438.

Grimsley, V. M., and G. Sadler. 1936. A bridge between the North and South. American Bee Journal 76:176–177.

Grobov, O. F. 1975. Mite fauna in *Apis* nest and its significance. Proceedings of the 25th International Apicultural Congress 25:366–369.

———. 1976. Varroa disease in honey bees. Apiacta (Bucharest) 11:145–148.

———. 1977. Varroasis in bees. *In* Varroasis: a honeybee disease. Apimondia, Bucharest.

Grobov, O. F., and O. I. Bondarenko. 1975. A study of the interaction between *Malpighamoeba mellificae* and *Nosema apis* in the honey bee. Trudy Vsesoyuznogo Ordena Lenina Instituta Eksperimental'noi Veterinarii 43:268–272.

Grobov, O. F., Y. A. Ivanov and M. Y. Shablii. 1983. Phenothiazine for the control of varroa disease and Braula infestations. Veterinarnaya Entomologiya i Akarologiya. Moscow, USSR. Kolos. 248-257.

Grobov, O. F., A. M. Smirnov, and G. D. Volkovskii. 1970. The efficiency of ethylene oxide and methyl bromide against *Nosema apis* spores [in Russian]. Veterinariya 6:66–67.

Gross, K. P., and F. Ruttner. 1970. Entwickelt *Nosema apis* Zander eine Resistenz gegenüber dem Antibiotikum Fumidil B? [Is *Nosema apis* Zander developing a resistance to the antibiotic Fumidil-B?]. Apidologie 1:401–421.

Grout, R. A., editor. 1946. The hive and the honeybee. Dadant & Sons, Hamilton, Illinois.

Grozdanic, S., and Z. Vasic. 1966. Biological investigation of some spider species of the family Thomisidae (Aranea) [in Croatian]. Glaznik Muzeum Beograde, Series B 21:71–88.

Gruszka, J. 1987. Honey-bee tracheal mites: are they harmful? American Bee Journal 127:653–654.

Guerra, J. C., Jr., L. S. Gonçalves, and D. De Jong. 1994. Remoção diferencial de crias de operárias infestadas pelo ácaro Varroa jacobsoni, por operárias de colônias de abelhas africanizadas, italianas e hêbridas [Differential removal of worker brood infested by the mite Varroa jacobsoni by workers of Africanized and Italian bees, and hybrids between the two.] Anais do IV Congreso Iberolatinoamericano de Apicultura y I Foro Expo-comercial Internacional de Apicultura. Rio Cuarto, Córdoba, Argentina.

Guilhon, H. 1945. Un nouveau parastieism apiare en France. C.R. Academie Agri. France 31(11): 548-550.

Guilloux, J. 1969. Protection contre les pics-verts [Protection against green woodpeckers]. Abeilles et Fleurs 191:8.

Gunst, K. W., J. L. Hilldrup, and G. C. Llewellyn. 1978. Aflatoxin production on Apis mellifera artificially inoculated with Aspergillus flavus and Aspergillus parasiticus. Journal of Apicultural Research 17:81–83.

Gunther, C. E. M. 1951. A mite from a beehive on Singapore Island (Acarina: Laelaptidae). Proceedings of the Linnean Society of New South Wales 76:155–157.

Guzmán, L. I. de, T. E. Rinderer, and G. T. Delatte. 1993. Preliminary evaluation of four stocks of Apis mellifera on their tolerance to Varroa jacobsoni. In Asian apiculture. L. J. Connor et al., editors. Wicwas Press, Cheshire, Connecticut.

Guzmán-Novoa, E., A. Sánchez, and T. García. 1994. Susceptibilidad de colonias de abejas melíferas Europeas y africanizadas a la infestación y reproducción del ácaro Varroa jacobsoni [Susceptibility of European and africanized honey bee colonies to the infestation and reproduction of the mite Varroa Jacobsoni]. In Memorias del XIV Congreso Panamericano de Ciencias Veterinarias, Acapulco, México. 2 pp.

Guzmán-Novoa, E., and A. Zozaya-Rubio. 1984. The effects of chemotherapy on the level of infestation and production of honey in colonies of honey bees with acariosis. American Bee Journal 124:669–672.

Hackett, K. J. 1980. A study of chalkbrood disease and viral infection of the alfalfa leafcutting bee. Ph.D. dissertation, University of California, Berkeley.

Hackett, K. J., and G. O. Poinar, Jr. 1973. The ability of Neoaplectana carpocapsae Weiser (Steinernematidae: Rhabditoidea) to infect adult honeybees (Apis mellifera, Apidae: Hymenoptera). American Bee Journal 113:100.

Haeseler, V. 1977. Der Bienenwolf Philanthus triangulum Fabricius in Nordwestdeutschland (Hym., Sphecidae) [The beewolf Philanthus triangulum Fabricius in northwestern Germany]. Allgemeine Deutsche Imkerzeitung 11:289–292.

Hajsig, M. 1959. On yeasts from the intestinal content of bees [in Serbo-Croatian]. Veterinarski Arhiv 29:145–155.

Hale, P. J., and D. M. Menapace. 1980. Effect of time and temperature on the viability of Ascosphaera apis. Journal of Invertebrate Pathology 36:429–430.

Hall, H. G., and K. Muralidharan. 1989. Evidence from mitochondrial DNA that African honey bees spread as continuous maternal lineages. Nature 339:211–213.

Hameed, S. F., and R. L. Adlakha. 1973. Scarabaeid beetles: a new menace to beekeepers in Kulu Valley. Indian Bee Journal 35:48.

Hamilton, W. J. 1939. American mammals: their lives, habits and economic relations. McGraw-Hill, New York.

Hamlin, J. C., W. D. Reed, and M. E. Phillips. 1931. Biology of the Indianmeal moth on dried fruits in California. U.S. Department of Agriculture Technical Bulletin number 242.

Hammer, O., and E. Karmo. 1947. Nosema und Honigertrag. Schweizerische Bienen-Zeitung 70:190–194.

Hänel, H. 1983. Effect of JH-III on the reproduction of *Varroa jacobsoni*. Apidologie 14:137–142.

———. 1986. Effect of juvenile hormone (III) from the host *Apis mellifera* (Insecta: Hymenoptera) on the neurosecretion of the parasitic mite *Varroa jacobsoni* (Acari: Mesostigmata). Experimental and Applied Acarology 2:257–271.

Hänel, H., and N. Koeniger. 1986. Possible regulation of the reproduction of the honey bee mite *Varroa jacobsoni* (Mesostigmata: Acari) by a host's hormone: juvenile hormone III. Journal of Insect Physiology 32:791–798.

Hanko, J. 1978. Varroa disease in Czechoslovakia. Apiacta (Bucharest) 13:177.

Hansen, H. 1983. The extent of the varroa infestation in Scandinavia. *In* Varroa *jacobsoni* Oud. affecting honey bees: present status and needs. R. Cavalloro, editor. Balkema, Rotterdam.

Hansen, H. 1984a. Methods for determining the presence of the foulbrood bacterium *Bacillus larvae* in honey. Danish Journal of Plant and Soil Science 88:325–328.

———. 1984b. The incidence of the foulbrood bacterium *Bacillus larvae* in honeys retailed in Denmark. Danish Journal of Plant and Soil Science 88:329–336.

Hanson, F. R., and T. E. Eble. 1949. An antiphage agent isolated from *Aspergillus* sp. Journal of Bacteriology 58:527–529.

Hansson, F. 1932. Oleander and other poisonous plants for bees. Bees and Honey 13:245.

Haragsim, O. 1965. Potemnici (*Tribolium* sp.) jako skudci pylovych zásob [*Tribolium* species as pests of pollen supplies]. Vedecke Prace Vyskumneho Ustavu Vcelar CSAZV 4:61–65.

Haragsim, O., K. Samsinak, and E. Vobrázková. 1978. The mites inhabiting the bee-hives in CSR. Zeitschrift für Angewandte Entomologie 87:52–67.

Harbo, J. R. 1976. The effect of insemination on the egg-laying behavior of honey bees. Annals of the Entomological Society of America 69:1036–1038.

———. 1977. Survival of honey bee spermatozoa in liquid nitrogen. Annals of the Entomological Society of America 70:257–258.

———. 1980. Mosaic male honey bees produced by queens inseminated with frozen spermatozoa. Journal of Heredity 71:435–436.

———. 1990. Artificial mixing of spermatozoa from honeybees and evidence for sperm competition. Journal of Apicultural Research 29:151–158.

———. 1992. Breeding honey bees (Hymenoptera: Apidae) for more rapid development of larvae and pupae. Journal of Economic Entomology 85:2125–2130.

———. 1993. Field and laboratory tests that associate heat with mortality of tracheal mites. Journal of Apicultural Research 32:159–165.

Hardin, J. W., and J. M. Arena. 1969. Human poisoning from native and cultivated plants. Duke University Press, Durham, North Carolina.

Hargreaves, H. 1934. Report of the government entomologist for 1933. *In* Annual Report of the Department of Agriculture, Uganda, 1933 (1934). Part III.

Harizanis, P. C. 1991. Infestation of queen cells by the mite *Varroa Jacobsoni*. Apidologie 22:533–538.

Harris, W. F., and D. W. Filmer. 1947. Botanical investigation of pollen and nectar flora. New Zealand Journal of Science and Technology 29(3):134–143.

Harrison, C. J. 1979. Birds, their life, their ways, their world. Readers Digest Association, New York.

Harry, O. G., and L. H. Finlayson. 1976. The life-cycle, ultrastructure, and mode of feeding of the locust amoeba *Malpighamoeba locustae*. Parasitology 72:127–135.

Hartshorn, J. K. 1988. A wild wild life. California Farmer 268:8, 9, 34, 38.

Hartwig, A. 1983. Curing chalkbrood by ascocidin application. Proceedings of the 29th International Congress of Apiculture 29:233.

Hartwig, A., and A. Przelecka. 1971. Nucleic acids in the intestine of *Apis mellifica* infected with *Nosema apis* and treated with Fumagillin DCH: cytochemical and autoradiographic studies. Journal of Invertebrate Pathology 18:331–336.

Hase, A. 1926. Über Warmeentwicklung in Kolonien von Wachsmottenraupen [Concerning the development of heat in colonies of wax moth larvae]. Naturwissenschaften 14:995–997.

Haseman, L. 1961. How long can spores of American foulbrood live? American Bee Journal 101:298–299.

Haseman, L., and L. F. Childers. 1944. Controlling American foulbrood with sulfa drugs. University of Missouri Agricultural Experiment Station Bulletin 482.

Hassanein, M. H. 1951. Studies on the effect of infection with *Nosema apis* on the physiology of the queen honey bee. Quarterly Journal of Microscopical Science 92:225–231.

———. 1952a. Some studies on amoeba disease. Bee World 33:109–112.

———. 1952b. The effects of infection with *Nosema apis* on the pharyngeal salivary glands of the worker honey-bee. Proceedings of the Royal Entomological Society of London, Series A 27:22–27.

———. 1953. The influence of infection with *Nosema apis* on the activities and longevity of worker honey bee. Annals of Applied Biology 40:418–423.

Hassanein, M. H., and A. L. Abd El-Salaam. 1962a. Biological studies on the bee louse, *Braula coeca* Nitzsch. Bulletin de la Société Entomologique d'Egypte 46:87–95.

Hassanein, M. H., and A. L. Abd El-Salaam. 1962b. Effect of different diets on some biological aspects of the lesser wax moth, *Achroia grisella* Fab. Bulletin de la Société Entomologique d'Egypte 46:87–95.

Hassanein, M. H., M. M. Ibrahim, M. A. El-Banby, and A. F. M. ElArousy. 1969. Effect of different diets on some biological aspects of the lesser wax moth, *Achroia grisella* Fab. Bulletin de la Société Entomologique d'Egypte 53:567–572.

Hasselman, G. 1926. Observações sobre a biologia de *Braula coeca* [Observations on the biology of *Braula coeca*]. Bo. Institute of Brazil Sciences 2(4): 130.

Haydak, M. H. 1948. Beekeeping in Minnesota. University of Minnesota Agricultural Extension Bulletin 204.

———. 1963. Unwanted guests of honey bee colonies. American Bee Journal 103:129–131.

Haynes, W. C. 1972. The catalase test, an aid in the identification of *Bacillus larvae*. American Bee Journal 112:130–131.

Heath, L. A. F. 1982a. Development of chalk brood in a honeybee colony: a review. Bee World 63:119–130.

———. 1982b. Chalk brood pathogens: a review. Bee World 63:130–135.

———. 1985. Occurrence and distribution of chalk brood disease of honeybees. Bee World 66:9–15.

Heath, L. A. F., and B. M. Gaze. 1987. Carbon dioxide activation of spores of the chalkbrood fungus, *Ascosphaera apis*. Journal of Apicultural Research 26:243–246.

Heath, L. A. F., and P. Smithers. 1986. Biscuit beetles from chalk brood mummies. Bee World 67:12–14.

Hefinger, A. L. 1922. Circumventing ants in Florida. Gleanings in Bee Culture 50:35.

Heiduschka, A., and G. Kaufmann. 1913. Über die Säuren im Honig [The acids in honey]. Süddeutsche Apotheker-Zeitung 53:118–119.

Heinrich, C. 1956. American moths of the subfamily Phycitinae. U.S. National Museum Bulletin number 207.

Heinze-Gerhard, W. 1968. Mit Zitronensäure gegen die Septikämie [Citric acid for septicemia]. Nordwestdeutsche Imkerzeitung 20:123–124.

Hejtmanek, J. 1951. *Cryptophagus scanicus* in the hive [in Slovak]. Slovenski Cebelar 29:123–124.

Hellmich II, R. L., and T. E. Rinderer. 1991. Beekeeping in Venezuela. *In* The "African" honey bee. M. Spivak, D. J. C. Fletcher, and M. D. Breed, editors. Westview Press, Boulder, Colorado.

Henderson, C. 1986. Control of *Varroa jacobsoni* in honey bee (*Apis mellifera*) packages. M.S. thesis, Cornell University, Ithaca, New York.

———. 1988. Tests of chemical control agents for *Varroa jacobsoni* in honey bee packages. *In* Africanized honey bees and bee mites. G. R. Needham, R. E. Page, M. Delfinado-Baker, and C. E. Bowman, editors. Ellis Horwood, Ltd., Chichester, England.

Henderson, C. E., and R. A. Morse. 1990. Tracheal mites. *In* Honey bee pests, predators, and diseases. 2d ed. R. A. Morse and R. Nowogrodzki, editors. Cornell University Press, Ithaca, New York.

Hepburn, H. R. 1991. Incidence of the tachinid bee fly, *Rondaniooestrus apivorus*, in southern Africa. South African Bee Journal. 63(4): 86-87.

Herbert, E. W., Jr., D. J. Chitwood, and H. Shimanuki. 1985. New compounds with potential for the control of chalkbrood. American Bee Journal 125:430–431.

Herbert, E. W., Jr., D. J. Chitwood, and H. Shimanuki. 1986. The effect of a candidate compound on chalkbrood disease in New Jersey. American Bee Journal 126:258–259.

Herbert, E. W., Jr., H. Shimanuki, and D. A. Knox. 1977. Transmission of chalk brood disease of honeybees by infected queens, and worker brood and adults. Journal of Apicultural Research 16:204–208.

Herbert, E. W., Jr., H. Shimanuki, and J. C. Matthenius, Jr. 1988. An evaluation of menthol placement in hives of honey bees for the control of *Acarapis woodi*. American Bee Journal 128:185–187.

Herbert, E. W., Jr., P. C. Witherell, W. A. Bruce, V. J. Mladjan, and H. Shimanuki. 1989. Observations on Varroa-infested honey bee packages treated with Apistan and hived. American Bee Journal 129:799–801.

Herrod-Hempsall, W. 1931. The blind louse of the honey bee. Journal Ministry of Agriculture 37(12):1176-1184.

———. 1937. Beekeeping new and old, described with pen and camera, volume 2. British Bee Journal, London.

Hertig, M. 1923. The normal and pathological histology of the ventriculus of the honey bee, with special reference to infection with *Nosema apis*. Journal of Parasitology 9:109–140.

Hicheri, K. 1978. *Varroa jacobsoni* in Africa. Apiacta (Bucharest) 13:178.

Higo, K. 1983. Mass trapping of *Vespa xanthoptera* in an apiary [in Japanese, English summary]. Honeybee Science 4:123–126.

Hilger, A. 1904. Zur Kenntnis der im rechtsdrehenden Koniferenhonig vorkommenden Dextrine [The dextrins found in conifer honey]. Zeitschrift für Untersuchung der Nahrungs- und Genussmittel 8:110–126.

Hilldrup, J. A. L., T. Eadie, and G. C. Llewellyn. 1977. Fungal growth and aflatoxin production on apiarian substrates. Journal of the Association of Official Analytical Chemists 60:96–99.

Hilldrup, J. L., and G. C. Llewellyn. 1979. Acute toxicity of the mycotoxin aflatoxin B_1 in *Apis mellifera*. Journal of Apicultural Research 18:217–221.

Hilse, B. 1986. Survey of cane toads and their effect on the bee population. Australasian Beekeeper 87:263–264.

Hines, L. H. 1992. Selection for "tracheal mite-resistant" honey bees in southern Arizona—comments by the beekeeper. American Bee Journal 132:607.

Hinkle, E. 1975. A beautiful predator. Gleanings in Bee Culture 103:263.

Hirsch, M. S., and J. C. Kaplan. 1987. Antiviral therapy. Scientific American 256(4):76–85.

Hirschfelder, H. 1952. Zur Bekämpfung des Bienenwolfes (*Philanthus triangulum* F.) [Control of the bee-wolf wasp (*Philanthus triangulum* F.)]. Anzeiger für Schädlingskunde und Pflanzenschutz 25:122–123.

Hirschfelder, H., and H. Sachs. 1952. Recent research on the acarine mite. Bee World 33:201–209.

Hirst, S. 1921. On the mite (*Acarapis woodi*, Rennie) associated with Isle of Wight bee disease. Annals and Magazine of Natural History 7:509–519.

Hischier, J. 1962. Untersuchungen über Flagellaten im Darm der Honigbiene (*Apis mellifica* L.) [Investigation of flagellates in the intestine of the honey bee]. Inaugural dissertation, Veterinärmedizinische Fakultät, Universität Bern.

Hitchcock, J. D. 1948. A rare gregarine parasite of the adult honey bee. Journal of Economic Entomology 41:854–858.

———. 1956. Honey bee queens whose eggs all fail to hatch. Journal of Economic Entomology 49:11–14.

———. 1959. Poisoning of honey bees by death camas blossoms. American Bee Journal 99:418–419.

———. 1966. Transmission of sacbrood disease to individual honey bee larvae. Journal of Economic Entomology 59:1154–1156.

Hitchcock, J. D., and M. Christensen. 1972. Chalk brood disease of honey bees in the United States. American Bee Journal 112:248–249, 254.

Hoage, T. R., and W. C. Rothenbuhler. 1966. Larval honey bee response to various doses of *Bacillus larvae* spores. Journal of Economic Entomology 59:42–45.

Holldobler, B., and E. O. Wilson. 1990. The ants. Belknap Press, Cambridge, Massachusetts.

Holm, S. N. 1986. Breeding honeybees for resistance to chalkbrood disease. Thirtieth International Congress of Apiculture, Nagoya, Japan, 1985.

Holm, S. N., and J. P. Skou. 1972. Studies on trapping, nesting and rearing of some *Megachile* species (Hymenoptera, Megachilidae) and on their parasites in Denmark. Entomologica Scandinavica 3:169–180.

Holst, E. C. 1946. A simple field test for American foulbrood. American Bee Journal 86:14, 34.

Homann, H. 1933. Die Milben in gesunden Bienenstöcken [The mites in healthy beehives]. Zeitschrift für Wissenschaftliche Biologie, Parasitenkunde 6:350–415.

Honeybee Act, the Act of August 31, 1922 (Public number 293 - 67th Congress), entitled "An Act to regulate foreign commerce in the importation into the United States of the adult honeybee (*Apis mellifica*)."

Hood, W. M. 1994. Mammals and their effect on beekeeping—survey. Unpublished.

Hoppe, H., and W. Ritter. 1988. The influence of the Nasonov pheromone on the recognition of house bees and foragers by *Varroa jacobsoni*. Apidologie 19:165–172.

Hoppe, H., and W. Ritter. 1989. Comparative examinations for the control of Varroatosis by means of ethereal oils. Apidologie 20:522–523.

Hoppe, H., W. Ritter, and E. W. C. Stephen. 1989. The control of parasitic bee mites *Varroa jacobsoni*, *Acarapis woodi* and *Tropilaelaps clareae* with formic acid. American Bee Journal 129:739–842.

Horn, H. 1984. Zum Zusammenhang zwischen *Varroa jacobsoni* und Bakteriosen bei der Honigbiene [On the relation between *Varroa jacobsoni* and bacterioses in the honey bee]. Allgemeine Deutsche Imkerzeitung 18:328–329.

Hornitzky, M. 1986. The use of gamma radiation in the control of honey bee infections. Australasian Beekeeper 88:55–58.

Hornitzky, M. A. Z., P. Stace, and J. G. Boulton. 1989. A case of stonebrood in Australian honey bees (*Apis mellifera*). Australian Veterinary Journal 66:64.

Hornitzky, M. A. Z., and V. E. Taylor. 1983. Preparation of specific antisera to honeybee viruses by immunization with agar gel precipitates. Journal of Apicultural Research 22:261–263.

Hornitzky, M. A. Z., and P. A. Wills. 1983. Gamma radiation inactivation of *Bacillus larvae* to control American foul brood. Journal of Apicultural Research 22:196–199.

Howard, L. O. 1907. Report of the meeting of inspectors of apiaries, San Antonio, Texas, November 12, 1906. U.S. Department of Agriculture, Bureau of Entomology Bulletin 70.

Howard, W. R. 1900. New York bee disease, or black brood. Gleanings in Bee Culture 28:121–127.

Howes, F. N. 1948. Plants and beekeeping. Faber and Faber Ltd., London.

———. 1949. Poisoning from honey. Food Manufacturing 24:459–462.

Howes, T. M. 1979. On mice and entrance reducers. American Bee Journal 119:237–238.

Huggins, F. L. 1942. Yellow jasmine poisoning. American Bee Journal 82:132.

Hughes, A. M. 1976. The mites of stored food and houses. Technical Bulletin 9, Ministry of Agriculture, Fisheries and Food, London.

Huheey, J. E. 1980. Studies in warning coloration and mimicry. VIII. Further evidence for a frequency-dependent model of predation. Journal of Herpetology 14:223–230.

Hung, A. C. F., J. R. Adams, and H. Shimanuki. 1995. Bee parasitic mite syndrome (II): the role of Varroa mite and viruses. American Bee Journal 135:702–704.

Hussain, S. A., and S. Ali. 1980. Beehive predation by wasps (genus Vespa) and its possible benefit to honeyguides (Indicatoridae) in Bhutan. Journal of the Bombay Natural History Society 76:157–159.

Hussein, M. H. 1983. Field and laboratory tests on the effect of juvenile hormone analogue on the larvae of *Galleria mellonella* L. (Lep., Pyralidae). Zeitschrift für Angewandte Entomologie 95:249–253.

———. 1988. Studies on some pests of honeybee colonies in Dhofar, Oman. Proceedings 4th International Conference on Apiculture in Tropical Climates; Cairo, Egypt: 370-376.

———. 1989. Studies on some pests of honeybee colonies in Dhofar, Oman. Proceedings of the 4th International Conference on Apiculture in Tropical Climates, Cairo, 1988.

Hutson, R. 1927. Gaseous chlorine as a disinfectant for American foulbrood infected combs. Journal of Economic Entomology 20:516–519.

Hüttinger, E. 1974. Dickkopffliegen (Conopidae, Dip., Ins.), wenig beachtete Bienenparasiten [Thick-headed flies (Insecta, Diptera, Conopidae), little studied parasites of bees]. Bienenvater 95:102–103.

Huxtable, R. J. 1989. Human health implications of pyrrolizidine alkaloids and herbs containing them. *In* Toxicants of plant origin. Volume 1, Alkaloids. P. R. Cheeke, editor. CRC Press, Boca Raton, Florida.

Hyser, D. 1986. Tracheal mite, *Nosema*, and wintered honeybee colonies in Minnesota. Proceedings of the Honey Bee Tracheal Mite (*Acarapis woodi*, R.) Scientific Symposium, July 8–9, 1986, Saint Paul, Minnesota. American Association of Professional Apiculturists, Gainesville, Florida.

Ibay, L. A., and D. M. Burgett. 1989. Biology of the two external Acarapis species of honey bees: *Acarapis dorsalia* Morgenthaler and *Acarapis externus* Morgenthaler. American Bee Journal 129:816.

Ibrahim, R., and M. R. Yusoh. 1989. Control of wasps and weaver ants in *Apis cerana* apiaries in West Malaysia. Proceedings of the Fourth International Conference of Apiculture in Tropical Climates, Cairo, 1988.

Ibrahim, S. H. 1972. Observation on some insect pests of stored pollen pellets. Agricultural Research Review, Cairo 50:115–117.

———. 1984. A study on a dipterous parasite of honeybees. Zeitschrift für Angewandte Entomologie 97:124–126.

Idígoras Rubio, I. 1985. Los gregarinos: protozoos que parasitan el intestino de las abejas [Gregarines: protozoa which parasitize the intestines of honey bees]. Vida Apícola 14:36–37.

Ifantidis, M. D. 1983. Ontogenesis of the mite *Varroa jacobsoni* in worker and drone honeybee brood cells. Journal of Apicultural Research 22:200–206.

———. 1988. Some aspects of the process of *Varroa jacobsoni* entrance into honey bee (*Apis mellifera*) brood cells. Apidologie 19:387–396.

Imai, S., S. Fujioka, K. Nakanishi, M. Koreeda, and T. Kurokawa. 1967. Extraction of ponasterone A and ecdysterone from podocarpaceae and related plants. Steroids 10:557–565.

Imms, A. D. 1942. On *Braula coeca* Nitzsch and its affinities. Parasitology, 34: 88-100

Inglis, G. D., M. S. Goettel, and L. Sigler. 1993. Influence of microorganisms on alfalfa leafcutter bee (*Megachile rotundata*) larval development and susceptibility to Ascosphaera aggregata. Journal of Invertebrate Pathology 61:236–243.

Isack, H. A., and H.-U. Reyer. 1989. Honeyguides and honey gatherers: interspecific communication in a symbiotic relationship. Science 243:1343–1346.

Iselin, W. A. 1968. A dermestid, *Trogoderma glabrum*, infests comb honey. Annals of the Entomological Society of America 61:1621.

Ishay, J., H. Bitinsky-Salz, and A. Shulov. 1967. Contributions to the bionomics of the oriental hornet (*Vespa orientalis* Fab.). Israel Journal of Entomology 2:45–106.

Ishihara, R. 1969. The life cycle of *Nosema bombycis* as revealed in tissue culture cells of *Bombyx mori*. Journal of Invertebrate Pathology 14:316–320.

Ishikawa, K. 1968. Studies on the mesostigmatid mites associated with insects in Japan. Report of Research, Matsuyama Shinonome Junior College 3:198–218.

Israel Beekeepers' Association. 1949. Secretary's diary. Yalhoot Hamicheret 1:30–32. *From* Apicultural Abstracts 23/52.

Issa, M. R. C., D. De Jong, and L. S. Gonçalves. 1993. Reproductive strategies of the mite *Varroa jacobsoni* (Mesostigmata, Varroidae): influence of larva type and comb cell size on honey bee brood infestation rates. Brazilian Journal of Genetics 16:219–224.

Issa, M. R. C., and L. S. Gonçalves. 1982. Estudo da preferência do ácaro *Varroa jacobsoni* por zangões de abelhas *Apis mellifera* (Nota preliminar) [A study of the preference of the mite *Varroa jacobsoni* for drones of *Apis mellifera*]. Ciência e Cultura 34(7):738.

Issa, M. R. C., and L. S. Gonçalves. 1984. Study of the preference of the acarid *Varroa jacobsoni* for drones of Africanized honey bees. *In* Advances in invertebrate reproduction 3. W. Engels, editor. Elsevier Science Publishers, Amsterdam.

Issa, M. R. C., and L. S. Gonçalves. 1986. Technique for inducing oviposition in the mite *Varroa jacobsoni* under laboratory conditions. *In* Proceedings of the 20th International Apicultural Congress of Nagoya-Japan. Apimondia, Bucharest.

Jack, R. W. 1916. Parthenogenesis amongst the workers of the cape honey bee. Mr. G. W. Onions' experiments. Transactions of the Entomological Society, London 64:396–403.

Jadczak, A. 1994. The bear facts. Leaflet. Maine Department of Agriculture. 2 pp.

James, W. R. 1973. Know your poisonous plants. Naturegraph Publishers, Healdsburg, California.

Jay, E. G., G. E. Cantwell, G. C. Pearman, Jr., and J. V. Thompson. 1972. Control of the greater wax moth, *Galleria mellonella* (Linnaeus), in comb honey with carbon dioxide. Part II. Field studies. American Bee Journal 112:342, 344.

Jay, S. C. 1963. The development of honeybees in their cells. Journal of Apicultural Research 2:117–134.

———. 1968. Factors influencing ovary development of worker honeybees under natural conditions. Canadian Journal of Zoology 46:345–347.

Jay, S. C., and D. Dixon. 1982. Nosema disease in package honeybees, queens, and attendant workers shipped to western Canada. Journal of Apicultural Research 21:216–221.

Jay, S. C., and E. V. Nelson. 1973. The effects of laying worker honeybees (*Apis mellifera* L.) and their brood on the ovary development of other worker honeybees. Canadian Journal of Zoology 51:629–632.

Jaycox, E. R. 1960. Surveys for nosema disease of honey bees in California. Journal of Economic Entomology 53:95–98.

———. 1969. Beekeeping in Illinois. University of Illinois Cooperative Extension Service Circular 1000.

———. 1975. Skunks and bees. Bees and Honey Newsletter, March: 3–4.

Jean-Prost, P. and P. Médori. 1994. Apiculture. Know the Bee, manage the Apiary. Sixth ed. Intercept, U.K. 659pp.

Jefferson, T. 1788. Notes on the State of Virginia. Pritchard and Hall, Philadelphia.

Jeffree, E. P. 1956. Winter brood and pollen in honeybee colonies. Insectes Sociaux 3:417–422.

Jeffree, E. P., and M. D. Allen. 1956. The influence of colony size and of nosema disease on the rate of population loss in honey bee colonies in winter. Journal of Economic Entomology 49:831–834.

Jelinsski, M., and F. Wojtowski. 1984. *Melittobia acasta* Walker (Hym, Chalcidoidea, Eulphida) malo znany pasozyt czerwia pszczelego. [*Melittobia acasta*, a little known parasite of honeybee brood] Przeglad Zoologiczny 28:507–511.

Jenn, R. A. 1973. Ravages of the bee-eater. American Bee Journal 113:21.

Jesper, D. M. 1954. War on wasps. British Bee Journal 82:217.

Johannsmeier, M. F. 1978. Nature down the honey trail: toads as bee predators. South African Bee Journal 50(2):2–4.

————. 1980. Beetles that eat brood. South African Bee Journal 52(5):3–5.

Johansen, C. 1962. Impregnated foundation for waxmoth control. Gleanings in Bee Culture 90:682–684.

————. 1975. Bear protection for bees. Washington State Extension Publication EM4005.

————. 1977. Pesticides and pollution. Annual Review of Entomology 22:177–192.

Johansen, C., and D. Mayer. 1976. Alkali bees: their biology and management for alfalfa seed production in the Pacific Northwest. Pacific Northwest Extension Publication 155.

Johnsen, P. 1951. Mere om amøbesygen og en opfordring [More about amoeba disease, and an appeal]. Tidsskrift for Biavl 85:120–121.

————. 1952a. Adselgravere som goester i bistader [Burying beetles in hives]. Tidsskrift for Biavl 86:98.

————. 1952b. Vanskabte bier og hestekastanieforgiftning [Deformed bees and horse chestnut poisoning]. Nordisk Bitidskrift 4:44–47.

————. 1952c. Om nogle sprøjtmidler og bier [Some sprays on bees]. Erhvevsfrugtauleren 18:214–216.

————. 1953. Guldbasse og strepsiptér som skadedyr for honningbier [Chafers and stylops as enemies of the honeybee]. Nordisk Bitidskrift 5:98–100.

————. 1955. Honnigbiens fjender [Honey bee enemies]. Dansk Videnskabs, Copenhagen.

Jordan, R. 1936. Der Krankheitsverlauf bei Malpighaamöben- und Nosemaerkrankung beim Einzelauftreten der Parasitem im Bienenvolke [The course of amoeba and nosema disease with only one of the parasites in the colony]. Deutsche Imker 49:152–158.

————. 1957. Essigsäure zur Bekämpfung der Wachsmotte vor allem aber Entkeim nosemainfisierter Waben. Bienenvater 78:101–105.

————. 1961. Bienenschäden verursacht durch Eibenpollen [Injuries to bees from yew pollen]. Bienenvater 83:246–247.

Kaczmarek, S. 1984. Investigations on insects associated with hives in some areas of Northern Poland [in Polish]. Pszczelnieze Zeszyty Neukowe 28:117–133.

Kadymov, V. A. 1981. The hornet Vespa crabro in the Lenkoran zone of Azerbaijan [in Russian]. Izvestiva Akademii Nauk Azerbaidzhankoi SSR, Biologicheskikh Nauk 1981(3):65–69.

Kajikawa, K., and T. Nakane. 1986. Preventive measures for chalk brood: selection of the most effective drug and its application to diseased colonies [in Japanese, English summary]. Honeybee Science 7:69–74.

Kamburov, G., and M. Hajsig. 1963. O kvasnicama slicnim gljivicama iz larvi pcela uginulih od blage gnjiloce [Yeast-like organisms in larvae killed by European foulbrood]. Veterinarski Arhiv 33:121–123.

Kandemir, I., and A. Kence. 1995. Allozyme variability in a central Anatolian honeybee (Apis mellifera L.) population. Apidologie 26:503–510.

Kang, Y. B., D. S. Kim, and D. H. Jang. 1976. Experimental studies on the pathogenicity and developmental stages of Nosema apis [in Korean, English summary]. Korean Journal of Veterinary Research 16:1–25.

Kapil, R. P., and K. Aggarwal. 1986. Some observations on reproduction and seasonal trends in Euvarroa sinhai (Mesostigmata: Varroidae). In Abstracts of the Seventh International Congress of Acarology, Bangalore, India.

Kashkovskii, U. G. 1958. An enemy of the apiary [in Russian]. Prioroda (Moscow) 47:120.

Kaston, B. J. 1978. How to know the spiders. 3d ed. W. C. Brown, Dubuque, Iowa.

Kates, A. H., D. E. Davis, J. McCormack, and J. F. Miller. 1974. Poisonous plants of the Southern United States. Publication 8. Agricultural Extension Service, University of Tennessee, Knoxville.

Katznelson, H. 1950. Bacillus pulvifaciens (n.sp.), an organism associated with powdery scale of honeybee larvae. Journal of Bacteriology 59:153–155.

Katznelson, H., J. Arnott, and S. E. Bland. 1952. Preliminary report on the treatment of European foulbrood of honey bees with antibiotics. Scientific Agriculture 32:180–184.

Katznelson, H., and C. A. Jamieson. 1952. Control of nosema disease of honey bees with fumagillin. Science 115:70–71.

Kauffeld, N. M. 1973. Queen honey bees in colonies: susceptibility to nosema disease. American Bee Journal 113:12–14.

Kauffeld, N. M., J. L. Williams, T. Lehnert, and F. E. Moeller. 1972. Nosema control in package bee production: fumigation with ethylene oxide and feeding with fumagillin. American Bee Journal 112:297–299, 301.

Kawarabata, T., and R. Ishihara. 1984. Infection and development of *Nosema bombycis* (Microsporida: Protozoa) in a cell line of *Antheraea eucalyptii*. Journal of Invertebrate Pathology 44:52–62.

Kawiak, J. 1955. Reakcja podniebienia zaby Rana esculenta L. na uzadlenie pszczoly *Apis mellifera* L. [Reaction of the palate of the frog *Rana esculenta* to the sting of the honey bee *Apis mellifera* L.]. Zoologica Poloniae 6:209–215.

Kaya, H. K., and R. Gaugler. 1993. Entomopathogenic nematodes. Annual Review of Entomology 38:181–206.

Kaya, H. K., J. M. Marston, J. E. Lindegren, and Y.-S. Peng. 1982. Low susceptibility of the honey bee, *Apis mellifera* L., to the entomogenous nematode, *Neoaplectana carpocapsae* Weiser. Environmental Entomology 11:920–924.

Keats, F. 1970. Our cover picture. South African Bee Journal 42(5):1.

Keck, C. B. 1949. The disappearance of American foulbrood in Hawaii. American Bee Journal 89:514–515.

Kefuss, J. A. 1995. Honey bee hygienic behavior: France, Tunisia and Chile. Apidologie 26:325–327.

Kellner, N. 1980. Studie van de levenscyclus van *Nosema apis* Zander in de honingbij (*Apis mellifera* L.). Dissertation, Rijksuniversitet Gent, Belgium.

Kellner, N., and F. J. Jacobs. 1978. In hoevel tijd bereiken de sporen van *Nosema apis* Zander de ventriculus van de honingbij (*Apis mellifera* L.)? [How long do the spores of *Nosema apis* Zander take to reach the ventriculus of the honey bee (*Apis mellifera* L.)?]. Vlaams Diergeneeskundig Tijdschrift 47:252–259.

Kellog, C. B. 1941. Some characteristics of the Oriental honeybee, *Apis indica* F., in China. Journal of Economic Entomology 34:717–719.

Kellogg, C. R. 1959. Chinese-Japanese bees. In The ABC and XYZ of bee culture. A. I. Root, Medina, Ohio.

Kelly, H. L. 1925. Bees drunk on lime-tree nectar attacked by murderous wasps. American Bee Journal 65:69.

Kempff Mercado, N. 1952. Enimigos del apiario: algunas abejas indigenas [Enemies of the hive bee: some indigenous wild bees]. Campo (La Paz) 5:5–7. *From* Apicultural Abstracts 173/53.

———. 1957. Control de las abejas indigenas indeseables [Control of undesirable indigenous bees]. Gaceta del Colmenar 19:89–90.

Kent, R. B. 1979. Technical and financial aspects of fixed comb and movable frame beekeeping in Costa Rica. American Bee Journal 119:36–38, 43.

Kenward, F. 1932. Foul brood. British Bee Journal 60:290–291.

Kerr, W. E. 1957. Introdução de abelhas africanas no Brasil [Introduction of African bees into Brazil]. Brasil Apiculture 3:211–213. *From* Apicultural Abstracts 184/58.

———. 1967. Multiple alleles and genetic load in bees. Journal of Apicultural Research 6:61–64.

Kessler, W. 1987. The therapy of Braula infestation (*Braula coeca*) in honeybee colonies with Perizin. Veterinary Medical Review, German Federal Republic (1): 54-57.

Khachatourians, G. G. 1991. Physiology and genetics of entomopathogenic fungi. *In* Handbook of applied mycology. Volume 2, Humans, animals, and insects. D. K. Arora, L. Ajello, and K. G. Murkeyji, editors. Marcel Dekker, New York.

Kharizanov, A. and K. Radeva. 1979. [Syrphid flies] Rastitelna Zashchita 27(10): 27-29.

Khristov, G., and S. Mladenov. 1961. Honey in surgical practice: the antibacterial properties of honey [in Bulgarian]. Khirurgiya 14:937–946.

Khrust, I. I. 1978. Thermal treatment during varroatosis [in Russian]. Pchelovodstvo 1978(6):5–8.

Killion, E. E., and L. A. Lindenfelser. 1988. Observations on the honey bee tracheal mite in Illinois. *In* Africanized honey bees and bee mites. G. R. Needham, R. E. Page, M. Delfinado-Baker, and C. E. Bowman, editors. Ellis Horwood Ltd., Chichester, England.

King, A. 1987. Saskatchewan experiments with *Acarapis woodi* mite. American Bee Journal 127:651–652.

Kingsbury, J. M. 1964. Poisonous plants of the United States and Canada. Prentice- Hall, Englewood Cliffs, New Jersey.

———. 1965. Deadly harvest, a guide to poisonous plants. Holt, Rinehart, and Winston, New York.

Kirchner, W. H., and K. Sommer. 1992. The dance language of the honeybee mutant *diminutive wings*. Behavioral Ecology and Sociobiology 30:181–184.

Kirkor, S., and Z. Czyz. 1958. Przypadek zakaznego rozproszkowywania sie czerwiu [A case of infectious powdery scale in the honeybee]. Pszczelnicze Zeszyty Naukowe 2:67–70.

Kish, L. P. 1977. Comparação de adelha Italiana com a adelha africanizada quanto a resistencia à doença da cria [Comparison of Italian bees with Africanized bees for resistance to brood diseases]. Anais do 4 Congresso Brasileiro de Apicultura, 1976, Curitiba, PR, Brazil.

———. 1980. Spore germination of *Ascosphaera* spp. associated with the alfalfa leafcutting bee, *Megachile rotundata*. Journal of Invertebrate Pathology 36:125–128.

———. 1983. The effect of high temperatures on spore germination of *Ascosphaera aggregata*. Journal of Invertebrate Pathology 42:244–248.

Kish, L. P., N. A. Bowers, G. L. Benny, and J. W. Kimbrough. 1988. Cytological development of *Ascosphaera atra*. Mycologia 80:312–319.

Kish, L. P., and P. M. Panlasigui. 1985. Effects of selected chemical treatments on spore germination of *Ascosphaera aggregata* (Ascosphaerales). Environmental Entomology 14:424–426.

Kitaoka, S. 1983. Notes on previous *Varroa* infestations and the recent occurrence of deformed honey bees in Japan [in Japanese]. Honeybee Science 4:105–108.

Kitprasert, C. 1984. Biology and systematics of the parasitic bee mite, *Tropilaelaps clareae* Delfinado and Baker (Acarina: Laelapidae). M.S. thesis, Kasetsart University, Thailand. *From* Apicultural Abstracts 1341/85.

Kjer, K. M., D. W. Ragsdale, and B. Furgala. 1989. A retrospective and prospective overview of the honey bee tracheal mite, *Acarapis woodi* R., Parts 1 & 2. American Bee Journal 129:25–28 and 112–115.

Kleinschmidt, G. J., and F. Furguson. 1989. Honey bee protein fluctuations in the Channel Country of South West Queensland. Australasian Beekeeper 91:163–165.

Klöcker, A. 1907. Handbuch der technischen Mykologie [Handbook of technical mycology]. F. Lafar, editor. C. Griffin, London.

Kloft, W., and P. Schneider. 1969. Gruppenverteidigungsverhalten bei wildlebenden Bienen (*Apis cerana* Fabricius) in Afghanistan [Group defensive behaviour in *Apis cerana* Fabricius living wild in Afghanistan]. Naturwissenschaften 56:219.

Knepp, T. H. 1987. Build a bear fence. Gleanings in Bee Culture 115(7):400–402, 405.

Knowlton, G. F. 1944. Poisoning of honey bees. Utah Agricultural Experiment Station Mimeo #310. 11 pp.

———. 1947. Boxelder bug nymphs feeding on dead honeybees. Journal of Economic Entomology 40:915.

———. 1951. Corrodentia in bee hives. Bulletin of the Brooklyn Entomological Society 46:134.

Knox, D. A., H. Shimanuki, and D. M. Caron. 1976. Ethylene oxide plus oxytetracycline for the control of American foulbrood in honey bees. Journal of Economic Entomology 69:606–608.

Knutson, L. W. and W. L. Murphy. 1990. Insect: Diptera (Flies). *In* Honey bee pests, predators and diseases, 2nd ed. R. A. Morse and R. Nowogrodzki, eds. Comstock Publishing:121-134.

Koenig, J. P., G. M. Boush, and E. H. Erickson, Jr. 1986. Effect of type of brood comb on chalk brood disease in honeybee colonies. Journal of Apicultural Research 25:58-62.

Koeniger, G., N. Koeniger, and S. Tingek. 1994. Mating flights, number of spermatozoa, sperm transfer, and degree of polyandry in *Apis koschevnikovi*. Apidologie 25:224-238.

Koeniger, N. 1986. A new approach to chemotherapy of varroatosis. Arbeitsgemein-schaft der Institute für Bienenforschung E.V. Abstracts of the Varroa Workshop in Feldafing/Starnberg, West Germany.

Koeniger, N. 1987. Die ostliche Honigbiene und ihre Milbe *Varroa jacobsoni* [The eastern honey bee and its mite *Varroa jacobsoni*]. Imkerfreund 42:303-306.

Koeniger, N., and S. Fuchs. 1989. Eleven years with Varroa—experiences, retrospects and prospects. Bee World 70:148-149.

Koeniger, N., G. Koeniger, and M. Delfinado-Baker. 1983. Observations on mites of the Asian honeybee species (*Apis cerana, Apis dorsata, Apis florea*). Apidologie 14:197-204.

Koeniger, N., G. Koeniger, L. I. De Guzman, and C. Lekprayoon. 1993. Survival of Euvarroa sinhai Delfinado and Baker (Acari, Varroidae) on workers of *Apis cerana* Fabr., *Apis florea* Fabr. and *Apis mellifera* L. in cages. Apidologie 24:403-410.

Koeniger, N., G. Koeniger, S. Tingek, M. Mardan, and T. E. Rinderer. 1988. Reproductive isolation by different time of drone flight between *Apis cerana* Fabricius, 1793 and *Apis vechti* (Maa, 1953). Apidologie 19:103-106.

Koeniger, N., G. Koeniger, and N. H. P. Wijayagunasekara. 1981. Beobachtungen über die Anpassung von *Varroa jacobsoni* an ihren natürlichen Wirt *Apis cerana* in Sri Lanka [Observations on the adaptation of *Varroa jacobsoni* to its natural host *Apis cerana* in Sri Lanka]. Apidologie 12:37-40.

Koivulehto, K. 1976. *Varroa jacobsoni*—a new mite infesting honeybees in Europe. British Bee Journal 104:16-17.

Komeili, A. B. 1988. The impact of the Varroa mite on Iranian commercial beekeeping. American Bee Journal 128:423-429.

Komeili, A. B., and J. T. Ambrose. 1990. Biology, ecology and damage of tracheal mites on honey bees (*Apis mellifera*). American Bee Journal 130:193-198.

Komeili, A. B., and J. T. Ambrose. 1991. Electron microscope studies of the tracheae and flight muscles of noninfested, *Acarapis woodi* infested, and crawling honey bees (*Apis mellifera*). American Bee Journal 131:253-257.

Komissar, A. D. 1978. Device for heat treatment of varroatosis [in Russian]. Pchelovodstvo 1978(11):18-20.

Koover, C. J. 1949. An ant proof hive stand. Gleanings in Bee Culture 77:149.

Koptev, V. 1948. The reasons for the summer deaths of bees in Siberia [in Russian]. Pchelovodstvo 1948(10):45-47.

———. 1957. Laying workers and swarming [in Russian]. Pchelovodstvo 1957(6): 31-32.

Korpela, S., A. Aarhus, I. Fries, and J. Hansen. 1992. *Varroa jacobsoni* Oud. in cold climates: population growth, winter mortality and influence on the survival of honey bee colonies. Journal of Apicultural Research 31:157-164.

Korschgen, L. J., and D. L. Moyle. 1955. Food habits of the bullfrog in central Missouri farm ponds. American Midland Naturalist 54:332-341.

Kovac, H., and K. Crailsheim. 1988. Lifespan of *Apis mellifera* carnica Pollm. infested by *Varroa jacobsoni* Oud. in relation to season and extent of infestation. Journal of Apicultural Research 27:230-238.

Kowalska, M. 1984a. Wlasciwosci biochemiczne *Ascosphaera apis* i *Bettsia alvei* [Biochemical properties of *Ascosphaera apis* and *Bettsia alvei*]. Polskie Archiwum Weterynaryjne 24:7-15. (English summary)

———. 1984b. Wrazliwosc "in vitro" *Ascosphaera apis* na chemioterapeutyki [In vitro sensitivity of *Ascosphaera apis* to chemotherapy]. Polskie Archiwum Weterynaryjne 24:165-172. (English summary)

Kramer, J. P. 1960. Observation on the emergence of the microsporidian sporoplasm. Journal of Insect Pathology 2:433–439.

Kramer, V. 1902. Eine Novität [A novelty]. Schweizerische Bienen-Zeitung 25:322–323.

Kraus, B. 1990. Effects of honey-bee alarm pheromone compounds on the behaviour of *Varroa jacobsoni*. Apidologie 21:127–134.

———. 1991. Zwischenbericht zur Winterbehandlung mit Michsäure als Varroatosetherpeutikum [Preliminary information concerning winter treatment for varroa]. Apidologie 22:473–475.

———. 1992. Weitere Ergebnisse zur Michlsäurebehandlung als Varroatose Therapie [Further results concerning formic acid treatment for varroa disease therapy]. Apidologie 23:385–387.

Kraus, B., N. Koeniger, and S. Fuchs. 1994. Screening of substances for their effect on *Varroa jacobsoni*: attractiveness, repellency, toxicity, and masking effects of etheral oils. Journal of Apicultural Research 33:34–43.

Kraus, B., and R.E. Page, Jr. 1995. Effect of *Varroa jocobsoni* (Mesostigmata: Varroidae) on Feral *Apis mellifera* (Hymenoptera: Apidae) in California. Population Ecology 24: 1473-1480.

Kraus, B., and R. E. Page, Jr. 1995. Population growth of *Varroa jacobsoni* Oud. in Mediterranean climates of California. Apidologie 26:149–159.

Krieg, A. 1963. Rickettsia and rickettsiosis. *In* Insect pathology: an advanced treatise, volume 1. E. A. Steinhaus, editor. Academic Press, New York.

Krombein, K. V. 1981. Biosystematic studies of Ceylonese wasps, VIII: A monograph of the Philanthidae (Hymenoptera: Sphecoidea). Smithsonian Contributions to Zoology number 343. Smithsonian Institution Press, Washington, D.C.

Kropácová, S., and H. Haslbachová. 1969. The development of ovaries in worker honeybees in a queenright colony. Journal of Apicultural Research 8:57–64.

Kropácová, S., and H. Haslbachová. 1970. The development of ovaries in worker honeybees in queenright colonies examined before and after swarming. Journal of Apicultural Research 9:65–70.

Kropácová, S., and H. Haslbachová. 1971. The influence of queenlessness and of unsealed brood on the development of ovaries in worker honeybees. Journal of Apicultural Research 10:57–61.

Kshirsagar, K. K. 1971. A wasp trap to control predator wasps in the apiary. Indian Bee Journal 33:55–57.

———. 1981. Morphometric studies on *Apis cerana indica* F. workers. III. Effect of geographical position on morphometric characters. Indian Bee Journal 43:1–5.

Kshirsagar, K. K., and D. B. Mahindre. 1975. Some notes on bee-predator wasps in India. Indian Bee Journal 37:4–9.

Kubasek, K. J., T. E. Rinderer, and W. R. Lee. 1980. Isogenic sperm line maintenance in the honey bee. Journal of Heredity 71:278–280.

Kudo, R. 1920. Notes on *Nosema apis* (Zander). Journal of Parasitology 7:85–90.

Kulikov, N. S. 1961. Treating bees with fumagillin for nosema infection [in Russian]. Pchelovodstvo 1961(5):43–44.

Kulincevic, J. M., and T. E. Rinderer. 1986. Differential survival of honey bee colonies infected by *Varroa jacobsoni* and breeding for resistance. Proceedings of the International Apicultural Congress (Japan, 1985) 30:175–177.

Kulincevic, J. M., and T. E. Rinderer. 1988. Breeding honey bees for resistance to *Varroa Jacobsoni*: analysis of mite population dynamics. In Africanized honey bees and bee mites. G. R. Needham, R. E. Page, Jr., M. Delfinado-Baker, and C. E. Bowman, editors. Ellis Horwood, Chichester, England.

Kulincevic, J. M., T. E. Rinderer, and V. J. Mladjan. 1991. Effects of fluvalinate and amitraz on bee lice (*Braula coeca* Nitzsch) in honey bee (*Apis mellifera* L.) colonies in Yugoslavia. Apidologie 22(1): 43-47.

Kulincevic, J. M., T. E. Rinderer, V. J. Mladjan, and S. M. Buco. 1991. Control of *Varroa Jacobsoni* in honey bee colonies in Yugoslavia by fumigation with low doses of Fluvalinate or Amitraz. Apidologie 22:147–153.

Kulincevic, J. M., T. E. Rinderer, V. J. Mladjan, and S. M. Buco. 1992. Five years of bi-directional genetic selection for honey bees resistant and susceptible to *Varroa jacobsoni*. Apidologie 23:443–452.

Kulincevic, J. M., T. E. Rinderer, and D. J. Urosevic. 1988. Seasonality and colony variation of reproducing and non-reproducing *Varroa jacobsoni* females in Western honey bee (*Apis mellifera*) worker brood. Apidologie 19:173–179.

Kulincevic, J. M., and W. C. Rothenbuhler. 1975. Selection for resistance and susceptibility to hairless-black syndrome in the honeybee. Proceedings of a Meeting of the EC-Experts Group, Udine, Italy, 1988:171–174.

Kumar, R., N. R. Kumar, and O. P. Bhalia. 1993. A comparative study on the response of *Apis* species to predation by *Vespa* species. In Asian apiculture. L. J. Connor, T. Rinderer, H. A. Sylvester, and S. Wongsiri, editors. Wicwas Press, Cheshire, Connecticut.

Kumar, N. R., R. Kumar, J. Mbaya, and R. W. Mwangi. 1993. *Tropilaelaps clareae* found on *Apis mellifera* in Africa. Bee World 74:101–102.

Kumbhojkar, M. S. 1972. *Schizosaccharomyces slooffiae* Kumbhojkar—a new species of osmophilic yeast from India. Current Science 38:347–348.

———. 1978. *Saccharomyces locheadii* Kumbhojkar—a new species of osmophilic yeast from Indian honey. Biovigyanam 4:169–171.

———. 1981. New taxa of Indian osmophilic yeasts. Biovigyanam 7:97–105.

Kunchev, K., B. Purvulov, M. Duparinova, and K. Gurgulova. 1983. Outbreaks of severe mycoses amongst honeybees [in Bulgarian, English summary]. Veterinarna Sbirka 81:36–38.

Kunike, G. 1930. Zur Biologie der kleinen Wachsmotte, *Achroia grisella* Fabr. [The biology of the small wax moth, *Achroia grisella* Fabr.]. Zeitschrift für Angewandte Entomologie 16:315–356.

Kurczewski, F. E., and R. C. Miller, 1983. Nesting behavior of *Philanthus sanborni* in Florida (Hymenoptera: Sphecidae). Florida Entomologist 66:199–206.

Kuz'mina, L. A., N. G. Lopatina, and V. V. Ponomarenko. 1979a. Neurological effect of the brick mutation in the honey bee. Doklady Biological Sciences 245:789–791.

Kuz'mina, L. A., N. G. Lopatina, and V. V. Ponomarenko. 1979b. Kynurenine in hereditary disorders of the function of the nervous system and behavior of the honeybee. Doklady Biological Sciences 245:800–802.

Laere, O., van. 1977. Factors influencing the germination of *Nosema apis* spores. *In* Biological aspects of nosema disease. Symposium of bee biology and pathology. Merelbeke, Belgium, July 14–16. Apimondia, Bucharest.

Lahitte, J. D. 1986. Perizin - Amitraz [effectiveness in controlling *Varroa jacobsoni* and *Braula coeca*]. Revue Francaise d'Apiculture (No. 452): 258-259.

———. 1987. Study of Perizin and its effects [A propos du Perizin]. Sante de l'Abeille (No. 98): 54-61.

Laidlaw, H. H., Jr., and J. E. Eckert. 1962. Queen rearing. 2d ed. University of California Press, Berkeley.

Laidlaw, H. H., and M. A. el-Banby. 1962. Inhibition of yellow color in the honey bee *Apis mellifera* L. Journal of Heredity 53:171–173.

Laidlaw, H. H., M. A. el-Banby, and K. W. Tucker. 1965. Three wing mutants of the honey bee. Journal of Heredity 56:84–88.

Laidlaw, H. H., Jr., F. P. Gomes, and W. E. Kerr. 1956. Estimation of the number of lethal alleles in a panmictic population of *Apis mellifera* L. Genetics 41:179–188.

Laidlaw, H. H., Jr., and R. E. Page, Jr. 1984. Polyandry in honey bees (*Apis mellifera* L.): sperm utilization and intracolony genetic relationships. Genetics 108:985–997.

Laidlaw, H. H., and K. W. Tucker. 1964. Diploid tissue derived from accessory sperm in the honey bee. Genetics 50:1439–1442.

Laidlaw, H. H., and K. W. Tucker. 1965. Three mutant eye shapes in honey bees. Journal of Heredity 56:190–192.

Laigo, F. M., and R. A. Morse. 1968. The mite *Tropilaelaps clareae* in *Apis dorsata* colonies in the Philippines. Bee World 49:116–118.

Laigo, F. M., and R. A. Morse. 1969. Control of the bee mites *Varroa jacobsoni* Oudemans and *Tropilaelaps clareae* Delfinado and Baker with chlorobenzilate. Philippine Entomologist 1:144–148.

Lampe, K. F. 1985. AMA handbook of poisonous and injurious plants. American Medical Association, Chicago Review Press, Chicago.

Landerkin, G. B., and H. Katznelson. 1959. Organisms associated with septicemia in the honeybee, *Apis mellifera*. Canadian Journal of Microbiology 5:169–172.

Langhe, A. B., K. V. Natskii, and V. M. Tatsii. 1976. The varroa mite and the methods of controlling it [in Russian]. Pchelovodstvo 1976(3):16–20.

Langridge, D. F. 1961. Nosema disease of the honeybee and some investigations into its control in Victoria, Australia. Bee World 42:36–40.

———. 1966. Flagellated protozoa (Trypanosomidae) in the honey bee *Apis mellifera*, in Australia. Journal of Invertebrate Pathology 8:124–125.

Langridge, D. F., and R. B. McGhee. 1967. *Crithidia mellificae* n. sp., an acidophilic trypanosomatid of the honey bee *Apis mellifera*. Journal of Protozoology 14:485–487.

Langstroth, L. L. 1860. The hive and the honeybee. 3d ed. C. M. Saxton, Barker, New York.

L'Arrivée, J. C. M. 1963. The effect of sampling sites on the *Nosema* determination. Journal of Insect Pathology 5(3):349–355.

———. 1965a. Sources of nosema infection. American Bee Journal 105:246–248.

———. 1965b. Tolerance of honey bees to nosema disease. Journal of Invertebrate Pathology 7:408–413.

L'Arrivée, J. C. M., and R. Hrytsak. 1964. Coprological examination for nosematosis in queen bees. Journal of Insect Pathology 6:126–127.

Larsson, R. 1986. Ultrastructure, function, and classification of Microsporidia. Progress in Protistology 1:325–390.

Larsson, R. J. I., C. Bach de Roca, and M. Gaju-Ricart. 1992. Fine structure of an amoeba of the genus *Vahlkampfia* (Rhizopoda, Vahlkampfiidae), a parasite of the gut epithelium of the bristletail *Promesomachilis hispanica* (Microcoryphia, Machilidae). Journal of Invertebrate Pathology 59:81–89.

Latham, A. 1922. Some more about catalepsy in queens. American Bee Journal 62:163.

———. 1939. Bees and spiders. American Bee Journal 79:251.

Laurence, G. A. 1973. Pests of honey bees in Trinidad. Journal of the Agricultural Society of Trinidad and Tobago 73:461–463.

Laurence, G., and I. Mohammed. 1974. The bee louse (*Braula coeca*) in Trinidad and Tobago. Journal of the Agricultural Society of Trinidad and Tobago 74:378–379.

Lavie, P. 1951. Investigations on a mysterious micro-organism, *Acaromyces laviae*, which appears to be a parasite of *Acarapis woodi*. Proceedings of the 14th International Beekeeping Congress, paper 20.

———. 1952. First results in the experimental treatment of acarine disease by *Acaromyces*. Apiculteur 96(2)Sect. sci.:1–8. From Apicultural Abstracts 20/53.

Lawrence, W. B. 1986. Infant botulism and its relation to honey: a review. American Bee Journal 126:484–486.

Le Conte, Y., G. Arnold, J. Trouiller, C. Masson, B. Chappe, and G. Ourisson. 1989. Attraction of the parasitic mite *Varroa* to the drone larvae of honey bees by simple aliphatic esters. Science 245:638–639.

Le Conte, Y., C. Bruchou, K. Benhamouda, C. Gauthier, and J. M. Cornuet. 1994. Heritability of the queen brood post-capping stage duration in *Apis mellifera mellifera* L. Apidologic 24:513–519.

Le Conte, Y., and J. M. Cornuet. 1989. Variability of the postcapping stage duration of the worker brood in three different races of *Apis mellifera*. Proceedings of a meeting of the EC-experts group, Udine, Italy, 1988.

Lecq, T. 1933. Le bousier [The dung beetle]. Apiculteur 77:138–140.

Lee, D. C. 1963. The susceptibility of honey bees of different ages to infestation by *Acarapis woodi* (Rennie). Journal of Insect Pathology 5:11–15.

Lee, P. E., and B. Furgala. 1965a. Electron microscopy of sacbrood virus in situ. Virology 25:387–392.

Lee, P. E., and B. Furgala. 1965b. Sacbrood virus: some morphological features and nucleic acid type. Journal of Invertebrate Pathology 7:502–505.

Lee, P. E., and B. Furgala. 1967a. Viruslike particles in adult honey bees (*Apis mellifera* Linnaeus) following injection with sacbrood virus. Virology 32:11–17.

Lee, P. E., and B. Furgala. 1967b. Electron microscopic observations on the localization and development of sacbrood virus. Journal of Invertebrate Pathology 9:178–187.

Lehnert, T. 1978. Nematodes. *In* Honey bee pests, predators, and diseases. R. A. Morse, editor. Cornell University Press, Ithaca, New York.

Lehnert, T., A. S. Michael, and M. D. Levin. 1974. Disease survey of South American africanized bees. American Bee Journal 114:338.

Lehnert, T., and H. Shimanuki. 1973a. The dosage of ethylene oxide necessary to control *Nosema apis* Zander on honey combs at 100°F. American Bee Journal 113:296.

Lehnert, T., and H. Shimanuki. 1973b. Production of nosema-free bees in the South. American Bee Journal 113:381–382.

Lehnert, T., H. Shimanuki, and D. Knox. 1973. Transmission of nosema disease from infected workers of the honey bee to queens in mailing cages. American Bee Journal 113:413–414.

LeMaistre, W. G. 1934. The wax moth in comb honey. Canadian Bee Journal 42:117–119.

Lengler, S. 1977. Comparação da abelha italiana com a abelha africanizada quanto a resistência à doença da cria [Comparison of Italian and africanized bees for resistance to brood diseases]. Anais do IV Congresso Brasileiro de Apicultura, Curitiba, 1976. L. S. Gonçalves, editor. Paraná, Brazil.

Lensky, Y. 1964. Résistance des abeilles (*Apis mellifica* L. var. *ligustica*) à des températures élévées [Resistance of honey bees (*Apis mellifica* L. var. *ligustica*) to high temperatures]. Insectes Sociaux 11:293–299.

Leonard, F. W. 1983. Investigation on the fauna associated with pollen collected by honey bees, *Apis mellifera* L. M.S. thesis, University of Maryland, College Park.

Leonard, F. W., H. Shimanuki, and C. G. Reichelderfer. 1983. A survey of arthropod contamination of front-mounted pollen traps. American Bee Journal 123:872–873.

Lescure, J. 1966. Le comportement prédateur du crapaud commun (*Bufo bufo*) envers les abeilles [The predatory behavior of the common toad (*Bufo bufo*) toward bees]. Annales de l'Abeille 9:83–114.

Levin, M. D. 1960. A comparison of two methods of mass-marking foraging honey bees. Journal of Economic Entomology 53:696–698.

Levine, N. D., J. O. Corliss, F. E. G. Cox, G. Deroux, J. Grain, B. M. Honigberg, G. F. Leedale, A. R. Loeblich III, J. Lom, D. Lynn, E. G. Merinfeld, F. C. Page, G. Poljansky, V. Sprague, J. Vavra, and F. G. Wallace. 1980. A newly revised classification of the Protozoa. Journal of Protozoology 27:37–58.

Levine, N. D. 1982. Apicomplexa. In McGraw Hill synopsis and classification of living organisms. S. P. Parker, editor. McGraw-Hill, New York.

Lewis, L. F. and W. C. Rothenbuhler. 1961. Resistance to American foulbrood in honey bees: III. Differential survival of the two kinds of larvae from two-drone matings. Journal of Insect Pathology 3:197–215.

Leyrer, R. L., and R. E. Monroe. 1973. Isolation and identification of scent of the moth *Galleria mellonella*, and a reevaluation of its sex pheromone. Journal of Insect Physiology 19:2267–2271.

Lian, C. Z. 1980. Nosema disease in honeybee (*Apis cerana cerana*) [in Chinese]. Apiculture in China 4:15–16.

Lim, B. H., M. L. Lee, and K. S. Woo. 1989. Studies on the control of hornets (Vespa spp.) by means of feeding attractants. Korean Journal of Apiculture 4:19–33.

Lin, H., G. W. Otis, M. E. Nasr, and C. D. Scott-Dupree. 1992. Comparative tracheal mite resistance of Canadian and Buckfast bees. American Bee Journal 132:810.

Lindauer, M. 1956. Über die Verständigung bei indischen Bienen. Zeitschrift für Vergleichende Physiologie 38:521–557.

Linder, A. 1947. Über das Auswerten zahlenmässiger Angaben in der Bienenkunde. Beihefte zur Schwezerischen Bienen-Zeitung 2:77–138.

Linder, K. E. 1962. Ein Beitrag zur Frage der antimikrobiellen Wirkung der Naturhonige [The antimicrobial activity of natural honeys]. Zentralblatt für Bakteriologie, Parasitenkunde, Infektionskrankheiten und Hygiene 115:720–736.

Lindquist, E. E. 1968. An unusual new species of *Tarsonemus* (Acarina: Tarsonemidae) associated with the Indian honey bee. Canadian Entomologist 100:1002–1006.

———. 1986. The world genera of Tarsonemidae (Acari: Heterostigmata): a morphological, phylogenetic, and systematic revision, with a reclassification of family-group taxa in the Heterostigmata. Memoirs of the Entomological Society of Canada 136:1–517.

Linsley, E. G. 1943. The dried fruit moth breeding in nests of the mountain carpenter bee in California. Journal of Economic Entomology 36:122–123.

———. 1960. Ethology of some bee- and wasp-killing robber flies of southeastern Arizona and western New Mexico (Diptera: Asilidae). University of California Publications in Entomology 16:357–392, pls. 48–55.

Linsley, E. G., and J. W. MacSwain. 1941. The bionomics of *Ptinus californicus*, a depredator in the nests of bees. Southern California Academy of Science Bulletin 40:126–137.

Linsley, E. G., and J. W. MacSwain. 1952. Notes on the biology and host relationships of some species of *Nemognatha* (Coleoptera: Meloidae). Wasmann Journal of Biology 10:91–102.

Lipa, J. J. 1977. Microsporidian infections of *Galleria mellonella* (L.) (Lepidoptera, Galleriidae), with the description of a new species, *Nosema galleriae* n. sp. Acta Protozoologica 16:141–149.

Lipa, J. J., and O. Triggiani. 1992. A newly recorded neogregarine (Protozoa, Apicomplexa) parasite in honey bees (*Apis mellifera*) and bumble bees (*Bombus* spp.). Apidologie 23:533–536.

Liu, T. P. 1973. Effects of Fumidil B on the spores of *Nosema apis* and on lipids of the host cell as revealed by freeze-etching. Journal of Invertebrate Pathology 22:364–368.

———. 1984a. Virus-like cytoplasmic particles associated with lysed spores of *Nosema apis*. Journal of Invertebrate Pathology 44:103–105.

———. 1984b. Ultrastructure of the midgut of the worker honey bee *Apis mellifera* heavily infected with *Nosema apis*. Journal of Invertebrate Pathology 44:282–291.

———. 1985. Scanning electron microscope observations on the pathological changes of Malpighian tubules in the worker honey bee, *Apis mellifera*, infected by *Malpighamoeba mellificae*. Journal of Invertebrate Pathology 46:125–132.

———. 1986. Comparative fine structure of the corpus allatum from healthy and nosema-infected honeybees. Journal of Apicultural Research 25:163–169.

———. 1990a. The release of *Nosema apis* spores from the epithelium of the honeybee midgut. Journal of Apicultural Research 29:221–229.

———. 1990b. Ultrastructure of the flight muscle of worker honey bees heavily infested by the tracheal mite *Acarapis woodi*. Apidologie 21:537–540.

———. 1991a. Melanosis, a little known honey bee disease. American Bee Journal 131:707.

———. 1991b. Virus-like particles in the tracheal mite *Acarapis woodi* (Rennie). Apidologie 22:213–219.

———. 1992a. Fluvalinate and its after-effects. American Bee Journal 132:398.

———. 1992b. Oöcytes degeneration in the queen honey bee after infection by *Nosema apis*. Tissue & Cell 24:131–138.

————. 1992c. Tracheal mite virus—a possible biological control? American Bee Journal 132:532–533.

————. 1995a. A possible control of chalkbrood and nosema diseases of the honey bee with neem. American Bee Journal 135:195–198.

————. 1995b. A rapid differential staining technique for live and dead tracheal mites. American Bee Journal 135:265–266.

Liu, T. P., B. Mobus, and G. Braybrook. 1989a. Fine structure of hypopharyngeal glands from honey bees with and without infestation by the tracheal mite *Acarapis woodi* (Rennie). Journal of Apicultural Research 28:85–92.

Liu, T. P., B. Mobus, and G. Braybrook. 1989b. A scanning electron microscope study on the prothoracic tracheae of honeybees, *Apis mellifera* L. infested by the mite *Acarapis woodi* (Rennie). Journal of Apicultural Research 28:81–84.

Liu, T. P., and M. Nasr. 1992. Effects of formic acid treatment on the infestation of tracheal mites, *Acarapis woodi* (Rennie), in the honey bee, *Apis mellifera* L. American Bee Journal 132:666–668.

Liu, T. P., and M. E. Nasr. 1993. Preventive treatment of tracheal mites, *Acarapis woodi* (Rennie) with vegetable oil extender patties in the honey bee, *Apis mellifera* L. colonies. American Bee Journal 133:873–875.

Lloyd, J. M. 1986. Simplified laboratory diagnosis of American foulbrood disease. Journal of Apicultural Research 25:55–57.

Lobo, J. A., M. A. Del Lama, and M. A. Mestriner. 1989. Population differentiation and racial admixture in the Africanized honeybee (*Apis mellifera* L.). Evolution 43:794–802.

Lochhead, A. G. 1928. The etiology of European foul-brood. Science 67:159–160.

————. 1933. Factors concerned with the fermentation of honey. Zentralblatt für Bakteriologie, Parasitenkunde, Infektionskrankheiten und Hygiene 88:298–302.

————. 1937. The nitrate reduction test and its significance in the detection of Bacillus larvae. Canadian Journal of Research, Section C 15:79–86.

————. 1942. *Zygosaccharomyces nectarophilus* n. sp. and *Zygosaccharomyces rugosus* n. sp. Canadian Journal of Research 20:89–91.

Lochhead, A. G., and L. Farrell. 1930. Soil as a source of infection of honey by sugar-tolerant yeasts. Canadian Journal of Research 3:51–64.

Lochhead, A. G., and L. Farrell. 1931. The types of osmophilic yeasts found in normal honey and their relation to fermentation. Canadian Journal of Research 5:665–672.

Lochhead, A. G., and D. A. Heron. 1929. Microbiological studies of honey. I. Honey fermentation and its cause. II. Infection of honey by sugar-tolerant yeasts. Canada Department of Agriculture Bulletin number 116.

Lochhead, A. G., and N. B. McMaster. 1931. Yeast infection of normal honey and its relation to fermentation. Scientific Agriculture 11:351–360.

Lodder, J., editor. 1970. The yeasts. North-Holland Publishing, Amsterdam.

Lodesani, M., A. Pellacani, S. Bergomi, E. Carpana, T. Rabitti, and P. Lasagni. 1992. Residue determination for some products used against Varroa infestation in bees. Apidologie 23:257–272.

Loglio, G. 1993. Probable apparacion de resitencias al fluvalinato [The appearance of resistance to fluvalinate]. Vida Apicola 60:16–18.

Loh, P. C., M. A. Dow, and R. S. Fujioka. 1985. Use of the nitrocellulose enzyme immunosorbent assay for rapid, sensitive and quantitative detection of human enteroviruses. Journal of Virological Methods 12:225–234.

Lolin, M. 1977. Prvi primer varoatoze ugotovljen v nasi drzavi [The first case of varroa disease found in Czechoslovakia]. Slovenski Cebelar 79:15. From Apicultural Abstracts 1039L/79.

Lom, J. 1964. The occurrence of a *Crithidia*-species within the gut of the honeybee, *Apis mellifera* L. *In* Colloque international sur la pathologie des insectes et la lutte microbiologique, Paris. Institut Pasteur, Service de Parasitologie Végétale, Paris.

Lom, J. and J. Vávra. 1963. The mode of sporoplasm extrusion in microsporidian spores. Acta Protozoologica 1:81–89.

Lomholdt, O. 1975. The Sphecidae (Hymenoptera) of Fernoscandia and Denmark. Fauna Entomologica Scandinavica, volume 4, part 1.

Lommel, S. A., T. J. Morris, and D. E. Pinnock. 1985. Characterization of nucleic acids associated with Arkansas bee virus. Intervirology 23:199–207.

Londt, J. G. H. 1993. Afrotropical robber fly (Diptera: Asilidae) predators of honey bees, *Apis mellifera* Linnaeus (Hymenoptera: Apidae). African Entomology 1(2): 167–173.

Longo, S. 1980. La difesa degli alveari dalle vespe [Control of injurious wasps]. Apicoltore Moderno 71:109–112.

Lopatina, N. G., E. G. Chesnokova, L. Z. Dolotovskaya, A. V. Fel'cher, and J. Woyke. 1984. Influence of the snow mutation on the functional activity of the nervous system and on the behaviour of the honey bee. Soviet Genetics 19:1276–1283.

Lopatina, N. G., and L. Z. Dolotovskaya. 1985. Role of serotonin in the behavioral and neurobiological effects of the snow and snow mutations of the bee *Apis mellifera*. Journal of Evolutionary Biochemistry and Physiology 20:249–252.

Loper, G. M. 1995. Some attributes of the Africanized honey bees in southern Arizona: wing length, hygienic behavior, worker emergence time and brood nest temperatures. American Bee Journal 135:828.

Loper, G. M., G. D. Waller, D. Steffens, and R. M. Roselle. 1992. Selection and controlled natural mating: a solution to the honey bee tracheal mite problem. American Bee Journal 132:603–606.

Lord, W. 1994. Beekeeping in Jericho and Gaza. American Bee Journal 134:43–45.

Lord, W. G. 1980. Bees, bears and a new bear fence. Gleanings in Bee Culture 108:323–325.

Lord, W. G., and J. T. Ambrose. 1981a. Bear depredation of bee hives in the United States and Canada. American Bee Journal 121:811–815.

Lord, W. G., and J. T. Ambrose. 1981b. Black bear depredation of beehives in North Carolina, 1977–1979. American Bee Journal 121:421–423.

Lord, W. G., and S. K. Nagi. 1987. *Apis florea* discovered in Africa. Bee World 68:39–40.

Lotmar, R. 1936. Nosema-Infektion und ihr Einfluss auf due Entwicklung der Futtersaftdrüse. Schweizerische Biengen-Zeitung 59:33–36, 100–104.

———. 1939. Der Eiweiss-Stoffwechsel im Bienenvolk (*Apis mellifica*) während der Überwinterung. Landwirtschhaftisches Jahrbuch der Schweiz 54:34–70.

———. 1940. Beiträge zur Pathölogie des Bienendarmes C. Über den infektionsverlauf und die Vermehrung des Parasiten *Nosema apis* (Quantitative Untersuchungen). Landwirtschaftliches Jahrbuch der Schweiz 54:775–805.

———. 1943. Über den Einfluss der Temperatur auf den Parasiten *Nosema apis* [On the effect of temperature on the parasite *Nosema apis*]. Beihefte Schweizerische Bienen-Zeitung 1:261–284.

———. 1946. Über Flagellaten und Bakteria im Dünndarm der Honigbiene [Flagellates and bacteria of the small intestine of the honey bee]. Schweizerische Bienen-Zeitung Beihefte 2:49–76.

Loubès, C. 1979. Recherches sur la méiose chez les Microsporidies: Conséquences sur les cycles biologiques [Research on meiosis in microsporidians: consequences of biological cycles]. Journal of Protozoology 26:200–213.

Louveaux, J. 1983. Thoughts and suggestions on varroatosis. *In* Proceedings of a meeting of the European Community Experts' Group, Wageningen, Netherlands. R. Cavalloro, editor. Balkema, Rotterdam.

Louveaux, J., and P. Lavie. 1949. Un cas de toxicité des fleurs du tilleul argenté pour les abeilles [Toxicity of the flowers of white linden to bees]. L'Apiculteur 93(1):9–12.

Love, J. L. 1990. Toxic honey—a New Zealand story. Analytical Proceedings 27:87–89.

Lovell, J. H. 1926. Honey plants of North America. A. I. Root, Medina, Ohio.

———. 1935. Poisonous honey plants. Gleanings in Bee Culture 63:405–406.

Lozano, L. G., J. O. Moffett, B. Campos-P., M. Guillen-M., O. N. Perez-E., D. L. Maki, and W. T. Wilson. 1989. Tracheal mite *Acarapis woodi* (Rennie) (Acari: Tarsonemidae) infestations in the honey bee, *Apis mellifera* L. (Hymenoptera: Apidae) in Tamaulipas, Mexico. Journal of Entomological Science 24:40–46.

Lunder, R. 1972. Undersøkelse av kalkyngel i 1971 [Investigations on chalkbrood in 1971]. Birøkteren 88:55–60.

Lundie, A. E. 1940. The small hive beetle *Aethina tumida*. Union of South Africa Department of Agriculture and Forestry Science Bulletin number 220.

———. 1951 and 1952. The principal diseases and enemies of honeybees. South African Bee Journal 26(4):15–16; 26(5):13, 15; 26(6):13–14; 27(1): 9; 27(2):13, 15.

———. 1952. The lesser enemies and diseases of honey bees. South African Bee Journal 27(3):15–17.

———. 1965. A conopid parasite of the honey-bee discovered. South African Bee Journal 37(5):14.

Lupo, A., and D. Gerling. 1990. A comparison between the efficiency of summer treatments using formic acid and Taktic against *Varroa jacobsoni* in beehives. Apidologie 21:261–267.

M., W. H. 1930. Dodder vine fatal in Texas. American Bee Journal 70:483.

Ma, S. C. 1949. Enemies of the adult honey bee. Utah State Agricultural College, Agricultural Experiment Station, Logan, Utah, Mimeograph Series 356.

Maa, T. 1953. An inquiry into the systematics of Tribus Apidini, or honeybees (Hym.). Treubia 21:525–640.

Maassen, A. 1906. Die Aspergillusmykose der Bienen [The aspergillus mycosis of bees]. Mitteilungen aus der Kaiserlichen Biologischen Anstalt für Land- und Forstwirtschaft 2:30–31.

———. 1913. Weitere Mitteilungen über die seuchenhaften Brutkrankheiten der Bienen [Further communication on the epidemic brood diseases of bees]. Mitteilungen aus der Kaiserlichen Biologischen Anstalt für Land- und Forstwirtschaft 14:48–58.

———. 1916. Über Bienenkrankheiten [On bee diseases]. Mitteilungen aus der Kaiserlichen Biologischen Anstalt für Land- und Forstwirtschaft 16:51–58.

Mabuchi, T. 1982. Sterilizing effects of ethylene oxide gas on *Ascosphaera apis* [in Japanese, English summary]. Journal of the Japan Veterinary Medical Association 35:32–34.

MacDonald, J. F., R. D. Akre, and W. B. Hill. 1975. Nest associates of *Vespula atropilosa* and *V. pensylvanica* in southeastern Washington State. Journal of the Kansas Entomological Society 48:53–63.

MacDonald, J. L. 1971. Beekeeping in Taiwan, Republic of China. American Bee Journal 111:58–60.

MacDougall, R. S. 1914. Bee diseases and bee enemies. Transactions of the Highland and Agriculture Society of Scotland. 20pp.

Mace, H. 1927. Modern beekeeping. Wyman and Sons, London.

———. 1934. Pests. Bee World 15:100.

MacFarlane, R. P., and B. J. Donovan. 1989. *Melittobia* spp. as parasitoids of bumble and lucerne leafcutting bees and their control in New Zealand Proceedings of the 42nd New Zealand Weed and Pest Control Conference, DSIRO, Lincoln, New Zealand.

Macieira, O. J. D., J. Chaud-Netto, and A. M. Zanon. 1983. Oviposition rate and relative viability of descendants from couples of *Megaselia scalaris* (Diptera: Phoridae) reared in different experimental conditions. Revista Brasileira de Biologia 43(3): 223-228.

Mackensen, O. 1943. The occurrence of parthenogenetic females in some strains of honeybees. Journal of Economic Entomology 36:465–467.

———. 1951. Viability and sex determination in the honey bee (*Apis mellifera* L.). Genetics 36:500–509.

———. 1955. Further studies on a lethal series in the honey bee. Journal of Heredity 46:72–74.

———. 1958. Linkage studies in the honey bee. Journal of Heredity 49:99–102.

MacLean, G. J., and J. H. Davidson. 1970. Poisonous plants, a major cause of livestock disorders. Down to Earth 26(2):5–11.

Maeda, S., A. Mukai, N. Kosugi, and Y. Okada. 1962. The flavor components of honey [in Japanese, English summary]. Nihon Shokuhin Kogyo GakkaiShi/Journal of Food Science and Technology 9:270–274.

Maehr, D. S., and J. R. Brady. 1982. Florida black bear–beekeeper conflict: 1981 beekeeper survey. American Bee Journal 122:372–375.

Mages, L. 1956. Un nouveau parasite des abeilles [A new parasite of bees]. Gazette Apicole 57:86.

Maghrabi, H. A., and L. P. Kish. 1985a. Isozyme characterization of Ascosphaerales associated with bees. 1. *Ascosphaera apis, Ascosphaera proliperda* and *Ascosphaera aggregata*. Mycologia 77:358–365.

Maghrabi, H. A., and L. P. Kish. 1985b. Isozyme characterization of Ascosphaerales associated with bees. 2. *Ascosphaera major, Ascosphaera atra* and A*scosphaera asterophora*. Mycologia 77:366–372.

Maghrabi, H. A., and L. P. Kish. 1986. Isozyme characterization of Ascosphaerales associated with bees. III. *Bettsia alvei*. Mycologia 78:676–677.

Maghrabi, H. A., and L. P. Kish. 1987a. Morphological and isozyme survey of chalkbrood disease of the alfalfa leafcutting bee in the western United States. Mycologia 79:565–570.

Maghrabi, H. A., and L. P. Kish. 1987b. Isozyme characterization of Ascosphaerales associated with bees. IV. Analyses. Mycologia 79:519–523.

Mahadevan, V. 1951. Acanthaspis siva Dist.: a predator of the Indian honey bee. Indian Bee Journal 13:143, 147.

Maher, J. F. 1968. The classroom. American Bee Journal 108:290.

Majak, W., R. Neufeld, and J. Corner. 1980. Toxicity of *Astragalus miser v. serotinus* to the honeybee. Journal of Apicultural Research 19:196–199.

Majak, W., and M. A. Pass. 1989. Aliphatic nitrocompounds. *In* Toxicants of plant origin. Volume II, Glycosides. P. R. Cheeke, editor. CRC Press, Boca Raton, Florida.

Majno, G. 1975. The healing hand. Harvard University Press, Cambridge, Massachusetts.

Makarov, I. I. 1966. Spiders that kill bees [in Russian]. Pchelovodstvo 1966(7):29.

Maki, D. L., W. T. Wilson, J. Vargas-C., R. L. Cox, and H. D. Petersen. 1988. Effect of *Acarapis woodi* infestation on honey-bee longevity. In Africanized honey bees and bee mites. G. R. Needham, R. E. Page, Jr., M. Delfinado-Baker, and C. E. Bowman, editors. Ellis Horwood, Chichester, England.

Makings, P. 1958. The oviposition behaviour of *Achroia grisella* (Fabricius) (Lepidoptera: Galleriidae). Proceedings of the Royal Entomological Society of London, Series A, General Entomology 33:136–148.

Malone, L. A., H. A. Giacon, R. J. Hunapo, and C. A. McIvor. 1992. Response of New Zealand honey bee colonies to *Nosema apis*. Journal of Apicultural Research 31:135–140.

Malone, L. A., C. A. McIvor, and H. A. Giacon. 1993. A visual test for *Nosema apis* infections in honey bees. Journal of Apicultural Research 32:143–146.

Manino, A., and F. Marletto. 1984. Il systema enzimatico MDH in populazione di *Apis mellifera* L. della Valle d'Aosta [The MDH enzymatic system in *Apis mellifera* populations of the Aosta Valley]. Apicoltore Moderno 75:89–94.

Manley, R. O. B. 1948. Bee-keeping in Britain. Faber and Faber, London.

Marcangeli, J., L. Monetti, and N. Fernandez. 1992a. Malformations produced by *Varroa jacobsoni* on *Apis mellifera* in the province of Buenos Aires, Argentina. Apidologie 23:399–402.

Marcangeli, J. M., M. J. Eguaras, and N. A. Fernandez. 1992b. Reproduction of *Varroa jacobsoni* (Acari: Mesostigmata: Varroidae) in temperate climates of Argentina. Apidologie 23:57–60.

Marchetti, S., and M. D'Agaro. 1986. Perizin liquid, a systemic agent for the chemical control of varroatosis. Apicoltura, Italy (No. 2): 67–76.

Margulis, L., J. O. Corliss, M. Melkonian, and D. Chapman. 1990. Handbook of Protoctista. Jones and Bartlett, Boston.

Marin, M. 1978. World spread of varroa disease. Apiacta (Bucharest) 41:163–166.

Markosyan, A. A., L. A. Tumasyan and Zh. K. Markosyan. 1973. Features of the biology of *Braula* and its control. Veterinariya, Moscow, USSR. (No. 6): 68-69.

Marletto, F., A. Manino, and A. Patetta. 1990. Evoluzione della varroasi in alveari sotto-posti a periodica asportazione di covata maschile [Evolution of varroa infestations in colonies treated only with periodic drone removal]. Apicoltore Moderno 81:3–9.

Marques, L. F. P., and A. F. Nascimento, Jr. 1981. Comportamento predatorio do pseudo-escorpiao em relacao a abelha (*Apis mellifera*) [Predatory behavior of pseudoscorpions on *Apis mellifera*]. Department of Genetics, Faculty of Medicine of the University of São Paulo, Ribeirão Preto, São Paulo.

Marques, M. H., M. R. C. Issa, and D. De Jong. 1994. Estudo dos danos causados pelas abelhas africanizadas ao ácaro *Varroa jacobsoni* [A study of the damage to the mite *Varroa jacobsoni* caused by africanized bees]. Anais do IV Congresso Iberolatinoamericano de Apicultura y I Foro Expo-comercial internacional de Apicultura. Río Cuarto, Córdoba, Argentina.

Marston, N., and B. Campbell. 1973. Comparison of nine diets for rearing *Galleria mellonella*. Annals of the Entomological Society of America 66:132–136.

Marston, N., B. Campbell, and P. E. Boldt. 1975. Mass producing eggs of the greater wax moth, *Galleria mellonella* (L.). U.S. Department of Agriculture Technical Bulletin number 1510.

Martignoni, M. E. 1964. Pathophysiology in the insect. Annual Review of Entomology 9:179–206.

Martin, S. J. 1994. Ontogenesis of the mite *Varroa jacobsoni* Oud. in worker brood of the honey bee *Apis mellifera* L. under natural conditions. Experimental and Applied Acarology 18:87–100.

Martins, M. I., and M. H. Rosa. 1982. A contribution to the study of parasitic fauna from Beira Interior, Portugal. The main species collected in 1979-81 [Contribucao para o estudo da fauna parasitologica da regiao da Beira Interior, portugal. Especies mais frequentes, clohidas de 1979-1981]. Revista Portuguesa de Ciencias Veterinarias 77(462): 105-109.

Matheson, A. 1993a. World bee health report. Bee World 74:176–212.

———. 1993. World bee health report. International Bee Research Association; Cardiff, UK

———. 1994. What's in a name? Bee World 75:101–103.

———. 1995. World bee health update. Bee World 76:31–39.

———. 1996. World bee health update 1996. Bee World 77:45–51.

Matheson, A. G. 1984. Wasps: social species, description and control. Ministry of Agriculture and Fisheries, New Zealand, FFP 196. 4 pp.

———. 1991. Beekeeping: leading agricultural change in New Zealand, parts 1 & 2. Bee World 72:60–73, 117–130.

Matheson, A. G., K. Clapperton, H. Moller, and P. Alspach. 1989. The impact of wasps on New Zealand beekeeping—the 1986/87 wasp survey. New Zealand Beekeeper 203:28–31.

Mathis, M. 1975. The fly *Senotaninia tricuspis*: probable cause of the 'disappearing disease' of honey bees [La mouche *Senotaninia tricuspis* Meig., agent probable de la maladie de la disparition qui qteint les abeilles]. Comptes Rendus Hebdomadaires des Séances de l'Académie des Sciences, D. (Sciences Naturelles). 281(4): 287-88.

Matsuura, M. 1985. Biology of hornets—why do they attack honeybee colonies? Honeybee Science 6:53–64.

———. 1988. Ecological study on vespine wasps (Hymenoptera: Vespidae) attacking honey bee colonies. I. Seasonal changes in the frequency of visits to apiaries by vespine wasps and damage inflicted, especially in the absence of artificial protection. Applied Entomology and Zoology 23:428–440.

Matsuura, M., and S. F. Sakagami. 1973. A bionomic sketch of the giant hornet, *Vespa mandarinia*, a serious pest for Japanese apiculture. Journal of the Faculty of Science, Hokkaido University, Series VI, Zoology 19:125–162.

Matsuura, M., and S. Yamane. 1984. Biology of the Vespine wasps. Springer Verlag, Berlin.

Matuka, S. 1960. O nalzu amebne bolesti psela (amoebosis) u NR Bosni i Hercegovini [The first discovery of amoeba disease of bees in Bosnia and Hercegovina PR]. Veterinaria (Sarajevo) 9:713–714.

Matus, F., and I. Sarbak. 1974. A méhek költésmeszesedesénék (acosphaeriosis) elöfordulása hazánkban [Occurrence of chalkbrood disease in Hungary]. Magyar Allatorvosok Lapja 29:250–255.

Maule, W. L. 1972. Bears and bees: now we can have both. Gleanings in Bee Culture 100:329–330.

Maurizio, A. 1929. On *Pericystis alvei* Betts. Bee World 10:91–92.

———. 1931. *Gymnoascus setosus* Eidam, ein saprophytischer Pilz aus Bienenwaben [*Gymnoascus setosus* Eidam, a saprophytic fungus of bee combs]. Sitzungsberichte der Bemischen Botanischen Gesellschaft 1930:44–46.

———. 1934. Über die Kalkbrut (Pericystis-Mykose) der Bienen [On chalkbrood (pericystis-mycosis) of bees]. Archiv für Bienenkunde 15:165–193.

———. 1935. Beitrage zur Kenntnis der Pilzflora im Bienenstock. I. Die Pericystis-Infektion der Bienenlarven [Fungi in bee colonies. I. Pericystis infection of bee larvae]. Berichte der Schweizerischen Botanischen Gesellschaft 44:133–156.

———. 1941. Über ein Massensterben von Bienen, verursacht durch Pollen von *Ranunculus puberulus* Koch [A widespread dying off of bees due to pollen from *Ranunculus puberulus* Koch]. Schweizerische Naturforschende Gesellschaft, Verhandlungen 121:149–150.

———. 1943. Bienenschäden während der Lindentracht [Damage to bees during linden flow]. Schweizerische Bienen-Zeitung 66:376–380.

———. 1944. Die Hahnenfussvergiftung (Bettlacher Maikrankheit) im Früjahr 1944 [Poisoning due to buttercup (Bettlacher maysickness) in the spring of 1944]. Schweizerische Bienen-Zeitung 67:288–291.

———. 1945a. Giftige Bienenpflanzen [Poisonous bee plants]. Schweizerische Bienen-Zeitung, Beihefte 1:430–440.

———. 1945b. Trachtkrankheiten der Bienen. I. Vergiftungen bei einseitiger Tracht von Rosskastanien [Brood disease of bees. I. Poisoning due to exclusive foraging on horse chestnut]. Schweizerische Bienen-Zeitung, Beihefte 1:337–368.

———. 1953. Beinenvergiftungen mit pflanzenlichen Wirkstoffen [Bee poisoning due to plant substances]. Proceedings of the Seventh International Botanical Congress, Stockholm 7:190–191.

———. 1968. Les plantes toxiques [Poisonous plants]. *In* Traite de biologie de l'abeille. R. Chauvin, editor. Masson, Paris.

———. 1975. How bees make honey. *In* Honey: a comprehensive survey. E. Crane, editor. Crane, Russak, New York.

May, A. F. 1961. Beekeeping in southern Africa. Haum, Capetown.

———. 1969. Beekeeping. Haum, Capetown.

Mayer, D. F., R. D. Akre, A. L. Antonelli, and D. M. Burgett. 1987. Protecting honey bees from yellowjackets. American Bee Journal 127:693.

Mayer, D. F., J. D. Lunden, and C. W. Kious. 1988. Effects of dipping alfalfa leaf-cutting bee nesting materials on chalkbrood disease. Applied Agricultural Research 3:167–179.

Mayer, D. F., J. D. Lunden, and E. R. Milczky. 1990. Effects of fungicides on chalkbrood disease of alfalfa leafcutting bee. Applied Agricultural Research 5:223–226.

Mazeed, M. M. 1987. Controlling acarine mites with natural materials. Gleanings in Bee Culture 115:517–518, 520.

McDonald, A. 1993. World bee health report. Bee World 74:176–212.

McDonald, D. N. 1978. Diseases of the honeybee, *Apis mellifera* (Hymenoptera: Apidae), in British Columbia with special emphasis on nosema disease, *Nosema apis* (Sporozoa: Nosematidae), in the lower Fraser Valley. Pest Management Papers, #13, Simon Fraser University, Canada.

McEvoy, M. V., and B. A. Underwood. 1988. The drone and species status of *Apis laboriosa* (Hymenoptera: Apidae). Journal of the Kansas Entomological Society 61:246–249.

McGregor, S. E. 1976. Insect pollination of cultivated crop plants. U.S. Department of Agriculture, Agricultural Research Service, Agricultural Handbook number 496.

McKay, C. 1967. Dragonflies attack bees. Australasian Beekeeper 69:20.

McLaughlin, R. E. 1971. Use of protozoans for microbial control of insects. *In* Microbial control of insects and mites. H. D. Burges and N. W. Hussey, editors. Academic Press, New York.

McLellan, A. R. 1964. Fifteen years' bee disease in the east of Scotland. I. "Brood disease." Scottish Beekeeper 41:37–39.

McManus, W. R., and N. N. Youssef. 1984. Life cycle of the chalk brood fungus, *Ascosphaera aggregata*, in the alfalfa leafcutting bee, *Megachile rotundata*, and its associated symptomatology. Mycologia 76:830–842.

McRory, D. 1969. Report of bear damage. Canadian Beekeeping 1:6.

Medina Solis, J. A. 1980. Melaloncha [*Melaloncha ronnai*] and its effects on apiculture [La Melalonca y su efecto en la apicultura]. Gaceta del Colmenar 43(3): 112-114.

Meged, A. G. 1960. Increase the fight against apimyiasis [in Russian]. Pchelovodstvo 37(6):41–42.

Mehr, Z., D. M. Menapace, W. T. Wilson, and R. R. Sackett. 1976. Studies on the initiation and spread of chalkbrood within an apiary. American Bee Journal 116:266–268.

Mehr, Z. A., W. T. Wilson, and R. R. Sackett. 1978. Persistence of chalkbrood (*Ascosphaera apis*) in some North American honey bee colonies one year after infection. Apiacta (Bucharest) 13:99–102.

Meixner, M. D., W. S. Sheppard, A. Dietz, and R. Krell. 1994. Morphological and allozyme variability in honey bees from Kenya. Apidologie 25:188–202.

Meixner, M., W. S. Sheppard, and J. Poklukar. 1993. Asymmetrical distribution of a mitochondrial DNA polymorphism between two introgressing honey bee races. Apidologie 24:147–153.

Mel'nik, V. N., and A. I. Muravskaya. 1981. Drone brood combs and *Varroa jacobsoni* infestations [in Russian]. Veterinariya (Moscow) 1981(4):50–51. From Apicultural Abstracts 1333/82.

Melville, R., and H. A. Dade. 1944. Chalk brood attacking a wild bee. Nature 153:112.

Menapace, D. M. 1978. Chalkbrood infection and detection in colonies of honey bees, *Apis mellifera*. American Bee Journal 118:158–159.

Menapace, D., and P. Hale. 1981. Citral and a combination of sodium propionate and potassium sorbate did not control chalkbrood. American Bee Journal 121:889–891.

Menapace, D. M., and W. T. Wilson. 1979. Feeding oxytetracyclines as Terramycin® does not aggravate chalkbrood infections. Apidologie 10:167–174.

Menapace, D. M., and W. T. Wilson. 1980. *Acarapis woodi* mites found in honey bees from Colombia, South America. American Bee Journal 120:761–762, 765.

Mendoza, M. R. Q., S. G. Delgado, and H. R. D. Villasboa. 1987. Investigación comparativa de la incidencia del ácaro *Varroa jacobsoni* Oudemans sobre distintas razas y lineas de abejas *Apis mellifera* L. en el Paraguay [Comparison of the incidence of the mite *Varroa jacobsoni* Oudemans in different races and lines of honey bees, *Apis mellifera* L., in Paraguay]. Miscellaneous publication of the Departamento de Apicultura, Universidad Nacional de Asunción, San Lorenzo, Paraguay.

Merfield, F. G., and H. Miller. 1956. Gorilla hunter. Farrar, Straus, New York.

Message, D. 1979. Efeito de condiçoes ambientais no comportamento higienico em abelhas africanizadas, *Apis mellifera* [Effect of environmental conditions on the hygienic behavior of Africanized honey bees *Apis mellifera*]. Master's thesis, Departamento Genetica, Facultad de Medicina, Universidad São Paulo, Riberão Preto, S.P., Brazil.

———. 1985. Aspectos reprodutivos do ácaro *Varroa jacobsoni* e seus efeitos em colônias de abelhas africanizadas [Reproduction of the mite *Varroa jacobsoni* and its effects on colonies of Africanized honey bees]. Ph.D. dissertation, University of São Paulo, Ribeirão Preto, São Paulo.

Message, D., and L. S. Gonçalves. 1995. Effect of the size of worker brood cells of Africanized honey bees on infestation and reproduction of the ectoparasitic mite *Varroa jacobsoni* Oud. Apidologie 26:381–386.

Messing, R. H. 1991. The status of beekeeping in the Hawaiian Islands. Bee World 72:147–160.

Metcalf, C. L., W. P. Flint, and R. L. Metcalf. 1962. Destructive and useful insects. 4th ed. McGraw-Hill, New York.

Meyer, O. 1983. The Beekeeper's Handbook. A practical manual of bee management. Sterling, NY. 253pp.

Meyerhoff, G. 1954. Beobachtungen zur Eiablage von *Mutilla europaea* L. [Observations on the egg laying of *Mutilla europea* L.]. Archiv für Geflugelzucht und Kleintierkunde 3:136–141.

Michael, A. S. 1957. Droplet method for the observation of living unstained bacteria. Journal of Bacteriology 74:831–832.

——. 1964. Ethylene oxide. A fumigant for control of pests and parasites of the honey bee. Gleanings in Bee Culture 92:102–104.

Michailoff, A. S. 1928. Statistische Untersuchungen über Nosema an der Tulaer Versuchsstation für Bienenzucht. Archive für Bienenkunde 9:89–114.

Michener, C. D. 1974. The social behavior of the bees. Harvard University Press, Cambridge, Massachusetts.

——. 1975. The Brazilian bee problem. Annual Review of Entomology 20:399–416.

Mickel, C. E. 1928. Biological and taxonomic investigations on the Mutilid wasps. Bulletin 143, United States National Museum, U.S. Government Printing Office, Washington, D.C.

Milani, N. 1995. The resistance of *Varroa jacobsoni* Oud. to pyrethroids: a laboratory assay. Apidologie 26:415–429.

Miles, E. S. 1922. Catalepsy in queens; death from fright. American Bee Journal 62:162–163.

Miller, C. D. F. 1961. Taxonomy and distribution of nearctic *Vespula*. Canadian Entomologist 93: supplement 22.

Milne, C. P. 1976. Morphogenetic fate map of prospective adult structures of the honey bee. Developmental Biology 48:473–476.

——. 1977. Blastoderm fate map of the honey bee (*Apis mellifera* L.). Ph.D. dissertation, Ohio State University, Columbus.

——. 1983. Honey bee (Hymenoptera: Apidae) hygienic behavior and resistance to chalkbrood. Annals of the Entomological Society of America 76:384–387.

Milne, C. P., G. W. Otis, F. A. Eischen, and J. M. Dormaier. 1991. A comparison of tracheal mite resistance in two commercially available stocks of honey bees. American Bee Journal 131:713–718.

Milne, P. S. 1942. Wax moth and bald-headed brood. Bee World 23:13–14.

——. 1957. Acarine disease in *Apis indica*. Bee World 38:156.

Milum, V. G. 1935. Lesser vs. greater wax-moth. Gleanings in Bee Culture 63:662–666.

——. 1938. A larval mermithid, *Mermis subnigrescens* Cobb, as a parasite of the honeybee. Journal of Economic Entomology 31:460.

——. 1940a. Larval pests common to the nests of bumble bees and combs of the honeybee. Journal of Economic Entomology 33:81–83.

——. 1940b. Moth pests of honeybee combs. Gleanings in Bee Culture 68:424–428, 488–493.

——. 1953. *Vitula edmandsii* as a pest of honeybee combs. Journal of Economic Entomology 46:710–711.

Milum, V. G., and H. W. Geuther. 1935. Observations on the biology of the greater wax moth, *Galleria mellonella* L. Journal of Economic Entomology 28:576–578.

Minkov, S. G., and K. V. Moiseev. 1953. Experiments on the control of *Meloe* larvae [in Russian]. Pchelovodstvo 1953(5):53–54.

Mishra, R. C., and H. D. Kaushik. 1993. Green bee eater, *Merops orientalis*, an evaluation of possible methods to reduce honeybee predation. Pavo 31:113–116.

Mishra, R. C., J. Kumar, and J. K. Gupta. 1989. A new approach to the control of predatory wasps (*Vespa* spp.) of the honeybee (*Apis mellifera* L.). Journal of Apicultural Research 28:126–130.

Mitroiu, P., A. Popa, M. Serban, and C. Toma. 1966. Etude sur la flore "mycosique" intestinale des abeilles soumises aux traitements aux antibiotiques et sulfathiazol [A study of the intestinal mycotic flora of bees submitted to treatment with antibiotics and sulfathiazole]. Bulletin Apicole 9:43–66.

Mo, C.-F. 1971. Studies on the life-history of *Neocypholaelaps indica* Evans (Acarina, Ameroseiidae). New Asia College Academic Annual 13:97–107.

Moeller, F. E. 1956. The behavior of nosema-infected bees affecting their position in the winter cluster. Journal of Economic Entomology 49:743–745.

———. 1962. Nosema disease control in package bees. American Bee Journal 109:390–392.

———. 1967. A study of the incidence of nosema infection in overwintered colonies in Wisconsin. Proceedings of the 21st International Apicultural Congress, Preliminary Scientific Meetings, summary paper number 47.

———. 1972. Effects of emerging bees and of winter flights on nosema disease in honey bee colonies. Journal of Apicultural Research 11:117–120.

Moeller, F. E., and P. H. Williams. 1976. Chalkbrood research at Madison, Wisconsin. American Bee Journal 116:484, 486, 495.

Moffett, J. O., J. J. Lackett, and J. D. Hitchcock. 1969. Compounds tested for control of nosema in honey bees. Journal of Economic Entomology 62:886–889.

Moffett, J. O., and W. T. Wilson, 1971. The viability and infectivity of frozen nosema spores. American Bee Journal 111:55, 70.

Moffett, J. O., W. T. Wilson, R. L. Cox, and M. Ellis. 1988. Four formulations of amitraz reduced tracheal mite, *Acarapis woodi*, populations in honey bees. American Bee Journal 128:805–806.

Moffett, J. O., W. T. Wilson, A. Stoner, and A. Wardecker. 1978. Feeding commercially purchased pollen containing mummies caused chalkbrood. American Bee Journal 118:412–414.

Mohamed, M. A., and H. C. Coppel. 1983. Mass rearing of the greater wax moth, *Galleria mellonella* (Lepidoptera: Pyralidae), for small-scale laboratory studies. Great Lakes Entomologist 16:139–141.

Mohrig, W., and B. Messner. 1968. Lysozym als antibakerielles Agens im Bienenhong und Binengift [Lysosyme as the antibacterial agent in honey and venom]. Acta Biologica et Medica Germanica 21:85–95.

Molan, P. C. 1992a. The antibacterial activity of honey 1. The nature of the antibacterial activity. Bee World 73:5–18.

———. 1992b. The antibacterial activity of honey 2. Variation in the potency of the antibiotic activity. Bee World 73:59–76.

Molan, P. C., and K. M. Russell. 1988. Non-peroxide antibacterial activity in some New Zealand honeys. Journal of Apicultural Research 27:62–67.

Moller, H., and J. A. V. Tilley. 1989. Beech honeydew: seasonal variation and use by wasps, honey bees, and other insects. New Zealand Journal of Zoology 16:289–302.

Moosbeckhofer, R. 1991. Apistan und Bayvarol—Langzeitwirkung behandelter Waben [Apistan and Bayvarol—longtime effects in the comb]. Bienenvater 112:90–92.

Moosbeckhofer, R. 1992. Beobachtungen Zum Auftreten beschädigter Varroamilben im natürlichen Totenfall bei Völkern von *Apis mellifera* carnica [Observations on the occurrence of damaged *Varroa* mites in natural mite fall of *Apis mellifera carnica* colonies]. Apidologie 23:523–531.

Moosbeckhofer, R., M. Fabsicz, and A. Kohlich. 1988. Untersuchungen über die Abhängigkeit der Nachkommensrate von *Varroa jacobsoni* vom Befallsgrad der Bienenvölkern [Investigations on the correlation between rate of reproduction of *Varroa jacobsoni* and infestation rate of honeybee colonies]. Apidologie 19:181–207.

Moosbeckhofer, R., and A. Kohlich. 1989. Erfahrungen bei den Anwendung von Apistan-streifen am Institut für Bienenkunde [Experience with Apistan strips at the bee institute]. Bienenvater 110:221–228.

Moosbeckhofer, R., and A. Kohlich. 1990. Nachwirkung von Apistan nach der Entfernung der Streifen! [Continued activity of Apistan after removal of the strips]. Bienenvater 111:3–9.

Moreaux, R. 1949. Contribution à l'étude d'une nouvelle maladie des abeilles [Contribution to the study of a new bee disease]. Apiculteur 93:70–74.

Moretto, G. 1988. Efeito de diferentes regiões climáticas Brasileiras e de tipos raciais de abelhas *Apis mellifera* na dinâmica de populações do ácaro *Varroa jacobsoni* [A study of *Varroa jacobsoni* infestations in Africanized and hybrid (Italian/Africanized) honey bees in different climate regions of Brazil]. M. S. thesis, Faculty of Medicine of the University of São Paulo, Riberão Preto, São Paulo, Brazil.

————. 1993. Estudo de algumas variáveis relacionadas ao mecanismo de defensa de operárias de *Apis mellifera* á reproduçáo do ácaro *Varroa jacobsoni* [Study of some variables associated to a defense mechanism of *Apis mellifera* workers against varroatosis, and against the reproduction of the mite *Varroa jacobsoni*]. Ph.D. dissertation. Departamento Genetica, Facultad de Medicina, Universidad São Paulo, Riberão Preto, S.P., Brazil.

Moretto, G., L. S. Gonçalves, and D. De Jong. 1987. Effect of climate and bee race on the infestation level of *Varroa jacobsoni* in honey bee colonies in Brazil. In Abstracts of the International Conference on Africanized Honey Bees and Bee Mites. 10th Biosciences Colloquium, Ohio State University, Columbus.

Moretto, G. 1988. Efeito de diferentes regiões climáticas brasileiras e de ripos raciais de abelhas *Apis mellifera* na dinâminca de populações do ácaro *Varroa jacobsoni*. [A study of *Varroa jacobsoni* infestations in Africanized and Italian/Africanized honey bees in three climate regions of Brazil.] M.Sc. thesis. Faculdade de Medicina de Ribeirão Preto, Universidade de São Paul. 102 pp.

Moretto, G., L. S. Gonçalves, and D. De Jong. 1991a. Africanized bees are more efficient at removing *Varroa jacobsoni*—Preliminary data. American Bee Journal 131:434.

Moretto, G., L. S. Gonçalves, and D. De Jong. 1991b. The effects of climate and bee race on *Varroa jacobsoni* Oud. infestations in Brazil. Apidologie 22:197–203.

Moretto, G., L. S. Gonçalves, and D. De Jong. 1993. Heritability of Africanized and European honey bee defensive behavior against the mite *Varroa jacobsoni*. Brazilian Journal of Genetics 16:71–77.

Moretto, G., L. S. Gonçalves, and D. De Jong. 1995a. Analysis of the F_1 generation, descendants of Africanized bee colonies with differing defense abilities against the mite *Varroa jacobsoni*. Brazilian Journal of Genetics 18:177–179.

Moretto, G., L. S. Gonçalves, D. De Jong, and M. Z. Bichuette. 1991. The effects of climate and bee race on *Varroa jacobsoni* Oud. infestations in Brazil. Apidologie 22:197–203.

Moretto, G., A. Pillati, D. De Jong, L. S. Gonçalves, and F. L. Cassini. 1995b. Reduction of Varroa infestations in the State of Santa Catarina, in Southern Brazil. American Bee Journal 135:498–500.

Morgenthaler, O. 1920. Bienenkrankheiten im Jahre 1919 [Bee diseases in 1919]. Schweizerische Bienen-Zeitung 43:146–154.

————. 1926. Bienenkrankheiten im Jahre 1925 [Bee diseases in 1925]. Schweizerische Bienen-Zeitung 49:176–180, 220–224.

————. 1927. Eine neue Pilzkrankheit der Bienenlarven [A new fungus disease of bee larvae]. Schweizerische Bienen-Zeitung 50:486–487.

————. 1929. Problems of acarine disease of bees. Bee World 10:19–24.

————. 1930. New investigations on acarine disease. Bee World 11:49–50.

————. 1931a. An acarine disease experimental apiary in the Bernese Lake District and some results obtained there. Bee World 12:8–10.

————. 1931b. Bienenkrankheiten in Jahre 1930 [Bee diseases in 1930]. Schweizerische Bienen-Zeitung 54:254–267.

————. 1933. *Acarapis woodi* in queens. Bee World 14:81.

————. 1934. Krankheitserregende und harmlose Arten der Bienenmilbe *Acarapis*, zugleich ein Beitrag zum Species-Problem [Infectious and harmless types of the bee mite *Acarapis*, with a note on the problem of species determination]. Revue Suisse de Zoologie 41:429–446.

————. 1939. Die ansteckende Frühjahrsschwindsucht (Nosema-Amöben-Infektion) der Bienen [Infective spring dwindling (nosema-amoeba infection) of bees]. Schweizerische Bienen-Zeitung 62:86–92, 154–162, 205–215.

————. 1944. Das jahreszeitliche Auftreten der Bieneseuche [The seasonal appearance of bee epidemics]. Schweizerische Bienen-Zeitung, Beihefte 1:285–336.

Morison, G. D. 1931. Amoebic disease of bees in Great Britain. Bee World 12:56.

————. 1941. Insect visitors to bee hives. Scottish Beekeeper 17:72.

Moritz, R. F. A. 1983. Homogeneous mixing of honeybee semen by centrifugation. Journal of Apicultural Research 22:249–255.

————. 1985. Heritability of the post-capping stage in *Apis mellifera* and its relation to varroatosis resistance. Journal of Heredity 76:267–270.

————. 1988. A reevaluation of the two-locus model for hygienic behavior in honeybees. Journal of Heredity 79:257–262.

————. 1994. Selection for varroatosis resistance in honeybees. Parasitology Today 10: 236–238.

Moritz, R. F. A., and H. Hänel. 1984. Restricted development of the parasitic mite *Varroa jacobsoni* Oud. in the Cape honeybee, *Apis mellifera capensis* Esch. Zeitschrift für Angewandte Entomologie 97:91–95.

Moritz, R. F. A., and M. Jordan. 1992. Selection of resistance against *Varroa jacobsoni* across caste and sex in the honeybee (*Apis mellifera* L., Hymenoptera: Apidae). Experimental and Applied Acarology 16:345–353.

Moritz, R. F. A., and D. Mautz. 1990. Development of *Varroa jacobsoni* in colonies of *Apis mellifera capensis* and *Apis mellifera carnica*. Apidologie 21:53-58.

Morse, D. H. 1979. Prey capture by the crab spider *Misumena calycina* (Araneae, Thomisidae). Oecologia (Berlin) 39:309–320.

————. 1984. How crab spiders hunt at flowers. Journal of Arachnology 12:307–316.

Morse, D. H. 1985. Costs in a milkweed-bumble bee mutualism. American Naturalist 125: 903–905.

Morse, D. H., and R. S. Fritz. 1989. Milkweed pollinia and predation risk to flower-visiting insects by the crab spider *Misumena vatia*. American Midland Naturalist 121(1):188–193.

Morse, G. D. 1979. Standing on guard. Gleanings in Bee Culture 107:614–616.

Morse, R. A. 1955. Larval nematode recorded from honeybee *Apis mellifera* L. Journal of Parasitology 41:553.

————. 1961. The effect of Sevin on honey bees. Journal of Economic Entomology 54:566–568.

————. 1972. The complete guide to beekeeping. E. P. Dutton, New York.

————. 1975a. Beekeeping in the Dominican Republic. Gleanings in Bee Culture 103:223–224.

————. 1975b. Bees and beekeeping. Cornell University Press, Ithaca, New York.

————. 1976. Brood patterns. Gleanings in Bee Culture 104:415–417.

————. 1978. Comb honey production. Wicwas Press, Cheshire, Connecticut.

————. 1981a. Honeybee pests and predators. *In* Handbook of pest management in agriculture, volume 2. D. Pimentel, editor. CRC Press, Boca Raton, Florida.

————. 1981b. The fall of BIPP. Gleanings in Bee Culture 109:290–291.

————. 1987. The braulids. American Bee Journal 127: 429-431.

————. 1988. Preface. *In* Africanized honey bees and bee mites. G. R. Needham, R. E. Page, Jr., M. Delfinado-Baker, and C. E. Bowman, editors. Ellis Horwood, Chichester, England.

————. 1990a. The Cornell project to breed mite-resistant bees from queen honey bees imported from Great Britain. American Bee Journal 130:186.

————. 1990b. Importation of British queen honey bees into the United States. American Bee Journal 130:106–107, 116–120.

————. 1993. A survey of the effects of pesticides on honey bees. Unpublished.

Morse, R. A., D. M. Burgett, J. T. Ambrose, W. E. Connor, and R. D. Fell. 1973. Early introductions of African bees into Europe and the New World. Bee World 54:57–60.

Morse, R. A., G. C. Eickwort, and R. S. Jacobson. 1977. The economic status of an immigrant yellowjacket, *Vespula germanica* (Hymenoptera: Vespidae), in northeastern United States. Environmental Entomology 6:109–110.

Morse, R. A., and K. Flottum. 1990. The ABC & XYZ of Bee Culture. 40th edition, A.I. Root Co., Ohio. 519pp.

Morse, R. A., and N. E. Gary. 1961. Insect invaders of the honeybee colony. Bee World 42:179–181.

Morse, R. A., and T. Hooper. 1985. The illustrated encyclopedia of beekeeping. E. P. Dutton, New York.

Morse, R. A., and F. M. Laigo. 1968. Beekeeping in the Philippines. University of the Philippines Farm Bulletin 27.

Morse, R. A., and F. M. Laigo. 1969a. *Apis dorsata* in the Philippines. Philippine Association of Entomologists Monograph number 1.

Morse, R. A., and F. M. Laigo. 1969b. The Philippine spine-tailed swift, *Chaetura dubia* McGregor, as a honey bee predator. Philippine Entomologist 1:138–143.

Morse, R. A., D. Miksa, and J. A. Masenheimer. 1991. Varroa resistance in U.S. honey bees. American Bee Journal 131:433–434.

Morse, R. A., and R. Nowogrodzki, editors. 1990. Honey bee pests, predators, and diseases. 2d ed. Cornell University Press, Ithaca, New York.

Mossadegh, M. S. 1990. Development of *Euvarroa sinhai* (Acarina: Mesostigmata), a parasitic mite of *Apis florea* on *Apis mellifera* worker brood. Experimental and Applied Acarology 9:73–78.

Most, H. 1989. The pollen louse *Trogium pulsatorium* (Psocoptera) a harmless comensal in bee hives (in German). Bienenwelt 31:112.

Moure, J. S. 1946. Sobre a abelha tataira ou cagafogo. Chacaras Quintais 73:612–613.

Mraz, C. 1973. Chalk brood. Gleanings in Bee Culture 101:115, 129.

Muck, O. 1936. Bericht der amtlichen Untersuchungsstelle für ansteckende Bienenkrankheiten an der Tierärztlichen Hochschule über das Jahr 1935 [Report of the veterinary institute's federal research station for contagious diseases of bees for the year 1935]. Wiener Tierärtzliche Monatsschrift 23:168–173.

Muenscher, W. C. 1960. Poisonous plants of the United States. The Macmillan Company, New York.

Muller, C. L. 1948. Die Trachtkrankheiten der Honigbiene [Brood diseases of the honey bee]. Tierzucht 2:117–120.

Muller, D. 1928. Studien über ein neues Enzym Glykoseoxydase I. [Studies of a new enzyme, glucose oxidase]. Biochemische Zeitschrift 199:136–170.

Muresan, E., C. Duca, and Z. Papay. 1975. The study of some histochemical indices of the midgut, healthy and infected with *Nosema apis* Z., of the *Apis mellifera carpatica* bee. In 25th International Apicultural Congress of Apimondia, Munchen. Apimondia, Bucharest.

Murthy, V. A., and R. Venkataraman. 1985. Contribution to the biology of the pseudoscorpion *Ellingsenius indicus* Chamberlain. Indian Bee Journal 47(1–4):34–35.

Murthy, V. A., and R. Venkataraman. 1986. *Ellingsenius indicus* (Arachnida: Chelonethi) as a tool to the assessment of the settling nature of honey bee (*Apis mellifera indica*) in a new habitat. Indian Bee Journal 48(1–4):54–55.

Mussen, E. C., and B. Furgala. 1975. Benomyl ineffective against *Nosema apis*. American Bee Journal 115:478.

Mussen, E. C., and B. Furgala. 1977. Replication of sacbrood virus in larval and adult honeybees, *Apis mellifera*. Journal of Invertebrate Pathology 30:20–34.

Mussen, E. C., B. Furgala, and R. A. Hyser. 1975. Enzootic levels of nosema disease in the continental United States (1974). American Bee Journal 115:48–50.

Muthel, H. 1955. Ameisen überfallen einen Bienenstand und vernichten 26 Völker [Ants overwhelm an apiary and destroy 26 colonies]. Bienenzucht 8:202–203.

Muzaffar, N., and R. Ahmad. 1986. Studies on hornets attacking honeybees in Pakistan. Pakistan Journal of Agricultural Research 7:59–63.

Mwale, V. G. D. 1992. An examination of the pests and pathogens of the honeybee (Apis mellifera L.) in Malawi. Master's thesis, School of Pure and Applied Biology, University of Wales College of Cardiff, Cardiff.

Nascimento, C. B. 1971. Acarine disease in Brazil. Ciencia y Abejas 1:35–39. From Apicultural Abstracts 399L/73.

Nascimento, C. B., R. P. de Mello, M. W. dos Santos, R. V. do Nascimento, and D. J. de Souza. 1971. Ocorrencia de acariose em Apis mellifera no Brazil [Occurrence of acarine disease in Apis mellifera in Brazil]. Pesquisla Agropecuaria Brasileira Seção Veterinaria 6:57–60.

Nascimento, C. B., and D. J. de Souza. 1970. Myrmeleon januarius, a new enemy of bees. Congresso Brasileiro de Apicultura 1:186–189.

Nasr, M. E., G. W. Otis, C. D. Scott-Dupree, and O. Welsh. 1992. Potential uses of Canadian honey bees in breeding for resistance to tracheal mites. American Bee Journal 132:812.

Nazzi, F. 1992. Morphometric analysis of honey bees from an area of racial hybridization in northeastern Italy. Apidologie 23:89–96.

Neboiss, A. 1962. Notes on distribution and descriptions of new species (orders: Odonata, Plecoptera, Orthoptera, Trichoptera, and Coleoptera). Memoirs of the National Museum, Melbourne.

Needham, J. G., and H. B. Heywood. 1929. A handbook of the dragonflies of North America. C. C. Thomas, Springfield, Illinois.

Neilson, R. 1986. Housecleaning chickadees. American Bee Journal 126:527.

Nelson, D., P. Sporns, P. Kristiansen, P. Mills, and M. Li. 1993. Effectiveness and residue levels of 3 methods of menthol application to honey bee colonies for the control of tracheal mites. Apidologie 24:549–556.

Nelson, D. L. 1974. Bear damage and control. Canadian Beekeeper 4:67–69.

———. 1975. An evaluation: a cross between New Zealand and California honey bee stocks. American Bee Journal 115:228–229, 234.

Nelson, D. L., R. Barker, E. Bland, J. Corner, U. Soehngen, and J. L. Villeneuve. 1976. Chalk brood disease survey of honey bees in Canada, 1975. American Bee Journal 116:108–109.

Nelson, D. L., R. G. Barker, S. E. Bland, U. Soehngen, and J. Corner. 1977. Western Canada chalkbrood disease survey of honey bees, 1976. American Bee Journal 117:494–496. 505.

Nelson, D. L., and T. A. Gochnauer. 1982. Field and laboratory studies on chalkbrood disease of honey bees. American Bee Journal 122:29–34.

Nelson, E. K., and H. H. Mottern. 1931. Some organic acids in honey. Industrial and Engineering Chemistry 23:335.

Nester, E. W., C. E. Roberts, N. N. Pearsall, and B. J. McCarthy. 1978. Microbiology. Holt, Rinehart, and Winston, New York.

Neuhauser, H. 1985. Information über den Wirkstoff des Perizin® [Information about the active ingredient in Perizin]. Apidologie 16:217–219.

Newell, W., and T. C. Barber. 1913. The Argentine ant. U.S. Department of Agriculture, Bureau of Entomology Bulletin 122.

Newswander, K. T. 1977. Queen rearing. American Bee Journal 117:224–226.

Newton, D. C., and D. J. Michl. 1974. Cannibalism as an indication of pollen insufficiency in honeybees: ingestion or recapping of manually exposed pupae. Journal of Apicultural Research 13:235–241.

Nicholls, H. M. 1932. The bee louse. Tasmanian Journal of Agriculture 3(4): 164-165.

———. Diseases of adult bees. Tasmanian Journal of Agriculture 5:13–17.

Nielsen, R. A. 1971. Radiation biology of the greater wax moth, *Galleria mellonella* (L.): effects on developmental biology, bionomics, mating competitiveness, and F_1 sterility. Ph.D. dissertation, Utah State University, Logan.

Nielsen, R. A., and C. D. Brister. 1979. Greater wax moth: behavior of larvae. Annals of the Entomological Society of America 72:811–815.

Nielsen, R. A., and C. D. Brister. 1980. Induced genetic load in descendants of irradiated greater wax moths. Annals of the Entomological Society of America 73:460–467.

Nielsen, R. A., and D. Brister. 1977. The greater wax moth: adult behavior. Annals of the Entomological Society of America 70:101–103.

Nielsen, R. A., and G. E. Cantwell. 1973. The question of parthenogenesis in the greater wax moth. Journal of Economic Entomology 66:37–38.

Nielsen, R. A., and E. N. Lambremont. 1976. Radiation biology of the greater wax moth: inherited sterility and potential for pest control. U.S. Department of Agriculture Technical Bulletin number 1539.

Nielsen, D., R. E. Page Jr., and M. W. J. Crosland. 1994. Clinal variation and selection of MDH allozymes in honey bee populations. Experientia 50:867–871.

Nikiel, A. H. 1972. Invading carpenter ants in the beehive. Gleanings in Bee Culture 100:330.

Nixon, G. 1954. The world of bees. Hutchinson, London.

Nixon, M. 1982. Preliminary world maps of honeybee diseases and parasites. Bee World 63:23–42.

———. 1983. World maps of *Varroa jacobsoni* and *Tropilaelaps clareae*, with additional records for honeybee diseases and parasites previously mapped. Bee World 64:124–131.

Nogueiro-Neto, P. 1949. Notas biônomicas sôbre meliponineos (Hymenoptera, Apoidea) II. Sôbre a pilhagem [Bionomic notes on Meliponids II. On robbing]. Revue Brazil Biology 9(2):13–32.

———. 1953. A criação de abelhas indigenas sem ferrào (Meliponinae) [The care of indigenous stingless bees (Meliponinae)]. Chacaras e Quintais, São Paulo, Brazil.

Noonan, K. C., and S. A. Kolmes. 1989. Kin recognition of worker brood by worker honey bees, *Apis mellifera* L. Journal of Insect Behavior 2:473–485.

North, P. M. 1967. Poisonous plans and fungi. Blanford Press Ltd., London.

Nunamaker, R. A., C. E. Nunamaker, W. T. Wilson, and B. R. Francis. 1985. Virus-like particles in the fat body tissue of adult queen honey bees (*Apis mellifera*). Journal of Invertebrate Pathology 46:337–342.

Nunamaker, R. A., W. T. Wilson, M. Cal, L. Choate, and G. Tween. 1986. Beekeeping in Belize, Central America, with notes on diseases and parasites. Bee World 67:151–156.

Nye, M. J., R. W. Schuel, and S. E. Dixon. 1973. Gluconic acid in the food of the larval honey bee. Journal of Apicultural Research 12:9–15.

Nyein, M. M., and C. Zmarlicki. 1982. Control of mites in European bees in Burma. American Bee Journal 122:638–639.

Nyffeler, M. 1982. Field studies on the ecological role of the spiders as predators of insects in agroecosystems. Ph.D. dissertation, Swiss Federal Institute of Technology, Zurich.

Nyffeler, M., and G. Benz. 1978. Prey selection by the web spinners *Argiope bruennichi* (Scop.), *Araneus quadratus* cl. and *Agelena labyrinthica* (cl) on fallow and near Zurich, Switzerland. Anzeiger für Schädlingskunde, Pflanzenschutz, Umweltschutz 54:33–39.

Nyffeler, M., and R. G. Breene. 1990. Spiders associated with selected European hay meadows and the effects of habitat disturbance with the predation ecology of the crab spiders, *Xysticus* spp. (Araneae, Thomisidae). Journal of Applied Entomology 110:179–189.

Nyffeler, M., and R. G. Breene. 1991. Impact of predators upon honey bees (Hymenoptera, Apidae) by orb-weaving spiders (Araneae, Araneidae & Tetragnathidae) in grassland ecosystems. Journal of Applied Entomology 111:179–189.

Nyffeler, M., D. A. Dean, and W. L. Sterling. 1987. Predation by green Lyme Spider, *Peucetia viridens* (Aranaea: Oxyopidae), inhabiting cotton and woolly croton plants in east Texas. Environmental Entomology 16:355–359.

O'Brien, J. M., and R. E. Marsh. 1990. Vertebrate pests of beekeeping. Proceedings of the 14th Vertebrate Pest Conference, University of California, Davis.

O'Brien, R. D. 1967. Insecticides: action and metabolism. Academic Press, New York.

OConnor, B. M. 1979. Evolutionary origins of astigmatid mites inhabiting stored products. *In* Recent advances in acarology, volume 1. J. G. Rodriguez, editor. Academic Press, New York.

———. 1982. Evolutionary ecology of astigmatid mites. Annual Review of Entomology 27:385–409.

Oertel, E. 1961. Effect of yellow jessamine on honey bees. American Bee Journal 101:174–175.

———. 1963. Greater wax moth develops on bumble bee cells. Journal of Economic Entomology 56:543–544.

———. 1965. Gregarines found in several honey bee colonies. American Bee Journal 105:10–11.

———. 1967. Colony disturbance and nosema disease. Journal of Apicultural Research 6:119–120.

———. 1969. Losses caused by the greater wax moth. American Bee Journal 109:145.

Oguz, T. 1976. [*Braula schmitzi* Örösi-Pál (Diptera, Braulidae) as an ectoparasite of honey bees in Turkey] Yurdumuz arilarinda tesbit ettigimiz *Braula schmitzi* Örösi-Pál (Diptera, Braulidae). Veteriner Fakültesi Dergisi, Ankara Üniversitesi 23(3/4): 345-351.

Okada, I. 1956. Notes on the Japanese hornet as a serious pest of honeybees. Japanese Bee Journal 9:110–114.

———. 1959. Natural enemies of the Japanese honeybee [in Japanese]. Heredity (Japan) 13:22–26.

———. 1960. Notes on the habits of a hornet (honey bee enemy) *Vespa crabroformis* [in Japanese]. Japanese Bee Journal 13:116–119.

———. 1961. Notes on the habits of a giant hornet (*Vespa mandarina* Smith) as a noxious agent against honey bees. Bulletin of the Faculty of Agriculture, Tamagawa University 2:73–89.

———. 1980. Trapping giant hornets on sticky paper [in Japanese, English summary]. Honeybee Science 1:24–25.

———. 1984. Aggressive behavior of honeybees against hornets in Japan [in Japanese, English summary]. Honeybee Science 5:105–112.

Oku, K., N. Hara, and H. Sano. 1983. The occurrence of wing abnormalities in honeybees [in Japanese]. Honeybee Science 4:109–110.

Okumura, G. T. 1966. The dried-fruit moth (*Vitula edmandsae serratilineella* Ragonot), pest of dried fruits and honeycombs. Bulletin of the California Department of Agriculture 55:180–186.

Olberg, G. 1952. Über die Lebensweise des Bienenwolfes [The life history of the beewolf]. Hessische Biene 87:273–275.

———. 1959. Das Verhalten der solitären Wespen Mitteleuropas (Vespidae, Pompilidae, Sphecidae) [The behavior of solitary wasps of central Europe]. Deutscher Verlag der Wissenschaften, Berlin.

———. 1960. Spinnen als Feinde der Honigbiene [Spider enemies of honeybees]. Imkerfreund 15:9–11.

Oldroyd, B. P., T. E. Rinderer, and S. M. Buco. 1991a. Honey bees dance with their super-sisters. Animal Behaviour 42:121–129.

Oldroyd, B. P., T. E. Rinderer, and S. M. Buco. 1991b. Intracolonial variance in honey bee foraging behaviour: the effects of sucrose concentration. Journal of Apicultural Research 30:137–145.

Oldroyd, B. P., T. E. Rinderer, J. R. Harbo, and S. M. Buco. 1992. Effects of intracolonial genetic diversity on honey bee (Hymenoptera: Apidae) colony performance. Annals of the Entomological Society of America 85:335–343.

Oldroyd, B. P., W. S. Sheppard, and J. A. Stelzer. 1992. Genetic characterization of the bees of Kangaroo Island, South Australia. Journal of Apicultural Research 31:141–148.

Olszowy, D. R. 1977. Of bees, rhododendrons and honey. American Bee Journal 117:498–500, 505.

O'Neill, K. M. 1992. Patch-specific foraging by the robber fly *Megaphorus willistoni* (Diptera: Asilidae). Environmental Entomology 21(6): 1330-1340.

O'Neill, K. M. and W. P. Kemp. 1991. Foraging of *Stenopogon inquinatus* (Loew) (Diptera: Asilidae) on Montana rangeland sites. Pan-Pacific Entomology 67(3): 177-180.

Ono, M., I. Okada, and M. Sasaki. 1987. Heat production by balling in the Japanese honeybee, *Apis cerana japonica*, as a defensive behavior against the hornet *Vespa simillima xanthoptera* (Hymenoptera: Vespidae). Experientia 43:1031–1032.

Ono, M., E. Ohno, and M. Sasaki. 1991. Foraging and mating strategies of the hornet, *Vespa mandarinia japonica* Radoszkowski. Abstracts from the 22nd International Ethological Conference, Kyoto, Japan.

Ono, M., and M. Sasaki. 1993. Defensive strategy of *Apis cerana japonica* against attack by *Vespa* spp. *In* Asian apiculture. L. J. Connor, T. Rinderer, H. A. Sylvester, and S. Wongsiri, editors. Wicwas Press, Cheshire, Connecticut.

Ordetx, G. S., and D. Espina Perez. 1966. La apicultura en los tropicos [Apiculture in the tropics]. Bartolome Trucco, Mexico City.

Örösi-Pál, Z. 1929. [On laying workers.] *From* an abstract in Bee World 10:134.

———. 1934. Experiments on the feeding habits of the *Acarapis* mites. Bee World 15:93–94.

———. 1935. Die Altersimmunität der Honigbiene gegen die Milbe (*Acarapis woodi*) [Age-correlated immunity of the honey bee against mites]. Zeitschrift für Parasitenkunde 7:401–407.

———. 1936. Über die Melanosekrankheit der Honigbiene [Melanosis disease of the honey bee]. Zeitschrift für Parasitenkunde 9:125–139.

———. 1938. Report of scientific research on beekeeping. Deutscher Imkerführer 11:297–301.

———. 1938. Humpbacked Flies and the Honey Bee. Bee World 19(5):64-68.

———. 1939a. Dragonflies as enemies of the honey bee. Bee World 20:70–71.

———. 1939b. Eischwarzsucht und Melanosenkrankheit der Bienenkönigin [Egg melanosis and melanosis disease of the honey bee queen]. VII. Internationaler Kongress für Entomologie. Verhandlungen 3:1865–1871.

———. 1939c. Méhellenségek és a Köpu állatvilága [Enemies of honey bees and the fauna of the hive]. Országos Magyar méhészeti egyesület, Budapest.

———. 1963. Amoeba disease in the queen honey bee. Journal of Apicultural Research 2:109–111.

———. 1963a. Request for Bee Lice. Bee World 44(3): 116.

———. 1963b. [Bee Lice] Pchelovodstvo 40(6): 28-30.

———. 1966. Die Bienenlaus-Arten [A classification of the bee-lice]. Angewandte Parasitologie 7:138-171.

———. 1976. Identifying the bee louse *Braula* and the big mite in Asia. *In* Varroasis, a honeybee disease. Apimondia, Bucharest.

———. 1980. Bee lice in the Americas. American Bee Journal. 120(6): 438-440.

Otis, G. W. 1991. A review of the diversity of species within *Apis*. *In* Diversity in the genus *Apis*. D. R. Smith, editor. Westview Press, Boulder, Colorado.

Otis, G. W., J. B. Bath, D. L. Randall, and G. M. Grant. 1987. Studies on the honey bee tracheal mite (*Acarapis woodi*) (Acari: Tarsonemidae) during winter. Canadian Journal of Zoology 66:2122-2127.

Otis, G. W., G. Grant, D. Randall, and J. Bath. 1986. Summary of the tracheal mite project in New York. Proceedings of the Honey Bee Tracheal Mite (*Acarapis woodi* R.) Scientific Symposium, July 8–9, 1986, St. Paul, Minnesota. American Association of Professional Apiculturists, Gainesville, Florida.

Otis, G. W., and C. D. Scott-Dupree. 1992. Effects of *Acarapis woodi* on overwintered colonies of honey bees (Hymenoptera: Apidae) in New York. Journal of Economic Entomology 85:40–46.

Otte, E. 1973. Ein Beitrag zur Labordiagnose der bösartigen Faulbrut der Honigbiene unter besonderer Berücksichtigung der Immunofluoreszenzmethode [Contributions to the laboratory diagnosis of American foulbrood of the honeybee with particular reference to the fluorescent antibody technique]. Apidologie 4:331–339.

Otten, C. 1991. Factors and effects of a different distribution of *Varroa jacobsoni* between adult bees and bee brood. Apidologie 22:465–467.

Otten, C., and S. Fuchs. 1988. Individual differences in *Varroa jacobsoni* of preference for drone larvae to worker bee larvae. In European research on varroatosis control. R. Cavalloro, editor. A. A. Balkema, Rotterdam, The Netherlands.

Ottoboni, M. A. 1991. The dose makes the poison, a plain language guide to toxicology. 2d ed. Van Nostrand Reinhold, New York.

Oudemans, A. C. 1904. Note VIII. On a new genus and species of parasitic acari. Notes from the Leyden Museum 24:216–233.

Oytun, H. S. 1963. Ankara arilarinda görülen *Braula coeca* Nitzsch, 1818, avi biti (ari sineui) [The bee louse, *Braula coeca* Nitzsch, observed on honey bees at Ankara]. Ankara Üniveristy vet. Fak. Derg. 10(3-4): 317-321.

Özbek, H. 1982. Türkiye için önemli bir balarisi (*Apis mellifera* L.) avciböcegi *Philanthus triangulum abdelkader* Lep. (Hymenoptera: Sphecidae) [*Philanthus triangulum abdelkader*, an important honey bee (*Apis mellifera* L.) predator in Turkey]. Atatürk Universitesi Fen Fakültesi Dergisi 13(3-4):47–54.

Packard, A. S., Jr. 1869. The parasites of the honey-bee. American Naturalist 2:195–205.

Paddock, F. B. 1918. The beemoth or waxworm. Texas Agricultural Experiment Station Bulletin number 231.

———. 1926. The chronological distribution of the beemoth. Journal of Economic Entomology 19:136–141.

———. 1926. Control of bee diseases and pests.Extension Service Bulletin, Iowa College of Agriculture. #138, 16pp.

———. 1930. The beemoths. Journal of Economic Entomology 23:422–428.

Page, R. E., Jr., and E. H. Erickson, Jr. 1988. Reproduction by worker honey bees (*Apis mellifera* L.). Behavioral Ecology and Sociobiology 23:117–126.

Page, R. E., M. K. Fondrk, and G. E. Robinson. 1993. Selectable components of sex allocation in the honeybee (*Apis mellifera* L.). Behavioral Ecology 4:239–245.

Page, R. E., Jr., and N. E. Gary. 1990. Genotypic variation in susceptibility of honey bees, *Apis mellifera*, to infestation by tracheal mites, *Acarapis woodi*. Experimental and Applied Acarology 8:275–283.

Page, R. E., Jr., and H. H. Laidlaw, Jr. 1982a. Closed population honeybee breeding. 1. Population genetics of sex determination. Journal of Apicultural Research 21:30–37.

Page, R. E., Jr., and H. H. Laidlaw, Jr. 1982b. Closed population honeybee breeding. 2. Comparative methods of stock maintenance and selective breeding. Journal of Apicultural Research 21:38–44.

Page, R. E., Jr., and R. W. Marks. 1982. The population genetics of sex determination in honey bees: random mating in closed populations. Heredity 48:263–270.

Page, R. E., and G. E. Robinson. 1991. The genetics of the division of labour in honey bee colonies. Advances in Insect Physiology 23:117–169.

Paillot, A., S. Kirkor, and A.-M. Granger. 1949. L'Abeille: anatomie, maladies, ennemis [Bees: anatomy, diseases, enemies]. 3d ed. Editions de Trevoux, France.

Pain, J., and J. Maugenet. 1966. Recherches biochimiques et physiologiques sur le pollen emmagasiné par les abeilles [Biochemical and physiological study of pollen stored by honey bees]. Annales de l'Abeille 9:209–236.

Palmer-Jones, T. 1947a. A recent outbreak of honey poisoning. Part I. Historical and descriptive. New Zealand Journal of Science and Technology 29:107–114.

————. 1947b. A recent outbreak of honey poisoning. Part III. Toxicology of the poisonous honey and the antagonism of tutin, mellitoxin, and picrotoxin by barbiturates. New Zealand Journal of Science and Technology 29:121–125.

————. 1949. Diseases of bees in New Zealand. New Zealand Journal of Agriculture 79:483–486.

————. 1965. Poisonous honey overseas and in New Zealand. New Zealand Medical Journal 64:631–637.

————. 1968. Nectar from karaka trees poisonous to honey bees. New Zealand Journal of Agricultural Research 117:77.

————. 1970. Causes of adult honey bee mortality in New Zealand. New Zealand Beekeeper 32:16–20.

Palmer-Jones, T., and J. S. Line. 1962. Poisoning of honey bees by nectar from the karaka tree (*Corynocarpus laevigata* J. R. et G. Forst). New Zealand Journal of Agricultural Research 5:433–436.

Pandey, R. S. 1967. *Varroa jacobsoni*: a new mite infesting honeybee (*Apis indica*) colonies in India. Bee World 48:16.

Pankiw, P., and J. Corner. 1966. Transmission of American foul brood by package bees. Journal of Apicultural Research 5:99–101.

Papadopoulo, P. 1964. Enemies of bees (I). Rhodesia Agricultural Journal 61:114–115.

————. 1965. Enemies of bees (II). Rhodesia Agricultural Journal 62:2–3.

Park, O. W. 1936. Disease resistance and American foulbrood. American Bee Journal 76:12–15.

————. 1937. Testing for resistance to American foulbrood in honeybees. Journal of Economic Entomology 30:504–512.

Park, W. 1922. A new bee parasite from Mexico. American Bee Journal 62:260.

Parker, F. D. 1984. Effect of fungicide treatments on incidence of chalkbrood disease in nests of the alfalfa leafcutting bee (Hymenoptera: Megachilidae). Journal of Economic Entomology 77:113–117.

————. 1985. Effective fungicide treatment for controlling chalkbrood disease (Ascomycetes: Ascosphaeraceae) of the alfalfa leafcutting bee (Hymenoptera: Megachilidae) in the field. Journal of Economic Entomology 78:35–40.

————. 1987. Further studies on the use of fungicides for control of chalkbrood of the alfalfa leafcutting bee. Journal of Apicultural Research 26:144–149.

————. 1988. Influence of wood, paper, and plastic nesting units on efficacy of three candidate fungicides for control of chalkbrood in the alfalfa leafcutting bee (Hymenoptera: Megachilidae). Journal of Economic Entomology 8:789–795.

Parks, H. B. 1930. The least shrew, *Cryptotis parva*. Beekeepers Item 14:157–159.

Parmly, E. 1867. Egyptian bees. American Bee Journal 3:114.

Paterson, C. R. 1947. A recent outbreak of honey poisoning. Part IV. The source of the toxic honey: field observations. New Zealand Journal of Science and Technology 29:125–129.

Paterson, C. R., and A. W. Bennett. 1963. Methyl bromide control of wax moth in bee combs. New Zealand Journal of Agriculture 106:230–231, 233, 235–236.

Patetta, A. 1984 [Old and new problems of bee pathology in north-western Italy] Problemi di patologia apicola vecchi e nuovi dell'Italia nord-occidentale. *In* Passato e Presente dell Apicoltura Subalpina, Turin, Italy; Consigilio Nazionale delle Richerhe (1982): 219-225. Instituto di Entomologia Agraria e Apicoltura, Universitá di Torino.

Patwardhan, V. N., and J. W. White. 1973. Problems associated with particular foods. *In* Toxicants occurring naturally in foods. National Academy of Sciences, Washington, D.C.

Pech, J. 1966. Die Spitzmaus, ein Schädling für das Bienenvolk in Winter [The shrew, a winter pest of the bee colony]. Imkerfreund 21:315–316.

Pechhacker, H. 1980. Über den Aussagewert des Nosemabefundes aus dem Wintertotenfall. Bienen-Welt 22:35–37.

Peer, D. F. 1956. Multiple mating of queen honey bees. Journal of Economic Entomology 49:741–743.

Peer, D. F., and C. L. Farrar. 1956. The mating range of the honey bee. Journal of Economic Entomology 49:254–256.

Pellett, F. C. 1916. Productive bee-keeping. Lippincott's farm manuals. J. B. Lippincott, Philadelphia.

———. 1938. History of American beekeeping. Collegiate Press, Ames, Iowa.

———. 1947. American honey plants. 4th ed. Orange Judd Publishing, New York.

———. 1978. American honey plants. 5th ed., 3d printing. Dadant & Sons, Inc., Journal Printing Co., Carthage, Illinois.

Pence, R. J., and P. Lomax, 1973. Biodynamics of the excised honey bee abdomen. Insect World Digest 1:16–24.

Peng, G., J. C. Sutton, and P. G. Kevan. 1992. Effectiveness of honey bees for applying the biocontrol agent *Gliocladium roseum* to strawberry flowers to suppress *Botrytis cinerea*. Canadian Journal of Plant Pathology 14:117–129.

Peng, Y. 1988. The resistance mechanism of the Asian honey bee (*Apis cerana*) to the mite *Varroa jacobsoni*. In Africanized honey bees and bee mites. G. R. Needham, R. E. Page, Jr., M. Delfinado-Baker, and C. E. Bowman, editors. Ellis Horwood, Chichester, England.

Peng, Y., and M. E. Nasr. 1985. Detection of honeybee tracheal mites (*Acarapis woodi*) by simple staining techniques. Journal of Invertebrate Pathology 46:325–331.

Peng, Y.-S. C., Y. Fang, S. Xu, and L. Ge. 1987a. The resistance mechanism of the Asian honey bee, *Apis cerana* Fabr., to an ectoparasitic mite, *Varroa jacobsoni* Oudemans. Journal of Invertebrate Pathology 49:54–60.

Peng, Y. S., Y. Fang, S. Xu, L. Ge., and M. E. Nasr. 1987b. Response of foster Asian honey bee (*Apis cerana* Fabr.) colonies to the brood of European honey bee (*Apis mellifera* L.) infested with parasitic mite, *Varroa jacobsoni* Oudemans. Journal of Invertebrate Pathology 49:259–264.

Peng, Y.-S., M. E. Nasr, J. M. Marston, and Y. Fang. 1984. Digestion of Torula yeast, *Candida utilis*, by the adult honeybee, *Apis mellifera*. Annals of the Entomological Society of America 77:627–632.

Peng, Y.-S., and K-Y. Peng. 1979. A study on the possible utilization of immunodiffusion and immunofluorescence techniques as the diagnostic methods for American foulbrood of honeybees (*Apis mellifera*). Journal of Invertebrate Pathology 33:284–289.

Perepelova, L. I. 1927. Summer death of bees in 1927 [in Russian]. Opytnaya Paseka 2:250–252.

———. 1949. Effect of hellebore pollen on bees [in Russian]. *In* Works of the 27th session of the Veterinary Section of the Lenin Academy of Agricultural Sciences, Moscow.

Pérez Gómez, J. D. 1975. Biologia e citogenética de *Pseudohypocera kertetzi* (Enderlein, 1912) [Biology and cytogenetics of *Pseudohypocera kertetzi* (Enderlein, 1912)]. M.S. thesis, University of São Paulo, Ribeirão Preto, São Paulo, Brazil.

Perez-Mellado, V., and I. De La Riva. 1993. Sexual size dimorphism and ecology: the case of a tropical lizard, *Tropidurus melanopleurus* (Sauria: Tropiduridae). Copeia 1993:969–976.

Perkins, H. 1935. Adult bee losses apparently due to poisonous nectar. Bees and Honey 16:261.

Perott, D. C. F. 1975. Factors affecting use of mirex-poisoned protein baits for control of the European wasp (*Paravespula germanica*) in New Zealand. New Zealand Journal of Zoology 2:491–508.

Perret-Maisonneuve. 1925. Le pou des abeilles. L'Apiculture IXIX(2):34-37.

Perrins, C. M. 1940. The illustrated encyclopedia of birds: the definitive reference to birds of the world. Prentice-Hall, New York.

Perrins, G. 1933. "Death camas" cause of bee mortality. American Bee Journal 73:381.

Perry, R. 1970. Bears. Areo Publishing, New York.

Pettis, J. S. 1991. Biology and dispersal behavior of the honey bee tracheal mite *Acarapis woodi*. Ph.D. dissertation, Texas A&M University, College Station, Texas.

Pettis, J. S., R. L. Cox, and W. T. Wilson. 1988. Efficacy of fluvalinate against the honey bee tracheal mite, *Acarapis woodi*, under laboratory conditions. American Bee Journal 128:806.

Pettis, J. S., A. Dietz, and F. A. Eischen. 1989. Incidence rates of *Acarapis woodi* (Rennie) in queen honey bees of various ages. Apidologie 20:69–75.

Pettis, J. S., and T. Pankiw. 1994. Grooming behavior by the honey bee and tracheal mite dispersal. American Bee Journal 134:834–835.

Pettis, J. S., and W. T. Wilson. 1989. Reproduction of *Acarapis woodi* as related to seasonal host longevity. American Bee Journal 129:820.

Pettis, J. S., and W. T. Wilson. 1996. Life history of the honey bee tracheal mite, *Acarapis woodi*: (Acari: Tarsonemidae). Annals of the Entomological Society of America 89:368–374.

Pettis, J. S., W. T. Wilson, and F. A. Eischen. 1992. Nocturnal dispersal by female *Acarapis woodi* in honey bee (*Apis mellifera*) colonies. Experimental and Applied Acaralogy 15:99–108.

Pettis, J. S., W. T., Wilson, F. A. Eischen, and A. Suarez. 1987. Distribution of *Acarapis woodi* among rustic and modern honey bee hives in northeast Mexico. American Bee Journal 127:849.

Pettis, J. S., W. T. Wilson, H. Shimanuki, and P. D. Teel. 1991. Fluvalinate treatment of queen and worker honey bees (*Apis mellifera* L.) and effects on subsequent mortality, queen acceptance and supersedure. Apidologie 22:1–7.

Phadke, K. G., D. S. Bisht, and R. B. P. Sinha. 1966. Occurrence of the mite *Varroa jacobsoni* Oudemans in the brood cells of the honey bee, *Apis indica* F. Indian Journal of Entomology 28:411–412.

Phelan, L. P., A. W. Smith, and G. R. Needham. 1991. Mediation of host selection by cuticular hydrocarbons in the honey bee tracheal mite *Acarapis woodi* (Rennie). Journal of Chemical Ecology 17:463–473.

Philipsborn, H. von. 1952. Über Calciumoxalat in Pflanzenzellen [Calcium oxalate in plant cells]. Protoplasma 41:415–424.

Phillips, E. F. 1922. The occurrence of diseases of adult bees. U.S. Department of Agriculture Circular 218.

———. 1923. The occurrence of diseases of adult bees, II. U.S. Department of Agriculture Circular 287.

———. 1925. The bee-louse *Braula coeca* in the United States. U.S. Department of Agriculture Circular 334.

———. 1928. Beekeeping. 2d ed. MacMillan, New York.

Phillips, E. F., and G. F. White. 1912. Historical notes on the causes of bee diseases. U.S. Department of Agriculture, Bureau of Entomology Bulletin 98.

Phokedi, K. M. 1985. Apiculture and its problems in Botswana. Proceedings of the Third International (IBRA) Conference on Apiculture in Tropical Climates, Nairobi 1984:64–65.

Piek, T., and W. Spanjer. 1986. Chemistry and pharmacology of solitary wasp venoms. *In* Venoms of the Hymenoptera: biochemical, pharmacological and behavioral aspects. T. Piek, editor. Academic Press, London, England.

Pike, H. A. 1947. Beekeeping in Rhode Island. Rhode Island Department of Agriculture and Conservation Publication.

Pinnock, D. E., R. B. Coles, and B. Donovan. 1988. A new spore cyst fungus from New Zealand. Australian Systematic Botany 1:387–389.

Pinnock, D. E., and N. E. Featherstone. 1984. Detection and quantification of *Melissococcus pluton* infection in honeybee colonies by means of enzyme-linked immunosorbent assay. Journal of Apicultural Research 23:168–170.

Piskovoi, F. R., and Y. E. Kolonschchakov. 1961. The poisoning of bees with rhododendron nectar [in Russian]. Pchelovodstvo 1961(5):44.

Pitt, D. T. 1937. The beekeeping industry in New Jersey. New Jersey Department of Agriculture Circular 279.

Plachy, E. 1944. Studie über die bakterizide Wirkung des Naturhonigs (Blütenund Blatthonig) aus verschiedenenen Höhenlagen sowie einige Untersuchungen über die Eigenschaft der antibakteriellen Hemmungsstoffe (Inhibine) im Naturhonig [Study on the bactericidal effect of natural honey from various altitudes as well as some investigations of antibacterial inhibitors in natural honey]. Zentralblatt für Bakteriologie, Parasitenkunde, Infektionskrankheiten und Hygiene 106:401–419.

Plath, O. E. 1923. The bee-eating proclivity of the skunks. American Naturalist 57:570–574.

Plugge, P. C. 1891. Giftiger Honig von *Rhododendron ponticum* [Poisonous honey from *Rhododendron ponticum*]. Archiv der Pharmazie und Berichte der Deutschen Pharmazeutischen Gesellschaft 229:554–558.

Plurad, S. B., and P. A. Hartman. 1965. The fate of bacterial spores ingested by adult honey bees. Journal of Invertebrate Pathology 7:449–454.

Poinar, G. O., Jr. 1975. Entomogenous nematodes. E. J. Brill, Leiden, West Germany.

Poinar, G. O., Jr., and G. M. Thomas. 1984. Laboratory guide to insect pathogens and parasites. Plenum Press, New York.

Polson, A. 1971. New approaches to ultracentrifugation. *In* Methods in virology, volume 5. K. Maramarosch and H. Koprowski, editors. Academic Press, New York.

Poltev, V. I. [also transliterated as W. I. Poltew]. 1953. Effect of nosema and amoeba infections on the productivity of honey bee colonies in the Primorsky region [in Russian]. Pchelovodstvo 1953(2):46–48.

———. 1968. Robbing and drifting of honeybees [in Russian]. Pchelovodstvo 1968(7):42.

Poltev, V. I., and O. F. Grobov. 1967. Rickettsiosis of bees [in Russian]. Pchelovodstvo 1967(8):27.

Poltev, V. I., and E. V. Neshatayeva. 1969. La mélanose expérimentale des abeilles causée par le champignon *Aureobasidium pullulans* (De Bary) Arnaud [Experimental melanosis of bees caused by the fungus *Aureobasidium pullulans* (De Bary) Arnaud]. Bulletin Apicole 12:189–198.

Poltev, V. I., N. P. Smaragdova, and G. V. Snoz. 1976. Diagnosis of pollen and nectar toxicosis in bees based on pollen morphology. *In* Pollination of entomophilous agricultural crops by bees. R. B. Kozin, editor. Amerind, New Delhi (available from the U.S. Department of Agriculture).

Poltew, W. I. [also transliterated as V. I. Poltev]. 1956. Food toxicosis of bees and their diagnosis. Proceedings of the 16th International Beekeeping Congress 16:73–74.

Popa, A. 1980. Apiculture in Lebanon. American Bee Journal 120:366–367.

Popolizio, E. R., and L. A. Pailhe. 1973a. Storing combs in "wax-moth-safe" storage rooms. Proceedings of the 24th International Apicultural Congress, Buenos Aires, Argentina, October 14–20, 1973. Apimondia, Bucharest.

Popolizio, E. R., and L. A. Pailhe. 1973b. Storage of wax materials, combs in hive bodies and supers in the field. Proceedings of the 24th International Apicultural Congress, Buenos Aires, Argentina, October 14–20, 1973. Apimondia, Bucharest.

Porter, C. E. 1936. El piojo de las abehas *Braula coeca* Nitzsch 1818 [The bee louse, *Braula coeca* Nitzsch (a summary from the literature)]. Review of Chilean Hist. nat. 39: 141-144.

Potts, R., and A. K. Behrensmeyer. 1992. Late Cenozoic terrestrial ecosystems. In Terrestrial ecosystems through time. A. K. Behrensmeyer, J. D. Damuth, W. DiMichele, R. Potts, H.-D. Sues, and S. L. Wing, editors. University of Chicago Press.

Poulton, E. B. 1906 [published 1907]. Predaceous insects and their prey. Transactions of the Entomological Society of London 1906, Part 3:323–409.

Poutiers, F., A. Giauffret, and C. Vago. 1969. Action de différents virus et de rickettsies d'invertébrés en cultures cellulaires d'*Apis mellifera* (Hym.) [Action of different viruses and rickettsia of invertebrates in cell cultures of *Apis mellifera* (Hym.)]. Annales de la Société Entomologique de France, nouvelle série 5:1001–1007.

Pouvreau, A. 1967. Contribution à l'étude morphologique et biologique d'*Aphomia sociella* L. (Lepidoptera, Heteroneura, Pyralidoidea, Pyralidae), parasite des nids de bourdons (Hymenoptera, Apoidea, Bombus Latr.) [Contributions to the morphological and biological study of *Aphomia sociella* L., parasite of bumble bee nests]. Insectes Sociaux 14:57–72.

————. 1973. Les ennemis des bourdons. I. Etude d'une zoocénose: le nid de bourdons [Enemies of bumble bees. I. Studies of a zoocoenosis: the nest of bumblebees]. Apidologie 4:103–148.

Powell, G. L., and A. P. Russell. 1984. The diet of the eastern short-horned lizard (*Phrynosoma douglasi brevirostre*) in Alberta and its relationship to sexual size dimorphism. Canadian Journal of Zoology 62:428–440.

Powell, J. A., and G. I. Stage. 1962. Prey selection by robberflies of the genus *Stenopogon*, with particular observations on *S. engelhardti* Bromley (Diptera: Asilidae). Wasmann Journal of Biology 20:139–157.

Prater, S. H. 1965. The book of Indian animals. Bombay Natural History Society, Bombay.

Prell, H. 1926a. The amoeba-disease of adult bees: a little-noticed springtime disease. Bee World 8:10–13.

————. 1926b. Beiträge zur Kenntnis der Amöbenseuche der erwachsenen Konigbiene. Archiv für Bienenkunde 7:113–121.

————. 1926c. Die Amöbenseuche der erwachsenen Bienen. Eine wenig beachtete Früjahrskrankheit. Leipziger Bienenzeitung 41:28–31.

Prest, D. B., M. Gilliam, S. Taber III, and J. P. Mills. 1974. Fungi associated with discolored honey bee, *Apis mellifera*, larvae and pupae. Journal of Invertebrate Pathology 24:253–255.

Preston, R. J. 1966. North American trees. M.I.T. Press, Cambridge, Massachusetts.

Prica, M. 1937. Über die bakterizide Wirkung des Naturhonigs [On the antibacterial effect of natural honey]. Zeitschrift für Hygiene und Infektionskrankheiten 120:437–443.

Proctor, M. and P. Yeo. 1972. The pollination of flowers. Taplinger Publishing Company, New York.

Prökschl, H. 1953. Beiträge zur Kenntnis der Entwicklungsgeschichte von *Pericystis apis* Maassen [Contributions to the knowledge of development of *Pericystis apis* Maassen]. Archiv für Mikrobiologie 18:198–209.

Prosie, V. 1959. Furnica mare de padure: un dauntor pericules al albinelor [The redwood ant: a dangerous pest of bees]. Apicultura (Bucharest) 32:89.

Pryce-Jones, J. 1944. Some problems associated with nectar, pollen, and honey. Proceedings of the Linnean Society of London 155:129–174.

Public Law 87-539, 87th Congress, July 19, 1962.

Public Law 34-319, 94th Congress, June 25, 1976.

Puerta, F., J. M. Flores, M. Bustos, F. Padilla, and F. Campano. 1994. Chalkbrood development in honeybee brood under controlled conditions. Apidologie 25:540–546.

Puerta Puerta, F., F. Padilla Alvarez, M. Bustos Ruiz, P. Pellin Martinez, J. M. Flores Serrano, and M. Hermoso de Mendoza Salcedo. 1989. Contribución al estudio de la etiologica de la ascoferosis en *Apis mellifera* [Contribution to the study of the etiology of ascosphaerosis in *Apis mellifera*]. Revista Ibérica de Micología 6:17–24.

Punjabi, G. M. 1961. Some suggestions for the improvement of beekeeping. Indian Bee Journal 23:40–42.

Quicke, D. L. J. 1988. Notes on prey taken by some N. American spiders. British Journal of Entomology and Natural History 1:107–110.

Quintero, M. T., and I. Canales. 1987. Un caso de asosiación parasitaria de un coleoptero Meloida con *Apis mellifera* [A case of parasitic association of a meloid coleopteran with *Apis mellifera*]. Veterinaria, Mexico 18:135–138.

Ragsdale, D., and B. Furgala. 1986. Progress on new detection methods for *Acarapis woodi*. Proceedings of the Honey Bee Tracheal Mite (*Acarapis woodi*, R.) Scientific Symposium, July 8-9, 1986, Saint Paul, Minnesota. American Association of Professional Apiculturists, Gainesville, Florida.

Rajchev, R. G. 1988. A study on the nutritive regime of the bear (*Ursus arctos* L., 1758) on the southern slopes of the Central Balkan Range [Bulgaria]. Ekologiya, Bulgaria (21):17–24.

Ramanan, R. V., and S. Ghai. 1984. Observations on the mite *Neocypholaelaps indica* Evans and its relationship with the honey bee *Apis cerana* indica Fabricius and the flowering of *Eucalyptus* trees. Entomon 9:291–292.

Ramirez, B. W. and J. Malavasi. 1992. Conformation of the pretarsus of *Barula coeca* Nitzsch (Braulidae: Diptera), the bee louse. Bee Science 2(2): 106-107.

Ramirez, W. 1984. Biologia del género *Melaloncha* (Phoridae), moscas parasitoides de la abeja doméstica (*Apis mellifera* L.) en Costa Rica [Biology of the genus *Melaloncha* (Phoridae), parasitoids of the honey bee (*Apis mellifera* L.) in Costa Rica]. Revista de Biologia Tropical 32:25–28.

Ramsey, F. W. 1946. Chasing ants from hives. Gleanings in Bee Culture 74:546.

Randall, J. B. 1982. Prey records of the green lynx spider *Peucetia viridans* (Hentz) (Araneae, Oxyopidae). Journal of Arachnology 7:149154.

Rangarajan, A. V. 1964. *Protaetia aurichalcea* F. (Cetonidae), a new beetle feeding on pollen stores of *Apis indica* F. Indian Bee Journal 26:26.

Rank, G. H., F. P. Rank, and R. Watts. 1990. Chalkbrood (*Ascosphaera aggregata* Skou.) resistance of a univoltine strain of the alfalfa leafcutting bee, *Megachile rotundata* F. Journal of Applied Entomology 109:524–527.

Rao, G. V. 1968. A note on a simple device for protecting bee colonies from bears. Indian Bee Journal 30:11–12.

Rath, W. 1985. Vorversuche zur unterschiedlichen Kalkbrutresistenz bei züchterisch bearbeitetem Bienenmaterial [Preliminary tests on resistance to chalkbrood in bee colonies of selected lines]. Apidologie 16:220–222. (Abstract.)

———. 1991. Investigations on the parasitic mites *Varroa jacobsoni* Oud. and *Tropilaelaps clareae* Delfinado and Baker and their host *Apis cerana* Fabr. *Apis dorsata* Fabr., and *Apis mellifera* L. Ph.D. dissertation, Math.-Nat. Fakultat der Rheinischen Friedrich Wilhelms Universität, Bonn, Germany.

———. 1992. The key to *Varroa*: the drones of *Apis cerana* and their cell cap. American Bee Journal 132:329–331.

Rath, W., O. Boecking, and W. Drescher. 1995. The phenomena of simultaneous infestation of *Apis mellifera* in Asia with the parasitic mites *Varroa jacobsoni* Oud. and *Tropilaelaps clareae* Delfinado & Baker. American Bee Journal 135:125–127.

Rath, W., and M. Delfinado-Baker. 1991. Analysis of *Tropilaelaps clareae* populations from the debris of *Apis dorsata* and *Apis mellifera* in Thailand. In Proceedings of the International Symposium on Recent Research on Bee Pathology, September 5–7, 1990, Ghent, Belgium. W. Ritter, editor. Apimondia, Bucharest.

Rath, W., M. Delfinado-Baker, and W. Drescher. 1991. Observations on the mating behavior, sex ratio, phoresy and dispersal of *Tropilaelaps clareae* (Acari: Laelapidae). International Journal of Acarology 17:201–208.

Rath, W., and W. Drescher. 1990. Response of *Apis cerana* Fabr. colonies towards brood infested with *Varroa jacobsoni* Oud. and infestation rate of colonies in Thailand. Apidologie 21:311–321.

Rathmayer, W. 1962. Paralysis caused by the digger wasp Philanthus. Nature 196:1148–1151.

———. 1966. The effect of the poison of spider and digger wasps on their prey (Hymenoptera: Pompilidae, Sphecidae). Memorias do Instituto Butantan, International Symposium on Animal Venoms 33:651–658.

Ratnieks, F. L. W., M. A. Piery, and I. Cuadriello. 1991. The natural nest and nest density of the africanized honey bee (Hymenoptera, Apidae) near Tapachula, Chiapas, Mexico. The Canadian Entomologist 123:353–359.

Rau, P. 1946. The nests and the adults of colonies of *Polistes* wasps. Annals of the Entomological Society of America 39:11–27.

Raw, G. R. 1954. Enemies of bees. Bee World 35:159–160.

Rawson, J. H. B. 1963. Wasps. Bee Craft 45:14.

Rayment, T. 1917. Money in bees in Australasia. Whitcombe and Tombs, Melbourne.

————. 1926. A wicked neighbour. Australasian Beekeeper 28:46–49.

————. 1935. A cluster of bees. Endeavour Press, Sydney.

Rea, J. 1974. Some beekeeping observations in Ethiopia. Bee World 55:61–64.

Read, P. E. C., and B. J. Donovan. 1984. Leafcutting bee life history allocation details & management techniques, 3d ed. Entomology Division. DSIR, Lincoln, New Zealand.

Reed, S. E., and D. A. J. Tyrrell. 1974. Viruses associated with the healthy individual. *In* The normal microbial flora of man. F. A. Skinner and J. G. Carr, editors. Academic Press, London.

Rehm, S.-M., and W. Ritter. 1989. Sequence of the sexes in the offspring of *Varroa jacobsoni* and the resulting consequences for the calculation of the developmental period. Apidologie 20:339–343.

Reichstein, T., J. von Euw, J. A. Parsons, and M. Rothschild. 1968. Heart poisons in the monarch butterfly. Science 161:861–866.

Reid, G. M. 1984. Chalkbrood and half-moon in N. Z. Australasian Beekeeper 85:232–234.

Reinganum, C. 1968. Aberrant forms of the chronic bee-paralysis virus particle. Journal of Invertebrate Pathology 12:471–472.

Rennie, J. 1921. Isle of Wight disease in hive bees—acarine disease: the organism associated with the disease—*Tarsonemus woodi*, n. sp. Transactions of the Royal Society of Edinburgh 52:768–779.

Revell, I. L. 1960. Longevity of refrigerated nosema spores. Journal of Economic Entomology 53:1132–1133.

Reyes, O., F. 1983. A new record of *Pseudohypocera kerteszi*, a pest of honey bees in Mexico. American Bee Journal 123:119–120.

Rhoades, D. F., and J. C. Bergdahl. 1981. Adaptive significance of toxic nectar. American Naturalist 117:798–803.

Rhodes, J., and R. Goebel. 1986. Diseases of honey bees. The importance of correct diagnosis. Queensland Agricultural Journal, March–April 1986:71–74.

Ribbands, C. R. 1953. The behavior and social life of honeybees. Bee Research Association, London.

Richards, K. W. 1983. Impact of damage by the dried fruit moth, *Vitula edmandsae serratilineella* (Lepidoptera: Pyralidae), a pest of the alfalfa leafcutter bee, *Megachile rotundata* (Hymenoptera: Megachilidae). Canadian Entomologist 115:1549–1553.

————. 1984. Food preference, growth and development of larvae of the driedfruit moth, *Vitula edmandsae serratilineella* Ragonot. Journal of the Kansas Entomological Society 57:28–33.

————. 1985. Detection of a chalkbrood fungus, *Ascosphaera aggregata*, in larvae of the alfalfa leafcutter bee (Hymenoptera: Megachilidae) from western Canada. Canadian Entomologist 117:1143–1145.

Richter, A. A. von. 1912. Über einen osmophilen Organismus, den Hefepilz *Zygosaccharomyces mellis acidi* sp. n. [A new species of osmophilic yeast, *Zygosaccharomyces mellis acidi*]. Mykologisches Zentralblatt 1:67–76.

Riddiford, L. M., and J. W. Truman. 1978. Biochemistry of insect hormones and insect growth regulators. *In* Biochemistry of insects. M. Rockstein, editor. Academic Press, New York.

Riedel, S. M., and H. Shimanuki. 1965. A conopid fly parasite found in the honey bee. Journal of Invertebrate Pathology 8:272–273.

Riggs, T. 1924. Peculiar case of poisoning. Gleanings in Bee Culture 52:369–370.

Riley, C. V., and L. O. Howard. 1892. General notes. A honey bee enemy in California. Insect Life 4:343.

Rinaldi, A. J. M., L. A. Pailhe, and E. Popolizio. 1967. *Euphoria lurida* Fabr., un enemigo de la abeja [*Euphoria lurida* Fabr., an enemy of the bee]. Tucumán Universidad, Facultad de Agronomia, Tucumán, Argentina, Miscelanea 22.

————. 1970 *Eicherax ricnotes*, a bee-hunting fly in Tucuman, Argentina [*Eicherax ricnotes*, Engel moscardon cazador de abejas en Tucuman]. I deg Congresso Brasileiro de Apicultura. Florianopolis, SC, Brazil; DEMA, SC and ACA. 306-308.

———. 1971. A predatory bee fly in Tucuman, Argentina: *Eicherax ricnotes* [Un diptero cazador de abejas en Tucuman: El*Eicherax ricnotes* Engel. (Asilidae)]. Revista Agronomica del Noroeste Argentino 8(3/4): 451–455.

Rinderer, T. E., M. S. Blum, H. M. Fales, Z. Bian, T. H. Jones, S. M. Buco, V. A. Lancaster, R. G. Danka, and D. F. Howard. 1988. Nest plundering allomones of the fire bee *Trigona (Oxytrigona) mellicolor*. Journal of Chemical Ecology 14:495–501.

Rinderer, T. E., A. M. Collins, and M. A. Brown. 1983. Heritabilities and correlations of the honey bee: response to *Nosema apis*, longevity, and alarm response to isopentyl acetate. Apidologie 14:79–85.

Rinderer, T. E., and K. D. Elliott. 1977. Worker honey bee response to infection with *Nosema apis*: influence of diet. Journal of Economic Entomology 70:431–433.

Rinderer, T. E., L. I. Guzmán, J. M. Kulincevic, G. T. Delatte, L. D. Beaman, and S. M. Buco. 1993. The breeding, importing, testing and general characteristics of Yugoslavian honey bees bred for resistance to *Varroa jacobsoni*. American Bee Journal 133:197–200.

Rinderer, T. E., R. L. Hellmich II, R. G. Danka, and A. M. Collins. 1985. Male reproductive parasitism: a new factor in the Africanization of European honey bee populations. Science 228:1119–1121.

Rinderer, T. E., and R. L. Hellmich II. 1991. The process of Africanization. *In* The "African" honey bee. M. Spivak, D. J. C. Fletcher, and M. D. Breed, editors. Westview Press, Boulder, Colorado.

Rinderer, T. E., and W. C. Rothenbuhler. 1969. Resistance to American foulbrood in honey bees. X. Comparative mortality of queen, worker, and drone larvae. Journal of Invertebrate Pathology 13:81–86.

Rinderer, T. E., and W. C. Rothenbuhler. 1975. The fate and effect of hairs removed from honeybees with hairless-black syndrome. Journal of Invertebrate Pathology 26:305–308.

Rinderer, T. E., W. C. Rothenbuhler, and J. M. Kulincevic. 1975. Responses of three genetically different stocks of the honey bee to a virus from bees with hairless-black syndrome. Journal of Invertebrate Pathology 25:297–300.

Rinderer, T. E., J. A. Stelzer, B. P. Oldroyd, S. M. Buco, and W. L. Rubink. 1991. Hybridization between European and Africanized honey bees in the neotropical Yucatan peninsula. Science 253:309–311.

Rinderer, T. E., and H. A. Sylvester. 1978. Variation in response to *Nosema apis*, longevity, and hoarding behaviour in a free-mating population of the honey bee. Annals of the Entomological Society of America 71:372–374.

Ritter, W. 1981. Varroa disease of the honeybee *Apis mellifera*. Bee World 62:141–153.

———. 1988. *Varroa jacobsoni* in Europe, the tropics, and the subtropics. In Africanized honey bees and bee mites. G. R. Needham, R. E. Page, Jr., M. Delfinado-Baker, and C. E. Bowman, editors. Ellis Horwood, Limited, Chichester.

Ritter, W., and D. De Jong. 1984. Reproduction of *Varroa jacobsoni* O. in Europe, the Middle East and tropical South America. Zeitschrift für Angewandte Entomologie 98:55–57.

Ritter, W., E. Leclercq, and W. Koch. 1984. Observations des populions d'abeilles et de *Varroa* dans les colonies à différents niveaux d'infestation [Observations on bee and *Varroa* mite populations in infested honey bee colonies]. Apidologie 15:389–400.

Ritter, W., P. Michel, M. Bartholdi, and A. Schwendemann. 1990. Development of tolerance to *Varroa jacobsoni* in bee colonies in Tunisia. Proceedings of the International Symposium on Recent Research in Bee Pathology. W. Ritter, editor. Apimondia, Bucharest.

Ritter, W., and U. Schneider-Ritter. 1988. Differences in biology and means of controlling *Varroa jacobsoni* and *Tropilaelaps clareae*, two novel parasitic mites of *Apis mellifera*. In Africanized honey bees and bee mites. G. R. Needham, R. E. Page, Jr., M. Delfinado-Baker, and C. E. Bowman, editors. Ellis Horwood, Chichester, England.

Rivnay, E., and H. Bitinsky-Salz. 1949. The oriental hornet (*Vespula orientalis* L.): its biology in Israel. Bulletin of the Agricultural Research Station, Rehovot, number 52.

Roberts, D., and E. Smaelli. 1958. Control of wax moth on comb honey and stored bee combs. New Zealand Journal of Agriculture 97:464–468.

Roberts, D. M. 1968. Fatty acids in honeybees (*Apis mellifera*) infected with the protozoan *Nosema apis*. Journal of Invertebrate Pathology 11:234–236.

Roberts, D. W., and R. A. Humber. 1984. Entomopathogenic fungi. In Conference report on infection processes of fungi. D. W. Roberts and J. R. Aist, editors. Rockefeller Foundation, New York.

Roberts, W. C. 1944. Multiple mating of queen bees proved by progeny and flight tests. Gleanings in Bee Culture 72:255–259, 303.

Robineau-Desvoidy, J. B. 1836. Notice sur un nouvel ennemi de l'abeille domestique [Record of a new enemy of the honey bee]. Comptes Rendus Hebdomadaires des Séances de l'Académie des Sciences 3:689.

Robinson, F. A. 1961. Bees, bears and electric fences. Gleanings in Bee Culture 89:138–141.

———. 1963. Beekeeping among the bears. American Bee Journal 103:454–456.

———. 1965. A "bear-proof" electric fence. Florida Agricultural Extension Service Circular 289.

Robinson, F. A., and E. Oertel. 1950. Chlordane for control of Argentine ants. American Bee Journal 90:406–407.

Robinson, F. A., K. L. Thel, R. C. Littell, and S. B. Linda. 1986. Sampling apiaries for honey bee tracheal mite (*Acarapis woodi* Rennie): effects of bee age and colony infestation. American Bee Journal 126:193–105.

Robinson, G. E. 1981. *Pseudohypocera kerteszi* (Enderlein) (Diptera: Phoridae), a pest of the honey bee. Florida Entomologist 64:456–457.

———. 1982. A unique beekeeping enterprise in Colombia. Bee World 63:43–46.

———. 1985. Effects of a juvenile hormone analogue on honey bee foraging behaviour and alarm pheromone production. Journal of Insect Physiology 31:277–282.

Robinson, W. S. 1981. Beekeeping in Jordan. Bee World 62:91–97.

———. 1988. Beekeeping with *Apis cerana* in Sri Lanka. Bee World 69:125–130.

Robinson, W. S., R. Nowogrodzki, and R. A. Morse. 1989. The value of honey bees as pollinators of U.S. crops. American Bee Journal 129:411–423, 477–487.

Rodriguez-Navarro, A. 1968. Flora zimogena de las mieles españolas [The floral yeasts of Spanish honey]. Boletin del Instituto Nacional Investigaciones Agronomicas 28:33–42.

Rocha, M. T. and L. M. Delgado. 1986. *Senotainia tricuspis* in Portugal [*Senotainia tricuspis* em portugal]. Repositorio de Trabalhos do Laboratorio Nacional de Investigacao Veterinaria 18: 71-74.

Roff, C. 1966. Beating the toads. Australasian Beekeeper 67:266.

———. 1975. Honeybees, giant toads, and hive stands. Queensland Agricultural Journal 101:689–691.

Roff, C., and A. R. Brimblecombe. 1953. Pests of the beehive and honeybee. Queensland Agricultural Journal 77:1–8.

Roff, C., and A. R. Brimblecombe. 1963. Pests of honeybees and beehives. Queensland Agricultural Journal 89:540–545.

Ron, M., and C. Rosenthal. 1989. Genetic differences in the resistance to varroasis of bees in Israel. Proceedings of the 31st International Apicultural Congress, Warsaw, Poland (1987). Apimondia, Bucharest.

Ronna, A. 1936a. Observacões biológicas sobre dois dipteros parasitas *Apis mellifera* L. (Dipt., Phoridae, Sarcophagidae) [Biological observations on two dipteran parasites of *Apis mellifera* L.]. Revista de Entomologia (Rio de Janeiro) 6(1):1-9.

———. 1936b. Piolho ou pulga da abelha (*Braula coeca* Nitzsch) [The bee louse]. Rev. Dept. nac. Proc. anim. 3(1-6): 143-148.

———. 1937a. Novos dados sobre os habitos de *Melaloncha ronnai* Borgmeir (Diptera, Phoridae) endoparasita de *Apis mellifera* L. [New data on the habits of *M. ronnai* Borg., an endoparasite of *A. mellifera* L.]. Rev. Ent. 7(4): 409-413.

————. 1937b. Animaes inimigos da abelha domestica e de seus productors. Fauna das colmeias. [Animal pests of the honey bee and its products. The fauna of the beehive]. Rev. Dep. Prod. anim. 4(4-6): 47-112.

————. 1937c. *Melaloncha ronnai* Borg. 1935 (Phoridae) endoparastia de *A. mellifera* L. (Abelha domestica). Rev. Dep. Prod. Anim. 4(4-6): 113-126.

Root, A. I. 1966. The ABC and XYZ of bee culture. 33d ed. Revised by E. R. Root, assisted by H. H. Root and J. A. Root. A. I. Root Company, Medina, Ohio.

————. 1974. The ABC and XYZ of bee culture. 35th ed. Revised by E. R. Root, H. H. Root, and J. A. Root. A. I. Root Company, Medina, Ohio.

Root, A. I., and E. R. Root. 1919. The ABC and XYZ of bee culture. The A. I. Root Company, Medina, Ohio.

————. 1940. The ABC and XYZ of bee culture. The A. I. Root Company, Medina, Ohio.

Root, E. R., editor. 1899. *Apis dorsata* in the Philippines; a chance to get them to the United States. Gleanings in Bee Culture 27:228–229.

Root, L. C. 1880. Quinby's new bee-keeping. Orange Judd, New York.

————. 1919. Quinby's new beekeeping. Rev. ed. Orange Judd, New York.

Rosário Nunes, J. F., and G. C. Tordo. 1960. Prospecções e ensaios experimentalis apícolas em Angola [Prospects and apiculture investigations in Angola]. Estudo, Ensaios e Documentos 70. Junta de Investigações do Ultramar, Lisboa.

Rose, J. B., M. Christensen, and W. T. Wilson. 1984. *Ascosphaera* species inciting chalkbrood in North America and a taxonomic key. Mycotaxon 19:41–55.

Rose, R. I., and J. D. Briggs. 1969. Resistance to American foulbrood in honey bees. IX. Effects of honey-bee larval food on the growth and viability of *Bacillus larvae*. Journal of Invertebrate Pathology 13:74–80.

Rosebury, T. 1969. Life on man. Viking Press, New York.

Rosenkranz, P. 1986. Factors affecting *Varroa* reproduction in colonies of *Apis mellifera*: a comparison of European and Africanized honey bees. Arbeitsgemeinschaft der Institute für Bienenforschung E.V. Abstracts of the Varroa Workshop in Feldafing/Starnberg, West Germany.

Rosenkranz, P., and W. Engels. 1985. Konsequente Drohnenbrutentnahme, eine wirksame biotechnische Massnahme zur Minderung von Varroatose-Schäden an Bienenvölkern [The efficacy of drone brood removal on varroa infestations in bee colonies]. Allgemeine Deutsche Imkerzeitung 19:265–271.

Rosenkranz, P., and W. Engels. 1994. Infertility of *Varroa jacobsoni* females after invasion into *Apis mellifera* worker brood as a tolerance factor against varroatosis. Apidologie 25:402–411.

Rosenkranz, P., A. Rachinsky, C. Strambi, A. Strambi, B. Schricker, P. Röpstorf, and Z. L. P. Simões. 1990. Juvenile hormone titer in capped L5 larvae of various races of honeybees. Apidologie 20:524–526.

Rosenkranz, P., and M. Stürmer. 1992. Ernährungsabhängige Fertilität der Varroa-Weibchen in Arbeiterinnenbrut von *Apis mellifera* carnica und *Apis mellifera* capensis [Reduction in the fertility of varroa females in worker brood of *Apis mellifera* carnica and *Apis mellifera* capensis.]. Univ. Marie Curie, Sklodowska, Lublin-Poland Sectio DD 47:55–60.

Rosenkranz, P., N. C. Tewarson, A. Rachinsky, A. Strambi, C. Strambi, and W. Engels. 1993a. Juvenile hormone titer and reproduction of *Varroa jacobsoni* in capped brood stages of *Apis cerana* indica in comparison to *Apis mellifera* ligustica. Apidologie 24:375–382.

Rosenkranz, P., N. C. Tewarson, A. Singh, and W. Engels. 1993b. Differential hygienic behaviour towards *Varroa jacobsoni* in capped worker brood of *Apis cerana* depends on alien scent adhering to the mites. Journal of Apicultural Research 32:89–93.

Rossi, C. O., and M. R. Carranza. 1980. Momificación de las larvas de la abeja "*Apis mellifera*" L.—una nueva enfermedad en la Argentina causada por *Ascosphaera apis* (Maasen ex Claussen) Olive & Spiltoir [Mummification of honeybee (*Apis mellifera* L.) larvae—a disease new to Argentina, caused by *Ascophaera apis* (Maasen ex Claussen) Olive & Spiltoir]. Gaceta del Colmenar 42:235–237.

Rossi, J. 1959. Indagini microbiologiche sul tecc ethiopico [Microbiological analysis of Ethiopian mead]. Annali di Microbiologia 9:150–160.

Rothenbuhler, W. C. 1955. Hereditary aspects of gynandromorph occurrence in honey bees (*Apis mellifera* L.). Iowa State College Journal of Science 29:487–488.

———. 1957. Diploid male tissue as new evidence on sex determination in honey bees. Journal of Heredity 48:160–168.

———. 1958a. Genetics and breeding of the honey bee. Annual Review of Entomology 3:161–180.

———. 1958b. Progress and problems in the analysis of gynandromorphic honey bees. Proceedings of the 10th International Congress of Entomology 2:867–873.

———. 1964a. Behaviour genetics of nest cleaning in honey bees: I. Responses of four inbred lines to disease-killed brood. Animal Behavior 12:578–583.

———. 1964b. Behaviour genetics of nest cleaning in honey bees: IV. Responses of F₁ and backcross generations to disease-killed brood. American Zoologist 4:111–123.

Rothenbuhler, W. C., J. W. Gowen, and O. W. Park. 1952. Androgenesis with zygogenesis in gynandromorphic honeybees (*Apis mellifera* L.). Science 115: 637–638.

Rothenbuhler, W. C., J. W. Gowen, and O. W. Park. 1953. Allelic and linkage relationships of five mutant genes in honey bees. Journal of Heredity 44:251–253.

Rothenbuhler, W. C., J. M. Kulincevic, and W. E. Kerr. 1968. Bee genetics. Annual Review of Genetics 2:413–438.

Rothenbuhler, W. C., M. S. Polhemus, J. W. Gowen, and O. W. Park. 1949. Gynandromorphic honey bees. Journal of Heredity 40:308–311.

Rothenbuhler, W. C., and V. C. Thompson. 1956. Resistance to American foulbrood in honeybees. I. Differential survival of larvae of different genetic lines. Journal of Economic Entomology 49:470–475.

Roubaud, E. 1954. La thermogénèse chez les mites des abeilles [The thermogenesis of bee moths]. Comptes Rendus des Séances de l'Académie des Sciences (Paris) 238:1086–1088.

Roubik, D. W., and F. Reyes. 1984. African honey bees have not brought acarine mite infestations to Panama. American Bee Journal 124:665–667.

Roubik, D. W., B. H. Smith, and R. G. Carlson. 1987. Formic acid in caustic cephalic secretions of stingless bee, *Oxytrigona* (Hymenoptera: Apidae). Journal of Chemical Ecology 13:1079–1086.

Roussy, L. 1942. Parasites occasionnels de l'abeille: les grégarines [Occasional parasites of honey bees: the gregarines]. Gazette Apicole 43:98–99.

———. 1949. Intoxication mortelle des abeilles due au pollen gelé [Fatal intoxication of honey bees on gelle pollen]. Gazette Apicole 50:259.

———. 1962. Nouvelle contribution à l'étude du Pericystis apis (Mn) [New contribution to the study of *Pericystis apis* (Mn)]. Gazette Apicole 63:101–105.

———. 1975. Une morte parfumée: les fleurs qui tuent [A perfumed death: flowers that kill]. Gazette Apicole 76:4–5.

Rowe, C. A. 1966. How to do it. American Bee Journal 106:287.

Royce, L. A., G. W. Krantz, L. A. Ibay, and D. M. Burgett. 1988. Some observations on the biology and behavior of *Acarapis woodi* and *Acarapis dorsalis* in Oregon. In Africanized honey bees and bee mites. G. R. Needham, R. E. Page, Jr., M. Delfinado-Baker, and C. E. Bowman, editors. Ellis Horwood, Chichester, England.

Royce, L. A., and P. A. Rossignol. 1990a. Epidemiology of honey bee parasites. Parasitology Today 6:348–353.

Royce, L. A., and P. A. Rossignol. 1990b. Honey bee mortality due to tracheal mite parasitism. Parasitology 100:147–151.

Royce, L. A., and P. A. Rossignol. 1991. Sex bias in tracheal mite [*Acarapis woodi* (Rennie)] infestation of honey bees (*Apis mellifera* L.). BeeScience 1:159–161.

Royce, L. A., P. A. Rossignol, D. M. Burgett, and B. A. Stringer. 1991. Reduction of tracheal mite parasitism of honey bees by swarming. Philosophical Transactions of the Royal Society of London B 331:123–129.

Royce, L. A., B. A. Stringer, C. Kitprasert, D. M. Burgett, and P. A. Rossignol. 1993. Infestation of tracheal mites (Acari: Tarsonemidae) in feral and managed colonies of honey bees (Hymenoptera: Apidae). Journal of Economic Entomology 86:712–714.

Royds, T. F. 1918. The beasts, birds, and bees of Virgil. D. H. Blackwell, London.

Rozov, S. A. 1947. On the summer death of bees [in Russian]. Pchelovodstvo 1947(5):48–53.

Ruijter, A. de. 1982. Tobacco smoke can kill varroa mites. Bee World 63:138.

———. 1987. Reproduction of *Varroa jacobsoni* during successive brood cycles of the honey-bee. Apidologie 18:321–326.

Ruijter, A de and J. Calis. 1988. Distribution of female mites in honey bee worker brood cells of normal and manipulated depth (Acarina: Varroidae). Entomologia Generalis 14:107-109.

Ruiz-Argüeso, T., and A. Rodriguez-Navarro. 1973. Gluconic acid–producing bacteria from honey bees and ripening honey. Journal of General Microbiology 76:211–216.

Rust, R. W., and P. F. Torchio. 1991. Induction of *Ascosphaera* (Ascomycetes: Ascosphaerales) infections in field populations of *Osmia lignaria propinqua* Cresson (Hymenoptera: Megachilidae). Pan-Pacific Entomologist 67:251–257.

Rust, R. W., and P. F. Torchio. 1992. Effects of temperature and host developmental stage on *Ascosphaera torchioi* Youssef and McManus prevalence in *Osmia lignaria propinqua* Cresson (Hymenoptera: Megachilidae). Apidologie 23:1-9.

Ruttner, F. 1975. Races of bees. In The hive and the honey bee. Dadant & Sons, editors. Rev. ed. Dadant & Sons, Hamilton, Illinois.

———. 1983. Varroatosis in honeybees: extent of infestation and effects. In Proceedings of a meeting of the European Community Experts' Group, Wageningen, Netherlands. R. Cavalloro, editor. Balkema, Rotterdam.

———. 1988. Biography and taxonomy of honeybees. Springer-Verlag, Berlin.

———. 1990. The dark European honey bee. G. Beard and Son, Brighton.

———. 1992. Naturgesichte der Honigbienen [Natural vision in honey bees]. Ehrenwirth, München.

Ruttner, F., and H. Hänel. 1992. Active defense against *Varroa* mites in a Carniolan strain of honey bee (*Apis mellifera carnica* Pollmann). Apidologie 23:173–187.

Ruttner, F., N. Koeniger, and W. Ritter. 1980. Die Varroatose der Honigbiene. Brutstop und Brutentnahme [Stopping brood production and brood removal—to control varroa]. Allgemeine Deutsche Imkerzeitung 14:159–160.

Ruttner, F., N. Koeniger, and H. J. Veith. 1976. Queen Substance bei eierlegenden Arbeiterinnen der Honigbiene (*Apis mellifera* L.) [Queen substance in laying workers of the honeybee (*Apis mellifera* L.)]. Naturwissenschaften 63:434.

Ruttner, F., H. Marx, and G. Marx. 1984. Beobachtungen über eine mögliche Anpassung von *Varroa jacobsoni* an *Apis mellifera* L. in Uruguay [Observations about a possible adaptation of *Varroa jacobsoni* to *Apis mellifera* L. in Uruguay]. Apidologie 15:43–62.

Sabater Pi, G. 1960. Beitrag zur Biologie des flachland Gorillas [The biology of flatland gorillas]. Zeitschrift für Säugetierkunde 25:133–141.

Sacchetti, M. 1932. Nota presentata dall'Acc. Benedettino Prof. V. Peglion nella 12a sessione, 29 Maggio, 1932.

Sachs, H. 1952. Über das Verhalten und die Orientierung der Tracheenmilbe *Acarapis woodi* (Rennie 1921) auf Bienen [On the behavior and orientation of the tracheal mite *Acarapis woodi* (Rennie 1921) toward bees]. Zeitschrift für Bienenforschung 1:148–170.

Sackett, W. G. 1919. Honey as a carrier of intestinal diseases. Agricultural Experiment Station of the Colorado Agricultural College, Bulletin 252.

Sadov, A. V. 1980. The respiratory apparatus of the female *Varroa jacobsoni* mite [in Russian]. Veterinariya, Moscow, USSR 1980(11):43–47. From Apicultural Abstracts 1337/82.

Sadov, A. V., V. I. Poltev, V. P. Chumakov, and O. F. Grobov. 1980. The pretarsus of the female *Varroa* and the mechanism of its action [in Russian]. Veterinariya, Moscow, USSR 1980(2):36–39. From Apicultural Abstracts 637/81.

Sakagami, S. F. 1958. A bumblebee thieving from a honeybee hive. Journal of the Kansas Entomological Society 31:286.

————. 1959. Some interspecific relations between Japanese and European honeybees. Journal of Animal Ecology 28:51–68.

————. 1960. Preliminary report on the specific difference of behavior and other ecological characters between European and Japanese honeybees. Acta Hymenopterologica 1:171–198.

Sakagami, S. F., T. Matsumura, and K. Ito. 1980. *Apis laboriosa* in Himalaya, the little known world largest honey bee (Hymenoptera, Apidae). Insecta Matsumurana 19:47–77.

Sakai, T. 1989. Defensive behavior of honey bees against *Vespa simillima xanthoptera* attacking *Apis cerana* and *A. mellifera* colonies. Honeybee Science 10:73–78.

Sakai, T., and M. Ono. 1990. Comparative studies on the bionomics of *Apis mellifera* L. and *A. cerana japonica* Radoszkowski, I. Differences in the seasonal foraging activity. Tamaqawa Daiqaku Nogakuba Kenkys Hokoku (Bulletin of the Faculty of Agriculture, Tamaqawa University) 30:73–86.

Sakofski, F., N. Koeniger, and S. Fuchs. 1990. Seasonality of honey bee colony invasion by *Varroa jacobsoni* Oud. Apidologie 21:547–550.

Salin, N. K. 1973. Be more careful when oak is in bloom [in Russian]. Pchelovodstvo 1973(4):19.

Salmon, T. 1980. Skunks and their control. U.C. Apiaries, June–July 1980:2–3.

Sammataro, D., S. Cobey, B. H. Smith, and G. R. Needham. 1994. Controlling tracheal mites (Acari: Tarsonemidae) in honey bees (Hymenoptera: Apidae) with vegetable oil. Journal of Economic Entomology 87:910–916.

Samsinák, K., and O. Haragsim. 1972. The mite *Varroa jacobsoni* imported into Europe [in Czech]. Vcelarstvi 25:268–269.

Samsinák, K., E. Vobrázková, and O. Haragsim. 1978. *Melittiphis alvearius* Berlese, a little known bee mite. Journal of Apicultural Research 17:50–51.

Samsináková, A., S. Kálalová, and O. Haragsim. 1977. Effects of some antimycotics and disinfectants on *Ascosphaera apis* Maassen fungus in vitro. Zeitschrift für Angewandte Entomologie 84:225–232.

Sanford, M. T. 1994. Bees poisoned by plants. Apis 12(6): 1-3.

Santas, L. A., N. G. Emmanouel, and D. G. Papadopoulou-Batzaki. 1981. Preliminary observations from the use of two control methods for Varroa disease in Greece. Proceedings of the 28th International Congress of Apiculture 28:356–360.

Santis, L., and J. A. V. S. de Regalia. 1978. Hormigas recolectoras de polen en colmenas [Ants, collectors of pollen from bee colonies]. Ciencia y Abejas 3(12):43–46.

Savory, T. 1977. Arachnida. 2d ed. Academic Press, London.

Sawathum, A., and W. Ritter. 1995. Vergleichende Untersuchung bei *Ascosphaera apis* verschiedner Herkunft [Comparative study on *Ascosphaera apis* strains]. Apidologie 26:317–318.

Scarbrough, A. G. 1979. Predatory behavior and prey of *Diogmites missouriensis* Bromley in Arkansas (Diptera: Asilidae). Proceedings of the Entomological Society of Washington. 81(3): 391-400.

Schaller, G. B. 1963. The mountain gorilla: ecology and behavior. University of Chicago Press, Chicago.

Schaupp, W. C., Jr., and H. M. Kulman. 1992. Attack behavior and host utilization of *Coccygomimus disparis* (Hymenoptera: Ichneumonidae) in the laboratory. Environmental Entomology 21:401–408.

Schepartz, A. I. 1965a. The glucose oxidase of honey. II. Stereochemical substrate specificity. Biochimica et Biophysica Acta 96:334–336.

————. 1965b. The glucose oxidase of honey. III. Kinetics and stoichiometry of the reaction. Biochimica et Biophysica Acta 99:161–164.

————. 1966a. The glucose oxidase of honey. IV. Some additional observations. Biochimica et Biophysica Acta 118:637–640.

————. 1966b. Honey catalase: occurrence and some kinetic properties. Journal of Apicultural Research 5:167–176.

Schepartz, A. I., and M. H. Subers. 1964. The glucose oxidase of honey. I. Purification and some· general properties of the enzyme. Biochimica et Biophysica Acta 85:228–237.

Schepartz, A. I., and M. H. Subers. 1966. Catalase in honey. Journal of Apicultural Research 5:37–43.

Schiff, N. M., and W. S. Sheppard. 1993. Mitochondrial DNA evidence for the 19th century introduction of African honey bee into the United States. Experientia 49:530–532.

Schiff, N. M., and W. S. Sheppard. 1995. Genetic analysis of commercial honey bees (Hymenoptera: Apidae) from the southern United States. Journal of Economic Entomology 88:1216–1220.

Schiff, N. M., W. S. Sheppard, G. R. Loper, and H. Shimanuki. 1994. Genetic diversity of feral honey bee (Hymenoptera: Apidae) populations in the Southern United States. Annals of the Entomological Society of America 87:842–848.

Schiller, J. 1937. Über zwei tödliche Fälle von reiner Amöbeninfektion [Two lethal cases of pure amoeba infection]. Bienenvater 11:382–384.

Schmid-Hempel, P. and S. Durrer. 1991. Parasites, floral resources and reproduction in natural populations of bumble bees. Oikos 62(3): 342–350.

Schmidt, E. 1933. Über die Beute des *Dasypogon teutonus* L. (Dip., Asilid.) [On the prey of *Dasypogon teutonus* L. (Dip., Asilid.)]. Mitteilungen der Deutschen Entomologischen Gesellschaft 4:95–96.

Schmidt, J. O., P. J. Schmidt, and C. K. Starr. 1985. Investigating the giant honey bee, *Apis dorsata*, in Sabah. American Bee Journal 125:649–751.

Schmidt, P. J., and J. O. Schmidt. 1986. *Apis mellifera*, an unusual prey for a paper wasp, *Polistes major castaneicolor* (Hymenoptera). Entomological News 97:73–75.

Schmutz, E. M., and L. B. Hamilton. 1979. Plants that poison, an illustrated guide for the American Southwest. Northland Press, Flagstaff, Arizona.

Schneider, H. 1941. Untersuchungen über die Acarapis-Milben der Honigbiene. Die Flügel- und Hinterleibsmilbe [Research on the mite *Acarapis* and wing and hindgut mites of the honey bee]. Mitteilungen der Schweizerischen Entomologischen Gesellschaft 18:318–327.

———. 1951. Die Waldtrachtkrankheit [Forest-sickness]. Schweizerische Bienen-Zeitung 74:106–110.

Schneider, P. 1986. Einfluss der Varroa-Parasitierung während der Puppenentwicklung auf die Flugaktivität der Arbeitsbienen [The influence of *Varroa* infestation on drones and worker bees]. Apidologie 17:366–368.

Schneider, S. S., and L. C. McNally. 1992. Factors influencing seasonal absconding in colonies of the African honey bee, *Apis mellifera scutellata*. Insectes Sociaux 39:403–423.

Schoener, T. W. 1968. The *Anolis* lizards of Bimini: resource partitioning in a complex fauna. Ecology 49:704–726.

Schoener, T. W., J. B. Slade, and C. H. Stinson. 1982. Diet and sexual dimorphism in the very catholic lizard genus, *Leiocephalus*, of the Bahamas. Oecologia (Berlin) 53:160–169.

Schoenfield. 1878. Ein noch unbekannter Bienenfeind [A previously unknown bee enemy]. Eichstädtische Bienen-Zeitung 34:145–147.

Scholl, L. M. 1912. Texas beekeeping. Texas Department of Agriculture Bulletin, number 24, Austin, Texas.

Schousboe, C. 1986. The duration of closed cell stage in worker brood of Danish honeybees (*Apis mellifera* L.) in relation to increased resistance to the *Varroa* mite (*Varroa jacobsoni* Oud.). Tidsskr Planteavl 90:293–299.

Schubert, A. 1972. Bienenfeindliche Pflanzen [Plants harmful to bees]. Bienenpflege 8:171–172.

Schulz, A., N. Koeniger, and F. Ruttner. 1983. Drohnenbrut als Varroa-Falle [Drone brood as a trap for varroa mites]. Allgemeine Deutsche Imkerzeitung 17:52–54.

Schulz, A. E. 1984. Reproduktion und populationsentwicklung der parasitischen milbe *Varroa jacobsoni* Oud. in Abhängigkeit vom brutzyklus ihres wirtes *Apis mellifera* L. [Reproduction and population dynamics of the parasitic mite *Varroa jacobsoni* Oud. in correlation with the brood cycle of *Apis mellifera*]. Apidologie 5:401–419.

Schulz-Langner, E. 1957. 1957 ein Amöbenjahr! Wie lässt sich die Seuche unter Kontrolle bringen? [1957, an amoeba year! How shall the epidemic be brought under control?] Bienenzucht 10:147–149.

———. 1961. Kultivierung in vitro der in der Honigbiene Schmarotzenden Amöbe. Die Naturwissenschaft 48:137.

———. 1964. Der Entwicklungsgang der in den malpighischen Gefässen der Honigbiene lebenden Amöbe (*Malpighamoeba mellificae* Prell) in Kulturen und im Bienenkörper [The life cycle of *Malpighamoeba mellificae* Prell in culture and in the Malpighian tubules of the honey bee]. Zeitschrift für Bienenforschung 7:1–22.

———. 1966. Quantitativer Nachweis kleinster Saponinmengen durch Beachten der Hämolysedauer. Untersuchungen am Nektar der Rosskastanie, *Aesculus hippocastanum* [Quantitative determination of minute amounts of saponin by observation of the duration of hemolysis. Experiments on the nectar of horse chestnut, *Aesculus hippocastanum*]. Planta Medica 14:49–56.

———. 1967. Über den Trachtwert der Rosskastanie (*Aesculus hippocastanum*) unter besonderer Berücksichtigung des Saponingehaltes im Nektar [On the value of the nectar flow of horse chestnut (*Aesculus hippocastanum*), with special reference to its saponin content]. Zeitschrift für Bienenforschung 9:49–65.

Schwantes, U., and D. Eichelberg. 1984. Elektronmikroskopische Untersuchungen zum Parasitenbefall der Malpighischen Gefässe von *Apis mellifera* durch *Malpighamoeba mellificae* (Rhizopoda) [Electron microscopic investigations of parasitosis of Malpighian tubules of *Apis mellifera* by *Malpighamoeba mellificae* (Rhizopoda)]. Apidologie 15:435–450.

Scott, C. D., M. L. Winston, K. N. Slessor, G. G. S. King, and G. G. Grant. 1984. The biology and pheromone-based monitoring of the driedfruit moth, *Vitula edmandsae serratilineella* (Lepidoptera: Pyralidae). Canadian Entomologist 116:1007–1013.

Scott, H. E., F. B. Meacham, and J. F. Greene. 1970. Honey bees in North Carolina. North Carolina Agricultural Extension Service Circular 512.

Scott, P. M., B. B. Caldwell, and G. S. Wiberg. 1971. Grayanotoxins, occurrence and analysis in honey and a comparison of toxicities in mice. Food and Cosmetic Toxicology 9:179–184.

Scott-Dupree, C. D., and G. D. Otis. 1988. Parasitic mites of honey bees: to bee or not to bee. Highlights 11:24–28. Ontario Ministry of Agriculture, Canada.

Seal, D. W. A. 1957. Chalk brood disease of bees. New Zealand Journal of Agriculture 95:562.

Sears, W. A. 1947. Getting rid of ants. Gleanings in Bee Culture 75:596.

Sechrist, E. L. 1944. Honey getting. American Bee Journal, Hamilton, Illinois.

Sechrist, E. L., F. E. Todd, and G. H. Vansell. 1933. A search for a method of producing honey in the poisonous buckeye area of California. American Bee Journal 73:390–391.

Sedova, V. I. 1959. A study of the biology of *Philanthus triangulum* Fabr. in the Turkmen SSR [in Russian]. Trudy Instituta Zoologii i Parazitologii, Akademii Nauk Tashkent 4:20–33.

Seeley, T. D., R. H. Seeley, and P. Akratanakul. 1982. Colony defense strategies of the honey bees in Thailand. Ecological Monographs 52:43–63.

Séguy, E. 1927. Diptères (Brachycères), Asilidae. Faune de France (Paris) 17:11–90.

———. 1930. Un nouveau parastie d'abeille domestique. Encylopaedia Entomologique Series B II. Diptera 5: 169-170.

———. 1965. Le *Sarcophaga nigriventris* parasite de l'abeille domestique en Europe occidentale (Insecte, Diptère, Calliphoride) [*Sarcophaga nigriventris*, a parasite of the honey bee in western Europe (Insecta, Diptera, Calliphoridae)]. Bulletin du Muséum National d'Histoire Naturelle, series 2, 37:407–411.

Seifert, L. 1968. Die Honigbiene: ihre Krankheiten und Schädlinge [Honeybees: their diseases and pests]. Deutscher Landwirtschaftsverlag, Berlin.

Selders, H. 1933. Ant proof hive stands. Gleanings in Bee Culture 61:230.

Seltner, B. 1950. Meloëlarven als Bienenschädlinge [Larvae of *Meloe* as bee pests]. Imkerfreund 5:56–58.

Selwyn, H. H. 1970. Is the kingbird a suspect? Gleanings in Bee Culture 98:114.

Senegacnik, J. 1990. O zatiranju varoze z vodno emulzijo fluvalinata [Control of varroasis using an aqueous emulsion of fluvalinate]. Slovenski Cebelar 92:292–297.

Severin, H. C. 1937. *Zodion fulvitrons* Say (Dip. Conopidae), a parasite of the honey bee. Entomological News 48:243–244.

Severson, D. L., R. E. Page, and E. H. Erickson, Jr. 1986. Closed population breeding in honey bees: a report on its practical application. American Bee Journal 126:93–94.

Shabanov, M., and D. Georgiev. 1979. The role of mildew and mildew-affected honey in summer and autumn deaths of honeybee colonies [in Bulgarian, English summary]. Veterinarnomeditsinski Nauki 16:88–95.

Shaginyan, E. G. 1956. Poisoning of bees by the alkaloids in henbane [in Russian]. Pchelovodstvo 1956(11):45–46.

Shah, F. A. 1987. Twenty years of acarine mite in India. Gleanings in Bee Culture 115:517.

Shah, F. A., and T. A. Shah. 1981. Apis iridescent virus. Indian Bee Journal 43:44–45.

Shah, F. A., and T. A. Shah. 1986. Search for *Nosema apis* in Kashmir *Apis cerana*. Indian Bee Journal 48:55–56.

Shah, F. A., and T. A. Shah. 1987. Thai sacbrood disease of *Apis cerana*. British Bee Journal 115:237–240.

Shah, F. A., and T. A. Shah. 1991. *Vespa velutina*, a serious pest of honey bees in Kashmir. Bee World 72:161–164.

Shahid, M. 1992. Beekeeping in the north west frontier province of Pakistan. *In* Honeybees in mountain agriculture. L. R. Verma, editor. Westview Press, Boulder, Colorado.

Sharma, O. P., G. S. Dogra, and R. C. Mishra. 1979. Evaluation of methods of control of predatory wasps, *Vespa* spp. in apiaries. Indian Bee Journal 41:10–16.

Sharma, O. P., and D. Raj. 1988. Ecological studies on predatory wasps attacking Italian honeybee, *Apis mellifera* L., in Kangra Shivaliks. Indian Journal of Ecology 15:168–171.

Sharma, O. P., A. K. Thakur, and R. Garg. 1985. Control of wasps attacking bee colonies. Indian Bee Journal 47:27–29.

Shaw, D. E. 1990. The incidental collection of fungal spores by bees and the collection of spores in lieu of pollen. Bee World 71:158–176.

Shaw, D. E., and D. F. Robertson. 1980. Collection of *Neurospora* by honeybees. Transactions of the British Mycological Society 74:459–464.

Shaw, F. R. 1941. Bee poisoning: review of the more important literature. Journal of Economic Entomology 34:16–21.

———. 1946. Does the common horse chestnut possess materials toxic to bees? Gleanings in Bee Culture 74:471.

———. 1961. The cabinet beetle: a pest of stored combs. Gleanings in Bee Culture 89:604.

Shaw, F. R., and J. Weidhaas, Jr. 1956. Distribution and habits of the giant hornet in North America. Journal of Economic Entomology 49:275.

Shelly, T. E. 1979. A list of known robber flies of Pennsylvania (Diptera: Asilidae). Entomological News 90(2): 95-99.

Shepherd, M. W. 1892. Punic (or Tunisian) bees. Gleanings in Bee Culture 20:504.

Sheppard, W. S. 1988. A comparative study of enzyme polymorphism in United States and European honey bee (Hymenoptera: Apidae) populations. Annals of the Entomological Society of America 81:886–889.

———. 1989. A history of the introduction of honey bee races into the United States—I and II. American Bee Journal 129:617–619, 664–667.

Sheppard, W. S., and B. A. McPheron. 1986. Genetic variation in honey bees from an area of racial hybridization in Western Czechoslovakia. Apidologie 17:21–32.

Sheppard, W. S., T. E. Rinderer, J. A. Mazzoli, J. A. Stelzer, and H. Shimanuki. 1991a. Gene flow occurs between African- and European-derived honey bee populations in Argentina. Nature 349:782–784.

Sheppard, W. S., T. E. Rinderer, M. D. Meixner, H. R. Yoo, J. A. Stelzer, N. M. Schiff, S. M. Kamel, and R. Krell. 1966. *Hin*F1 variation in mitochondrial DNA of Old World honey bee races. Journal of Heredity 87:35–40.

Sheppard, W. S., A. E. E. Soares, and D. DeJong. 1991b. Hybrid status of honey bee populations near the historic origin of Africanization in Brazil. Apidologie 22:643–652.

Shestakova, R. A., and I. I. Chebotarev. 1971. Pollen toxicosis in bees [in Russian]. Veterinariia 8:77–78.

Shimamori, K. 1987. On the biology of *Apanteles galleriae*, a parasite of the two species of moths [in Japanese]. Honeybee Science 88:107–112.

Shimanuki, H. 1967. Ethylene oxide and control of American foulbrood: a progress report. American Bee Journal 107:290–291.

——. 1981. Controlling the greater wax moth—a pest of honey combs. U.S. Department of Agriculture Farmers' Bulletin number 2217.

Shimanuki, H., N. W. Calderone, and D. A. Knox. 1994. Parasitic mite syndrome: the symptoms. American Bee Journal 134:827–828.

Shimanuki, H., P. A. Hartman, and W. C. Rothenbuhler. 1965. In vitro growth studies of *Bacillus larvae* White. Journal of Invertebrate Pathology 7:437–441.

Shimanuki, H., E. W. Herbert, Jr., and D. A. Knox. 1984. High velocity electron beams for bee disease control. American Bee Journal 124:865–867.

Shimanuki, H., and D. A. Knox. 1964. Honey bee disease interactions: the impact of chalkbrood on other honey bee diseases. American Bee Journal 132:735–736.

Shimanuki, H., and D. A. Knox. 1988. Improved method for the detection of *Bacillus larvae* spores in honey. American Bee Journal 128:353–354.

Shimanuki, H., and D. A. Knox. 1989. Tracheal mite surveys. American Bee Journal 129:671–672.

Shimanuki, H., and D. A. Knox. 1991. Laboratory diagnosis of honey bee diseases. U.S. Department of Agriculture Agricultural Handbook 690.

Shimanuki, H., and D. A. Knox. 1991. Diagnosis of honey bee diseases. U.S. Department of Agriculture, Agricultural Research Service Handbook number 690.

Shimanuki, H., D. Knox, and D. De Jong. 1991. Bee diseases, parasites and pests. *In* The African honey bee. M. Spivak, D. J. C. Fletcher, and M. D. Breed, editors. Westview Press, Boulder, Colorado.

Shimanuki, H., D. A. Knox, M. Delfinado-Baker, and P. J. Lima. 1983. National honey bee mite survey. Apidologie 14:329–332.

Shimanuki, H., D. A. Knox, and M. F. Feldlaufer. 1992. Honey bee disease interactions: the impact of chalkbrood on other honey bee brood diseases. American Bee Journal 132:735–736.

Shimanuki, H., D. A. Knox, B. Furgala, D. M. Caron, and J. L. Williams. 1992. Diseases and pests of honey bees. *In* The hive and the honey bee. J. M. Graham, editor. Dadant & Sons, Hamilton, Illinois: 1083-1151

Shimanuki, H., D. A. Knox, and E. W. Herbert. 1970. Fumigation with ethylene oxide to control diseases of honey-bees. Journal of Economic Entomology 63:1062–1063.

Shimanuki, H., T. Lehnert, and D. Knox. 1973. Transmission of nosema disease from infected honey bee workers to queens in mating nuclei. Journal of Economic Entomology 66:777–778.

Shimanuki, H., T. Lehnert, D. A. Knox, and E. W. Herbert, Jr. 1969. Control of European foulbrood disease of the honey bee. Journal of Economic Entomology 62:813–814.

Shiner, D. 1983. Beekeeper discovers recycled pop bottles keep bears at bay. American Bee Journal 123:404.

Shoreit, M. N., and M. M. K. Bagy. 1995. Mycoflora associated with stonebrood disease in honeybee colonies in Egypt. Microbiological Research 150:207–211.

Showers, R. E., A. Jones, and F. E. Moeller. 1967. Cross-inoculation of the Bumble bee *Bombus feridus* with the microsporidian Nosema apis from the honey bee. Journal of Economic Entomology 60:744–777.

Sihag, R. C. 1982. Problem of the wax moth (*Galleria mellonella* L.) infestation on giant honey bee (*Apis dorsata* Fab.) colonies in Haryana. Indian Bee Journal 44:107–109.

———. 1991. Ecology of European honeybee (*Apis mellifera* L.) in semi-arid subtropical climates. 2. Seasonal incidence of diseases, pests, predators, and enemies. Korean Journal of Apiculture 6:16–26.

Silacek, D. L., and G. L. Miller. 1972. Growth and development of the Indian meal moth, *Plodia interpunctella* (Lepidoptera: Phycitidae), under laboratory mass-rearing conditions. Annals of the Entomological Society of America 65:1084–1087.

Simintzis, G. 1949. Larves de diptères du genre *Senotainia*, parasites thoraciques internes de l'abeille adulte [Diptera larvae of the genus *Senotainia* internal thoracic parasites of adult honey bees]. Revue Française d'Apiculture 11(48):13–16.

———. 1958. Pouvoir pathogène du *Senotainia tricuspis* Meig. pour les abeilles domestiques [The strength of *Senotainia tricuspis* Meig. as a disease vector for honey bees]. Revue du Médecine Vétérinaire 134:919–940.

Simintzis, G., and S. Fiasson. 1951. Les myiases des abeilles, en France [Apimyiasis in France]. Revue de Médecine Vétérinaire 102:351–361.

Simmons, P., and H. D. Nelson. 1975. Insects on dried fruits. U.S. Department of Agriculture Handbook 464.

Simonthomas, R. T. 1966. A method for breeding *Philanthus triangulum* F. (Sphecidae, Hymenoptera). Entomologische Berichten 26:114–116.

Simonthomas, R. T., and A. M. J. Simonthomas. 1972. Some observations on the behavior of females of *Philanthus triangulum* (F.) (Hymenoptera, Sphecidae). Tijdschrift voor Entomologie 115:123–139.

Simonthomas, R. T., and A. M. J. Simonthomas. 1978. A pest of the bee wolf in the apiculture of the Dakhla Oasis, Egypt. Pharmacological Laboratory, University of Amsterdam, Netherlands. 23 pp.

Simonthomas, R. T., and A. M. J. Simonthomas. 1980. *Philanthus triangulum* and its recent eruption as a predator of honeybees in an Egyptian oasis. Bee World 61:97–107.

Simonthomas, R. T., and R. L. Veenendaal. 1978. Observations on the behavior underground of *Philanthus triangulum* (Fabricius) (Hymenoptera, Sphecidae). Entomologische Berichten 38:3–8.

Singh, G. 1972. Defensive behavior of *Apis indica* F. (hill strain) against predatory hornets in Kashmir. Indian Bee Journal 34:65–69.

Singh, J. N., and T. V. Venkataraman. 1948. Pseudoscorpions in bee hives in India. Indian Bee Journal 10:6.

Singh, S. 1957. Acarine disease in the Indian honey bee (*Apis indica* F.). Indian Bee Journal 19:27–28.

———. 1959. The honeybees of India. In The ABC and XYZ of bee culture. 31st ed. A. I. Root, editor. A. I. Root, Medina, Ohio.

———. 1962. Beekeeping in India. Indian Council of Agricultural Research, New Delhi.

Singh, Y. 1975. Nosema in Indian honey bee (*Apis cerana indica*). American Bee Journal 115:59.

Sipos, E., A. Kerpely, and E. Zamory. 1960. Magyar mezek inhibin vizsgalata [Investigation of the inhibine effect of Hungarian honey]. Elelmiszervizgalati Kozlemenyek 6:81–89.

Sistrank, T. W. 1956. Bee defense against ants. Gleanings in Bee Culture 84:681.

Skaife, S. H. 1921a. A Tachinid Parasite of the Honey Bee. South African Journal of Science, Johannesburg, 17(2):196-200.

———. 1921b. On *Braula coeca* Nitzsch, a dipterous parasite of the honey bee. Transactions of the Royal Society of South Africa 10:41–48.

———. 1930. Insect pests of the hive. I. The tachinid parasite. Bee World 11(9): 106-7.

Skou, J. P. 1972. Ascosphaerales. Friesia 10:1–24.

———. 1975. Two new species of *Ascosphaera* and notes on the conidial state of *Bettsia alvei*. Friesia 11:62-74.

———. 1982. *Ascosphaera asterophora* species nova. Mycotaxon 14:149–159.

———. 1985. Notes on habitats, morphology and taxonomy of spore cyst fungi (Ascosphaerales). Apiacta (Bucharest) 20:105–108, 117.

———. 1988a. More details in support of the class Ascosphaeromycetes. Mycotaxon 31:191–198.

———. 1988b. Japanese species of *Ascosphaera*. Mycotaxon 31:173–190.

———. 1992. A series of xerophilic *Chrysosporium* species. Mycotaxon 43:237–259.

Skou, J. P., and K. Hackett. 1979. A new, homothallic species of *Ascosphaera*. Friesia 11:265–271.

Skou, J. P., and S. N. Holm. 1980. Occurrence of melanosis and other diseases in the queen honeybee, and risk of their transmission during instrumental insemination. Journal of Apicultural Research 19:133–143.

Skou, J. P., and S. N. Holm. 1989. *Ascosphaera tenax* species nova and a variant of *Ascosphaera aggregata*. Mycotaxon 35:211–218.

Skou, J. P., and J. King. 1984. *Ascosphaera osmophila* sp. nov., an Australian spore cyst fungus. Australian Journal of Botany 32:225–231.

Slabezki, Y., H. Gal, and Y. Lensky. 1991. The effect of fluvalinate application in bee colonies on population levels of *Varroa jacobsoni* and honey bees (*Apis mellifera* L.) and on residues in honey and wax. Bee Science 1:189–195.

Smirnov, A. M. 1970. How to control Braula infestations. Pchelovodstvo 90(1): 21-24.

Smirnov, A. M. 1978. Research results obtained in USSR concerning aetiology, parthenogenesis, epizootiology, diagnosis and control of *Varroa jacobsoni* in bees. Apiacta (Bucharest) 13:149–162.

Smirnov, A. M., and K. S. Chernov. 1976. Beekeeping in Japan [in Russian]. Pchelovodstvo 1976(2):27–28; (3):36–38; (4):29–30; (5):45–47.

Smirnov, A. M., and S. N. Luganskil 1987. Parasitism of honeybees by *Senotainia*. Veterinariya, Moscow, USSR (No. 6): 43-44.

Smirnova, N. I. 1954. A bacteriophage against *Bacterium larvae*: its use in diagnosis, prophylaxis, and therapeutics of American foulbrood [in Russian]. Thesis summary, Veterinary Institute of the Department of Higher Education, Leningrad.

Smith, A. W. 1990. Population dynamics and chemical ecology of the honey bee tracheal mite (Acari: Tarsonemidae), in Ohio. Ph.D. dissertation, Ohio State University, Columbus.

Smith, A. W., G. R. Needham, and R. E. Page, Jr. 1987. A method for the detection and study of live honey bee tracheal mites (*Acarapis woodi* Rennie). American Bee Journal 127:433–434.

Smith, A. W., G. R. Needham, R. E. Page, Jr., and M. K. Fondrk. 1991a. Dispersal of the honeybee tracheal mite, *Acarapis woodi* (Acari: Tarsonemidae) to old winter bees. Bee Science 1:95–99.

Smith, A. W., R. E. Page, and G. R. Needham. 1991b. Vegetable oil disrupts the dispersal of tracheal mite *Acarapis woodi* (Rennie) to young host bees. American Bee Journal 131:44–46.

Smith, D. R., M. F. Palopoli, B. R. Taylor, L. Garnery, J. M. Cornuet, M. Solignac, and W. M. Brown. 1991. Geographical overlap of two mitochondrial genomes in Spanish honeybees (*Apis mellifera iberica*). Journal of Heredity 82:96–100.

Smith, D. R., O. R. Taylor, and W. M. Brown. 1989. Neotropical Africanized honey bees have African mitochondrial DNA. Nature 339:213–215.

Smith, F. G. 1953. Beekeeping in the tropics. Bee World 34:233–245.

———. 1960. Beekeeping in the tropics. Longmans, Green, London.

Smith, I. B., Jr. 1978. The bee louse, *Braula coeca* Nitzsch, its distribution and biology on honey bees. M.S. Thesis, University of Maryland, USA. viii + 111 pp. Apiary Inspection, Dept. Agriculture, Annapolis, MD 21401, USA.

Smith, I. B., and D. M. Caron. 1984a. Distribution of the bee louse, *Braula coeca*, in Maryland and worldwide. American Bee Journal 125:294–296.

Smith, I. B., and D. M. Caron. 1984b. Distribution of the bee louse *Braula coeca* Nitzsch in honey bee colonies and its preferences among workers, queens and drones. Journal of Apicultural Research 23:171–176.

Smith, K. G. V. 1966. The larva of *Theocophora occidensis*, with comments upon the biology of Conopidae (Diptera). Journal of the Zoological Society of London 149:263–276.

Smith, K. M. 1967. Insect pathogenic viruses. *In* Methods of virology. K. Maramorosch and H. Koprowski, editors. Academic Press, New York.

Smith, M. 1969. The British amphibians and reptiles. 4th ed. Collins, London.

Smith, M. R. 1965. House-infesting ants of the eastern United States: their recognition, biology, and economic importance. ARS, USDA Technical Bulletin number 1326.

Smith, M. V. 1964. Honey bee importations from Europe. American Bee Journal 104:134–135.

Smith, T. L. 1941. Some notes on the development and regulation of heat among *Galleria* larvae. Proceedings of the Arkansas Academy of Science 1:29–33.

Snodgrass, R. E. 1956. Anatomy of the honey bee. Cornell University Press, Ithaca, New York.

Soares, A. E. E. 1980. A mutation preventing bees from stinging. American Bee Journal 120:834–835.

———. 1981a. Split-sting: a new honeybee character. Journal of Apicultural Research 20:140–142.

———. 1981b. Chartreuse-limao: first eye mutation induced by gamma radiation with ^{60}Co in the honeybee. Journal of Apicultural Research 20:137–139.

Sols, A., E. Cadenas, and F. Alvarado. 1960. Enzymatic basis of mannose toxicity in honey bees. Science 131(3306):297–298.

Sommerville, L. 1986. Amitraz: safe and effective acaricide for the control of varroa. Arbeitsgemeinschaft der Institute für Bienenforschung E. V. Abstracts of the Varroa Workshop in Feldafing/Starnberg, West Germany.

Sonan, J. 1927. Specific names and observations on some Formosan wasps [in Japanese]. Transactions of the Natural History Society of Formosa 89:121–138.

Spangler, H. G. 1984. Attraction of female lesser wax moths (Lepidoptera: Pyralidae) to male-produced and artificial sounds. Journal of Economic Entomology 77:346–349.

———. 1985. Sound production and communication by the greater wax moth (Lepidoptera: Pyralidae). Annals of the Entomological Society of America 78:54–61.

Spangler, H. G., and S. Taber III. 1970. Defensive behavior of honeybees toward ants. Psyche 77:184–189.

Spangler, H. G., and A. Takessian. 1983. Sound perception by two species of wax moths (Lepidoptera: Pyralidae). Annals of the Entomological Society of America 76:94–97.

Spiltoir, C. F. 1955. Life cycle of *Ascosphaera apis* (*Pericystis apis*). American Journal of Botany 42:501–508.

Spiltoir, C. F., and L. S. Olive. 1955. A reclassification of the genus *Pericystis* Betts. Mycologia 47:238–244.

Spivak, M., and M. Gilliam. 1991. New ideas on the role of hygienic behaviour in disease resistance in honey bees. American Bee Journal 131:782.

Spivak, M., and M. Gilliam. 1993. Facultative expression of hygienic behaviour of honey bees in relation to disease resistance. Journal of Apicultural Research 32:147–157.

Spivak, M., G. A. Reuter, R. Metton, and J. Breyfogle. 1994. Honey bee hygienic behavior and tolerance to *Varroa jacobsoni*. American Bee Journal 134:836–837.

Staal, G. B. 1967. Plants as a source of insect hormones. Proceedings of the Koninklijke Nederlandse Akademie van Wetenschappen, Series C 70:409–418.

Stacy, G. W. 1955. Bees and *Tilia petiolaris*. Journal of the Royal Horticultural Society 80:329.

Stairs, G. R. 1978. Effects of a wide range of temperatures on the development of *Galleria mellonella* and its specific baculovirus. Environmental Entomology 7:297–299.

Stanford, M. T. 1994. Bees poisoned by plants? Apis 12(6):1–3. University of Florida, Gainesville.

Stanger, W., L. Foote, H. H. Laidlaw, R. W. Thorp, N. E. Gary, and L. H. Watkins. 1971. Fundamentals of California beekeeping. California Agricultural Experiment Station Extension Service Manual 42.

Stanley, R. G., and H. F. Linskins. 1974. Pollen: biology, biochemistry, management. Springer-Verlag, New York.

Starnes, H. T. 1946. Ants. American Bee Journal 86:329.

Stearman, A. M. 1981. Working the "Africans" in eastern Bolivia. American Bee Journal 121:28–35, 43–44.

Steche, W. 1965. Zur Ontogonie von *Nosema apis* Zander im Mitteldarm der Arbeitsbienen [On the ontogeny of *Nosema apis* Xander in the midgut of the worker honey bee]. Bulletin Apicole de Documentation Scientifique et Technique et d'Information 8:181–212.

Steche, W., and T. Held. 1981. Rasterelektronenmikroskopische Untersuchungen über die Ontogenie von *Nosema apis* Zander [Scanning electron microscope studies of the ontogenesis of *Nosema apis* Zander]. Apidologie 12:185–207.

Steiner, J. 1988. Sex discrimination based on external structures in nymphal and adult *Varroa jacobsoni* mites (Acarina: Varroidae). Entomologia Generalis 14:133–138.

——— 1993. Verteilung von *Varroa jacobsoni* im drohnenfreien Binenvolk (*Apis mellifera carnica*) [Distribution of *Varroa jacobsoni* within a drone-free honey bee colony (*Apis mellifera carnica*)]. Apidologie 24:45–50.

———. 1995. Leg formation in the honeybee broodmite *Varroa jacobsoni* (Acarina: Varroidae). Entomologia Generalis 19:179–183.

Steiner, J., F. Dittmann, P. Rosenkranz, and W. Engels. 1994. The first gonocycle of the parasitic mite (*Varroa jacobsoni*) in relation to preimaginal development of its host, the honey bee (*Apis mellifera carnica*). Invertebrate Reproduction and Development 25:175–183.

Steiner, J., M. R. C. Issa, A. H. Buriolla, E. Engels, L. S. Gonçalves, and W. Engels. 1984. Varroatose na Alemanha e no Brasil [Varroa disease in Germany and in Brazil]. Apicultura no Brasil 1(5):35–37.

Steiner, J., S. D. G. Pompolo, C. S. Takahashi, and L. S. Gonçalves. 1982. Cytogenetics of the acarid *Varroa jacobsoni*. Revista Brasileira de Genetica 5:841–844.

Steiner, J., Diehl, P.A. and M. Vlimant. 1995. Vitellogenesis in *Varroa jacobsoni*, a parasite of honey bees. Experimental and Applied Acarology 19:411-422.

Steinhaus, E. A. 1946. Insect microbiology. Comstock Publishing, Ithaca, New York.

———. 1949a. Principles of insect pathology. McGraw-Hill, New York.

———. 1949b. Nomenclature and classification of insect viruses. Bacteriological Reviews 13:203–223.

———. 1963. Insect pathology, volume 2. Academic Press, New York.

———. 1967. Principles of insect pathology. Hafner Publishing, New York.

Steinhaus, E. A., and G. A. Marsh. 1962. Report of the diagnosis of diseased insects. Hilgardia 33:349–479.

Stejskal, M. 1955. Gregarines found in the honey-bee *Apis mellifera* Linnaeus in Venezuela. Journal of Protozoology 2:185–188.

———. 1958. Correlations between bee diseases and atmospheric conditions in Venezuela. Proceedings of the 17th International Beekeeping Congress 17:634–640.

———. 1962. Duft als 'Sprache' der tropischen Bienen [Odor in the language of tropical bees]. Südwestdeut Imker 49:271.

———. 1965. Gregarines parasitizing honey bees. American Bee Journal 105:374–375.

———. 1966. Amibiasis. American Bee Journal 106:292–293.

———. 1967. El piojo de las abejas (*Braula coeca* Nitzsch) y la "Glosiella mesonella" in Venezuela [The bee louse (*Braula coeca* Nitzsch) and *Glosiella mesonella* in Venezuela]. Apicultura Venezolana 2(1):12.

———. 1969. *Apiomerus crassipes*, parasito de las abejas *Apis mellifica* L. [*Apiomerus crassipes*, a parasite of the honey bee, *Apis mellifera* L.]. Oriente Agropecuario 1:125–134.

———. 1974. *Arrhenosphaera cranei*, gen. et sp. nov., a beehive fungus found in Venezuela. Journal of Apicultural Research 13:39–45.

————. 1976a. *Labyrinthula apis* sp. nov., fungus, parásito de las abejas *Apis mellifera* L. [*Labyrinthula apis* sp. nov., a fungus parasitizing the honeybee *Apis mellifera* L.]. Acta Botánica Venezuelica 11:385–396.

————. 1976b. *Endomycopsis apis*, sp. nov., fungus, parásito de las abejas *Apis mellifica* L. en Venezuela [*Endomycopsis apis* sp. nov., a fungal parasite of the honeybee *Apis mellifera* L. in Venezuela]. Turrialba 26:274–278.

Stephen, W. A. 1957. Chunk comb honey. Pennsylvania Beekeeper 32:1–2, 4–7, 9, 11–12.

————. 1968. Mites: a beekeeping problem in Vietnam and India. Bee World 49:119–120.

Stephen, W. P., and B. L. Fichter. 1990a. Chalkbrood (*Ascosphaera aggregata*) resistance in the leafcutting bee Megachile rotundata. I. Challenge of selected lines. Apidologie 21:209–219.

Stephen, W. P., and B. F. Fichter. 1990b. Chalkbrood (Ascosphaera aggregata) resistance in the leafcutting bee *Megachile rotundata*. II. Random matings of resistant lines to wild type. Apidologie 21:221–231.

Stephen, W. P., J. D. Vandenberg, and B. L. Fichter. 1981. Etiology and epizootiology of chalkbrood in the leafcutting bee, *Megachile rotundata* (Fabricius), with notes on *Ascosphaera* species. Agricultural Experiment Station, Oregon State University, Corvallis, Oregon, Station Bulletin number 653.

Stephen, W. P., J. D. Vandenberg, B. L. Fichter, and G. Lahm. 1982. Inhibition of chalkbrood spore germination in vitro (*Ascosphaera aggregata*: Ascosphaerales). Agricultural Experiment Station, Oregon State University, Corvallis, Oregon, Station Bulletin number 656.

Stephens, H. A. 1980. Poisonous plants of the central United States. Regents Press of Kansas, Lawrence, Kansas.

Stepp, W. E. 1935. Tanzy makes ants "skidoo." Gleanings in Bee Culture 63:342–343.

Stinson, E. E., M. H. Subers, J. Petty, and J. W. White. 1960. The composition of honey. V. Separation and identification of the organic acids. Archives of Biochemistry and Biophysics 89:6–12.

Stomfay-Stitz, J., and S. D. Kiminos. 1960. Ueber bakteriostatische Wirkung des Honigs [The bacteriostatic action of honey]. Zeitschrift für Lebensmitteluntersuchung und -forschung 133:304–309.

Storer, T. I., and G. H. Vansell. 1935. Bee eating proclivities of the striped skunk. Journal of Mammalogy 16:118–121.

Storer, T. I., G. H. Vansell, and B. O. Moses. 1938. Protection of mountain apiaries from bears by use of electric fence. Journal of Wildlife Management 2:172–178.

Stoya, W. G. 1985. Methansäurerückstände im Bienenhonig nach Behandlung der Bienenvölker mit dieser Säure im Rahmen der Varroatose-Bekämpfung [Residues in honey following treatment with formic acid in the course of controlling varroatosis]. Apidologie 16:216–217.

Strick, H., and G. Madel. 1986. The ectoparasitic bee mite *Varroa jacobsoni* Oud. as a carrier of bacteria to its host, the honey bee *Apis mellifera* L. Arbeitsgemeinschaft der Institute für Bienenforschung E.V. Abstracts of the Varroa Workshop in Feldafing/Starnberg, West Germany.

Stringer, B. A. 1989. Wasps—the honeydew thieves of New Zealand. American Bee Journal 129:465–467.

Struble, D. L., and K. W. Richards. 1983. Identification of sex pheromone components of the female driedfruit moth, *Vitula edmandsae serratilineella*, and a blend for attraction of male moths. Journal of Chemical Ecology 9:785–801.

Studier, H. 1958. The sterilization of American foulbrood by irradiation with gamma rays. American Bee Journal 98:192.

Sturtevant, A. P. 1926. The sterilization of American foulbrood combs. U.S. Department of Agriculture Circular 284.

————. 1932. Relation of commercial honey to the spread of American foulbrood. Journal of Agricultural Research 45:257–285.

————. 1936. Quantitative demonstration of the presence of spores of *Bacillus larvae* in honey contaminated by contact with American foulbrood. Journal of Agricultural Research (Washington, D.C.) 52:697–704.

Sturtevant, A. P., G. F. Knowlton, J. D. Hitchcock, G. H. Vansell, E. C. Holst and W. P. Nye. 1941. A report of investigations of the extent and causes of heavy losses of adult honeybees in Utah. U.S. Department of Agriculture, Bureau of Entomology and Plant Quarantine Mimeo E-545.

Sturtevant, A. P., and I. L. Revell. 1953. Reduction of *Bacillus larvae* spores in liquid food of honey bees by action of the honey stopper, and its relation to the development of American foulbrood. Journal of Economic Entomology 46:855–860.

Subba Rao, P. V., M. Mohanasundaram, and T. R. Subramaniam. 1972. *Odontomantis micans* Stal (Mantidae:Dictyoptera) as an enemy of *Apis indica* F., the Indian honey bee in Coimbatore. Indian Bee Journal 34:72.

Subbiah, M. S., and V. Mahadevan. 1957a. The life history and biology of the reduviid *Acanthaspis siva* Dist: a predator of Indian honey bee, *Apis indica*. Indian Journal of Veterinary Science 27:117–122.

Subbiah, M. S., and V. Mahadevan. 1957b. *Vespa cincta* Fabr., a predator of the hive bee and its control. Indian Journal of Veterinary Science 27:153–154. *From* Apicultural Abstracts 33/61.

Subbiah, M. S., V. Mahadevan, and R. Janakiraman. 1957. A note on the occurrence of an arachnid *Ellingsenius indicus* Chamberlain infesting bee hives in South India. Indian Journal of Veterinary Science 27:155–156.

Subers, M. H., A. I. Schepartz, and R. P. Koob. 1966. Separation and identification of some organic phosphates in honey by column and paper chromatography. Journal of Apicultural Research 5:49–57.

Subhapradha, C. K. 1957. A hive stand of proved utility. Indian Bee Journal 19:150–151.

————. 1961. Enemies and pests of bees encountered in this area. Indian Bee Journal 23:16–19.

Sugden, E. A., and A. M. Collins. 1990. Burglar bars exclude vertebrates, minimize bait hive loss. Proceedings of the American Bee Research Conference, October 1990. American Bee Journal 130:814–815.

Suhayda, J. 1968. A japánakácot látogáto méhek pusztulása [Mortality of bees on *Sophora japonica*]. Méhészet (Budapest) 16:208. From Apicultural Abstracts 171L/70.

Suire, J. 1931. Contribution à l'étude de *Braula coeca* Nitzsch [in French]. Rev. Zool. agri. 30(6-7): 85-89, 101-114.

Sulimanovic, D., D. Grbic and I. Tomac. 1983. Use of Apiakaridim [smoke] to control ectoparasites [Varroa and Braula] of bees [Primjena Apiakaridim u suzbijanju ektoparazita pcela]. Veterinarski Arhiv 53 (supplement): S67-S68.

Sulimanovic, D., F. Ruttner, and H. Pechhacker. 1982. Studies on the biology of reproduction in *Varroa jacobsoni* [in Japanese]. Honeybee Science 3:109–112.

Sulimanovic, D., F. Ruttner, M. Spitzer, H. Pechhacker, and I. Tomac. 1986. Reduced fertility of *Varroa jacobsoni* on worker brood. Arbeitsgemeinschaft der Institute für Bienenforschung E.V. Abstracts of the Varroa Workshop in Feldafing/Starnberg, West Germany.

Surface, H. A. 1913. Pests of domestic animals, households and buildings, bush fruits and lawn plants. Bi-monthly Zoology Bulletin, Division of Zoology, Pennsylvania Department of Agriculture 3(1), 30pp.

Sutherland, M. D. 1992. New Zealand toxic honey—the actual story. Analytical Proceedings 29:112–115.

Sutherland, M. D., and T. Palmer-Jones. 1947a. A recent outbreak of honey poisoning. II. The toxic substances of the poisonous honey. New Zealand Journal of Science and Technology 29:129–133.

Sutherland, M. D., and T. Palmer-Jones. 1947b. A recent outbreak of honey poisoning. V. The source of the toxic honey—laboratory investigations. New Zealand Journal of Science and Technology 29:129–133.

Sutter, G. R., W. C. Rothenbuhler, and E. S. Raun. 1968. Resistance to American foulbrood in honey bees. VII. Growth of resistant and susceptible larvae. Journal of Invertebrate Pathology 12:25–28.

Swadener, S. O., and T. R. Yonke. 1973a. Immature stages and biology of *Apiomerus crassipes* (Hemiptera: Reduviidae). Annals of the Entomological Society of America 66:188–196.

Swadener, S. O., and T. R. Yonke. 1973b. Immature stages and biology of *Sinea complexa*, with notes on four additional reduviids (Hemiptera: Reduviidae). Journal of the Kansas Entomological Society 46:123–136.

Sychevskaya, V. I. 1956. On myiasis of bees in Tadzhikistan [in Russian]. Izv. Akademii nauk tadzhikistan SSR Otd. estest. Nauk 13: 159–165.

Sylvester, H. A. 1982. Electrophoretic identification of Africanized honeybees. Journal of Apicultural Research 21:93–97.

Szabo, T. I. 1986. Mating distance of the honey bee in north-western Canada. Journal of Apicultural Research 25:227–233.

———. 1989. The capping scratcher: a tool for detection and control of *Varroa jacobsoni*. American Bee Journal 129:402–403.

Szabo, T. I., and D. T. Heikel. 1987a. Fumigation with SO$_2$ to control dried fruit moth in honeybee combs. Bee World 68:37–38.

Szabo, T. I., and D. T. Heikel. 1987b. Effect of dry fumagillin feeding on spring *Nosema* spore counts in overwintered colonies. American Bee Journal 127:210–211.

Tabarly, O. 1962. Mycoses (nourissements et antibiotiques) [Fungal diseases (feeding and antibiotics)]. Bulletin Apicole 5:105–111.

Tabarly, O., and E. Monteira. 1961. Influence de la teneur en eau du miel sur le développement des mycoses [Influence of the water content of honey on the development of fungal diseases]. Bulletin Apicole 4:31–44.

Taber, S., III. 1954. The frequency of multiple mating of queen honey bees. Journal of Economic Entomology 47:955–998.

———. 1955a. Evidence of binucleate eggs in the honey bee. Journal of Heredity 46:156.

———. 1955b. Sperm distribution in the spermathecae of multiply-mated queen honey bees. Journal of Economic Entomology 48:522–525.

———. 1979. A population of feral honey bee colonies. American Bee Journal 119:842–847.

———. 1982. Bee behavior: determining resistance to brood diseases. American Bee Journal 122:422–425.

———. 1986. Breeding bees resistant to chalkbrood disease. American Bee Journal 126:823–825.

———. 1990. Breeding acarine resistant bees—second generation. American Bee Journal 130:115–116.

———. 1993. Choosing an apiary site. Bee Culture 121:285–286.

Taber, S., and M. Gilliam. 1987. Breeding honey bees for resistance to diseases. Korean Journal of Apiculture 2:15–20.

Taber, S., III, R. Sackett, and J. Mills. 1975. A possible control for chalk brood disease. American Bee Journal 115:20.

Taber, S., III, and J. Wendel. 1958. Concerning the number of times queen bees mate. Journal of Economic Entomology 51:786–789.

Takatori, K., and I. Tanaka. 1982. *Ascosphaera apis* isolated from chalk brood in honey bees. Japanese Journal of Zootechnical Science 53:89–92.

Takeuchi, K., and T. Sakai. 1986. Parasitic ecology of Varroa mite in honeybee colonies and its annual control scheme [in Japanese]. Bulletin of the Faculty of Agriculture of Tamagawa University 26:75–88.

Takeuchi, K., I. Suzuki, and A. Inoue. 1989. Occurrence and control of chalk brood disease in Japan. Thirty-second International Congress of Apiculture of Apimondia, Rio de Janeiro, Brazil.

Tal, J., and T. Attathom. 1993. Insecticidal potential of the insect parvovirus GmDNV. Archives of Insect Biochemistry and Physiology 22:345–356.

Tamasko, M., J. Finley, W. Harkness, and E. Rajotte. 1993. A sequential sampling scheme for detecting the presence of tracheal mite (*Acarapis woodi*) infestations in honey bee (*Apis mellifera* L.) colonies. Pennsylvania State University, College of Agricultural Sciences, Bulletin 871.

Tanada, Y., and H. K. Kaya. 1993. Insect pathology. Academic Press, San Diego.

Tanaka, M., T. Watanabe, T. Tawara, S. Hanaki, K. Uchiyama, M. Tominaga, and R. Inaji. 1984. An experiment to protect honeybees from chalk brood disease [in Japanese, with English summary]. Honeybee Science 5:117–120.

Tarr, H. L. A. 1937. Studies on American foul brood of bees. I. The relative pathogenicity of vegetative cells and endospores of *Bacillus larvae* for the brood of the bee. Annals of Applied Biology 24:377–384.

———. 1938. Studies on European foul brood of bees. IV. On the attempted cultivation of *Bacillus pluton*, the susceptibility of individual larvae to inoculation with this organism and its localization within its host. Annals of Applied Biology 25:815–821.

Taylor, F. 1934. Beekeeping for the beginner. 15. Diseases and enemies of bees. Farming in South Africa #2: 71-78.

Tewarson, N. C. 1983. Nutrition and reproduction in the ectoparasitic honey bee (*Apis* sp.) mite *Varroa jacobsoni*. Ph.D. dissertation, Fakultät für Biologie der Eberhard-Karls, University of Tübingen, Germany.

Tewarson, N. C., and W. Engels. 1982. Undigested uptake of non-host proteins by *Varroa jacobsoni*. Journal of Apicultural Research 21:222–225.

Tewarson, N. C., and A. Singh. 1986. *Varroa jacobsoni* in the colonies of the eastern honey bee *Apis cerana indica* in North India infected with sacbrood virus. Arbeitsgemeinschaft der Institute für Bienenforschung E.V. Abstracts of the Varroa Workshop in Feldafing/Starnberg, West Germany.

Tewarson, N. C., A. Singh, and W. Engels. 1992. Reproduction of *Varroa jacobsoni* in colonies of *Apis cerana indica* under natural and experimental conditions. Apidologie 23:161–171.

Thiem, H. 1932. Die Bienenwolf-Plage (*Philanthus triangulum* F.) im Kaligebiet der Werra und ihre Bekampfung [The beewolf (*Philanthus triangulum* F.) plague in the potash district of the Werra River, and its control]. Deutsche Bienenzucht in Theorie und Praxis 40:173–186.

———. 1935. Der Bienenwolf, gefährlicher Bienenschädling [The beewolf, a dangerous pest to honey bees]. Kranke Pflanze 12:112–115.

Thoenes, S. C. 1993. Fatal attraction of certain large-bodied native bees to honey bee colonies. Journal of the Kansas Entomological Society 66:210–213.

Thoenes, S. C., and S. L. Buchmann. 1992. Colony abandonment by adult honey bees: a behavioural response to high tracheal mite infestation? Journal of Apicultural Research 31:167–168.

Thomas, C. R. 1960. The European wasp (*Vespula germanica* Fab.) in New Zealand. New Zealand Department of Scientific and Industrial Research, Information Series 27:1–74.

Thomas, G. M., and A. Luce. 1972. An epizootic of chalk brood, *Ascosphaera apis* (Maassen ex Claussen) Oliver and Spiltoir in the honey bee, *Apis mellifera* L. in California. American Bee Journal 112:88–90.

Thomas, G. M., and G. O. Poinar, Jr. 1973. Report of diagnoses of diseased insects, 1962–1972. Hilgardia 42:261–359.

Thompson, A., K. Anno, M. L. Wolfrom, and M. Inatome. 1954. Acid reversion products from D-glucose. Journal of the American Chemical Society 76:1309–1311.

Thompson, C. R. 1990. Ants that have pest status in the United States. *In* Applied myrmecology. R. K. Vander Meer, K. Jaffe, and A. Cedeno, editors. Westview Press, Boulder, Colorado.

Thomson, S. V., D. R. Hansen, K. M. Flint, and J. D. Vandenberg. 1992. Dissemination of bacteria antagonistic to *Erwinia amylovora* by honey bees. Plant Disease 76:1052–1056.

Thompson, V. C., and W. C. Rothenbuhler. 1957. Resistance to American foulbrood in honey bees. II. Differential protection of larvae by adults of different genetic lines. Journal of Economic Entomology 50:731–737.

Thorpe, H. S. 1967. Scoundrel of the bee hives. American Bee Journal 107:288–289.

Thrasyvoulou, A. T., and N. Pappas. 1988. Contamination of honey and wax with malathion and coumaphos used against *Varroa* mite. Journal of Apicultural Research 27:55–61.

Thrybom, B., and I. Fries. 1991. Development of infestations by *Varroa jacobsoni* in hybrid colonies of *Apis mellifera* monticola and *Apis mellifera ligustica*. Journal of Apicultural Research 30:151–155.

Thurber, P. F. 1975. The bear facts. Gleanings in Bee Culture 103:112–113, 131, 152–153.

Tiedemann, V. O. 1958. *Vitula serratilineella* Ragonot, a North American microlepidopteran becoming naturalized in Europe. Zeitschrift der Wiener Entomologischen Gesellschaft 43:282–286.

Tingek, S., M. Mardan, T. E. Rinderer, N. Koeniger, and G. Koeniger. 1988. Rediscovery of *Apis vechti* (Maa, 1953): the Saban honey bee. Apidologie 19:97–102.

Tirgari, S., H. Polous, H. Foroughi, F. Fakhari, and M. A. Nabavi. 1969. On the occurrence and biology of the little bee (*Micrapis florea* F.) and the first account of its migration (Hymenoptera, Apidae). *In* Proceedings of the Second National Congress of Entomology and Pest Control of Iran. Mimeo.

Tissot, A. N., and F. A. Robinson. 1954. Some unusual insect nests. Florida Entomologist 37:73–92.

Tomac, I., D. Grbic, B. Bukovic, A. Erceg, and D. Sulimanovic. 1983. Chalk and stone brood in Yugoslavia. Proceedings of the 29th International Congress of Apiculture 29:274.

Tomasec, I. 1957. O djelovanju nekih antibiotika na pcele [On the effect of certain antibiotics on the honey bee]. Veterinarski Arhiv 27:71–80.

Tomasko, M. 1989. Live aid. Gleanings in Bee Culture 117:619–620.

Tomaszewska, B. 1979. Badania nad zachowaniem sie niektórych skladników hemolimfy pszczó zarazonych sporowcem pszczelim (*Nosema apis* Z.) [Alterations in the levels of some components of haemolymph in honey bees infected with *Nosema apis* Z.]. Pszczelnicze Zeszyty Naukowe 23:181–207.

Tomsik, B., E. Lisy, J. Svoboda, and J. Hejtmanek. 1955. Vcelarstvi [Beekeeping]. Nakladatelstvi Ceskoslovenske Akademie Ved, Prague.

Torchio, P. F. 1971. The biology of *Anthophora (Micranthophora) peritomae* Cockerell (Hymenoptera: Apoidea, Anthophoridae). Contributions in Science 206:1–14.

———. 1992. Effects of spore dosage and temperature on pathogenic expressions of chalkbrood syndrome caused by *Ascosphaera torchioi* within larvae of *Osmia lignaria propinqua* (Hymenoptera: Megachilidae). Environmental Entomology 21:1086–1091.

Toschkov, A. von, Ts. Valerianov, and A. Tomov. 1970. Die Immunofluores- zenzmethode und die schnelle und spezifische Diagnostik der amerikanischen Faulbrut bei der Bienenbrut [Method of fluorescent immunization; rapid and specific diagnosis of American foulbrood]. Bulletin Apicole 13:13–14.

Toumanoff, C. [also transliterated as Toumanov]. 1939. Les ennemis des abeilles [The enemies of bees]. Imprimerie d'Extrême-Orient, Hanoi.

———. 1951. Les maladies des abeilles [The diseases of bees]. Revue Française d'Apiculture numéro spécial 68.

Toumanoff, K. [also transliterated as Toumanov]. 1928. Aspergillusmycosis of bees. Bee World 9:187–188.

Townsend, G. F. 1975. Processing and storing liquid honey. *In* Honey: a comprehensive survey. E. Crane, editor. Crane, Russak, New York.

Townsend, G. F., P. W. Burke, and M. V. Smith. 1965. Bee diseases and pests of the apiary. Ontario Agricultural College, Department of Agriculture Publication 429.

Trembly, A., and M. Burgett. 1979. Controlled release fumigation of the greater wax moth. Journal of Economic Entomology 72:616–617.

Tribe, G. D. Drones caught in a spider's web within a drone congregation area. South African Bee Journal 61:110-111.

Tribe, G. D., A. P. du Toit, N. J. van Rensburg and N. F. Johannsveier. 1989. The collection and release of *Tetrastichus* sp. (Eulopidae) as a biological control agent for *Drosophila flavohirta* Malloch (Drosophilidae) in South Africa. Journal of the Entomology Society of Southern Africa. 52(1): 181-182.

Trouiller, J., G. Arnold, B. Chappe, Y. Le Conte, A. Billion, and C. Masson. 1994. The kairomonal esters attractive to the *Varroa jacobsoni* mite in the queen brood. Apidologie 25:314–321.

Tryasko, V. V. 1969. Study of spontaneous female parthenogenesis in honeybees. Proceedings of the 22nd International Beekeeping Congress 22:599.

Tryon, H. 1919. Report of the entomologist and vegetable pathologist. Queensland Annual Report, Dept. Agriculture and Stock for the Year 1918-1919:37-49.

Tsivilev, I. V. 1968. Veterinary phenothiazine for the control of varrosis and braullosis of bees [in Russian]. Veterinariya, 45(6): 53-54b.

Tucker, K. W. 1958. Automictic parthenogenesis in the honey bee. Genetics 43:299–316.

———. 1980. Tests for linkage and other interactions in the honey bee. Journal of Heredity 71:452–454.

———. 1986. Visible mutants. *In* Bee genetics and breeding. T. E. Rinderer, editor. Academic Press, Orlando, Florida.

Tucker, K. W., and H. H. Laidlaw, Jr. 1966. The potential for multiplying a clone of honey bees by androgenesis. Journal of Heredity 57:213–214.

Tucker, K. W., and H. H. Laidlaw. 1967. Honey bee drones with jet black bodies. Journal of Heredity 58:184–185.

Turner, M. 1979. Diet and feeding phenology of the green lynx spider *Peucetia viridans* (Araneae: Oxyopidae). Journal of Arachnology 7:149–154.

Tutkun, E. 1982. Research on the definition, bioecology, economic importance and a new control method of *Merops apiaster*, harmful to honeybees in Central Anatolia. Bitki Koruma Bulteni 22:148–159.

Tyne, J., and A. J. Berger. 1976. Fundamentals of ornithology. John Wiley & Sons, New York.

Tysset, C., and Y. de Rautlin de la Roy. 1974. Essais sur l'étude des levures "osmophiles," agents des fermentations des miels récoltés en France [An investigation of the osmophilic yeasts as the agents of fermentation of French honey]. Bulletin de l'Association des Diplomes de Microbiologie de la Faculté de Pharmacie de Nancy numéro 134.

Underwood, B. A. 1986. The natural history of *Apis laboriosa* Smith in Nepal. M.S. thesis, Cornell University, Ithaca, New York.

———. 1992. Notes on the orange-rumped honeyguide *Indicator xanthonotus* and its association with the Himalayan honey bee *Apis laboriosa*. Journal of the Bombay Natural History Society 130:290–295.

Urata, U. 1954. Pollination requirements of macadamia. Hawaii Agricultural Experiment Station Technical Bulletin 22.

U.S. Food and Drug Administration. 1969. Poison honey. Food and Drug Administration Papers 3:29.

USSR, Chief Veterinary Directorate, Gosagroprom. 1988. Identification of infestation of honeybees by *Zodion* sp. (Conopidae). Veterinariya, Moscow, USSR (No. 7): 76.

Vago, C. 1968. Non-inclusion virus diseases of invertebrates. *In* Current topics in microbiology and immunology, 42: insect viruses. K. Maramorosch, editor. Springer-Verlag, New York.

Van Buren, N. W. M., A. G. H. Mariën, and H. H. W. Velthuis. 1993. The effectiveness of systemic agents used to control the mite, *Varroa jacobsoni*, in colonies of the honey bee, *Apis mellifera* depends on food distribution patterns. Apidologie 24:33–43.

Van Buren, N. W. M., A. G. H. Mariën, H. H. W. Velthuis, and R. C. H. M. Oudemans. 1992. Residues in beeswax and honey of Perizin, an acaricide to combat the mite *Varroa jacobsoni* Oudemans (Acari: Mesostigmata). Environmental Entomology 21:861–865.

Vandenberg, J. D. 1990. Nematodes. *In* Honey bee pests, predators, and diseases. R. A. Morse and R. Nowogrodzki, editors. Cornell University Press, Ithaca, New York.

————. 1992a. Bioassay of the chalkbrood fungus *Ascosphaera aggregata* on larvae of the alfalfa leafcutting bee, *Megachile rotundata*. Journal of Invertebrate Pathology 60:159–163.

————. 1992b. Chalkbrood susceptibility among larvae of the alfalfa leafcutting bee, *Megachile rotundata*, from different source populations. Journal of Invertebrate Pathology 60:213–214.

————. 1994. Chalkbrood susceptibility among larvae of the alfalfa leafcutting bee (Hymenoptera: Megachilidae) reared on different diets. Journal of Economic Entomology 87:350–355.

Vandenberg, J. D., B. L. Fichter, and W. P. Stephen. 1980. Spore load of *Ascosphaera* species on emerging adults of the alfalfa leafcutting bee, *Megachile rotundata*. Applied and Environmental Microbiology 39:650–655.

Vandenberg, J. D., and M. S. Goettel. 1995. Chalkbrood susceptibility among alfalfa leafcutting bee (Hymenoptera: Megachilidae) larvae reared at different temperatures. Journal of Economic Entomology 88:810–814.

Vandenberg, J. D., and H. Shimanuki. 1990a. Application methods for *Bacillus thuringiensis* used to control larvae of greater wax moth (Lepidoptera: Pyralidae) on stored beeswax combs. Journal of Economic Entomology 83:766–771.

Vandenberg, J. D., and H. Shimanuki. 1990b. Isolation and characterization of *Bacillus coagulans* associated with half-moon disorder of honey bees. Apidologie 21:233–241.

Vandenberg, J. D., and H. Shimanuki. 1990c. Viability of *Bacillus thuringiensis* and its efficacy for larvae of the greater wax moth (Lepidoptera: Pyralidae) following storage of treated combs. Journal of Economic Entomology 83:760–765.

Vandenberg, J. D., and W. P. Stephen. 1982. Etiology and symptomatology of chalkbrood in the alfalfa leafcutting bee, *Megachile rotundata*. Journal of Invertebrate Pathology 39:133–137.

Vandenberg, J. D., and W. P. Stephen. 1983a. Pathogenicity of *Ascosphaera* species for larvae of *Megachile rotundata*. Journal of Apicultural Research 22:57–63.

Vandenberg, J. D., and W. P. Stephen. 1983b. Pathogenesis of chalkbrood in the alfalfa leafcutting bee, *Megachile rotundata*. Apidologie 14:333–341.

van der Vecht, J. 1957. The Vespinae of the Indo-Malayan and Papuan areas (Hymenoptera, Vespidae). Leyden Rijks Musejm van Natuurlijke Historie. Zoologische Verhandelingen 34:1–83.

Van de Sande, M. W. C. M., J. A. F. de Oliveira, and P. B. V. Ribeiro. 1989 Apimyíasis—ocorrência de *Melaloncha ronnai* Borgmeier, 1935 (Diptera, Phoridae), associada à mortalidade de abelhas africanizadas (*Apis mellifera*) em Santa Catarine e Minas Gerais [Apimyasis—the phorid fly *Melaloncha ronnai* Borgmeier, 1935 (Diptera, Phoridae) associated with Africanized bee mortality in Santa Catarina and Minas Gerais]. *In* Proceedings of the Seventh Brazilian Congress on Apiculture, 1986.

Van Duzee, M. C. 1934. Conopidae from North Dakota and the Rocky Mountain region. Annals of the Entomological Society of America 27:315–323.

Van Emden, F. I. 1944. Keys to the Ethiopian Tachinidae. I. Phasiinae. Proceedings of the Zoological Society of London 114:389–436.

Van Emelen, A. 1921. Combate aos Piolhos das Abelhas [Measures against *Braula coeca*]. Chacaras e Quintaes, S. Paulo, XXIII (1):51.

Van Handel, E. 1971. Mannose metabolism: a comparison between the honey bee and the mosquito. Comparative Biochemistry and Physiology 38B(1):141–145.

Vansell, G. H. 1925. Buckeye poisoning in California. American Bee Journal 65:575–578.

————. 1926a. Buckeye poisoning of the honey bee. California Agricultural Experiment Station Circular 301.

————. 1926b. The California buckeye and its relation to the hive bee. Journal of Economic Entomology 19:133–136.

————. 1935. Western plants poisonous to bees. Bees and Honey 16:303–304.

————. 1936. Lesser wax moth found in Oregon and California. Journal of Economic Entomology 29:1176.

Vansell, G. H., and T. I. Storer. 1930. Skunks caught with the goods. American Bee Journal 70:339.

Vansell, G. H., and F. E. Todd. 1932. Resistance of hybrid honeybees to a plant poison in California. Journal of Economic Entomology 25:503–506.

Vansell, G. H., and W. G. Watkins. 1933. A plant poisonous to adult bees. Journal of Economic Entomology 26:168–170.

Vansell, G. H., and W. G. Watkins. 1934. Adult bees found dying on spotted loco. Journal of Economic Entomology 27:636–637.

Vansell, G. H., and W. G. Watkins. 1936. The distribution of California buckeye in the south-central Sierra Nevada counties, in relation to honey production. U.S. Department of Agriculture, Bureau of Entomology and Plant Quarantine Mimeo E-397.

Vansell, G. H., W. G. Watkins, and L. F. Hosbrook. 1940. The distribution of California buckeye in the Sierra Nevada in relation to honey production. California Agricultural Experiment Station Report.

Van Tyne, J., and A. J. Berger. 1976. Fundamentals of ornithology. John Wiley & Sons, New York.

Varanda, E. A., C. S. Takahashi, S. E. Soares, and L. A. de Oliveira Campos. 1984. Considerations on the courtship and copulation behavior, sex ratio and life cycles of *Melittobia hawaiiensis* (Hym. Eulophidae). Revista Brasileira de Genetica 7:65–72.

Varis, A. L., B. V. Ball, and M. Allen. 1992. The incidence of pathogens in honey bee (*Apis mellifera* L.) colonies in Finland and Great Britain. Apidologie 23:133–137.

Varitchak, B. 1933. L'évolution nucléaire dans le sac sporifère *de Pericystis apis* Maassen et sa signification pour la phylogénie des Ascomycètes [Nuclear development in the spore sac of *Pericystis apis* Maassen and its significance for the phylogeny of Ascomycetes]. Le Botaniste 25:343–391.

Vaughn, J. L. 1974. Virus and rickettsial diseases. *In* Insect diseases, volume 1. G. E. Cantwell, editor. Marcell Dekker, New York.

Vavruch, I. 1952. Chromatograficka studie vceliho medu [Investigation of honey by chromatography]. Chemicke Listy 46:116–117.

Vecchi, M. A. 1959. La microflora dell'ape mellifica [Microflora of the honey bee]. Annali di Microbiologia ed Enzimologia 9:73–86.

————. 1976. Sulla patologia dell'ape regina (*Apis mellifica* L.) [On the pathology of the queen honey bee (*Apis mellifera* L.)]. Edizioni l'Apicoltore d'Italia, Rome.

Vecchi, M. A., and G. Giordani. 1968. Chemotherapy of acarine disease. I. Laboratory tests. Journal of Invertebrate Pathology 10:390–416.

Vecht, J. 1957. The Vespinae of the Indo-Malayan and Pauan areas (Hymenoptera, Vespidae). Zoologische Verhandelingen 34, Rijksmuseum van Natuurlijke Historie, Leyden.

Veeresh, G. K. 1990. Pest ants of India. *In* Applied myrmecology. R. K. Vander Meer, K. Jaffe, and A. Cedeno, editors. Westview Press, Boulder, Colorado.

Veitch, R. 1936. Report of the chief entomologist. Annual Report of the Department of Agriculture and Stock, Queensland, for the year 1935–1936.

Velthoen, A. A. 1947. Kastanjevergiftiging bij bijen [Chestnut-poisoning in bees]. Tijdschrift voor Diergeneeskunde 72:18–21.

Verdcourt, B. 1969. Common poisonous plants of East Africa. Collins, London.

Vergara, C., A. Dietz, and A. Perez de Leon. 1993. Female parasitism of European honey bees by Africanized honey bee swarms in Mexico. Journal of Apicultural Research 32:34–40.

Verheijen-Voogd, C. 1959. How worker bees perceive the presence of their queen. Zeitschrift für Vergleichende Physiologie 41:527–582.

Verma, S., and F. Ruttner. 1983. Cytological analysis of the thelytokous parthenogenesis in the Cape honeybee (*Apis mellifera capensis* Escholtz). Apidologie 14:41–57.

Verma, S. K., and N. K. Joshi. 1988. *Syntretomorpha szyboi* Papp.—a new hymenopterous parasite of Indian hive bee, *Apis cerana indica* Fabr. Indian Bee Journal 50:39–40.

Vesely, V., and M. Peroutka. 1984. Bewertung der Methode zur radikalen Eindämmung der Varroatose [An evaluation of a method for the radical control of Varroa disease]. Apidologie 15:379–388.

Vidano, C., and G. Onore. 1971. Le mediterranea *Potosia opaca* (Fabricius) (Coleoptera, Scarabaeidae), cetonino dannoso agli alveari [The Mediterranean *Potosia opaca* (Fabricius) (Coleoptera, Scarabaeidae), a chafer injurious to bee colonies]. Apicoltore Moderno 62:169–182.

Villeneuve, J. 1916. A new species of tachino-oestrid from South Africa. Annals of the South African Museum 15:465–468.

Villumstad, E. 1970. Overvintring og vårutveckling av bifolk på tavler som ble bygget ut under høstforinga [Wintering and spring development of colonies on combs drawn out during autumn feeding]. Birøkteren 86:36–43.

———. 1980. Eine Methode zur Nosema-Prophylaxe aus Norwegen [A method for Nosema prophylaxis in Norway]. Allgemeine Deutsche Imkerzeitung 14:38–40.

Vincens, F. 1923. Sur l'aspergillomycose des abeilles [On aspergillus mycosis of bees]. Comptes Rendus de l'Académie des Sciences (Paris) 177:540–542.

Visscher, P. K. 1989. A quantitative study of worker reproduction in honey bee colonies. Behavioral Ecology and Sociobiology 25:247–254.

Vitt, L. J. 1993. Ecology of isolated open-formation *Tropidurus* (Reptilia: Tropiduridae) in Amazonian lowland rain forest. Canadian Journal of Zoology 71:2370–2390.

Vitt, L. J., and T. E. Lacher, Jr. 1981. Behavior, habitat, diet, and reproduction of the iguanid lizard *Polychrus acutirostris* in the Caatingas of northeastern Brazil. Herpetologica 37:53–63.

Vitzthun, H. G. 1930. Investigations of the causes of May-sickness. Bee World 11:14–15.

Volcinschi, T. 1969. Measures for controlling the bee louse (*Braula coeca*) [Masuri pentru combaterea paduchelui albinelor]. Apicultura, Rumania 22 (9): 25-26.

Voller, A., D. E. Bidwell, and A. Bartlett. 1976. Enzyme immunoassays in diagnostic medicine. Theory and practice. Bulletin of the World Health Organization 53:55–65.

von Posern, H. 1988. The synthetic comb, a new weapon to fight the Varroa mite? American Bee Journal 128:698–702.

Vyslouzil, L. 1984. Arthropod parasites in bee-hives and diagnosis of *Varroa jacobsoni* Oud. (in Czechoslovak). Sbornik Vedeckych Praci Ustredniho Statniho Veterinarniho Ustavu 14:56–57.

Wafa, A. K. 1956. Ecological investigations on the activity of the oriental hornet *Vespa orientalis*. Bulletin of the Faculty of Agriculture, Cairo University number 103.

Wafa, A. K., and S. G. Sharkawi. 1972. Contribution to the biology of *Vespa orientalis* Fab. (Hymenoptera, Vespidae). Bulletin de la Société Entomologique d'Egypte 56:219–226.

Wagner, R. E., and D. A. Rierson. 1969. Yellow jacket control by baiting. Journal of Economic Entomology 62:1192–1197.

Wagner, S. 1866. [Review of an account by Dr. Barton of poisonous honey plants.] American Bee Journal 2:14.

Wali-ur-Rehman, and M. I. Chaudhry. 1980. Efficacy of microbial insecticides against wax moth, *Galleria mellonella* (L.). Pakistan Journal of Zoology 12: 148–150.

Walker, A. K., N. K. Joshi, and S. K. Verma. 1990. The biosystematics of *Syntretomorpha szaboi* Papp. (Hymenoptera: Braconidae: Euphorinae) attacking the Oriental honey bee, *Apis cerana* Fabricius (Hymenoptera: Apidae), with a review of braconid parasitoids attacking bees. Bulletin of Entomological Research 80:79–83.

Walker, E. P., F. Warnick, S. E. Hamlet, K. I. Lange, M. A. Davis, H. E. Uible, and P. F. Wright. 1975. Mammals of the world. 3d ed. Johns Hopkins University Press, Baltimore.

Wallace, A. R. 1895. Natural selection and tropical nature. Macmillan, London.

Wallace, F. G. 1966. The trypanosomatid parasites of insects and arachnids. Experimental Parasitology 18:124–193.

Waller, G. D., E. W. Carpenter, O.A. Ziehl. 1972. Potassium in onion nectar and its probable effect on attractiveness of onion flowers to honey bees. Journal of the American Society or Horticultural Science 97:535-539.

Waller, G. D., and L. H. Hines. 1990. A search for tracheal mite resistance in Arizona honey bees. American Bee Journal 130:818.

Wallner, A. 1990. Beobachtungen naturlicher *Varroa*-Abwehrreaktionen in meinen Bienenvolkern [Observations of honey bee natural Varroa-defensive reactions]. Imerfreund 9:4-5.

———. 1993. Mein Weg in der Varroa-resistenzzüchtung [My method of developing Varroa-resistance]. Bienenvater 114:107-108.

Wallner, K. 1990. Beobachtungen natürlicher *Varroa*-abwehrreaktionin in meinen Bienenvölkern [Observation of natural defense against varroa in my beehives]. Bienenvater 111:299-300.

———. 1992a. Diffusion varroazider Wirkstoffe aus dem Wachs in den Honig [Diffusion of varroa-treatment chemicals from the wax to the honey]. Apidologie 23:387-389.

———. 1992b. The residues of p-dichlorobenzene in wax and honey. American Bee Journal 132:538-541.

Walshaw, E. R. 1967. The raiders. American Bee Journal 107:14-15.

Walton, G. M., and G. M. Reid. 1976. The 1975 New Zealand European wasp survey. New Zealand Beekeeper 38(2):26-30.

Wang, D.-I. 1971. Ultrastructural changes in the hypopharyngeal glands of worker honey bees infected by *Nosema apis*. Journal of Invertebrate Pathology 17:308-320.

Wang, Der-I., and F. E. Moeller. 1969. Histological comparisons of the development of hypopharyngeal glands in healthy and nosema-infected worker honey bees. Journal of Invertebrate Pathology 14:135-142.

Wang, Der-I., and F. E. Moeller. 1970. The division of labor and queen attendance behavior of nosema-infected worker honey bees. Journal of Economic Entomology 63:1539-1541.

Wang, Der-I., and F. E. Moeller. 1971. Ultrastructural changes in the hypopharyngeal glands of worker honey bees infected by *Nosema apis*. Journal of Invertebrate Pathology 17:308-320.

Waren, L. O., and P. Huddleston. 1962. Life history of the greater wax moth, *Galleria mellonella* L., in Arkansas. Journal of the Kansas Entomological Society 35:212-216.

Warnecke, B., and H. Duisberg. 1958. Die bakteriostatische (inhibitorische) Wirkung des Honigs [Bacteriostatic (inhibitory) action of honey]. Zeitschrift für Lebensmitteluntersuchung und -forschung 107:340-344.

Warren, L. O., and P. Huddleston. 1962. Life history of the greater wax moth, *Galleria mellonella* L., in Arkansas. Journal of the Kansas Entomological Society 35:212-216.

Warren, P. 1975. The bear-bee yard conflict in British Columbia. British Columbia Department of Agriculture, Victoria.

Watson, E. B. 1922. The food habits of wasps (*Vespa*). Bulletin of Chambers Horticultural Society of London 1:26-31.

Weatherston, J., and J. E. Percy. 1968. Studies of physiologically active arthropod secretions. I. Evidence for a sex pheromone in female *Vitula edmandsae* (Lepidoptera: Phycitidae). Canadian Entomologist 100:1065-1070.

Weaver, J. 1979. Living with bears. Bee aware. Penn State Cooperative Extension Service Newsletter 4(8):1-4.

Weaver, R. S., Jr. 1969. Pampered package bees. American Bee Journal 109:49.

Webster, R. L. 1947. Beekeeping in Washington. Washington Extension Service Bulletin 289. Revised.

Wedenig, B. 1987. *Apis cerana, Varroa jacobsoni,* and *Tropilaelaps clareae* found in Papua New Guinea. International Bee Research Association Newsletter for Beekeepers in Tropical and Subtropical Countries 1987(10):5.

Wedmore, E. B. 1932. A manual of beekeeping. Edward Arnold, London.

Weems, H. V., Jr. 1983. Beelouse, *Braula coeca* Nitzsch (Diptea: Braulidae). Entomology Circular, Division of Plant Industry, Florida Department of Agriculture and Consumer Services (No. 252): 2pp.

Wegner, A. M. R. 1949. A remarkable observation on the Indian honeybee versus the yellow-throated marten from Java. Treubia 20:31–33.

Weidner, E., W. Byrd, A. Scarbrough, J. Pleshinger, and D. Sibley. 1984. Microsporidian spore discharge and the transfer of polaroplast organelle membrane into plasma membrane. Journal of Protozoology 31:195–198.

Weinberg, K. P., and G. Madel. 1985. The influence of the mite *Varroa jacobsoni* Oud. on the protein concentration and the hemolymph volume of the brood of worker bees and drones of the honey bee *Apis mellifera* L. Apidologie 16:421–436.

Welch, H. E. 1963. Nematode infections. *In* Insect pathology: an advanced treatise. E. A. Steinhaus, editor. Academic Press, New York.

Westbrooks, R. G. 1986. Poisonous plants of Eastern North America. University of South Carolina Press, Durham.

Western, T. 1989. Improved ant isolator. South African Bee Journal 61:105–107.

Whitcomb, W., Jr. 1936. The wax moth and its control. U.S. Department of Agriculture Circular 386.

Whitcomb, W. H., H. Exline, and R. C. Hunter. 1963. Spiders of the Arkansas cotton field. Annals of the Entomological Society of America 56:653–660.

Whitcomb, W. H., M. Hite, and R. Eason. 1966. Life history of the green lynx spider *Pencetia viridans* (Araneida: Oxyopidae). Journal of the Kansas Entomological Society 39:259–267.

White, A. J. 1968. Opossums too [letter to the editor]. American Bee Journal 108:265.

White, B. 1993. Exotic honey bee disease found in Queensland. Australasian Beekeeper 94:345–346.

White, G. F. 1906. The bacteria of the apiary, with special reference to bee diseases. U.S. Department of Agriculture Technical Series number 14.

———. 1907. The cause of American foul brood. U.S. Department of Agriculture, Bureau of Entomology Circular 94.

———. 1912. The cause of European foul brood. U.S. Department of Agriculture Circular 157.

———. 1913. Sacbrood, a disease of bees. U.S. Department of Agriculture, Bureau of Entomology Circular 169.

———. 1914. Destruction of germs of infectious bee diseases by heating. U.S. Department of Agriculture Bulletin 92.

———. 1917. Sacbrood. U.S. Department of Agriculture Bulletin 431.

———. 1919. Nosema disease. U.S. Department of Agriculture Bulletin 780.

———. 1920a. American foulbrood. U.S. Department of Agriculture Bulletin 809.

———. 1920b. European foulbrood. U.S. Department of Agriculture Bulletin 810.

White, J. W. 1975a. Composition of honey. *In* Honey: a comprehensive survey. E. Crane, editor. Crane, Russak, New York.

———. 1975b. Physical characteristics of honey. *In* Honey: a comprehensive survey. E. Crane, editor. Crane, Russak, New York.

White, J. W., and I. Kushnir. 1967. The enzymes of honey: examination by ion exchange chromatography, gel filtration and starch-gel electrophoresis. Journal of Apicultural Research 6:69–89.

White, J. W., M. L. Riethof, M. H. Subers, and I. Kushnir. 1962a. Composition of American honeys. U.S. Department of Agriculture Technical Bulletin 1261.

White, J. W., and M. H. Subers. 1963. Studies on honey inhibine. 2. A chemical assay. Journal of Apicultural Research 2:93–100.

White, J. W., and M. H. Subers. 1964a. Studies on honey inhibine. 3. Effect of heat. Journal of Apicultural Research 3:45–50.

White, J. W., and M. H. Subers. 1964b. Studies on honey inhibine. 4. Destruction of the peroxide accumulation system by light. Journal of Food Science 29:819–828.

White, J. W., M. H. Subers, and A. I. Schepartz. 1962b. The identification of inhibine. American Bee Journal 102:430–431.

White, J. W., M. H. Subers, and A. I. Schepartz. 1963. The identification of inhibine, the antibacterial factor in honey, as hydrogen peroxide, and its origin in a honey glucose oxidase system. Biochimica et Biophysica Acta 73:57–70.

White, P. B. 1921. The pathology of Isle of Wight Disease in hive bees. Transactions of the Royal Society of Edinburgh 52:755–764.

Whitefoot, L. O., and B. F. Detroy. 1968. Pollen: milling and storing. American Bee Journal 108:138, 140.

Whiting, P. W. 1943. Multiple alleles in complementary sex determination. Genetics 28:365–382.

Wiese, H. 1977. Apiculture with Africanized bees in Brazil. *In* African bees: taxonomy, biology and economic use. Proceedings of an Apimondia International Symposium, 17–19 November 1976. D. J. C. Fletcher, editor. Pretoria, South Africa.

Wieting, J., and H.-J. Ferenz. 1991. Behavioral study on the invasion of honey bee brood by the mite *Varroa jacobsoni* on wax combs and ANP combs. American Bee Journal 131:117–118.

Will, L. 1978. I solved the skunk problem. Gleanings in Bee Culture 106:367.

Wille, H. 1962. Septikämien und Mischinfektionen [Septicemias and mixed infections]. Schweizerische Bienen-Zeitung 85:222–226, 280–285.

———. 1964a. Bekämpfung der Kalkbrut [Control of chalkbrood]. Schweizerische Bienen-Zeitung 87:381.

———. 1964b. Weitere Untersuchungen an wenig bekannten Krankheitsformen der erwachsenen Honigbiene [Further investigations of little-known forms of adult honey bee diseases]. Schweizerische Bienen-Zeitung 87:18–28.

———. 1966. Neue Erkenntnisse über krankhafte Zustände im Bienenvolk [New knowledge about disease conditions in honey bee colonies]. Bienenvater 87:3–13.

———. 1967. Mischinfektionen in der Honigbiene (*Apis mellifica* L.) nach Ermittlungen in schweizerischem Material der Jahre 1965/66 [Mixed infections in Swiss honey bees found in 1965–1966]. Zeitschrift für Bienenforschung 9:150–171.

Wille, H., and L. Pinter. 1961. Untersuchungen über bakterielle Septikämie der erwachsenen Honigbiene in der Schweiz [Investigations on bacterial septicemia of the adult honey bee in Switzerland]. Bulletin Apicole 4:141–180.

Williams, C. E. 1990. Late winter foraging by honeybees (Hymenoptera: Apidae) at sapsucker drill holes. The Great Lakes Entomologist 23:29–32.

Williams, H. 1985. Beekeeping in Tennessee. Agricultural Extension Service Bulletin 697, University of Tennessee.

Williams, I. H. 1979. Beekeeping in the Sultanate of Oman. American Bee Journal 119:510–511, 532, 535.

Williams, J. L. 1973. Fumagillin-treated extender patties ineffective for *Nosema* control in nuclei. American Bee Journal 113:58–59.

———. 1976. Status of the greater wax moth, *Galleria mellonella* (L.), in the United States beekeeping industry. American Bee Journal 116:524–526.

Wilson, C. A., and L. L. Ellis. 1966. A new technique for the detection of nosema in apiaries. American Bee Journal 106:131.

Wilson, H. F., and G. E. Marvin. 1929. On the occurrence of yeasts which may cause the spoilage of honey. Journal of Economic Entomology 22:513–517.

Wilson, W. T. 1974. Residues of oxytetracycline in honey stored by *Apis mellifera*. Environmental Entomology 3:674–676.

Wilson, W. T., and M. Alzubaidy. 1975. The role of the adult queen bee (*Apis mellifera*) in the epizootiology of American foulbrood disease. Proceedings of the 25th International Apicultural Congress 25:372–374.

Wilson, W. T., J. R. Baxter, A. M. Collins, R. L. Cox, and D. Cardoso-T. 1993. Formic acid fumigation for control of tracheal mites in honey bee colonies. BeeScience 3:26–32.

Wilson, W. T., J. R. Baxter, W. L. Rubink, C. Garza-Q., and A. M. Collins. 1995. Parasitic mite population changes during Africanization of managed honey bee colonies in Mexico. American Bee Journal 135:833.

Wilson, W. T., and J. W. Brewer. 1974. Beekeeping in the Rocky Mountain Region. Colorado State University Cooperative Extension Service WRP-12.

Wilson, W. T., and A. M. Collins. 1993. Formic acid or amitraz for spring or fall treatment of *Acarapis woodi*. American Bee Journal 133:871.

Wilson, W. T., and A. M. Collins. 1994. Economic impact and chemical control of *Acarapis woodi* in North America. Proceedings of the 5th International Conference on Apiculture in Tropical Climates. Trinidad-Tobago, West Indies.

Wilson, W. T., R. Cox, and D. Maki. 1986. Research related to the honey bee tracheal mite in southern Texas and northern Mexico. *In* Proceedings of the honey bee tracheal mite (*Acarapis woodi* R.) scientific symposium. St. Paul, Minnesota. American Association of Professional Apiculturists, Gainesville, Florida.

Wilson, W., T., R. L. Cox, and J. O. Moffett. 1990. Menthol-grease board: a new method of administering menthol to honey bee colonies. American Bee Journal 130:409–412.

Wilson, W. T., J. R. Elliot, and J. D. Hitchcock. 1971. Antibiotic extender patties for control of American foulbrood. Journal of Apicultural Research 10:143–147.

Wilson, W. T., J. R. Elliot, and J. D. Hitchcock. 1973a. Treatment of American foulbrood with antibiotic extender patties and antibiotic paper packs. American Bee Journal 113:341–344.

Wilson, W. T., J. R. Elliott, and J. J. Lackett. 1970. Antibiotic treatments that last longer. American Bee Journal 110:348, 351.

Wilson, W. T., M. Ellis, and A. M. Collins. 1994. Citral fumigation and three methods of feeding apitol for tracheal mite control. American Bee Journal 134:839.

Wilson, W. T., D. L. Maki, J. Vargas C., R. L. Cox, and W. L. Rubink. 1988a. Honey bee swarm frequency and tracheal mite infestation levels in bait hives along the U.S./Mexico border. American Bee Journal 128:811–812.

Wilson, W. T., J. O. Moffett, R. L. Cox, D. L. Maki, H. Richardson, and R. Rivera. 1988b. Menthol treatment for *Acarapis woodi* control in *Apis mellifera* and the resulting residues in honey. *In* Africanized honey bees and bee mites. G. R. Needham, R. E. Page, Jr., M. Delfinado-Baker, and C. E. Bowman, editors. Ellis Horwood, Chichester, England.

Wilson, W. T., and R. T. Nunamaker. 1982. The infestation of honey bees in Mexico with *Acarapis woodi*. American Bee Journal 122:503–505, 508.

Wilson, W. T., and R. T. Nunamaker. 1983. The incidence of *Nosema apis* in honeybees in Mexico. Bee World 64:132–136.

Wilson, W. T., and R. T. Nunamaker. 1985. Further distribution of *Acarapis woodi* in Mexico. American Bee Journal 125:107–111.

Wilson, W. T., G. H. Rose, J. R. Elliot, and J. J. Lackett. 1973b. Antibiotic paper packs: another method of prolonging drug treatment in honey bee colonies. American Bee Journal 113:338–340.

Winston, M. L., G. G. Grant, K. Slessor, and J. Corner. 1981. The moth *Vitula edmandsae*: a pest of honeybee combs. Bee World 62:108–110.

Winston, M. L., G. W. Otis, and O. R. Taylor, Jr. 1979. Absconding behavior of the africanized honeybee in South America. Journal of Apicultural Research 18:85–94.

Wistreich, G. A., and M. D. Lechtman. 1980. Microbiology. Macmillan, New York.

Witherell, P. C. 1973. Behavior of the honey bee (*Apis mellifera* L.) mutant "diminutive-wing." Ph.D. dissertation, University of California, Davis.

Witherell, P. C., and W. A. Bruce. 1990. *Varroa* mite detection in beehives: Evaluation of sampling methods using tobacco smoke, fluvalinate smoke, amitraz smoke and ether-roll. American Bee Journal 130:127–129.

Witherell, P. C., and E. W. Herbert. 1988. Evaluation of several possible treatments to control Varroa mite *Varroa jacobsoni* (Oud.) on honey bees in packages. American Bee Journal 128:441–445.

Wolcott, O. N. 1922. Insect parasite introduction in Porto Rico. Journal of Department of Agriculture, Porto Rico, 6(1):5-20.

Wongsiri, S., P. Polnurak, and H. A. Sylvester. 1986. Enemies of *Apis mellifera* in a rubber plantation in Thailand. Proceedings of the 30th International Congress of Apiculture, Nagoya, Japan.

Woodrow, A. W. 1941a. Susceptibility of honeybee larvae to American foulbrood. Gleanings in Bee Culture 69:148–151, 190.

———. 1941b. Behavior of honeybees toward brood infected with American foulbrood. American Bee Journal 81:363–366.

———. 1942. Susceptibility of honeybee larvae to individual inoculations with spores of *Bacillus larvae*. Journal of Economic Entomology 35:892–895.

———. 1948. Drying and storing pollen trapped from honey bee colonies. American Bee Journal 88:124–125.

Woodrow, A. W., and E. C. Holst. 1942. The mechanisms of colony resistance to American foulbrood. Journal of Economic Entomology 35:327–330.

Woodward, L. 1985. Poisonous plants, a color field guide. Hippocrene Books, New York.

Worden, S. H. 1917. Caged bees and queen murdered by ants. Gleanings in Bee Culture 45:795.

World Trade Organization. 1995. The results of the Uruguay round of multilateral trade negotiations; the legal texts. World Trade Organization, Geneva, Switzerland.

Woyke, J. 1962a. The hatchability of "lethal" eggs in a two sex-allele fraternity of honeybees. Journal of Apicultural Research 1:6–13.

———. 1962b. Geneza powstawania niezwyklych pszczó [The origin of unusual bees]. Pszczelnicze Zeszyty Naukowe 6:49–63.

———. 1963a. Drone larvae from fertilized eggs of the honeybee. Journal of Apicultural Research 2:19–24.

———. 1963b. What happens to diploid drone larvae in a honeybee colony. Journal of Apicultural Research 2:73–75.

———. 1967. Diploid drone substance: cannibalism substance. Proceedings of the 21st International Apicultural Congress, Preliminary Scientific Meetings, summary paper number 10.

———. 1976a. Population genetic studies on sex alleles in the honeybee using the example of the Kangaroo Island bee sanctuary. Journal of Apicultural Research 15:105–123.

———. 1976b. Brood rearing efficiency and absconding in Indian honeybees. Journal of Apicultural Research 15:133–143.

———. 1977. Cannibalism and brood-rearing efficiency in the honeybee. Journal of Apicultural Research 16:84–94.

———. 1980. Effect of sex allele homo-heterozygosity on honeybee colony populations and on their honey production 1. Favourable development conditions and unrestricted queens. Journal of Apicultural Research 19:51–63.

———. 1984a. Increases in life span, unit honey productivity and honey surplus with fumagillin treatment of honey bees. Journal of Apicultural Research 23:209–212.

———. 1984b. Survival and prophylactic control of *Tropilaelaps clareae* infesting *Apis mellifera* colonies in Afghanistan. Apidologie 15:421–434.

———. 1985a. Further investigations into control of the parasite bee mite *Tropilaelaps clareae* without medication. Journal of Apicultural Research 24:250–254.

———. 1985b. *Tropilaelaps clareae*, a serious pest of *Apis mellifera* in the tropics, but not dangerous for apiculture in temperate zones. American Bee Journal 125:497–499.

———. 1986. Sex determination. In Bee genetics and breeding. T. E. Rinderer, editor. Academic Press, Orlando, Florida.

———. 1987a. Comparative population dynamics of *Tropilaelaps clareae* and *Varroa jacobsoni* mites on honeybees. Journal of Apicultural Research 26:196–202.

————. 1987b. Infestation of honeybee (*Apis mellifera*) colonies by the parasitic mites *Varroa jacobsoni* and *Tropilaelaps clareae* in South Vietnam and results of chemical treatment. Journal of Apicultural Research 26:64–67.

————. 1987c. Length of successive stages in the development of the mite *Tropilaelaps clareae* in relation to honeybee brood age. Journal of Apicultural Research 26:110–114.

————. 1989. Change in shape of *Tropilaelaps clareae* females and the onset of egg laying. Journal of Apicultural Research 28:196–200.

————. 1993. Practical control method of the parasitic bee mite *Tropilaelaps clareae*. American Bee Journal 133:510–511.

————. 1994a. Repeated egg laying by females of the parasitic honeybee mite *Tropilaelaps clareae* Delfinado and Baker. Apidologie 25:327–330.

————. 1994b. *Tropilaelaps clareae* females can survive for four weeks when given open bee brood of *Apis mellifera*. Journal of Apicultural Research 33:21–25.

Woyke, J., and W. Skowronek. 1974. Spermatogenesis in diploid drones of the honeybee. Journal of Apicultural Research 13:183–190.

Wright, A. H., and A. A. Wright. 1949. Handbook of frogs and toads. Comstock Publishing, Ithaca, New York.

Wright, A. H., and A. A. Wright. 1970. Handbook of frogs and toads of the United States and Canada. 3d ed. Cornell University Press, Ithaca, New York.

Wright, M. 1944. Some random observations on dragonfly habits with notes on their predaceousness on bees. Journal of the Tennessee Academy of Science 19:295–301.

Wright, P. H. 1952. The control of ants by DDT. Gleanings in Bee Culture 80:342.

Wyborn, M. H., and D. M. McCutcheon. 1987. A comparison of dry and wet fumagillin treatments for spring *Nosema* disease suppression of overwintered colonies. American Bee Journal 127:207–209.

Yabuki, A. T., and C. da Cruz-Landim. 1977. Nota sobre infestação dupla do intestino de *Apis mellifera* por *Nosema* e *Gregarina* [A note on simultaneous infestation of the midgut of *Apis mellifera* by *Nosema* and *Gregarina*]. Revista de Agricultura (Piracicaba) 52(4):253–258.

Yacobson, B. A., D. Elad, K. Rosenthal, I. Kramer, I. Slovecky, and H. Efrat. 1991. A recent chalkbrood outbreak in Israel: attempts at therapeutic intervention. American Bee Journal 131:786.

Yakobsen, B. A., S. Rosen, and A. Hadani. 1986. The occurrence and distribution of varroasis in apiaries in Israel. American Bee Journal 126:120–121.

Yakobson, B., S. Pothichot, and S. Wongsiri. 1992. Possible transfer of *Nosema apis* from *Apis mellifera* to *Apis cerana*. In Abstracts of papers of the International Conference on Asian Honey Bees and Bee Mites, Bangkok, 9–14 February 1992; Bee Biology Research Unit, Department of Biology, Chulalongkorn University, Bangkok, Thailand.

Yang, R. S. H., and M. R. Kare. 1968. Taste response of a bird to constituents of arthropod defensive secretions. Annals of the Entomological Society of America 61:781–782.

Youssef, N. N., and W. A. Brindley. 1989. Effectiveness of Botran and DPX 965 as growth inhibitors of *Ascosphaera aggregata* (Ascosphaeraceae) in the alfalfa leafcutting bee (Hymenoptera: Megachilidae). Journal of Economic Entomology 82:1335–1338.

Youssef, N. N., and D. M. Hammond. 1971. The fine structure of the developmental stages of the microsporidian *Nosema apis*. Tissue & Cell 3:283–294.

Youssef, N. N., and W. R. McManus. 1985. Captan: a promising fungicide for management of chalkbrood diseases in the alfalfa leafcutting bee. Journal of Economic Entomology 78:428–431.

Youssef, N. N., and W. R. McManus. 1991. In vitro culture of *Ascosphaera aggregata* (Ascosphaeraceae), a pathogen of the alfalfa leafcutting bee *Megachile rotundata* (Apidae). Journal of Invertebrate Pathology 58:335–347.

Youssef, N. N., W. R. McManus, and P. F. Torchio. 1985. Cross-infectivity potential of *Ascosphaera* spp. (Ascomycetes: *Ascosphaera*) on the bee *Osmia lignaria propinqua* Cresson (Megachilidae: *Osmia*). Journal of Economic Entomology 78:227–231.

Youssef, N. N., C. F. Roush, and W. R. McManus. 1984. In vivo development and pathogenicity of *Ascosphaera proliperda* (Ascosphaeraceae) to the alfalfa leafcutting bee, *Megachile rotundata*. Journal of Invertebrate Pathology 43:11–20.

Zaitsev, V. F. 1976a. The bee flies (Diptera: Bombyliidae) of Iraq. Entomological Review 55(3): 131-134.

————. 1976b. Parasitic bee flies [Bombyliidae]. Zashchita Rastenii No. 11: 34-35.

Zakatimova, Z. I. 1948. False hellebore and the summer death of bees in the Primorsk Region [in Russian]. Pchelovodstvo 1948(7):47–49.

Zander, E. 1909a. Tierische Parasiten als Krankenheitserreger bei der Biene. Münchener Bienenzeitung 31:196–204.

————. 1909b. Tierische Parasiten als Krankheitserreger bei der Biene [Animal parasites as disease-producers in bees]. Leipziger Bienenzeitung 24:147–150, 164–166.

————. 1909c. Tierische Parasiten als Krankheitserreger bei der Biene [Animal parasites as disease-producers in bees]. Bienenvater 41:233–240.

————. 1911. Krankheiten und Schädlinge der erwachsenen Bienen [Diseaes and injuries of adult bees]. 2d ed. Verlagsbuchhandlung Eugen Ulmer, Stuttgart.

————. 1919. Die Brutkrankheiten und ihre Bekampfung [The brood diseases and their control]. *In* Handbuch der Bienenkunde in Einzeldarstellungen, volume 1. 2d ed. Eugen Ulmer, Stuttgart.

————. 1944. Krankheiten und Schädlinge der erwachsenen Bienen [Diseases and parasites of adult bees]. *In* Handbuch der Bienenkunde in Einzeldarstellungen, volume 2. 4th ed. Eugen Ulmer, Stuttgart.

Zappi-Recordati, A. 1971. Apicoltura moderna [Modern apiculture]. 4th ed. Ramo Editoriale degli Agricoltori, Rome.

Zhavnenko, V. M. 1971. Indirect method of immunofluorescence in the diagnosis of foul brood (American and European) [in Russian]. Veterinariya 1971(8):109–111.

Zherebkin, M. V. 1976. Resistance of the honeybee to nosema in relation to chimozin. Apiacta 11:5–9.

Zimina, L. V. 1968. New data concerning the parasitic relationships of conopids (Dip.). Zoologicheskii Zhurnal Moskau 47:780–781.

————. 1977. *Physocephala truncata* (Diptera: Conopidae), a parasite of honeybees in Tuva ASSR [USSR]. Zoologicheskii Ayhurnall 52(11): 1732-1733.

Zivihinovic, S. 1936. Insects preying upon the honeybee. Beograd.

Zozaya-R., J. A., E. Tanus-S., and E. Guzmán-N. 1982. Mexicans report on acarine mite survey. The Speedy Bee 10:16.

INDEX

abnormalities, effects of, 403-423

Acarapis woodi, 4, 13, 18, 19, 21, 28, 29, 35, 158, 248-249, 250, 253-277, 294, 483, 508-509, 536, 547-548, 557, 559-560, 592-602,
    see tracheal mites

acarine disease, 255, 536

acetic acid fumigation, 72, 75, 131, 504

Achroia grisella, 121, 129, 136-138, 412

Achromobacter (Bacterium) eurydice, 46, 48, 51

acid foulbrood, 535

acidity of honey, 457-458

acute paralysis, 23-24

acute paralysis virus, 14, 15, 23-24, 294, 296

Africanized bees, 3, 97, 166, 196, 199, 289, 291, 293, 295, 301, 302 303, 306, 363, 485

Africanized bees, varroa resistance, 485-491

Africanized honey bees, 97, 127, 169, 171, 264, 300, 301, 305, 306, 393, 474, 485, 486, 520, 521, 523, 531, 560, 565

Agamomermis sp., 115

Agriculture Bee Research Laboratory, 8

alfalfa leafcutting bee, 3, 82, 86, 91, 92, 95, 97, 135, 136, 139

aluminum phosphide, 131, 507, 576, 577

American badger, 367

American foulbrood, 4, 5, 6, 23, 35-45, 50, 85, 91, 95, 96, 124, 277, 296, 410, 471, 472, 476, 534, 540-541, 564-569, 573, 575, 586-602

American foulbrood, control of, 277, 495-503

American foulbrood disease, 35-45, 53, 157, 464, 474, 475, 476, 495, 496, 540-541

American foulbrood, resistance to, 477-480

amitraz, 151, 262, 274, 275, 276, 307, 312

amoeba disease, 73-75, 535, 546-547, 586-602

amphibians, 331, 336

Anacardiaceae, 428-429

Anagasta kuehniella, 121, 140
    see Mediterranean flour moth

Animal and Plant Health Inspection Service, 558

antibacterial activity of honey, 457

antibiotic extender patties, 5, 44, 277, 499-500, 504

antibiotic systems, 455-466

ants, 165-175, 335, 364

ants, wasps, and bees, 163-200, 510
    see Hymenoptera

Apanteles galleriae, 129, 130, 138

APHIS, 270, 271, 383, 558, 559, 560

Aphomia sociella, 121, 135, 139

apimyiasis, 155

Apis andreniformis, 282, 323, 518

Apis cerana, 6, 22, 23, 25, 28, 29, 62, 63, 97, 121, 128, 149, 166, 169, 171, 173, 175-177, 179, 181, 182, 184, 190, 191, 195, 198, 208, 213, 220, 223, 236, 244, 246, 247, 248, 262, 264, 272, 281, 282, 283, 290, 291, 293, 302, 305, 319, 320, 321, 324, 334, 335, 349, 366, 374, 408-409, 474, 476, 485-491, 516-518, 533, 557, 587

Apis dorsata, 121, 128, 129, 172, 175-177, 181, 182, 198, 248, 264, 281, 282, 323, 324, 326, 327, 343, 348, 349, 355, 374, 516-518

Apis florea, 6, 62, 102, 129, 166, 171-172, 175-177, 178, 182, 198, 244, 248, 282, 516-518,

Apis iridescent virus, 15, 28-29

Apis koschenikovi, 282, 321

Apis laboriosa, 152, 282, 327, 348, 356, 517

Apis mellifera, 23, 24, 25, 29, 62, 63, 73, 97, 121, 151, 171, 175-177, 178, 179, 181, 182, 184, 189, 191, 196, 197, 198, 218, 220, 222, 223, 244, 245, 246, 248, 255, 262, 264, 281, 282, 283, 286, 289, 291, 293, 299, 302, 305, 306, 320, 321, 322, 323, 324, 326, 334, 344, 349, 403, 462, 485-491, 508, 516-518, 519-533, 560, 562

Apis mellifera capensis, 301, 417, 487, 523, 525
    see Cape honey bee

Apis mellifera carnica, 480, 483, 487, 523

Apis mellifera ligustica, 488, 520, 523, 525-526, 528, 530, 532

Apis mellifera mellifera, 488, 489, 523, 525-526, 527, 528, 530

Apistan, 151, 276, 287, 299, 313, 314-318, 508, 571, 576, 587

Apitol, 274, 275, 312, 313

Apocynaceae, 429

Apodidae, 341-342, 348-349

Arachnida, 225-237, 241-250

Argentine ants, 166-167, 173

Aristotle, 16, 341, 520

Arkansas bee virus, 15, 29

armadillos, 399, 511

army ants, 166

Arrhenosphaera cranei, 83, 89

arsenicals, 551

Ascosphaera aggregata, 82, 83, 86, 87, 91, 92

Ascosphaera apis, 81-99, 535, 544, 559-560, 592-602
    see chalkbrood

Ascosphaera atra, 82

Ascosphaera fimicola, 82

Ascosphaera osmophila, 83

Ascosphaera proliperda, 82, 92

Asian honey bees, 248, 281, 516-518, 591

Asiatic black bear, 374

Asilidae, 152-153

Aspergillus flavus, 101-105, 545

Aspergillus fumigatus, 71, 101-105, 545
    see fumagillin

Aspergillus niger, 102, 104, 105, 108, 109, 545

assassin bugs, 213

Asteraceae, 429

autumn collapse, 537

Azalea, 434-435

baboons, 346, 348, 363, 364, 367, 511

Bacillus alvei, 45, 46, 50, 51, 534, 535, 542

Bacillus apisepticus, 51, 52

www.ingramcontent.com/pod-product-compliance
Lightning Source LLC
Chambersburg PA
CBHW081424270326
41932CB00019B/3093